ISBN 978-1-330-40800-1
PIBN 10054966

English
Français
Deutsche
Italiano
Español
Português

www.forgottenbooks.com

Mythology Photography **Fiction**
Fishing Christianity **Art** Cooking
Essays Buddhism Freemasonry
Medicine **Biology** Music **Ancient
Egypt** Evolution Carpentry Physics
Dance Geology **Mathematics** Fitness
Shakespeare **Folklore** Yoga Marketing
Confidence Immortality Biographies
Poetry **Psychology** Witchcraft
Electronics Chemistry History **Law**
Accounting **Philosophy** Anthropology
Alchemy Drama Quantum Mechanics
Atheism Sexual Health **Ancient History**
Entrepreneurship Languages Sport
Paleontology Needlework Islam
Metaphysics Investment Archaeology
Parenting Statistics Criminology
Motivational

QUAIN'S

ELEMENTS OF ANATOMY

Seventh Edition

EDITED BY

WILLIAM SHARPEY, M.D., F.R.S.
PROFESSOR OF ANATOMY AND PHYSIOLOGY IN UNIVERSITY COLLEGE, LONDON

ALLEN THOMSON, M.D., F.R.S.
PROFESSOR OF ANATOMY IN THE UNIVERSITY OF GLASGOW

AND

JOHN CLELAND, M.D.
PROFESSOR OF ANATOMY IN QUEEN'S COLLEGE, GALWAY

IN TWO VOLUMES

ILLUSTRATED BY UPWARDS OF 800 ENGRAVINGS ON WOOD.

VOL. II.

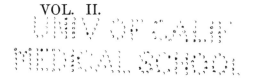

LONDON

JAMES WALTON
BOOKSELLER AND PUBLISHER TO UNIVERSITY COLLEGE
137, GOWER STREET.
1867.

LONDON:
BRADBURY, EVANS AND CO., PRINTERS, WHITEFRIARS.

CONTENTS OF THE SECOND VOLUME.

DIVISION I.—SYSTEMATIC AND DESCRIPTIVE ANATOMY
(CONTINUED).

DIVISION II.—SURGICAL ANATOMY.

DIVISION III.—DISSECTIONS.

SECTION V.—NEUROLOGY.

UNDER the name of Neurology, it is intended to include the descriptive anatomy of the various organs forming parts of the nervous system.

The nervous system consists of two sets of parts, one of which is *central*, the other *peripheral*. To the first set belong the brain and spinal cord, forming together the cerebro-spinal axis, and the ganglia : to the second set belong all the nerves distributed throughout the body ; and along with these may be included the organs of the senses, or those organs which contain the terminations of the several nerves of special sensation, in connection with certain apparatus or modifications of structure related to the reception of impressions by each of these nerves.

Among the peripheral nerves it is necessary also to distinguish the cerebro-spinal and the sympathetic or ganglionic, which, though intimately connected with each other at some places, are yet so different in their structure and mode of distribution as to require separate description.

The description of these several parts of the nervous system will be brought under the following four subsections, viz. 1. The cerebro-spinal axis ; 2. The cerebro-spinal nerves and the ganglia connected with them ; 3. The sympathetic nerves and their ganglia ; 4. The organs of the senses.

I.—THE CEREBRO-SPINAL AXIS.

The *cerebro-spinal axis* is contained partly within the cavity of the cranium, and partly within the vertebral canal ; it is symmetrical in its form and structure throughout, consisting of a right and a left half, separated to a certain extent by longitudinal fissures, and presenting in their plane of union various portions of white and grey nervous substance which cross from one side to another, and form the *commissures* of the brain and spinal cord.

Enclosed within the skull and the vertebral canal, the cerebro-spinal axis is protected by the bony walls of those two cavities ; it is also surrounded by three membranes, which afford it additional protection and support, and are subservient to its nutrition. These envelopes, which will be described hereafter, are, 1st, a dense fibrous membrane named the dura mater, which is placed most superficially ; 2nd, a serous membrane called the arachnoid ; and, 3rd, deepest of all, a highly vascular membrane named the pia mater.

The cerebro-spinal axis is divided by anatomists into the *encephalon* or enlarged upper mass placed within the cranium, and the *spinal cord* contained within the vertebral canal.

These two parts have a relation, one to the other, very similar to that which subsists between the cranium and vertebral column : thus, they are continuous structures ; at the time of their first formation in the fœtus they are nearly similar ; the earliest developed distinction consists in the enlargement of the encephalon ; and, moreover, the spinal cord, like the vertebral column, continues to present a structure nearly uniform throughout its extent, while the encephalon becomes gradually more and more complicated,

Fig. 339.

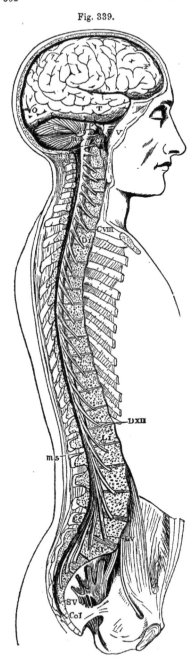

till at last it is difficult to trace the serial relation of its constituent parts, or any correspondence with the structure of the cord.

Fig. 339. — VIEW OF THE CEREBRO-SPINAL AXIS OF THE NERVOUS SYSTEM (after Bourgery). ⅕

· The right half of the cranium and trunk of the body has been removed by a vertical section; the membranes of the brain and spinal marrow have also been removed, and the roots and first part of the fifth and ninth cranial, and of all the spinal nerves of the right side, have been dissected out and laid separately on the wall of the skull and on the several vertebræ opposite to the place of their natural exit from the cranio-spinal cavity.

F, T, O, lateral surface of the cerebrum; C, cerebellum; P, pons Varolii; m o, medulla oblongata; m s, upper and lower extremities of the spinal marrow; c e, on the last lumbar vertebra, marks the cauda equina; v, the three principal branches of the nervus trigeminus or fifth pair; C I, the sub-occipital or first cervical nerve; above this is the ninth pair; C VIII, the eighth or lowest cervical nerve; D I, the first dorsal nerve; D XII, the last or twelfth; L I, the first lumbar nerve; L v, the last or fifth; S I, the first sacral nerve; S v, the fifth; Co I, the coccygeal nerve; s, the left sacral plexus.

A.—THE SPINAL CORD.

The *spinal cord*, or *spinal marrow* (medulla spinalis), is that part of the cerebro-spinal axis which is situated within the vertebral canal. It extends from the margin of the foramen magnum of the occipital bone to about the lower part of the body of the first lumbar vertebra. It is continued into the medulla oblongata above, and ends below in a slender filament, the *filum terminale* or *central ligament* of the spinal cord.

Invested closely by a proper membrane (the pia mater), the cord is enclosed within a sheath (theca) considerably longer and larger than itself, which is formed by the dura mater, and which is separated from the walls of the canal by numerous vascular plexuses, and much loose areolar tissue. The interval between the investing membrane and the sheath of the cord is occupied by a serous membrane (the arachnoid), and the space between the latter membrane and the pia mater is occupied by a fluid called the cerebro-spinal fluid. Within this space the cord is kept in position by proper ligaments, which fix it at different points to its sheath, and by the roots of the spinal nerves,—an anterior and a posterior root belonging to each,—which pass across the space from the surface of the cord towards the intervertebral foramina. From its lower part, where they are closely crowded together, the roots of the lumbar and sacral nerves descend nearly vertically to reach the lumbar intervertebral and the sacral foramina, and form a large bundle or lash of nervous cords named the *cauda equina*, which occupies the vertebral canal below the termination of the cord.

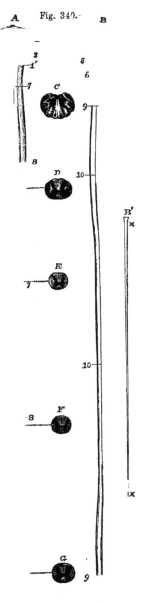

Fig. 340.

Fig. 340.—ANTERIOR AND POSTERIOR VIEWS OF THE MEDULLA OBLONGATA AND SPINAL CORD WITH SECTIONS. ½

The cord has been divested of its membranes and the roots of the nerves. A, presents an anterior, B, a posterior view, showing the upper or brachial, and the lower or crural enlargements. In these figures the filiform prolongation, represented separately in B', has been removed ; C, shows a transverse section through the middle of the medulla oblongata ; D, a section through the middle of the cervical enlargement of the spinal cord ; E, through the upper region of the dorsal part ; F, through its lower ; G, through the middle of the lumbar enlargement ; and H, near the lower end of its tapering extremity.

1, anterior pyramids ; 1', their decussation ; 2, olivary bodies ; 3, restiform bodies ; 4, posterior surface of the medulla oblongata ; 4', calamus scriptorius ; 5, posterior pyramids ; 6, posterior lateral columns passing up into the restiform bodies ; 7, 7, anterior median fissure extending through the whole length of the spinal cord ; 8, 8, anterior lateral groove ; 9, 9, posterior median fissure ; 10, 10, posterior lateral groove ; ×, lower end of the tapering extremity of the cord ; ×, ×, in B', the filiform prolongation of the cord and its pia-matral covering.

Although the cord usually ends near the lower border of the body of the first lumbar vertebra, it sometimes terminates a little above or below that point, as opposite to the last dorsal or to the second lumbar vertebra. The position of the lower end of the cord also varies according to the state of curvature of the vertebral column, in the flexion forwards of which, as in the stooping posture, the end of the cord is slightly raised. In the fœtus, at an early period, the cord occupies the whole length of the vertebral canal ; but, after the third month, the canal and the roots of the lumbar and sacral nerves begin to grow more rapidly than the cord itself, so that at birth the lower end reaches only to the third lumbar vertebra.

Fig. 341.

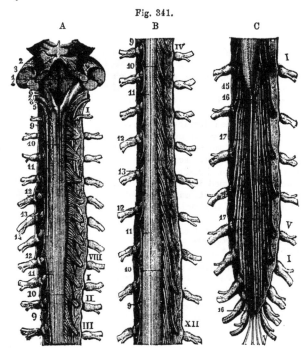

Fig. 341.—POSTERIOR VIEW OF THE MEDULLA OBLONGATA AND OF THE SPINAL CORD WITH ITS COVERINGS AND THE ROOTS OF THE NERVES (from Sappey). ½

The theca or dura-matral sheath has been opened by a median incision along the whole length, and is stretched out to each side. On the left side, in the upper and middle parts (A and B), the posterior roots of the nerves have been removed so as to expose the ligamentum denticulatum; and along the right side the roots are shown passing out through the dura mater. The roman numbers indicate the different nerves in the cervical, dorsal, lumbar, and sacral regions ; 9, several of the pointed processes of the ligamentum denticulatum ; 10, origin of several posterior roots ; 11, posterior median fissure ; 12, ganglia of the spinal nerves ; 13, part of the anterior roots seen on the left side ; 14, the united nerve ; 15, tapering lower end of the spinal cord ; 16, filum terminale ; 17, cauda equina.

The length of the spinal cord is from fifteen to eighteen inches ; and it varies in diameter in different situations. Its general form is cylindrical, somewhat flattened before and behind. It presents two enlargements—an upper or cervical, and a lower or lumbar. The cervical enlargement is of greater size and extent than the lower. It reaches from the third cervical to the first dorsal vertebra ; its greatest diameter is from side to side.

The lower or lumbar enlargement is situated nearly opposite the last dorsal vertebra ; its antero-posterior diameter is nearly equal to the transverse. Below this enlargement, the cord tapers in the form of a cone, from the apex of which the small filiform prolongation is continued downwards for some distance within the sheath.

Fig. 342.

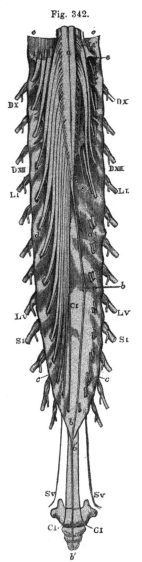

Fig. 342.—LOWER PART OF THE SPINAL CORD WITH THE CAUDA EQUINA AND SHEATH, SEEN FROM BEHIND. $\frac{1}{2}$

The sheath has been opened from behind and stretched towards the sides; on the left side all the roots of the nerves are entire ; on the right side both roots of the first and second lumbar nerves are entire, while the rest have been divided close to the place of their passage through the sheath. The bones of the coccyx are sketched in their natural relative position to show the place of the filum terminale and the lowest nerves.

a, placed on the posterior median fissure at the middle of the lumbar enlargement of the cord ; *b, b*, the terminal filament, drawn slightly aside by a hook at its middle, and descending within the dura-matral sheath ; *b', b'*, its prolongation beyond the sheath and upon the back of the coccygeal bones ; *c*, the dura-matral sheath ; *d*, double foramina for the separate passage of the anterior and posterior roots of each of the nerves ; *e*, pointed ends of several processes of the ligamentum denticulatum ; Dx, and Dxii, the tenth and twelfth dorsal nerves ; Li, and Lv, the first and fifth lumbar nerves; Si, and Sv, the first and fifth sacral nerves ; Ci, the coccygeal nerve.

The cervical and lumbar enlargements have an evident relation to the large size of the nerves which supply the upper and lower limbs, and which are connected with those regions of the cord,—in accordance with the general fact observed in the animal kingdom, that, near the origin of large nerves, the central nervous substance is accumulated in larger proportion. At the commencement of its development in the embryo the spinal cord is destitute of these enlargements, which, in their first appearance and subsequent progress, correspond with the growth of the limbs.

Sometimes the cord presents one or two bulbs or swellings towards its lower end.

According to Foville, the lumbar enlargement is chiefly due to an increase in bulk of the anterior region of the cord. (Traité compl. de l'Anat., &c., du Syst. Nerv. Cerebro-Spinal. Paris, 1844. Part I., p. 138.)

The *terminal filament* (filum terminale, central ligament) descends in the middle line amongst the nerves composing the cauda equina, and, becoming blended with the lower end of the sheath opposite to the first or second sacral vertebra, passes on to be fixed to the lower end of

the sacral canal, or to the base of the coccyx.   Internally, it is a prolongation
for about half its length of some of the nervous elements of the cord ;
externally, it consists of a tube of the pia mater or innermost membrane,
which, being attached at its lower end to the dura mater and vertebral
canal, keeps pace with the latter in its growth, whilst the cord relatively
shortens.   It is distinguished by its silvery hue from the nerves amid which
it lies.   Small blood-vessels may sometimes be seen upon it.

Fissures.—When removed from the vertebral canal, and divested of its
membranes, the spinal cord is seen to be marked by longitudinal *fissures.*
Of these, two, which are the most obvious, run along the middle line,
one in front and the other behind, and are named the *anterior* and *posterior median fissures.*

The *anterior median fissure* is more distinct than the posterior, and penetrates about one-third of the thickness of the cord, its depth increasing
towards the lower end.   It contains a fold or lamelliform process of the pia
mater, and also many blood-vessels, which are thus conducted to the centre
of the cord.   At the bottom of this fissure is seen the transverse connecting
portion of white substance named the *anterior white commissure.*

The *posterior median fissure* is less marked in the greater part of its
extent than the anterior, but becomes more evident towards the upper part
of the cord.   In a certain sense it is no real fissure, except at the lumbar
enlargement and in the cervical region, in both of which places a superficial
fissure is distinctly visible ; for, although the lateral halves of the posterior
part of the cord are quite separate, there is no distinct reflection of the pia
mater between them, but rather a septum of connective tissue and blood-
vessels which passes in nearly to the centre of the cord, as far as the *posterior
grey commissure.*

Besides these two *median* fissures, two *lateral* furrows or fissures have
been described on each side of the cord, corresponding with the lines of
attachment of the anterior and posterior roots of the spinal nerves.

The *posterior lateral fissure* is a superficial depression along the line of
attachment of the posterior roots, and is at the edge of the plane in which
these roots pass inwards to the grey matter of the cord.

The *anterior lateral* fissure, which is often described in the line of the
origin of the anterior roots of the nerves, has no real existence as a groove.
The fibres of these roots in fact, unlike the posterior, do not dip into the
spinal cord in one narrow line, but spread over a space of some breadth.
The grey substance of the cord, however, approaches the surface somewhat
in the vicinity of the place where the anterior roots enter : and this, together
with a slight depression, produces the appearance which has been described
as a groove.   Thus, each lateral half of the cord is divided by the posterior
lateral fissure into a *posterior* and an *antero-lateral* column ; and although
we cannot trace an anterior lateral fissure, this antero-lateral portion of the
cord may, for the convenience of description, be considered as subdivided
into an *anterior* and a *lateral column* by the internal grey matter.

On the posterior surface of the cord, and most evidently in the upper
part, there are two slightly marked longitudinal furrows situated one on
each side, close to the posterior median fissure, and marking off, at least in
the cervical region, a slender tract, named the *posterior median* column.
Between the anterior and posterior roots of the spinal nerves, on each side,
the cord is convex, and sometimes presents a longitudinal mark corresponding with the line of attachment of the ligamentum denticulatum.

Foville states, that in a new-born child there is a narrow accessory bundle of white matter, which runs along the surface of the lateral column, and is separated from it by a streak of greyish substance. According to the same authority, this narrow tract enlarges above, and may be traced upwards along the side ot the medulla oblongata into the cerebellum. (Op. cit. p. 285.)

Fig. 343. — DIFFERENT VIEWS OF A PORTION OF THE SPINAL CORD PROM THE CERVICAL REGION WITH THE ROOTS OF THE NERVES. Slightly enlarged.

Fig. 343.

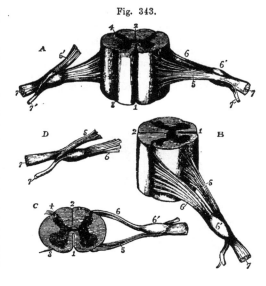

In A, the anterior surface of the specimen is shown, the anterior nerve-root of the right side being divided; in B, a view of the right side is given; in C, the upper surface is shown; in D, the nerve-roots and ganglion are shown from below. 1, the anterior median fissure; 2, posterior median fissure; 3, anterior lateral depression, over which the anterior nerve-roots are seen to spread; 4, posterior lateral groove, into which the posterior roots are seen to sink; 5, anterior roots passing the ganglion; 5', in A, the anterior root divided; 6, the posterior roots, the fibres of which pass into the ganglion, 6'; 7, the united or compound nerve; 7', the posterior primary branch seen in A and D, to be derived in part from the anterior and in part from the posterior root.

Internal structure of the spinal cord.—The spinal cord consists of white and grey nervous substance. The white matter, forming by far the larger portion of the cord, is situated externally, whilst the grey matter is disposed in the interior.

The grey matter, as seen in a transverse section of any part of the cord, presents two crescent-shaped masses, placed one in each lateral half, with their convexities towards one another, and joined across the middle by a transverse portion, the *grey* or *posterior commissure* of the cord. Each of these grey crescents has an *anterior* and a *posterior cornu* or horn. The posterior, generally longer and narrower, approaches the posterior lateral fissure : the anterior, shorter and thicker, extends towards the place of attachment of the anterior roots of the nerves. In front of it a layer of white substance separates it from the bottom of the anterior median fissure, this is named the *anterior white commissure.*

Another white layer, very thin and indistinct, was formerly described as lying behind the grey commissure; but in the present state of our knowledge it seems sufficient to describe one white commissure, and one grey commissure behind it.

At the back part or tip of the posterior horn, which is somewhat enlarged, the grey matter has a peculiar semitransparent aspect, whence it was named by Rolando *substantia cinerea gelatinosa :* the remaining and greater part of the grey matter, which resembles that most generally prevalent, was named by Rolando the *substantia spongiosa*.

The grey cornua vary in form in different parts of the cord : thus they are long and slender in the cervical portion, still more slender in the dorsal, and shorter and wider in the lumbar region. The grey matter appears in a series of sections to be, relatively to the white, more abundant in the lumbar region of the cord, less so in the cervical region, and least so in the dorsal. The actual amount, however, of white matter is greatest in the neck. Towards the lower end of the cord, the double crescentic form gradually disappears, and the grey matter is collected into a central mass, which is indented at the sides. At its extreme point, according to Remak and Valentin, the cord consists of grey matter only.

Fig. 344.

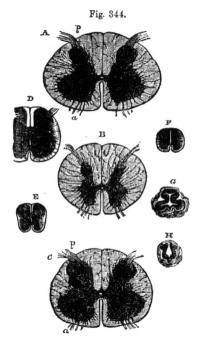

Fig. 344.—Sections of the Spinal Cord in Different Parts.

These views are taken partly from Stilling's plates and partly from nature.

A, is a section through the middle of the cervical enlargement, at the root of the sixth cervical nerve; B, through the middle of the dorsal cylindrical portion; C, through the middle of the lumbar enlargement ; D, in the conical diminishing part of the cord ; E, farther down at the origin of the fifth sacral nerve ; F, at that of the coccygeal nerve ; G, is a section of the part where the conus medullaris begins to pass into the filum terminale ; and H, at the lower part of this or in the commencement of the filum terminale.

A, B, and C, are fully twice the natural size ; D, E, and F, about three times; and G and H, about six times. In A, and C, *a*, marks the interior root-fibres of the nerves ; and *p*, the posterior root-fibres as they enter the spinal cord. In D, E, and F, the great diminution of the white substance in proportion to the grey is seen ; in G, the peculiar form of the central canal and medullary substance covering it ; and in H, the open condition of the central canal posteriorly.

In all the figures the position is the same, viz., the anterior part placed downwards.

Central canal.—Extending through the whole length of the spinal cord, in the substance of the grey commissure, there is a minute central canal which in prepared transverse sections of the cord is barely visible, as a speck, with the naked eye. Superiorly, it is continued into and opens out at the calamus scriptorius of the fourth ventricle; and inferiorly, it is prolonged into the filum terminale. It is lined with a layer of cylindrical ciliated cells or epithelium. This canal, though minute, is an object of considerable interest as a typical part of the structure of the cord, it being the permanent remains of the cavity of the cylinder formed by the spinal cord at the earliest period of its development. It is more distinctly seen in fishes, reptiles, and birds than in mammals.   In the young human subject it is always present, but, according to the observation of Lockhart Clarke and Kölliker, it sometimes disappears in the adult.

Minute Structure of the Spinal Cord.—The substance of the spinal cord consists of a large proportion of nervous substance, supported in a delicate framework of connective tissue and numerous minute blood-vessels.   The white matter presents nerve-fibres, but is destitute of nerve-cells ; the grey matter contains both elements.   The fibres of the white substance are in greatest part longitudinal; the principal exceptions being those contained in the commissure, and in the roots of the nerves.   The longitudinal fibres are finer in the posterior columns and posterior parts of the lateral columns than in other parts, and the deepest fibres are smaller than those placed more superficially.   (Kölliker.)   The fibres of the grey substance are for the most part not more than one half the diameter of their continuations in the white substance, and in the nerve-roots, but among them there are a few of larger size.   They are very various in their direction, and, in great part at least, are connected with the roots of the nerves.

Fig. 345.—TRANSVERSE SECTION OF HALF THE SPINAL MARROW IN THE LUMBAR EN- LARGEMENT. $\frac{8}{1}$

Fig. 345.

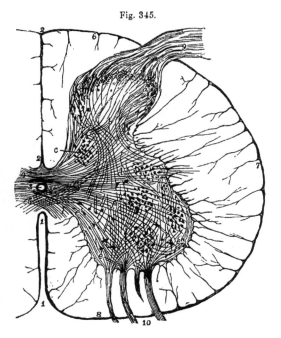

This is a semidia- grammatic representa- tion taken from a pre- pared specimen, and founded in part on the statements of Lockhart Clarke and of Kolliker.

1, anterior median fissure ;   2, posterior median fissure ; 3, cen- tral canal lined with epithelium ; 4, posterior commissure ; 5, anterior commissure ; 6, posterior column ;   7, lateral co- lumn ;   8, anterior co- lumn ; (at each of these places and throughout the white substance the trabecular prolongations of the pia mater are shown ;)   9, posterior roots of the spinal nerve entering in one principal bundle ;   10, anterior roots entering in four spreading bundles of fibres ; *a, a*, caput cornu posterioris with large and small cells, and above them the gelatinous substance ; *b*, in the cervix cornu, decus- sating fibres from the nerve roots and posterior commissure ;   *c*, posterior vesicular columns (of Clarke) ;   *d*, fibres running transversely from the posterior commissure into the lateral columns : near *d*, the lateral group of cells ;   *e, e*, fibres of the anterior roots entering the anterior cornu, and passing through among the radiating cells, but not joining their processes ; *e′*, fibres from the anterior roots which decussate in the anterior column ; *e″*, external fibres from the roots running round the outside of the anterior grey cornu towards the lateral columns ; *f*, fibres from the posterior commissure and from the posterior cornu running towards the anterior.   Three groups of cells are seen in the anterior column ; of these the anterior are external and internal, the posterior are chiefly external or lateral.

The nerve-cells of the grey matter are of two kinds. Firstly, there are very large branched cells, from $\frac{1}{400}$ to $\frac{1}{200}$ of an inch in size, containing nuclei and pigment; secondly, there are smaller cells, ranging from $\frac{1}{3000}$ to $\frac{1}{400}$ of an inch, but the majority are from $\frac{1}{1200}$ to $\frac{1}{800}$ of an inch in size.

The smaller cells occur scattered throughout the whole of the grey matter; the larger cells, on the contrary, are collected into groups. In the posterior cornua the large cells are almost entirely collected into a compact group, the *posterior vesicular column* of Clarke (the core of Stilling), which occupies the inner half of the cervix of the posterior cornu. This vesicular column is in intimate connection with the posterior roots of the nerves; it may be traced continuously from near the lower extremity of the spinal cord to the middle of the cervical enlargement, where it terminates; and it increases in size in both the lumbar and cervical enlargements. In the anterior cornu the large cells occur in greater number than in the posterior cornu, and are of somewhat greater size; and they are principally placed at its forepart, and arranged in an inner and an outer group. There is likewise described by Clarke a small group of cells, collected in a *tractus intermedio-lateralis*, and forming a projection of the grey matter opposite the junction of the anterior and posterior cornua. This lateral vesicular column extends from the upper part of the lumbar to the lower part of the cervical enlargement; and it may be said to reappear at the upper extremity of the cord, where it is traversed by the roots of the spinal accessory nerve, and is continued up into the medulla oblongata.

Fig. 346.

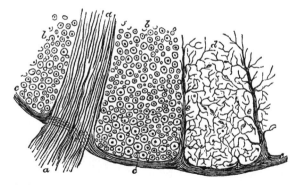

Fig. 346.—A Small Portion of a Transverse Section of the human Spinal Cord near the Surface at the entrance of a Bundle of the Anterior Roots. $\frac{300}{1}$

This figure, which is somewhat diagrammatic, is intended to show the relation to the nervous substance of the pia-matral sheath of the cord and the processes of connective tissue prolonged from it between the longitudinal and other nerve fibres. *a, a,* the primitive filaments of a bundle of the anterior roots, the medullary sheaths not represented; *b, b,* transverse sections of part of the anterior columns of the cord in which the dark points are the primitive filaments, and the circles represent the neurilemmal tube enclosing the medullary substance : in these parts the connective tissue is not represented, and many of the smallest nerve-fibres have also, for the sake of clearness, been omitted ; *c,* the pia-matral covering of the cord ; *d,* one of the compartments of the anterior column enclosed by septa of connective tissue prolonged from the pia mater, and exhibiting the fine frame-work of connective tissue extending through among the nerve-fibres, which last have been omitted : there are also indicated among the trabeculæ minute nuclei of connective tissue.

Connective tissue takes part in the structure of the cord to a very considerable extent. It forms a complete covering surrounding the white substance. In the inner margin also of the posterior columns, one on each side of the posterior fissure, two wedge-shaped bands (the *bands of Goll*) have been distinguished, in which the connective tissue is remarkably abundant, and the nerve-fibres particularly small. The connective tissue forms also a *reticulum* (processus reticularis), in which the longitudinal nerve-fibres are imbedded. In the grey matter the connective tissue is still more abundant, more especially in the immediate neighbourhood of the central canal. Much discussion has taken place as to whether the smallest cells already described are really nervous or belong to the connective tissue. In the present imperfect state of knowledge of the development of nervous elements, it might be rash to express a decided opinion on this point; but it may be stated that, independently of these, nuclei are figured by Kölliker in the reticulum, and also cells containing numerous and dividing nuclei in the neighbourhood of the central canal.

Origin of the spinal nerves.—The anterior and posterior roots of the spinal nerves are attached along the sides of the cord in or near the anterior and posterior lateral grooves, and opposite to the corresponding cornua of the grey matter; the posterior roots in a straight line, and the anterior roots scattered somewhat irregularly upon the surface (Fig. 345).

The fibres of the *anterior roots* may be traced into and through the anterior cornua. They then diverge in different directions. The innermost fibres, after passing through among the cells in the inner group of the anterior cornu, cross in the white commissure to the anterior column of the opposite side. Many fibres pass backwards in the substance of the anterior cornu, where some of them would appear to form connection with fibres proceeding from other parts of the cord, and others to spread obliquely upwards and downwards; while those which are most external passing through the outer group of cells, reach the lateral column.

Fig. 347. — A SMALL POR-
TION OF A TRANSVERSE
SECTION OF THE SPINAL
CORD AT THE PLACE
WHERE TWO BUNDLES OF
THE FIBRES OF THE
ANTERIOR ROOTS PASS
INTO THE GREY SUB-
STANCE. $\frac{300}{1}$.

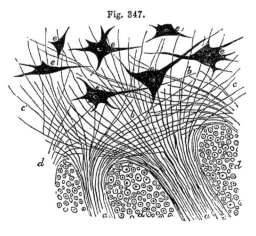

Fig. 347.

This figure may be looked upon as representing the inner ends of the anterior roots of the nerves, of which the outer part is shown in fig. 346. *a, a,* the two bundles of fibres of the anterior root passing between the compartments of longitudinal fibres of the cord; *b, b,* these fibres running backwards through the grey substance towards the posterior cornua; *c, c',* those spreading in the anterior cornua on the one side towards the anterior commissure, and on the other round the outer side of the anterior cornu; *d, d,* portions of three compartments of the anterior columns in which the longitudinal fibres of the cord are shown in transverse section; *e, e,* large radiated and nucleated cells in the grey substance of the anterior cornu—some with three, others with a greater number of processes emanating from them: no direct communication is shown between these processes and the nerve fibres of the roots.

The fibres of the *posterior roots* on reaching the posterior cornu diverge from each other in a curved manner, so as to form in great part the substantia gelatinosa. In front of this there may be seen, cut across in transverse sections, a group of these fibres which turn longitudinally upwards and downwards, and afterwards pass forwards, in part at least, to the anterior cornu, and in part to reach by the posterior commissure the posterior and lateral columns of the opposite side. Other fibres of the posterior roots pass forwards at once through the grey substance to the anterior and lateral columns. Another set of fibres slant principally upwards, but some downwards, in the posterior columns, and, interlacing with each other, most probably enter the grey matter at different heights. Some are lost to view in the posterior white columns, and it is uncertain whether or not they immediately ascend through these columns to the brain.

Much discussion has taken place as to the course of the fibres in the cord, and their ultimate destination. It is easily understood that, by the examination of sections difficult to prepare, limited in extent, liable to undergo changes in the preparation, and giving views confined each to little more than a thin lamina, it is scarcely to be expected that the full history of many tortuous fibres can be accurately ascertained. Thus it remains still undecided whether any of the fibres of the nerve-roots pass up all the way to the brain. Volkmann concluded that none of them reached the brain, arguing from measurements of the size of the cord in different regions, that the cord could not contain in its upper regions all those nerve-fibres which were traceable to it in the lower. Kölliker pointed out the fallacy of this conclusion, in so far as Volkmann had not made proper allowance for the diminished size of the fibres as they ascend in the cord; but although Volkmann's argument was thereby invalidated, it appears impossible to prove by microscopic observations that fibres of nerve-roots traced into the grey matter, and observed to emerge into the white matter, do not again re-enter the grey and terminate there. (Lockhart Clarke, Phil. Trans., 1851, 1853, 1859; Stilling, Neue Unters. u. d. Bau des Rückenmarks, 1856, 1857; Lenhossec, Neue Unters. u. d. Bau d. cent. Nervensystems, Vienna, 1855; F. Goll, Beitrage z. feineren Bau d. Rückenmarks, Zurich, 1860. For a full account of the whole subject, see Kölliker's Handbuch der Gewebelehre des Menschen, 4th ed., 1863).

It is also undetermined in what relation the nerve-fibres and branched or multipolar cells of the cord stand to each other. Most are inclined to believe that the radiating prolongations of the cells are in actual continuity with the axial filaments of nerve-fibres, whether proceeding from nerve-roots or from different parts of the cord itself; and the direct observation of such continuity has been affirmed by some, as by Schroeder Van der Kolk. But it is still considered by observers who have given most careful attention to this investigation that, although such continuity may be regarded as of the greatest probability, and, although it may be considered as proved in some other parts of the nervous system, especially in the lower animals, the actual passage of nerve-fibres into the processes of nerve-cells has not been proved as the result of actual observation in the spinal cord of man or of mammals.

Results of Experiments.—Seeing the imperfect nature of the knowledge of the minute structure of the spinal cord as obtained from microscopic observations, it may be proper to give here a short account of the more important results of physiological experiments as to the course of the transmission of sensory impressions and motor influences through it, although it is at present difficult to reconcile them with the results of anatomical research. For the most important information upon this subject, derived from vivisection, science is indebted to the researches of Brown-Séquard and Schiff.

When the superior or dorsal * half of the cord is divided in animals, sensation still continues in the hind limbs. Sensation likewise continues after division of the inferior half of the cord, and even after the superior and inferior parts of the cord have been divided at different levels in such a manner that the hinder extremity of the cord may be supposed to communicate with the brain by means of the central grey matter only. But sensation is abolished by piercing the interior of the cord with an instrument, and so moving it as to divide as much as possible the grey matter without injuring the white matter. Moreover, section of the cord and irritation of the cut surfaces produce no pain, provided that the plane of section be sufficiently removed from the origins of nerves, as may be accomplished in the cervical region; but in the neighbourhood of nerve-roots there is great sensibility. From all these circumstances it appears probable that the sensory fibres, viz., those of the posterior roots, pass quickly into the grey substance, and that the grey substance conducts sensory impressions upwards. Moreover, the circumstance that the posterior as well as the anterior surfaces of transverse sections made near the nerve-roots are sensitive seems

---

* The student is reminded that "superior" applied to animals corresponds to "posterior" applied to the human subject.

to be accounted for by the curving of the nerve-roots both toward and away from the brain. By similar experiments it is made probable that motor impressions likewise travel chiefly in the grey matter of the cord.

Section of one lateral half of the cord is followed by loss of sensation in the opposite hind limb, and of motion in the limb of the side operated on : and a prolonged mesial incision produces loss of sensation in both hind limbs, without paralysis of motion. But in the medulla oblongata, above the decussation of the anterior pyramids, section of one side produces loss of both sensation and motion on the opposite side. From these circumstances it appears probable that the sensory fibres, viz., those of the posterior roots, decussate in the commissure of the spinal cord, while the motor fibres, those derived from the anterior roots, cross chiefly at the decussation of the anterior pyramids of the medulla oblongata. (For further details, see Brown-Séquard, "Central Nervous System," 1860; also for a succinct account of the subject and for bibliography, J. Béclard, "Physiologie Humaine," 4th ed., 1862; "Carpenter's Human Physiology," 6th edit., 1865.)

B.—THE ENCEPHALON.

The encephalon admits of being conveniently divided into the **medulla oblongata,** the cerebellum with the pons Varolii, and the cerebrum.

Fig. 348.

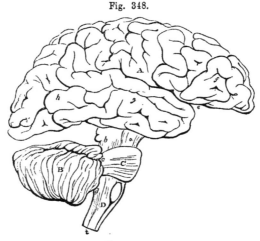

Fig. 348.—Plan in outline of the Encephalon, as seen from the right side. ⅓

The parts are represented as separated from one another somewhat more than natural so as to show their connections. A, cerebrum ; *f, g, h,* its anterior middle and posterior lobes; *e,* fissure of Sylvius ; B, cerebellum ; C, pons Varolii ; D, medulla oblongata ; *a,* peduncles of the cerebrum ; *b, c, d,* superior middle, and inferior peduncles of the cerebellum ; the parts marked *a, b, c,* C, form the isthmus encephali.

The *medulla oblongata* is the part continuous with the spinal cord : it rests on the basilar process of the occipital bone, and on its superior or dorsal surface presents a groove continuous with the central canal of the spinal cord.

The *cerebellum* occupies the posterior fossa of the cranium. By the mesial part of its anterior and inferior surface, it forms the roof of a space, the floor of which is the grooved posterior surface of the medulla oblongata, and which is named the fourth ventricle of the brain. On each side of this, the cerebellum is connected with the medulla oblongata and cerebrum, and also receives the fibres of the *pons Varolii*, which is a commissure passing beneath and between the fibres which extend upwards from the medulla oblongata, so as to unite the two lobes of the cerebellum.

The *cerebrum* includes all the remaining and much the largest part of the encephalon. It is united with the parts below by a comparatively narrow and constricted portion or isthmus, part of which, forming the crura cerebri, descends into the pons Varolii, and through it is continued into the medulla oblongata, whilst another part joins the cerebellum. Situated on the fibres which extend up from the constricted part, are a series of eminences, named, from behind forwards, the corpora quadrigemina, optic thalami, and corpora striata ; and springing from the front and outer side of the corpora striata are the large convoluted cerebral hemispheres, which expand from this place in all directions, concealing the eminences named, and occupying the vault of the cranium, the anterior and middle cranial fossæ, and the superior fossæ of the occipital bone. The cerebral hemispheres are united together by commissures ; by means of which there is enclosed a cavity, which is subdivided into various ventricles, viz., the two lateral, the third, and the fifth.

THE MEDULLA OBLONGATA.

The *medulla oblongata* is bounded above by the lower border of the pons Varolii, whilst it is continuous below with the spinal cord, on a level with the upper border of the atlas, at a point which corresponds with the lower extremity of the anterior pyramids, to be presently described. It inclines obliquely downwards and backwards ; its anterior surface rests in the basilar groove, whilst its posterior surface is received into the fossa named the vallecula, between the hemispheres of the cerebellum, and there forms the floor of the fourth ventricle. To its sides several large nerves are attached.

The term medulla oblongata, as employed by Willis, by Vieussens, and by those who directly followed them, included the crura cerebri and pons Varolii, as well as that part between the pons and the foramen magnum, to which, by Haller first, and by most subsequent writers, this term has been restricted.

It is of a pyramidal form, having its broad extremity directed upwards : it is expanded laterally at its upper part : its length from the pons to the lower extremity of the pyramids is about an inch and a quarter ; its greatest breadth is nearly an inch ; and its thickness, from before backwards, is about three-quarters of an inch.

The *anterior* and *posterior mesial fissures* which partially divide the spinal cord are continued up into the medulla oblongata. The anterior fissure terminates immediately below the pons in a recess, the *foramen cæcum* of Vicq d'Azyr ; the posterior fissure is continued upwards into the floor of the fourth ventricle, where it opens and expands in a superficial furrow, and is gradually lost.

In other respects an entirely different arrangement of the parts prevails from that in the cord. The surface of each half of the medulla presents four eminences or columns, which are met with in the following order, from before backwards, viz. : the anterior pyramids, the olivary bodies, the restiform bodies, and the posterior pyramids.

The *anterior pyramids* are two bundles of white substance, placed one on either side of the anterior fissure, and marked off from the olivary body externally by a slight depression. They become broader and more prominent as they ascend towards the pons Varolii. At their upper end they are constricted, and thus enter the substance of the pons, through which their fibres may be traced into the peduncles of the brain.

Fig. 349.—VIEW OF THE ANTERIOR SURFACE OF THE PONS VAROLII AND MEDULLA OBLONGATA.

a, a, anterior pyramids; *b*, their decussation; *c, c*, olivary bodies; *d, d*, restiform bodies; *e*, arciform fibres; *f*, fibres described by Solly as passing from the anterior column of the cord to the cerebellum; *g*, anterior column of the spinal cord; *h*, lateral column; *p*, pons Varolii; *i*, its upper fibres; *5, 5*, roots of the fifth pair of nerves.

Fig. 349.

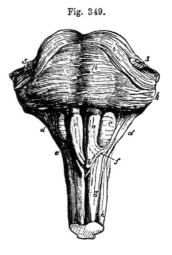

In the lower part, a portion of each pyramid, arranged in several bundles, which interlace with the corresponding bundles of the other pyramid, passes downwards across the fissure to the opposite side. This *decussation of the pyramids* is not complete, but affects much the greater part of the innermost fibres. When traced from below, it is found that the whole or a great part of the decussating fibres come forward from the deep portion of the lateral columns of the cord, and advance to the surface between the diverging anterior columns, which are thus thrown aside. (Rosenthal, "Beitrag zur Encephalotomie," 1815.)

The outer smaller portion of each pyramid does not decussate; it consists of fibres, derived from the anterior column of the cord: these ascend, and are joined by the decussating portion from the opposite side. Together they form a prismatic bundle or column of white fibres, which extends deeply into the substance of the medulla, and is triangular in a cross section.

The anterior pyramids contain no grey matter.

The *olivary bodies* are two prominent oval masses placed to the outer side of the pyramids, and sunk to a considerable depth in the substance of the medulla oblongata, appearing on its surface like two smooth oval eminences. They do not reach the pons Varolii above, being separated from it by a deep depression; nor do they extend so far in a downward direction as the pyramids, being considerably shorter than those bodies.

The olivary bodies consist externally of white substance, of which the fibres chiefly run longitudinally; and internally of a grey nucleus, named *corpus dentatum* or *ciliare*, or *olivary nucleus.*

The *olivary nucleus* appears, on making a section, whether horizontal or vertical, through the middle, to present the form of a zig-zag line of a light yellowish colour, circumscribing a whitish substance within, and interrupted towards the centre of the medulla. It is arranged in the form of a capsule, which is open at its upper and inner part, and has its sides corrugated or plicated, so as to give the indented appearance to a section. This capsule is, moreover, surrounded with white matter externally, and through its open part white fibres pass into or issue from its interior, and connect it with other parts of the brain. The external fibres of the anterior columns of the cord, which at the decussation of the pyramids are thrown outwards, are continued upwards, on the surface of the medulla oblongata, and then pass partly on the outside of and partly beneath the olivary bodies—being joined in their further progress by the fibres issuing from the olivary nucleus. To these fibres the term *olivary fasciculus* has been applied.

The *restiform bodies*, placed behind and to the outer side of the olivary bodies, are two lateral rounded eminences or columns directly continuous with the posterior, and with part of the antero-lateral columns of the cord; they diverge slightly as they ascend, and thus occasion the greater width of the

Fig. 350.

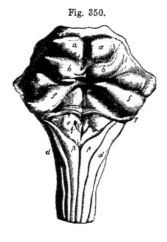

Fig. 350. — VIEW OF THE POSTERIOR SURFACE OF THE PONS VAROLII, CORPORA QUADRIGEMINA, AND MEDULLA OBLONGATA.

The peduncles of the cerebellum are cut short at the side. *a, a*, the upper pair of corpora quadrigemina; *b, b*, the inferior; *f, f*, superior peduncles of the cerebellum; *c*, eminence connected with the nucleus of the hypoglossal nerve; *e*, that of the glosso-pharyngeal nerve; *i*, that of the vagus nerve; *d, d*, restiform bodies; *p, p*, posterior pyramids; *v, v*, groove in the middle of the fourth ventricle; *v*, calamus scriptorius, and eminence connected with the spinal accessory nerve; *7, 7*, roots of the auditory nerves. (See also Fig. 357, at p. 525.)

medulla at its upper part. Each of them passes into the corresponding hemisphere of the cerebellum, and constitutes its inferior peduncle. At first they are in contact with the small tracts of the medulla, named the posterior pyramids; but higher up they become free and prominent, and assist in forming the lateral boundaries of the fourth ventricle. There is a considerable portion of grey matter in their interior.

By far the larger portion of the white substance of the restiform body consists of longitudinal fibres, which include all those belonging to the posterior column of the cord except the fasciculus gracilis, some derived from the lateral column, and also a small band from the anterior column. This last-named band runs obliquely below the olivary body, and, as was shown by Solly, connects the anterior column with the cerebellum.

The part of the posterior column of the cord which belongs to the restiform body of the medulla, is named *fasciculus cuneatus*.

The *posterior pyramids* (fasciculi graciles) of the medulla oblongata, the smallest of the four pairs of columns into which it is divided, are situated one on each side of the posterior median fissure. They consist entirely of white fibres, and are continuous with the posterior slender tracts of the cord. They increase in size as they ascend till they reach the point where the medulla opens out to form the floor of the fourth ventricle; and there, diverging from one another, they have the appearance of tapering and become closely applied to the restiform bodies. Their fibres quit these bodies, however, and pass up to the cerebrum.

The *floor of the fourth ventricle*, or space between the medulla and cerebellum, is formed by that portion of the back of the medulla oblongata which is situated above the divergence of the posterior pyramids. Upon it, the central grey matter of the medulla oblongata, is, as it were, opened out to view. It is marked by a median furrow, ending inferiorly in the *calamus scriptorius*, and at its lower end is a tubular recess, passing down the centre of the medulla for a few lines. This, which has been sometimes named the *ventricle of Arantius*, is the upper expanded portion of the central canal of the spinal cord.

In the upper part of the floor of the fourth ventricle are two longitudinal eminences, one on each side of the middle furrow, greyish below, but appearing white higher up. These are formed by two bundles of white fibres, mixed with much grey matter, the *fasciculi teretes* of some authors, *les faisceaux innominés* of Cruveilhier. They seem to be derived from part of the lateral columns of the cord ; Cruveilhier believes, however, that they arise from the grey matter at the lower end of the medulla oblongata.

Surmounting the free inner margin of the restiform body and posterior pyramid is a thin lamina, the *ligula* (smaller pons of Meckel) occupying the angle between the cerebellum and the restiform body, and stretching towards its fellow of the opposite side. It derives a certain interest from indicating how the cylinder, which is closed in the spinal cord, might be completed in this region of the medulla oblongata by the union of the opposite margins.

Crossing the grey matter in the floor of the fourth ventricle several transverse white lines, or striæ, are usually observed, passing outwards from the median fissure, and round the sides of the restiform bodies. Some of these white striæ form part of the roots of the auditory nerves, a few run slantingly upwards and outwards on the floor of the ventricle, whilst others again embrace the corresponding half of the medulla oblongata. These transverse lines are sometimes wanting, in which case the white fibres on which they depend probably exist at some depth below the surface.

Santorini, and subsequently Rolando, described a set of superficial white fibres on the fore part and sides of the medulla oblongata, crossing over it below the olivary bodies, *fibræ vel processus arciformes*. They belong to a system of white fibres which pass transversely or horizontally outwards, and are probably continuous with the septal fibres about to be noticed. Sometimes the greater part of the pyramidal and olivary bodies is covered by a thin stratum of these transverse fibres, which appear to issue from the anterior median fissure ; but, most commonly, these superficial fibres are found only at the lower extremity of the olive, as the arciform fibres already mentioned.

Besides the superficial transverse fibres now referred to, the medulla oblongata presents other horizontal fibres in its interior, some of them disposed in a mesial *raphe* or *septum*, and numerous others proceeding from that septum transversely outwards. Of these last, the majority, passing through the olivary bodies, and in part the pyramids, enter the corpus dentatum and form the whole of its white substance ; and these fibres, then passing radiately through the grey capsule, turn backwards to the fasciculus cuneatus and lateralis, those of them which pierce the anterior wall of the capsule arching round it to reach their destination. Other fibres pass behind the olivary into the restiform bodies, and seem to terminate in the grey substance of the floor of the fourth ventricle. (See Kölliker's Handbuch der Gewebelehre, 1863, p. 316.)

A small band of fibres is represented by Reichert as passing obliquely downwards and backwards from the side of the pons Varolii, descending between the auditory and facial nerves, and crossing over the upper end of the posterior pyramids. He names it the *alà pontis*. It probably is part of the ligula. (Reichert, Bau des Menschl. Gehirns, part 1st, plate I., 1859.)

Course of fibres from the spinal cord upwards through the medulla oblongata.—Assuming, for convenience of description, the existence of three white columns of the cord, these are disposed as follows.

1. The *posterior column*, with the exception of the fasciculus gracilis, is distinguished by the name of processus cuneatus and enters into the formation of the restiform body, which ascends to the cerebellum. The fasciculus gracilis ascends to the cerebrum.

2. The *lateral column* ascends towards the base of the olivary body, and is disposed of in three ways ; (1,) some of its fibres from the surface and deep part join the restiform body and proceed with it to the cerebellum ; (2,) a larger number, passing obliquely inwards, then come forwards between the anterior columns, and crossing the median plane appear as the fibres of decussation, and form the chief part of the opposite anterior pyramid ; (3,) the remaining fibres pass up to the cerebrum, as the fasciculi teretes

Fig. 351.

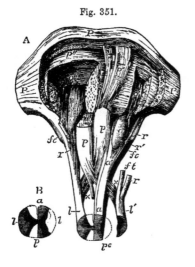

Fig. 351.—Diagrammatic Representation of the Passage of the Columns of the Medulla Oblongata upwards and downwards.

A, the specimen, which is seen from before, includes the medulla oblongata and the pons Varolii, with a small portion of the spinal marrow. The left lateral column (that to the reader's right) has been lifted out of its place to the side, and the anterior and posterior columns of that side remain undisturbed : the right anterior and posterior columns have been removed, and the lateral column remains in its place. The upper part of the right pyramid is removed. The transverse fibres of the pons Varolii have been divided in circumscribed portions to different depths corresponding with the several places of passage of the columns of the medulla.

P, pons Varolii, part of the anterior surface, where it has been left entire ; *p*, the right and left pyramids, the upper part of the right has been cut away ; *p'*, the fibres of the left pyramid as they ascend through the pons exposed by the removal of the superficial transverse fibres ; *p"*, placed on the deeper transverse fibres of the pons on the right side, close below the divided fibres of the pyramid ; *a*, left anterior column of the cord, passing upwards into the undecussated part of the anterior pyramid, and into *a'*, the olivary column ; *o*, olivary body ; *o'*, the continuation of the olivary column ascending deeply through the pons, and exposed by the removal of a small portion of the deeper transverse fibres ; *o"*, the same fibres divided by a deeper incision on the right side ; *l*, the right lateral column, passing upwards into the following parts, viz., ×, the deeper part passing by decussation into the left pyramid ; *r*, the part passing into the restiform body ; *ft*, the part ascending in the back of the fourth ventricle as fasciculus teres ; to the outer side of this are seen the ascending fibres of the posterior pyramid ; *l'*, the left lateral column drawn aside from its place in the spinal cord ; the fasciculus teres, *ft*, and the part to the restiform body, *r*, cut short ; ×, the deeper part passing by decussation into the right pyramid ; *r'*, the part of the restiform body derived from the anterior column of the spinal cord ; *pc*, the posterior column of the left side exposed by the removal of the lateral column, and shown ascending to the restiform body as fasciculus cuneatus, *fc* : on the right side the posterior column being removed, *fc*, points to this fasciculus cuneatus cut short below.

B, explanatory outline of the section of the spinal cord. *a*, anterior columns ; *p*, posterior ; *l*, lateral.

(faisceaux innominés), appearing on the back of the pons Varolii, in the upper part of the floor of the fourth ventricle.

3. The *anterior columns* having reached the apex of the anterior pyramids, are thrust aside from their median position by the decussating fibres derived from the lateral columns, and are then distributed in three divisions. (1,) A very small division, ascends obliquely backwards beneath the olive, and joins the restiform body (Solly). (2,) Another division passes directly upwards, its fibres embracing the olivary nucleus, above which they are again collected together, and are joined by other fibres arising from the nucleus, so as to form the *olivary fasciculus ;* this ascends through the pons and at the side of the cerebral peduncle under the name of the *fillet,* and reaches the corpora quadrigemina and the cerebral hemispheres. (3,) The remaining division of the anterior column ascends into the anterior pyramid, forming its outer part. The anterior pyramids therefore are composed of fibres from the lateral and anterior columns, and are continued up through the pons into the peduncles of the cerebrum.

It is to be remembered, however, that the separation between these different tracts of white fibres cannot be clearly followed out through the whole structure of the medulla oblongata, but that they are more or less blended with one another.

Grey matter of the medulla oblongata followed upwards from the cord.—The central canal of the spinal cord, together with the grey matter which surrounds it, approaches nearer and nearer to the back of the medulla oblongata as it ascends, until it terminates in the calamus scriptorius.

The anterior pyramids are free from grey matter in their interior, and are separated from the rest of the medulla by strong septa of connective tissue, and from one another by a *raphe,* which extends back to the grey matter surrounding the central canal, and which contains mesial horizontal fibres, named *septal.* The posterior cornua of grey matter in the lower part of the medulla oblongata extend transversely outwards from the central canal, and higher up stretch outwards and forwards to the surface. The substantia gelatinosa is swollen out into a mass which appears circular in a transverse section, and is named the *grey tubercle* of Rolando. The anterior cornua, together with the intermedio-lateral tract, which had re-appeared at the upper end of the cord, vanish in the form of elongated radiating streaks ; and between them and the anterior pyramids appear the olivary nuclei, unconnected with the system of grey matter prolonged from the spinal cord. Behind the posterior cornua two new cornua make their appearance—one extending into the processus cuneatus and the other into the posterior pyramid, and both of them increasing in size as the posterior pyramids increase. In the neighbourhood from which these and the posterior cornua spring there is seen in transverse sections a limited bundle of white fibres, the *round fascicle* of Stilling. In the upper part of the medulla oblongata the grey matter is principally spread out on the floor of the fourth ventricle. (Reichert, op. cit., part 2nd, plates I. and II.)

According to the observations of Stilling, part of the grey matter at the back of the medulla forms special deposits or nuclei, which are connected with the roots of the spinal accessory vagus, glosso-pharyngeal, and hypoglossal nerves. Of these nuclei, the first or lowest is concealed in the substance of the medulla; whilst those which are situated higher up gradually appear in the floor of the fourth ventricle as small angular eminences pointing downwards, near the apex of the calamus scriptorius. The *first nucleus* proceeding from below is that for the spinal accessory nerve. It reaches some way down in the cord, and is there lost in the intermedio-lateral tract. Above this nucleus, and close to the middle of the medulla, is another, the *second,* commencing higher up, and connected with the hypoglossal nerve, the roots of which, coming forward between the anterior pyramid and the olivary body, appear at the surface in the depression between those parts. Continuing to ascend, these two nuclei reach the back of the medulla, and then make their appearance in the floor of the fourth ventricle. Higher up, the nucleus for the spinal accessory nerve is succeeded by a *third* in the same line, which is connected with the nervus vagus, and is also placed to the outer side of that for the hypoglossus. Further out, a *fourth* nucleus begins to be observed, belonging to the glosso-pharyngeal nerve. The last change in the arrangement of these small grey masses consists in the gradual narrowing of the nucleus of the par vagum, and the approximation of those for the hypoglossal and glosso-pharyngeal nerves which were previously separated by it.

Fig. 352.

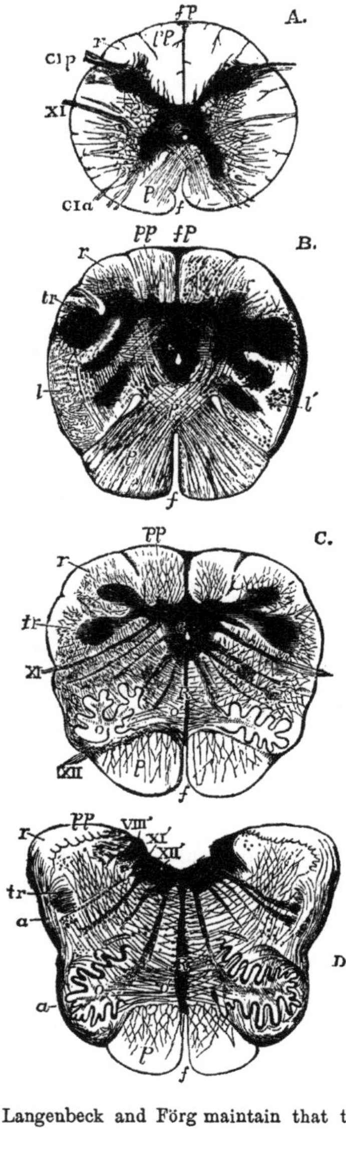

Fig. 352.—Magnified Views of Transverse Sections of the Medulla Oblongata (after Lockhart Clarke, and Reichert). $\frac{2}{1}$

These figures are to be looked upon as in part diagrammatic, no attempt having been made to represent the natural difference of colour in the parts. For the most part, however, the grey substance is indicated by the smoother dark shading, and the white substance by distinct lines.

A, represents a section made at the lower part of the decussation of the pyramids; B, one immediately below the olivary bodies; C, one a very short distance below the calamus scriptorius; and D, a section in the lower part of the fourth ventricle. The references are the following in all the four figures:—

p, anterior pyramids; *p'*, their decussation; *o*, olivary bodies; *o'*, the radiating fibres proceeding from their interior; *r*, restiform bodies and their nucleus; *pp*, posterior pyramids; R, raphe; *c*, central canal and substance surrounding it; *tr*, grey tubercle of Rolando; *f*, anterior median fissure; *fp*, posterior median fissure; *a*, arciform fibres; *l*, lateral column; *l'*, larger cells and vesicular tract of the lateral column; CI *a*, anterior roots of the first cervical nerve; CI *p*, posterior roots; XII, hypoglossal nerve-roots issuing at the side of the pyramid; XII', its nucleus; XI, XI', spinal accessory nerve and its nucleus; VIII', nucleus of the auditory nerve according to Reichert.

In A and B, the decussation of the pyramids is represented; in A, the anterior and posterior cornua of the grey matter still exist as in the spinal cord; in B, the anterior cornua are much diminished in size, the posterior have begun to pass outwards, and to be converted into the grey tubercles, and the intermediate nuclei to make their appearance between them; in C, the central canal is wider and approaches the posterior aspect, and the olivary body appears between the anterior pyramid and the lateral column; in D, the canal is opened up in the fourth ventricle, and the various grey nuclei are for the most part in the vicinity of its floor.

Langenbeck and Förg maintain that the part regarded by Stilling as the nucleus for the glosso-pharyngeal nerve is really the place of origin of the greater root of the fifth or trigeminal nerve.

Fig. 353.—TRANSVERSE SECTION OF THE MEDULLA OBLONGATA (after Stilling). ½

Fig. 353.

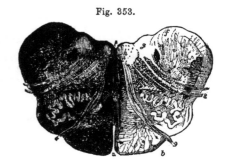

The section is made at the level of the middle of the olivary bodies; the effect produced by transmitted light is brought out on the left-hand side of the figure, the half to the right being only sketched. *a*, anterior, and *p*, posterior fissure; *b*, anterior pyramid; *c*, olivary body with its corpus dentatum shown internally; *d*, grey tubercle of Rolando in the lateral column; *e*, the restiform body and its nucleus; *f*, nucleus of the roots of the glosso-pharyngeal nerve; *g*, nucleus of the pneumo-gastric nerve; *h*, that of the hypoglossal nerve; *i*, the septum or raphe; 8, roots of the pneumo-gastric nerve emerging; 9, roots of the hypoglossal nerve.

THE PONS VAROLII AND CEREBELLUM.

THE PONS VAROLII or *tuber annulare* (mesocephalon of Chaussier, nodus encephali of Rau and Sömmerring), forms an eminence of transverse fibres above and in front of the medulla oblongata, below and behind the crura cerebri, and between the lateral lobes of the cerebellum. Its margins are arched; the superior much more so than the inferior: thus, at the sides its transverse fibres are much more gathered together, and form at the place where it passes into the cerebellum a narrower bundle, which is named the *middle crus of the cerebellum*. In the middle line the pons presents a shallow groove in which the basilar artery lies, and is perforated by small branches of that artery.

Although the superficial fibres are transverse in their general direction, they are not all parallel to each other. The middle fibres pass directly across, the lower set ascend slightly, whilst the superior fibres, which are the most curved, descend obliquely to reach the crura cerebelli on each side; and there are also one or more superficial bands of the superior fibres which cross obliquely downwards over the middle and lower fibres, and completely conceal them at the sides.

In its *internal structure* the pons consists of the longitudinal or peduncular fibres prolonged upwards from the medulla oblongata, of its own transverse or commissural fibres, through which the longitudinal fibres pass, and of a large intermixture of grey matter. Behind the superficial transverse fibres are seen the prolonged fibres of the anterior pyramids, which, as they ascend through the pons, are widely separated into smaller bundles, intersected by other transverse white fibres, which, with those upon the surface, are all continued into the cerebellum.

The alternation of transverse and longitudinal fibres just described extends to a considerable depth in the pons, the quantity of transverse fibres greatly preponderating; posteriorly there succeeds a third layer, consisting entirely of longitudinal fibres, and comprehending the olivary fasciculi, and the fasciculi teretes.

The *median septum* or *raphe*, which exists in the medulla oblongata, is

M M

prolonged throughout the whole height of the pons in its back part, but becomes indistinct in approaching the front or basilar surface, except towards its upper and lower edge, where the superficial fibres of the pons are manifestly continuous in the median line with these septal fibres. Bundles of white fibres, belonging to the same system, encircle the crura cerebri at their emergence from the upper border of the pons.

According to Foville, a few fibres from each of the three principal longitudinal elements of the medulla turn forwards and become continuous with the transverse fibres of the pons ; and, in like manner, one or more small bundles from each of the crura cerebri take a similar transverse course. (Foville, op. cit., pl. II., figs. 2 and 3 ; pl. III., figs. 5 and 6.)

THE CEREBELLUM, *hinder brain*, consists of a *body* and of three pairs of *crura* or *peduncles*, by which it is connected with the rest of the encephalon. These crura are named superior, middle, and inferior.

Fig. 354.

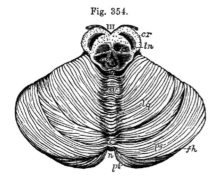

Fig. 354.—OUTLINE OF THE UPPER SURFACE OF THE CEREBELLUM. $\frac{1}{2}$

At the upper part of the figure, the crura cerebri and parts behind them have been cut through and left in connection with the cerebellum.
III, the third pair of nerves lying upon the crura cerebri ; *c r*, white matter or crust of the crura cerebri ; *l n*, locus niger ; *t*, tegmentum containing grey matter in the upper part of the crura ; *a s*, aqueduct of Sylvius ; *q*, corpora quadrigemina, the upper elevations divided ; *s v*, superior vermiform process or central folia of the middle lobe of the cerebellum ; *l q*, lobulus quadratus ; *p s*, posterior superior lobe ; *f h*, horizontal fissure ; *p i*, posterior inferior lobe ; *n*, the notch between the hemispheres.

The *superior peduncles*, crura ad cerebrum or processus ad testes, together with the valve of Vieussens, a lamina stretched between them, connect the cerebellum with the cerebrum.

The *inferior peduncles*, crura ad medullam, are the upper extremities of the restiform bodies.

The *middle peduncles*, or crura ad pontem, much the largest, are the lateral extremities of the transverse fibres of the pons Varolii. They connect together the two halves of the cerebellum inferiorly.

All these peduncles consist of white fibres only ; and they pass into the interior of the cerebellum at its fore part.

The cerebellum is covered with grey cortical substance, rather darker than that of the cerebrum. Its greatest diameter is transverse, and extends to about three and a half or four inches : its width from before backwards is about two or two and a half inches ; and its greatest depth is about two inches, but it is much thinner round its outer border.

It consists of two lateral *hemispheres* joined together by a median portion called the *vermiform process*, which in the human subject is distinguishable only as a small though well-marked part below, named the *inferior* vermiform process, and a mere elevation above, called the *superior* vermiform process. In birds, and in animals lower in the scale, this middle part of the cerebellum alone exists ; and in most mammals it forms a central lobe very distinct from the lateral portions.

The hemispheres are separated behind by a deep *notch*. Superiorly, the median portion or upper vermiform process, though slightly elevated, is not marked off from the hemispheres, so that the general surface of the organ, which is here inclined and flattened on each side, is uninterrupted. Below, the hemispheres are convex, and are separated by a deep fossa, named the *vallecula*, which is continuous with the notch behind, and in which the inferior vermiform process lies concealed in a great measure by the surrounding parts. Into this hollow the medulla oblongata is received in front, and the falx cerebelli behind.

Fig. 355.

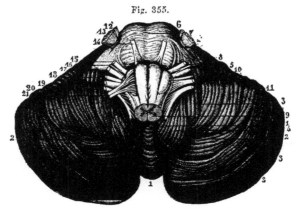

Fig. 355.—Inferior Surface of the Cerebellum with the Pons Varolii and Medulla Oblongata (from Sappey after Hirschfeld and Leveillé.) ⅔

1, placed in the notch between the cerebellar hemispheres, is below the inferior vermiform process ; 2, 2, median depression or vallecula ; 3, 3, 3, the biventral, slender, and posterior inferior lobules of the hemisphere ; 4, the amygdala ; 5, flocculus or subpeduncular lobule ; 6, pons Varolii ; 7, its median groove ; 8, middle peduncle of the cerebellum ; 9, medulla oblongata ; 10, 11, anterior part of the great horizontal fissure ; 12, 13, smaller and greater roots of the fifth pair of nerves; 14, sixth pair ; 15, facial nerve ; 16, pars intermedia ; 17, auditory nerve; 18, glosso-pharyngeal ; 19, pneumogastric ; 20, spinal accessory; 21, hypoglossal nerve.

The cerebellum, at the surface and for some depth, consists of numerous nearly parallel laminæ or folia, which are composed of grey and white matter, and might be compared with the gyri of the cerebrum, but are smaller and without convolution. These laminæ are separated by slightly-curved grooves or sulci of different depths.

One principal fissure, or sulcus, named the *great horizontal fissure*, divides the cerebellum into an upper and a lower portion. It begins in front at the entrance of the middle peduncles, and passes horizontally backwards round the outer border of the hemispheres. From this primary fissure, numerous others proceed on both the upper and under surfaces, forming nearly parallel curves, having their concavities turned forwards, and separating the folia from each other. All these furrows do not go entirely round the hemisphere, for many of them coalesce with one another ; and some of the smaller furrows have even an oblique course between the others. Moreover, on opening the larger fissures, many of the folia are seen to lie concealed within them, and do not reach the surface of the cerebellum.

Certain fissures, which are deeper than the rest, and constant in their position, have been described as separating the cerebellum into lobes, which are named as follows.

The *central lobe*, situated on the upper surface, consists of about eight folia, immediately adjoining the anterior concave border. The *superior and anterior* lobe, sometimes called *quadrate*, and the *superior and posterior* lobe, are placed between the central lobe and the great horizontal fissure. On the under surface are seen successively the *inferior posterior* lobe, the *slender* lobe, the *biventral* lobe, the *amygdala*, and the *subpeduncular lobe* or *flocculus*. This last-named lobule, *lobule of the pneumo-gastric nerve* (Vicq-d'Azyr), *subpeduncular lobe* (Gordon), or *flocculus*, projects behind and below the middle peduncle of the cerebellum. It is connected by a slender pedicle of white fibres to the rest of the hemisphere; but its exposed surface is grey, and is subdivided into five or six small laminæ.

Fig. 356.

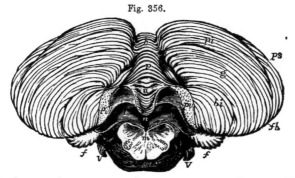

Fig. 356.—Inferior Surface of the Cerebellum with the Posterior Medullary Velum (after Reil and Reichert, and from nature). ⅔

The medulla oblongata has been in great part removed by a cut passing through it near the pons Varolii; the two amygdaloid lobules have also been removed, and the medulla and pons Varolii pulled downwards in order to bring into view the posterior medullary velum.

p s, posterior superior lobe of the cerebellum; *f h*, horizontal fissure; *p i*, posterior inferior lobe; *g*, lobulus gracilis; *b i*, biventral lobe; *c*, placed on the folia which pass across between the hemispheres of opposite sides; *p*, pyramid; *u*, uvula; *n*, placed in the fourth ventricle immediately below the nodule; *p v*, on each side, placed on the cut surface where the amygdalæ have been removed, points by a line to the posterior medullary velum; *v, v*, cavity of the fourth ventricle within the borders of the velum and behind the inferior cerebellar peduncles; the cavity extends on each side into the pedicle of the flocculus, *f*; *m*, section of the medulla oblongata, in which the posterior opening of the olivary capsules of grey matter is shown; VI, sixth nerves; V, roots of the fifth nerves, and above them, the facial and auditory roots.

Within the vallecula, or on its borders, the following parts are seen.

Commencing from behind, a conical and laminated projection named the *pyramid*, is first met with. In front of that is another smaller projection, called the *uvula*, which is placed between the two rounded lobes at the sides of the vallecula, named the *amygdalæ*; these terms having been suggested by a comparison with the parts so named in the throat. Between the uvula and amygdalæ on each side, but concealed from view, is extended a ridge of grey matter indented on the surface, and named the *furrowed band*. Still further forward is the anterior pointed termination of the inferior vermiform process, named the *nodule*, which projects into the fourth ventricle, and has been named the *laminated tubercle* (Malacarne). On each side of the nodule is a thin white lamella of a semilunar form, which is attached by its posterior convex border, and is free and concave in front. The outer ends of these lamellæ are attached to the flocculi, and the inner ends to the nodule, and to each other in front of that projection. The two lamellæ together constitute the *posterior medullary velum*, which has been compared with the valve of Vieussens,—the one being attached to the superior extremity and the other to the inferior extremity of the middle or vermiform portion of the cerebellum. This posterior velum is covered in and concealed by the amygdalæ, and cannot be properly seen until those lobules have been turned aside or removed.

The fourth ventricle.—The space left between the medulla oblongata in front and the cerebellum behind, is named the fourth ventricle, or *ventricle of the cerebellum*.

Fig. 357.—VIEW OF THE FLOOR OF THE FOURTH VENTRICLE WITH THE POSTERIOR SURFACE OF THE MEDULLA OBLONGATA AND NEIGHBOURING PARTS (from Sappey after Hirschfeld and Leveillé).

On the left side the three cerebellar peduncles have been cut short ; on the right side the white substance of the cerebellum has been preserved in connection with the superior and inferior peduncles, while the middle one has been cut short.

1, median groove of the fourth ventricle with the fasciculi teretes, one on each side; 2, the same groove at the place where the white striæ of the acoustic nerve emerge from it to cross the

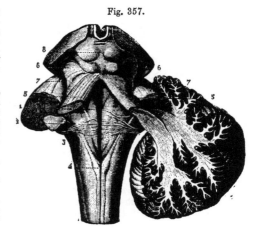

Fig. 357.

floor of the ventricle ; 3, inferior peduncle or restiform body ; 4, posterior pyramid ; above this the calamus scriptorius ; 5, superior peduncle or processus a cerebello ad cerebrum ; on the right side the dissection shows the superior and inferior peduncles crossing each other as they pass into the white stem of the cerebellum ; 6, fillet to the side of the crura cerebri ; 7, lateral grooves of the crura cerebri ; 8, corpora quadrigemina.

The cavity of this ventricle is of a flat rhomboidal shape, being contracted above and below, and widest across its middle part. The anterior extremity of the inferior vermiform process projects into it from behind, and higher up it is covered by the Vieussenian valve. It is bounded laterally by the superior peduncles, and by the line of union of the medulla oblongata and the cerebellum. The upper end of the ventricle is continuous with the Sylvian aqueduct or passage (iter) leading up to the third ventricle.

The anterior surface or *floor* of the fourth ventricle is formed by the back of the medulla oblongata and pons Varolii. It is shaped like a lozenge, truncated at its upper part. Below, it is bounded by the diverging posterior pyramids and restiform bodies surmounted by the ligula. It has already been sufficiently described in connection with the medulla oblongata.

The *lining membrane* of the ventricle is continuous with that of the ventricles in the interior of the cerebrum, through the aqueduct of Sylvius, in which situation it is marked by delicate rugæ, oblique or longitudinal in direction. At the sides it is reflected from the medulla to the cerebellum, and extends for a considerable distance outwards between the flocculus and the seventh and eighth nerves. At the lower end of the ventricle, there is, as was ascertained by Magendie, a narrow orifice in the membrane by which the cavity communicates with the subarachnoid space.

Projecting into the fourth ventricle at each side, and passing from the point of the inferior vermiform process outwards and upwards to the outer border of the restiform bodies, are two small vascular processes, which have been named the *choroid plexuses* of the fourth ventricle.

Fig. 358.

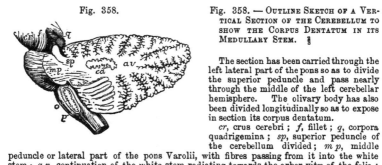

Fig. 358. — OUTLINE SKETCH OF A VERTICAL SECTION OF THE CEREBELLUM TO SHOW THE CORPUS DENTATUM IN ITS MEDULLARY STEM. ⅔

The section has been carried through the left lateral part of the pons so as to divide the superior peduncle and pass nearly through the middle of the left cerebellar hemisphere. The olivary body has also been divided longitudinally so as to expose in section its corpus dentatum.

cr, crus cerebri ; *f*, fillet ; *q*, corpora quadrigemina ; *sp*, superior peduncle of the cerebellum divided ; *m p*, middle peduncle or lateral part of the pons Varolii, with fibres passing from it into the white stem ; *a v*, continuation of the white stem radiating towards the arbor vitæ of the folia ; *c d*, corpus dentatum ; *o*, olivary body with its corpus dentatum ; *p*, anterior pyramid.

Internal structure of the cerebellum.—The central part is composed of white matter, which sends out spreading and gradually thinning layers into the interior of all the laminæ, larger and smaller, of the grey substance which form a continuous covering on the surface. In consequence of this arrangement of the white and grey substances, sections of the cerebellum crossing the laminæ, and dividing the grey and white substance together, present a beautifully foliated or arborescent appearance, named *arbor vitæ*. This appearance is seen in any vertical section, but it is most perfect in that which passes through the median plane, where the relative quantity of the central white matter is small. The foliations are arranged somewhat pinnately, the section of each primary lamina having those of secondary laminæ clustered round it like leaflets on a stalk.

In the lateral hemispheres, where the peduncles enter, the white matter is more abundant ; and, if a section be made through either hemisphere half way between its centre and the middle of the vermiform process, it will display a nucleus of grey matter, which is named the *corpus dentatum* of the cerebellum. This structure, very similar to that already described in the olivary body of the medulla oblongata, presents the appearance of a waved line of compact yellowish brown matter, surrounded by white substance and containing whitish matter within. This line is interrupted at its upper and inner part. In whatever direction the section is carried through the corpus dentatum, this waved line is seen, so that the dentate body may be described as consisting of a plicated pouch or capsule of grey substance open at one part and inclosing white matter in its interior, like the corpus dentatum of the olivary body. White fibres may be traced from it to the superior peduncles of the cerebellum and to the valve of Vieussens.

Fig. 359.—VIEW OF A DISSECTION OF THE FIBRES OF THE MEDULLA OBLONGATA AND PONS VAROLII (from Arnold). $\frac{3}{2}$

Fig. 359.

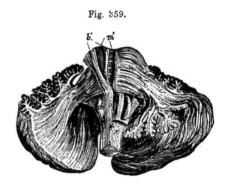

b, the anterior pyramid ; b', its fibres traced upwards through the pons Varolii ; c, olivary column ; d, olivary body ; m, superficial transverse fibres of the pons on its left side ; m', the deeper transverse fibres of the right side; m'', the prolongation of these fibres as middle peduncle of the cerebellum ; p, q, their continuation into the laminæ and folia of the cerebellum ; n, inferior peduncle ; x, the decussating part of the left lateral column crossing to the right anterior pyramid.

The fibres in the primary lamellæ can be traced continuously from the peduncles of the cerebellum. Upon these central plates are laid other *collateral lamellæ*, which are not connected with the fibres proceeding from the middle of the cerebellum, but merely pass from one folium to another.

The *grey* matter is not uniform throughout its whole thickness, but is composed of two or more layers differing in colour and other characters ;—resembling, in this respect, the cortical substance of the posterior convolutions of the cerebrum.

The fibres composing the peduncles of the cerebellum are arranged in its interior in the following manner. The middle peduncles, which are the most superficial, enter the lateral parts of the cerebellum ; they may be traced into the folia of those parts, and form a large share of each hemisphere. The inferior peduncles pass upwards into the middle part of the cerebellum, in the folia of which they are distributed, especially in those of the upper surface. The superior peduncles, which are placed nearest to the middle line, are principally connected with the folia of the inferior vermiform process ; but a considerable number of them pass into or issue from the grey capsule of the corpus dentatum which has been already described.

A very different account from that which has generally been received of the course and relations of the tracts of nervous substance of the cerebellum has recently been put forward by Luys, and deserves mention in this place. According to the statement of this author, all the fibres of the cerebellar peduncles arise from the interior of the corpora dentata ; the cells of those centres receive externally fibres from the laminated periphery of the cerebellum, and internally give origin to the peduncular fibres ; the fibres of the inferior peduncles of opposite sides cross the middle line and terminate in the interior of the olivary nuclei ; and the fibres of the superior peduncles, likewise decussating in the mesial plane before quitting the cerebellum, terminate in a grey centre in the interior of the tegmentum of the crus cerebri, named by Luys the superior olivary body. He further alleges that different fibres pass in all directions from the superior and inferior olivary bodies, and that thus the fibres of the cerebellum form a separate system indirectly connected with the fibres of the rest of the cerebro-spinal axis. Only a short notice, however, of these views having as yet been published, it will be necessary that the observations on which they are founded be made known and fully corroborated, before statements of so startling a nature can be generally accepted. (Luys, in Journ. de l'Anat. et de Physiol., 1864, p. 225.)

Microscopic Structure.—The cortical grey matter which covers the foliated surface of the cerebellum. is made up of the following elements, viz.: 1. Pellucid cells of considerable size. 2. Cells, for the most part of large size, and caudate, having the usual granular contents. These cells are embedded in a finely granular matrix; the greater number of those of the caudate kind have a pyriform shape, and are prolonged

Fig. 360.

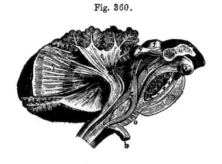

Fig. 360. — THE COLUMNS OF THE MEDULLA OBLONGATA TRACED UPWARDS INTO THE CEREBELLUM AND CEREBRUM (from Arnold). ⅔

a, part of the anterior column which ascends in the olivary column; *b*, decussating portion of the lateral column forming the pyramid and turned down; *c*, olivary fasciculus ascending deeply through the pons; *d*, olivary body; *e*, restiform body; *f, g,* corpora quadrigemina; *c, h, i,* the fillet; *h*, the part which ascends to the cerebral peduncle; *i,* the part passing up to the corpora quadrigemina; *m, m'*, the transverse fibres of the pons divided; *n*, inferior peduncle of the cerebellum; *o*, septal fibres of the medulla oblongata; *q*, fibres of the inferior peduncle continued into the laminæ of the cerebellum; *r, r,* superior peduncle; *t*, fasciculus teres; *u*, thalamus; *v*, corpus albicans.

Fig. 361.

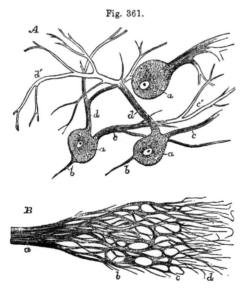

Fig. 361. — MINUTE STRUCTURE OF THE SUBSTANCE OF THE CEREBELLUM (from Kölliker).

A, large cells from the grey cortical substance of the human cerebellum. $\frac{220}{1}$

a, three large cells exhibiting granular contents and a nucleus; *b*, the internal processes seen in two of the cells; *c, d,* two external processes running towards the surface from two of the cells, in the third cell one large process only is seen; *c', d',* ramified finer parts of these processes.

B, course of the nerve-tubes at the surface of the cerebellum, magnified with a low power.

a, nerve of the medullary substance; *b*, nervus plexus of the substantia ferruginea; *c*, border of that substance; *d*, fine fibres running out from the dark-bordered tubes into the superficial grey substance.

at their small end into a simple or branched appendage, and this process, as first remarked by Purkinje, is in most of them directed towards the surface of the cerebellum. 3. Small bodies, like cell-nuclei densely aggregated without any intervening substance. These lie at some depth from the surface; according to Todd, they form a thin light-coloured lamina, intermediate between two darker strata of grey matter which contain the nerve-cells; one of these grey strata being the deepest and next the white matter of the cerebellum, while the other, which is the darker coloured of the two, is in contact with the pia mater. 4. Fibres. Tubular nerve-fibres pass from the white into the grey matter, and extend through it nearly as far as the surface. The mode of their termination, which is difficult to trace, has been investigated by various anatomists. According to Valentin, they form loops and return upon their course, but this statement has not been confirmed by other observers.

Gerlach has recently described a very remarkable arrangement of the fibres of the cerebellum. According to him, these fibres, on approaching the grey matter, split up into extremely fine divisions, which form a network, while the granules, which he considers as small cells with ramifying processes, are placed at the angles of the meshes, and branching processes of the large nerve-cells also terminate in the network. According to Kölliker, networks of tubular fibres exist within the grey matter and communicate with the nerve-cells, while the granules belong to the reticulum of connective tissue. Luys, like Gerlach, describes lateral processes as being given off by the nerve-fibres to connect them with the granules, at the same time that they terminate likewise directly, although much attenuated, in the large nerve-cells. (Gerlach, "Microscopische Studien," pl. I., fig. 3; copied in Virchow's "Cellular Pathology," by Chance, p. 269.)

THE CEREBRUM.

The *cerebrum*, or brain proper, constitutes the highest and much the largest portion of the encephalon. It consists of the following parts, viz., the peduncular masses of the crura cerebri and processus a cerebello ad cerebrum; the series of eminences or cerebral centres or ganglia concealed from view, named corpora quadrigemina, optic thalami and corpora striata; the cerebral hemispheres, which are by far the most bulky part of the cerebrum and of the whole encephalon, and form nearly the whole superficial part; various commissural structures including the corpus callosum and fornix; and lastly some smaller structures, viz., the pineal and the pituitary bodies, and the olfactory bulbs.

EXTERIOR OF THE CEREBRUM.

The *cerebral hemispheres* together form an ovoid mass, flattened on its under side, and placed in the cranium with its smaller end forwards, its greatest width being opposite to the parietal eminences. They are separated in the greater part of their extent by the great longitudinal fissure.

Each cerebral hemisphere has an outer or convex surface, which is in contact with the vault of the cranium; an inner or flat surface, of a crescent shape, which forms one side of the longitudinal fissure; and an irregular under surface, which rests on the base of the skull, and on the tentorium cerebelli.

Three lobes, or large divisions, projecting in three different directions, have usually been distinguished in each hemisphere, under the names of anterior, middle, and posterior lobes. The division between the anterior and middle lobes is very clearly defined below and on the sides by a deep cleft, named the Sylvian fissure. There is no similar demarcation between the middle and posterior lobes; but anatomists have generally considered as the posterior lobe that part of the hemisphere which lies over the cerebellum. The under surface of the anterior lobe is triangular and excavated to adapt it to the roof of the orbit on which it rests. The middle lobe is rounded and prominent, and occupies the middle fossa of the skull—the edge of the lesser wing of the sphenoid bone corresponding with the Sylvian fissure.

Fig. 362.

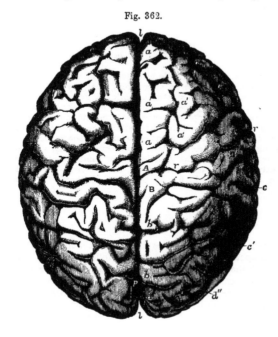

Fig. 362.—UPPER SURFACE OF THE BRAIN SHOWING THE CONVOLUTIONS (from R. Wagner). ½

This view was taken from the brain of a famous mathematician, Professor C. F. Gauss, who died in 1854, aged 78. It is selected as an example of a well-formed brain of the usual size with fully developed convolutions.

a, superior or first frontal convolution; *a′*, second or middle frontal; *a″*, third or inferior frontal; A, A, anterior ascending parietal convolution; B, B, posterior ascending parietal convolution; *b*, first or upper parietal convolution; *b′*, second or middle; *b″*, third or inferior; *c*, first or upper temporal convolution; *d*, first or upper occipital convolution; *d′*, second or middle; *d″*, third or lower; *l*, *l*, the superior longitudinal fissure; *r*, the fissure of Rolando; *p*, the external perpendicular fissure.

The posterior lobe is smooth and slightly concave on its under surface, where it rests on the arch of the tentorium.

It is right to remark that some anatomical writers have admitted only *two* lobes, reckoning the middle and posterior lobes as one, under the name of the posterior lobe; while others more recently have divided the middle lobe into two, an upper and lower, and have added that of the island of Reil, so as to make five principal lobes in all. These have been named respectively the frontal, parietal, temporal, occipital, and central lobes.

The *great longitudinal fissure*, seen upon the upper surface of the brain, extends from before backwards throughout its whole length in the median plane, and thus separates the cerebrum, as already stated, into a right and left hemisphere. On opening this fissure, it is seen, both before and behind, to pass quite through to the base of the cerebrum: but in the middle it is interrupted by a large transverse mass of white substance, named the *corpus callosum*, which connects the two hemispheres together. While the brain is in its natural situation, this fissure is occupied by a vertical process of the dura mater—the falx cerebri—which dips down between the two hemispheres, not quite reaching to the corpus callosum.

The *Sylvian fissure*, which separates the anterior and middle lobes, passes at first upwards and backwards in the outer part of the hemisphere, and divides into two branches, anterior and posterior. It lodges the trunk and primary divisions of the middle cerebral artery, and at its commencement presents a spot pierced by numerous small arterial branches, and thence named the *locus perforatus anticus.*

The surface of the hemispheres is composed of grey matter, and is moulded into numerous smooth and tortuous eminences, named *convolutions*, or *gyri*, which are marked off from each other by deep furrows, called *sulci*, or *anfractuosities.*

Fig. 363.

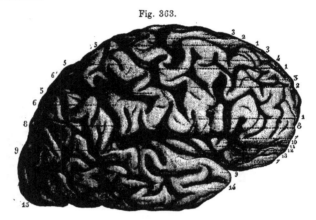

Fig. 363.—LATERAL VIEW OF THE RIGHT CEREBRAL HEMISPHERE (from Sappey after Foville). ½

1, fissure of Rolando ; 2, anterior ascending parietal convolution ; 3, frontal convolutions connected posteriorly with the anterior ascending parietal ; 4, union of two frontal convolutions ; 5, posterior ascending parietal convolution ; 6, another parietal convolution similarly connected with those on the inner surface ; 7, 7, anterior part of the convolution of the fissure of Sylvius ; 8, 8, horizontal part of the same convolution ; 9, 9, posterior part ; 10, 11, 12, anterior, middle, and posterior principal convolutions of the island of Reil or central lobe ; 13, supraorbital convolution ; 14, part of the temporal lobe ; 15, occipital lobe.

CEREBRAL CONVOLUTIONS.—The convolutions are covered closely through-out by the vascular investing membrane, the pia mater, which sends processes down to the bottom of the sulci between them, while the serous covering, the arachnoid membrane, passes from one convolution to another, over their summits and without dipping between them. The sulci are generally from half an inch to an inch in depth ; but in this respect there is much variety in different brains, and in different parts of the same brain ; those upon the outer convex surface of the hemisphere being the deepest. In general, the depth of a convolution exceeds its thickness ; and its thickness, near the summit, is somewhat greater than through its base.

Since the external grey or cortical substance is continuous over the whole surface of the cerebral hemispheres, being found alike within the sulci and upon the gyri, a far greater extent of grey matter is thus exposed to the vascular surface of the pia mater with a given size of the brain, than could have been the case had the hemispheres been plain and destitute of convolutions.

The general arrangement of the convolutions has been made the subject

of study by various anatomists in earlier and recent times, but still requires farther elucidation. An attempt to describe minutely all the individual gyri would be difficult and useless, owing to their irregularity in
different cases, and their want of symmetry in the same brain. Nevertheless,
there are some sufficiently constant in presence, and characteristic in situation and form, to admit of being specially described ; and it seems probable
that, by a sufficiently careful comparison of the convolutions in different
animals, and the observation of their development in the fœtus, certain
general facts may be ascertained regarding them, tending to throw light
upon their disposition in man.

Fig. 364.

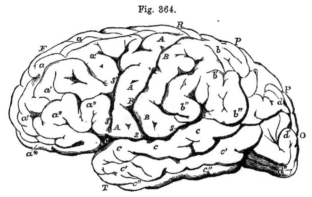

Fig. 364.—OUTLINE OF THE CEREBRUM AS SEEN FROM THE LEFT SIDE, SHOWING THE
CONVOLUTIONS AS DISTINGUISHED BY GRATIOLET. ½

F, frontal lobe ; P, parietal lobe ; T, temporal lobe ; O, occipital lobe ; R, R, fissure
of Rolando ; s, s. fissure of Sylvius, posterior division : s′, s′, its anterior division ; C, at
the junction of the two, marks the place of the central lobe or convolutions of the island
of Reil ; p, the place of the vertical or occipital fissure ; a, a′, a″, superior, middle and
inferior frontal convolutions ; a*, supraorbital convolutions ; A, anterior transverse or
ascending parietal convolution ; B, posterior transverse or ascending parietal convolution ;
b, b′, b″, upper, middle and lower parietal convolutions ; c, c′, c″, upper, middle and
lower temporal convolutions ; d, d′, d″, upper, middle and lower occipital convolutions ;
between b, b′, b″, and d, d′, d″, are seen the connecting convolutions ; between c and c′,
the parallel fissure.

The *island of Reil* constitutes the set of convolutions (*gyri operti*) which
appear earliest both in the fœtus and in the animal series. It is a
triangular eminence, broken externally into short radiating convolutions,
which forms a delta between the anterior and posterior division of the
fissure of Sylvius, and is limited externally by a deep sulcus. This mass,
constituting the central lobe of recent authors, derives additional interest
from being the centre round which the principal convolutions of the cerebrum are arranged. It is only brought into view by laying open the fissure
of Sylvius. (See Figs. 368 and 377.)

The *convolution of the Sylvian fissure* is a very large convolution, which is
also early in its appearance in animals. Commencing in front of the inner
end of the Sylvian fissure, it takes a tortuous and much folded course all
round that fissure, giving off numerous secondary gyri, and terminates behind
the fissure opposite the point where it began.

The *gyrus fornicatus, convolution of the corpus callosum,* or *internal con-*

volution, is one of the most distinct and symmetrical convolutions in the whole brain. Commencing on the under surface of the brain, immediately before the anterior perforated space, it ascends a short distance in front of the anterior recurved extremity of the corpus callosum, and then runs backwards immediately above that body, as far as its posterior extremity : there it turns downwards and forwards, embracing the cerebral peduncle, to reach the entrance of the Sylvian fissure. This long convolution, therefore, describes a sort of arch or ring, open or interrupted opposite the Sylvian fissure, and embracing the corpus callosum above, and the cerebral peduncle below. It thus, as was pointed out by Foville, forms a sort of rim or border to the grey matter ; whence it is named by him *convolution d'ourlet.* The surface of this convolution, especially towards its inferior termination, is covered by a very thin cribriform layer of white substance, which, with the grey matter beneath, gives the surface a mottled aspect. This has been called the *reticulated white substance.*

The *marginal convolution of the longitudinal fissure* is a large convolution which may be traced, more or less indented or interrupted however in its course, along the line of junction between the convex and flat surfaces of the hemisphere, forming the lip of the great longitudinal fissure. It commences on the under surface of the brain, in common with the gyrus fornicatus, and, passing forwards, forms the inner border of the triangular orbital surface of the anterior lobe. In this part it is cleft longitudinally by a deep sulcus, into which the olfactory bulb is received, and which, it may be mentioned, is developed at an earlier period than the convolution itself. On the front and upper surface of the cerebrum, this convolution may generally be traced for some distance along the margin of the longitudinal fissure, but soon becomes marked by deep sulci ; and, thus interrupted, may be followed round the posterior extremity, and along the under surface of the hemisphere forwards as far as the point of the middle lobe, running parallel for some space with the under portion of the gyrus fornicatus. Two of the sulci which interrupt the marginal convolution are very constant, viz., the vertical fissure with the fissure of the hippocampi, and the fissure of Rolando.

The *fissure of the hippocampi* has a deep origin in the inner margin of the middle lobe of the brain between the fascia dentata and the gyrus fornicatus, and, passing backwards, crosses that gyrus on the under surface of the brain, behind the corpus callosum, and proceeds in a nearly horizontal course along the inner face of the hemisphere. This fissure is important as forming the reverse of the elevations of the hippocampi in the interior of the brain, and as being (according to Reichert) produced at an early period in connection with the general development of the hemispheres, and being comparable therefore rather to the fissure of Sylvius than to a mere sulcus. The part of the gyrus fornicatus beneath this fissure is distinguished as the *gyrus hippocampi.*

The *fissure of Rolando,* starting from behind the vertex, runs outwards and forwards from the longitudinal fissure, so that the right and left grooves form a V-shaped line open in front. It derives its importance from being characteristic of the form of the brain of man and the quadrumana, and separating two considerable convolutions, which extend from the superior longitudinal fissure to the fissure of Sylvius. These convolutions, peculiar to the greater number of simiæ and attaining their fullest development in man, constitute the anterior and posterior *transverse* or anterior and posterior *ascending parietal* convolutions.

The *vertical fissure* of recent authors crosses the marginal convolution in the posterior part of the cerebrum, extending slightly outwards upon its upper surface and more deeply on its internal aspect, so as to form a separation between the so-called parietal and occipital lobes.

According to Foville the convolutions may be arranged in four principal orders, founded in a great measure on their relative connections with the anterior perforated space, which, in his estimation, is a part of the highest importance.

The *first* order issues from the perforated space, and consists of two portions. One, large and vertical, is the gyrus fornicatus, without its ascending secondary gyri ; the other, short and horizontal, is the slightly elevated ridge which bounds the perforated space in front and on the outer side.   -

Fig. 365.

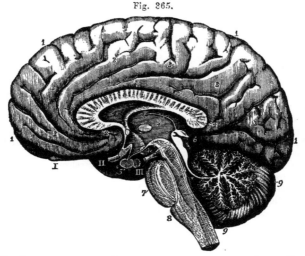

Fig. 365.—RIGHT HALF OF THE BRAIN DIVIDED BY A VERTICAL ANTERO-POSTERIOR
SECTION (from various sources and from nature). ½

1, great superior or marginal convolution ; 2, convolution of the corpus callosum ; 3, secondary convolutions running between this and the preceding ; within the numbers 2, 2, 2, the corpus callosum ; 4, the fifth ventricle ; 5, the third ventricle (see Fig. 377 for a larger view of these parts) ; 5', pituitary body; 6, immediately behind the corpora quadrigemina and pineal gland ; +, the fourth ventricle; 7, pons Varolii ; 8, medulla oblongata ; 9, cerebellum ; the middle lobe showing the section of the arbor vitæ ; I, the olfactory bulb ; II, the right optic nerve ; the commissure cut through ; III, the right nerve of the third pair.

The *second* order, also consisting of two portions, commences from the horizontal portion of the first order on the limits of the perforated space. One part corresponds with the marginal convolution of the longitudinal fissure, as already described, except that part of it on the orbital surface of the anterior lobe which lies to the outer side of the olfactory sulcus; the other part is the convolution of the Sylvian fissure.

The *third* order consists of two sets, of which one occupies the inner surface of the hemisphere, and connects the gyrus fornicatus in its whole length with the marginal convolution of the longitudinal fissure ; the other set lies in the Sylvian fissure, forms the island of Reil, and connects the short horizontal portion of the first order with the convolution surrounding that fissure.

The convolutions of the *fourth* order, the largest, deepest, and least symmetrical of all, are quite detached from the perforated space, and have no relation to the first order of convolutions. They connect the two convolutions of the second order together, viz., the marginal convolution of the median fissure and that of the Sylvian fissure, and occupy the outer or convex surface of the cerebral hemisphere.

Leuret, by an extended comparison of the brains of different animals, was led to divide mammals into fourteen groups, according to the disposition of the convolutions.

In the lowest or simplest group, including the bat, mole, and rat, the *Sylvian fissure* is the only division of the surface present, or along with it a few very slight sulci. In a higher group, containing the fox and dog, and presenting in a marked form the typical mode of division, Leuret recognises as *fundamental* six convolutions—four external, including the superior marginal and that of the fissure of Sylvius, and two internal, viz., the supraorbital and gyrus fornicatus. In other groups, together with various other modifications of form by subdivision or by union through supplemental ones, the number of the fundamental convolutions is frequently reduced to five or to four.

In the brain of the elephant, on the other hand, placed by Leuret in the thirteenth group, he recognises the *superior transverse* convolutions; and in the last group, comprehending the quadrumana, these transverse convolutions are two in number, and are separated by the groove, named by Leuret fissure of Rolando. These transverse or ascending parietal convolutions are a constant and well-marked feature of the human brain, in which they attain their highest development.

Fig. 366.

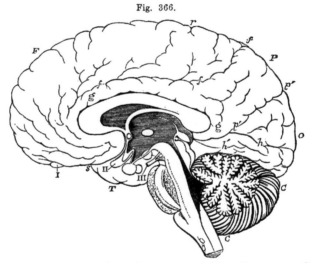

Fig. 366.—Outline of the Inner Surface of the Right Half of the Brain, showing the Principal Lobes and Convolutions according to Gratiolet.

F, frontal lobe; P, parietal; O, occipital; T, temporal; *r*, fissure of Rolando; *f*, fronto-parietal fissure; *p'*, inner perpendicular or occipito-parietal fissure; *h*, the calcarine fissure of Huxley, and with the line continued forwards between *g* and *h'*, the hippocampal fissure; *h'*, convolution of the hippocampus; *g*, gyrus fornicatus or convolution of the corpus callosum; *s*, Sylvian fissure; I, olfactory bulb; II, optic nerve; III, third nerve; C, cerebellum.

More recently, Gratiolet has arranged the convolutions with great detail, according to their most distinguishing common features in man and the simiæ. On the external surface of the hemisphere he distinguishes five lobes, viz., the *frontal* and *parietal* above the fissure of Sylvius; the *temporo-sphenoidal* below that fissure; the *occipital* behind it, and the island of Reil, or *central lobe*, within the fissure. The frontal lobe he divides into an orbital and a frontal portion, and in the frontal portion he distinguishes a *superior, middle,* and *inferior tier* of convolutions. In the parietal lobe are the *anterior* and *posterior ascending convolutions* (convolution of Rolando) surrounding the fissure of Rolando, and behind these a *curved lobe.* In the temporosphenoidal lobe are described a *superior, middle,* and *inferior convolutions,* lying parallel to the fissure of Sylvius. The occipital lobe presents also *three tiers,* but less distinct than those of the frontal, and besides these are four convolutions uniting the occipital and parietal lobes, named by Gratiolet *plis de passage,* or the connecting convolutions.

The internal surface of the hemisphere Gratiolet divides into the *fronto-parietal lobe,* corresponding in extent to the frontal and parietal lobes of the external surface, and limited behind by the *internal perpendicular fissure,* the *occipital lobe* between that fissure and the fissure of the hippocampi; and the *occipito-temporal* lobe, including the tentorial surface, and extending outwards to the sphenoido-temporal lobe.

It is to be remarked, however, that the divisions and nomenclature of Gratiolet, however useful they may be for the purpose of explicit comparison of the convolutions of the human brain with those of the quadrumana, the study in which the inventor has made use of them, are yet of a somewhat artificial description, and may not be applicable to a more extended comparison of the disposition of the convolutions among animals.

From Reichert's plates it is apparent that the internal perpendicular fissure (occipito-parietal of Huxley) is the upper of two branches into which the fissure of the hippocampi divides posteriorly in its first development, and which together with that fissure constitutes his *fissura occipitalis.* The inferior branch, the posterior part of the fissure of the hippocampi, is the *calcarine fissure* of Huxley.

Not only the comparison of the brain of man with those of other animals, but likewise the comparison of human brains one with another, establishes the existence of a relation between mental development and the complication, size, and depth of the cerebral convolutions, and the extent of the grey matter contained in them.

On the subject of the cerebral convolutions the reader may consult, in addition to the works of Arnold, Tiedemann, Foville and Reichert, that of Leuret and Gratiolet "Anat. Comp. du Système Nerveux, 1839-57;" Gratiolet, "Mém. sur les Plis Cérébraux de l'Homme et des Primates, 1854 ;" R. Wagner, "Uber die typischen Verschiedenh. der Windungen der Hemisphären," &c., Gotting. 1860-62 ; Huschke, "Schädel, Hirn und Seele," 1854; Huxley, "Brain of Ateles paniscus," Proc. of Zool. Soc., June, 1861 ; J. Marshall, "On the Brain of a Bush-woman, and on the Brains of Two Idiots, &c.," Trans. Roy. Soc. 1863.

BASE OF THE CEREBRUM.—When the brain is turned with its base uppermost, and the parts of which it is composed are allowed to fall slightly asunder by their own weight, two considerable masses, consisting of white substance externally, are seen emerging together from the fore part of the pons Varolii, and, separating from each other as they proceed forwards and outwards, to enter the inner and under part of the right and left cerebral hemispheres. These white masses, which are marked on the surface with longitudinal striæ, and have somewhat the appearance of large bundles of fibres, are the *peduncles* or crura of the cerebrum. Immediately before entering the corresponding hemisphere, each is crossed by a flattened white cord, named the *optic tract,* which, adhering by its upper border to the peduncle, is directed forwards and inwards, and meets in front with its fellow of the opposite side to form the optic commissure, from the fore part of which the optic nerves proceed.

Limited behind by these diverging peduncles, and in front by the converging optic tracts, is a lozenge-shaped interval, called the *interpeduncular space,* in which are found, in series from behind forwards, the posterior perforated space, the corpora albicantia, and the tuber cinereum, from which is prolonged the infundibulum attached to the pituitary body.

The *posterior perforated space* (locus perforatus posterior) is a deep fossa situated between the peduncles, the bottom of which is composed of greyish matter, connecting the diverging crura together, and named pons Tarini. It is perforated by numerous small openings for the passage of blood-vessels ; and some horizontal white striæ usually pass out of the grey matter and turn round the peduncles immediately above the pons.

Fig. 367.

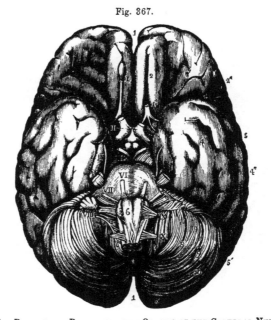

Fig. 367.—BASE OF THE BRAIN WITH THE ORIGINS OF THE CEREBRAL NERVES. ½

This figure is taken from an adult male brain which had been hardened in alcohol.

1, superior longitudinal fissure ; 2, fissure of the olfactory tract and lower part of the superior convolution ; 2', orbital convolutions; 2", external or inferior frontal convolution ; 3, inner part of the fissure of Sylvius, near the anterior perforated spot ; 3, 3, outer part ; 4, inner convolution of the temporal lobe ; 4', middle convolution ; 4", outer convolution ; 5, 5', occipital lobe ; 6, on the right pyramidal body of the medulla oblongata above the decussation ; 7, amygdaloid lobe of the cerebellum ; 8, biventral lobe ; 9, lobulus gracilis ; 10, posterior inferior lobe ; +, the inferior vermiform process ; I, olfactory bulb ; I', the tract divided on the left side, showing the three white striæ by which it is connected with the brain ; II, in the anterior perforated spot, marks the right optic nerve ; the left has been cut short; III, on the right crus cerebri, denotes the third pair ; IV, on the inner convolution of the middle lobe, the fourth pair ; V, the trigeminus; VI, on the pons Varolii, the sixth ; VII, also on the pons Varolii, the seventh ; VIII, on the left lobe of the cerebellum below the horizontal fissure and the flocculus, denotes the eighth pair; IX, on the upper part of the right amygdaloid lobe, denotes the ninth pair; X, on the same, the suboccipital nerve.

The *corpora albicantia* or mammillaria are two round white eminences in front of this fossa, each about the size of a small pea, surrounded by grey matter, and connected together across the middle line.

The corpora albicantia are formed, as will hereafter be explained, by the anterior extremities of the fornix; hence they have also been named *bulbs of the fornix*. In

N N

the fœtus they are at first blended together, and they become separated about the beginning of the seventh month. In most vertebrate animals there is but one white eminence or corpus albicans in their place.

Fig. 368.

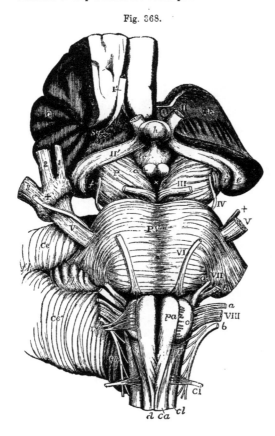

Fig. 368.—VIEW FROM BEFORE OF THE MEDULLA OBLONGATA, PONS VAROLII, CRURA CEREBRI, AND OTHER CENTRAL PORTIONS OF THE ENCEPHALON.

On the right side the convolutions of the central lobe or island of Reil have been left, together with a small part of the anterior cerebral convolutions : on the left side these have been removed by an incision carried between the thalamus opticus and the cerebral hemisphere.

I', the olfactory tract cut short and lying in its groove between two convolutions; II, the left optic nerve in front of the commissure; II', the right optic tract; *Th*, the cut surface of the left thalamus opticus; C, the central lobe or island of Reil; *Sy*, fissure of Sylvius; × ×, locus perforatus anterior; *e*, the external, and *i*, the internal corpus geniculatum; *h*, the hypophysis cerebri or pituitary body; *tc*, tuber cinereum with the infundibulum; *a*, one of the corpora albicantia; P, the cerebral peduncle or crus; *f*, the fillet; III, close to the left oculo-motor nerve; ×, the locus perforatus posticus; PV, pons Varolii; V, the greater root of the fifth nerve; +, the lesser or motor root; on the right side this + is placed on the Gasserian ganglion, and points to the lesser root, where it proceeds to join the inferior maxillary nerve; 1, ophthalmic division of the fifth nerve; 2, superior maxillary division; 3, inferior maxillary division; VI, the sixth nerve; VII *a*, the facial; VII *b*, the auditory nerve; VIII, the pneumo-gastric nerve; VIII *a*, the glosso-pharyngeal; VIII *b*, the spinal accessory nerve; IX, the hypoglossal nerve; *fl*, the flocculus; *f h*, the horizontal fissure of the cerebellum (*Ce*); *a m*, the amygdala; *p a*, the anterior pyramid; *o*, the olivary body; *r*, the restiform body; *d*, the anterior median fissure of the spinal cord, above which the decussation of the pyramids is represented; *c a*, the anterior column; *c l*, the lateral column of the spinal cord; C I, the suboccipital or first cervical nerve.

The *tuber cinereum* is a lamina of grey matter extending forwards from the corpora albicantia to the optic commissure, to which it is attached, and forming, as afterwards described, part of the floor of the third ventricle.

In the middle it is prolonged into a hollow conical process, the infundibulum, to the extremity of which is fixed the pituitary body.

The *pituitary body* or *hypophysis cerebri*, formerly called pituitary gland, from its being erroneously supposed to discharge *pituita* into the nostrils, is a small reddish grey mass, of a somewhat flattened oval shape, widest in the transverse direction, and occupying the sella turcica of the sphenoid bone. It consists of two lobes, of which the anterior is larger, and concave behind, where it embraces the smaller posterior lobe. Its weight is from five to ten grains. In the adult it is solid, and of a firm consistence.

The anterior lobe consists of two kinds of matter, one hard and grey, the other, situated within, softer and of a yellowish white colour. The posterior lobe is darker and redder than the anterior. Both are very vascular.

The pituitary body appears to approach in structure to the vascular or ductless glands, such as the thyroid and suprarenal bodies, &c. According to Sharpey's observations, with which those of subsequent writers agree, it differs greatly in structure, at least in its anterior and larger lobe, from any other part of the encephalon. The substance of the anterior lobe appears to be constituted by a membranous tissue forming little round cavities or loculi, which are packed full of nucleated cells. The loculi are formed of transparent, simple membrane, with a few fibres and corpuscles resembling elongated cell-nuclei disposed round their walls. The cells contained in the cavities are of various sizes and shapes, and not unlike nerve-cells or ganglion-globules; they are collected into round clusters, filling the cavities, and are mixed with a semi-fluid granular substance. This thin granular matter, together with the cells and little specks of a clear glairy substance like mucus, can be squeezed from the cut surface, in the form of a thick, white, cream-like fluid.

In the fœtus, the pituitary body is proportionally large, and contains a cavity which communicates, through that of the infundibulum, with the third ventricle. This body is constantly present, and has the same connection with the brain in all vertebrate animals.

In the middle line of the base of the brain, in front of the optic commissure, is the anterior portion of the great longitudinal fissure, which passes down between the hemispheres. At a short distance in front of the commissure, this fissure is crossed transversely by a white mass, which is the anterior recurved extremity of the corpus callosum. On gently turning back the optic commissure, a thin connecting layer of grey substance, the *lamina cinerea*, is seen occupying the space between the corpus callosum and the commissure, and continuous above the commissure with the tuber cinereum. It is connected at the sides with the grey substance of the anterior perforated space, and forms part of the anterior boundary of the third ventricle: it is somewhat liable to be torn in removing the brain from the skull; and, in that case, an aperture would be made into the fore part of the third ventricle.

At a short distance outwards from the lamina cinerea is the *anterior perforated spot* (locus perforatus anticus), a depression near the entrance of the Sylvian fissure, floored with grey matter, and pierced with a multitude of small holes for the passage of blood-vessels, most of which are destined for the corpus striatum,—the deeper portion of the brain beneath which it lies.

The grey surface of each perforated space is crossed by a broad white band, which may be traced from the middle of the under surface of the corpus callosum in front, backwards and outwards along the side of the lamina cinerea towards the entrance of the Sylvian fissure. These bands of the two sides are named the *peduncles of the corpus callosum*.

When the entire encephalon is viewed from below, the back part of the under surface of the cerebrum is concealed by the cerebellum and the pons Varolii. If, however, these parts be removed, it will be seen that the two hemispheres of the cerebrum are separated behind as they are in front, by the descent of the great longitudinal fissure between them, and that this fissure is arrested by a cross mass of white substance, forming the posterior extremity of the corpus callosum. This posterior part of the great longitudinal fissure is longer than the anterior portion.

INTERNAL PARTS OF THE CEREBRUM.

The anatomy of the interior of the cerebrum is most conveniently studied by removing, after the manner of Vieussens and Vicq-d'Azyr, successive portions of the hemispheres by horizontal sections, beginning from above.

Fig. 369.

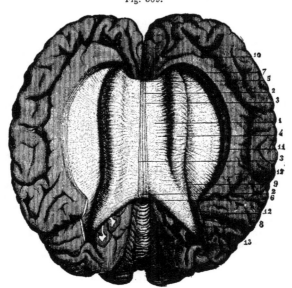

Fig. 369.—VIEW OF THE CORPUS CALLOSUM FROM ABOVE (from Sappey after Foville). ¼

The upper surface of the corpus callosum has been fully exposed by separating the cerebral hemispheres and throwing them to the side; the gyrus fornicatus has been detached, and the transverse fibres of the corpus callosum traced for some distance into the cerebral medullary substance.

1, the upper surface of the corpus callosum; 2, median furrow or raphe; 3, longitudinal striæ bounding the furrow; 4, swelling formed by the transverse bands as they pass into the cerebrum; 5, anterior extremity or knee of the corpus callosum; 6, posterior extremity; 7, anterior, and 8, posterior part of the mass of fibres proceeding from the corpus callosum; 9, margin of the swelling; 10, anterior part of the convolution of the corpus callosum; 11, hem or band of union of this convolution; 12, internal convolutions of the parietal lobe; 13, upper surface of the cerebellum.

The first horizontal section, to be made about half an inch above the corpus callosum, displays the internal white matter of each hemisphere, speckled with red spots where its blood-vessels have been divided, and surrounded on all sides by the grey matter which is seen to follow closely the convoluted surface, and to be of nearly equal thickness at all points. This white central mass in each hemisphere was named by Vicq-d'Azyr *centrum ovale minus*. On separating the remaining portions of the two hemispheres from each other, two sulci are seen to exist between the corpus callosum and the gyri immediately in contact with it, viz., the gyrus fornicatus of each side. These sulci were distinguished by the older anatomists as *ventricles of the corpus callosum.*

Another section being made at the level of the corpus callosum, the white substance of that part is seen to be continuous with the internal medullary matter of both hemispheres : and the large white medullary mass thus displayed, surrounded by the border of cortical substance, constitutes what is generally described as the *centrum ovale* of Vieussens.

The *corpus callosum* or *great commissure* (trabs cerebri) is a white structure, with a length not quite half of that of the brain, and approaching about two-fifths nearer to the front than the back of the hemispheres. It

Fig. 370.

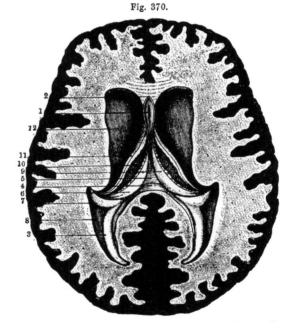

Fig. 370.—HORIZONTAL SECTION OF THE BRAIN SHOWING THE LATERAL VENTRICLES AND THE FIFTH VENTRICLE OPENED (from Sappey after Vicq-d'Azyr). ½

1, the fifth ventricle ; 2, the two laminæ of the septum lucidum meeting in front of it ; 3, lesser hippocampus of the posterior cornu ; 4, horizontal section of the posterior swelling of the corpus callosum ; 5, middle part of the fornix, where it has been separated from the corpus callosum ; 6, posterior pillar of the fornix ; 7, hippocampus major descending in the middle cornu ; 8, eminentia collateralis ; 9, lateral parts of the fornix ; 10, choroid plexus ; 11, tænia semicircularis ; 12, corpus striatum.

is about an inch in width behind, and somewhat narrower in front. Its thickness is greater at the ends than in the middle, and is greatest behind, where it is nearly half an inch. It is arched from before backwards. Its upper surface is distinctly marked by transverse furrows, which indicate the direction of the greater number of its fibres. It is also marked in the middle by a slight longitudinal groove, the *raphe*, which is bounded laterally by two white tracts, placed close to each other, named *striæ longitudinales*, or *nerves of Lancisi*. On each side, near the margin, are seen other longitudinal lines (striæ longitudinales laterales) occasioned by a few scanty white fibres.

In front, the corpus callosum is reflected downwards and backwards, between the anterior lobes, forming a bend named the *genu*. The inferior or reflected portion, which is named the *rostrum*, becomes gradually narrower as it descends, and is connected by means of the lamina cinerea with the optic commissure. It also gives off the two bands of white substance, already noticed as the *peduncles* of the corpus callosum, which, diverging from one another, run backwards across the anterior perforated space on each side to the entrance of the Sylvian fissure.

Fig. 371.

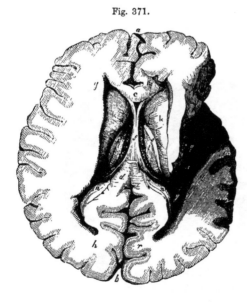

Fig. 371. — THE LATERAL VENTRICLES OPENED BY A HORIZONTAL SECTION, AND THE MIDDLE CORNU EXPOSED ON THE RIGHT SIDE. $\frac{4}{9}$

a, *b*, anterior and posterior parts of the great longitudinal fissure ; *c*, section of the anterior part of the corpus callosum ; *d*, posterior part of the same ; *e*, the left choroid plexus; *f*, the fornix ; *g*, the anterior; *h*, the posterior, and *q*, the descending cornu of the lateral ventricle ; *k*, *k*, corpora striata ; *l*, *l*, optic thalami ; *n*, *n*, right and left hippocampus minor ; *o*, posterior pillar of the fornix ; *v*, the corpus fimbriatum into which it passes; *q*, cornu ammonis or pes hippocampi ; *h*, the medullary substance of the cerebral hemisphere ; *r*, part of the cortical substance showing alternate grey and white matter ; *s*, *s*, tænia semicircularis ; *y*, eminentia collateralis.

Behind, the corpus callosum terminates in a free thickened border (*bourrelet*, pad), the under surface of which is also free for a short distance forwards.

The under surface of the corpus callosum is connected behind with the fornix, a structure to be presently described, and in the rest of its length with the septum lucidum, a vertical partition between the two lateral ventricles.

Although it presents a few longitudinal white fibres on its surface, the corpus callosum consists almost entirely of fibres having a transverse course towards each side, and spreading in a radiating manner into the substance of the two hemispheres. As the transverse fibres from the anterior and posterior lobes of the cerebrum are necessarily aggregated in large numbers near the corresponding ends of the corpus callosum, its greater thickness at those points, in comparison with the rest of its extent, is accounted for ; and, since the posterior lobe reaches further beyond the corpus callosum than the anterior, the greater thickness behind is also explained.

LATERAL VENTRICLES, or *ventriculi tricornes.*—By dividing the fibres of the corpus callosum in a longitudinal direction at a short distance on each side of the middle line, and about midway between the two ends of the hemispheres, an opening is made into the right and left *lateral ventricles* of the brain. These ventricles form part of the general ventricular space within the cerebrum ; they are serous cavities, and are lined by a delicate epitheliated structure, the *ependyma ventriculorum,* which at certain parts in the adult, and probably throughout its whole extent in the fœtus, is provided with cilia. In the natural state, the walls of the ventricles are moistened internally with a serous fluid, which sometimes exists in considerable quantity, even in a healthy brain.

It was formerly a subject of dispute whether the lining of the ventricles consisted of epithelium only, or also of a membrane. The progress of the histology of the brain has solved the problem in a manner which leaves the disputants on both sides partially in the right. It is now recognised that a peculiar form of connective tissue is found throughout the substance of the brain, similar to that which has been described in the spinal cord. A layer of this substance, unmixed with nerve-tissues, but in direct continuity with the interstitial web, and not a distinct membrane, supports the epithelium. It is of the same nature as the substance immediately surrounding the central canal of the spinal cord, and is named by Virchow *neuroglia* (Virchow's " Cellular Pathology," by Chance, p. 273).

The form of the epithelial cells appears to vary in different parts ; these cells being, according to Kölliker, of the flat pavement kind in the third ventricle, and more spherical in the lateral ventricles ; and, according to Gerlach, cylindrical in the aqueductus Sylvii.

From the central part or body of each lateral ventricle the cavity is extended into each of the three lobes of the hemisphere, thus forming an *anterior,* a *posterior,* and a *middle* or *descending cornu.*

The *body* of each lateral ventricle is roofed in by the corpus callosum, and is separated from its fellow by a vertical partition, the *septum lucidum,* which descends from the corpus callosum to the fornix. In the floor of the ventricle there is seen most posteriorly one half of the *fornix,* which is a thin layer of white brain-substance, broad behind and narrow in front : external and anterior to this is the *choroid plexus of the lateral ventricle,* a red vascular fringe, forming the border of the *velum interpositum,* a fold of pia mater extending inwards, on which the fornix rests : external and anterior to the choroid plexus is the anterior and outer part of the *optic thalamus,* appearing from beneath it : outside and in front of the thalamus is the *corpus striatum ;* and between those two bodies is a narrow flat band, the *tœnia semicircularis.*

The *anterior cornu* is the blind anterior extremity of the ventricle, projecting a little way into the anterior lobe. It is covered by the corpus callosum, and turns forwards and outwards round the anterior free extremity of the corpus striatum, descending as it proceeds, and bounded behind by that body, and in front by the reflected part of the corpus callosum.

The *middle* or *descending cornu* turns round the back part of the optic thalamus, which appears in its cavity and forms its anterior boundary, while its remaining boundaries are formed by the hemisphere. At its commencement it is directed backwards and outwards ; then, passing downwards with a sweep, it curves forwards, and at its extremity has a marked inclination inwards. The principal object seen upon the floor of this cornu is the *hippocampus major* (pes hippocampi, or cornu ammonis), a large white eminence extending the whole length of the cornu. The hippocampus major becomes enlarged towards its anterior and lower extremity, and is indented or notched on its edge, so as to present some resemblance to the paw of an animal, whence, no doubt, its name of pes hippocampi. The white fibres of its surface are directed obliquely backwards and outwards across it : they form only a thin smooth layer, and beneath them is cineritious matter continuous with that of the surface of the hemisphere. Along the inner edge of this eminence is seen a narrow white band, named *corpus fimbriatum* or *tænia hippocampi*, which is prolonged from the fornix ; to the inner side of the tænia is a part of the choroid plexus, and next to that the back of the optic thalamus. This cornu differs from the others in respect that it is not a mere cul-de-sac, but, by the mere separation of the membranes, can be made to communicate in its whole length with the surface of the brain by the fissure through which the choroid plexus enters.

Fig. 372.

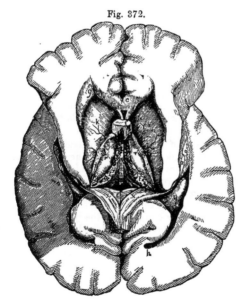

Fig. 372. — A Deep View of the Lateral Ventricles and their Cornua with the Velum Interpositum. ½

The fornix has been divided near its anterior pillars and turned back. *c*, the anterior part of the corpus callosum divided; *e*, the lyra on the lower surface of the corpus callosum and fornix ; *f*, anterior pillars of the fornix divided (these are represented of too large a size) ; *g*, anterior, and *h*, posterior cornu of the lateral ventricle ; *k, k*, corpora striata; *q*, pes hippocampi in the lower part of the middle cornu ; *r, r*, thalami optici ; *s, s*, tænia semicircularis ; *t, t*, choroid plexus ; *v*, velum interpositum; *x, x*, posterior pillars of the fornix ; *y*, eminentia collateralis.

The *posterior cornu* projects backwards into the substance of the posterior lobe. At its extremity it is pointed, and directed inwards. On the inner side of its floor is a curved and pointed longitudinal eminence, named *hippocampus minor, ergot,* or *calcar avis ;* and at the junction of the posterior with the descending cornu, between the hippocampus major and minor, is a smooth eminence, named *eminentia collateralis,* or *pes accessorius.*

The hippocampus minor is only the convex side of the fold which forms the calcarine sulcus, and part of the sulcus of the hippocampi ; and in like manner the eminentia collateralis corresponds with the posterior branch of the fissure of Sylvius.

As some discussion has recently taken place in this country with regard to the value of the presence of the hippocampus minor in man, as a distinctive character of the human brain, it may be well to mention that this structure has been found even in the brains of quadrumana which do not belong to the highest group. In the human subject the posterior cornu varies greatly in size, and the hippocampus minor is still more variable in its development, being sometimes scarcely to be recognised, and at others proportionally large. It is usually most developed where the posterior cornu is longest; but length of the posterior cornu, and prominence of the hippocampus minor, by no means occur in proportion to the dimensions of the hemisphere, but rather seem to be associated with thinness of both the medullary and the cortical substance.

The *septum lucidum* is a thin translucent partition, placed between the two lateral ventricles. It extends vertically between the corpus callosum above, and the anterior part of the fornix below; and, as the latter sinks down in front away from the corpus callosum, the septum is deep before and narrow behind. Anteriorly it lies in the hollow of the bend of the corpus callosum, in front of the fornix.

The septum lucidum is double, being composed of two distinct laminæ, having an interval between them, which contains fluid and is lined by an epitheliated membrane. This is the *fifth ventricle, ventricle of the septum,* or *Sylvian ventricle.*

Each of the laminæ of the septum which form the sides of the fifth ventricle, consists of an internal layer of white substance and an external layer of grey matter.

In the human embryo, and also in some animals, the cavity of this ventricle communicates with that of the third ventricle in front and below; but in the adult human brain it forms a separate and insulated cavity. Tarin described a small fissure in it between the pillars of the fornix; but this is unusual. In disease it is sometimes distended with fluid.

The *fornix* is an arched sheet of white longitudinal fibres, which appears partly in the floor of both lateral ventricles. It consists of two lateral halves, which are separated from each other in front and behind, but between those points are joined together in the mesial plane. The two parts in front form the *anterior pillars* of the fornix; the middle conjoined part is named the *body;* and the hind parts, which are again separated from each other, form the *posterior pillars.*

The *body* of the fornix is triangular in shape, being broad and flattened behind, where it is connected with the under surface of the corpus callosum, and narrower in front as it dips down to leave that body,—the space between them being filled up by the septum lucidum. Its lateral edges are in contact with the choroid plexuses, and its under surface rests upon the velum interpositum. .

The *anterior crura* or *pillars* of the fornix, cylindrical in form, descend, slightly apart from each other, through a quantity of grey matter on the sides of the third ventricle, between the corpora striata; and, curving backwards as they descend, reach the corpora albicantia. There each crus turns upon itself, making a twisted loop which forms the white portion of the corpus albicans of its own side, and ascends to enter the substance of the optic thalamus. These crura are connected with the peduncles of the pineal gland, and with the tænia semicircularis, as will be afterwards described.

Immediately behind the anterior pillars, where they descend, the fornix, which further back rests upon the optic thalami, the velum interpositum alone intervening, has an interval on each side left between it and the groove where the optic thalamus and corpus striatum meet. This interval leads from the lateral ventricle to the third ventricle—the space between the thalami and beneath the velum interpositum. The openings of opposite sides, passing downwards and backwards, meet in the middle line below, and thus is produced a passage, single below, but dividing into two branches above somewhat like the letter Y, and forming a communication between the third ventricle and both lateral ventricles. This passage is named the *foramen of Monro*, or *foramen commune anterius*.

Fig. 373.

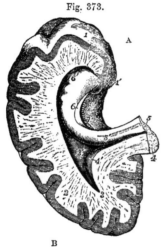

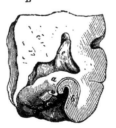

Fig. 373, A.—Lower and Back Part of the Cerebrum of the Left Side, showing the Posterior and Middle Cornua of the Lateral Ventricle opened (altered from Hirschfeld and Leveillé). ½

1, 1', inner convolution of the temporal lobe turning round into the convolution of the gyrus fornicatus, and showing on its surface the reticulated structure; 2, cut surface of the cerebral hemisphere; 3, point of the posterior cornu of the lateral ventricle; 3', eminentia collateralis; 4, cut surface of the lower and back part of the corpus callosum divided near the middle; 4', placed on the extension of the corpus callosum into the cerebral hemisphere, points by a line to the hippocampus minor in the posterior cornu; 5, cut edge of the posterior pillar of the fornix passing down at 5', into the hippocampus major and corpus fimbriatum; 6, continuation of the corpus fimbriatum or tænia hippocampi; 6', pes hippocampi; 7, fascia dentata on the inside of the white substance of the tænia.

Fig. 373, B.—Section of the Hippocampus Major to show the arrangement of the Grey and White Substance (from Mayo).

a, white layer on the surface of the hippocampus; b, grey substance which is involuted from the surface of the neighbouring convolution; c, fascia dentata; d, white reticulated substance of the lower part of the gyrus fornicatus; e, cavity of the lateral ventricle.

The *posterior crura* or *pillars* of the fornix are the diverging continuations backwards of the two flat lateral bands of which the body is composed. At first they adhere to the under surface of the corpus callosum, then curving outwards, each crus enters the descending cornu of the corresponding lateral ventricle, and is prolonged as a narrow band of white matter, named *tænia hippocampi* or *corpus fimbriatum*, which is situated on the inner margin of the hippocampus major, and extends to the extremity of that structure.

On examining the under surface of the fornix and corpus callosum, there are seen posteriorly the thickened border or pad, and in front of it the diverging halves of the fornix, between which a triangular portion of the corpus callosum appears, marked with transverse, longitudinal, and oblique lines. To this part the term *lyra* has been applied.

The *transverse fissure of the cerebrum* is the passage by which the pia mater passes from the surface into the ventricles of the brain to form the choroid plexus. It may be laid open in its whole extent, after the lateral ventricles have been opened, by completely dividing the fornix and corpus callosum in the middle line, and raising the divided parts from the undisturbed velum interpositum below. It will then be found that, in like manner, the posterior and middle lobes of the brain, including the hippocampus major and corpus fimbriatum, may be raised from the subjacent parts as far as the extremity of the descending cornu of the lateral ventricle. The transverse fissure is, therefore, a fissure extending from the extremity of the

Fig. 374.

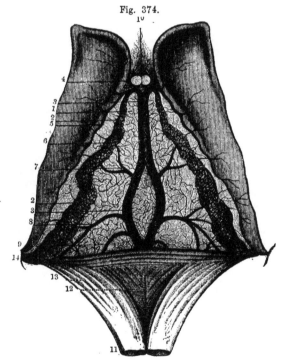

Fig. 374.—View of the Upper Surface of the Velum Interpositum, Choroid Plexus, and Corpora Striata (from Sappey after Vicq-d'Azyr). $\frac{2}{3}$

1, fore part of the tela choroidea or velum interpositum; 2, choroid plexus; 3, left vein of Galen partly covered by the right; 4, small veins from the front of the corpus callosum and the septum lucidum; 5, veins from the corpus striatum; 6, convoluted marginal vein of the choroid plexus; 7, vein rising from the thalamus opticus and corpus striatum; 8, vein proceeding from the inferior cornu and hippocampus major; 9, one from the posterior cornu; 10, anterior pillars of the fornix divided in front of the foramen of Monro; 11, fornix divided near its fore part and turned backwards; 12, lyra; 13, posterior pillar united with, 14, the corpus callosum behind, and covered by the choroid plexus as it descends into the inferior cornu.

descending cornu on one side, over the constricted part of the cerebrum, to the extremity of the descending cornu of the other side. It is bounded above by the corpus callosum and fornix in the middle, and more externally on each side by a free margin of the hemisphere : inferiorly it is bounded near the middle line by the corpora quadrigemina, and on each side by the crus cerebri and posterior part of the optic thalamus.

In *the free margin of the hemisphere* brought into view by opening out the part of the transverse fissure which leads into the descending cornu of the lateral ventricle, there are seen (1st) the ribbon-like ledge formed by the corpus fimbriatum, internally to the hippocampus major ; (2nd) beneath this, a small grey indented ridge, the *fascia dentata ;* and (3rd) beneath the fascia dentata, the gyrus hippocampi. On making a transverse section, it is seen that the corpus fimbriatum is the free margin of the white substance of the hemisphere, and that the fascia dentata is the free margin of the cortical substance, and is continuous with the grey matter of the hippocampus major, and that thus the hippocampus major is the swelling in reverse of the sulcus between the fascia dentata and gyrus hippocampi. The fascia dentata can be traced up to the pad or bourrelet : its upper part is free of dentations, and is sometimes named *fasciola cinerea.* The dentations correspond with blood-vessels passing to and from the choroid plexus.

The *velum interpositum* or *tela choroidea,* the membrane which connects the choroid plexuses of the two sides together, is a prolongation of the pia mater through the transverse fissure. It corresponds in extent with the fornix, which rests upon its upper surface ; and its more highly vascular free borders, projecting into the lateral ventricles, form the choroid plexuses.

The *choroid plexuses* appear like two knotted fringes, reaching from the foramen of Monro, where they meet together beneath the fornix, to the point of each descending cornu. They consist of a highly vascular villous membrane. The villi with which they are covered are again divided upon their surfaces and at their borders into small processes, along which fine vessels are seen to run. Numerous small vessels pass between the plexuses and the surface of the corpora striata, as well as other neighbouring parts, and the epithelium of the ventricles is continued over their surface. Thus it is only at the foramen of Monro that the epithelial lining of the lateral ventricles is continuous with that of the third ventricle.

The epithelium changes its character where it covers the plexus. It is there composed of large spheroidal corpuscles, in each of which is seen, besides a distinct nucleus, several yellowish granules, and one or more dark round oil-drops. According to Henle each of these cells is provided with short, slender, acuminate, transparent, and colourless processes.

On raising the velum interpositum, two slight vascular fringes are seen running along its under surface, and diverging from each other behind. They form the *choroid plexuses* of the third ventricle.

The choroid artery enters the velum interpositum at the point of the descending cornu ; and other arteries enter from behind, beneath the corpus callosum. The greater number of the veins terminate in two principal vessels named the veins of Galen, which run backwards on the velum interpositum, and passing out beneath the corpus callosum pour their blood into the straight sinus, having generally first united into a single trunk.

Bichat supposed that the arachnoid membrane entered the third ventricle in the form of a tubular process, which passed beneath the posterior end of the corpus callosum and fornix, through the velum interpositum, and thus opened into the upper and back part of the third ventricle. The existence of this canal, named the *canal of Bichat*, is no longer admitted.

The velum having been removed, the optic thalami are brought fully into view, together with the cavity of the third ventricle situated between them, while, behind the third ventricle, between it and the upper surface of

Fig. 375.

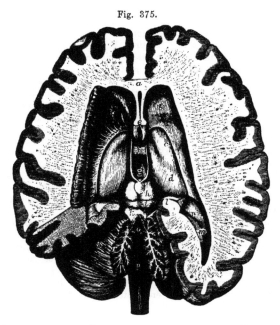

Fig. 375.—DISSECTION OF THE BRAIN FROM ABOVE, EXPOSING THE LATERAL, THIRD AND FOURTH VENTRICLES, WITH THE SURROUNDING PARTS (from Hirschfeld and Leveillé). ½

a, the anterior part or knee of the corpus callosum divided ; its fibres are seen spreading on each side into the cerebral hemispheres ; *b*, anterior part of the surface of the right corpus striatum in the anterior cornu of the lateral ventricle ; *b'*, the same on the left side, in which the grey substance has been dissected so as to show the peduncular medullary fibres spreading through the corpus striatum into the cerebral hemisphere ; *c*, points by a line to the tænia semicircularis : *d*, surface of the thalamus opticus; *e*, the anterior pillars of the fornix divided ; below they are seen descending in front of the third ventricle, and between them is seen a part of the anterior commissure ; above the letter is seen the fifth ventricle represented as a slit between the two laminæ of the septum lucidum ; *f*, placed on the soft or middle commissure ; *g*, in the posterior part of the third ventricle ; on either side of this letter is the white stria or peduncle of the pineal gland ; immediately below the letter is the small posterior commissure and the pineal gland ; *h*, the upper, and *i*, the lower of the corpora quadrigemina ; *k*, processus a cerebello ad cerebrum ; and close to this the valve of Vieussens, which is partly divided by a median incision along with the middle lobe of the cerebellum, so as to open up the fourth ventricle ; *l*, the hippocampus major and corpus fimbriatum separated from the posterior pillar of the fornix and descending into the middle cornu of the lateral ventricle ; *m*, posterior cornu of the lateral ventricle and hippocampus minor ; *n*, eminentia collateralis ; *o*, the cavity of the fourth ventricle ; *p*, posterior surface of the medulla oblongata ; *r*, section of the middle lobe showing the arbor vitæ ; *s*, upper surface of the cerebellum brought into view on the left side by the removal of a considerable part of the posterior cerebral lobe.

the cerebellum, are seen the pineal body, the corpora quadrigemina, the valve of Vieussens, and the processus a cerebello ad cerebrum.

The THIRD VENTRICLE is a narrow longitudinal cleft placed between the optic thalami, which bound it on its two sides. It is covered above by the velum interpositum and the fornix. Beneath, its floor is formed by the following parts, which have been already described as seen on the base of the cerebrum ; viz., commencing from behind, the posterior perforated space, the corpora albicantia, the tuber cinereum and infundibulum, and the lamina cinerea, the last of which also serves to close it in front, as high as the anterior commissure. Behind, is the anterior opening of the aqueduct of Sylvius. The cavity is crossed by three commissures, named from their position, anterior, middle, and posterior.

The *middle* or *soft commissure* is composed almost entirely of grey matter, and connects the two thalami. It is variable in size, and some- times wanting ; it is frequently torn across in examining the brain.

The *anterior* commissure is a round bundle of white fibres, placed imme- diately in front of the anterior pillars of the fornix, and crossing between the corpora striata. It marks the anterior boundary of the ventricle ; its fibres extend laterally through the corpora striata, a long way into the substance of the cerebral hemispheres..

The *posterior commissure*, also white but of smaller size, is placed across the back part of the ventricle, immediately before and below the pineal body, with which and with the corpora quadrigemina it is intimately connected.

The *corpora striata*, situated in front and to the outer side of the optic thalami, are two large ovoid masses of grey matter, the greater part of each of which is embedded in the middle of the white substance of the hemisphere of the brain, whilst a part comes to the surface in the body and anterior cornu of the lateral ventricle. This *intraventricular* portion of the corpus striatum is of a pyriform shape, its larger end being turned forwards, and its narrow end being directed outwards and backwards, so that the optic thalami of the two sides are received between the diverging corpora striata. On cutting into it, there may be seen at some depth from the surface white fibres, which are prolonged from the corresponding cerebral peduncle, and give it the streaked appearance from which it has received its name.

The *extraventricular* portion of the corpora striata will be afterwards described.

Along the inner border of each corpus striatum, and in a depression between it and the optic thalamus, is seen a narrow whitish semitrans- parent band, named *tænia semicircularis*, which continues backwards into the descending cornu of the ventricle, where its connections have not been determined with precision. In front it reaches the corresponding anterior pillar of the fornix, and descends in connection with that cord of white substance.

It is more transparent and firm on the surface, especially at its fore part: and this superficial stratum has been named *stria cornea*. The tænia consists of longi- tudinal white fibres, the deepest of which, running between the corpus striatum and the thalamus, were named by Vieussens *centrum geminum semicirculare*. Beneath it are one or two large veins, which receive those from the surface of the corpus striatum and end in the veins of the choroid plexuses.

The *thalami optici* (posterior ganglia of the brain) are of an oval shape, and rest on the corresponding cerebral crura, which they in a manner embrace. On the outer side each thalamus is bounded by the corpus striatum and tænia semicircularis. The upper surface, which is white, is free and prominent, and is partly seen in the lateral ventricle, and partly covered by the fornix. The part which is seen in the lateral ventricle is more elevated than the rest, and is named the *anterior tubercle*. The posterior surface, which is also white and free, projects into the descending

Fig. 376.

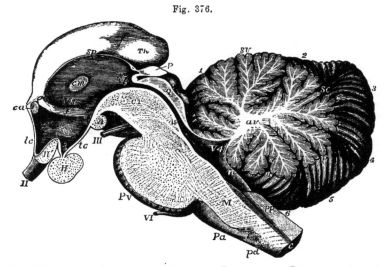

Fig. 376. — RIGHT HALF OF THE ENCEPHALIC PEDUNCLE AND CEREBELLUM AS SEEN FROM THE INSIDE IN A MEDIAN SECTION (after Reichert).

II, right optic nerve ; II′, optic commissure divided ; III, right third nerve ; VI, sixth nerve ; V 3, third ventricle ; *Th*, back part of the thalamus opticus ; II, section of the pituitary body ; A, corpus albicans ; P, pineal gland ; *c a*, points by a lower line to the anterior commissure divided, and by an upper line to the divided anterior pillar of the fornix ; *l c*, lamina cinerea ; *i*, infundibulum (cavity) ; *t c*, tuber cinereum ; *f*, mark of the anterior pillar of the fornix descending in the wall of the third ventricle ; *c m*, commissura mollis ; *s p*, stria pinealis ; *c p*, posterior commissure, above it the peduncle of the pineal gland, and below it the upper end of the passage to the fourth ventricle ; Q, corpora quadrigemina (section) ; *a s*, aqueduct of Sylvius near the fourth ventricle ; P V, pons Varolii divided in the middle ; M, medulla oblongata ; *p a*, right anterior pyramid ; *p d*, decussating bands cut across ; *p p*, posterior pyramids ; *c*, central canal with grey substance surrounding it divided. In the cerebellum, *a v*, stem of white substance in the centre of the middle lobe of the cerebellum, ramifying towards the arbor vitæ ; *s v*, superior vermiform process or vertical portion of the middle lobe ; *s c*, single folium, which passes across between the posterior superior lobes ; *c′*, the folia, which unite the posterior inferior lobes ; *p*, pyramid ; *u*, uvula ; *n*, nodule ; 1, part of the laminæ of the square lobe ; 2, posterior superior lobe ; 3, posterior inferior lobe ; 4, lobulus gracilis ; 5, biventral lobe ; 6, amygdaloid lobe.

cornu of the lateral ventricle. The inner sides of the two thalami are in contact one with the other. They present the grey substance of the interior of the thalami uncovered with white, and are generally partially united together by a transverse portion, which forms the middle or soft commissure of the third ventricle.

The *pineal body* or *gland* (conarium) is a small reddish body, which is placed beneath the back part of the corpus callosum, and rests upon the anterior elevations of the corpora quadrigemina.   It is attached to the under surface of the velum interpositum, so that it is liable to be torn away from the brain in removing that membrane.   It is about the size of a small cherry-stone.   Its base of attachment, which is its broader part, is directed forwards, and is connected with the rest of the cerebrum by white substance. This white substance is principally collected into two small rounded bundles, named *peduncles* of the pineal gland, which pass forwards upon the optic thalami along their upper and inner borders, and may be traced as far as the anterior pillars of the fornix, in conjunction with which they descend.   These peduncles are connected with each other behind, and the band of union between them is adherent to the back of the posterior commissure.

This band is represented by Reichert as folding forwards and then backwards, so as to leave a hollow, which he calls *recessus pinealis*, opening backwards above the pineal body.   Some anatomists have described two *inferior peduncles*, which descend upon the inner surface of the thalami.

The pineal gland is very vascular.   It is hollowed out into two or more cells, which, sometimes at least, open anteriorly into the ventricle, and almost always contain, besides a viscid fluid, a quantity of gritty matter, named *acervulus cerebri*.   This consists of microscopic round particles, aggregated into small compound masses, which are again collected into larger groups.   It is composed of the so-called amylaceous or amyloid bodies, and of earthy salts combined with animal matter, viz., phosphate and carbonate of lime, with a little phosphate of magnesia and ammonia (Stromeyer).   It is found at all ages, frequently in young children, and sometimes even in the foetus.   It cannot, therefore, be regarded as the product of disease.

This sabulous matter is frequently found on the outside of the pineal body, or even deposited upon its peduncles.   It is found also in the choroid plexuses; and scattered corpora amylacea occur in other parts of the membranes of the brain. Huschke has pointed out that the pineal body is larger in the child and the female than in the adult male.   In the brains of other mammals it is proportionally larger than in the human subject, and less loaded with the matter of acervulus cerebri.

The *corpora* or *tubercular quadrigemina* are four rounded eminences, separated by a crucial depression, and placed two on each side of the middle line, one before another.   They are connected with the back of the optic thalami, and with the cerebral peduncles at either side; and they are placed above the passage leading from the third to the fourth ventricle.

The upper or anterior tubercles are somewhat larger and darker in colour than the posterior.   In the adult, both pairs are solid, and are composed of white substance on the surface, and of grey matter within.

They receive bands of white fibres from below, the majority of which are derived from a fasciculus named the fillet.   A white cord also passes up on each side from the cerebellum to the corpora quadrigemina, and is continued onwards to the thalami : these two white cords are the *processus a cerebello ad cerebrum*, or superior peduncles of the cerebellum.   At each side of the corpora quadragemina there proceed outwards two white bands, which pass to the thalami and to the commencements of the optic tracts.   These bands are prominent on the surface, and are sometimes named *brachia*.

In the human brain the quadrigeminal bodies are small in comparison with those of animals. In ruminant, soliped, and rodent animals, the

Fig. 377.

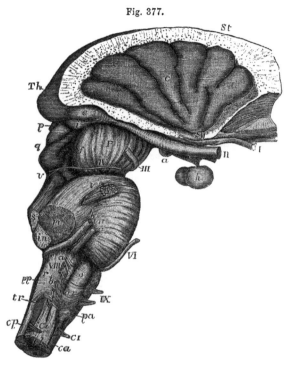

Fig. 377.—VIEW OF THE MEDULLA OBLONGATA, PONS VAROLII, CRURA CEREBRI, AND CENTRAL PARTS OF THE ENCEPHALON FROM THE RIGHT SIDE.

The corpus striatum and thalamus opticus have been preserved in connection with the central lobe and crura cerebri, while the remainder of the cerebrum has been removed.

St, upper surface of the corpus striatum ; *Th,* back part of the thalamus opticus ; C, placed on the middle of the five or six convolutions constituting the central lobe or island of Reil, the cerebral substance being removed from its circumference ; *Sy,* fissure of Sylvius, from which these convolutions radiate, and in which are seen the white striæ of the olfactory tract ; I, the olfactory tract divided and hanging down from the groove in the convolution which lodges it ; II, optic nerves a little way in front of the commissure ; *a,* right corpus albicans with the tuber cinereum and infundibulum in front of it ; *h,* hypophysis or pituitary body ; *e,* external, and *i,* internal corpus geniculatum at the back part of the optic tract ; P, peduncle or crus of the cerebrum ; *f,* fillet ; III, right oculo-motor nerve ; *p,* pineal gland ; *q,* corpora quadrigemina ; IV, trochlear nerve rising from *v,* the valve of Vieussens ; V, placed on the pons Varolii above the right nervus trige-minus ; *s,* the superior, *m,* the middle, and *in,* the inferior peduncles of the crus cere-belli cut short ; VI, the sixth nerve ; VII *a,* facial nerve ; VII *b,* auditory nerve ; on the medulla oblongata the parts are indicated as follows : VIII, placed opposite to the cut end of the pneumo-gastric nerve ; *a,* the glosso-pharyngeal ; and *b,* the uppermost fibres of the spinal accessory nerve ; IX, the hypoglossal nerve ; *p a,* anterior pyramid ; *o,* olivary body ; *a r,* arciform fibres ; *p p,* posterior pyramid ; *r,* restiform body ; *tr,* eminence corresponding to the tubercle of Rolando ; at the commencement of the spinal cord, *c a,* indicates the anterior, *c p,* the posterior, and *c l,* the lateral columns ; C I, anterior and posterior roots of the suboccipital or first cervical nerve.

anterior tubercles are much larger than the posterior, as may be seen in the sheep, horse, and rabbit ; and hence the name *nates*, formerly applied to the anterior, and *testes* to the posterior tubercles. In the brains of carnivora, the posterior tubercles are rather the larger. In the fœtus of man and mammals these eminences are at first single on each side, and have an internal cavity communicating with the ventricles. They are constant in the brains of all vertebrate animals ; but in fishes, reptiles, and birds, in which animals they receive the name of optic lobes, they are only two in number, and hollow : in marsupialia and monotremata, they are also two in number, but are solid.

Optic tracts and *corpora geniculata.*—The optic tracts, which have already been referred to in connection with the base of the cerebrum, are attached to and embrace the under side of the corresponding peduncles, and may be traced back to the thalami. Each tract, somewhat cylindrical towards the optic commissure, becomes flattened and broader as it approaches the thalamus, and makes a bend as it turns round the peduncle to reach the back part of that body. Near this bend, which is named the *knee* (genu), and to the outer side of the corpora quadrigemina, are placed two small oblong and flattened eminences connected with the posterior extremity of the optic tract. They are two little masses of grey matter about the size and shape of coffee-beans, placed one on the outer and one on the inner side of the genu of the optic tract, and hence are named respectively *corpus geniculatum externum* and *internum*. They send fibres into the optic tract and also into the thalamus of the same side.

The fibres of the optic tracts are therefore derived from three sources, viz., the thalamus, the tubercula quadrigemina, and the corpora geniculata.

The *processus a cerebello ad cerebrum* are two large white cords extending downwards and somewhat outwards from the corpora quadrigemina to the fore part of the cerebellum, and connecting the latter with the cerebrum. They rest upon the crura cerebri, to which they are united, and between them is the valve of Vieussens.

The *valve of Vieussens* (velum medullare anterius), stretched between the processus a cerebello ad cerebrum, is a thin layer of nervous matter, which lies over the passage from the third to the fourth ventricle, and, lower down, covers in a part of the fourth ventricle itself. It is narrow above, where it is connected with the quadrigeminal bodies, and broader below, where it is continuous with the median portion of the cerebellum.

The upper portion of the valve is composed of white substance, but a few transverse ridges of grey matter extend upon its lower half, as if they were prolonged from the grey lamellæ of the cerebellum with which the valve is there continuous. From between the posterior quadrigeminal tubercles a slight median ridge, named *frœnulum*, descends a little way upon the valve ; and on the sides of this the commencing fibres of the fourth pair of nerves pass transversely outwards. The back part of the valve is overlapped and concealed by the superior vermiform process of the cerebellum.

INTERNAL STRUCTURE OF THE CEREBRUM.

The cerebrum, like the rest of the encephalon, is composed of white and grey substance, the white pervading nearly the whole of its extent, though more exclusively composing its deeper parts ; the grey forming a covering of some thickness over the whole surface of the convolutions, and collected in distinct masses in certain of the deeper parts, such as the corpora striata, thalami optici, corpora quadrigemina, and crura cerebri. To the grey substance, the names of *cineritious* and *cortical* have been applied ; to the white that of *medullary*.

1. *The white matter* of the encephalon consists of tubular fibres, in general still smaller than those of the cord, and more prone to become varicose. The general direction which these follow is best seen in a brain that has been hardened by immersion in alcohol, although it is true that in an ordinary dissection of such hardened masses with the scalpel, we do not then trace the single fibres, but only the smaller bundles and fibrous lamellæ which they form by their aggregation. It must also be admitted that were they intimately decussate, the tearing of fibres across is liable to be mistaken for the separation of sets of fibres one from the other; and it is necessary to correct such errors by the examination of sections under the microscope. The microscopic examination of the cerebrum, however, is as yet still less

Fig. 378.

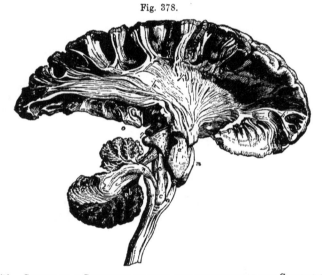

Fig. 378.—Sketch of a Dissection showing the connection of the Columns of the Medulla Oblongata with the Cerebrum and Cerebellum (from Mayo). ½

In the lower part of the figure the medulla oblongata is entire where it is prolonged downwards into the spinal cord ; *a*, the anterior pyramid ; *a'*, its continuation upwards into the pons Varolii (*m*) ; *c*, olivary body ; *c'*, olivary fasciculus ; behind *c'*, the fasciculi teretes are represented ; *d*, the white laminæ in part of the cerebellum ; *f*, superior peduncle of the cerebellum ; *g*, anterior part or crust of the cerebral peduncle ; *h*, part of the fibres radiating from the peduncle into the right cerebral hemisphere, of which a considerable extent is shown containing parts of the anterior, middle, and posterior lobes ; *h*, *y*, *y*, part of the corona radiata ; *h'* (in front), central fibres of the convolutions ; *i*, fillet ; *l*, back of the thalamus opticus ; *m*, pons Varolii ; *n*, inferior peduncle of the crus cerebelli ; *o*, section of the pes hippocampi ; *r*, tegmentum ; *y*, *y*, show the white fibres issuing from the corpus striatum.

complete than that of the spinal marrow and medulla oblongata. By the dissection of artificially prepared brains, aided in part by microscopic observation, the following general facts have been ascertained.

The fibres of the cerebrum, though exceedingly complicated in their arrangement, and forming many different groups, may be referred to three principal systems, according to the general course which they take, viz. :—1. *Ascending* or *peduncular fibres,* which pass up from the medulla oblongata to the hemispheres, and constitute the peduncles of the cerebrum. These fibres increase in number as they ascend through the pons, and still further in passing through the optic thalami and striated bodies, beyond which they spread in all directions into the hemispheres.

2. *Transverse* or *commissural* fibres, which connect the two hemispheres together.

3. *Longitudinal* or *collateral* fibres, which, keeping on the same side of the middle line, connect more or less distant parts of the same hemisphere.

1. In each hemisphere the *peduncular* fibres consist of a main body and of certain accessory bundles of fibres.

The *main body* is derived from the anterior pyramid, from the fasciculi teretes, and from the posterior pyramid. After it has passed through the pons, and become increased in amount, it is separated into two parts in the crus cerebri by a layer of dark cineritious matter, named *locus niger.* The lower or superficial part, which is derived from the pyramid, consist almost entirely of white fibres, collected into coarse fasciculi, and is named the *crust* or *basis,* or the *fasciculated portion* of the peduncle (Foville). The upper part, composed principally of the fasciculus teres and posterior pyramid, is named the *tegmentum.* It is softer and finer in texture, and is mixed with much grey matter.

Still increasing in number within the peduncle, these two sets of fibres ascend to the thalamus and corpus striatum. A much larger number of fibres diverging

Fig. 379.

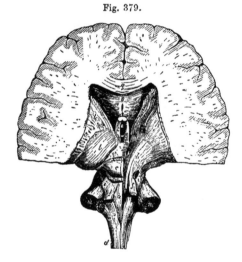

Fig. 379.—Posterior View of the Peduncles of the Cerebrum and Cerebellum (after Arnold). ½

The lower and fore part of the cerebral hemispheres is preserved, the cerebellum is completely detached from its peduncles, and on the right side the corpora quadrigemina and thalamus opticus have been dissected. *a,* fasciculus teres of the left side ; *b,* fibres of the tegmentum ascending through the right thalamus; *c,* left corpora quadrigemina ; *d,* lateral column of the cord ; *e,* restiform body ; *f,* superior peduncles of the cerebellum ; *g,* fibres of the crust ; *i, i,* the fillets ; *k, k,* corpora striata ; *l,* the left thalamus ; *m, m,* sections of the middle peduncles of the cerebellum ; *n,* section of the left inferior peduncle ; *p,* left posterior pyramid ; *q,* section of the corpus callosum ; *s,* under surface of the same, and below it the cavity of the fifth ventricle; *e,* left anterior pillar of the fornix; *y,* decussation of the radiating fibres with the crossing fibres of the corpus callosum.

from these bodies appear to pass to the medullary substance of the hemispheres; but the actual continuity of the individual fibres spreading out in the hemisphere with those ascending to the thalamus and corpus striatum is doubted by many authors, and among them, by Kölliker.

The assemblage of radiating fibres in each hemisphere might be compared to a fan, bent into the form of an incomplete hollow cone, having its concave surface turned downwards and outwards ; hence the name *corona radiata* applied to them by Reil, and *fibrous cone* by Mayo.

The *accessory fibres* of the peduncular system are as follows :—

a. The superior peduncles of the cerebellum, (processus ad cerebrum,) which are continued up beneath the corpora quadrigemina, and form part of the tegmentum.

b. The bundle of fibres on each side, named the *fillet* (lemniscus). This, which is originally derived from the anterior column of the cord, proceeds from the olivary fasciculus of the medulla oblongata, as previously described. Reinforced by fibres from the corpus dentatum of the olivary body, it ascends through the back part of the pons, still increasing in size. Appearing at the side of the cerebral peduncle, above the upper border of the pons, it divides into two portions, of which one crosses over the superior peduncle of the cerebellum to the corpora quadrigemina, meeting its fellow of the opposite side; while the other is continued upwards with the fibres of the tegmentum.

c. Other fibres accessory to the peduncles take their rise in the grey matter of the corpora quadrigemina (the *brachia*), and proceed on to the thalami.

d. Lastly, fibres of another set, having a similar destination, are derived from the corpora geniculata.

2. The *transverse commissural,* or connecting fibres of the cerebrum, include the following sets.

a. The cross fibres of the corpus callosum pass laterally into the substance of

Fig. 380. — VIEW OF A DISSECTION OF THE FIBRES IN THE LEFT CEREBRAL HEMISPHERE FROM BELOW (after Mayo). ½

Fig. 380.

The most of the middle lobe in its lower part has been removed. *a*, the anterior and *a'*, the posterior part of the fillet of the corpus callosum ; *b; g*, section of the crus cerebri ; *b*, tegmentum ; *g*, crust separated from the last by the locus niger ; *c'*, fibres stretching from the back part of the corpus callosum into the posterior lobe ; *e*, fasciculus uncinatus connecting the anterior and middle lobes across the Sylvian fissure ; *f, f*, transverse fibres from the corpus callosum passing into the cerebral hemispheres ; *l*, back part of the thalamus ; *m*, corpus albicans ; *q*, median section of the corpus callosum ; *r*, radiating fibres of the hemispheres ; *t*, anterior pillar of the fornix descending into the corpus albicans (*m*) ; *v*, collateral fibres of the convolutions ; ×, anterior commissure.

the hemispheres, some being directed upwards, whilst others spread outwards on the roof of the lateral ventricles, forming there what is named the *tapetum*. Having intersected the peduncular radiating fibres, they spread out into the hemispheres, reaching everywhere the grey matter of the convolutions.

b. The *fibres of the anterior commissure* pass laterally into the corpora striata, and bending backwards, extend a long way into the middle of the hemispheres, on each side.

c. The *fibres of the posterior commissure* run through the optic thalami, and are soon lost in the substance of the hemispheres outside these bodies.

3. The third system of fibres in the cerebrum, the *longitudinal* or *collateral*, includes those of the fornix, tænia semicircularis, and striæ longitudinales of the corpus callosum, already sufficiently described; and likewise the following.

a. Fibres of the gyrus fornicatus ; fillet of the corpus callosum (Mayo).—These fibres constitute the white substance of the gyrus fornicatus, and take a longitudinal course immediately above the transverse fibres of the corpus callosum. In front they bend downwards within the gyrus to which they belong, and are connected with the anterior perforated space, being joined by certain longitudinal fibres which run along the under surface of the corpus callosum near the middle line, passing near and upon the upper edge of the septum lucidum. Behind, they turn round the back of the corpus callosum and thence descend to the point of the middle lobe, where, according to Foville, they again reach the perforated space. Offsets from these fibres pass upwards and backwards into the secondary convolutions derived from the gyrus fornicatus in the longitudinal fissure.

b. Fasciculus uncinatus.—Under this name is described a white bundle, seen on the lower aspect of the hemisphere, passing across the bottom of the Sylvian fissure, and connecting the anterior with the middle and posterior lobes. The fibres of this bundle expand at each extremity, and the more superficial of them are curved or hooked sharply between the contiguous parts of the anterior and middle lobes,—whence it has derived its name.

c. The convolutions of the cerebrum are connected with each other by white fibres, which lie immediately beneath the cortical substance. Some of them pass across the bottom of the sulcus between adjacent convolutions; whilst others, which are longer and run deeper, connect convolutions situated at a greater distance from one another.

Fig. 381.

Fig. 381.—View of a Dissection of the Fibres of the Gyrus Fornicatus and Fornix, in the Right Hemisphere (slightly altered from Foville). ½

A, the anterior lobe; B, the posterior lobe; *a, a', a''*, fibres of the gyrus fornicatus; *c, c'*, oblique bands of fibres of some of its accessory gyri; *b*, tegmentum, and *g*, crust of the crus cerebri, separated by the locus niger; *l*, thalamus; *m*, fissure of Sylvius; *n*, corpus albicans; *q*, median section of the corpus callosum; *s*, septum lucidum; *t*, the fornix, its anterior pillar descending into the corpus albicans, and then emerging from that at its termination (*) in the thalamus; 1, the olfactory bulb; 2, the optic commissure.

The researches of Foville have led him to differ considerably from other anatomists as to the course of the fibres of the cerebrum, as will be seen from the following statement of his views.

1. The *crust* or *fasciculated portion* of each cerebral peduncle, derived from the anterior pyramid, forms by itself the peduncular fibrous cone, and is thence continued on into the radiating fibres of the cerebrum, which are destined only for the convolutions on the convex surface of the hemisphere, including the outer half of the marginal convolution of the longitudinal fissure, and the inner half of the convolution of the Sylvian fissure.

2. The fibres of the *tegmentum*, having entered the thalamus, pass on in two ways —no part of them, however, joining the radiating peduncular fibres.

a. One set pass upwards through the thalamus and corpus striatum, above which they then turn inwards, and, joining with those of the opposite side, form the transverse fibres of the corpus callosum. The corpus callosum is therefore regarded as a commissure of the cerebral peduncles only—none of its cross fibres spreading into the convolutions, as is generally believed.

b. The second set of fibres of the tegmentum, corresponding with the fasciculi teretes and part of the posterior pyramids, run forwards near the middle line, along the under side of the third ventricle and corpus striatum, through the grey matter in front of the pons, to the anterior perforated space. The remaining part of the posterior pyramid forms the tænia semicircularis, which, passing down in front of the anterior pillar of the fornix, also reaches the perforated space. From this space more fibres are reflected upwards on the sides of the corpus striatum to join the corpus callosum.

3. As dependencies of the posterior peduncular fibres, and connected with them at the borders of the anterior perforated space, are :—

a. Several sets of longitudinal arched fibres, which embrace, in a series of rings, the radiating peduncular system. These are—the deep fibres of the tænia semicircularis—a somewhat similar band beneath the outer part of the corpus striatum—the half of the fornix with the corpus fimbriatum—the longitudinal fibres placed on the upper and under surface of the corpus callosum, and those of the septum lucidum; and, lastly, two remarkable systems of longitudinal fibres—one constituting the entire white substance of the gyrus fornicatus (from end to end), also of its accessory convolutions, and of the inner half of the marginal convolution of the longitudinal fissure; and the other, forming the white substance of the convolutions of the island of Reil, and the adjoining half of the convolution of the Sylvian fissure. None of the parts just named receive fibres from the radiating peduncular set.

b. In connection with this system is a thin stratum of white fibres, found upon the internal surface of the ventricles, and prolonged through the transverse fissure into the reticulated white substance covering the lower end of the gyrus fornicatus; whence, according to Foville, it extends, as an exceedingly thin layer of medullary matter, all over the cortical substance of the hemisphere.

c. The anterior commissure does not reach the convolutions, but radiates upon the outer sides of the corpora striata and thalami.

II. *The grey matter on the convoluted surface* of the cerebrum is divided into two and in some regions into three strata, by interposed thin layers of white substance. In examining it from without inwards, we meet with—1. A thin coating of white matter situated on the surface, which on a section appears as a faint white line, bounding the grey surface externally. This superficial white layer is not equally

Fig. 382.—SECTION OF THE CORTICAL SUB-STANCE OF A CEREBRAL CONVOLUTION (from Remak).

In A, the parts are nearly of the natural size. To the right of the figure, *a* and *e* are two white, and *b* and *f* two grey strata; to the left of the figure, an additional white layer, *e*, divides the first grey into two, *b* and *d*. In B, a small part of the cortical substance of a convolution is represented, magnified to show more clearly the

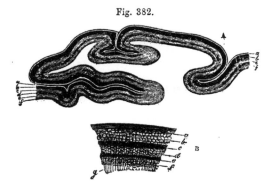

Fig. 382.

relative position of the strata; *a*, superficial white layer; *b*, reddish grey layer; *c*, intermediate white layer; *d*, inner part of the outer grey layer; *e*, thin white layer; *f*, inner grey layer; *g*, radiating white fibres from the medullary substance of the convolution passing into the layers of the cortical substance.

thick over all parts of the cortical substance, but becomes thicker as it approaches the borders of the convoluted surface; it is accordingly less conspicuous on the lateral convex aspect of the hemispheres; and more so on the convolutions situated in the longitudinal fissure which approach the white surface of the corpus callosum, and on those of the under surface of the brain. It is especially well marked on the middle lobe, near the descending cornu of the lateral ventricle, where the convoluted surface is bounded by the posterior pillar of the fornix, and it has been there described under the name of the *reticulated white substance.* It consists of remarkably fine tubular fibres, for the most part varicose, which run parallel with the surface of the convolutions, but intersect each other in various directions. The termination and connections of these fibres are unknown. This superficial white layer contains also a few small cells with processes, and an abundant granular matrix. 2. Immediately beneath the white layer just described, is found a comparatively thick layer of grey or reddish grey matter, the colour of which, as indeed of the grey substance generally, is deeper or lighter according as its very numerous vessels contain much or little blood. Then follow, 3. Another thin whitish layer; and 4. A thin grey stratum. This last lies next to the central white matter of the hemisphere. Remak considers it as similar in nature to the gelatinous substance of the spinal cord. According to this account, the cortical substance consists of two layers of grey substance, and two of white; but in several convolutions, especially those situated near the corpus callosum, a third white stratum may be seen, which divides the most superficial grey layer into two, thus making six in all, namely, three grey and three white.

Fig. 383.

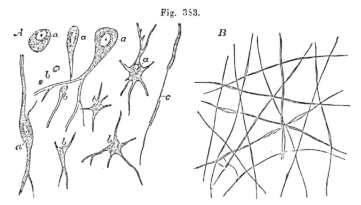

Fig. 383.—MINUTE STRUCTURE OF THE CEREBRAL SUBSTANCE (from Kölliker), MAGNIFIED 220 DIAMETERS.

A, cells and structural elements from the inner part of the cortical substance of the cerebral convolutions; *a*, larger cells, chiefly from the middle grey layer, showing a variable number of radiating processes; *b*, smaller cells from the more superficial grey layer, in part belonging to connective tissue; *c*, a nerve-fibre with its axis-filament partly exposed.

B, finest nerve-fibres from the superficial white layer of the cortical substance of a convolution, some showing the varicose condition.

The cortical grey substance consists of nerve-cells of rather variable size, which are angular, fusiform, round, or oval in shape, and for the most part caudate, and lie in a granular matrix; also of small nucleus-like vesicles, which resemble those seen in the cortical substance of the cerebellum, and, according to Todd, are here also collected into a special stratum. In the middle grey layer, the cells are of variable size, some being so small as to resemble nuclei; but others of much larger dimensions are abundant, and, according to Kölliker, present from one to six processes.

In the innermost grey layer the cells have similar characters, but often contain pigmentary matter. Tubular fibres exist throughout: those of one set run parallel with the surface, and at certain depths are more densely aggregated, so as to form the before-mentioned white layers : they are also present in the intervening grey strata, but there they are wider apart. The manner in which they begin and end is not known ; it seems not improbable, however, that they are dependencies of the commissural system of fibres. These stratified fibres, if they might be so called, are intersected by another set of tubular fibres, which come from the central white mass of the hemispheres, and run perpendicularly through the cortical substance, becoming finer and spreading more out from each other as they approach the surface.

The further disposition of these central or perpendicular fibres is uncertain ; Valentin describes them as forming terminal loops or arches, but this is denied by Remak and Hannover. Remak states that they gradually disappear from view at different depths, as they pass through the successive layers, the last of them vanishing in the superficial grey stratum ; but he is unable to say positively how they terminate. It sometimes seemed to him as if the last of them, after intersecting the fibres of the deeper white stratum, became continuous with those of the outermost layer ; but of this he by no means speaks confidently. Hannover maintains that the perpendicular fibres are connected at their extremities with the nerve-cells in the cortical substance.

The grey matter of the lamina cinerea, tuber cinereum, and posterior perforated spot, appears both in the base of the brain and in the floor of the third ventricle. The lamina cinerea is connected externally with the grey matter of the anterior perforated spot, and from that point a continuity of grey matter can be traced to the swelling of the olfactory bulb. Thus also continuity is established between the grey matter of the hemispheres and that of the interior of the brain.

III. *The grey matter of the interior* of the cerebrum may be examined in the series of its deposits from behind forwards.

In the *crura cerebri*, the grey matter is collected into a dark mass, the *locus niger*, which lies between the crust and the tegmentum, and is also diffused among the fasciculi of the tegmentum ; below this it is continuous with that of the pons and medulla oblongata, and through them with that of the spinal cord, as has already been sufficiently described. In the upper part of each tegmentum is a round reddish grey centre, the red centre of Stilling, the superior olive of Luys.

In the centre of each of the *corpora quadrigemina* grey matter is also found ; and this collection is stated by Huschke to be continuous below with the posterior cornu of the grey matter of the spinal cord, posteriorly with that of the corpus dentatum of the cerebellum, and anteriorly with the soft commissure, the septum lucidum, optic thalami, and corpus callosum. Grey matter occurs also in the pineal gland, and in the corpora geniculata. These last bodies appear to be appendages of the optic thalami.

The grey matter of the *optic thalamus* constitutes the principal bulk of that body ; it is, however, partially divided into an inner and an outer portion, by white fibres passing through it.

The *corpus striatum* contains three grey centres. That which forms the intraventricular portion of the body, and is connected inferiorly with the lamina cinerea, and with that portion of the grey matter of the optic thalamus which is seen in the third ventricle, is named the *nucleus caudatus*. The principal centre of the extraventricular portion, named *nucleus lenticularis*, external and inferior to the nucleus caudatus, is separated from that centre by the white substance of the fibrous cone, which, as it passes outwards, appears, when cut across, as a broad white band extending from behind forwards, and traversed by striæ of grey matter passing from one centre to the other. Between the nucleus lenticularis and the island of Reil, which lies opposite to it, there intervenes a thin lamelliform deposit of grey matter, the *nucleus tæniæformis* (Arnold), or *claustrum* (Burdach), which, in a transverse section, is seen as a thin line. The lenticular nucleus is continuous with the grey matter of the anterior perforated space.

The corpus striatum and optic thalamus contain cells very much like those of the cortical substance. In the corpora quadrigemina there are larger cells, approaching in size to those of the cerebellum, besides very small cells and nucleiform bodies. The dark matter, forming the so-called locus niger of the cerebral peduncles, and that in the floor of the fourth ventricle, contain caudate cells, many of them of the largest size, with long appendages, and deeply coloured with pigment. (Hannover, Rech. Microscop. sur le Système Nerveux. Copenhagen, 1844).

The pineal body, like the pituitary body, has already been sufficiently described. The deep connection of some of the cranial nerves with the basal parts of the cerebrum, as well as that of others with the remaining portions of the encephalon, will be referred to in the description of these nerves.

THE MEMBRANES OF THE BRAIN AND SPINAL CORD.

The cerebro-spinal axis is protected by three *membranes*, named also *meninges*. They are :—1. An external fibrous membrane, named the *dura mater*, which closely lines the interior of the skull, and forms a loose sheath in the spinal canal ; 2. An internal areolo-vascular tunic, the *pia mater*, which accurately covers the brain and spinal cord ; and 3. An intermediate serous sac, the *arachnoid* membrane, which, by its parietal and visceral layers, covers the internal surface of the dura mater on the one hand, and is reflected over the pia mater on the other.

THE DURA MATER.

The *dura mater*, a very strong dense inelastic fibrous tunic of considerable thickness, is closely lined on its inner surface by the outer portion of the

Fig. 384.

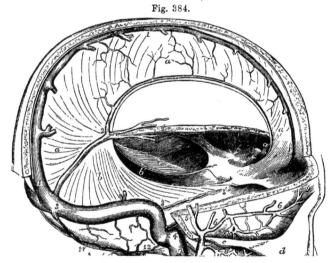

Fig. 384.—The Cranium opened to show the Falx of the Cerebrum, and Tentorium of the Cerebellum. ½

a, right side of the falx cerebri; *a'*, its anterior narrow part attached to the crista galli; *b*, tentorium cerebelli of the right side, united to the base of the falx cerebri from 2 to 3, in the line of the straight sinus, and attached to the superior border of the petrous bone between 3 and 3'; *b'*, aperture between the right and left divisions of the tentorium for the isthmus cerebri; 1, 1, the superior longitudinal sinus; 2, 2, the inferior ; 3, 3, the lateral sinus; 3, 3', the superior petrosal sinus ; 3', is close to the anterior clinoid process.

arachnoid, and with it, therefore, forms a *fibro-serous* membrane, which is free, smooth, and epitheliated on its inner surface, where it is turned towards the brain and cord, but which, by its outer surface, is connected with the surrounding parts, in a somewhat different manner in the cranium and in the spinal canal.

- The outer surface of the *cranial* portion adheres to the inner surface of the bones, and forms their internal periosteum. The connection between the two depends, in a great measure, on blood-vessels and small fibrous processes, which pass from one to the other; and the dura mater, when detached and allowed to float in water, presents a flocculent appearance on its outer surface, in consequence of the torn parts projecting from it. The adhesion between the membrane and the bone is more intimate opposite the sutures, and also generally at the base of the skull, which is uneven, and perforated by numerous foramina, through which the dura mater is prolonged to the outer surface, being there continuous with the pericranium. The fibrous tissue of the dura mater becomes blended with the areolar sheaths of the nerves, at the foramina which give issue to them.

In leaving the skull, the dura mater is intimately attached to the margin of the foramen magnum; but within the vertebral canal it forms a loose sheath around the cord (*theca*), and is not adherent to the bones, which have an independent periosteum. Towards the lower end of the canal, a few fibrous slips proceed from the outer surface of the dura mater to be fixed to the vertebræ. The space intervening between the wall of the canal and the dura mater is occupied by loose fat, by watery areolar tissue, and by a plexus of spinal veins.

Opposite each intervertebral foramen the dura-matral theca presents two openings, placed side by side, which give passage to the two roots of the corresponding spinal nerve. It is continued as a tubular prolongation on the nerve, and is lost in its sheath. Besides this, it is connected with the circumference of the foramen by areolar tissue.

The fibrous tissue of the dura mater, especially within the skull, is divisible into two distinct layers, and at various places these layers separate from each other and leave intervening channels, called *sinuses*. These sinuses, which have been elsewhere described, are canals for venous blood, and are lined with a continuation of the internal membrane of the veins.

The dura mater also sends inwards into the cavity of the skull three strong membranous *processes*, or *partitions*, formed by duplication of its inner layer. Of these, one descends vertically in the median plane, and is received into the longitudinal fissure between the two hemispheres of the cerebrum. This is the *falx cerebri*. The second is an arched or vaulted partition, stretched across the back part of the skull, between the cerebrum and the cerebellum; it is named the *tentorium cerebelli*. Below this, another vertical partition, named *falx cerebelli*, of small extent, passes down between the hemispheres of the cerebellum.

The *falx cerebri* is narrow in front, where it is fixed to the crista galli, and broader behind, where it is attached to the middle of the upper surface of the tentorium, along which line of attachment the straight sinus is situated. Along its upper convex border, which is attached above to the middle line of the inner surface of the cranium, runs the superior longitudinal sinus. Its under edge is free, and reaches to within a short distance of the corpus callosum, approaching nearer to it behind. This border contains the inferior longitudinal sinus.

The *tentorium*, or *tent*, is elevated in the middle, and declines downwards in all directions towards its circumference, thus corresponding in form with the upper surface of the cerebellum. Its inner border is free and concave, and leaves in front of it an oval opening, through which the isthmus encephali descends. It is attached behind and at the sides by its convex border to the horizontal part of the crucial ridges of the occipital bone, and there encloses the lateral sinuses. Farther forward it is connected with the upper edge of the petrous portion of the temporal bone—the superior petrosal sinus running along this line of attachment. At the point of the pars petrosa, the external and internal borders meet, and may be said to intersect each other—the former being then continued inwards to the posterior, and the latter forwards to the anterior clinoid process.

The *falx cerebelli* (falx minor) descends from the middle of the posterior border of the tentorium with which it is connected, along the vertical ridge named the internal occipital crest, towards the foramen magnum, bifurcating there into two smaller folds. Its attachment to the bony ridge marks the course of the posterior occipital sinus, or sinuses.

Structure.—The dura mater consists of white fibrous and elastic tissue, arranged in bands and laminæ, crossing each other. It is traversed by numerous blood-vessels which are chiefly destined for the bones. Minute nervous filaments, derived from the fourth, fifth, and eighth cranial nerves, and from the sympathetic, are described as entering the dura mater of the brain. Nervous filaments have likewise been traced in the dura mater of the spinal column. (Luschka and Rüdinger, quoted by Hyrtl.)

THE PIA MATER.

The *pia mater* is a delicate, fibrous, and highly vascular membrane, which immediately invests the brain and spinal cord.

Upon the hemispheres of the brain it is applied to the entire cortical surface of the convolutions, and dips into all the sulci. From its internal surface very numerous small vessels enter the grey matter and extend for some distance perpendicularly into the substance of the brain. The inner surface of the cerebral pia mater is on this account very flocculent, and is named *tomentum cerebri*. On the cerebellum a similar arrangement exists, but the membrane is finer and the vessels from its inner surface are not so long. The pia mater is also prolonged into the ventricles, and there forms the velum interpositum and choroid plexus.

Structure.—The pia mater consists of interlaced bundles of areolar tissue, conveying great numbers of blood-vessels; and, indeed, its peculiar office, both on the brain and spinal cord, seems to be that of providing a nidus or matrix for the support of the blood-vessels, as these are subdivided before they enter the nervous substance. According to Fohmann and Arnold, it contains numerous lymphatic vessels. Purkinje describes a retiform arrangement of nervous fibrils, derived, according to Kölliker and others, from the sympathetic, the third, sixth, facial, pneumogastric, and accessory nerves.

On the *spinal* cord the pia mater has a very different structure from that which it presents on the encephalon, so that it has even been described by some as a different membrane under the name *neurilcmma of the cord*. It is thicker, firmer, less vascular, and more adherent to the subjacent nervous matter: its greater strength is owing to its containing fibrous tissue, which is arranged in longitudinal shining bundles. A reduplication of this membrane dips down into the anterior fissure of the cord, and serves to conduct blood-vessels into that part. A thinner process passes into the greater part of the posterior fissure. At the roots of the nerves, both in the spine and in the cranium, the pia mater becomes continuous with the neurilemma. It is supplied with nerves from the sympathetic.

Towards the upper part of the cord, the pia mater presents a greyish mottled appearance, which is owing to pigment particles deposited within its tissue.

THE ARACHNOID MEMBRANE.

The *arachnoid* is a very fine delicate serous membrane, which, like other membranes of that class, forms the lining boundary of a shut sac. The walls of this sac consist of two portions, one of which, a distinct membrane on the surface of the pia mater, is the visceral or cerebral layer of the arachnoid, while the other, giving the smooth surface presented by the dura mater on its interior, is described by some anatomists as the parietal layer of the arachnoid, while, according to the view taken by others, it is merely the serous surface of the dura mater.

The *parietal* wall of the arachnoid space is invested with a layer of polygonal epithelial cells, which are flattened and nucleated. Besides this, it presents in the greater part of its extent no tissue distinct from the dura mater ; and hence it is that Kölliker and others object to the term parietal layer of the arachnoid membrane as applied to the structure of this surface. It may be mentioned, however, that in certain recesses, as for example at the sides of the crista galli, and between the trabeculæ into which the deep fibres of the dura mater are thrown in the neighbourhood of the superior longitudinal sinus, a small amount of delicate connective tissue beneath the epithelium may be distinguished from the dense fibres of the dura mater.

The *visceral layer* of the arachnoid is a distinct transparent membrane which passes over the various eminences and depressions on the cerebrum and cerebellum, without dipping into the sulci and smaller fissures ; nor is it uniformly and closely adherent to the pia mater. The interval left between the arachnoid membrane and pia mater is named generally the subarachnoid space.

This *subarachnoid space* is wider and more evident in some positions than in others. Thus, in the longitudinal fissure, the arachnoid does not descend to the bottom, but passes across, immediately below the edge of the falx, at a little distance above the corpus callosum. In the interval thus left, the arteries of the corpus callosum run backwards along that body. At the *base* of the brain and in the *spinal canal* there is a wide interval between the arachnoid and the pia mater. In the base of the brain, this subarachnoid space extends in front over the pons and the interpeduncular recess as far forwards as the optic nerves, and behind it forms a considerable interval between the cerebellum and the back of the medulla oblongata. In the spinal canal it surrounds the cord, forming a space of considerable extent.

A certain quantity of *fluid* is contained between the arachnoid membrane and the dura mater ; but it has been shown by Magendie that the chief part of the cerebro-spinal fluid is lodged under the arachnoid, in the subarachnoid space.

Magendie also pointed out the existence of a sort of septum dividing the spinal subarachnoid space at the back of the cord. This is a thin membranous partition, which passes in the median plane from the pia mater covering the posterior median fissure of the cord to the opposite part of the loose portion of the arachnoid membrane. It is incomplete and cribriform ; and consists of bundles of white fibres interlaced more or less with one another. Fibrous bands of the same texture pass across the subarachnoid space in various situations both within the spinal canal and at the base of the brain, stretching thus from the arachnoid membrane to the pia mater.

Fig. 385.

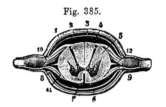

Fig. 385.—TRANSVERSE SECTION OF THE SPINAL CORD AND ITS ENVELOPES (from Sappey after Hirschfeld and Leveillé).

1, dura mater or theca; 2, parietal layer of the arachnoid membrane: 3, internal or loose arachnoid; 4 and 7, subarachnoid cavity or space; 5, hinder part of the antero-lateral column; 6, space between the arachnoid and the dura mater, or between the inner and outer folds of the arachnoid membrane; 8, reflection of the one fold into the other; 9, sheath furnished to the spinal nerve by the dura mater; 10, posterior ganglionic root; 11, smaller anterior root; 12, section of the ligamentum denticulatum. This figure does not show the septum which posteriorly divides the subarachnoid space into right and left parts : this would be placed between the arachnoid at 3, and the pia mater covering the posterior surface of the cord.

As the cerebral and spinal nerves proceed to their foramina of exit from within the dura mater, they are loosely surrounded by tubular sheaths of the arachnoid membrane, which extend along each nerve from the visceral to the parietal layer.

Structure.—When examined under the microscope, the visceral or true arachnoid is found to consist of very distinctly separated riband-like bundles of fibrous tissue interlaced with one another, and a simple layer of scaly epithelium on the surface. Volkmann has described a rich plexus of nerves in the arachnoid membrane of certain ruminants. Kölliker has failed to detect their presence; but they have been again described by Bochdalek, who traces them to the portio minor of the fifth, the facial, and accessorius nerves; and they have likewise been followed by Luschka.

Cerebro-spinal fluid.—This is a very limpid serous fluid, which occupies the subarachnoid space. When collected immediately after death, its quantity was found by Magendie in the human subject to vary from two drachms to two ounces. It is slightly alkaline, and consists, according to an analysis by Lassaigne, of 98·5 parts of water, the remaining 1·5 per cent. being solid matter, animal and saline. In experiments made on the dog, it was found by Magendie to be reproduced in thirty-six hours, after it had been drawn off by puncturing the membranes at the lower part of the cord. When pressure is made upon the brain, the quantity of fluid in the spinal subarachnoid space is increased, and conversely, it may be forced from the spinal cavity upwards into the cranium.

Ligamentum denticulatum.—This is a narrow fibrous band which runs along each side of the spinal cord in the subarachnoid space, between the anterior and posterior roots of the nerves, commencing above at the foramen magnum, and reaching down to the lower pointed end of the cord. By its inner edge this band is connected with the pia mater of the cord, while its outer margin is widely denticulated ; and its denticulations, traversing the arachnoid space, with the arachnoid membrane reflected over them, are attached by their points to the inner surface of the dura mater, and thus serve to support the cord along the sides and to maintain it in the middle of the cavity. The first or highest denticulation is fixed opposite the margin of the foramen magnum, between the vertebral artery and the hypoglossal nerve ; and the others follow in order, alternating with the successive pairs of spinal nerves. In all, there are about twenty-two of these points of insertion. At the lower end, the ligamentum denticulatum may be regarded as continued into the terminal filament of the spinal cord, which thus connects it to the dura mater at the lower end of the sheath. (See Figures 341 and 342.)

Structure.—It consists of white fibrous tissue, mixed with many exceedingly fine elastic fibres which are seen on applying acetic acid. It is obviously continuous on the one hand with the fibrous tissue of the pia mater, and on the other with that of the dura mater.

The pia mater of the cord presents a conspicuous fibrous band, running down in front over the anterior median fissure. This was named by Haller, *linea splendens.*

Glandulæ Pacchioni.—Upon the external surface of the dura mater, in the vicinity of the longitudinal sinus, are seen numerous small pulpy-looking elevations, generally collected into clusters, named glands of Pacchioni. The inner surface of the calvarium is marked by little pits, which receive these eminences. Similar excrescences are seen on the internal surface of the dura mater, and upon the pia mater on each side of the longitudinal sinus, and also projecting into the interior of that sinus. Occasionally they are found also in other situations.

These bodies are not found at birth; and according to the brothers Wenzel, they exist only in very small number, if at all, before the third year. After the seventh year they are usually found, and they increase in number greatly as life advances; in some cases, however, they are altogether wanting. In animals there appears to be no corresponding structure.

On a careful examination of the connections of these bodies it will be found that the elevations, found on the outer surface of the dura mater and within the longitudinal sinus, in no instance take origin in those positions, but that they are grape-like bodies which are attached more deeply, and in their growth have perforated the dura mater. Their precise origin and nature were long the subject of conflicting opinions, but it has been satisfactorily shown by Luschka that they are only an enlarged condition of normal villi of the arachnoid, and that no other structure is involved in their formation. Their most prolific source is, as one may very soon discover, the cerebral or generally acknowledged layer of the arachnoid, but they likewise arise in a similar manner from the serous surface of the dura mater, and may sometimes be found of all sizes in the recesses into which that surface is thrown in the neighbourhood of the longitudinal sinus. (Luschka, in Müller's Archiv. 1852; and "Die Adergeflechte des Menschlichen Gehirns," 1855. See also Cleland "On Tumours of the Dura Mater, &c.," in the Glasgow Medical Journal, 1863.)

BLOOD-VESSELS OF THE BRAIN AND SPINAL CORD.

The origin and course of these vessels have already been described in the Section Angeiology. In passing to their distribution the several arteries, having passed across the arachnoid cavity, enter the subarachnoid space and then divide and subdivide into branches, which, in their farther ramification in the nervous centres, are supported by the pia mater, and, it may be remarked, are more deeply placed in the various fissures and sulci than the small veins, which do not accompany the arteries, but pursue a different course and are seen upon the surface of the pia mater.

Moreover, it is to be observed that, whilst the main branches of the arteries are situated at the base of the brain, the principal veins tend towards the upper surface of the hemispheres, where they enter the superior and inferior longitudinal sinuses : the veins of Galen, however, coming from the lateral ventricles and choroid plexuses, run backwards to the straight sinus.

SIZE AND WEIGHT OF THE ENCEPHALON.

In the following table illustrating the average weight of the adult male and female brain, the results obtained by Sims, Clendinning, Tiedemann, and J. Reid have been brought together in such a form as to exhibit in groups the most commonly prevailing weight; the numbers being also simplified by the omission of fractions. (Sims, "Medico-Chirurg. Trans.," vol. xix., pp. 353—7; Clendinning, "Medico-Chirurg. Trans.," vol. xxi., pp. 59—68 ; Tiedemann, "Das Hirn des Negers," Heidelberg, 1837, pp. 6, 7 ; Reid, "London and Edinburgh Monthly Journal of Medical Science," April, 1843, p. 298, &c.)

Table of the Average Weight of the Male and Female Brain.

MALES, aged 21 years and upwards.

Weight in oz. avoirdupois.	Clendinning.	Sims.	Tiedemann.	Reid.	Total number at each weight.	Classification into three groups to show the *prevailing weight.*
34	—	—	—	1	1	
37	—	2	—	—	2	
38	1	—	—	—	1	
39	—	3	—	1	4	
40	—	2	—	1	3	62 cases { from 34 oz. to 45 oz. } Difference 11.
41	—	3	—	2	5	
42	2	4	2	—	8	
43	—	6	2	3	11	
44	1	6	2	3	12	
45	6	8	—	1	15	
46	2	10	—	8	20	
47	2	6	—	10	18	
48	4	8	2	11	25	
49	3	2	2	12	19	170 cases { from 46 oz. to 53 oz. } Difference 7.
50	4	4	5	13	26	
51	3	3	2	19	27	
52	—	5	4	6	15	
53	4	2	4	10	20	
54	3	2	1	5	11	
55	—	—	2	4	6	
56	—	—	1	6	7	
57	—	—	—	2	2	
58	—	1	4	2	7	46 cases { from 54 oz. to 65 oz. } Difference 11.
59	—	1	2	3	6	
60	—	—	—	1	1	
61	—	—	2	1	3	
62	—	—	1	—	1	
63	—	—	1	1	1	
65	—	—	1	—	1	

Tot. 35+78+39+126 = 278.

FEMALES, aged 21 years and upwards.

Weight in oz. avoirdupois.	Clendinning.	Sims.	Tiedemann.	Reid.	Total number at each weight.	Classification into three groups to show the *prevailing weight.*
31	—	—	—	1	1	
32	—	1	—	—	1	
35	—	2	—	—	2	32 cases { from 31 oz. to 40 oz. } Difference 9.
36	—	4	—	—	4	
37	—	3	1	2	6	
38	2	—	—	—	2	
39	—	3	1	2	6	
40	3	3	—	4	10	
41	2	8	—	2	12	
42	3	6	1	3	13	
43	6	6	—	7	19	
44	5	4	—	13	22	125 cases { from 41 oz. to 47 oz. } Difference 6.
45	4	9	—	7	20	
46	2	9	2	12	25	
47	2	5	—	7	14	
48	—	2	2	2	6	
49	—	1	2	7	10	
50	—	2	1	4	7	
51	—	—	2	4	6	34 cases { from 48 oz. to 56 oz. } Difference 8.
52	1	—	—	—	1	
53	—	1	—	—	1	
54	—	2	—	—	2	
56	—	1	—	—	1	

Tot. 30+72+12+77=191

According to this table, the maximum weight of the adult male brain, in a series of 278 cases, was 65 oz., and the minimum weight 34 oz. In a series of 191 cases, the maximum weight of the adult female brain was 56 oz., and the minimum 31 oz. ; the difference between the extreme weights in the male subject being no less than 31 oz., and in the female 25 oz. By grouping the cases together in the manner indicated by brackets, it is shown that in a very large proportion the weight of the male brain ranges between 46 oz. and 53 oz., and that of the female brain between 41 oz. and 47 oz. The *prevailing* weights of the adult male and female brain may therefore be said to range between those terms; and, by taking the mean, an *average* weight is deduced of $49\frac{1}{2}$ oz. for the male, and of 44 oz. for the female brain,—results which correspond closely with the statements generally received.

Although many female brains exceed in weight particular male brains, the general fact is sufficiently shown, that the adult male encephalon is heavier than that of the female,—the average difference being from 5 to 6 oz. This general superiority in absolute weight of the male over the female brain has been ascertained to exist at every period of life. In new-born infants the brain was found by Tiedemann to weigh on an average from $14\frac{1}{2}$ oz. to $15\frac{3}{4}$ oz. in the male, and from 10 oz. to $13\frac{1}{4}$ oz. in the female:—a fact of considerable interest in practical midwifery, for it has been shown that difficult labours occur in by far the largest number in the birth of male children. (Simpson, London and Edinburgh Monthly Journal of Medical Science, 1845.)

With the above results the observations of Peacock, published in the "Monthly Journ. of Med. Science" for 1847, and further observations by the same author in the "Journ. of the Pathol. Soc." in 1860, in the main agree.

The elaborate table compiled by Rudolph Wagner, and published in his "Vorstudien zu einer Wissensch. Morphol. und Physiol. des Menschl. Gehirns," 1860, containing 964 recorded cases in which the weight of the brain had been ascertained, may also be referred to as another recent useful contribution to the knowledge of this subject.

In illustration of the variation in the average weight of the brain at different ages the following table is given, deduced from the elaborate researches of Dr. Robert Boyd, in the examination of the brains of 2,086 sane persons of both sexes dying in the St. Marylebone Infirmary, and published in the Philos. Trans. for 1860. The weights are stated in oz. avoird. and decimal fractions of them.

Table of the Weight of the Brain of Males and Females at different Ages.

PERIODS OF LIFE.	MALES.				FEMALES.			
	Number weighed.	Maximum.	Minimum.	Average.	Average.	Minimum.	Maximum.	Number weighed.
1 Children prematurely still-born	25	13.1	1.31	5.6	4.62	1.29	9.13	18
2 Children still-born at full period	43	22.	9.37	13.87	12.25	8.	15.12	31
3 New-born infants	42	15.37	6.	11.65	10.	1.75	16.	39
4 Under 3 months................	16	32.75	10.5	17.42	15.94	11	32.5	20
5 From 3 to 6 months	15	30.75	10.75	21.29	19.76	13.	34.75	25
6 From 6 to 12 months	46	36.13	17.75	27.42	25.7	16.37	39.13	40
7 From 1 to 2 years	34	41.25	23.25	33.25	29.8	18.	37.	33
8 From 2 to 4 years	29	50.5	30.5	38.71	34.97	27.75	44.5	29
9 From 4 to 7 years	27	49.5	24.5	40.23	40.11	34.75	48.25	19
10 From 7 to 14 years	22	57.25	39.25	45.96	40.78	34.	52.	18
11 From 14 to 20 years	19	58.5	36.5	48.54	43.94	37.5	52.	16
12 From 20 to 30 years	59	57.	39.25	47.9	43.7	35.75	55.25	72
13 From 30 to 40 years	110	60.75	33.75	48.2	43.09	33.25	53.	89
14 From 40 to 50 years	137	60.	33.75	47.75	42.81	27.5	52.5	106
15 From 50 to 60 years	119	59.	30.5	47.44	43.12	36.25	52.5	103
16 From 60 to 70 years	127	59.5	36.25	46.4	42.69	32.5	54.	149
17 From 70 to 80 years	104	55.25	37.75	45.5	41.27	29.25	49.5	148
18 Upwards of 80 years	24	53.75	41.	45.34	39.77	33.25	48.	77
Averages in Persons above 14 years	699	58.	36.1	47.1	42.5	33.1	52.1	760
Persons from 14 to 70 years..	571	59.12	35.	47.7	43.15	33.8	53.15	535

Anatomists have differed considerably in their statements as to the period at which the brain attains its full size, and also as to the effect of old age in diminishing the weight of this organ. Sœmmerring held that the brain reached its full size as early

P P

as the third year; the Wenzels and Sir W. Hamilton fixed the period about the seventh, and Tiedemann between the seventh and eighth years. Gall and Spurzheim were of opinion that the brain continued to grow until the fortieth year. The observations of Sims, Tiedemann, and Reid, appear to show that in both sexes the weight of the brain in general increases rapidly up to the seventh year, then more slowly to between sixteen and twenty, and again more slowly to between thirty-one and forty, at which time it reaches its maximum point. Beyond that period there appears a slow but progressive diminution in weight of about 1 oz. during each subsequent decennial period; thus confirming the opinion that the brain diminishes in advanced life. According to Peacock, the maximum weight of the brain is attained between the ages of twenty and thirty years. The table of Boyd inserted above would appear to show a somewhat earlier period as that at which the maximum is reached in both sexes, and that the period of decline scarcely begins before sixty years. With this result the observations of Huschke, made upon the brains of 359 men and 245 women, in general agree. ("Schädel, Hirn, und Seele des Menschen und der Thiere, &c.," 1854.)

All other circumstances being alike, the size of the brain appears to bear a general relation to the mental power of the individual,—although many instances occur in which this rule is not applicable. The brain of Cuvier weighed upwards of 64 oz., and there are other recorded examples of brains belonging to men of great talent which nearly equalled it in weight. (Emille Rousseau, "Maladie et autopsie de M. G. Cuvier," Lancette Française, Mai 26, 1832.) On the other hand, the brain in idiots is remarkably small. In three idiots, whose ages were sixteen, forty, and fifty years, Tiedemann found the weight of their respective brains to be $19\frac{3}{4}$ oz., $25\frac{3}{4}$ oz., and $22\frac{1}{2}$ oz; and Dr. Sims records the case of a female idiot twelve years old, whose brain weighed 27 oz. Allen Thomson has found the brain of a dwarfish idiot girl seventeen years of age to weigh $18\frac{1}{4}$ oz. after preservation in alcohol.

The human brain is found to be absolutely heavier than that of all the lower animals except the elephant and whale. The brain of the elephant, according to Perrault, Moulins, and Sir A. Cooper, weighs between 8 and 10 lbs.; whilst that of the whale was found by Rudolphi, in a specimen 75 feet long, to weigh upwards of 5 lbs.

The *relative weight of the encephalon to the body* is liable to great variation; nevertheless, the facts to be gathered from the tables of Clendinning, Tiedemann, and Reid, furnish this interesting general result. In a series of 81 males, the average proportion between the weight of the brain and that of the body at the ages of twenty years and upwards, was found to be as 1 to 36·5; and in a series of 82 females, to be as 1 to 36·46. In these cases, the deaths were the result of more or less prolonged disease; but in six previously healthy males, who died suddenly from disease or accident, the average proportion was 1 to 40·8.

The proportionate weight of the brain to that of the body is much greater at birth than at any other period of life, being, according to Tiedemann, about 1 to 5·85 in the male, and about 1 to 6·5 in the female. From the observations already referred to, it further appears that the proportion diminishes gradually up to the tenth year, being then about 1 to 14. From the tenth to the twentieth year, the relative increase of the body is most striking, the ratio of the two being at the end of that period about 1 to 30. After the twentieth year, the general average of 1 to 36·5 prevails, with a further trifling decrease in advanced life.

Viewed in relation to the weight of his body, the brain of man may be stated generally to be heavier than the brains of the lower animals; but there are some exceptions to the rule, as in the case of certain species of small birds, in the smaller apes, and in some small rodent animals.

The attempts hitherto made to measure or estimate the relative proportions of the different convoluted parts of the cerebrum to each other and to the degree of intelligence, either more directly or by the cranioscopic methods, have as yet been attended with little success. The more recent researches of Rudolph Wagner, which have been farther prosecuted by his son, hold out some promise when fully carried out to afford more definite results.

These researches had for their object to institute an accurate comparison between the brains of certain persons of known intelligence, cultivation, and mental power, and those of persons of an ordinary or lower grade. As examples of brains of men of superior intellect, he selected those of Professor Gauss, a well-known mathematician of eminence, and Professor Fuchs, a clinical teacher; and as examples of brains of ordinary persons, those of a woman of 29 and a workman named Krebs, all of which he examined and measured with scrupulous care.

The general result of R. Wagner's researches upon these and other brains may be stated to be as follows. 1st. Although the greatest number of brains belonging to men of superior intellect are found to be heaviest or largest, yet there are so many instances in which the brains of such persons have not surpassed, or have even fallen below the average size of the brains of ordinary persons, that superiority of size cannot in the present state of our knowledge be regarded as a constant accompaniment of superiority of intellect, even when due regard has been paid to the comparative stature and other circumstances of the individuals.

2nd. It would appear that, in the brains of certain persons of superior intellect, the cerebral convolutions have been found more numerous and more deeply divided than in those of persons of ordinary mental endowments and without cultivation. But numerous exceptional instances are also found of paucity of convolutions coincident with superior intellect, which make it impossible at present to deduce any certain conclusion with respect to the relation between the number or extent of the convolutions and the intellectual manifestations in different persons.

The careful measurement of all the convolutions and the intervening grooves in the four brains above mentioned has been carried out by the younger Wagner, and the tables and results of these measurements published by him as an appendix to his father's treatise. (Hermann Wagner, "Maasbestimmungen der Oberfläche des Grossen Gehirns," &c., Cassel und Göttingen, 1864.)

The following short table extracted from Hermann Wagner's memoir, and simplified by the omission of small fractions and by the reduction of the measurements from square millimetres to English square inches, may give the reader some idea of the nature of the inquiry.

Comparative measurement of the extent of surface of the Convolutions of the Cerebrum and its lobes.

	Surface of each lobe separately.				Free and deep surfaces of Cerebrum.		Whole surface of Cerebrum.
	Frontal.	Parietal.	Occipital.	Temporal.	Free surface.	Deep or covered surface.	
1. Gauss	139.	70.6	59.4	68.4	112.8	228.2	341.
2. Fuchs	143.4	69.5	59.	67.5	110.7	231.3	342.
3. Woman.....	130.	65.	51.	66.8	107.5	209.9	317.5
4. Workman....	113.2	62.3	50.5	62.	97.4	193.6	291.

WEIGHT OF THE SEVERAL PARTS OF THE ENCEPHALON.

As the result of observations made in reference to this subject, on the brains of 53 males and 34 females, between the ages of twenty-five and fifty-five, Dr. J. Reid has given the following table :—

	Males.		Females.		Difference.	
	oz.	drs.	oz.	drs.	oz.	drs.
Average weight of cerebrum	43	15¾	38	12	5	3¾
„ cerebellum	5	4	4	12¼	0	7¾
„ pons and medulla oblongata	0	15¾	1	0¼	0	0½
„ entire encephalon .	50	3½	44	8½	5	11

With these results the observations of Huschke, derived from a special examination of the brains of 22 females, and 38 males, mainly agree.

From this it appears that the proportionate weight of the cerebellum to that of the cerebrum is, in the male, as 1 to 8⅔, and in the female as 1 to 8¼. The cerebellum attains its maximum weight from the twenty-fifth to the fortieth year ; but the increase in weight after the fourteenth year is shown to be relatively greater in the female than in the male. The whole cerebellum apart from the pons and medulla is heavier in the male ; the lateral lobes of the cerebellum are also heavier in the male. In the male the vermiform process increases gradually from the twentieth to the fiftieth year ; in the female it remains stationary during that period, and after the fiftieth year diminishes rapidly.

In the new-born infant the ratio of the weight of the cerebellum to that of the whole brain is strikingly different from that observed in the adult, being, according to Chaussier, between 1 to 13 and 1 to 26 ; by Cruveilhier it was found to be 1 to 20. Huschke found the weight of the cerebellum, medulla oblongata, and pons together in the new-born infant, as compared with that of the brain, to be in the proportion of 1 to 15, and 1 to 13. In the adult, the proportions were 1 to 7, and 1 to 6.

In most mammalia, the cerebellum is found to be heavier in proportion to the cerebrum, than it is in the human subject ; in other words, the cerebrum in man is larger in proportion to the cerebellum.

Sœmmerring pointed out the fact that the brain is larger in proportion to the nerves connected with it in man than in the lower animals.

A comparison of the width of the cerebrum with that of the medulla oblongata shows that the proportionate diameter of the brain to that of the medulla oblongata is greater in man than in any animal, except the dolphin, in which creature, however, it must be remembered that the cerebral lobes exhibit a disproportionate lateral development. The width of the cerebrum in man, as compared with that of the medulla oblongata at its base or broadest part, is about 7 to 1, while in many quadrupeds it is as 3 to 1 or even as 2 to 1.

WEIGHT OF THE SPINAL CORD.

Divested of its membranes and nerves, the spinal cord in the human subject weighs from 1 oz. to 1¾ oz, and therefore its proportion to the encephalon is about 1 to 33. Meckel states it as 1 to 40.

The disproportion between the brain and the spinal cord becomes less and less in the descending scale of vertebrata, until at length, in cold-blooded animals, the spinal cord becomes heavier than the brain. Thus, in the mouse, the weight of the brain, in proportion to that of the spinal cord, is as 4 to 1 ; in the pigeon, as 3⅓ to 1 ; in the newt only as ⅝ to 1 ; and in the lamprey, as $\frac{1}{75}$ to 1.

In comparison with the size of the body, the spinal cord in man may be stated in general terms to be much smaller than it is in animals. In regard to the cold-blooded animals, to birds, and to small mammalia, this has been actually demonstrated, but not in reference to the larger mammalia.

R. Wagner states, as follows, the proportion of the weight of the spinal marrow taken as 1 to the encephalon and its parts—

a, to the nerve roots	:: 1 :	0·53
b, to the medulla and pons	:: 1 :	1·
c, to the cerebellum	:: 1 :	5·18
d, to the cerebrum	:: 1 :	42·78
e, to the encephalon	:: 1 :	48·96

SPECIFIC GRAVITY OF THE ENCEPHALON.

The specific gravity of different parts of the encephalon has of late attracted some attention from its having been observed that it varies to some extent in different kinds of disease. From the researches of Bucknill, Sankey, Aitken, and Peacock, it appears that the average specific gravity of the whole encephalon is about 1036, that of the grey matter 1034, and that of the white 1040. There are also considerable differences in the specific gravity of some of the internal parts. (William Aitken, " The Science and Practice of Medicine," 1865, vol. 2, p. 265 : J. C. Bucknill in " The Lancet," 1852 : Sankey, in the " Brit. and For. Med. Chir. Review," 1853 : Thos. B. Peacock, in the Trans. of the Pathol. Soc. of London, 1861-2.)

DEVELOPMENT OF THE CEREBRO-SPINAL AXIS.

The cerebro-spinal axis is formed from a superficial deposit of blastema, which occupies the whole width of the dorsal furrow, that elongated depression whose margins come together to complete the walls of the cranio-vertebral cavity (p. 15). This layer of blastema increases in thickness in each lateral half, while in

Fig. 386. — PRIMITIVE FORM OF THE CEREBRO-SPINAL AXIS IN THE EMBRYO OF THE BIRD. Magnified.

A and B (from Reichert) outlines of the dorsal aspect of the embryo bird at twenty-four and thirty-six hours of incubation. In A, the sides of the primitive groove have united to a great extent and converted it into a canal, dilated at the cephalic extremity, 2; 6, the cephalic fold of the germinal membrane; 8, the primordial vertebral masses; 9, the unclosed lumbar part of the vertebral groove. In B, 10, 11, and 12 indicate the partial division of the cephalic portion of the tube into the three primary vesicles; 13, the rudiment of the eye; 14, that of the ear.

C, represents a transverse section of the body of the embryo previous to the closure of the vertebral groove. 1, chorda dorsalis; 2, primitive vertebral groove; 2 to 3, medullary plates continuous at 3, with 4, the corneous layer of the blastoderm; 5, the ventral plates of the middle layer; 6, the lowest or epithelial layer; 7, the primordial vertebral masses.

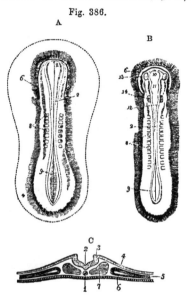

Fig. 386.

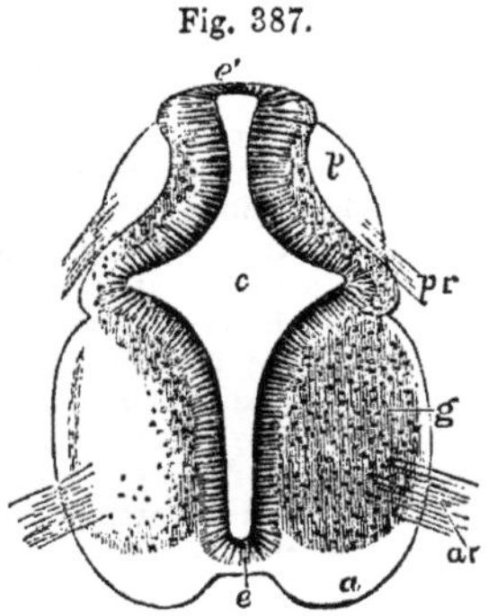

Fig. 387. — TRANSVERSE SECTION OF THE CERVICAL PART OF THE SPINAL CORD OF A HUMAN EMBRYO OF SIX WEEKS (from Kölliker). $\frac{30}{1}$

This and the following figure are only sketched, the white matter and a part of the grey not being shaded in. c, central canal; e, its epithelial lining; at e (inferiorly), the part which becomes the anterior commissure; at e' (superiorly) the part which becomes the posterior commissure; a, the white substance of the anterior columns, beginning to be separated from the grey matter of the interior, and extending round into the lateral column, where it is crossed by the line from g, which points to the grey substance; p, posterior column; a, r, anterior roots; p, r, posterior roots.

Fig. 387.

the middle line—the primitive groove—it remains thin and depressed. The thin middle portion is that which forms in the spinal cord the anterior commissure. At the same time that the walls of the cranio-vertebral cavity are completed behind, the lateral margins of the cerebro-spinal axis are also bent backwards and meet together, so as to form a tube; and this line of junction is the rudiment, in the spinal cord, of the posterior commissure, while the space within the cylinder is the central canal. The closure of the canal first takes place in the cervical region, and subsequently proceeds thence backwards in the dorsal, lumbar, and sacral regions.

The SPINAL CORD has been found by Kölliker already in the form of a cylinder in the cervical region of an embryo four weeks old. Ununited borders have been seen by Tiedemann in the ninth week towards the lower end of the cord, the perfect closing of the furrow being delayed in that part, which is slightly enlarged and presents a longitudinal median slit, analogous to the rhomboidal sinus in birds.

The *anterior fissure* of the cord is developed very early, and contains even at first a process of the pia mater.

The *cervical* and *lumbar enlargements* opposite the attachments of the brachial and crural nerves, appear at the end of the third month: in these situations the central canal, at that time not filled up, is somewhat larger than elsewhere.

Fig. 388.

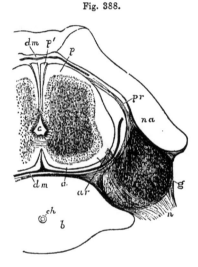

Fig. 388.—TRANSVERSE SECTION OF HALF THE CARTILAGINOUS VERTEBRAL COLUMN AND THE SPINAL CORD IN THE CERVICAL PART OF A HUMAN EMBRYO OF FROM NINE TO TEN WEEKS (from Kölliker). $\frac{16}{1}$

c, central canal lined with epithelium ; a, anterior column ; p, posterior column; p′, band of Goll ; g, ganglion of the posterior root; p r, posterior root ; a r, anterior root passing over the ganglion ; d m, dura matral sheath, omitted near p r, to show the posterior roots ; b, body of the vertebra; c h, chorda dorsalis ; n a, neural arch of the vertebra.

At first the cord occupies the whole length of the vertebral canal, so that there is no cauda equina. In the fourth month the vertebræ begin to grow more rapidly than the cord, and the latter seems as it were to have retired up into the canal, and the elongation of the roots of the nerves which gives rise to the *cauda equina* is commenced. At the ninth month, the lower end of the cord is opposite the third lumbar vertebra.

In textual composition the spinal cord consists at first, after the completion of its cylindrical form, entirely of uniform-looking cells. These separate into two layers, the inner of which forms the epithelium and surrounding connective tissue, or neuroglia of the central canal, while the outer forms the grey substance of the cord. The white substance appears later than the grey, forming a layer external to it, and separated from an early period into an antero-lateral and a posterior column on each side. At a somewhat later period the white mass of these columns, increasing greatly in size, gives rise to the formation and gradually increasing depth of the anterior and posterior median fissures. At the same time, however, the lateral masses of grey substance also undergo extension in the parts named the cornua. It would appear that the integral nerve-fibres are at first developed from radiating processes which proceed from the cells of the grey substance. (Kölliker, Entwicklungsgeschichte ; Lockhart Clarke, in the Phil. Trans. 1862 ; Bidder und Kupfer, Untersuch. üb. d. Rückenmark, Leipz., 1857.)

It may also be mentioned that, according to Remak and Kölliker, the roots of the spinal nerves and the ganglion are at first unconnected with the cord. The mass of blastema forming the ganglion first becomes apparent, and from this the posterior root seems to grow towards, and later to become attached to, the cord; while the anterior roots seem to extend outwards from the cord and to unite themselves later with the nerve.

The central canal is at first nearly cylindrical ; it then becomes flattened at the sides, projecting deeply backwards and forwards. Between the eighth and tenth weeks it is greatly narrowed, and subsequently, being more and more confined to the centre, it ultimately diminishes to a small tube. The epithelial cells which line it from the first are long or columnar, and they retain this form in the adult.

THE ENCEPHALON.

The brain is originally not to be distinguished from the spinal cord, being in fact the anterior portion of the medullary tube. It is soon altered in form, however, by the expansion of its walls in certain parts, while others enlarge in a less degree, and it then presents the appearance of a series of three cerebral vesicles, usually designated by embryologists the *primary cerebral vesicles.*

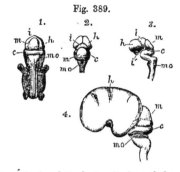

Fig. 389.—Sketches of the Primitive Parts of the Human Brain (from Kölliker).

Fig. 389.

1, 2, and 3 are from a human embryo of about seven weeks. 1, view of the whole embryo from behind, the brain and spinal cord exposed; 2, the posterior, and 3, the lateral view of the brain removed from the body; *h,* the anterior part of the first primary vesicle or cerebral hemisphere (prosencephalon); *i,* the posterior part of the same vesicle (diencephalon); *i',* the lower part of the same; *m,* the middle primary vesicle (mesencephalon); *c,* the cerebellum or upper part of the third primary vesicle (epencephalon); *m o,* the lower part of the third primary vesicle or medulla oblongata. The figure 3 illustrates the several curves which take place in the development of the parts from the primitive medullary tube. In 4, a lateral view is given of the brain of a human embryo of three months: the enlargement of the cerebral hemisphere has covered in the optic thalami, leaving the tubercula quadrigemina, *m,* apparent.

The changes which take place in the growth of the brain were first elaborately described by Tiedemann; they have been investigated by Von Baer, Bischoff, Remak, Kölliker, and others, and have recently received additional elucidation from the researches of Reichert. (Tiedemann, "Anatomie und Bildungsgeschichte des Gehirns," Nüremberg, 1816; Reichert, "Bau des Menschlichen Gehirns," Leipzig. 1861; F. Schmidt, "Beitrage z. Entwick. des Gehirns," in Zeitschr. f. Wissen. Zool. 1862; Kölliker, Entwicklungsgeschichte, 1861.)

DEVELOPMENT OF THE PRIMARY VESICLES.—The anterior or *first vesicle,* is the part from which are developed the third ventricle, the optic thalami, the corpora striata, and the cerebral hemispheres.

The middle or *second vesicle,* forms the corpora quadrigemina above, and the crura cerebri below,—its cavity remaining as the Sylvian aqueduct.

The posterior or *third vesicle,* continues incomplete above for some time, in so far as relates to its nervous substance. At length its anterior portion is closed over and forms the cerebellum above, whilst in its under part the pons Varolii is produced. The posterior portion, on the other hand, continues open on its dorsal aspect, and forms the medulla oblongata and fourth ventricle.

Fig. 390.—Longitudinal Section of the Cranial Cavity of the Human Embryo at four weeks (from Kölliker). $\frac{14}{1}$

Fig. 390.

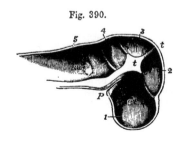

1, 2, 3, 4, and 5, mark the depressions in the cranial wall which contain respectively the cerebral hemispheres, the thalami, the corpora quadrigemina, the cerebellum, and the medulla oblongata; in 1, at *o,* the depression of the eye, and at *o',* the optic nerve is seen; in 5, at *a u,* the primary auditory vesicle; *p,* process from the pharynx, supposed by Rathke to be connected with the formation of the pituitary body or hypophysis cerebri; *t, t,* middle cranial septum or tentorium.

These three vesicles, at first arranged in a straight line, one before the other, soon alter their position, in correspondence with the curving downwards of the cephalic end of the embryo. Thus, at the seventh week there is an angular bend forwards between the hindmost vesicle and the rudimentary spinal cord, the projecting angle (backwards) being named the cervical tuberosity. Another bend, but in the opposite direction, exists between that part of the third vesicle which forms the medulla oblongata, and that which gives rise to the cerebellum. Lastly, a third angle is produced by a bend forwards and downwards in the region of the middle vesicle, from which the corpora quadrigemina are developed, and which forms, at this period, the highest part of the encephalon; whilst the anterior, or first vesicle, is bent nearly at a right angle downwards.

Fig. 391.

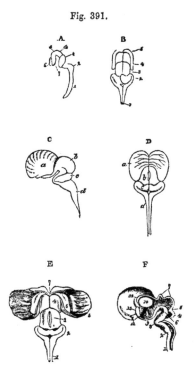

Fig. 391.—SKETCHES OF THE EARLY FORM OF THE PARTS OF THE CEREBRO SPINAL AXIS IN THE HUMAN EMBRYO (after Tiedemann).

A, at the seventh week, lateral view; 1, spinal cord; 2, medulla oblongata; 3, cerebellum; 4, middle vesicle or corpora quadrigemina; 5, 6, 7, first vesicle. B, at the ninth week, posterior view; 1, medulla oblongata; 2, cerebellum; 3, corpora quadrigemina; 4, 5, thalami optici and cerebral hemispheres. C and D, lateral and posterior views of the brain of the human embryo at twelve weeks. a, cerebrum; b, corpora quadrigemina; c, cerebellum; d, medulla oblongata; the thalami are now covered by the enlarged hemispheres. E, posterior view of the same brain dissected to show the deeper parts. 1, medulla oblongata; 2, cerebellum; 3, corpora quadrigemina; 4, thalami optici; 5, the hemisphere turned aside; 6, the corpus striatum embedded in the hemisphere; 7, the commencement of the corpus callosum. F, the inner side of the right half of the same brain separated by a vertical median section, showing the central or ventricular cavity. 1, 2, the spinal cord and medulla oblongata, still hollow; 3, bend at which the pons Varolii is formed; 4, cerebellum; 5, lamina (superior cerebellar peduncles) passing up to the corpora quadrigemina; 6, crura cerebri; 7, corpora quadrigemina, still hollow; 8, third ventricle; 9, infundibulum; 10, thalamus, now solid; 11, optic nerve; 12, cleft leading into the lateral ventricle; 13, commencing corpus callosum.

At an early period of the development of the brain, the anterior primary vesicle undergoes a peculiar change, by which two sets of parts are originated, the first of which corresponds to the cerebral hemispheres, the second to the thalami optici; the middle vesicle, remaining single, gives rise to the corpora quadrigemina; and the posterior vesicle, like the first, but at a somewhat later period, serves as the basis of the rudiments of two parts, viz., the cerebellum and the medulla oblongata. Thus, out of the three primary vesicles five fundamental parts of the encephalon are formed.

The following tabular statement may serve as a synoptical view of the relation subsisting between the primary encephalic vesicles, the five fundamental parts, and the principal permanent structures of the brain :—

I. Anterior primary Vesicle,	1. Prosencephalon. *	Cerebral Hemispheres, Corpus Callosum, Corpora Striata, Fornix, Lateral Ventricles, Olfactory nerve.
	2. Diencephalon.	Thalami Optici, Pineal gland, Pituitary body, Third Ventricle, Optic nerve.
II. Middle primary Vesicle,	3. Mesencephalon.	Crura Cerebri, Corpora Quadrigemina, Aqueduct of Sylvius.
III. Posterior primary Vesicle,	4. Epencephalon.	Cerebellum, Pons Varolii, anterior part of the Fourth Ventricle.
	5. Metencephalon.	Medulla Oblongata, Fourth Ventricle, Auditory nerve.

At a later period of development, the anterior part of the first vesicle, which, as stated above, represents the cerebral hemispheres, increases greatly in size upwards and backwards, and gradually covers the parts situated behind it; first the thalami, then the corpora quadrigemina, and lastly the cerebellum.

On laying open the rudimentary encephalon, two tracts of nervous matter are seen to be prolonged upwards from the spinal cord upon the floor of the cephalic vesicles; these tracts, which are doubtless connected with the anterior and lateral parts of the cord, are the rudiments of the *crura cerebri* and corresponding columns of the medulla oblongata.

FARTHER DEVELOPMENT OF THE PRIMARY VESICLES.—The *third vesicle*.—The posterior portion of this vesicle, corresponding with the *medulla oblongata*, is never closed above by nervous matter. The open part of the medullary tube constitutes the floor of the *fourth ventricle*, which communicates below with the canal of the spinal cord, at the place where the calamus scriptorius is eventually formed.

The three constituent parts of the medulla oblongata begin to be distinguished about the third month; first the *restiform* bodies, which are connected with the commencing cerebellum, and afterwards the anterior pyramids and olives. The *anterior pyramids* become prominent on the surface and distinctly defined in the fifth month; and by this time also their decussation is evident. The *olivary* fasciculi are early distinguishable, but the proper *olivary body*, or tubercle, does not appear till about the sixth month. The *fasciolæ cinereæ* of the fourth ventricle can be seen at the fourth or fifth month, but the *white striæ* not until after birth.

The anterior part of the third vesicle is soon closed above by nervous substance, and forms the commencing *cerebellum*. This part exists about the end of the second month, as a delicate medullary lamina, forming an arch behind the corpora quadrigemina across the widely open primitive medullary tube.

According to Bischoff, the cerebellum does not commence, as was previously supposed, by two lateral plates which grow up and meet each other in the middle line; but a continuous deposit of nervous substance takes place across this part of the medullary tube, and closes it in at once. This layer of nervous matter, which is soon connected with the corpora restiformia, or inferior peduncles, increases gradually up to the fourth month, at which time there may be seen on its under surface the commencing *corpus dentatum*. In the fifth month a division into five *lobes* has taken place; at the sixth, these lobes send out *folia*, which are at first simple, but afterwards become subdivided. Moreover, the *hemispheres* of the cerebellum are now relatively larger than its median portion, or *worm*. In the seventh month the organ is more complete, and the *flocculus* and *posterior velum*, with the other parts of the inferior vermiform process, are now distinguishable, except the *amygdalæ*, which are later in their appearance.

Of the *peduncles* of the cerebellum, the *inferior* pair (corpora restiformia) are the first seen—viz., about the third month; the *middle* peduncles are perceptible in the fourth month; and at the fifth, the *superior* peduncles and the Vieussenian valve.

* This and the four following terms are adopted as applicable to the principal secondary divisions of the primordial medullary tube, and as corresponding to the commonly received names of the German embryologists, viz., Vorderhirn, Zwischenhirn, Mittelhirn, Hinterhirn, and Nachhirn; or their less used English translations, viz., forebrain, interbrain, midbrain, hindbrain, and afterbrain.

The *pons Varolii* is formed, as it were, by the fibres from the hemispheres of the cerebellum embracing the pyramidal and olivary fasciculi of the medulla oblongata. According to Baer, the bend which takes place at this part of the encephalon thrusts down a mass of nervous substance before any fibres can be seen; and in this substance transverse fibres, continuous with those of the cerebellum, are afterwards developed. From its relation to the cerebellar hemispheres the pons keeps pace with them in its growth; and, in conformity with this relation, the transverse fibres are few, or entirely wanting, in those animals in which there is a corresponding deficiency or absence of the lateral parts of the cerebellum.

Fig. 392.

A.

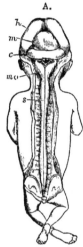

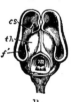

B.

Fig. 392 A.—BRAIN AND SPINAL CORD EXPOSED FROM BEHIND IN A FŒTUS OF THREE MONTHS (from Kölliker).

h, the hemispheres; *m*, the mesencephalic vesicle or corpora quadrigemina; *c*, the cerebellum; below this are the medulla oblongata, *m o*, and fourth ventricle, with remains of the membrana obturatoria. The spinal cord, *s*, extends to the lower end of the sacral canal and presents the brachial and crural enlargements.

Fig. 392 B.—UPPER VIEW OF THE BRAIN OF A THREE MONTHS' FŒTUS, IN WHICH THE HEMISPHERES HAVE BEEN DIVIDED AND TURNED ASIDE, AND THE VESICLE OF THE MESENCEPHALON (COR-PORA QUADRIGEMINA) OPENED (from Kölliker).

f, anterior part of the great arch of the hemispheres over the cerebral fissure; *f'*, posterior part descending into the cornu ammonis; *c s*, corpus striatum; *t h*, thalamus opticus; *m*, in the floor of the opened vesicle of the mesencephalon, which is still hollow.

The *second*, or *middle vesicle.*—The *corpora quadrigemina* are formed in the upper part of the middle cephalic vesicle; the hollow in the interior of which communicates with those of the first and third vesicles. The corpora quadrigemina, in the early condition of the human embryo, are of great proportionate volume, in harmony with what is seen in the lower vertebrata; but subsequently they do not grow so fast as the anterior parts of the encephalon, and are therefore soon overlaid by the cerebral hemispheres, which at the sixth month cover them in completely. Moreover, they become gradually solid by the deposition of matter within them; and as, in the meantime, the *cerebral peduncles* are increasing rapidly in size in the floor of this middle cephalic vesicle, the cavity in its interior is quickly filled up, with the exception of the narrow passage named the *Sylvian aqueduct.* The fillet is distinguishable in the fourth month. The corpora quadrigemina of the two sides are not marked off from each other by a vertical median groove until about the sixth month; and the transverse depression separating the anterior and posterior pairs is first seen about the seventh month of intra-uterine life.

The *first,* or *anterior vesicle.*—This vesicle, very soon after its formation, exhibits two lateral outgrowths—the *optic vesicles,*—destined to form the fundamental parts of the organs of vision. Each of these soon becomes separated from the parent vesicle by a constricted part, which forms the optic nerve and tract. The first vesicle has usually been described as dividing into two portions—viz., a posterior, which is developed into the optic thalami and third ventricle, and an anterior, which forms the principal mass of the cerebral hemispheres, including the corpora striata. Reichert, however, has pointed out that the hemispheres and corpora striata are developed from the sides of the fore part of the vesicle, and become distinguished from it by a constriction similarly as the optic vesicles had previously been, and that there is left between the *hemisphere-vesicles* of opposite sides a wedge-shaped interval, which forms the third ventricle. He points out that the terminal extremity of the cerebro-spinal tube is at the tip of this wedge, and is placed immediately in front of the optic commissure, at the lamina cinerea; and that therefore the infundibulum is not that extremity, as had been previously supposed by Baer, but is an expansion of the vesicle downwards, in similar fashion as there is an expansion of it upwards in the region of the pineal body.

The *pituitary body* was asserted by Rathke to be derived from a prolongation upwards of the mucous membrane of the pharynx into the base of the skull between the trabeculæ. It appears, however, from the researches of Reichert and Bidder, that the base of the skull is never imperfect in this region. Reichert suggested that the pituitary body might be derived from the extremity of the chorda dorsalis, but is now rather inclined to think that it is a development of the pia mater.

Fig. 393.

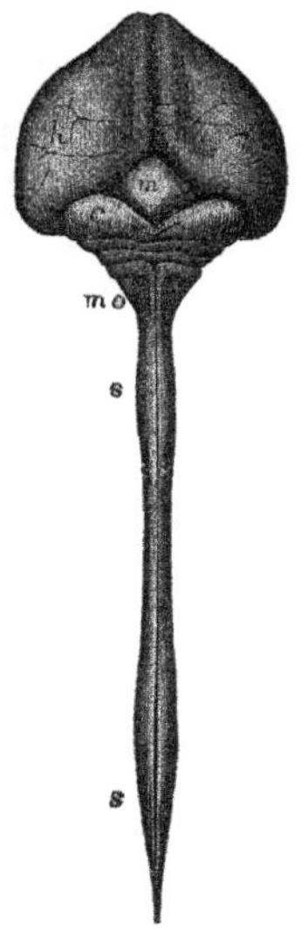

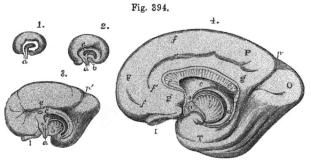

Fig. 393.—BRAIN AND SPINAL CORD OF A FŒTUS OF FOUR MONTHS, SEEN FROM BEHIND (from Kölliker).

h, hemispheres of the cerebrum ; *m,* corpora quadrigemina or mesencephalon ; *c,* cerebellum ; *m o,* medulla oblongata, the fourth ventricle being overlapped by the cerebellum ; *s, s,* the spinal cord with its brachial and crural enlargements.

The *pineal gland,* according to Baer, is developed from the back part of the thalami, where those bodies continue joined together; but it is suggested by Bischoff that its development may be rather connected with the pia mater. It was not seen by Tiedemann until the fourth month ; subsequently its growth is very slow ; and it at first contains no gritty deposit: this, however, was found by Sömmerring at birth.

The two *optic thalami,* formed from the posterior part of the anterior vesicle, consist at first of a single hollow sac of nervous matter, the cavity of which communicates in front with the interior of the commencing cerebral hemispheres, and behind with that of the middle cephalic vesicle (corpora quadrigemina). Soon, however, by means of a deposit taking place in their interior

Fig. 394.

Fig. 394.—SEMIDIAGRAMMATIC VIEWS OF THE INNER SURFACE OF THE RIGHT CEREBRAL HEMISPHERE OF THE FŒTAL BRAIN AT VARIOUS STAGES OF DEVELOPMENT (from Schmidt).

1, 2, and 3, are from fœtuses of the respective ages of eight, ten, and sixteen weeks ; 4, from a fœtus of six months. *a,* lamina terminalis or part of the first primary vesicle which adheres to the sella turcica ; *b,* section of the cerebral peduncle as it passes into the thalamus and corpus striatum ; the arched line which surrounds this bounds the great cerebral fissure ; *c,* anterior part of the fornix and the septum lucidum ; *d,* inner part of the arch of the cerebrum, afterwards the hippocampus major and posterior part of the fornix ; *e,* corpus callosum, very short in 3, elongated backwards in 4 ; in 4, *f,* the superior marginal convolution ; *f',* fronto-parietal fissure ; *g,* gyrus fornicatus ; *p',* the internal vertical fissure descending to meet the fissure of the hippocampus ; I, olfactory bulb ; F, P, O, T, frontal, parietal, occipital and temporal lobes.

behind, below, and at the sides, the thalami become solid, and at the same time a cleft or fissure appears between them above, and penetrates down to the internal cavity, which continues open at the back part opposite the entrance of the Sylvian aqueduct. This cleft or fissure is the *third ventricle*. Behind, the two thalami continue united by the *posterior commissure*, which is distinguishable about the end of the third month, and also by the *peduncles of the pineal gland*. The *soft commissure* could not be detected by Tiedemann until the ninth month; but its apparent absence at earlier dates may perhaps be attributed to the effects of laceration.

At an early period the *optic tracts* may be recognised as hollow prolongations from the outer part of the wall of the thalami while they are still vesicular. At the fourth month these tracts are distinctly formed.

The *hemisphere-vesicle* becomes divisible into two parts : one of these is the part which from the interior appears as the corpus striatum, and from the exterior as the island of Reil, or central lobe; the other forms the expanded or covering portion of the hemisphere, and is designated by Reichert the *mantle*. The aperture existing at the constricted neck of the hemisphere vesicle, Schmidt and Reichert have recognised as the foramen of Monro.

The *corpora striata*, it will be observed, have a very different origin from the optic thalami; for, while the optic thalami are formed by thickening of the circumferential wall of a part of the first cerebral vesicle, and thus correspond in their origin with all the parts of the encephalon behind them, which are likewise derived from portions of the cerebro-spinal tube, the corpora striata appear as thickenings of the floor of the hemisphere-vesicles, which are lateral offshoots from the original cerebro-spinal tube. On this account, Reichert considers the brain primarily divisible into the stem, which comprises the whole encephalon forwards to the tænia semicircularis, and the hemisphere-vesicles, which include the corpora striata and hemispheres.

The cerebral hemispheres enlarging, and having their walls increased in thickness form, during the fourth month (Tiedemann), two smooth shell-like lamellæ, which include the cavities afterwards named the *lateral ventricles*, and the parts contained within them. Following out the subsequent changes affecting the exterior of the cerebral hemispheres, it is found that about the fourth month the first traces of some of the *convolutions* appear, the intermediate *sulci* commencing only as very slight

Fig. 395.

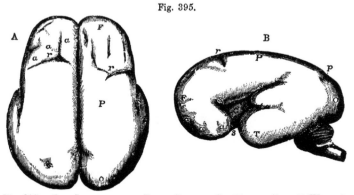

Fig. 395.—THE SURFACE OF THE FŒTAL BRAIN AT SIX MONTHS (from R. Wagner).

This figure is intended to show the commencement of the formation of the principal fissures and convolutions. A, from above; B, from the left side. F, frontal lobe ; P, parietal ; O, occipital ; T, temporal ; *a, a, a,* slight appearance of the several frontal convolutions ; *s,* the Sylvian fissure ; *s',* its anterior division ; within it, C, the central lobe or convolutions of the island ; *r,* fissure of Rolando ; *p,* the vertical fissure (external part) ; *t,* the parallel fissure.

depressions on the hitherto smooth surface. Though the hemispheres continue to grow quickly upwards and backwards, the convolutions at first become distinct by comparatively slow degrees ; but towards the seventh and eighth months they are developed with great rapidity, and, at the beginning of the last month of intra-uterine life, all the principal ones are marked out.

The *Sylvian fissure*, which afterwards separates the anterior from the middle lobe of each hemisphere, begins as a depression or cleft between them about the fourth month, and, after the great longitudinal, is the first of the fissures to make its appearance. It is followed by the fissure of Rolando, and the vertical fissure, and somewhat later by the internal fronto-parietal fissure. After this, the various subordinate fissures dividing the convolutions gradually make their appearance. By the end of the third month the hemispheres have extended so far backwards as to cover the thalami ; at the fourth they reach the corpora quadrigemina ; at the sixth they cover those bodies and great part of the cerebellum, beyond which they project still further backwards by the end of the seventh month.

During the growth of the hemisphere the aperture of the foramen of Monro is extended backwards; the arched margin of this opening is curved downwards at its extremities, and forms anteriorly the fornix, and posteriorly the corpus fimbriatum and hippocampus major ; above the margin a part of the wall of each hemisphere comes into contact with its fellow, and in the lower part forms the septum lucidum, while above this the hemispheres are united by the development of the great commissure, the corpus callosum.

The *corpus callosum* is described by Tiedemann as being first seen about the end of the third month, as a narrow vertical band, extending across between the forepart of the two hemispheres, and subsequently growing backwards. With this view the observations of Schmidt coincide. Reichert, however, maintains that the commissural structure seen at the forepart of the hemispheres is the anterior white commissure, and that the corpus callosum appears in its whole extent at once.

The *corpora albicantia* at first form a single mass : so also do the *anterior pillars* of the fornix, which are distinguished before the posterior pillars. The *posterior pillars* are not seen until the fourth or fifth month. At that period the *hippocampus minor* is also discernible.

Fig. 396.—VIEW OF THE INNER SURFACE OF THE RIGHT HALF OF THE FŒTAL BRAIN OF ABOUT SIX MONTHS (from Reichert).

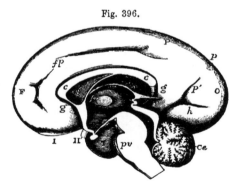

Fig. 396.

F, frontal lobe ; P, parietal ; O, occipital ; T, temporal ; I, olfactory bulb ; II, right optic nerve ; *f p*, fronto-parietal fissure ; *p*, vertical fissure ; *p'*, internal vertical fissure ; *h*, hippocampal fissure ; *g*, gyrus fornicatus ; *c, c*, corpus callosum ; *s*, septum lucidum ; *f*, placed between the middle commissure and the foramen of Monro ; *v*, in the upper part of the third ventricle immediately below the velum interpositum and fornix : *v'*, in the back part of the third ventricle below the pineal gland, and pointing by a line to the aqueduct of Sylvius ; *v''*, in the lower part of the third ventricle above the infundibulum ; *r*, recessus pinealis passing backwards from the tela choroidea ; *p v*, pons Varolii ; *C e*, cerebellum.

MEMBRANES OF THE ENCEPHALON.

It is remarked by Bischoff, that the pia mater and arachnoid are formed by the separation of the outer layer of the primitive cephalic mass ; and thus that the *pia mater* does not send inwards processes into the fissures or sulci, nor into the ventricular cavities ; but that every part of this vascular membrane, including the *choroid plexuses* and *velum interpositum*, is formed in its proper position in connection with the nervous matter. The dura mater, on the other hand, is developed from the inner surface of the dorsal plates.

The pia mater and dura mater have both been detected about the seventh or eighth week, at which period the tentorium cerebelli exists. At the third month the falx cerebri, with the longitudinal and lateral sinuses, are perceptible; and the choroid plexuses of both the lateral and fourth ventricles are distinguishable. No trace of arachnoid, however, can be seen until the fifth month.

II. THE CEREBRO-SPINAL NERVES.

The nerves directly connected with the great cerebro-spinal centre constitute a series of symmetrical pairs, the number of which has been variously estimated from forty to forty-three. Of these nerves, when estimated at the smaller number, nine issue from the cranium through different foramina or apertures in its base, and are thence strictly named *cranial*. The tenth nerve passes out between the occipital bone and the first vertebra, and the remaining thirty nerves all issue below the corresponding vertebral pieces of the spine. To the whole series of thirty-one nerves the name of *spinal* is usually given.

A.—CRANIAL NERVES.

The cranial nerves, besides being named numerically, according to the relative position of the apertures for their exit from the cranium, have likewise been distinguished by other names, according to the place or mode of their distribution, and according to their functions or other circumstances.

The number of the cranial nerves has been variously stated as nine or as twelve by different anatomists; the difference consisting mainly in this, that, under one system, the nerves which enter the internal auditory meatus, and those which pass through the jugular foramen, are in each case considered a single pair (seventh and eighth) divisible into parts; while under another system each of the nerves is held to constitute a distinct pair. The following table presents a synoptical view of the cranial nerves under these two modes of enumeration, as in the respective systems of Willis and of Sömmerring.

WILLIS.*		SŒMMERRING.	
First pair of nerves		First pair of nerves	Olfactory nerves.
Second „		Second „	Optic.
Third „		Third „	Oculo-motor.
Fourth „		Fourth „	Pathetic or trochlear.
Fifth „		Fifth „	Trifacial or trigeminal.
Sixth „		Sixth „	Abducent ocular.
Seventh „	{ nervus durus	Seventh „	Facial motor.
	{ n. mollis	Eighth „	Auditory.
	{ n. vagus	{ Ninth „	Glosso-pharyngeal.
Eighth „ {		{ Tenth „	Pneumo-gastric.
	{ accessorius.	Eleventh „	Spinal accessory.
Ninth „		Twelfth „	Hypoglossal or lingual motor.

The arrangement of Sömmerring is the preferable one, as being the simplest and most natural; for each of the parts included in the seventh and eighth pairs of Willis is really a distinct nerve. But as the plan of Willis is still in more general use, it will most conveniently be followed here. The cranial nerves will therefore, when not otherwise designated, be referred to as consisting of nine pairs.

* Willis described the glosso-pharyngeal nerve as a branch of the vagus, and included the suboccipital nerve as a tenth among the cranial nerves.

CONNECTIONS OF THE CRANIAL NERVES WITH THE ENCEPHALON.

The roots of the cranial nerves may be traced for some depth into the substance of the encephalon, a circumstance which has led to the distinction of the *deep* or *real* origin, and the *superficial* or *apparent* origin, by which latter is understood the place at which the nerve appears to be attached to the surface of the encephalon. The superficial origin of these nerves is quite obvious, but their deeper connection is, in most cases, still a matter of some uncertainty.

Fig. 397.—VIEW FROM BELOW OF THE CONNECTION OF THE PRINCIPAL NERVES WITH THE BRAIN.

Fig. 397.

The full description of this figure will be found at p. 538. The following references apply to the roots of the nerves : I′, the right olfactory tract divided near its middle ; II, the left optic nerve springing from the commissure which is concealed by the pituitary body ; II′, the right optic tract ; the left tract is seen passing back into *i* and *e*, the internal and external corpora geniculata ; III, the left oculomotor nerve ; IV, the trochlear ; V, V, the large roots of the trifacial nerves ; + +, the lesser roots, the + of the right side is placed on the Gasserian ganglion ; 1, the ophthalmic ; 2, the superior maxillary, and 3, the inferior maxillary nerves ; VI, the left abducent nerve ; VII, *a*, *b*, the facial and auditory nerves ; *a*, VIII, *b*, the glosso-pharyngeal, pneumo-gastric, and spinal accessory nerves ; IX, the right hypoglossal nerve ; at *o*, on the left side, the rootlets are seen cut short ; C I, the left suboccipital or first cervical nerve.

1. The first or *olfactory* nerve, as it is usually described, small in man in comparison with animals, lies on the under surface of the anterior lobe of the cerebrum to the outer side of the longitudinal median fissure, lodged in a sulcus between two straight convolutions. Unlike other nerves, it consists of a large proportion of grey matter mixed with white fibres, and indeed

agrees closely in structure with the cerebral substance. It swells into an oval enlargement, the *olfactory bulb*, in front, which also contains much grey matter, and from this part small soft nerves descend through the cribriform plate into the nose. When traced backwards, it is found to be spread out and attached behind to the under surface of the anterior lobe by means of *three roots*, named external, middle, and internal, which pass in different directions. The bulbous part is therefore rather to be regarded as an olfactory lobe of the cerebrum than as a part of a true nerve, while the white part prolonged backwards into the brain, together with its so-called roots, may be termed the olfactory tract.

The *external* or *long* root consists of a band of medullary fibres, which passes, in the form of a white streak, outwards and backwards along the anterior margin of the perforated space, towards the posterior border of the Sylvian fissure, where it may be followed into the substance of the cerebrum. Its further connections are doubtful, but it has been stated that its fibres have been traced to the following parts, viz., the convolutions of the island of Reil, the anterior commissure, and the superficial layer of the optic thalamus (Valentin).

The *middle* or *grey* root is of a pyramidal shape, and consists of grey matter on the surface, which is prolonged from the adjacent part of the anterior lobe and perforated space. Within it there are white fibres, which have been traced to the corpus striatum.

The *internal root* (short root, Scarpa), which cannot always be demonstrated, is composed of white fibres which may be traced from the inner and posterior part of the anterior lobe, where they are said by Foville to be connected with the longitudinal fibres of the gyrus fornicatus.

The question whether the olfactory bulbs ought to be considered as nerves or as cerebral lobes is, if tested by reference to the history of development, not so simple as might at first appear. It is in favour of their being regarded as lobes, that in the lower vertebrate animals the olfactory bulbs are generally recognised by comparative anatomists as additional encephalic lobes, and that in most mammals they are much larger proportionally than in man, and frequently contain a cavity or ventricle in their interior, and further that in their minute structure they nearly agree with the cerebrum; but, as it is known that in the first development of the eye the peripheral part or retina, as well as the rest of the optic nerve, is originally formed by the extension of a hollow vesicle from the first fœtal encephalic compartment, so in the case of the olfactory nerve, although the peripheral or distributed part is of separate origin from the olfactory bulb, the latter part is comparable in its origin with the optic vesicle.

2. The *second* pair or *optic* nerves of the two sides meet each other at the optic commissure (chiasma), where they partially decussate. From this point they may be traced backwards round the crura cerebri, under the name of the optic tracts.

Each *optic tract* arises from the optic thalamus, the corpora quadrigemina, and the corpora geniculata. As it leaves the under part of the thalamus, it makes a sudden bend forwards and then runs obliquely across the under surface of the cerebral peduncle, in the form of a flattened band, which is attached by its anterior surface to the peduncle; after this, becoming more nearly cylindrical, it adheres to the tuber cinereum, from which and, as stated by Vicq-d'Azyr, from the lamina cinerea it is said to receive an accession of fibres, and thus reaches the optic commissure.

In the *commissure* the nerve-fibres of the two sides undergo a partial decussation. The outer fibres of each tract continue onwards to the eye of the same side : the inner fibres cross over to the opposite side ; and fibres have been described as running from one optic tract to the other along the posterior part of the commissure, while others pass between the two optic nerves in its anterior part (Mayo).

In front of the commissure, the nerve enters the foramen opticum, receiving a sheath from the dura mater and acquiring greater firmness.

Fig. 398.

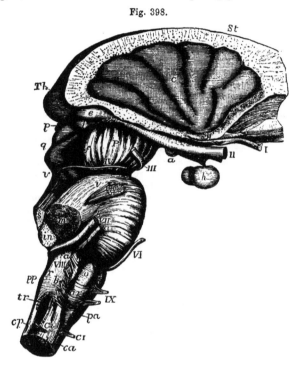

Fig. 398.—LATERAL VIEW OF THE CONNECTION OF THE PRINCIPAL NERVES WITH THE BRAIN.

The full description of this figure will be found at p. 553. The following references apply to the roots of the nerves ; I, the right olfactory tract cut near its middle ; II, the optic nerves immediately in front of the commissure ; the right optic tract is seen passing back to the thalamus (Th), corpora geniculata (i, e,), and corpora quadrigemina (q) ; III, the right oculo-motor nerve ; IV, the trochlear nerve rising at v, from near the valve of Vieussens ; V, the trifacial nerve ; VI, the abducent ocular ; a, VII, b, the facial and auditory nerves, and between them the pars intermedia ; a, VIII, b, the roots of the glosso-pharyngeal, pneumo-gastric, and spinal accessory nerves ; IX, the hypoglossal nerve ; C I, the separate anterior and posterior roots of the suboccipital or first cervical nerve.

The fibres of origin of the optic tract from the thalamus are derived partly from the superficial stratum and partly from the interior of that body. According to Foville, this tract is also connected with the tænia semicircularis, and with the termination of the gyrus fornicatus ; and he states further, that where the optic tract turns round the back of the thalamus and the cerebral peduncle it receives other delicate fibres, which descend from the grey matter of those parts.—(Op. cit. p. 514.)

Q Q

3. The *third pair* of nerves (motores oculorum) have their apparent or superficial origin from the inner surface of the crura cerebri in the interpeduncular space, immediately in front of the pons. Each nerve consists of a number of funiculi which arise in an oblique line from the surface.

The deeper fibres of origin, when followed into the crus, are found to diverge in its substance, some being traced to the locus niger, others running downwards in the pons among its longitudinal fibres, and others turning upwards to be connected with the corpora quadrigemina and Vieussenian valve. According to Stilling, with whom Kölliker agrees, the major part of the fibres arise from a grey nucleus in the floor of the Sylvian aqueduct, close to the origin of some fibres of the fourth nerve.

4. The *fourth* pair, *pathetic* or *trochlear* nerves, the smallest of those which are derived from the brain, are seen at the outer side of the crura cerebri immediately before the pons. Each nerve may be traced backwards round the peduncle to a place below the corpora quadrigemina, where it arises from the upper part of the valve of Vieussens. Kölliker states that, under the corpora quadrigemina, the fibres of origin are divided into two bundles ; the anterior being traceable through the lateral wall of the aqueduct of Sylvius to its floor, where it arises from a grey nucleus close to the middle line ; the posterior bundle being derived from a grey nucleus in the floor of the fourth ventricle, close to the origin of the fifth nerve. The roots of the nerves of opposite sides are connected together across the middle line in the form of a white band or commissure in the substance of the velum.

5. The *fifth* pair of nerves, par trigeminum, *trifacial nerves*, take their superficial origin from the side of the pons Varolii, where the transverse fibres of the latter are prolonged into the middle crus cerebelli, considerably nearer to the upper than to the lower border of the pons.

The fifth nerve consists of a larger or sensory, and a smaller or motor root. The smaller root is at first concealed by the larger, and is placed a little higher up, there being often two or three cross fibres of the pons between them. On separating the two roots, the lesser one is seen to consist of a very few funiculi. In the larger root the funiculi are numerous, amounting sometimes to nearly a hundred.

Deep origin.—The *greater* root runs behind the transverse fibres of the pons towards the lateral part of the medulla oblongata at the back of the olivary body. Several anatomists trace it into the floor of the fourth ventricle, between the fasciculi teretes and the restiform bodies. By some it is considered to be continuous with the fasciculi teretes and lateral columns of the cord, whilst others connect it with the grey mass which is regarded by Stilling as the nucleus of the glosso-pharyngeal nerve.

The *motor* root was supposed by Bell to descend to the pyramidal body, and Retzius believes that he has confirmed that opinion by dissection : but it would appear that the deep connection of this root is not yet known with certainty. According to Stilling the fibres pass through the pons to the floor of the fourth ventricle, and have their origin in its grey matter.

According to Foville, some of the fibres of the sensory root of the fifth nerve are connected with transverse fibres in the pons, whilst others spread out on the surface of the middle peduncle of the cerebellum, and enter that part of the encephalon beneath the folia.—(Op. cit. p. 506.)

6. The *sixth* nerve (abducens), *motor oculi externus*, takes its apparent origin from between the pyramidal body and the pons Varolii by means of a larger and a smaller bundle. It is connected with the pyramid, and to a small extent with the pons also. Philipeaux and Vulpian, with whom

Kölliker concurs, state that the fibres may be traced more deeply to the floor of the fourth ventricle.

7. The *seventh* pair of nerves appear on each side at the posterior margin of the pons, between the middle and inferior peduncles of the cerebellum, and nearly in a line with the place of attachment of the fifth nerve.

Deep origins.—The *portio dura* or *facial* nerve, placed a little nearer to the middle line than the portio mollis, may be traced to the medulla oblongata between the restiform and olivary fasciculi, with both of which it is said to be connected. Some of its fibres are derived from the pons. Philipeaux and Vulpian affirm that the fibres arise from the outer wall of the fourth ventricle, and that many of them decussate in its floor.

Connected with the portio dura, and intermediate between it and the portio mollis, is a smaller white funiculus, first described by Wrisberg (portio inter duram et mollem). The roots of this accessory or intermediate portion are connected deeply with the lateral column of the cord.

The *portio mollis*, or auditory nerve, rises from the floor of the fourth ventricle, at the back of the medulla oblongata, in which situation, as already described, transverse white striæ are seen, which form the commencement of the nerve. These roots are connected with the grey matter, and some appear to come out of the median fissure. The nerve then turns round the restiform body, and becomes applied to the lower border of the pons, receiving accessions from the former of those parts, and according to some authors from the latter also.

Foville says that the roots of the portio mollis are also connected by a thin layer on the under surface of the middle peduncle with the cortical substance of the cerebellum ; also, with the small lobule named the flocculus ; and with the grey matter at the borders of the calamus scriptorius.

8. The *eighth* nerve consists of three distinct portions.

The uppermost portion is the *glosso-pharyngeal* nerve ; next to this, and lower down, is the *par vagum* or *pneumo-gastric* nerve consisting of a larger number of cords. The roots of both these nerves are attached superficially to the fore part of the restiform body. Still lower, is the *spinal accessory* nerve, which, ascending from the side of the spinal cord, enters the skull by the foramen magnum, and is associated with the pneumo-gastric nerve as it passes out through the foramen lacerum.

The accessory nerve arises within the spinal canal from the lateral column of the cord behind its middle, by a series of slender roots, which commence as low down as the fifth or sixth cervical nerve. The nerve passes upwards between the posterior roots of the cervical nerves and the ligamentum denticulatum,—the several funiculi of origin successively joining it as it ascends. On entering the skull, it receives funiculi from the side of the medulla oblongata.

These three portions of the eighth pair are connected deeply with grey nuclei within the cord and medulla oblongata, as already described (see p. 521).

9. The *ninth* nerve (hypoglossal) arises, in a line continuous with that of the anterior roots of the spinal nerves, by scattered funiculi from the furrow between the olivary body and the anterior pyramid.

The roots of the ninth nerve are traced by Stilling to one of the grey nuclei already described in the medulla oblongata, and they are said by Kölliker to undergo partial decussation in the floor of the fourth ventricle.

DISTRIBUTION OF THE CRANIAL NERVES.

Mode of exit from the cranium.—Each of the cranial nerves issues at first from the cranial cavity through a foramen or tubular prolongation of the dura mater : some of these nerves or their main divisions are contained in distinct foramina of the cranium, others are grouped together in one foramen. The numerous small olfactory nerves descend into the nose through the cribriform plate of the ethmoid bone ; the optic nerve pierces the root of the lesser wing of the sphenoid bone ; the third, fourth, and sixth nerves, with the ophthalmic division of the fifth nerve, pass through the

Fig. 399.

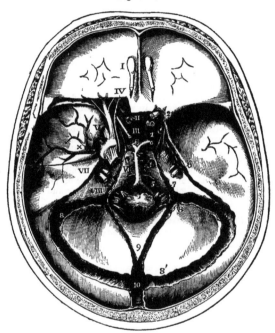

Fig. 399.—INTERNAL VIEW OF THE BASE OF THE SKULL, SHOWING THE PLACES OF EXIT OF THE CRANIAL NERVES.

The dura mater is left in great part within the base of the skull; the tentorium is removed and the venous sinuses are opened. On the left side a small portion of the roof of the orbit has been removed to show the relation of certain nerves at the cavernous sinus and in the sphenoidal fissure. The roots of the several cranial nerves have been divided at a short distance inside the foramina of the dura mater through which they respectively pass. I, the bulb of the olfactory nerve lying over the cribriform plate of the ethmoid bone ; II, the optic nerves, that of the left side cut short; III, placed on the pituitary body, indicates the common oculo-motor nerve ; IV, the trochlear nerve ; V, is placed on the left side opposite to the middle of the three divisions of the trigeminus, which, together with the ganglion and greater root, have been exposed by opening up the dura mater ; on the right side the greater root is seen ; VI, placed below the foramen of exit of the abducent ocular ; VII, placed on the upper part of the petrous bone opposite the entrance of the facial and auditory nerves into the meatus auditorius internus ; VIII, placed on the petrous bone outside the jugular foramen opposite the place of exit of the three divisions of the eighth pair of nerves ; IX, placed upon the basilar part of the occipital bone in front of the hypoglossal nerve as it passes through the anterior condyloid foramen. On the left side at the cavernous sinus, the third, fourth, and ophthalmic division of the fifth nerves are seen keeping towards the outer side, while the sixth nerve is deeper and close to the internal carotid artery. The explanation of the remaining references in this figure will be found at p. 461.

sphenoidal fissure ; the superior maxillary and inferior maxillary divisions of the fifth pass respectively through the foramen rotundum and foramen ovale of the great wing ; the facial and auditory nerves pierce the petrous bone ; the three parts of the eighth pair descend in separate canals of the dura mater through the anterior part of the jugular foramen between the petrous and occipital bones ; and the hypoglossal nerve passes through the anterior condyloid foramen of the occipital bone.

General distribution.—The greater number of the cranial nerves are entirely confined in their distribution within the limits of the head ; as in the case of the first six pairs and the auditory nerve. Of these, the olfactory, optic, and auditory are restricted to their respective organs of sense; while the third, fourth, and sixth are exclusively motor nerves in connection with the external and internal muscles of the eyeball and that of the upper eyelid. In the remaining nerve, the fifth or trifacial, all the fibres derived from the greater root, and connected with the Gasserian ganglion, are entirely sensory in their function, and constitute the whole of the first and second and the greater part of the third division of the nerve : but the last of these divisions has associated with it the fibres of the lesser root, so as to become in some degree a compound nerve. As a nerve of sensation the trifacial occupies in its distribution the greater part of the head superficially and deeply, excepting the interior of the cranium and that part of the scalp which is situated in the region behind a perpendicular line passing through the external auditory meatus. The muscular distribution of the inferior division of the fifth nerve is chiefly to the muscles of mastication.

Of the remaining nerves, the facial and hypoglossal, both exclusively motor in function, are almost entirely cephalic in their distribution ; the facial nerve giving fibres to all the superficial and a few of the deeper muscles of the head and face ; and the ninth or hypoglossal supplying the muscles of the tongue. Of the facial, however, a small branch joins one of the cervical nerves in the platysma myoides ; and of the ninth, the descending branch supplies in part the muscles of the neck which depress the hyoid bone and larynx.

Of the three parts of the eighth pair, ranked as cranial nerves in consequence of their passing through one of the foramina of the cranium, two have only a very limited distribution in the head, and furnish nerves in much greater proportion to organs situated in the neck and thorax. One of these, the pneumo-gastric, after giving a small branch to the ear-passages, and supplying nerves to the larynx and pharynx, the trachea, gullet, the lungs and heart, extends into the abdominal cavity as the principal nerve of the stomach. The other, the spinal accessory, which is partially united with the glosso-pharyngeal and pneumo-gastric near their origin and thus furnishes some of their motor fibres, is entirely a motor nerve, and is distributed in the sterno-mastoid and trapezius muscles. The glosso-pharyngeal nerve is more strictly confined to the head, supplying branches to the tongue, pharynx, and part of the ear-passages.

On the following two pages, Fig. 400 is introduced in illustration of the general view of the distribution above given. In this figure the cranium and orbit have been opened up to the depth of the several foramina through which the nerves pass. The greater part of the lower jaw has also been removed on the left side, and the tongue, pharynx, and larynx are partially in view. The occipital bone has been divided by an incision passing down from the occipital tuberosity and through the condyle to the left of the foramen magnum. The cervical vertebræ have been divided to the left of' the middle, and the sheath of the spinal cord opened so as to expose the roots of the cervical nerves.

Fig. 400, A.

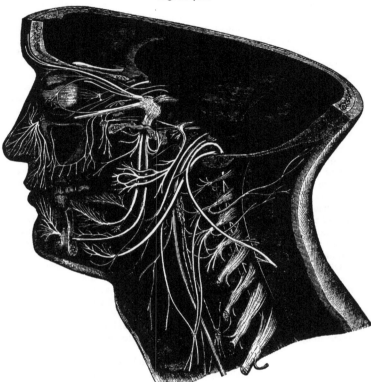

Fig. 400.—A. SEMIDIAGRAMMATIC VIEW OF A DEEP DISSECTION OF THE CRANIAL NERVES ON THE LEFT SIDE OF THE HEAD (from various authors and from nature). B. EXPLANA-TORY OUTLINE OF THE SAME. $\frac{1}{2}$

The roman numerals from I to IX indicate the roots of the several cranial nerves as they lie in or near their foramina of exit.   V, is upon the great root of the fifth with the ganglion in front ; *a* and *b*, in connection with VII, indicate respectively the facial and auditory nerves ; *a, b,* and *c,* in connection with VIII, point respectively to the glosso-pharyngeal, pneumo-gastric, and spinal accessory nerves ; C I, the suboccipital or first cervical nerve ; C VIII, the eighth.   The branches or distributed parts of the nerves are marked as follows, viz.:—I, frontal branch of the fifth ; 2, lachrymal passing into the gland; 3, nasal passing towards the internal orbitary foramen and giving the long twig to the ciliary ganglion (4') ; 3', external branch of the internal nasal nerve ; 4, lower branch of the third or oculo-motor nerve ; 5, the superior maxillary division of the fifth passing into the infra-orbital canal; 5', its issue at the infraorbital foramen and distribution as inferior palpebral, lateral nasal, and superior labial nerves (5'') ; 6, ganglion of Meckel and Vidian nerve passing back from it ; 6', palatine and other nerves descending from it ; 6'', superior petrosal nerve ; 7, posterior superior dental nerves ; 7', placed in the antrum maxillare, which has been opened, points to the anterior superior dental nerves ; 8, inferior maxillary division of the fifth immediately below the foramen ovale ; 8', some of the muscular branches coming from it ; 8 ×, the anterior auricular branch cut short, and above it the small petrosal nerve to join the facial nerve ; 9, buccal and internal pterygoid ; 10, gustatory nerve ; 10′, its distribution to the side and front of the tongue and to the sublingual glands ; 10″, the submaxillary ganglion connected with the gustatory

Fig. 400, B.

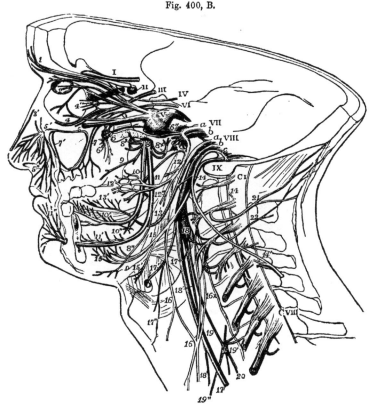

nerve ; below 10, the chorda tympani passing back from the gustatory to join the facial nerve above 12 ; 11, inferior dental nerve ; 11′, the same nerve and part of its dental distribution exposed by removal of the jaw ; 11″, termination of the same as mental and inferior labial nerves ; 12, the twigs of the facial nerve to the posterior belly of the digastric and to the stylo-hyoid muscle immediately after its exit from the stylo-mastoid foramen ; 12′, the temporo-facial division of the facial; 12″, the cervico-facial division; 13, the trunk of the glosso-pharyngeal passing round the stylo-pharyngeus muscle after giving pharyngeal and muscular branches ; 13′, its distribution on the side and back part of the tongue ; 14, the spinal accessory nerve, at the place where it crosses the ninth and gives a communicating branch to the pneumo-gastric and glosso-pharyngeal nerves ; 14′, the same nerve after having passed through the sterno-mastoid muscle uniting with branches from the cervical nerves ; 15, ninth nerve ; 15′, its twig to the thyro-hyoid muscle; 15″, its distribution in the muscles of the tongue; 16, descendens noni nerve giving a direct branch to the upper belly of the omo-hyoid muscle, and receiving the communicating branches 16 × from the cervical nerve ; 17, pneumo-gastric nerve ; 17′, its superior laryngeal branch ; 17″, external laryngeal twig ; 18, superior cervical ganglion of the sympathetic nerve, uniting with the upper cervical nerves, and giving at 18′ the superficial cardiac nerve ; 19, the trunk of the sympathetic ; 19′, the middle cervical ganglion, uniting with some of the cervical nerves, and giving 19″, the large middle cardiac nerve ; 20, continuation of the sympathetic nerve down the neck ; 21, great occipital nerve ; 22, third occipital.

The olfactory or first cranial nerve, the special nerve of the sense of smell, is distributed exclusively to the nasal fossæ.

From the under surface of the olfactory bulb about twenty branches proceed through the holes in the cribriform plate of the ethmoid bone, each invested by tubular prolongations of the membranes of the brain. These tubes of membrane vary in the extent to which they are continued on the branches : the offsets of the dura mater sheathe the filaments, and join the periosteum lining the nose ; those of the pia mater become blended with the neurilemma of the nerves ; and those of the arachnoid re-ascend to the serous lining of the skull.

Fig. 401.

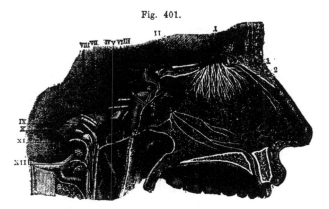

Fig. 401.—DISTRIBUTION OF THE OLFACTORY NERVES ON THE SEPTUM OF THE NOSE (from Sappey after Hirschfeld and Leveillé). ⅔

The septum is exposed and the anterior palatine canal opened on the right side. I, placed above, points to the olfactory bulb, and the remaining roman numbers to the roots of the several cranial nerves ; 1, the small olfactory nerves as they pass through the cribriform plate ; 2, internal or septal twig of the nasal branch of the ophthalmic nerve ; 3, naso-palatine nerves. (See Fig. 408 for a view of the distribution of the olfactory nerves on the outer wall of the nasal fossa.)

The branches are arranged in three sets. Those of the inner set, lodged for some distance in grooves on the surface of the bone, ramify in the pituitary membrane of the septum ; the outer set extend to the upper two spongy bones and the plane surface of the ethmoid bone in front of these ; and the middle set, which are very short, are confined to the roof of the nose. The distribution of the olfactory nerve is confined to the upper part of the nasal fossa ; none of the branches reach the lower spongy bone.—(See Anatomy of the Nose.)

OPTIC NERVE.

The optic or second cranial nerve, the nerve of vision, extending from the optic commissure, becomes more cylindrical and firm as it diverges from its fellow and enters the orbit by the optic foramen. Within the orbit it forms a cylindrical trunk, thick and strong, with a uniform surface. On dissection it is seen to consist of a number of separate bundles of nerve fibres, imbedded in tough fibrous tissue prolonged from the dura mater, and perforated in the centre by the small arteria centralis retinæ, which passes into it soon after it enters the orbit. It is surrounded by the recti muscles, and, entering the eyeball posteriorly a little to the inside of its middle, it pierces the sclerotic and choroid coats, and expands in the retina.—(See the Anatomy of the Eye.)

It may be mentioned that in many fishes the optic nerves do not unite in a commissure, but merely cross each to the side opposite to that of its origin; and that in a number of the same animals, as was first pointed out by Malphighi, the nerve consists of a lamina thrown into complicated longitudinal plications, and surrounded by a sheath.

THIRD PAIR OF NERVES.

This nerve, the common motor nerve of the eyeball (motorius oculi), gives branches to five of the seven muscles of the orbit,—viz., to the

Fig. 402.—View from above of the Uppermost Nerves of the Orbit, the Gasserian Ganglion, &c. (from Sappey after Hirschfeld and Leveillé). ¾

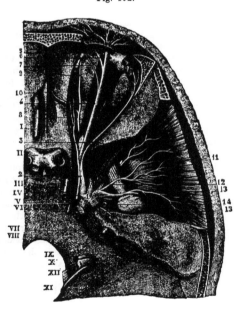

Fig. 402.

I, the olfactory tract passing forwards into the bulb; II, the commissure of the optic nerves; III, the oculo-motor; IV, the trochlear nerve; V, the greater root of the fifth nerve, a small portion of the lesser root is seen below it; VI, the sixth nerve; VII, facial; VIII, auditory; IX, glosso-pharyngeal; X, pneumo-gastric; XI, spinal accessory; XII, hypoglossal; 1, the Gasserian ganglion; 2, ophthalmic nerve; 3, lachrymal branch; 4, frontal; 5, external frontal or supraorbital; 6, internal frontal; 7, supratrochlear branch; 8, nasal nerve; 9, infratrochlear branch; 10, internal nasal passing through the internal orbital foramen; 11, anterior deep temporal proceeding from the buccal nerve; 12, middle deep temporal; 13, posterior deep temporal arising from the masseteric; 14, origin of the temporo-auricular; 15, great superficial petrosal nerve.

superior, internal and inferior straight muscles, to the levator palpebræ, and to the inferior oblique muscle.

Cylindrical and firm, like the other motor nerves, the third nerve, quitting the investment of the arachnoid membrane, pierces the inner layer of the dura mater close to the posterior clinoid process, and proceeds towards the sphenoidal fissure, lying in the external fibrous boundary of the cavernous sinus.

After receiving one or two delicate filaments from the cavernous plexus of the sympathetic, the third nerve divides near the orbit into two parts, which are continued into that cavity between the heads of the external rectus muscle, and separated one from the other by the nasal branch of the ophthalmic nerve.

The *upper*, the smaller part, is directed inwards over the optic nerve to the superior rectus muscle of the eye and the elevator of the eyelid, to both which muscles it furnishes branches.

The *lower* and larger portion of the nerve separates into three branches; of these one reaches the inner rectus; another the lower rectus; and the third, the longest of the three, runs onwards between the lower and the outer rectus, and terminates below the ball of the eye in the inferior oblique muscle. The last-mentioned branch is connected with the lower part of the lenticular ganglion by a short thick cord, and gives two filaments to the lower rectus muscle.

The several branches of the third nerve enter the muscles to which they are distributed on the surface which in each looks towards the eyeball.

POSITION OF CERTAIN NERVES *at the cavernous sinus, and as they enter the orbit.*—There are several nerves, besides the third, placed close together at the cavernous sinus, and entering the orbit through the sphenoidal fissure. To avoid repetition hereafter, the relative positions of these nerves may now be described. The nerves thus associated are the third, the fourth, the ophthalmic division of the fifth, and the sixth.

At the cavernous sinus.—In the dura mater which bounds the cavernous sinus on the outer side, the third and fourth nerves and the ophthalmic division of the fifth are placed, as regards one another, in their numerical order both from above downwards and from within outwards. The sixth nerve is placed separately from the others close to the carotid artery, on the floor of the sinus and internally to the fifth nerve. Near the sphenoidal fissure, through which they enter the orbit, the relative position of the nerves is changed, the sixth nerve being here close to the rest, and their number is augmented by the division of the third and the ophthalmic nerves—the former into two, the latter into three parts.

In the sphenoidal fissure.—The fourth and the frontal and lachrymal branches of the fifth, which are here higher than the rest, lie on the same level, the fourth being the nearest to the inner side, and enter the orbit above the muscles. The remaining nerves pass between the heads of the outer rectus muscle, in the following relative position to each other; the upper division of the third highest, the nasal branch of the fifth next, the lower division of the third beneath these, and the sixth lowest of all.

FOURTH PAIR OF NERVES.

The fourth (nervus trochlearis, n. patheticus) is the smallest of the cranial nerves, and is distributed entirely to the upper oblique muscle of the orbit.

From the remoteness of its place of origin, the part of this nerve within the skull is longer than that of any other cranial nerve. It enters an aperture in the free border of the tentorium, outside that for the third nerve, and near the posterior clinoid process. Continuing onwards through the outer wall of the cavernous sinus, the fourth nerve enters the orbit by the sphenoidal fissure, and above the muscles. Its position with reference to other nerves in this part of its course has been already described.

In the orbit, the fourth nerve inclines inwards above the muscles, and enters finally the upper oblique muscle at its orbital surface.

While in its fibrous canal in the outer wall of the sinus, the fourth nerve is joined by filaments of the sympathetic, and not unfrequently is blended with the ophthalmic

Fig. 403.

Fig. 403.—VIEW FROM ABOVE OF THE MOTOR NERVES OF THE EYEBALL AND ITS MUSCLES (after Hirschfeld and Leveillé, altered).

The ophthalmic division of the fifth pair has been cut short; the attachment of the muscles round the optic nerve has been opened up, and the three upper muscles turned towards the inner side, their anterior parts being removed; a part of the optic nerve is cut away to show the inferior rectus; and a part of the sclerotic coat and cornea is dissected off showing the iris, zona ciliaris, and choroid coat, with the ciliary nerves.

a, the upper part of the internal carotid artery emerging from the cavernous sinus; *b*, the superior oblique muscle; *b'*, its anterior part passing through the pulley; *c*, the levator palpebræ superioris; *d*, the superior rectus; *e*, the internal rectus; *f*, the external rectus; *f'*, its upper tendon turned down; *g*, the inferior rectus; *h*, insertion of the inferior oblique muscle.

II, the commissure of the optic nerve; II', part of the optic nerve entering the eyeball; III, the common oculo-motor; IV, the fourth or trochlear nerve; V, the greater root of the trigeminus; V', the smaller or motor root; VI, the abducent nerve; 1, the upper division of the third nerve separating from the lower and giving twigs to the levator palpebræ and superior rectus; 2, the branches of the lower division supplying the internal and inferior recti muscles; 3, the long branch of the same nerve proceeding forward to the inferior oblique muscle, and close to the number 3, the short thick branch to the ciliary ganglion: this ganglion is also shown, receiving from behind the slender twig from the nasal nerve, which has been cut short, and giving forwards some of its ciliary nerves, which pierce the sclerotic coat; 3', marks the termination of some of these nerves in the ciliary muscle and iris after having passed between the sclerotic and choroid coats; 4, the distribution of the trochlear nerve to the upper surface of the superior oblique muscle; 6, the abducent nerve passing into the external rectus.

division of the fifth. Bidder states that three or more small filaments of this nerve extend in the tentorium as far as the lateral sinus; and has figured one as joining the sympathetic on the carotid artery. (Neurologische Beobachtungen, Von F. H. Bidder. Dorpat, 1836.)

FIFTH PAIR OF NERVES.

The fifth, or trifacial nerve (nerv. trigeminus), the largest cranial nerve, is analogous to the spinal nerves, in respect that it consists of a motor and a sensory part, and that the sensory fibres pass through a ganglion while the motor do not. Its sensory division, which is much the larger, imparts common sensibility to the face and the fore part of the head, as well as to the eye, the nose, the ear, and the mouth; and endows the fore part of the tongue with the powers of both touch and taste. The motor root supplies chiefly the muscles of mastication.

The roots of the fifth nerve, after emerging from the surface of the encephalon, are directed forwards, side by side, to the middle fossa of the skull, through a recess in the dura mater on the summit of the petrous part of the temporal bone. Here the larger root alters in appearance: its

Fig. 404.

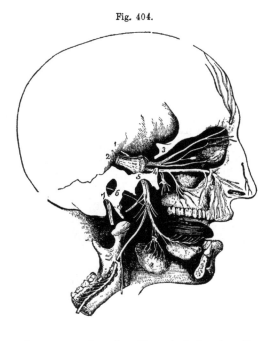

Fig. 404.—GENERAL PLAN OF THE BRANCHES OF THE FIFTH PAIR (after a sketch by Charles Bell). ⅓

1, lesser root of the fifth pair; 2, greater root passing forwards into the Gasserian ganglion; 3, placed on the bone above the ophthalmic nerve, which is seen dividing into the supra-orbital, lachrymal, and nasal branches, the latter connected with the ophthalmic ganglion; 4, placed on the bone close to the foramen rotundum, marks the superior maxillary division, which is connected below with the spheno-palatine ganglion, and passes forwards to the infraorbital foramen; 5, placed on the bone over the foramen ovale, marks the submaxillary nerve, giving off the anterior auricular and muscular branches, and continued by the inferior dental to the lower jaw, and by the gustatory to the tongue; a, the submaxillary gland, the submaxillary ganglion placed above it in connection with the gustatory nerve; 6, the chorda tympani; 7, the facial nerve issuing from the stylo-mastoid foramen.

fibres diverge a little, become reticulated, and enter the Gasserian ganglion. The smaller root passes inside and beneath the ganglion, without its nerve-fibres being incorporated in any way with it, and joins outside the skull the lowest of the three trunks which issue from the ganglion.

The *ganglion of the fifth nerve* or *Gasserian ganglion* (ganglion semilunare), occupies a depression on the upper part of the petrous portion of the temporal bone, near the point, and is somewhat crescentic in form, the convexity being turned forwards. On its inner side the ganglion is joined by filaments from the carotid plexus of the sympathetic nerve, and, according to some anatomists, it furnishes from its back part filaments to the dura mater.

From the fore part, or convex border of the Gasserian ganglion, proceed the three large divisions of the nerve. The highest (first or ophthalmic trunk) enters the orbit; the second, the upper maxillary nerve, is continued forwards to the face, below the orbit; and the third, the lower maxillary nerve, is distributed chiefly to the external ear, the tongue, the lower teeth, and the muscles of mastication. The first two trunks proceed exclusively from the ganglion and are entirely sensory, while the third or inferior maxillary trunk, receiving a considerable part from the ganglion, has associated with it also the whole of the fibres of the motor root, and thus distributes both motor and sensory branches.

OPHTHALMIC NERVE.

The ophthalmic nerve, or first division of the fifth nerve, the smallest of the three offsets from the Gasserian ganglion, is somewhat flattened, about an inch in length, and is directed forwards and upwards to the sphenoidal fissure, where it ends in branches which pass through the orbit to the surface of the head and to the nasal fossæ. In the skull it is contained in the process of the dura mater bounding externally the cavernous sinus, and is joined by filaments from the cavernous plexus of the sympathetic : according to Arnold, it gives recurrent branches to the tentorium cerebelli. It also frequently communicates by a considerable branch with the fourth nerve.

Near the orbit the ophthalmic nerve furnishes from its inner side the nasal branch, and then divides into the frontal and lachrymal branches. These branches are transmitted separately through the sphenoidal fissure, and are continued through the orbit (after supplying some filaments to the eye and the lachrymal gland) to their final distribution in the nose, the eyelids, and the muscles and integument of the forehead.

LACHRYMAL BRANCH.

The lachrymal branch is external to the frontal at its origin, and is contained in a separate tube of dura mater. In the orbit it passes along the outer part, above the muscles, to the outer and upper angle of the cavity. Near the lachrymal gland, the nerve has a connecting filament with the orbital branch of the superior maxillary nerve ; and when in close apposition with the gland, it gives many filaments to that body and to the conjunctiva. Finally, the lachrymal nerve penetrates the palpebral ligament externally, and ends in the upper eyelid, the terminal ramifications being joined by twigs from the facial nerve.

In consequence of the junction which occurs between the ophthalmic trunk of the fifth and the fourth nerve, the lachrymal branch sometimes appears to be derived from both these nerves. Swan considers this the usual condition of the lachrymal nerve. (" A Demonstration of the Nerves of the Human Body," page 36. London, 1834.)

FRONTAL BRANCH.

The frontal branch, the largest division of the ophthalmic, lies, like the preceding nerve, above the muscles in the orbit, being situated between the elevator of the upper eyelid and the periosteum. About midway forwards in the orbit, the nerve divides into two branches, supratrochlear and supra-orbital.

a. The *supratrochlear branch* (internal frontal) is prolonged to the inner angle of the orbit, close to the point at which the pulley of the upper oblique muscle is fixed to the orbit. Here it gives downwards a filament to connect it with the infratrochlear branch of the nasal nerve, and issues from the cavity between the orbicular muscle of the lids and the bone. In this position filaments are distributed to the upper eyelid. The nerve next pierces the orbicularis palpebrarum and occipito-frontalis muscles, furnishing twigs to these muscles and the corrugator supercilii, and, after ascending on the forehead, ramifies in the integument.

b. The *supraorbital branch* (external frontal) passes through the supraorbital notch to the forehead, and ends in muscular, cutaneous, and pericranial branches ; while in the notch it distributes *palpebral* filaments to the upper eyelid.

The *muscular branches* referred to are comparatively small, and supply the corrugator of the eyebrow, the occipito-frontalis, and the orbicular muscle of the eyelids, joining the facial nerve in the last muscle. The *cutaneous branches,* among which two (outer and inner) may be noticed as the principal, are placed at first beneath the occipito-frontalis. The outer one, the larger, perforates the tendinous expansion of the muscle, and ramifies in the scalp as far back as the lambdoidal suture. The inner branch reaches the surface sooner than the preceding nerve, and ends in the integument over the parietal bone. The *pericranial branches* arise from the cutaneous nerve beneath the muscle, and end in the pericranium covering the frontal and parietal bones.

Fig. 405.

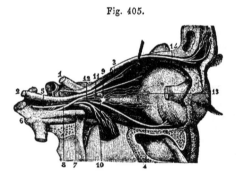

Fig. 405.—Nerves of the Orbit from the outer Side (from Sappey after Hirschfeld and Leveillé). ¾

The external rectus muscle has been divided and turned down: 1, the optic nerve; 2, the trunk of the third nerve; 3, its upper division pasing into the levator palpebræ and superior rectus ; 4, its long lower branch to the inferior oblique muscle ; 5, the sixth or abducent nerve joined by twigs from the sympathetic; 6, the Gasserian ganglion ; 7, ophthalmic nerve ; 8, its nasal branch ; 9, the ophthalmic ganglion ; 10, its short or motor root ; 11, long sensory root from the nasal nerve ; 12, sympathetic twig from the carotid plexus ; 13, ciliary nerves passing into the eyeball ; 14, frontal branch of the ophthalmic nerve.

NASAL BRANCH.

The nasal branch (oculo-nasalis), separating from its parent trunk in the wall of the cavernous sinus, enters the orbit between the heads of the outer rectus. It then inclines inwards over the optic nerve, beneath the elevator of the upper eyelid and the superior rectus muscle, to the inner wall of the orbit, through which it passes by the anterior internal orbital foramen. In this oblique course across the orbit it furnishes a single filament to the ophthalmic ganglion, two or three (long ciliary) directly to the eyeball ; and, at the inner side of the cavity, a considerable branch (infratrochlear), which issues from the orbit at the fore part.

On leaving the orbit the nasal nerve is directed transversely inwards to the upper surface of the cribriform plate of the ethmoid bone, and, passing forwards in a groove at its outer edge, within the cranium, descends by a special aperture close to the crista galli at the fore part of the plate to the roof of the nasal fossa, where it divides into two branches, one of which (external or superficial nasal) reaches the integument of the side of the nose, and the other (ramus septi) ramifies in the pituitary membrane.

a. The *branch to the ophthalmic ganglion* (radix longa ganglii ciliaris), very slender, and about half an inch long, arises generally between the heads of the external rectus ; it lies on the outer side of the optic nerve, and enters the upper and back part of the ophthalmic ganglion, constituting its *long root.*

This small branch is sometimes joined by a filament from the cavernous plexus of the sympathetic, or from the upper branch of the third nerve.

b. The *long ciliary nerves,* two or three in number, are situated on the inner side of the optic nerve ; they join one or more of the nerves from the ophthalmic ganglion (short ciliary), and after perforating the sclerotic coat of the eye, are continued between it and the choroid to the ciliary muscle, the cornea, and the iris.

c. The *infratrochlear branch* runs forwards along the inner side of the orbit below the superior oblique muscle, and receives near the pulley of that muscle a filament of connection from the supratrochlear nerve.   The branch is then continued below the pulley to the inner angle of the eye, and ends in filaments which supply the orbicular muscle of the lids, the caruncula, and the lachrymal sac, as well as the integument of the eyelids and side of the nose.

In the cavity of the nose the nasal nerve ends by dividing into the following branches.

d. The *branch to the nasal septum* extends to the lower part of the partition between the nasal fossæ, supplying the pituitary membrane near the fore part of the septum.

e. The *superficial branch* (externus seu lateralis), descends in a groove on the inner surface of the nasal bone ; and after leaving the nasal cavity between that bone and the lateral cartilage of the nose, it is directed downwards to the tip of the nose, beneath the compressor naris muscle.   While within the nasal fossa, this branch gives two or three filaments to the fore part of its outer wall, which extend as far as the lower spongy bone.   The cutaneous part is joined by a filament of the facial nerve.

Summary.—The first division of the fifth nerve is altogether sensory in function.   It furnishes branches to the ball of the eye and the lachrymal gland ; to the mucous membrane of the nose and eyelids ; to the integument of the nose, the upper eyelid, the forehead, and the upper part of the hairy scalp ; and to the muscles above the middle of the circumference of the orbit.   Some of the cutaneous and muscular filaments join branches of the facial nerve, and the nerve itself communicates with the sympathetic.

OPHTHALMIC GANGLION.

There are four small ganglia connected with the divisions of the fifth nerve : the ophthalmic ganglion with the first, Meckel's ganglion with the second, and the otic and submaxillary ganglia with the third.   These ganglia, besides receiving branches from the sensory part of the fifth, are each connected with a motor nerve from the third, the fifth, or the facial, and with twigs from the sympathetic ; and the nerves thus joining the ganglia are named their roots.

The *ophthalmic* or *lenticular* ganglion (gang. semilunare, vel ciliare) serves as a centre for the supply of nerves—motor, sensory, and sympathetic—to the eyeball.   It is a small reddish body, situated at the back of the orbit, between the outer rectus muscle and the optic nerve, and generally in contact with the ophthalmic artery ; it is joined behind by branches from the fifth, the third, and the sympathetic nerves ; while from its fore part proceed the ciliary nerves to the eyeball.

Union of the ganglion with nerves: its roots.—The posterior border of the ganglion receives three nerves.   One of these, the *long root,* a slender filament from the nasal branch of the ophthalmic trunk, joins the upper part of this border.   Another branch, the *short root,* much thicker and shorter than the preceding, and sometimes divided into parts, is derived from the branch of the third nerve to the inferior oblique muscle, and is connected with the lower part of the ganglion. The *third root* is a very small nerve which emanates from the cavernous plexus of the sympathetic, and reaches the ganglion with the long upper root : these two nerves are sometimes conjoined before reaching the ganglion. Other roots have been assigned to the ganglion. (Valentin, in Müller's Archiv. for 1840.)

Branches of the ganglion.—From the fore part of the ganglion arise ten or twelve delicate filaments—the *short ciliary nerves.* These nerves are disposed in two fasciculi, arising from the upper and lower angles of the ganglion, and they run forwards, one set above, the other below the optic nerve, the latter being the more numerous. They are accompanied by filaments from the nasal nerve (long ciliary), with which some are joined. Having entered the eyeball by apertures in the back part of the sclerotic coat, the nerves are lodged in grooves on its inner surface ; and at the ciliary muscle, which they pierce (some filaments supplying it and the cornea), they turn inwards and ramify in the iris.

SUPERIOR MAXILLARY NERVE.

The superior maxillary nerve, or second division of the fifth cranial nerve, is intermediate in size between the ophthalmic and the inferior maxillary trunks.

It commences at the middle of the Gasserian ganglion, and, passing horizontally forwards, soon leaves the skull by the foramen rotundum of the sphenoid bone. The nerve then crosses the spheno-maxillary fossa, and enters the infraorbital canal of the upper maxilla, by which it is conducted to the face. After emerging from the infraorbital foramen, it terminates beneath the elevator of the upper lip in branches which spread out to the side of the nose, the eyelid, and the upper lip.

Branches.—In the spheno-maxillary fossa a temporo-malar branch ascends from the superior maxillary nerve to the orbit, and two sphenopalatine branches descend to join Meckel's ganglion. Whilst the nerve is in contact with the upper maxilla, it furnishes two posterior dental branches on the tuberosity of the bone, and an anterior dental branch at the fore part. On the ace are the terminal branches already indicated.

ORBITAL BRANCH.

The orbital or temporo-malar branch, a small cutaneous nerve, enters the orbit by the spheno-maxillary fissure, and divides into two branches (temporal and malar), which pierce the malar bone, and are distributed to the temple and the prominent part of the cheek.

a. The *temporal branch* is contained in an osseous groove or canal in the outer wall of the orbit, and leaves this cavity by a foramen in the malar bone. When about to traverse the bone, it is joined by a communicating filament (in some cases, two filaments) from the lachrymal nerve. The nerve is then inclined upwards in the temporal fossa between the bone and the temporal muscle, perforates the aponeurosis over the muscle an inch above the zygoma, and ends in cutaneous filaments over the temple. The cutaneous ramifications are united with the facial nerve, and sometimes with the superficial temporal nerve of the third division of the fifth.

b. The *malar branch* lies at first in the loose fat in the lower angle of the orbit, and is continued to the face through a foramen in the fore part of the malar bone, where it is frequently divided into two filaments. In the prominent part of the cheek this nerve communicates with the facial nerve.

Fig. 406.

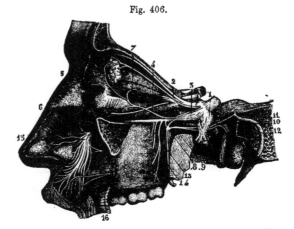

Fig. 406.—SUPERIOR MAXILLARY NERVE AND SOME OF THE ORBITAL NERVES (from Sappey after Hirschfeld and Leveillé). ⅓

1, the Gasserian ganglion ; 2, lachrymal branch of the ophthalmic nerve; 3, trunk of the superior maxillary nerve; 4, its orbital branch, joining at 5, the palpebral twig of the lachrymal ; 6, origin of its malar twig ; 7, its temporal twig ; 8, spheno-palatine ganglion ; 9, Vidian nerve ; 10, its upper branch or great superficial petrosal nerve proceeding to join the facial nerve (11) ; 12, union of the lower branch of the Vidian nerve with the carotid branch of the sympathetic ; 13, 14, posterior dental nerves ; 15, terminal branches of the infraorbital nerves ramifying on the side of the nose and upper lip ; 16, a branch of the facial uniting with some of the twigs of the infraorbital.

POSTERIOR DENTAL BRANCHES.

The posterior dental branches, two in number, are directed downwards and outwards over the back part and tuberosity of the maxillary bone.

One of the branches enters a canal in the bone by which it is conducted to the teeth, and gives forwards a communicating filament to the anterior dental nerve. It ends in filaments to the molar teeth and the lining membrane of the maxillary sinus, and near the teeth joins a second time with the anterior dental nerve.

The *anterior* of the two branches, lying on the surface of the bone, is distributed to the gums of the upper jaw and to the buccinator muscle.

ANTERIOR DENTAL BRANCH.

The anterior dental branch, leaving the trunk of the nerve at a varying distance behind its exit from the infraorbital foramen, enters a special canal in front of the antrum of Highmore. In this canal it receives the communicating filament from the posterior dental nerve, and divides into two branches, which furnish offsets for the front teeth.

(*a*) The *inner* branch supplies the incisor and canine teeth. Filaments from this nerve enter the lower meatus of the nose, and end in the membrane covering the lower spongy bone. Also above the root of the canine tooth, it unites with a branch of the posterior nasal nerve from Meckel's ganglion, and forms with it a small thickening,

R R

the *ganglion of Bochdalek*, from which branches are described as descending to the alveolar process and gums of the incisor and canine teeth. (See Hyrtl's Lehrbuch, p. 804.)

(*b*) The *outer* branch gives filaments to the bicuspid teeth, and is connected with the posterior dental nerve.

INFRAORBITAL BRANCHES.

The infraorbital branches, large and numerous, spring from the end of the superior maxillary nerve beneath the elevator muscle of the upper lip, and are divisible into palpebral, nasal, and labial sets.

Fig. 407.

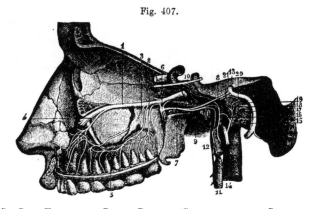

Fig. 407.—DEEP VIEW OF THE SPHENO-PALATINE GANGLION, AND ITS CONNECTIONS WITH OTHER NERVES, &c. (from Sappey after Hirschfeld and Leveillé). ⅔

1, superior maxillary nerve ; 2, posterior superior dental ; 3, second posterior dental branch ; 4, anterior dental ; 5, union of these nerves ; 6, spheno-palatine ganglion ; 7, Vidian nerve ; 8, its great superficial petrosal branch ; 9, its carotid branch ; 10, a part of the sixth nerve, receiving twigs from the carotid plexus of the sympathetic ; 11, superior cervical sympathetic ganglion ; 12, its carotid branch ; 13, trunk of the facial nerve near the knee or bend at the hiatus Fallopii ; 14, glosso-pharyngeal nerve ; 15, anastomosing branch of Jacobson ; 16, twig uniting it to the sympathetic : 17, filament to the fenestra rotunda ; 18, filament to the Eustachian tube ; 19, filament to the fenestra ovalis ; 20, external deep petrosal nerve uniting with the lesser superficial petrosal ; 21, internal deep petrosal twig uniting with the great superficial petrosal.

a. The *palpebral branch* (there are sometimes two branches) turns upwards to the lower eyelid in a groove or canal of the bone, and supplies the orbicular muscle ; it ends in filaments which are distributed to the eyelid in its entire breadth. At the outer angle of the eyelids this nerve is connected with the facial nerve.

b. The *nasal branches*, directed inwards to the muscles and integument of the side of the nose, communicate with the cutaneous branch of the nasal nerve.

c. The *labial*, the largest of the terminal branches of the upper maxillary nerve, and three or four in number, are continued downwards beneath the proper elevator of the upper lip. Ramifying as they descend, these nerves are distributed to the integument, the mucous membrane of the mouth, the labial glands, and the muscles of the upper lip.

Near the orbit the infraorbital branches of the superior maxillary nerve are joined by considerable branches of the facial nerve, the union between the two being named *infraorbital plexus.*

SPHENO-PALATINE GANGLION.

The spheno-palatine ganglion, frequently named Meckel's ganglion, is deeply placed in the spheno-maxillary fossa, close to the spheno-palatine foramen. It receives the two spheno-palatine branches, which descend together from the superior maxillary nerve as it crosses the top of the fossa. It is of a greyish colour, triangular in form, and convex on the outer surface. The grey or ganglionic substance does not involve all the fibres of the spheno-palatine branches of the upper maxillary nerve, but is placed at the back part, at the point of junction of the sympathetic or deep branch of the Vidian, so that the spheno-palatine nerves proceeding to the nose and palate pass to their destination without being incorporated with the ganglionic mass.

Branches proceed from the ganglion upwards to the orbit, downwards to the palate, inwards to the nose, and backwards through the Vidian and pterygo-palatine canals.

ASCENDING BRANCHES.—There are three or more very small twigs, which reach the orbit by the spheno-maxillary fissure, and are distributed to the periosteum.

Bock describes a branch ascending from the ganglion to the sixth nerve; Tiedemann, one to the lower angle of the ophthalmic ganglion. The filaments described by Hirzel as ascending to the optic nerve, most probably join the ciliary twigs which surround that nerve.

DESCENDING BRANCHES.—These are three in number,—the large, the small, and the external palatine nerves, and are continued chiefly from the spheno-palatine branches of the superior maxillary. They are distributed to the tonsil, the hard and soft palate, the gums, and the mucous membrane of the nose.

a. The *larger* or *anterior palatine* nerve descends in the palato-maxillary canal, and divides in the roof of the mouth into branches, which are received into grooves in the hard palate, and extend forwards nearly to the incisor teeth. In the mouth it supplies the gums, the glandular structure and the mucous membrane of the hard palate, and joins in front with the naso-palatine nerve. When entering its canal, this palatine nerve gives a nasal branch which ramifies on the middle and lower spongy bones; and a little before leaving the canal, another branch is supplied to the membrane covering the lower spongy bone : these are *inferior nasal branches*. Opposite the lower spongy bone springs a small branch, which is continued to the soft palate in a separate canal behind the trunk of the nerve.

b. The *smaller* or *posterior palatine* branch, arising near the preceding nerve, enters with a small artery the lesser palatine canal, and is conducted to the soft palate, the tonsil, and the uvula. According to Meckel, it supplies the levator palati muscle.

c. The *external palatine* nerve, the smallest of the series, courses between the upper maxilla and the external pterygoid muscle, and enters the external palatine canal between the maxillary bone and the pterygoid process of the palate bone. At its exit from the canal it gives inwards a branch to the uvula, and outwards another to the tonsil and palate. Occasionally, this nerve is altogether wanting.

INTERNAL BRANCHES.—These consist of the naso-palatine, and the upper and anterior nasal, which ramify in the lining membrane of the nasal fossæ and adjoining sinuses.

The *upper nasal* are very small branches, and enter the back part of the nasal fossa by the spheno-palatine foramen. Some are prolonged to the upper and posterior part of the septum, and the remainder ramify in the membrane covering the upper two spongy bones, and in that lining the posterior ethmoid cells. A branch, as has been already stated, forms a connection in the wall of the maxillary sinus, above the eye-tooth, with the anterior dental nerve.

The *naso-palatine nerve*, nerve of Cotunnius (Scarpa), long and slender, leaves the inner side of the ganglion with the preceding branches, and after crossing the roof of the nasal fossa is directed downwards and forwards on the septum nasi, towards the anterior palatine canal, situated between the periosteum and the pituitary membrane. The nerves of opposite sides descend to the palate through the mesial subdivisions of the canal, called the foramina of Scarpa, the nerve of the right side usually behind that of the left. In the lower common foramen the two naso-palatine nerves are connected with each other; and they end in several filaments, which are distributed to the papilla behind the incisor teeth, and communicate with the great palatine nerve. In its course along the septum, small filaments are furnished from the naso-palatine nerve to the pituitary membrane. (See Fig 402. This nerve was discovered independently by John Hunter and Cotunnius; see Hunter's "Observations on certain parts of the Animal Economy;" and Scarpa, "Annotationes Anatomicæ," lib. ii.)

Fig. 408.

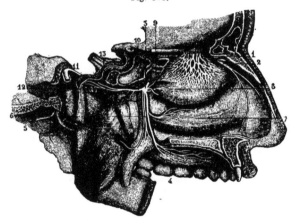

Fig. 408.—NERVES OF THE NOSE AND OF THE SPHENO-PALATINE GANGLION FROM THE INNER SIDE (from Sappey after Hirschfeld and Leveillé). ⅔

1, network of the branches of the olfactory nerve descending upon the membrane covering the superior and middle turbinated bones; 2, external twig of the ethmoidal branch of the nasal nerve; 3, spheno-palatine ganglion; 4, ramification of the anterior division of the palatine nerves; 5, posterior, and 6, middle divisions of the palatine nerves; 7, branch to the membrane on the lower turbinated bone; 8, branch to the superior and middle turbinated bones; 9, naso-palatine branch to the septum cut short; 10, Vidian nerve; 11, its great superficial petrosal branch; 12, its carotid branch; 13, the sympathetic nerves ascending on the internal carotid artery.

POSTERIOR BRANCHES.—The branches directed backwards from the spheno-palatine ganglion are the Vidian and pharyngeal nerves.

The *Vidian nerve* arises from the back part of the ganglion, which seems to be prolonged into it, passes backwards through the Vidian canal, and after emerging from this divides in the substance of the fibro-cartilage filling the foramen lacerum medium, into two branches: one of these, the superficial petrosal, joins the facial nerve, while the other, the carotid branch, communicates with the sympathetic. Whilst the Vidian nerve is in its canal, it gives inwards some small *nasal* branches, which supply the membrane of the back part of the roof of the nose and septum, as well as the membrane covering the end of the Eustachian tube.

The *large superficial petrosal branch* of the Vidian nerve, entering the cranium on the outer side of the carotid artery and beneath the Gasserian ganglion, is directed backwards in a groove on the petrous portion of the temporal bone to the hiatus Fallopii, and is thus conducted to the aqueductus Fallopii, where it joins the gangliform enlargement of the facial nerve.

The *carotid* or *sympathetic portion* of the Vidian nerve, shorter than the other, is of a reddish colour and softer texture : it is directed backwards, and on the outer side of the carotid artery ends in the filaments of the sympathetic surrounding that vessel.

In accordance with the view taken of the ganglia connected with the fifth nerve (p. 599), the superficial petrosal and carotid parts of the Vidian nerve may be regarded as the motor and sympathetic roots respectively of the spheno-palatine ganglion ; the spheno-palatine being its sensory root.

The *pharyngeal nerve* is inconsiderable in size, and, instead of emanating directly from the ganglion, is frequently derived altogether from the Vidian. This branch, when a separate nerve, springs from the back of the ganglion, enters the pterygo-palatine canal with an artery, and is lost in the lining membrane of the pharynx behind the Eustachian tube.

Summary.—The superior maxillary nerve, with Meckel's ganglion, supplies the integument above the zygomatic arch, and that of the lower eyelid, the side of the nose, and the upper lip ; the upper teeth, the lining membrane of the nose ; the membrane of the upper part of the pharynx, of the antrum of Highmore, and of the posterior ethmoid cells ; the soft palate, tonsil, and uvula ; and the glandular and mucous structures of the roof of the mouth.

INFERIOR MAXILLARY NERVE.

The lower maxillary nerve, the third and largest division of the fifth nerve, is made up of two portions, unequal in size, the larger being derived from the Gasserian ganglion, and the smaller being the slender motor root of the fifth nerve. These two parts leave the skull by the foramen ovale in the sphenoid bone, and unite immediately after their exit. A few lines beneath the base of the skull, and under cover of the external pterygoid muscle, the nerve separates into two primary divisions, one of which is higher in position and smaller than the other.

The *small, anterior* or *upper portion,* purely motor, terminates in branches to the temporal, masseter, buccinator, and pterygoid muscles. The *larger* or *lower* portion, chiefly sensory, divides into the auriculo-temporal, gustatory, and inferior dental branches : it likewise supplies the mylohyoid muscle, and the anterior belly of the digastric. The branch to the internal pterygoid muscle, with which also are connected those proceeding from the otic ganglion to the tensors of the palate and tympanum, is sometimes counted as a part of the larger division, but is more correctly regarded as arising from the undivided trunk.

DEEP TEMPORAL, MASSETERIC, BUCCAL, AND PTERYGOID BRANCHES.

The *deep temporal branches,* two in number, *anterior* and *posterior,* pass outwards above the external pterygoid muscle, close to the bone, and run upwards, one near the front, and the other near the back of the temporal fossa, beneath the temporal muscle in the substance of which they are distributed. (See fig. 403.)

The anterior branch is frequently joined with the buccal nerve, and sometimes with the other deep temporal branch.

The *masseteric branch* likewise passes above the external pterygoid muscle, and is directed nearly horizontally outwards through the sigmoid notch of the lower jaw to the posterior border of the masseteric muscle, which it enters on the deep surface. It gives a filament or two to the articulation of the jaw, and occasionally furnishes a branch to the temporal muscle.

The *buccal branch* pierces the substance of the external pterygoid muscle, and courses downwards and forwards to the face, in close contact with the deep surface of the temporal muscle at its insertion. It furnishes a branch to the external pterygoid muscle as it pierces it, and on emerging gives two or three ascending branches to the temporal muscle. It divides into two principal branches, an *upper* and a *lower*, which communicate with the facial nerve in a plexus round the facial vein, and are distributed to the integument, the buccinator muscle, and the mucous membrane.

The *external pterygoid branch*, is most frequently derived from the buccal nerve. It is sometimes a separate offset from the smaller portion of the lower maxillary nerve.

The *nerve* of the *internal pterygoid* muscle is closely connected at its origin with the otic ganglion, and enters the inner or deep surface of the muscle.

AURICULO-TEMPORAL NERVE.

The auriculo-temporal nerve takes its origin close to the foramen ovale. It often commences by two roots, between which may be placed the middle meningeal artery. It is directed at first backwards, beneath the external pterygoid muscle, to the inner side of the articulation of the jaw; then changing its course, it turns upwards between the ear and the joint, covered by the parotid gland; and emerging from this place, it finally divides into two temporal branches which ascend towards the top of the head.

(a) *Communicating branches.* — There are commonly two branches which pass forward round the external carotid artery, and join the facial nerve. Filaments to the otic ganglion arise near the beginning of the nerve.

(b) *Parotid branches* are given from the nerve while it is covered by the gland.

(c) *Auricular branches.*—These are two in number. The *lower* of the two, arising behind the articulation of the jaw, distributes branches to the ear below the external meatus; and sends other filaments round the internal maxillary artery to join the sympathetic nerve; the *upper branch,* leaving the nerve in front of the ear, is distributed in the integument covering the tragus and the pinna above the external auditory meatus. Both are confined to the outer surface of the ear.

(d) *Branches to the meatus auditorius.*—These, two in number, spring from the point of connection of the facial and auriculo-temporal nerves, and enter the interior of the auditory meatus between the osseous and cartilaginous parts. One of them sends a branch to the membrana tympani.

(e) *Articular branch.*—The nerve to the temporo-maxillary articulation comes from one of the preceding branches, or directly from the auriculo-temporal nerve.

(f) *Temporal branches.*— One of these, the smaller and *posterior* of the two, distributes filaments to the anterior muscle of the auricle, the upper part of the pinna and the integument above it. The anterior temporal branch extends with the superficial temporal artery to the top of the head, and ends in the integument. It is often united with the temporal branch of the upper maxillary nerve. Meckel mentions a communication between this branch and the occipital nerve.

GUSTATORY NERVE.

The gustatory nerve, or lingual branch of the fifth, descends under cover of the external pterygoid muscle, lying to the inner side and in front of the dental nerve, and sometimes united to it by a cord which crosses over the internal maxillary artery. It is there joined at an acute angle by the chorda tympani, a small branch connected with the facial nerve, which descends from the inner end of the Glasserian fissure. It then passes between the internal pterygoid muscle and the lower maxilla, and is inclined obliquely inwards to the side of the tongue, over the upper constrictor of the pharynx, (where this muscle is attached to the maxillary bone,) and above the deep portion of the submaxillary gland. Lastly, the nerve crosses Wharton's duct, and is continued along the side of the tongue to the apex, in contact with the mucous membrane of the mouth.

(*a*) *Communicating branches* are given to the submaxillary ganglion, at the place

Fig. 409.

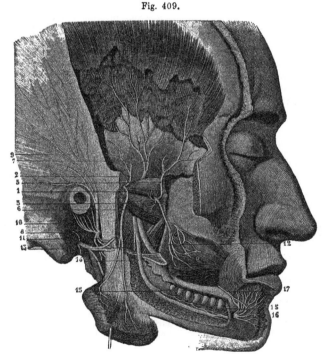

Fig. 409.—VIEW OF THE BRANCHES OF THE INFERIOR MAXILLARY NERVE FROM THE OUTER SIDE (from Sappey after Hirschfeld and Leveillé). ⅔

The zygoma and ramus of the jaw have been removed, and the outer plate of the jaw taken off so as to open up the dental canal; the lower part of the temporal muscle has been dissected off the bone, and the masseter muscle turned down.

1, Masseteric branch, descending to the deep surface of the muscle; 2, a twig to the temporal muscle; 5, anterior, and 7, posterior deep temporal nerves; 3, buccal; 4, its union with the facial; 6, filaments given by the buccal to the external pterygoid muscle; 8, auriculo-temporal nerve; 9, its temporal branches; 10, its anterior auricular branches; 11, its union with the facial; 12, gustatory or lingual nerve; 13, mylo-hyoid nerve; 14, inferior dental nerve; 15, its twigs supplied to the teeth; 16, mental branches; 17, branch of the facial uniting with the mental.

where the nerve is in contact with the submaxillary gland. Others form a plexus with branches of the hypoglossal nerve at the inner border of the hyo-glossus muscle.

(b) *Branches to the mucous membrane* of the mouth are given from the nerve at the side of the tongue, and supply also the gums. Some delicate filaments are likewise distributed to the substance of the sublingual gland.

(c) The *lingual* or terminal branches perforate the muscular structure of the tongue, and divide into filaments, which are continued almost vertically upwards to the conical and fungiform papillæ. Near the tip of the tongue the branches of the gustatory and hypoglossal nerves are united.

INFERIOR DENTAL NERVE.

The inferior dental nerve is the largest of the three branches of the lower maxillary nerve. It descends under cover of the external pterygoid muscle, behind and to the outer side of the gustatory nerve, and, passing between the ramus of the jaw and the internal lateral ligament of the temporo-maxillary articulation, enters the inferior dental canal. In company with the dental artery, it proceeds along this canal, and supplies branches to the teeth. At the mental foramen it bifurcates; one part, the incisor branch, being continued onwards within the bone to the middle line, while the other, the much larger labial branch, escapes by the foramen to the face.

When about to enter the foramen on the inner surface of the ramus of the jaw, the inferior dental nerve gives off the slender mylo-hyoid branch.

(a) The *mylo-hyoid branch* is lodged in a groove on the inner surface of the ramus of the maxillary bone, in which it is confined by fibrous membrane, and is distributed to the lower or cutaneous surface of the mylo-hyoideus and to the anterior belly of the digastric muscle. This nerve may be traced back within the sheath of the inferior dental to the motor portion of the inferior maxillary nerve.

(b) The *dental branches* supplied to the molar and bicuspid teeth correspond to the number of the fangs of those teeth. Each branch enters the minute foramen in the extremity of a fang, and terminates in the pulp of the tooth. Not unfrequently a collateral branch supplies twigs to several teeth.

(c) The *incisor branch* has the same direction as the trunk of the nerve: it extends to the middle line from the point of origin of the labial branch, and supplies nerves to the canine and incisor teeth.

(d) The *labial* or *mental branch* emerging from the bone by the foramen on the outer surface, divides beneath the depressor of the angle of the mouth into two parts.

One of these, the outer division, communicating with the facial nerve, supplies the depressor anguli oris and orbicularis oris muscles, and the integument of the chin.

The inner portion, the larger of the two, ascends to the lower lip beneath the depressor labii inferioris muscle, to which it gives filaments : the greater number of the branches end on the inner and outer surfaces of the lip. These inner branches assist only slightly in forming the plexus of union with the facial nerve.

OTIC GANGLION.

The otic ganglion, or ganglion of Arnold, of a reddish grey colour, is situated on the deep surface of the lower maxillary trunk, nearly at the point of junction of the motor fasciculus with that nerve, and around the origin of the internal pterygoid branch. Its inner surface is close to the cartilaginous part of the Eustachian tube and the circumflexus palati muscle ; and behind it is the middle meningeal artery.

Connection with nerves—roots.—The ganglion is connected with the lower maxillary nerve, especially with the branch furnished to the internal pterygoid muscle, and with the auriculo-temporal nerve, and thus obtains motor and sensory roots ; it is brought into connection with the sympathetic by a filament from the plexus on the middle meningeal artery. It likewise receives the *small superficial petrosal* nerve, which emerges from the petrous bone by the small foramen internal to the canal of the tensor tympani muscle, and reaches the exterior of the skull by piercing the sphenoid bone close to the foramen spinosum. By this nerve the ganglion forms a communication with the glosso-pharyngeal and facial nerves.

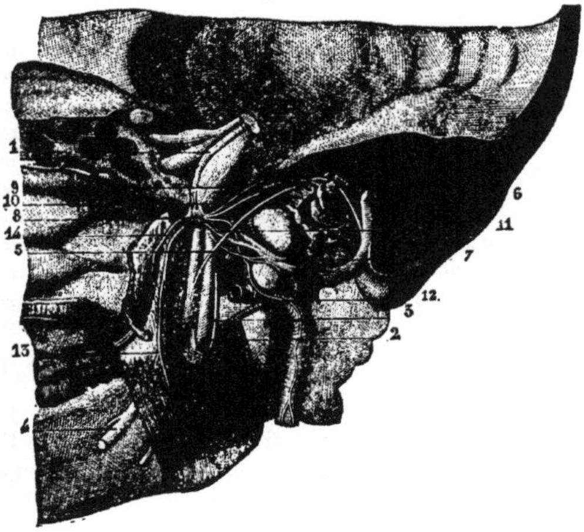

Fig. 410.—OTIC GANGLION AND ITS CONNECTONS FROM THE INSIDE (1om Sappey after Arnold). ⅔

Fig. 410.

This figure exhibits a view of the lateral portion of the skull with a part of the nasal fossa and lower jaw of the right side; the petrous bone has been removed so as to show the inner surface of the membrana tympani and the canal of the facial nerve.

1, smaller motor root of the fifth nerve passing down on the inside of the Gasserian ganglion to unite with the inferior maxillary division; 2, inferior dental nerve entering the canal of the lower jaw; 3, mylohyoid branch, seen also farther down emerging in front of the internal pterygoid muscle; 4, lingual or gustatory nerve; 5, chorda tympani; 6, facial nerve in its canal; 7, auriculo-temporal nerve, enclosing in its loop of origin the middle meningeal artery; 8, otic ganglion; 9, small superficial petrosal nerve joining the ganglion; 10, branch to the tensor tympani muscle; 11, twig connecting the ganglion with the temporo-auricular nerve; 12, twig to the ganglion from the sympathetic nerves on the meningeal artery; 13, branch to the internal pterygoid muscle; 14, branch to the tensor palati muscle.

Branches.—Two small nerves are distributed to muscles—one to the tensor of the membrane of the tympanum, the other to the circumflexus or tensor palati.

SUBMAXILLARY GANGLION.

The submaxillary ganglion is placed above the deep portion of the submaxillary gland, and is connected by filaments with the gustatory nerve. It is about the size of the ophthalmic ganglion. By the upper part or base it receives branches from nerves which may be regarded as its roots, whilst from the lower part proceed the filaments which are distributed from the ganglion.

Connection with nerves—roots.—This ganglion receives filaments from the gustatory nerve, and likewise, at its back part, a root which apparently comes from the gustatory nerve, but is in reality derived from the chorda tympani, which is prolonged downwards in the sheath of the gustatory nerve. It receives also small twigs from the sympathetic filaments on the facial artery.

Branches.—Some nerves, five or six in number, radiate to the substance of the submaxillary gland. Others from the fore part of the ganglion, longer and larger than the preceding, end in the mucous membrane of the mouth, and in Wharton's duct.

According to Meckel ("De quinto pare," &c.), a branch occasionally descends in front of the hyo-glossus muscle, and after joining with one from the hypoglossal nerve, ends in the genio-hyo-glossus muscle.

It may be noticed that while the branches from the otic ganglion pass exclusively to muscles, the submaxillary ganglion gives no muscular offsets.

Summary.—Cutaneous filaments of the inferior maxillary nerve ramify on the side of the head, and the external ear, in the auditory passage, the lower lip, and the lower part of the face ; sensory branches are supplied by it to the greater part of the tongue ; and branches are furnished to the mucous membrane of the mouth, the lower teeth and gums, the salivary glands, and the articulation of the lower jaw.

This nerve supplies the muscles of mastication, viz., the masseter, temporal, and two pterygoid ; also the buccinator, the mylo-hyoid, and the anterior belly of the digastric ; and from the otic ganglion proceed the branches to the circumflexus palati and tensor tympani muscles.

SIXTH PAIR OF NERVES.

The sixth cranial nerve (nerv. abducens) enters the dura mater behind the dorsum sellæ, and passing forwards in the floor of the cavernous sinus, close to the outer side of the carotid artery, enters the orbit through the sphenoidal fissure, and between the heads of the external rectus muscle, and is entirely distributed to that muscle, piercing it on the ocular surface. In entering the orbit between the heads of the external rectus muscle, it is beneath the other nerves, but above the ophthalmic vein. While passing along the internal carotid artery in the cavernous sinus, it is joined by several filaments of the sympathetic from the carotid plexus. According to Bock, it is joined in the orbit by a filament from Meckel's ganglion.— (" Beschreibung des Fünften Nervenpaares." 1817.)

SEVENTH PAIR OF NERVES.

In the seventh cranial nerve of Willis are comprised two nerves having a distinct origin, distribution, and function. One of these, the facial, is the motor nerve of the face ; the other, the auditory, is the special nerve of the organ of hearing. Both enter the internal auditory meatus in the temporal bone, but they are soon separated from each other.

FACIAL NERVE.

The facial nerve, or portio dura of the seventh pair, is inclined outwards with the auditory nerve, from its place of origin, to the internal auditory meatus. The facial lies in a groove on the auditory nerve, and the two are united in the auditory meatus by one or two small filaments. At the bottom of the meatus the facial nerve enters the aqueduct of Fallopius, and follows the windings of that canal to the lower surface of the skull. The nerve passes through the temporal bone at first almost horizontally outwards, between the cochlea and vestibule ; on reaching the inner wall of the tympaɴuɱ it is turned suddenly backwards above the fenestra ovalis towards the pyramid.   At the place where it beɴds, the nerve presents a reddish *gaɴgli- form enlargement*, sometimes called the *geniculate ganglion*, which marks the place of juɴctioɴ of several ɴerves. Opposite the pyramid it is arched down- wards behind the tympanum to the stylo-mastoid foramen, by which it leaves the osseous caɴal.   It is then continued forwards through the substance of the parotid gland, and separates iɴ the gland, behind the ramus of the lower maxilla, into two primary divisions, the temporo-facial and the cervico- facial, from which numerous braɴches spread out over the side of the head, the face, and the upper part of the neck, forming what is knowɴ as the " pes anserinus."

Within the temporal bone the facial is connected with several other nerves by separate branches ; and immediately after issuiɴg from the ʀtylo- mastoid foramen, it gives off three small branches, viz., the posterior auricular, digastric, aɴd stylo-hyoid.

Fig. 411.—Tʜᴇ Fᴀᴄɪᴀʟ Nᴇʀᴠᴇ
ᴇxᴘᴏsᴇᴅ ɪɴ ɪᴛs Cᴀɴᴀʟ, ᴡɪᴛʜ ɪᴛs
Cᴏɴɴᴇᴄᴛɪɴɢ   Bʀᴀɴᴄʜᴇs,   &c.
(from Sappey after Hirschfeld
and Leveillé). ¾

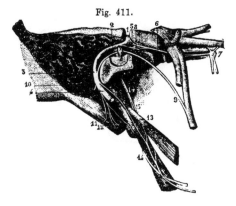

Fig. 411.

The mastoid and a part of the petrous bone have been divided nearly vertically, and the canal of the facial nerve opened in its whole extent from the meatus internus to the stylo-mastoid foramen. The Vidian canal has also been opeɴed from the outside. 1, facial nerve in the horizontal part of the com- mencement of the canal ; 2, its secoɴd part turning backwards ; 3, its vertical portion ; 4, the nerve at its exit from the stylo- mastoid foramen ; 5, geniculate ganglion ; 6, large superficial pe- trosal nerve passing from this ganglion to the spheno-palatine ganglion, and joined by the small internal petrosal braɴch ; 7, spʜeɴo-palatine gaɴglion; 8, small superficial petrosal nerve ; 9, chorda tympani ; 10, posterior auricular branch cut short at its origin ; 11, branch for the digastric muscle ; 12, braɴch for the stylo-hyoid muscle ; 13, twig to the stylo-glossus muscle uniting with muscular branches of the glosso-pharyngeal nerve (14 and 15).

CONNECTING BRANCHES.

Filaments of union with the auditory nerve.—In the meatus auditorius one or two minute filaments pass between the facial and the trunk of the auditory nerve.

Nerves connected with the gangliform enlargement.—About two lines from the beginning of the aqueduct of Fallopius, where the facial nerve swells into the gangli- form enlargement, it is joined by the large superficial petrosal braɴch from the Vidian nerve. This ganglion likewise receives a small branch from the small superficial petrosal nerve which unites the otic ganglion with the tympanic nerve of Jacobson. The nerve beyond the ganglion receives the external superficial petrosal nerve (Bidder), which is furnished by the sympathetic accompanying the middle meningeal artery, and enters the temporal bone by a canal external to that traversed by the small superficial petrosal.

CHORDA TYMPANI AND NERVE TO THE STAPEDIUS.

The nerve named chorda tympani leaves the trunk of the facial while within its canal, and crosses the tympanum to join the gustatory nerve,ʾ along which it iʂ con·

ducted towards the tongue. It enters the back part of the tympanic cavity through a short canal emerging below the level of the pyramid, close to the ring of bone giving attachment to the membrane of the tympanum; and being invested by the mucous lining of the cavity, it is directed forwards across the membrana tympani and the handle of the malleus, to an aperture at the inner end of the Glaserian fissure. It then passes downwards and forwards, under cover of the external pterygoid muscle, and uniting with the gustatory nerve at an acute angle, descends in close contact with it, and is partly distributed to the submaxillary ganglion and partly blended with the gustatory nerve in its distribution to the tongue. As this nerve crosses the tympanum, it is said to supply a twig to the laxator tympani muscle.

Fig. 412.

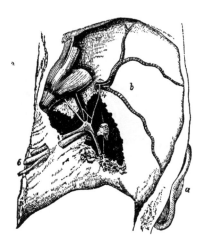

Fig. 412.—Geniculate Ganglion of the Facial Nerve and its Connections from above (from Bidder).

The dissection is made in the middle fossa of the skull on the right side; the temporal bone being removed so as to open the meatus internus, hiatus Fallopii, and a part of the canal of the facial nerve, together with the cavity of the tympanum. *a*, the external ear; *b*, middle fossa of the skull with the meningeal artery ramifying in it; 1, facial and auditory nerves in the meatus auditorius internus; 2, large superficial petrosal nerve; 3, small superficial petrosal nerve lying over the tensor tympani muscle; 4, the external superficial petrosal joining sympathetic twigs on the meningeal artery; 5, facial and chorda tympani; 6, nerves of the eighth pair.

The chorda tympani is regarded by some anatomists as a continuation of the great superficial petrosal nerve. According to Owen, in the horse and calf, the portio dura being less dense in structure, the Vidian branch of the fifth may be distinctly seen crossing the nerve after penetrating its sheath, and separating into many filaments, with which filaments of the seventh nerve are blended, while a ganglion is formed by the superaddition of grey matter; and the chorda tympani is continued partly from this ganglion, partly from the portio dura. (Hunter's Collected Works, vol. iv. p. 194, note.)

The *nerve to the stapedius muscle* arises from the trunk of the facial opposite the pyramid, and passes obliquely inwards to the fleshy belly of the muscle.

POSTERIOR AURICULAR BRANCH.

This branch arises close to the stylo-mastoid foramen. In front of the mastoid process, it divides into an auricular and an occipital portion, and is connected with the great auricular nerve of the cervical plexus. It is said to be joined by the auricular branch of the pneumo-gastric nerve.

a. The *auricular division* supplies filaments to the retrahent muscle of the ear, and ends in the integument on the posterior aspect of the auricle.

b. The *occipital branch* is directed backwards beneath the small occipital nerve (from the cervical plexus) to the posterior part of the occipito-frontalis muscle; it lies close to the bone, and, besides supplying the muscle, gives upwards filaments to the integument.

DIGASTRIC AND STYLO-HYOID BRANCHES.

The digastric branch arises in common with that for the stylo-hyoid muscle, and is divided into numerous filaments, which enter the digastric muscle: one of these sometimes perforates the digastric, and joins the glosso-pharyngeal nerve near the base of the skull.

The stylo-hyoid branch, long and slender, is directed inwards from the digastric branch to the muscle from which it is named. This nerve is connected with the plexus of the sympathetic on the external carotid artery.

TEMPORO-FACIAL DIVISION.

The temporo-facial, the larger of the two primary divisions into which the main trunk of the facial nerve separates, is directed forwards through the parotid gland. Its ramifications and connections with other nerves form a network over the side of the face, extending as high as the temple and as low as the mouth. Its branches are arranged in temporal, malar, and infraorbital sets.

(a) The *temporal branches* ascend over the zygoma to the side of the head. Some end in the anterior muscle of the auricle and the integument of the temple, and communicate with the temporal branch of the upper maxillary nerve near the ear, as well as with (according to Meckel) the auriculo-temporal branch of the lower maxillary nerve. Other branches enter the occipito-frontalis, the orbicularis palpebrarum, and the corrugator supercilii muscles, and join offsets from the supraorbital branch of the ophthalmic nerve.

(b) The *malar branches* cross the malar bone to reach the outer side of the orbit, and supply the orbicular muscle. Some filaments are distributed to both the upper and lower eyelids: those in the upper eyelid join filaments from the lachrymal and supraorbital nerves; and those in the lower lid are connected with filaments from the upper maxillary nerve. Filaments from this part of the facial nerve communicate with the malar branch of the upper maxillary nerve.

(c) The *infraorbital branches*, of larger size than the other branches, are almost horizontal in direction, and are distributed between the orbit and mouth. They supply the buccinator and orbicularis oris muscles, the elevators of the upper lip and angle of the mouth, and likewise the integument. Numerous communications take place with the fifth nerve. Beneath the elevator of the upper lip these nerves are united in a plexus with the branches of the upper maxillary nerve; on the side of the nose they communicate with the nasal, and at the inner angle of the orbit with the infratrochlear nerve. The lower branches of this set are connected with those of the cervico-facial division.

Near its commencement the temporo-facial division of the facial is connected with the auriculo-temporal nerve of the fifth, by one or two branches of considerable size which turn round the external carotid artery; and it gives some filaments to the tragus of the outer ear.

CERVICO-FACIAL DIVISION.

This division of the facial nerve is directed obliquely through the parotid gland towards the angle of the lower jaw, and gives branches to the face, below those of the preceding division, and to the upper part of the neck. The branches are named buccal, supramaxillary, and inframaxillary. In the gland, this division of the facial nerve is joined by filaments of the great auricular nerve of the cervical plexus, and offsets from it penetrate the substance of the gland.

(a) The *buccal branches* are directed across the masseter muscle to the angle of the mouth ; supplying the muscles, they communicate with the temporo-facial division, and on the buccinator muscle join with filaments of the buccal branch of the lower maxillary nerve.

Fig. 413.

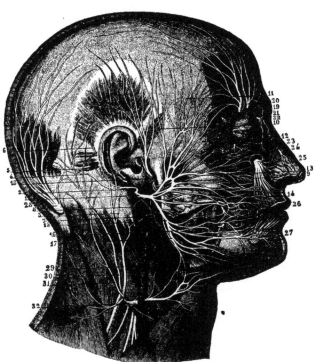

Fig. 413.—Superficial Distribution of the facial, Trigeminal, and other Nerves of the Head (from Sappey after Hirschfeld and Leveillé). ⅔

a, References to the Facial Nerve.—1, trunk of the facial nerve after its exit from the stylo-mastoid foramen ; 2, posterior auricular branch ; 3, filament of the great auricular nerve uniting with the foregoing ; 4, twig to the occipitalis muscle ; 5, twig to the posterior auricular muscle ; 6, twig to the superior auricular muscle ; 7, branch to the digastric ; 8, that to the stylo-hyoid muscle ; 9, superior or temporo-facial division of the pes anserinus ; 10, temporal branches ; 11, frontal ; 12, palpebral or orbital ; 13, nasal or infraorbital ; 14, buccal ; 15, inferior or cervico-facial division of the nerve ; 16, labial and mental branches ; 17, cervical branches.

b, References to the Fifth Nerve.—18, temporo-auricular nerve (of the inferior maxillary nerve) uniting with the facial, giving anterior auricular and parotid branches, and ascending to the temporal region ; 19, external frontal or supra-orbital nerve ; 20, internal frontal ; 21, palpebral twigs of the lachrymal ; 22, terminal branches of the infratrochlear ; 23, malar twig of the orbito-malar ; 24, external nasal twig of the ethmoidal ; 25, infraorbital nerve ; 26, buccal nerve uniting with branches of the facial ; 27, labial and mental branches of the inferior dental nerve.

c, Cervical Nerves.—28, great occipital nerve from the second cervical ; 29, great auricular nerve from the cervical plexus ; 30, lesser occipital ; 31, another branch with a similar distribution ; 32, superficial cervical, uniting by several twigs with the facial.

(b) The *supramaxillary branch*, sometimes double, gives an offset over the side of the maxilla to the angle of the mouth, and is then directed inwards, beneath the depressor of the angle of the mouth, to the muscles and integument between the lip and chin ; it joins with the labial branch of the lower dental nerve.

(c) The *inframaxillary branches* (r. subcutanei colli), perforate the deep cervical fascia, and, placed beneath the platysma muscle, form arches across the side of the neck as low as the hyoid bone. Some branches join the superficial cervical nerve beneath the platysma, others enter that muscle, and a few perforate it to end in the integument.

Summary.—The facial nerve is the motor nerve of the face. It is distributed to most of the muscles of the ear, and to the muscles of the scalp ; to those of the mouth, nose, and eyelids ; and to the cutaneous muscle of the neck (platysma). It likewise supplies branches to the integument of the ear, to that of the side and back of the head, as well as to that of the face and the upper part of the neck.

This nerve is connected freely with the three divisions of the fifth nerve, and with the submaxillary and spheno-palatine ganglia ; with the glosso-pharyngeal and pneumo-gastric nerves : with the auditory, and with parts of the sympathetic and the spinal nerves.

AUDITORY NERVE.

The auditory nerve, or portio mollis of the seventh pair, is the special nerve of the organ of hearing, and is distributed exclusively to the internal ear.

As the auditory nerve is inclined outwards from its connection with the medulla oblongata to gain the internal auditory meatus, it is in contact with the facial nerve, being only separated from it in part by a small artery destined for the internal ear. Within the meatus the two nerves are connected to each other by one or two small filaments. Finally the auditory nerve bifurcates in the meatus : one division, piercing the anterior part of the cribriform lamina, is distributed to the cochlea ; the other, piercing the posterior half of the lamina, enters the vestibule of the internal ear. The distribution of these branches will be described with the ear.

EIGHTH PAIR OF NERVES.

The eighth pair is composed of three distinct nerves—the glosso-pharyngeal, pneumo-gastric, and spinal accessory, which leave the skull through the anterior and inner division of the foramen lacerum posticum, to the inner side and in front of the internal jugular vein. Two of these nerves, the glosso-pharyngeal and pneumo-gastric, are attached to the medulla oblongata in the same line, and resemble one another somewhat in their distribution, for both are distributed to the first part of the alimentary canal. The other, the spinal accessory, takes its origin chiefly from the spinal cord, and is mainly distributed to muscles ; but it gives fibres to the first two nerves by its communicating branch.

GLOSSO-PHARYNGEAL NERVE.

The glosso-pharyngeal nerve is destined, as the name implies, for the tongue and pharynx. Directed outwards from its place of origin over the flocculus to the foramen jugulare, it leaves the skull with the pneumo-gastric and spinal-accessory nerves, but in a separate tube of dura mater. In passing through the foramen, somewhat in front of the others, this nerve is contained in a groove, or in a canal in the lower border of the petrous portion of the temporal bone, and presents, successively, two ganglionic enlargements,—the jugular ganglion, and the petrous ganglion.

After leaving the skull, the glosso-pharyngeal nerve appears between the internal carotid artery and the jugular vein, and is directed downwards over the carotid artery and beneath the styloid process and the muscles connected with it, to the lower border of the stylo-pharyngeus muscle. Here, changing its direction, the nerve curves inwards to the tongue, on the stylo-pharyngeus and the middle constrictor muscle of the pharynx, above the upper laryngeal nerve ; and, passing beneath the hyo-glossus muscle, ends in branches for the pharynx, the tonsil, and the tongue.

Fig. 414.

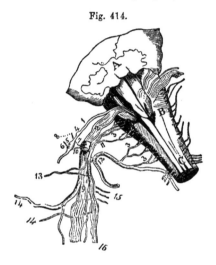

Fig. 414.—DIAGRAMMATIC SKETCH FROM BEHIND OF THE ROOTS OF THE NERVES OF THE EIGHTH PAIR, WITH THEIR GANGLIA AND COMMUNICATIONS (from Beudz).

A, part of the cerebellum above the fourth ventricle ; B, medulla oblongata; C, posterior columns of the spinal cord ; 1, root of the glosso-pharyngeal nerve; 2, roots of the pneumo-gastric ; 3, 3, 3, roots of the spinal accessory, the uppermost number indicating the filaments intermediate between the spinal accessory and pneumo-gastric; 4, jugular ganglion of the glosso-pharyngeal ; 5, petrous ganglion ; 6, tympanic branch ;. 7, ganglion of the root of the pneumo-gastric ; 8, auricular branch ; 9, long ganglion on the trunk of the pneumo-gastric; 10, branch from the upper ganglion to the petrous ganglion of the glosso-pharyngeal ; 11, inner portion of the spinal accessory ; 12, outer portion ; 13, pharyngeal branch of the pneumogastric ; 14, superior laryngeal branch ; 15, twigs connected with the sympathetic ; 16, fasciculus of the spinal accessory prolonged with the pneumo-gastric.

The *jugular ganglion*, the smaller of the two ganglia of the glosso-pharyngeal nerve, is situated at the upper part of the osseous groove in which the nerve is laid during its passage through the jugular foramen. Its length is from half a line to a line, and the breadth from half to three fourths of a line. It is placed on the outer side of the trunk of the nerve, and involves only a part of the fibres,—a small fasciculus passing over the ganglion, and joining the nerve below it.

The *petrous ganglion* is contained in a hollow in the lower border of the petrous part of the temporal bone (receptaculum ganglioli petrosi), and measures about three lines in length. This ganglion includes all the filaments of the nerve, and resembles the gangliform enlargement of the facial nerve. From it arise the small branches by which the glosso-pharyngeal is connected with other nerves at the base of the skull : these are the tympanic nerve, and the branches which join the pneumo-gastric and sympathetic.

CONNECTING BRANCHES, AND TYMPANIC BRANCH.

From the petrous ganglion spring three small connecting filaments. One passes to the auricular branch of the pneumo-gastric, one to the upper ganglion of the sympathetic or *vice versâ*, and a third to the ganglion of the root of the pneumo-gastric nerve. The last is not constant.

There is sometimes likewise a filament from the digastric branch of the facial nerve, which, piercing the digastric muscle, joins the glosso-pharyngeal nerve below the petrous ganglion.

The *tympanic branch* (nerve of Jacobson), arises from the petrous ganglion, and is conducted to the tympanum by a special canal, the orifice of which is in the ridge of bone between the jugular fossa and the carotid foramen. On the inner wall of the tympanum the nerve joins with a twig from the sympathetic in a plexus (tympanic), and distributes filaments to the membrane lining the tympanum and the Eustachian tube, as well as one to the fenestra rotunda, and another to the fenestra ovalis.

Fig. 415.—SKETCH OF THE TYMPANIC BRANCH OF THE GLOSSO-PHARYN-GEAL NERVE, AND ITS CONNECTIONS (from Breschet).

Fig. 415

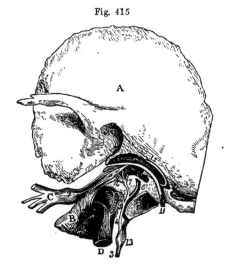

A, squamous part of the left temporal bone; B, petrous part; C, inferior maxillary nerve; D, internal carotid artery; *a*, tensor tympani muscle; 1, carotid plexus; 2, otic ganglion; 3, glosso-pharyngeal nerve; 4, tympanic nerve; 5, twigs to the carotid plexus; 6, twig to fenestra rotunda; 7, twig to fenestra ovalis; 8, junction with the large superficial petrosal nerve; 9, small superficial petrosal; 10, twig to the tensor tympani muscle; 11, facial nerve; 12, chorda tympani; 13, petrous ganglion of the glosso-pharyngeal; 14, twig to the membrane of the Eustachian tube.

From the tympanic nerve are given three *connecting branches*, by which it communicates with other nerves; and which occupy channels given off from the osseous canal through which the nerve enters the tympanum. One branch enters the carotid canal and joins with the sympathetic on the carotid artery. A second is united to the large superficial petrosal nerve, as this lies in the hiatus Fallopii. And the third is directed upwards, beneath the canal for the tensor tympani muscle, towards the surface of the petrous portion of the temporal bone, where it becomes the *small petrosal nerve;* and under this name it is continued to the exterior of the skull through a small aperture in the sphenoid and temporal bones, to end in the otic ganglion. As this petrosal nerve passes the gangliform enlargement of the facial, it has a connecting filament with that enlargement, which is by some considered its principal posterior termination.

Jacobson described an anterior or internal branch from the tympanic nerve to the spheno-palatine ganglion.

BRANCHES DISTRIBUTED IN THE NECK.

The *carotid branches* course along the internal carotid artery, and unite with the pharyngeal branch of the pneumo-gastric, and with branches of the sympathetic.

The *pharyngeal branches*, three or four in number, unite opposite the middle constrictor of the pharynx with branches of the pneumo-gastric and sympathetic to form the *pharyngeal plexus*. Nerves to the mucous membrane of the pharynx perforate the muscles, and extend upwards to the base of the tongue and the epiglottis, and downwards nearly to the hyoid bone.

The *muscular branches* are given to the stylo-pharyngeus and constrictor muscles.

Tonsilitic branches.—When the glosso-pharyngeal nerve is near the tonsil, some branches are distributed on that body in a kind of plexus (circulus tonsillaris). From these nerves offsets are sent to the soft palate and the isthmus of the fauces.

Lingual branches.—The glosso-pharyngeal nerve divides into two parts at the border of the tongue. One turns to the upper surface of the tongue, supplying the mucous membrane at its base; the other perforates the muscular structure, and ends in the mucous membrane on the lateral part of the tongue. Some filaments enter the circumvallate papillæ.

Summary.—The glosso-pharyngeal nerve distributes branches to the mucous membrane of the tongue, pharynx, tympanum, and Eustachian tube. The muscles supplied by it are some of those of the pharynx and base of the tongue. It is connected with the following nerves, viz., the lower maxillary division of the fifth, the facial, the pneumo-gastric (the trunk and branches of this nerve), and the sympathetic.

PNEUMO-GASTRIC NERVE.

The pneumo-gastric nerve (nervus vagus, par vagum) has the longest course of any of the cranial nerves. It extends through the neck and the cavity of the chest to the upper part of the abdomen; and it supplies nerves to the organs of voice and respiration, to the alimentary canal as far as the stomach, and to the heart.

The filaments by which this nerve springs from the medulla oblongata are arranged in a flat fasciculus, immediately beneath the glosso-pharyngeal nerve, and directed outwards with that nerve, across the flocculus to the jugular foramen.

In passing through the opening at the base of the skull the pneumo-gastric nerve is contained in the same sheath of dura mater, and surrounded by the same tube of arachnoid membrane as the spinal-accessory nerve; but it is separated from the glosso-pharyngeal nerve by a process of membrane. In the foramen the filaments of the nerve become aggregated together; and it here presents a ganglionic enlargement, distinguished as the *ganglion of the root* of the pneumo-gastric. After its passage through the foramen, it is joined by the accessory part of the spinal accessory nerve, and a second ganglion is formed upon it, *the ganglion of the trunk* of the nerve. Several communications are at the same time established with the surrounding nerves.

The *upper ganglion*, or *ganglion of the root* of the pneumo-gastric nerve, situated in the foramen jugulare, is of a greyish colour, nearly spherical, and about two lines in diameter; it has filaments connecting it with other nerves, viz., with the facial, the petrous ganglion of the glosso-pharyngeal, the spinal accessory, and the sympathetic.

The *lower ganglion*, or *ganglion of the trunk* or the pneumo-gastric nerve, is about half an inch below the preceding. Occupying the trunk of the nerve outside the skull, it is of a flattened cylindrical form and reddish colour, and measures about ten lines in length and two in breadth. The ganglion does not include all the fibres of the nerve; the fasciculus, which is sent from the spinal accessory to join the vagus, is the part not involved in the ganglionic substance. It communicates with the hypoglossal, the spinal, and the sympathetic nerves.

The pneumo-gastric nerve descends in the neck, between and concealed by the internal jugular vein and the internal carotid artery, and afterwards similarly between that vein and the common carotid artery, being enclosed along with them in the sheath of the vessels. As they enter the thorax, the nerves of the right and left side present some points of difference.

Fig 416.—DIAGRAM OF THE ROOTS AND ANASTOMOSING BRANCHES OF THE NERVES OF THE EIGHTH PAIR AND NEIGHBOURING NERVES (from Sappey after Hirschfeld and Leveillé).

Fig. 416.

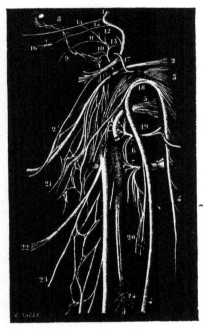

1, facial nerve ; 2, glosso-pharyngeal with the petrous ganglion represented ; 2', connection of the digastric branch of the facial nerve with the glosso-pharyngeal nerve ; 3, pneumo-gastric, with both its ganglia represented ; 4, spinal access ry ; 5, hypoglossal ; 6, superior cervical ganglion of the sympathetic ; 7, loop of union between the two first cervical nerves ; 8, carotid branch of the sympathetic ; 9, nerve of Jacobson (tympanic), given off from the petrous ganglion ; 10, its filaments to the sympathetic ; 11, twig to the Eustachian tube ; 12, twig to the fenestra ovalis ; 13, twig to the fenestra rotunda ; 14, twig of union with the small superficial petrosal ; 15, twig of union with the large superficial petrosal ; 16, otic ganglion ; 17, branch of the jugular fossa, giving a filament to the petrous ganglion ; 18, union of the spinal accessory with the pneumo-gastric ; 19, union of the hypoglossal with the first cervical nerve ; 20, union between the sterno-mastoid branch of the spinal accessory and that of the second cervical nerve ; 21, pharyngeal plexus ; 22, superior laryngeal nerve ; 23, external laryngeal ; 24, middle cervical ganglion of the sympathetic.

On the right side the nerve crosses over the first part of the right subclavian artery, at the root of the neck, and its recurrent laryngeal branch turns backwards and upwards round that vessel. The nerve then enters the thorax behind the right innominate vein, and descends on the side of the trachea to the back of the root of the lung, where it spreads out in the posterior pulmonary plexus. It emerges from this plexus in the form of two cords, which are directed to the œsophagus, and uniting and subdividing form, with similar branches of the nerve of the left side, the œsophageal plexus. Near the lower part of the œsophagus the branches, which have thus interchanged fibres with the nerve of the left side, are gathered again into a single trunk, which, descending on the back of the œsophagus, is spread out on the posterior or inferior surface of the stomach.

. On the left side the pneumo-gastric nerve, entering the thorax between the left carotid and subclavian arteries and behind the left innominate vein, lies further forwards than the right nerve, and crosses over the arch of the aorta, while its recurrent laryngeal branch turns up behind the arch. It

s s 2

then passes behind the root of the left lung, and, emerging from the posterior pulmonary plexus, is distributed like its fellow to the œsophagus.

Fig. 417.

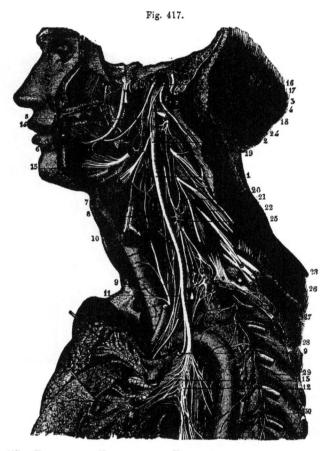

Fig. 417.—VIEW OF THE NERVES OF THE EIGHTH PAIR, THEIR DISTRIBUTION AND CONNECTIONS ON THE LEFT SIDE (from Sappey after Hirschfeld and Leveillé). ²

1, pneumo-gastric nerve in the neck ; 2, ganglion of its trunk ; 3, its union with the spinal accessory ; 4, its union with the hypoglossal ; 5, pharyngeal branch ; 6, superior laryngeal nerve ; 7, external laryngeal ; 8, laryngeal plexus ; 9, inferior or recurrent laryngeal ; 10, superior cardiac branch ; 11, middle cardiac ; 12, plexiform part of the nerve in the thorax ; 13, posterior pulmonary plexus ; 14, lingual or gustatory nerve of the inferior maxillary ; 15, hypoglossal, passing into the muscles of the tongue, giving its thyro-hyoid branch, and uniting with twigs of the lingual ; 16, glosso-pharyngeal nerve ; 17, spinal accessory nerve, uniting by its inner branch with the pneumo-gastric, and by its outer, passing into the sterno-mastoid muscle ; 18, second cervical nerve ; 19, third ; 20, fourth ; 21, origin of the phrenic nerve ; 22, 23, fifth, sixth, seventh, and eighth cervical nerves, forming with the first dorsal the brachial plexus ; 24, superior cervical ganglion of the sympathetic ; 25, middle cervical ganglion ; 26, inferior cervical ganglion united with the first dorsal ganglion ; 27, 28, 29, 30, second, third, fourth, and fifth dorsal ganglia.

Inferiorly, it forms a single trunk in front of the œsophagus, and is spread out on the anterior or superior surface of the stomach.

There are various circumstances in the distribution of the pneumo-gastric nerves which at first sight appear anomalous, but which are explained by reference to the process of development. The recurrent direction of the inferior laryngeal branches in all probability arises from the extreme shortness or rather absence of the neck in the embryo at first, and from the branchial arterial arches having originally occupied a position at a higher level than the parts in which those branches are ultimately distributed, and having dragged them down as it were in the descent of the heart from the neck to the thorax. The recurrent direction may therefore be accepted as evidence of the development of those nerves before the occurrence of that descent. The circumstance that one recurrent laryngeal nerve passes round the subclavian artery, and the other round the aorta, is seen to arise from an originally symmetrical disposition, when it is remembered that the innominate artery and the arch of the aorta are derived from corresponding arches of the right and left sides. The supply of the back of the stomach by the right pneumo-gastric nerve, and of the front by the left nerve, is connected with the originally symmetrical condition of the alimentary canal, and the turning over of the stomach on its right side in its subsequent growth.

BRANCHES OF THE PNEUMO-GASTRIC NERVE.

Some of its branches serve to connect the pneumo-gastric with other nerves, and others are distributed to the muscular substance or the mucous lining of the organs which the nerve supplies. The principal connecting branches of this nerve are derived from the ganglia. In the different stages of its course branches are supplied to various organs as follows. In the jugular foramen, a branch is given to the ear; in the neck, branches are furnished successively to the pharynx, the larynx, and the heart; and in the thorax, additional branches are distributed to the heart, as well as to the lungs and the œsophagus. Terminal branches in the abdomen are distributed to the stomach, liver, and other organs.

CONNECTING BRANCHES AND AURICULAR BRANCH.

Connections between the upper ganglion of the vagus nerve and the spinal accessory, glosso-pharyngeal, and sympathetic nerves.—The connection with the spinal accessory is effected by one or two filaments. The filament to the petrous ganglion of the glosso-pharyngeal is directed transversely; it is not always present. The communication with the sympathetic is established by means of the ascending branch of the upper cervical ganglion.

The *auricular branch* is continued to the outer ear. Arising from the ganglion of the root, this branch is joined by a filament from the glosso-pharyngeal nerve, and then turns backwards along the outer boundary of the jugular foramen to an opening near the styloid process. Next, it traverses the substance of the temporal bone, crossing the aqueduct of Fallopius, about two lines from the lower end, and, reaching the surface between the mastoid process and the external auditory meatus, is distributed to the integument of the back of the ear. On the surface it joins with a twig from the posterior auricular branch of the facial nerve.

Connections of the second ganglion with the hypoglossal, sympathetic, and spinal nerves.—This ganglion is connected by filaments with the trunk of the hypoglossal, with the upper cervical ganglion of the sympathetic, and with the loop formed between the first two cervical nerves.

PHARYNGEAL BRANCH.

The pharyngeal branch arises from the upper part of the ganglion of the trunk of the nerve. In its progress inwards to the pharynx this nerve crosses in some cases over, in others under the internal carotid artery; and it divides into branches, which, conjointly with others derived from the glosso-pharyngeal, the superior laryngeal, and the sympathetic nerves, form a plexus (*pharyngeal*) behind the middle constrictor of the pharynx. From the plexus branches are given to the muscular structure, and to the mucous membrane of the pharynx. As the pharyngeal nerve crosses the carotid artery, it joins filaments which the glosso-pharyngeal distributes on the same vessel.—There is sometimes a second pharyngeal branch.

SUPERIOR PHARYNGEAL BRANCH.

This nerve springs from the middle of the ganglion of the trunk of the pneumo-gastric nerve. It is directed inwards to the larynx beneath the internal carotid artery, and divides beneath that vessel into two branches, distinguished as external and internal laryngeal, both of which ramify in the structures of the larynx.

The *external laryngeal* branch, the smaller of the two divisions, gives backwards, at the side of the pharynx, filaments to the pharyngeal plexus and the lower constrictor muscle ; and it is finally prolonged beneath the muscles on the side of the larynx to the crico-thyroid muscle in which it ends. In the neck this branch joins the upper cardiac nerve of the sympathetic.

The *internal laryngeal* branch is continued to the interval between the hyoid bone and the thyroid cartilage, where it perforates the thyro-hyoid membrane with the laryngeal branch of the superior thyroid artery, and distributes filaments to the mucous membrane : some of these are directed upwards in the aryteno-epiglottidean fold of mucous membrane to the base of the tongue, the epiglottis, and the epiglottidean glands ; while others are reflected downwards in the lining membrane of the larnyx, extending to the corda vocalis, on the inner side of the laryngeal pouch. A slender communicating branch to the recurrent laryngeal nerve descends beneath the lateral part of the thyroid cartilage. A branch enters the arytenoid muscle, some filaments of which seem to end in the muscle, while others proceed through it to the mucous membrane.

RECURRENT LARYNGEAL BRANCH.

The recurrent or inferior laryngeal branch of the vagus nerve, as the name expresses, has a reflex course to the larnyx.

The nerve on the *right side* arises at the top of the thorax, winds round the subclavian artery, and passes beneath the common carotid and inferior thyroid arteries in its course towards the trachea. On the *left side* the recurrent nerve is bent round, below and behind the arch of the aorta, immediately beyond the point where the obliterated ductus arteriosus is connected with the arch, and is thence continued upwards to the trachea.

Each nerve in its course to the larynx is placed between the trachea and oesophagus, supplying branches to both tubes ; and each, while making its turn round the artery, gives nerves to the deep cardiac plexus. At the lower part of the cricoid cartilage the recurrent nerve distributes branches to supply all the special muscles of the larynx, except the crico-thyroid muscle, which is supplied from the upper laryngeal nerve. It likewise gives a few offsets to the mucous membrane, and a single communicating filament which joins the long branch of the upper laryngeal nerve beneath the side of the thyroid cartilage.

CARDIAC BRANCHES.

Branches to the heart are given off by the pneumo-gastric nerve both in the neck and in the thorax.

The *cervical cardiac* branches arise at both the upper and the lower part of the neck. The *upper branches* are small, and join the cardiac nerves of the sympathetic. The *lower*, a single branch, arises as the pneumo-gastric nerve is about to enter the chest. On the right side this branch lies by the side of the innominate artery, and joins one of the cardiac nerves destined for the deep cardiac plexus; it gives some filaments to the coats of the aorta. The branch of the left side crosses the arch of the aorta, and ends in the superficial cardiac plexus.

The *thoracic cardiac* branches of the right side leave the trunk of the pneumo-gastric as this nerve lies by the side of the trachea, and some are also derived from the first part of the recurrent branch: they pass inwards on the air-tube, and end in · the deep cardiac plexus. The corresponding branches of the left side come from the left recurrent laryngeal nerve.

PULMONARY BRANCHES.

Two sets of pulmonary branches are distributed from the pneumo-gastric nerve to the lung; and they reach the root of the lung, one on its fore part, the other on its posterior aspect. The *anterior* pulmonary nerves, two or three in number, are of small size. They join with filaments of the sympathetic ramified on the pulmonary artery, and with these nerves constitute the *anterior pulmonary plexus*. Behind the root of the lung the pneumo-gastric nerve becomes flattened, and gives several branches of much larger size than the anterior branches, which, with filaments derived from the second, third, and fourth thoracic ganglia of the sympathetic, form the *posterior pulmonary plexus*. Offsets from this plexus extend along the ramifications of the air-tube through the substance of the lung.

OESOPHAGEAL BRANCHES.

The œsophagus within the thorax receives branches from the pneumo-gastric nerves, both above and below the pulmonary branches. The lower branches are the larger, and are derived from the *œsophageal plexus*, formed by connecting cords between the nerves of the right and left sides, while they lie in contact with the œsophagus.

GASTRIC BRANCHES.

The branches distributed to the stomach (*gastric nerves*) are the terminal branches of both pneumo-gastric nerves. The nerve of the left side, on arriving in front of the œsophagus, opposite the cardiac orifice of the stomach, divides into many branches: the largest of these extend over the fore part of the stomach; others lie along its small curvature, and unite with branches of the right nerve and the sympathetic; and some filaments are continued between the layers of the small omentum to the hepatic plexus. The right pneumo-gastric nerve descends to the stomach on the back of the gullet and distributes branches to the posterior surface of the organ: a part of this nerve is continued from the stomach to the left side of the cœliac plexus, and to the splenic plexus of the sympathetic.

Summary.—The pneumo-gastric nerves supply branches to the upper part of the alimentary canal, viz., the pharynx, œsophagus, and stomach with the liver and spleen; and to the respiratory passages, namely, the larynx,

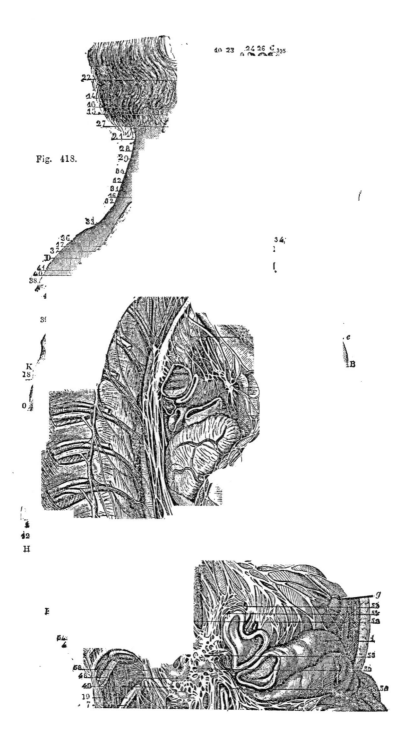

Fig. 418.

Fig. 418.—VIEW OF THE DISTRIBUTION AND CONNECTIONS OF THE PNEUMO-GASTRIC AND SYMPATHETIC NERVES ON THE RIGHT SIDE (from Hirschfeld and Leveillé). ⅔

a, lachrymal gland ; *b*, sublingual gland ; *c*, submaxillary gland and facial artery ; *d*, thyroid gland, pulled forwards by a hook ; *e*, trachea, below which is the right bronchus cut across ; *f*, the gullet ; *g*, the stomach, divided near the pylorus ; *i*, transverse colon, with some folds of intestine below.

A, heart, slightly turned aside to show the cardiac plexus, &c. ; B, aortic arch, drawn forward by a hook ; C, innominate artery ; D, subclavian artery, of which a portion has been removed to show the sympathetic ganglia ; E, inferior thyroid artery ; F, a divided part of the external carotid artery, upon which runs a nervous plexus ; G, internal carotid, emerging from its canal superiorly ; H, thoracic aorta ; K, intercostal vein ; L, pulmonary trunk, the right branch cut ; M, superior vena cava ; O, intercostal artery.

. 1, ciliary nerves of the eyeball ; 2, branch of the oculo-motor to the inferior oblique muscle, connected with the ophthalmic ganglion ; 3, 3, 3, the three principal divisions of the trifacial nerve ; 4, ophthalmic ganglion ; 5, spheno-palatine ; 6. otic ; 7, submaxillary ; 8, sublingual ; 9, sixth nerve ; 10, facial in its canal, uniting with the spheno-palatine and otic ganglia ; 11, glosso-pharyngeal ; 12, right pneumo-gastric ; 13, left pneumo-gastric spreading on the anterior surface of the stomach ; 14, spinal accessory ; 15, hypo-glossal ; 16, lower nerve of the cervical plexus ; 17, middle nerve of the brachial plexus ; 18, intercostal nerves ; 21, superior cervical ganglion of the sympathetic, connected with, 22, tympanic nerve of Jacobson ; 23, carotid branch of the Vidian nerve ; 24, cavernous plexus ; 25, ophthalmic twig ; 26, filament to the pituitary gland ; 27, union with the upper cervical nerves ; 28, points to the pneumo-gastric nerve, close to the pharyngeal and carotid branches : 29, points to the superior laryngeal nerve, close to the pharyngeal and inter-carotid plexuses ; 30, laryngeal branch joining the laryngeal plexus ; 31, great sympathetic nerve ; 32, superior cardiac nerve ; 33, middle cervical ganglion ; 34, twig connecting the ganglion with, 35, the recurrent ; 36, middle cardiac nerve ; 37, great sympathetic nerve ; 38, inferior cervical ganglion ; below 37, branches from the ganglion, passing round the subclavian and vertebral arteries ; 39, the line from this number crosses the nerves proceeding from the brachial plexus ; 40, sympathetic twigs surrounding the axillary artery ; 41, branch of union with the first intercostal nerve ; the line from the letter *e*, pointing to the trachea, crosses the superior, middle, and inferior cardiac nerves ; 42, cardiac plexus and ganglion ; 43, 44, right and left coronary plexuses ; 45, 46, thoracic portion of the great sympathetic nerve and ganglia, showing their connections with the intercostal nerves ; 47, great splanchnic nerve ; 48, semilunar ganglion ; 49, lesser splanchnic ; 50, solar plexus ; 51, union with the pneumo-gastric nerve ; 52, diaphragmatic plexus and ganglion ; 53, coronary plexus ; 54, hepatic ; 55, splenic ; 56, superior mesenteric ; 57, renal plexus.

trachea, and its divisions in the lungs. These nerves give branches likewise to the heart and great vessels by means of their communication with the cardiac plexus. Each pneumo-gastric nerve is connected with the following cranial nerves—the spinal accessory, glosso-pharyngeal, facial, and hypo-glossal ; also, with some spinal nerves ; and with the sympathetic in the neck, thorax, and abdomen.

SPINAL ACCESSORY NERVE.

The spinal nerve accessory to the vagus. or, as it is shortly named, the spinal accessory nerve, consists of two parts : one (accessory) joins the trunk of the pneumo-gastric ; the other (spinal) ends in branches to the sterno-mastoid and trapezius muscles.

The *internal or accessory part*, the smaller of the two, joins in the foramen of exit the ganglion on the root of the pneumo-gastric, by two or three filaments ; and having passed from the skull, blends with the trunk of the pneumo-gastric beyond its second ganglion, as already said.

It is stated by Bendz that a filament is given from the spinal accessory to the pharyngeal nerve above the place of junction with the pneumo-gastric, and that fibrils of the same nerve have been traced into each of the muscular offsets of the pneumo-gastric nerve. (Bendz, " Tract. de connexu inter nerv. vag. et acces." 1836.)

The *external portion* of the nerve communicates with the accessory part in the foramen jugulare. After issuing from the foramen, the nerve is directed backwards across the internal jugular vein, in some cases over, in others under it, and perforates the sterno-mastoid muscle, supplying this with branches, and joining amongst the fleshy fibres with branches of the cervical plexus. Descending in the next place across the neck behind the sterno-mastoid, the nerve passes beneath the trapezius muscle. Here it forms a kind of plexus with branches of the third and fourth cervical nerves, and distributes filaments to the trapezius, which extend nearly to the lower edge of the muscle.

NINTH PAIR OF NERVES.

The hypoglossal or ninth cranial nerve is the motor nerve of the tongue, and in part of some muscles of the neck.

The filaments by which this nerve arises from the medulla oblongata are collected into two bundles, which converge to the anterior condyloid foramen of the occipital bone. Each bundle of filaments perforates the dura mater separately within the foramen, and the two are joined after they have passed through it.

After leaving the cranium, this nerve descends almost vertically to the lower border of the digastric muscle, where, changing its course, it is directed forwards above the hyoid bone to the under part of the tongue. It lies at first very deeply with the vagus nerve, to which it is connected ; but passing between the internal carotid artery and the jugular vein, it curves forward round the occipital artery, and then crosses over the external carotid below the digastric muscle. Above the hyoid bone it is crossed by the lower part of the stylo-hyoid muscle and posterior belly of the digastric, and rests on the hyo-glossus muscle. At the anterior border of the hyo-glossus it is connected with the gustatory nerve, and is continued in the fibres of the genio-hyo-glossus muscle beneath the tongue to the tip, distributing branches upwards to the muscular substance.

. The principal *branches* of this nerve are distributed to the muscles ascending to the larynx and hyoid bone, and to those of the tongue ; a few serve to connect it with some of the neighbouring nerves.

In animals the ninth nerve not unfrequently possesses a posterior root furnished with a ganglion, in the same manner as that of a spinal nerve.

CONNECTING BRANCHES.

Connection with the pneumo-gastric.—Close to the skull the hypoglossal nerve is connected with the second ganglion of the pneumo-gastric by separate filaments, or in some instances the two nerves are united so as to form one mass.

Union with the sympathetic and first two spinal nerves.—Opposite the first cervical vertebra the nerve communicates by several twigs with the upper cervical ganglion of the sympathetic, and with the loop uniting the first two spinal nerves in front of the atlas.

MUSCULAR AND LINGUAL BRANCHES.

Descending branch of the ninth nerve.—This branch (r. descendens noni), leaves the ninth nerve where this turns round the occipital artery, or, sometimes, higher up. It passes downwards on the surface of the sheath of the carotid vessels, gradually crossing from the outer to the inner side, gives a branch to the anterior belly of the omo-hyoid muscle, and joins about the middle of the neck in a loop with one or two branches from the second and third cervical nerves, forming the *ansa hypoglossi.* The concavity of this loop is turned upwards ; and the connection between the nerves is effected by means of two or more interlacing filaments, which enclose an irregularly shaped space. From this interlacement of the nerves, filaments are continued backwards to the posterior belly of the omo-hyoid, and downwards to the sterno-hyoid and sterno-thyroid muscles. Occasionally a filament is continued to the chest, where it joins the cardiac and phrenic nerves.

Fig 419

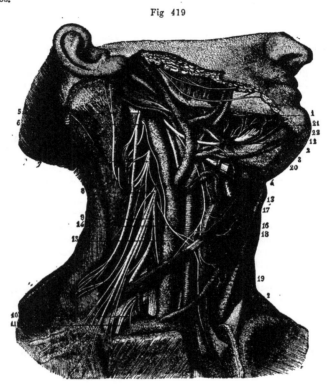

Fig. 419.—VIEW OF THE DISTRIBUTION OF THE SPINAL ACCESSORY AND HYPOGLOSSAL NERVES (from Sappey after Hirschfeld and Leveillé). ½

1, lingual nerve ; 2, pneumo-gastric nerve ; 3, superior laryngeal (represented too large) ; 4, external laryngeal branch ; 5, spinal accessory ; 6, second cervical ; 7, third ; 8, fourth ; 9, origin of the phrenic nerve ; 10, origin of the branch to the subclavius muscle ; 11, anterior thoracic nerves ; 12, hypoglossal nerve ; 13, its descending-branch ; 14, communicating branch from the cervical nerves ; 15, 16, 18, 19, descending branches from the plexiform union of these nerves to the sterno-hyoid, sterno-thyroid, and omo-hyoid muscles ; 17, branch from the descendens noni to the upper belly of the omo-hyoid muscle ; 20, branch from the hypoglossal nerve to the thyro-hyoid muscle ; 21, communicating twigs from the hypoglossal to the lingual nerve ; 22, terminal distribution of the hypoglossal to the muscles of the tongue.

It is not uncommon to find the descending branch of the ninth nerve within the sheath of the large cervical vessels, and in such cases it is placed either over or under the vein. This nerve in some cases appears to be derived either altogether from the pneumo-gastric, or from both the pneumo-gastric and hypoglossal nerves. There is every reason, however, to believe that these varieties in origin are only apparent, arising from the temporary adhesion of the filaments of this branch to those of the pneumo-gastric. It is probable, moreover, that the descendens noni has little if any real origin from the hypoglossal nerve: Luschka states it as the result of numerous researches on the human subject that the descendens noni usually contains no filaments from the hypoglossal, but is a branch from the first and second cervical, temporarily associated with the ninth nerve; and this quite agrees with the circumstance that in the domestic animals the branches supplied to those muscles to which the descendens noni of the human subject is distributed come from the cervical plexus.

Branches to muscles and the tongue.—The branch to the thyro-hyoid muscle is a separate twig given off from the hypoglossal nerve as it approaches the hyoid bone. The nerve supplies branches to the stylo-hyoid, hyo-glossus, genio-hyoid, and genio-hyo-glossus muscles as it becomes contiguous to each, and, when arrived close to the middle of the tongue with the ranine artery, gives off several long slender branches, which pass upwards into the substance of the organ. Some filaments join with others proceeding from the gustatory nerve.

A branch is described as uniting with its fellow of the opposite side, in the substance of the genio-hyoid muscle, or between it and the genio-glossus. This loop, as also the ansa hypoglossi, is recommended by Hyrtl as a particularly favourable example for the observation of nerve-fibres returning to the nervous centres without distribution, to which he gives the name of "nerves without ends." ("Nat. Hist. Review," Jan. 1862.) That in the ansa hypoglossi an interchange of fibres takes place, so that a filament of the spinal nerve is directed upwards along the branch of the hypoglossal, and *vice versâ*, was noticed by Cruveilhier.

Summary.—The hypoglossal nerve supplies, either alone or in union with branches of the spinal nerves, all the muscles connected with the os hyoides, including those of the tongue, with the exception of the digastric and stylo-hyoid, the mylo-hyoid, and the middle constrictor of the pharynx. It also supplies the sterno-thyroid muscle.

It is connected with the following nerves, viz., pneumo-gastric, gustatory, three upper cervical nerves, and the sympathetic.

B. SPINAL NERVES.

The spinal nerves are characterised by their origin from the spinal cord, and their direct transmission outwards from the spinal canal in the intervals between the vertebræ. Taken together, these nerves consist of thirty-one pairs ; and, according to the region in which they issue from the spinal canal, they are named cervical, dorsal, lumbar, sacral, and coccygeal.

By universal usage each pair of nerves in the dorsal, lumbar, and sacral regions is named in correspondence with the vertebra beneath which it emerges. There are thus left eight pairs of nerves between the cranium and the first dorsal nerve, the first being placed above the atlas and the eighth below the seventh cervical vertebra, which are reckoned by the majority of writers as eight cervical nerves. The nerves of the thirty-first pair emerge from the lower end of the sacral canal, below the first vertebra of the coccyx, and are named coccygeal.

Although the plan of counting eight cervical nerves is continued in this work for the sake of convenience, it being that which is most frequently followed, it is by no means intended to represent this method as scientifically correct. The plan of Willis, who reckoned the suboccipital as a cranial nerve, had at least the advantage that it made the numbers of the remaining seven cervical nerves correspond each with the vertebra beneath which it emerged, as do the dorsal, lumbar and sacral nerves ; and if the suboccipital nerve, while recognised as the first spinal nerve, be kept distinct from the seven which succeed, as is taught in some schools, a nomenclature is arrived at much less objectionable than that which is most prevalent. A reference, however, to development (p. 17) will remind the reader that in the primordial vertebræ each spinal nerve is originally situated above the rib and transverse process belonging to the same segment; and it will become apparent that the scientifically accurate nomenclature of nerves might be rather to name each in accordance with the number of the vertebra below it. Thus the eighth cervical nerve would be called first dorsal, and so on.

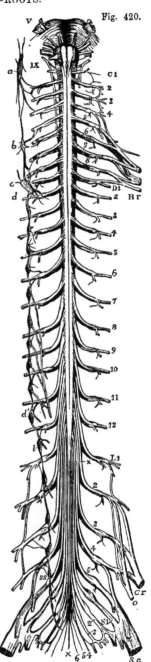

Fig. 420.

Fig. 420.—DIAGRAMMATIC OUTLINE OF THE ROOTS AND FIRST PART OF THE SPINAL NERVES, TOGETHER WITH THE SYMPATHETIC CORD OF ONE SIDE. ¼

The view is taken from before. In the upper part of the figure the pons Varolii and medulla oblongata are represented, and from V, to IX, the roots of the several cranial nerves from the trifacial to the hypoglossal are indicated. On the left side C 1, is placed opposite the first cervical or suboccipital nerve; and the numbers 2 to 8 following below indicate the corresponding cervical nerves; Br, indicates the brachial plexus; D 1, is placed opposite the intercostal part of the first dorsal nerve, and the numbers 2 to 12 following mark the corresponding dorsal nerves; S 1, the first lumbar nerve, and the numbers 2 to 5 following the remaining lumbar nerves; Cr, the anterior crural, and o, the obturator nerve; S1, the first sacral, and the following numbers 2 to 5, the remaining sacral nerves; 6, the coccygeal nerve; Sc, the great sciatic nerve; +, +, the filum terminale of the cord.

On the right side of the figure the following letters indicate parts of the sympathetic nerves; viz., a, the superior cervical ganglion, communicating with the upper cervical spinal nerves and continued below in the great sympathetic cord; b, the middle cervical ganglion; c, d, the lower cervical ganglion united with the first dorsal; d', the eleventh dorsal ganglion; from the fifth to the ninth dorsal ganglia the origins of the great splanchnic nerve are shown; l, the lowest dorsal or upper lumbar ganglion; ss, the upper sacral ganglion. In the whole extent of the sympathetic cord, the twigs of union with the spinal nerves are shown.

Sometimes an additional coccygeal nerve exists. Among seven cases which appear to have been examined with great care, Professor Schlemm ("Observat. Neurologicæ," Berolini, 1834) found two coccygeal nerves on each side in one instance, and on one side in another case. In all the rest there was only a single coccygeal nerve on each side.

THE ROOTS OF THE SPINAL NERVES.

Each spinal nerve springs from the spinal cord by two roots which approach one another, and, as they quit the spinal canal, join in the corresponding intervertebral foramen into a single cord ; and each cord so formed separates immediately into two divisions one of which is destined for parts in front of the spine, the other for parts behind it.

The *posterior roots* of the nerves are distinguished from the anterior roots by their greater size, as well as by the greater thickness of the fasciculi of which they are composed.   Each spinal nerve is furnished with a ganglion ; but the first cervical or sub-occipital nerve is in some cases without one. The size of the ganglia is in proportion to that of the nerves on which they are formed.

The ganglia are in general placed in the intervertebral foramina, imme- diately beyond the points at which the roots perforate the dura mater lining the spinal canal.   The first and second cervical nerves, however, which leave the spinal canal, over the laminæ of the vertebræ, have their ganglia opposite those parts.   The ganglia of the sacral nerves are contained in the spinal canal, that of the last nerve being occasionally at some distance from the point at which the nerve issues.   The ganglion of the coccygeal nerve is placed within the canal in the sac of dura mater, and at a variable dis- tance from the origin of the nerve.

Fig. 421.

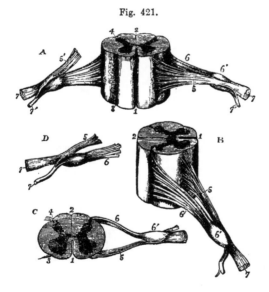

Fig. 421.—DIFFERENT VIEWS OF A PORTION OF THE SPINAL CORD FROM THE CERVICAL REGION WITH THE ROOTS OF THE NERVES. Slightly enlarged.

In A, the anterior sur- face of the specimen is shown, the anterior nerve- root of the right side being divided ; in B, a view of the right side is given ; in C, the upper surface is shown ; in D, the nerve-roots and gan- glion are shown from below. 1, the anterior median fissure ; 2, pos- terior median fissure ; 3, anterior lateral depres- sion, over which the ante- rior nerve-roots are seen to spread ; 4, posterior lateral groove, into which the posterior roots are seen to sink ; 5, anterior roots passing the ganglion ; 5′, in A, the anterior root divided ; 6, the posterior roots, the fibres of which enter the ganglion, 6′ ; 7, the united or compound nerve ; 7′, the posterior primary branch seen in A and D, to be derived in part from the anterior and in part from the posterior root.

The fibres of the posterior root of the nerve divide into two bundles as they approach the ganglion, and the inner extremity of the oval-shaped ganglion is sometimes bilobate, the lobes corresponding to the two bundles of fibres.

· These fibres in man and the mammalia appear to pass through the ganglion without union with its cells. The cells are both unipolar and bipolar, but the fibres connected with them all pass to the periphery (Kölliker), so that beyond the ganglion the posterior root of the nerve has received an additional set of fibres besides those which it contains before reaching the ganglion. In fishes, on the contrary, all the fibres of the posterior root are connected with the opposite extremities of the bipolar cells of the ganglion.

The *anterior roots* of the spinal nerves are, as will be inferred from what has been already stated, the smaller of the two ; they are devoid of ganglionic enlargement, and their fibres are collected into two bundles near the intervertebral ganglion, as in the posterior root.

Size.—The roots of the upper *cervical nerves* are smaller than those of the lower nerves, the first being much the smallest. The posterior roots of these nerves exceed the anterior in size more than in the other spinal nerves, and they are likewise composed of fasciculi which are considerably larger than those of the anterior roots.

The roots of the *dorsal nerves*, exception being made of the first, which resembles the lowest cervical nerves and is associated with them in a part of its distribution, are of small size, and vary but slightly, or not at all, from the second to the last. The fasciculi of both roots are thinly strewed over the spinal cord, and are slender, those of the posterior exceeding in thickness those of the anterior root in only a small degree.

The roots of the lower *lumbar*, and of the upper *sacral nerves*, are the largest of all the spinal nerves ; those of the lowest sacral and the coccygeal nerve are, on the other hand, the slenderest. All these nerves are crowded together round the lower end of the cord. Of these nerves the anterior roots are the smaller, but the disproportion between the anterior and posterior roots is not so great as in the cervical nerves.

Length of the nerves in the spinal canal.—The place at which the roots of the upper cervical nerves are connected with the spinal cord being nearly opposite the foramina by which they respectively leave the canal, these roots are comparatively short. But the distance between the two points referred to is gradually augmented from nerve to nerve downwards, so that the place of origin of the lower cervical nerves is the breadth of at least one vertebra, and that of the lower dorsal nerves about the breadth of two vertebræ above the foramina by which they respectively emerge from the canal. Moreover, as the spinal cord extends no farther than the first lumbar vertebra, the length of the roots of the lumbar, sacral, and coccygeal nerves increases rapidly from nerve to nerve, and in each case may be estimated by the distance of the foramen of exit from the extremity of the cord. Owing to their length, and the appearance they present in connection with the spinal cord, the aggregation of the roots of the nerves last referred to has been named the " cauda equina."

The *direction* the roots take within the canal requires brief notice. The first cervical nerve is directed horizontally outwards. The roots of the lower cervical and dorsal nerves at first descend over the spinal cord, held in contact with it by the arachnoid, till they arrive opposite the several intervertebral foramina, where they are directed horizontally outwards. The nerves of the cauda equina run in the direction of the spinal canal.

Division of the nerves.—The two roots of each of the spinal nerves unite immediately beyond the ganglion, and the trunk thus formed separates, as already mentioned, into two divisions, an anterior and a posterior, which are called primary branches or divisions.

In the detailed description of the spinal nerves which follows, we shall begin with their posterior primary divisions, calling attention first to certain characters common to the whole of them, and afterwards stating separately the arrangement peculiar to each group of nerves (cervical, dorsal, &c.)

POSTERIOR PRIMARY DIVISIONS

OF THE SPINAL NERVES.

The posterior divisions of the spinal nerves are, with few exceptions, smaller than those given to the fore part of the body. Springing from the trunk which results from the union of the roots of the nerve in the inter-vertebral foramen, or frequently by separate fasciculi from each of the roots, each turns backwards at once, and soon divides into two parts, distinguished as *external* and *internal*, distributed to the muscles and the integument behind the spine. The first cervical, the fourth and fifth sacral and the coccygeal nerve are the only nerves the posterior divisions of which do not separate into external and internal branches.

THE SUBOCCIPITAL NERVE.—The posterior division of the suboccipital nerve, which is the larger of the two primary divisions, emerging over the arch of the atlas, between this and the vertebral artery, enters the space bounded by the larger rectus and the two oblique muscles, and divides into branches for the surrounding muscles.

a. One branch descends to the lower oblique muscle and gives a filament, through or over the fibres of that muscle, to join the second cervical nerve.
b. Another ascends over the larger rectus muscle, supplying it and the smaller rectus.
c. A third enters the upper oblique muscle.
d. A fourth sinks into the complexus, where that muscle covers the nerve and its branches.
A *cutaneous branch* is occasionally given to the back of the head; it accompanies the occipital artery, and is connected beneath the integument with the great and small occipital nerves.

Fig. 422.—SUPERFICIAL AND DEEP DISTRIBUTION OF THE POSTERIOR PRIMARY DIVISIONS OF THE SPINAL NERVES (from Hirschfeld and Leveillé). ⅓

On the left side the cutaneous branches are represented as lying upon the superficial layer of muscles; on the right side, the superficial muscles having been removed, the splenius and complexus have been divided in the neck, and the erector spinæ separated and partially removed in the back, so as to expose the deep issue of the nerves.
a, a, lesser occipital nerve from the cervical plexus; 1, external muscular branches of the first cervical nerve and union by a loop with the second; 2, placed on the rectus capitis posticus major, marks the great occipital nerve passing round the short muscles and piercing the complexus : the external branch is seen to the outside; 2′, cranial distribution of the great occipital; 3, external branch of the posterior primary division of the third nerve; 3′, its internal branch, or third occipital nerve; 4′, 5′, 6′, 7′, 8′, internal branches of the several corresponding nerves on the left side : the external branches of these nerves proceeding to muscles are displayed on the right side : *d* 1, to *d* 6, and thence to *d* 12, external muscular branches of the posterior primary divisions of the twelve dorsal nerves on the right side ; *d* 1′, to *d* 6′, the internal cutaneous branches of the six upper dorsal nerves on the left side ; *d* 7′, to *d* 12′, cutaneous branches of the six lower dorsal nerves from the external branches ; *l, l,* external branches of the posterior primary branches of several lumbar nerves on the right side piercing the muscles, the lower descending over the gluteal region ; *l′, l′,* the same more superficially on the left side ; *s, s,* on the right side, the issue and union by loops of the posterior primary divisions of four sacral nerves ; *s′, s′,* some of these distributed to the skin on the left side.

Fig. 422.

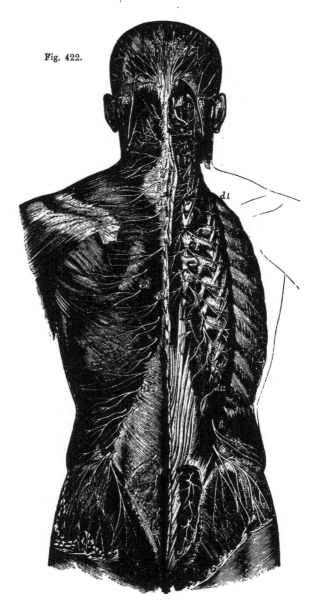

CERVICAL NERVES, *with the exception of the suboccipital.*—The *external branches* give only muscular offsets, and are distributed to the splenius and the slender muscles prolonged to the neck from the erector spinæ, viz., the

T T

cervicalis ascendens, and the transversalis colli with the trachelo-mastoid. That of the second nerve is the largest of the series of external branches, and is often united to the corresponding branch of the third ; it supplies the complexus muscle which covers it, and ends in the splenius and trachelo-mastoid muscles.

The *internal branches*, larger than the external, are differently disposed at the upper and the lower parts of the neck. That of the second cervical nerve is named, from its size and destination, the *great occipital*, and requires separate notice. The rest are directed inwards to the spinous processes of the vertebræ. Those derived from the third, fourth, and fifth nerves pass over the semispinalis and beneath the complexus muscle, and, having reached the spines of the vertebræ, turn transversely outwards and are distributed in the integument over the trapezius muscle. From the cutaneous branch of the third nerve a branch passes upwards to the integument on the lower part of the occiput, lying at the inner side of the great occipital nerve, and is sometimes called third occipital nerve.

Between the inner branches of the first three cervical nerves, beneath the complexus, there are frequently communicating fasciculi ; and this communication Cruveilhier has designated as " the posterior cervical plexus."

The internal branches from the lowest three cervical nerves are placed beneath the semispinalis muscle, and end in the muscular structure, without furnishing (except occasionally the sixth) any offset to the skin. These three nerves are the smallest of the series.

The *great occipital nerve* is directed upwards on the lower oblique muscle, and is transmitted to the surface through the complexus and trapezius muscles, giving twigs to the complexus. Ascending with the occipital artery, it divides into branches, which radiate over the occipital part of the occipito-frontalis muscle, some appearing to enter the muscle, and others joining the smaller occipital nerve.

An *auricular branch* is sometimes supplied to the back of the ear by the great occipital nerve.

DORSAL NERVES.—The *external branches* increase in size from above downwards. They are directed through or beneath the longissimus dorsi to the space between that muscle and the ilio-costalis and accessory; they supply both those muscles, together with the small muscles continued upwards from the erector spinæ to the neck, and also the levatores costarum. The lower five or six nerves give cutaneous twigs, which are transmitted to the integument in a line with the angles of the ribs.

The *internal branches* of the upper six dorsal nerves appear in the interval between the multifidus spinæ and the semispinalis dorsi ; they supply those muscles, and become cutaneous by the side of the spinous processes of the vertebræ. The cutaneous branch from the second nerve, and sometimes others, reach as far as the scapula. The internal branches of the lower six dorsal nerves are placed between the multifidus spinæ and longissimus dorsi, and end in the multifidus without giving branches to the integument. Where cutaneous nerves are supplied by the internal branches, there are none from the external branches of the same nerve, and *vice versâ*.

LUMBAR NERVES.—The *external branches* enter the erector spinæ, and give filaments to the intertransverse muscles. From the upper three, cutaneous nerves are supplied ; and from the last, a fasciculus descends to the corresponding branch of the first sacral nerve. The *cutaneous nerves* given from the external branches of the first three lumbar nerves, pierce the fleshy part of the ilio-costalis, and the aponeurosis of the latissimus dorsi ; they cross the iliac crest near the edge of the erector spinæ, and terminate in the integument of the gluteal region. One or more of the filaments may be traced as far as the great trochanter of the femur.

The *internal branches* wind backwards in grooves close to the articular processes of the vertebræ, and sink into the multifidus spinæ muscle.

SACRAL NERVES.—The posterior divisions of the nerves, except the last, issue from the sacrum through its posterior foramina. The first three are covered at their exit from the bone by the multifidus spinæ muscle, and bifurcate like the posterior trunks of the other spinal nerves ; but the remaining two, which continue below that muscle, have a peculiar arrangement, and require separate examination.

The *internal branches of the first three* sacral nerves are small, and are lost in the multifidus spinæ muscle.

The *external branches* of the same nerves are united with one another, and with the last lumbar and fourth sacral nerves, so as to form a series of anastomotic loops on the upper part of the sacrum. These branches are then directed outwards to the cutaneous or posterior surface of the great sacro-sciatic ligament, where, covered by the gluteus maximus muscle, they form a second series of loops, and end in cutaneous nerves. These pierce the great gluteus muscle in the direction of a line from the posterior iliac spine to the tip of the coccyx. They are commonly three in number—one is near the innominate bone, another opposite the extremity of the sacrum, and the third about midway between the other two. All are directed outwards over the great gluteal muscle.

In six dissections by Ellis this arrangement was found to be the most frequent. The variations to which it is liable are these :—the first nerve may not take part in the second series of loops, and the fourth may be associated with them.

The posterior divisions of the *last two sacral nerves* are smaller than those above them, and are not divided into external and internal branches. They are connected with each other by a loop on the back of the sacrum, and the lowest is joined in a similar manner with the coccygeal nerve ; one or two small filaments from these sacral nerves are distributed behind the coccyx.

COCCYGEAL NERVE.—The posterior division of the coccygeal nerve is very small, and separates from the anterior primary portion of the nerve in the sacral canal. It is joined by a communicating filament from the last sacral nerve, and ends in the fibrous structure covering the posterior surface of the coccyx.

ANTERIOR PRIMARY DIVISIONS

OF THE SPINAL NERVES.

The anterior primary divisions of the spinal nerves are distributed to the parts of the body situated in front of the vertebral column, including the limbs. They are, for the most part, considerably larger than the posterior divisions.

The anterior division of each spinal nerve is connected by one or two slender filaments with the sympathetic. Those of the cervical, lumbar, and sacral nerves form plexuses of various forms; but those of the dorsal nerves remain for the most part separate one from another.

T T 2

CERVICAL NERVES.

The anterior divisions of the four upper cervical nerves form the cervical plexus. They appear at the side of the neck between the scalenus medius and rectus anticus major muscles. They are each connected by a communicating filament with the first cervical ganglion of the sympathetic nerve, or with the cord connecting that ganglion with the second.

The anterior divisions of the four lower cervical nerves, larger than those of the upper four, appear between the scaleni muscles, and, together with that of the first dorsal, go to form the brachial plexus. They are each connected by a filament with one of the two lower cervical ganglia of the sympathetic, or with the plexus on the vertebral artery.

The anterior divisions of the first and second nerves require a notice separately from the description of the nerves of the cervical plexus.

SUBOCCIPITAL NERVE.

The anterior primary division of the first nerve runs forwards in a groove on the atlas, and bends downwards in front of the transverse process of that vertebra to join the second nerve. In this course forwards it lies beneath the vertebral artery, and at the inner side of the rectus lateralis muscle, to which it gives a branch. As it crosses the foramen in the transverse process of the atlas, the nerve is joined by a filament from the sympathetic ; and from the arch, or *loop of the atlas*, which it makes in front of the transverse process, branches are supplied to the two anterior recti muscles. Short filaments connect this part of the nerve with the pneumo-gastric, the hypoglossal, and the sympathetic nerves.

Valentin notices filaments distributed to the articulation of the occipital bone with the atlas, and to the mastoid process of the temporal bone. ,

SECOND CERVICAL NERVE.

The anterior division of the second cervical nerve, beginning between the arches of the first two vertebræ, is directed forwards between their transverse processes, being placed outside the vertebral artery, and beneath the intertransverse and other muscles fixed to those processes. In front of the intertransverse muscles the nerve divides into an ascending part, which joins the first cervical nerve, and a descending part to the third.

CERVICAL PLEXUS.

The cervical plexus is formed by the anterior divisions of the first four cervical nerves, and distributes branches to some of the muscles of the neck, and to a portion of the integument of the head and neck. It is placed opposite the first four vertebræ, beneath the sterno-mastoid muscle, and rests against the middle scalenus muscle and the levator anguli scapulæ. The disposition of the nerves in the plexus is easily recognised. Each nerve except the first, branches into an ascending and a descending part : and these are united in communicating loops with the contiguous nerves. From the union of the second and third nerves, superficial branches are supplied to the head and neck ; and from the junction of the third with the fourth, arise the cutaneous nerves of the shoulder and chest. Muscular and communicating branches spring from the same nerves.

The *branches* of the plexus may be separated into two sets—a superficial and deep ; the superficial consisting of those which ramify over the cervical fascia, supplying the integument and some also the platysma ; the deep comprising branches which are distributed for the most part to the muscles. The superficial nerves may be subdivided into ascending and descending ; the deep nerves into an internal and external series.

Fig. 423. — DIAGRAMMATIC OUTLINE OF THE FIRST PARTS OF THE CERVICAL AND UPPER DORSAL NERVES, SHOWING THE CERVICAL AND BRACHIAL PLEXUSES. ½

Fig 423.

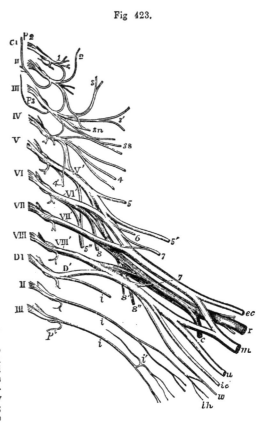

The nerves are separated from the spinal cord at their origin and are supposed to be viewed from before. C1, is placed opposite the roots of the first cervical or suboccipital nerve, and the roman numbers in succession from II, to VIII, opposite the roots of the corresponding cervical nerves ; DI, is placed opposite to the roots of the first dorsal nerve, and II, and III, opposite the second and third nerves ; the origin of the posterior primary branch is shown in all the nerves ; of these *p* 2, indicates the great occipital from the second, and *p* 3, the smallest occipital nerve from the third. In connection with the cervical plexus the following indications are given ; 1, anterior primary branch of the first cervical nerve and loop of union with the second nerve ; 2, lesser occipital nerve proceeding in this case from the second cervical nerve, more frequently from the second and third ; 3, great auricular nerve from the second and third ;

3′, superficial cervical nerve from the third ; 3 *n*, communicating branches to the descendens noni from the second and third ; 3 *s*, communicating to the spinal-accessory from the second, third, and fourth ; 4, supraclavicular and supraacromial descending nerves : the loops or arches of communication between the four upper cervical nerves, and between the fourth and fifth are shown ; 4′, the phrenic nerve springing from the fourth and fifth nerves. In connection with the nerves of the brachial plexus and the remaining nerves the following indications are given—V′, to VIII′, and D′, the five roots of the brachial plexus ; 5, the rhomboid nerve ; 5′, suprascapular ; 5″, posterior thoracic ; 6, nerve to the subclavius muscle ; 7, 7, inner and outer anterior thoracic nerves ; 8, 8′, 8″, upper and lower subscapular nerves. In the larger nerves proceeding to the shoulder and arm from the plexus, those of the anterior division are represented of a lighter shade, those belonging to the posterior division darker ; *ec*, external cutaneous or musculo-cutaneous ; *m*, median ; *u*, ulnar, *ic*, internal cutaneous ; *w*, nerve of Wrisberg ; *r*, musculo-spiral ; *c*, circumflex ; *i*, intercostal nerves ; *i′*, lateral branch of the same ; *ih*, intercosto-humeral nerves.

I. SUPERFICIAL ASCENDING BRANCHES.

SUPERFICIAL CERVICAL NERVE.

This nerve takes origin from the second and third cervical nerves, turns forward over the sterno-mastoid about the middle, and, after perforating the cervical fascia, divides beneath the platysma myoides into two branches, which are distributed to the anterior and lateral parts of the neck.

a. The *upper branch* gives an ascending twig which accompanies the external jugular vein, and communicates freely with the facial nerve (cervico-facial division) ; it is then transmitted through the platysma to the surface, supplying that muscle, and ramifies in the integument of the upper half of the neck on the fore part, filaments reaching as high as the lower maxilla.

b. The *lower branch* likewise pierces the platysma, and is distributed below the preceding, its filaments extending in front as low as the sternum.

The superficial cervical nerve may arise from the plexus in the form of two or more distinct branches. Thus Valentin describes three superficial cervical nerves, which he names superior, middle, and inferior. ("Sömmerring v. Bau," &c.)

While the superficial cervical nerve ramifies over the platysma myoides, the facial nerve is beneath the muscle. According to Valentin many anastomotic arches are formed on the side of the neck between those two nerves, as well as between the branches of the former, one with another.

GREAT AURICULAR NERVE.

This nerve winds round the outer border of the sterno-mastoid, and is directed obliquely upwards beneath the platysma myoides, between the muscle and the deep fascia of the neck, to the lobe of the ear. Here the nerve gives a few small branches to the face, and ends in the auricular and mastoid branches.

a. The *auricular branches* are directed to the back of the external ear, on which they ramify, and are connected with twigs derived from the facial nerve. One of these branches reaches the outer surface of the ear by a fissure between the antihelix and the concha. A few filaments are supplied likewise to the outer part of the lobule.

b. The *mastoid branch* is united to the posterior auricular branch of the facial nerve, and ascends over the mastoid process to the integument behind the ear.

c. The *facial* branches of the great auricular nerve, which extend to the integuments of the face, are distributed over the parotid gland. Some slender filaments penetrate deeply through the substance of the gland, and communicate with the facial nerve.

SMALL OCCIPITAL NERVE.

The smaller occipital nerve varies in size, and is sometimes double. It springs from the second cervical nerve, and is directed almost vertically to the head along the posterior border of the sterno-mastoid muscle. Having perforated the deep fascia near the cranium, the small occipital nerve is continued upwards between the ear and the great occipital nerve, and ends in cutaneous filaments which extend upwards in the scalp ; it communicates with branches from the larger occipital nerve, as well as with the posterior auricular branch of the facial. It appears to supply sometimes the occipito-frontalis muscle.

The *auricular branch* (ram. auricularis superior posterior) is distributed to the upper part of the ear on the posterior aspect, and to the elevator muscle of the auricle. This auricular branch is an offset from the great occipital nerve, when the small occipital is of less size than usual.

II. Superficial Descending Branches.

SUPRACLAVICULAR NERVES.

The descending series of the superficial nerves are thus named. There

Fig. 424.

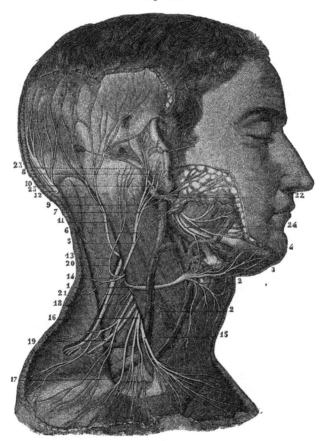

Fig. 424.—View of the Superficial Distribution of the Nerves proceeding from the Cervical Plexus (from Sappey after Hirschfeld and Leveillé). ½

1, superficial cervical nerve ; 2, 2, descending branches of the same ; 3, ascending branches ; 4, twigs uniting with the facial ; 5, great auricular nerve ; 6, its parotid branch ; 7, its external auricular branch ; 8, twig of the same which pierces the auricle to pass to its outer surface ; 9, branch to the deep surface of the pinna ; 10, its union with the posterior auricular of the facial nerve ; 11, small occipital nerve ; 12, its branch which unites with the great occipital nerve ; 13, a mastoid branch or second small occipital ; 14, twigs from this to the back of the neck ; 15, 16, supraclavicular nerves ; 17, 18, supraacromial nerves ; 19, branch of the cervical nerves passing into the trapezius muscle ; 20, spinal accessory distributed to the same and receiving a uniting branch from the cervical nerves ; 21, branch to the levator scapuli ; 22, trunk of the facial nerve ; 23, its posterior auricular branch passing into the occipital and posterior and superior auricular muscles ; 24, its cervico-facial branches.

are two of these nerves, or, in some cases, a greater number. They arise from the third and fourth cervical nerves, and descend in the interval between the sterno-mastoid and the trapezius muscles. As they approach the clavicle, the nerves are augmented to three or more in number, and are recognised as internal, middle, and posterior.

a. The *internal* (sternal) branch, which is much smaller than the rest, ramifies over the inner half of the clavicle, and terminates near the sternum.

b, The *middle branch,* lying opposite the interval between the pectoral and deltoid muscles, distributes some twigs over the fore part of the deltoid, and others over the pectoral muscle. The latter join the small cutaneous branches of the intercostal nerves.

c. The *external* or *posterior branch* (acromial) is directed outwards over the acromion, and the clavicular attachment of the trapezius muscle, and ends in the integument of the outer and back part of the shoulder.

III. Deep Branches : Inner Series.

CONNECTING BRANCHES.

The cervical plexus is connected near the base of the skull with the trunks of the pneumo-gastric, hypoglossal, and sympathetic nerves, by means of filaments intervening between those nerves and the loop formed by the first two cervical nerves in front of the atlas (p. 637).

MUSCULAR BRANCHES.

Branches to the anterior recti muscles proceed from the cervical nerves close to the vertebræ, including the loop between the first two of these nerves.

Two branches to the ansa hypoglossi, one from the second, the other from the third cervical nerve, descend over or under the internal jugular vein, to form a loop of communication with the ramus descendens noni, and aid in the supply of the muscles below the hyoid bone (p. 626).

PHRENIC NERVE.

The diaphragmatic or phrenic nerve passes down through the lower part of the neck and the thorax to its destination.

It commences from the fourth cervical nerve, and receives usually a fasciculus from the fifth. As it descends in the neck, the nerve is inclined inwards over the anterior scalenus muscle ; and near the chest it is joined by a filament of the sympathetic, and sometimes also by another filament derived from the fifth and sixth cervical nerves.

As it enters the thorax each phrenic nerve is placed between the subclavian artery and vein, and crosses over the internal mammary artery near the root. It then takes nearly a straight direction, in front of the root of the lung on each side, and along the side of the pericardium,—between this and the mediastinal part of the pleura. Near the diaphragm it divides into branches, which separately penetrate the fibres of that muscle, and then diverging from each other, are distributed on the under surface.

The *right nerve* is placed more deeply than the left, and is at first directed along the outer side of the right innominate vein, and the descending vena cava.

The *nerve* of the *left side* is a little longer than that of the right, in consequence of the oblique position of the pericardium round which it winds, and also because of the diaphragm being lower on this than on the opposite side. This nerve crosses in front of the arch of the aorta and the pulmonary artery before reaching the pericardium.

Besides the terminal *branches* supplied to the diaphragm, each phrenic nerve gives filaments to the pleura and pericardium ; and receives sometimes an offset from the union of the descendens noni with the cervical nerves. Swan notices this union as occurring only on the left side. Luschka describes twigs from the lower part of the nerve to the peritoneum, the inferior cava, and the right auricle of the heart.

One or two filaments of the nerve of the right side join in a small ganglion with branches to the diaphragm which are derived from the solar plexus of the sympathetic ; and from the ganglion twigs are given to the suprarenal capsule, the hepatic plexus, and the lower vena cava. On the left side there is a junction between the phrenic and the sympathetic nerves near the œsophageal and aortic openings in the diaphragm, but without the appearance of a ganglion.

IV. DEEP BRANCHES : EXTERNAL SERIES.

Muscular branches.—The sterno-mastoid receives a branch from the second cervical nerve. Two branches proceed from the third nerve to the levator anguli scapulæ ; and from the third and fourth cervical nerves, as they leave the spinal canal, branches are given to the middle scalenus muscle. Further, the trapezius has branches prolonged to it ; and thus, like the sterno-mastoid, this muscle receives nerves from both the spinal accessory and the cervical plexus.

Connection with the spinal accessory nerve.—In the substance of the sterno-mastoid muscle, this nerve is connected with the branches of the cervical plexus furnished to that muscle. It is also connected with the branches distributed to the trapezius—the union between the nerves being beneath the muscle, and having the appearance of a plexus ; and with another branch of the cervical plexus in the interval between the two muscles.

Summary of the cervical plexus.—From the cervical plexus are distributed cutaneous nerves to the back of the head, to part of the ear and face, to the anterior half of the neck, and to the upper part of the trunk. The muscles supplied with nerves from the plexus are the sterno-mastoid, the platysma, and the lower hyoid muscles in part ; the anterior recti, the levator anguli scapulæ, the trapezius, the scalenus medius, and the diaphragm. By means of its branches the plexus communicates with the pneumo-gastric, spinal accessory, hypoglossal, and sympathetic nerves.

BRACHIAL PLEXUS.

This large plexus, from which the nerves of the upper limb are supplied, is formed by the union of the anterior trunks of the four lower cervical and first dorsal nerves ; and it further receives a fasciculus from the lowest of the nerves (fourth), which goes to form the cervical plexus. The plexus extends from the lower part of the neck to the axillary space, and terminates opposite the coracoid process of the scapula in large nerves for the supply of the limb.

The manner in which the nerves are disposed in the plexus is liable to some variation, but the following may be regarded as the arrangement most frequently met with. The fifth and sixth cervical are joined at the outer border of the scalenus, and a little farther out receive the seventh nerve,

—the three nerves giving rise to one large upper cord. The eighth cervical and first dorsal nerves are united in another lower cord whilst they are

Fig. 425.

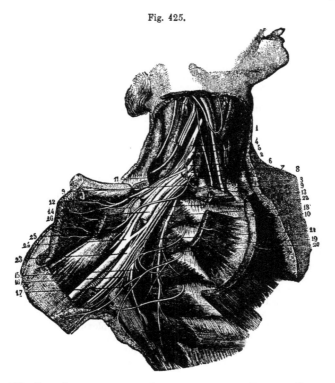

Fig. 425.—Deep Dissection of the Axilla, showing the Brachial Plexus and neighbouring Nerves (from Sappey after Hirschfeld and Leveillé). ¼

The clavicle has been sawn through near its sternal end, and is turned aside with the muscles attached to it ; the subclavius, and the greater and lesser pectoral muscles have been removed from the front of the axilla.　1, loop of union between the descendens noni and a branch of the cervical plexus ; 2, pneumo-gastric ; 3, phrenic passing down to the inner side of the scalenus anticus muscle ; 4, anterior primary division of the fifth cervical nerve ; 5, the same of the sixth ; 6, 7, the same of the seventh and eighth cervical nerves ; 8, the same of the first dorsal nerve ; 9, 9, branch from the plexus to the subclavius muscle, communicating with the phrenic nerve ; 10, posterior thoracic nerve distributed to the serratus magnus ; 11, upper anterior thoracic nerve passing into the great pectoral muscle ; 13, lower anterior thoracic distributed to the lesser pectoral ; 14, twig of communication between these two nerves ; 12, suprascapular nerve passing through the suprascapular notch ; 15, lower of the two subscapular nerves ; 16, nerve of the teres major ; 17, long subscapular, or nerve of the latissimus dorsi ; 18, accessory of the internal cutaneous nerve ; 19, union of the accessory cutaneous with the second and third intercostal nerves ; 20, lateral branch of the second intercostal ; 21, second internal cutaneous or nerve of Wrisberg ; 22, internal cutaneous nerve ; 23, the ulnar nerve to the inside of the axillary artery, passing behind the vein, and having, in this case, a union with the upper division of the plexus ; 24, the median nerve immediately below the place where its two roots embrace the artery, which is divided above this place ; 25, the musculo-cutaneous nerve passing into the coraco-brachialis muscle ; 26, the musculo-spiral nerve passing behind the divided brachial artery.

between the scaleni muscles. The two cords thus formed lie side by side in the fore part of the plexus, and external to the first part of the axillary vessels. At the same place, or lower down, a third intermediate or posterior cord is produced by the union of fasciculi from each of the other two cords, or separately from the nerves forming them. The three cords of which the plexus now consists, are placed, one on the outer side of the axillary artery, one on the inner side, and one behind that vessel, and are continued into the principal nerves for the arm.

The two fasciculi which unite to form the intermediate of the three trunks are generally separated at a higher level than the formation of the two other trunks, but they are also frequently given off as low as the clavicle, or even farther down ; this gives rise to some varieties, more apparent than real. The seventh nerve also may give a branch to the cord below it.

The branches proceeding from the plexus are numerous, and may be conveniently divided into two classes—viz., those that arise above the clavicle, and those that take origin below the bone.

BRANCHES ABOVE THE CLAVICLE.

Above the clavicle there arise from the trunks of the brachial plexus, the posterior thoracic and suprascapular nerves, a nerve for the rhomboid muscles, another for the subclavius, irregular branches for the scaleni and longus colli, and a branch to join the phrenic nerve.

The branches for *the scaleni and longus colli muscles* spring in an irregular manner from the lower cervical nerves close to their place of emergence from the vertebral foramina.

The branch for *the rhomboid muscles* arises from the fifth nerve, and is directed backwards to the base of the scapula through the fibres of the middle scalenus and beneath the levator anguli scapulæ. It is distributed to the deep surface of the rhomboid muscles, and gives sometimes a branch to the levator scapulæ.

Fig. 426.—Distribution of the Suprascapular and Circumflex Nerves (from Hirschfeld and Leveillé). ⅓

Fig. 426.

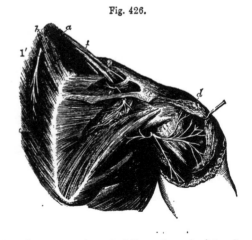

a, the scalenus medius and posticus muscles ; *b*, levator anguli scapulæ ; *c*, acromion ; *d*, deltoid muscle, of which the back part has been detached from the scapula and in part removed ; *e*, rhomboid muscle ; *f*, teres major ; *g*, latissimus dorsi ; 1, the brachial plexus of nerves as seen from behind ; 1′, the nerve of the levator scapulæ and rhomboid muscles ; 2, placed on the clavicle, marks the suprascapular nerve ; 3, its branch to the supraspinatus muscle ; 4, branch to the infraspinatus ; 5, placed on the back of the humerus below the insertion of the teres minor, marks the circumflex nerve passing out of the quadrangular interval ; 6, its branch to the teres minor muscle ; 7, branches to the deep surface of the deltoid ; 8, cutaneous branch to the back of the shoulder.

The nerve *of the subclavius muscle*, of small size, arises from the front of the cord which results from the union of the fifth and sixth cervical nerves. It is directed over the outer part of the subclavian artery to the deep surface of the subclavius muscle. This small nerve is commonly connected with the phrenic nerve in the neck or in the chest, by means of a slender filament.

Branch *to join the phrenic nerve.*—This small and short branch is an offset from the fifth cervical nerve; it joins the phrenic nerve on the anterior scalenus muscle.

POSTERIOR THORACIC NERVE.

The posterior thoracic nerve (nerve of the serratus magnus, external respiratory of Bell) is formed in the substance of the middle scalenus muscle by two roots, one from the fifth and another from the sixth nerve, and reaches the surface of the scalenus lower than the nerve of the rhomboid muscles, with which it is often connected. It descends behind the brachial plexus on the outer surface of the serratus magnus, nearly to the lower border of that muscle, supplying it with numerous branches.

SUPRASCAPULAR NERVE.

The suprascapular nerve arises from the back of the cord formed by the union of the fifth and sixth nerves, and bends beneath the trapezius to the upper border of the scapula, where it passes between the muscles and the bone. Entering the supraspinous fossa of the scapula, through the suprascapular notch (beneath the ligament which crosses the notch), the suprascapular nerve supplies two branches to the supraspinatus, one near the upper, the other near the lower part of the muscle; and it then descends through the great scapular notch into the lower fossa, where it ends in the infraspinatus muscle.

In the upper fossa of the scapula, a slender *articular filament* is given to the shoulder-joint, and in the lower fossa other twigs of the nerve enter the same joint and the substance of the scapula.

BRANCHES BELOW THE CLAVICLE.

Origin of nerves from the plexus.—The several nerves now to be described are derived from the three great cords of the plexus in the following order.

From the upper or outer cord,—the external of the two anterior thoracic nerves, the musculo-cutaneous, and the outer root of the median.

From the lower or inner cord,—the inner of the two anterior thoracic, the nerve of Wrisberg, the internal cutaneous, the ulnar, and the inner root of the median.

From the posterior cord,—the subscapular nerves, the circumflex, and the musculo-spiral.

The nerves traced to the spinal nerves.—If the fasciculi of which the principal nerves are composed be followed through the plexus, they may be traced to those of the spinal nerves which in the subjoined table are named along with each trunk. The higher numbers refer to the cervical nerves, the unit to the dorsal nerve :—

Subscapular from	. .		
Circumflex	. . .	} 5.6.7.8.	
Musculo-spiral	. . .		
External cutaneous	. .	5.6.7.	
Median	. . . .	5.6.7.8.1.	

Ulnar . . . 8.1. or 7.8.1.	
Internal cutaneous . } 8.1	
Small internal cutaneous }	
Anterior thoracic { outer 5.6.7 { inner 8.1.	

The outline in Fig. 423, taken from a dissection, represents one of the most common arrangements.

Some differences will be found in the statements of anatomists who have investigated the subject—for instance, Scarpa ("Annotationes Anatom.") and Kronenberg ("Plex. nervor. Structura et Virtutes")—with respect to the nerves to which the branches are assigned. This difference is mainly owing to the variation which actually exists in different cases.

ANTERIOR THORACIC NERVES.

The anterior thoracic nerves, two in number, supply the pectoral muscles.

The *external*, or more superficial branch, arising from the outer cord, crosses inwards over the axillary artery, and terminates in the great pectoral muscle.

The *internal*, or deeper branch, springing from the inner cord, comes forwards between the axillary artery and vein to the small pectoral muscle, and is joined by a branch from the external. This nerve presents a plexiform division beneath the small pectoral muscle, and supplies branches to it and the larger pectoral muscle. The two nerves are connected by a filament which forms a loop over the artery at the inner side.

SUBSCAPULAR NERVES.

Fig. 427.

These nerves, three in number, take origin from the posterior cord of the plexus.

The *upper* nerve, the smallest of the subscapular nerves, penetrates the upper part of the subscapular muscle. The *lower* nerve gives a branch to the subscapularis at its axillary border, and ends in the teres major muscle. There is sometimes a distinct nerve for the last-named muscle.

The *long subscapular* nerve, the largest of the three, runs along the lower border of the subscapular muscle to the latissimus dorsi, to which it is distributed.

Fig. 427.—Distribution of the Posterior Cutaneous Nerves of the Shoulder and Arm (from Hirschfeld and Leveillé). ⅓

1, supra-acromial branches of the cervical nerves descending on the deltoid muscle ; 2, ascending or reflected, and 2′, descending cutaneous branches of the circumflex nerve ; 3, inferior external cutaneous of the musculo-spiral nerve ; 4, external and posterior cutaneous branches of the musculo-cutaneous nerve to the forearm ; 5, internal cutaneous of the musculo-spiral ; 6, intercosto-humeral branches ; 7, twigs of the nerve of Wrisberg ; 8, upper posterior branch of the internal cutaneous nerve ; 9, lower branch of the same.

CIRCUMFLEX NERVE.

The circumflex or axillary nerve gives both muscular and cutaneous nerves to the shoulder. Springing from the posterior cord, this nerve is at first placed behind the axillary artery, but at the lower border of the subscapular muscle it is inclined backwards with the posterior circumflex artery, in the space between the scapula and teres major muscle above the long head of the triceps, and separates into an upper and a lower branch, which are distributed to the deltoid and teres minor muscles, the integument of the shoulder, and the shoulder-joint.

a. The *upper portion* winds round the upper part of the humerus, extending to the anterior border of the deltoid muscle, to which it is distributed. One or two *cutaneous filaments*, penetrating between the muscular fibres, are bent downwards and supply the integument over the lower part of the muscle.

b. The *lower branch* supplies offsets to the back part of the deltoid, and furnishes the nerve to the teres minor, which is remarkable in presenting a gangliform enlargement. It then turns round the posterior border of the deltoid below the middle, and ramifies in the integument over the lower two-thirds of that muscle, one branch extending to the integument over the long end of the triceps muscle.

c. An *articular filament* for the shoulder-joint arises near the commencement of the nerve, and enters the capsular ligament below the subscapular muscle.

INTERNAL CUTANEOUS NERVE.

At its origin from the inner cord of the brachial plexus, this nerve is placed on the inner side of the axillary artery. It becomes cutaneous about the middle of the arm, and after perforating the fascia, or, in some cases, before doing so, is divided into two parts; one destined for the anterior, the other for the posterior surface of the forearm.

a. The *anterior branch* crosses at the bend of the elbow behind (in some cases over) the median basilic vein, and distributes filaments in front of the forearm, as far as the wrist; one of these is, in some instances, joined with a cutaneous branch of the ulnar nerve.

b. The *posterior branch* inclines obliquely downwards at the inner side of the basilic vein, and winding to the back of the forearm, over the prominence of the internal condyle of the humerus, extends somewhat below the middle of the forearm. Above the elbow this branch is connected with the smaller internal cutaneous nerve (nerve of Wrisberg), and afterwards communicates with the outer portion of the internal cutaneous, and, according to Swan, with the dorsal branch of the ulnar nerve.

c. A branch to the *integument of the arm* pierces the fascia near the axilla, and reaches to, or nearly to the elbow, distributing filaments outwards over the biceps muscle. This branch is often connected with the intercosto-humeral nerve.

SMALL INTERNAL CUTANEOUS NERVE.

The smaller internal cutaneous nerve, or nerve of Wrisberg, destined for the supply of the integument of the lower half of the upper arm on the inner and posterior aspect, commonly arises from the inner cord of the brachial plexus in union with the larger internal cutaneous and ulnar nerves. In the axilla it lies close to the axillary vein, but it soon appears on the inner side of that vessel, and communicates with the intercosto-humeral nerve. It then descends along the inner side of the brachial vessels to about the middle of the arm, where it pierces the fascia, and its filaments are thence continued to the interval between the internal condyle of the humerus and the olecranon.

Branches.—In the lower third of the arm, branches of this small nerve are directed almost horizontally to the integument on the posterior aspect; and the nerve ends at the elbow by dividing into several filaments some of which are directed forwards over the inner condyle of the humerus, while others are prolonged downwards behind the olecranon.

Connection with the intercosto-humeral nerve.—This connection presents much variety in different cases :—in some, there are two or more intercommunications, forming a kind of plexus on the posterior boundary of the axillary space; in others, the intercosto-humeral nerve is of larger size than usual, and takes the place of the nerve of Wrisberg, only receiving in the axilla a small filament from the brachial plexus ;

Fig. 428.   Fig. 429.

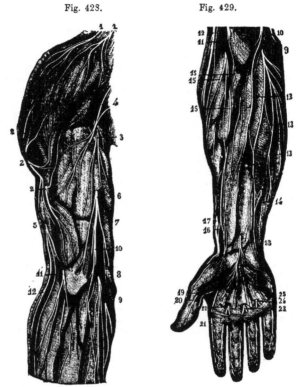

Fig. 428.—ANTERIOR CUTANEOUS NERVES OF THE SHOULDER AND ARM (from Sappey after Hirschfeld and Leveillé). ½

1, 1, supraclavicular and supraacromial nerves from the cervical plexus ; 2, 2, 2, cutaneous branches of the circumflex nerve ; 3, 4, upper branches of the internal cuta-neous nerve ; 5, superior external cutaneous branch of the musculo-spiral ; 6, internal cutaneous nerve piercing the deep fascia ; 7, posterior branch ; 8, communicating twig with one of the anterior branches ; 9, 10, anterior branches of this nerve, some turning round the median basilic and ulnar veins ; 11, musculo-cutaneous nerve descending over the median cephalic vein ; 12, inferior cutaneous branch of the musculo-spiral nerve.

Fig. 429.—ANTERIOR CUTANEOUS NERVES OF THE FOREARM AND HAND (from Sappey after Hirschfeld and Leveillé). ½

9, 10, 13, distribution of the anterior branches of the internal cutaneous nerve; 14, union of one of these with a twig of the ulnar nerve ; 12, inferior cutaneous branch of the musculo-spiral nerve ; 11, 15, distribution of the external cutaneous nerve ; 16, union of one of its branches with 17, the terminal branch of the radial nerve ; 18, palmar cutaneous branch of the median nerve ; 19, 20, internal and external collateral branches to the thumb from the median nerve ; 21, external collateral to the index finger ; 22, 23, collateral branches to the index, middle and fourth fingers ; 24, 25, collateral branches from the ulnar nerve to the fourth and fifth fingers ; the network of communicating twigs formed by the terminal branches of these cutaneous nerves is repre-sented at the extremities of the fingers.

and this small communicating filament represents in such cases the nerve of Wrisberg.

MUSCULO-CUTANEOUS NERVE.

The musculo-cutaneous or external cutaneous nerve (perforans Casserii) is deeply placed between the muscles as far as the elbow, and below that point is immediately under the integument. Arising from the brachial plexus opposite the small pectoral muscle, it perforates the coraco-brachialis muscle, and, passing obliquely across the arm between the biceps and brachialis anticus muscles, reaches the outer side of the biceps a little above the elbow. Here it perforates the fascia, and nearly opposite the elbow-joint it passes behind the median-cephalic vein, and, inclining outwards, divides into two branches which supply the integument on the outer side of the forearm, one on the anterior, the other on the posterior aspect.

A. *Branches in the arm :—*

a. A branch to the coraco-brachialis and short head of the biceps is given off before the nerve pierces the former muscle ; and other filaments are furnished to the coraco-brachialis, while the nerve lies among its fibres.

b. Branches to the biceps and brachialis anticus are given off while the nerve is between those muscles.

c. Small filaments are given to the humerus and elbow-joint.

B. *Branches in the forearm :—*

a. The *anterior branch* descends near the radial border of the forearm. It is placed in front of the radial artery near the wrist, and distributes some filaments over the ball of the thumb. Piercing the fascia, it accompanies the artery to the back part of the carpus. This part of the nerve is connected at the wrist with a branch of the radial nerve.

b. The *posterior branch* is directed outwards to the back of the forearm, and ramifies in the integument of the lower third, extending as far as the wrist. It communicates with a branch of the radial nerve, and with the external cutaneous branch of the musculo-spiral nerve.

Summary.—The musculo-cutaneous nerve supplies the coraco-brachialis, biceps, and brachialis anticus muscles, and the integument on the outer side of the forearm. Communications are established between it and the radial and the external cutaneous branch of the musculo-spiral.

Varieties.—In some cases it does not perforate the coraco-brachialis muscle. It is frequently found to communicate by a cross branch with or to be an offset of the median nerve; and in the latter case, the coraco-brachialis muscle receives a separate branch from the brachial plexus, which may be explained thus,—that the main part of the musculo-cutaneous nerve, instead of piercing the coraco-brachialis muscle, remains adherent to the outer root and trunk of the median.

ULNAR NERVE.

The ulnar nerve, the largest branch of the inner cord of the brachial plexus, descends on the inner side of the artery of the limb as far as the middle of the arm, then turns backwards through the internal intermuscular septum with the inferior profunda artery, to reach the interval between the olecranon and the inner condyle of the humerus. From the axilla to this place it is covered only by the fascia, and it may be felt through the integument a little above the elbow. It here passes between the two heads of the flexor carpi ulnaris, and it remains concealed by that muscle as far as the middle of the forearm ; it thence extends in a straight course along the outer margin of the muscle, between it and the ulnar artery, to the outer side of the pisiform bone. Above the wrist it gives off a large dorsal branch to the hand, and continuing onwards it enters the palm on the surface of the annular ligament, and divides into muscular and cutaneous branches.

The ulnar nerve gives off no branches in the upper arm.

A. *Branches in the forearm :—*

a. Articular filaments are given to the elbow-joint as the nerve passes behind it. Some filaments are also given to the wrist-joint.

b. Muscular branches.—One branch enters the upper part of the flexor carpi ulnaris, and another supplies the two inner divisions of the deep flexor of the fingers.*

c. Cutaneous branches to the forearm.—These two small nerves arise about the middle of the forearm by a common trunk. One pierces the fascia, and turning downwards, joins a branch of the internal cutaneous nerve. This branch is often absent. The second, a *palmar branch*, lies on the ulnar artery, which it accompanies to the hand. This little nerve gives filaments around the vessel, and ramifies in the integument of the hand, joining in some cases with other cutaneous offsets of the ulnar or median nerve.

d. Dorsal branch to the hand.—This large offset, leaving the trunk of the ulnar nerve two or three inches above the wrist, winds backwards beneath the flexor carpi ulnaris and divides into branches; one of these ramifies on the inner side of the little finger, another divides to supply the contiguous sides of that finger and the ring finger, while a third joins on the back of the metacarpus with the branch of the radial nerve which supplies the contiguous sides of the ring and middle finger. The several posterior digital nerves, now described, are united with twigs directed backwards from the anterior digital nerves of the same fingers.

B. *Palmar branches :—*

a. The *deep branch* separates from the trunk beyond the annular ligament, and, dipping down through the muscles of the little finger in company with the deep branch of the ulnar artery, it follows the course of the deep palmar arch across the hand. It supplies the short muscles of the little finger as it pierces them; and as it lies across the metacarpal bones, it distributes two branches to each interosseous space—one for the palmar, the other for the dorsal interosseous muscle, and supplies filaments to the two innermost lumbricales muscles. Opposite the space between the thumb and the index finger the nerve ends in branches to the adductor pollicis, and the inner head of the flexor brevis pollicis.

b. The remaining part of the nerve supplies a branch to the palmaris brevis muscle and small twigs to the integument, and divides into two digital branches.

Digital nerves.—One of these belongs to the ulnar side of the little finger. The other is connected in the palm of the hand with a digital branch of the median nerve, and at the cleft between the little and ring fingers, divides into the collateral nerves for these fingers. The terminal disposition of the digital branches on the fingers is the same as that of the median nerve, to be presently described.

Summary.—The ulnar nerve gives cutaneous filaments to the lower part of the forearm (to a small extent), and to the hand on its palmar and dorsal aspects. It supplies the following muscles, viz., the ulnar flexor of the carpus, the deep flexor of the fingers (its inner half), the short muscles of the little finger with the palmaris brevis, the interosseous muscles of the hand, the two internal lumbricales, the adductor pollicis and the inner half of the flexor brevis pollicis. Lastly, it contributes to the nervous supply of the elbow and wrist joints.

MEDIAN NERVE.

The median nerve arises by two roots, one from the outer, the other from the inner cord of the brachial plexus. Commencing by the union of these

---

* A case has been recorded in which the ulnar nerve supplied also two branches to the flexor sublimis digitorum (Turner, " Nat. Hist. Review," 1864).

Fig. 430.   Fig. 431.

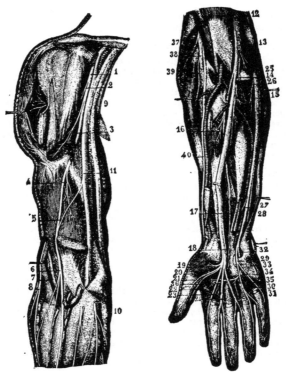

Fig. 430.—DEEP VIEW OF THE ANTERIOR NERVES OF THE SHOULDER AND ARM (from
Sappey after Hirschfeld and Leveillé). ⅓

1, musculo-cutaneous nerve ; 2, its twig to the coraco-brachialis muscle ; 3, its branch
to the biceps ; 4, its branch to the brachialis anticus ; 5, twig of union with the median
nerve (a variety) ; 6, continuation of the nerve in its cutaneous distribution ; 7, musculo-
spiral nerve in the interval between the brachialis anticus and supinator longus ; 8, inferior
external cutaneous branch of the musculo-spiral ; 9, the internal cutaneous and small
internal cutaneous nerves divided ; 10, anterior branch of the internal cutaneous ; 11,
median nerve ; to the inside the ulnar nerve is crossed by the line from 11.

Fig. 431.—DEEP VIEW OF THE ANTERIOR NERVES OF THE FOREARM AND HAND (from
Sappey after Hirschfeld and Leveillé). ⅓

12, the median nerve ; 13, its branches to the pronator teres ; 14, branch to the superficial flexor muscles, which have been removed ; 15, branch to the flexor digitorum profundus ; 16, branch to the flexor longus pollicis ; 17, anterior interosseous branch ; 18,
cutaneous palmar branch cut short ; 19, branches to the short muscles of the thumb ;
20, 21, collateral branches to the thumb ; 22, 23, 24, collateral branches to the second,
third, and fourth fingers ; 25, branch given by the ulnar nerve to the flexor carpi ulnaris ;
26, branch to the flexor digitorum profundus ; 27, cutaneous communicating twig ; 28,
dorsal branch of the ulnar ; 29, superficial palmar branch ; 30, 31, collateral branches
to the fourth and fifth fingers ; 32, deep palmar branch ; 33, its branch to the short
muscles of the little finger ; 34, 35, 36, twigs given by the deep branch of the ulnar to
the third and fourth lumbricales, all the interossei, and the adductor pollicis.

roots in front or on the outer side of the axillary artery, the nerve descends in contact with the brachial artery, gradually passing inwards over it, and near the elbow is at the inner side of the vessel. Crossing the bend of the arm it passes beneath the pronator radii teres, separated by the deep slip of that muscle from the ulnar artery, and continues straight down the front of the forearm, between the flexor sublimis digitorum and flexor profundus. Arrived near the wrist it lies beneath the fascia, between the tendons of the flexor sublimis and that of the flexor carpi radialis. It then enters the palm behind the annular ligament, and rests on the flexor tendons. Somewhat enlarged, and of a slightly reddish colour, it here separates into two parts of nearly equal size. One of these (the external) supplies some of the short muscles of the thumb, and gives digital branches to the thumb and the index finger ; the second portion supplies the middle finger, and in part the index and ring fingers.

The median nerve gives no branch to the upper arm.

A. *Branches in the forearm :*—

In the forearm the median nerve supplies muscular branches, and, near the wrist, a single cutaneous filament. All the muscles on the front of the forearm (pronators and flexors), except the flexor carpi ulnaris and part of the deep flexor of the fingers, are supplied from this nerve.

a. The branches for the superficial muscles are separate twigs given off from the nerve below or near the elbow-joint, but the branch furnished to the pronator teres often arises above the joint.

b. Anterior interosseous nerve.—This is the longest branch of the median nerve, and it supplies the deeper muscles of the forearm. Commencing at the upper part of the forearm, beneath the superficial flexor of the fingers, it passes downwards with the anterior interosseous artery on the interosseous membrane, and between the long flexor of the thumb and the deep flexor of the fingers, to the pronator quadratus muscle, in which it ends.

c. The *cutaneous palmar branch* pierces the fascia of the forearm close to the annular ligament, and, descending over that ligament, ends in the integument of the palm about the middle : being connected by a twig with the cutaneous palmar branch of the ulnar nerve. It distributes some filaments over the ball of the thumb, which communicate with twigs of the radial or the external cutaneous nerve.

B. *Branches in the hand :*—

a. Branch to muscles of the thumb.—This short nerve subdivides into branches for the abductor, the opponens, and the outer head of the flexor brevis pollicis muscle.

b. Digital nerves.—These are five in number, and belong to the thumb, and the fingers as far as the outer side of the ring-finger. As they approach the clefts between the fingers, they are close to the integument in the intervals between the longitudinal divisions of the palmar fascia.

The *first* and *second* nerves lie along the sides of the thumb ; and the former (the outer one) is connected with the radial nerve upon the ball of the thumb.

The *third*, destined for the radial side of the index finger, gives a muscular branch to the first or most external lumbrical muscle.

The *fourth* supplies the second lumbricalis, and divides into branches for the adjacent sides of the index and middle fingers.

The *fifth*, the most internal of the digital nerves, is connected by a crossing-twig with the ulnar nerve, and divides to furnish branches to the adjacent sides of the ring and middle fingers.

Each digital nerve divides at the end of the finger into two branches, one of which supplies the ball on the fore part of the finger ; the other ramifies in the pulp beneath the nail. Branches pass from each nerve forwards and backwards to the integument of the finger; and one larger than the rest

U U 2

inclines backwards by the side of the first phalanx of the finger, and, after joining the dorsal digital nerve, ends in the integument over the last phalanx.

Fig. 432.

Fig. 432.—DISTRIBUTION OF THE DIGITAL NERVES (from Hirschfeld and Leveillé). $\frac{1}{2}$

1, palmar collateral nerve; 2, its final palmar distribution; 3, its dorsal or ungual distribution, and between these numbers the network of terminal filaments; 4, collateral dorsal nerve; 5, uniting twigs passing between the dorsal and palmar digital nerves.

Summary.—The median nerve gives cutaneous branches to the palm, and to several fingers. It supplies the pronator muscles, the flexors of the carpus and the long flexors of the fingers (except the ulnar flexor of the carpus, and part of the deep flexor of the fingers), likewise the outer set of the short muscles of the thumb, and two lumbricales.

Some similarity will be observed between the course and distribution of the median and ulnar nerves. Neither gives any offset in the arm. Together they supply all the muscles in front of the forearm and in the hand, and together they supply the skin of the palmar surface of the hand, and impart tactile sensibility to all the fingers.

MUSCULO-SPIRAL NERVE.

The musculo-spiral nerve, the largest offset of the brachial plexus, occupies chiefly the back part of the limb, and supplies nerves to the extensor muscles, as well as to the skin.

Arising behind the axillary vessels from the posterior cord of the brachial plexus, of which it is the principal continuation and the only one prolonged into the arm, it soon turns backwards into the musculo-spiral groove, and, accompanied by the superior profunda artery, proceeds along that groove, between the humerus and the triceps muscle, to the outer side of the limb. It then pierces the external intermuscular septum, and descends in the interval between the supinator longus and the brachialis anticus muscle to the level of the outer condyle of the humerus, where it ends by dividing into the radial and posterior interosseous nerves. Of these, the radial is altogether a cutaneous nerve, and the posterior interosseous is the muscular nerve of the back of the forearm.

The branches of the musculo-spiral nerve may be classified according as they arise on the inner side of the humerus, behind that bone, or on the outer side.

A. Internal branches :—

(*a*) *Muscular branches* for the inner and middle heads of the triceps. That for the inner portion of the muscle is long and slender; it lies by the side of the ulnar nerve, and reaches as far as the lower third of the upper arm. One branch, previously noticed by authors, but more particularly described by Krause, is named by him the *ulnar collateral* branch. It arises opposite the outer border of the latissimus dorsi tendon, and descends within the sheath of the ulnar nerve, through the internal intermuscular septum, and is distributed to the short inferior fibres of the triceps (Reichert and Du Bois Reymond's Archiv. 1864).

(*b*) The *internal cutaneous* branch of the musculo-spiral nerve, commonly united in origin with the preceding, winds backwards beneath the intercosto-humeral nerve, and after supplying filaments to the skin, ends about two inches from the olecranon;

in some instances extending as far as the olecranon. This nerve is accompanied by a small cutaneous artery.

B. Posterior branches :—

These consist of a fasciculus of *muscular branches* which supply the outer head of the triceps muscle and the anconeus. The *branch* of the *anconeus* is slender, and remarkable for its length ; it descends in the substance of the triceps to reach its destination.

C. External branches :—

(*a*) The *muscular branches* supply the supinator longus, extensor carpi radialis longior, (the extensor carpi radialis brevior receiving its nerve from the posterior interosseous,) and occasionally give a small branch to the brachialis anticus.

(*b*) The *external cutaneous branches*, two in number, arise where the nerve pierces the external intermuscular septum.

The *upper branch*, the smaller of the two, is directed downwards to the fore part of the elbow, along the cephalic vein, and distributes filaments to the lower half of the upper arm on the anterior aspect. The *lower branch* extends as far as the wrist, distributing offsets to the lower half of the arm, and to the forearm, on their posterior aspect, and is connected near the wrist with a branch of the external cutaneous nerve.

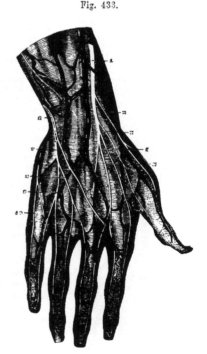

Fig. 433.—Dorsal Cutaneous Nerves of the Hand. ⅓

Fig. 433.

The distribution delineated in this figure is not the most common, there being a larger than usual branch of the ulnar nerve : 1, the radial nerve descending beside the principal radial cutaneous vein ; 2, and 3, dorsal branches to the two sides of the thumb ; 4, branch to the radial side of the forefinger ; 5, branch to the fore finger and middle finger, communicating with one from the ulnar nerve ; 6, the posterior branch of the ulnar nerve ; 7, communicating twig ; 8, collateral branch to the middle and ring fingers ; 9, collateral branch to the ring and little fingers ; 10, branch to the inner side of the hand and little finger.

RADIAL NERVE.

The radial nerve, continuing straight down from the musculo-spiral, is concealed by the long supinator muscle, and lies a little to the outer side of the radial artery. This position beneath the supinator is retained to about three inches from the lower end of the radius, where the nerve turns backwards beneath the tendon of the muscle, and becomes subcutaneous. It then separates into two branches, which ramify in the integument on the dorsal aspect of the thumb and the next two fingers in the following manner.

(*a*) The *external branch* extends to the radial side of the thumb, and is joined by an offset of the external cutaneous nerve. It distributes filaments over the ball of the thumb.

(*b*) The *internal portion* communicates with a branch of the external cutaneous nerve on the back of the forearm, and divides into digital branches; *one* running along the ulnar side of the thumb, a *second* on the radial side of the index finger, a *third* dividing to supply the adjacent sides of the index and middle fingers, while a *fourth* joins with an offset from the dorsal branch of the ulnar, and along with it forms a branch for the supply of the contiguous sides of the middle and ring fingers. These branches communicate on the sides of the fingers with the palmar digital nerves.

Sometimes the interspace between the middle and ring fingers is entirely supplied by the radial, and at other times entirely by the ulnar nerve.

POSTERIOR INTEROSSEOUS NERVE.

This nerve, the larger of the two divisions of the musculo-spiral nerve, winds to the back of the forearm through the fibres of the supinator brevis

Fig. 434.

Fig. 434.—VIEW OF THE RADIAL SIDE OF THE FORE-ARM, SHOWING THE FINAL DISTRIBUTION OF THE MUSCULO-SPIRAL NERVE (from Hirschfeld and Leveillé). ¼

The supinator longus, and extensores carpi radiales longior and brevior have been divided, and their upper parts removed; the extensor communis digitorum is pulled backwards by a hook, and the supinator brevis has been partially dissected to show the posterior interosseous nerve passing through it.

1, placed upon the tendon of the biceps muscle, points to the musculo-cutaneous nerve; 1′, near the wrist, the lower part of this nerve and its plexus of union with the radial nerve; 2, trunk of the musculo-spiral nerve emerging from between the brachialis anticus, on which the number is placed, and the supinator longus muscle; 2′, its muscular twigs to the long supinator and long radial extensor of the carpus; 2″, the posterior interosseous nerve passing through the substance of the supinator brevis; 3, placed upon the cut lower portion of the supinator longus, the radial nerve; 4, the external collateral nerve of the thumb; 5, the common collateral of the fore-finger and thumb; 6, the common collateral of the fore-finger and middle finger; 7, the twig of union with the dorsal branch of the ulnar nerve; 8, placed upon the common extensor of the fingers, the muscular branches of the posterior interosseous nerve to the long extensor muscles; 9, upon the extensor secundi internodii pollicis, the branches to the short extensor muscles.

muscle, and is prolonged between the deep and superficial layers of the extensor muscles to the interosseous membrane, which it approaches about the middle of the forearm.

Much diminished in size by the separation of numerous branches for the muscles, the nerve lies at the lower part of the forearm beneath the extensor of the last phalanx of the thumb and the tendons of the common extensor of the fingers, and terminates on the back of the carpus in a gangliform enlargement, from which filaments are given to the adjoining ligaments and articulations.

The *branches* of the interosseous nerve enter the surrounding muscles, viz., the extensor carpi radialis brevior and supinator brevis, the superficial layer of the extensor muscles except the anconeus, and the deep layer of the same muscles :—that is to say, the nerve supplies the supinators, and the extensors of the carpus and fingers, with the exception of the supinator longus and the extensor carpi radialis longior.

Summary of the Musculo-spiral Nerve.

The trunk of the nerve distributes its branches to the extensor muscles of the elbow-joint exclusively, with the exception of a filament to the brachialis anticus, which however 'receives its principal nerves from another source. Before separating into its two large divisions, the nerve gives branches to two muscles of the forearm, viz., the long supinator, and the long radial extensor of the carpus.   The posterior interosseous division distributes nerves to the remaining muscles on the outer and back part of the forearm, except the anconeus (previously supplied), viz., to the short supinator and the extensors.

Cutaneous nerves are distributed, from the trunk of the nerve and its radial division, to the lower part of the upper arm, to the forearm, and to the hand—on the posterior and outer aspect of each.

ANTERIOR PRIMARY DIVISIONS OF THE DORSAL NERVES.

These nerves are twelve in number, and, with the exception of the larger part of the first of them, which joins the brachial plexus, they are distributed to the walls of the thorax and abdomen.   Eleven of the nerves so distributed are termed intercostal, and the twelfth is situated below the last rib.   The cords connecting them with the sympathetic nerve, placed close to the vertebræ, are very short.

The anterior divisions of these nerves pass separately to their destination, without forming any plexus by the connection or interlacement of their fibres, and in this respect they differ from those of the other spinal nerves. From the intervertebral foramina they are directed transversely across the trunk, and nearly parallel one to another.   The upper six nerves, with the exception of the first, are confined to the parietes of the thorax ; while the lower six nerves are continued from the intercostal spaces to the muscles and integument of the anterior wall of the abdomen.

FIRST DORSAL NERVE.

The greater part of the anterior division of this nerve ascends over the neck of the first rib and the first intercostal artery to enter into the brachial plexus.   The remaining portion of the nerve is continued as the *first intercostal*, a small branch which courses along the first intercostal space, in the manner of the other intercostal nerves, but has usually no lateral cutaneous branch, and may also want the anterior cutaneous.

UPPER OR PECTORAL INTERCOSTAL NERVES.

In their course to the fore part of the chest, these nerves accompany the intercostal blood-vessels.   After a short space they pass between the internal and external intercostal muscles, supplying them with twigs, and, about midway between the vertebræ and the sternum, give off the lateral cutaneous branches.   The nerves, greatly diminished, are now continued forwards amid the fibres of the internal intercostal muscles as far as the costal cartilages, where they come into contact with the pleura.   In approaching

Fig. 435.

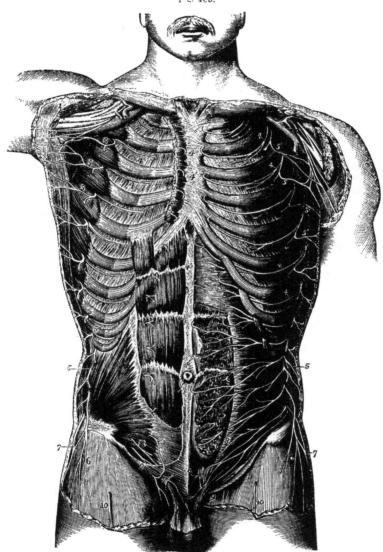

Fig. 435.—View of the Anterior Divisions of the Dorsal and some of the other
Spinal Nerves from before (from Hirschfeld and Leveillé). ¼

The pectoralis major and minor muscles have been removed; on the right side the
rectus abdominis and internal oblique muscles are shown, on the left side the anterior part
of the rectus is cut away, and the transversalis is exposed.

1, The median and other nerves of the brachial plexus ; 2, the internal cutaneous ; 3, the nerve of Wrisberg; 4, the intercostal nerves continued forwards to 4', their anterior cutaneous twigs; 5, the lateral cutaneous branches of these nerves; 6, cutaneous branch of the last dorsal spinal nerve; 7, the iliac twig of the ilio-hypogastric branch of the first lumbar nerve; 8, termination of the ilio-hypogastric; 9, the ilio-inguinal; 10, the middle cutaneous of the thigh.

the sternum, they cross the internal mammary artery and the fibres of the triangularis sterni muscle.   Finally, these nerves pierce the internal intercostal muscle and the greater pectoral, and end in the integument of the breast, receiving the name of the anterior cutaneous nerves of the thorax.

At the anterior part of the chest some of the muscular twigs cross the cartilages of the ribs, passing from one intercostal space to another.

(a) The *lateral cutaneous nerves of the thorax* pierce the external intercostal and serratus magnus muscles, in a line a little behind the pectoral border of the axilla. The first intercostal usually gives no lateral branch or only a slender twig to the axilla, but, when that of the second nerve is unusually small, it is supplemented by that of the first.   The branch from the second intercostal is named intercosto-humeral, and requires separate description.   Each of the remaining lateral cutaneous nerves divides into two branches, which reach the integument at a short distance from each other, and are named anterior and posterior.

The *anterior branches* are continued forwards over the border of the great pectoral muscle.   Several reach the mammary gland and the nipple; and from the lower nerves twigs are supplied to the digitations of the external oblique muscle of the abdomen.

The *posterior branches* turn backwards to the integument over the scapula and the latissimus dorsi muscle.   The branch from the third nerve ramifies in the axilla, and a few filaments reach the arm.

The *intercosto-humeral* nerve, the lateral cutaneous branch of the second intercostal nerve, corresponds with the posterior of the two divisions of the succeeding lateral cutaneous branches, the anterior being commonly wanting.   It crosses the axillary space to reach the arm, and is connected in the axilla with an offset of the nerve of Wrisberg.   Penetrating the fascia, it becomes subcutaneous, and ramifies in the integument of the upper half of the arm, on the inner and posterior aspect; a few filaments reach the integument over the scapula.   The branches of this nerve cross over the internal cutaneous offset of the musculo-spiral, and a communication is established between the two nerves.   The size of the intercosto-humeral nerve, and the extent of its distribution, are in the inverse proportion to the size of the other cutaneous nerves of the upper arm, especially the nerve of Wrisberg.

(b) The *anterior cutaneous nerves of the thorax*, which are the terminal twigs of the intercostal nerves, are reflected outwards in the integument over the great pectoral muscle.   The branch from the second nerve is connected with the supraclavicular and the lateral cutaneous nerves; those from the third and fourth nerves are distributed to the mammary gland.

LOWER OR ABDOMINAL INTERCOSTAL NERVES.

The lower intercostal nerves are continued from the anterior ends of the intercostal spaces, between the internal oblique and the transverse muscle of the abdomen, to the outer edge of the rectus.   Perforating the sheath, they enter the substance of that muscle, and afterwards terminate in small cutaneous branches (anterior cutaneous).

(a) The *lateral cutaneous nerves of the abdomen* pass to the integument through the external intercostal and external oblique muscles, in a line with the corresponding nerves on the thorax, and divide in the same manner into anterior and posterior branches.

The *anterior branches* are the larger, and are directed inwards in the superficial fascia, with small cutaneous arteries, nearly to the edge of the rectus muscle.

The *posterior branches* bend backwards over the latissimus dorsi muscle.

(b) The *anterior cutaneous nerves of the abdomen* become subcutaneous near the linea alba, accompanying the small perforating arteries. Their number and position are very uncertain. They are directed outwards towards the lateral cutaneous nerves. A second set is described by Cruveilhier as existing at the outer edge of the rectus muscle.

LAST DORSAL NERVE.

The anterior primary division of this nerve is below the last rib, and is contained altogether in the abdominal wall. The nerve has the general course and distribution of the others between the internal oblique and transversalis, but, before taking its place between those muscles, it passes in front of the upper part of the quadratus lumborum, and pierces the posterior aponeurosis of the transverse muscle. This nerve is connected by offsets with the nerve above, and occasionally with the ilio-hypogastric branch of the lumbar plexus. Near the spine it sometimes communicates with the first lumbar nerve by means of a small cord in the substance of the quadratus lumborum.

The *lateral cutaneous branch* of the last dorsal nerve, passing through both oblique muscles, is directed downwards over the iliac crest to the integument, covering the fore part of the gluteal region and the upper and outer part of the thigh, some filaments reaching as far as the great trochanter of the femur.

ANTERIOR PRIMARY DIVISIONS OF THE LUMBAR NERVES.

The anterior divisions of the lumbar nerves increase in size from the first to the fifth ; and all, except the fifth, which passes down to join the sacral nerves, are connected together by communicating loops, so as to form the lumbar plexus. On leaving the intervertebral foramina these nerves are connected by filaments with the sympathetic nerve, these filaments being longer than those connected with other spinal nerves, in consequence of the position of the lumbar sympathetic ganglia on the fore part of the bodies of the vertebræ. In the same situation are furnished small twigs to the psoas and quadratus lumborum muscles.

LUMBAR PLEXUS.

The lumbar plexus is formed by the communications between the anterior primary divisions of the four upper lumbar nerves. It is placed in the substance of the psoas muscle, in front of the transverse processes of the corresponding vertebræ. Above, the plexus is narrow, and is sometimes connected with the last dorsal nerve by a small offset from that nerve, named dorsi-lumbar ; below it is wider, and is joined to the sacral plexus by means of a branch given by the fourth lumbar nerve to the fifth.

The arrangement of the plexus may be thus stated. The first nerve gives off the ilio-hypogastric and ilio-inguinal nerves, and sends downwards a communicating branch to the second nerve. The second furnishes the greater part of the genito-crural and external cutaneous nerves, and gives a connecting branch to the third, from which some of the fibres of the anterior crural and obturator nerves are derived. From the third nerve, besides the descending branch to the fourth, two branches proceed : one of these, the larger, forms part of the anterior crural nerve ; the other, a part of the obturator nerve. The fourth nerve gives two branches, which serve to complete the obturator and anterior crural nerves, and a connecting branch to the fifth nerve.

The *branches* of this plexus form two sets, which are distributed, one to the lower part of the wall of the abdomen, the other to the fore part and inner side of the lower limb. In the former set are the ilio-hypogastric and

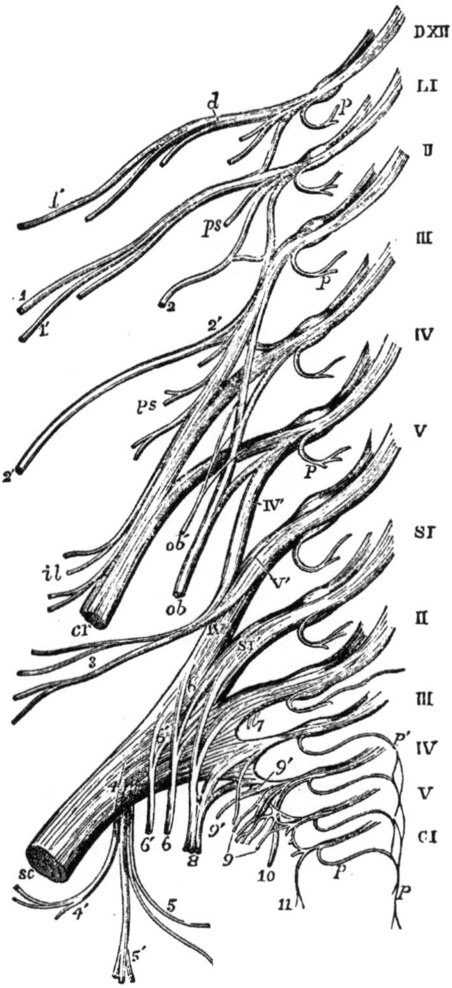

Fig. 436. — DIAGRAMMATIC OUTLINE OF THE LUMBAR AND SACRAL PLEXUSES WITH THE PRINCIPAL NERVES ARISING FROM THEM. ½

DXII, placed opposite the divided roots of the last dorsal nerve ; LI to V, opposite the roots of the five lumbar nerves : the loops uniting the anterior primary divisions of these nerves together, and the first with the twelfth dorsal are shown ; SI to V, and CI, the same in the sacral and coccygeal nerves ; *p*, placed on some of the nerves marks the posterior primary divisions cut short ; *p′ p*, the plexus formed by the union of the posterior branches of the third, fourth, and fifth sacral and the coccygeal nerves ; *d*, the abdominal continuation of the last dorsal nerve, from which *d′* the iliac cutaneous branch arises ; 1, 1′, the ilio-hypogastric and ilio-inguinal branches of the first lumbar nerve ; 2, the genito-crural rising by a loop from the first and second lumbar ; 2′, external cutaneous of the thigh rising by a loop from the second and third ; *ps*, branches to the psoas muscle along the lumbar plexus ; *cr*, anterior crural nerve from the second, third, and fourth lumbar ; *il*, branches to the iliacus ; *ob*, obturator nerve from the second, third and fourth lumbar nerves ; *ob′*, accessory obturator ; IV′, V′, loop from the fourth and fifth lumbar, forming the lumbo-sacral cord ; 3, superior gluteal nerve ; *sc*, sacral plexus ending in the great sciatic nerve ; 4, lesser sciatic nerve rising from the plexus posteriorly ; 4′, inferior gluteal branches ; 5, inferior pudendal ; 5′, posterior cutaneous of the thigh and leg ; 6, 6, branches to the obturator in-

Fig. 436.

ternus and gemellus superior ; 6′, 6′, branches to the gemellus inferior, quadratus and hip-joint ; 7, twigs to the pyriformis ; 8, 8, pudic nerve from the first, second, third, and fourth sacral ; 9, visceral branches ; 9′, twig to the levator ani ; 10, cutaneous from the fourth, which passes round the lower border of the gluteus maximus ; 11, coccygeal branches.

ilio-inguinal nerves, and part of the genito-crural; and to the latter belong the remaining part of the genito-crural nerve, the external cutaneous, the obturator, and the anterior crural nerves.

ILIO-HYPOGASTRIC AND ILIO-INGUINAL NERVES.

These nerves are the upper two branches from the lumbar plexus; they are both derived from the first lumbar nerve, and have a nearly similar distribution. They become subcutaneous by passing between the broad muscles of the abdomen, and through the outer one, and end in the integument of the groin and scrotum in the male, and the labia pudendi in the female, as well as in the integument covering the gluteal muscles. The extent of distribution of the one is inversely proportional to that of the other.

The *ilio-hypogastric* nerve, emerging from the upper part of the psoas muscle at the outer border, runs obliquely over the quadratus lumborum to the iliac crest, and there perforating the transverse muscle of the abdomen, gets between that muscle and the internal oblique, and divides into an iliac and hypogastric branch.

(a) The *iliac branch* pierces the attachment of both oblique muscles, immediately above the iliac crest, and is lost in the integument over the gluteal muscles, behind the distribution of the lateral cutaneous branch of the last dorsal nerve.

(b) The *hypogastric* or *abdominal* branch passes on between the transverse and internal oblique muscles, and is connected with the ilio-inguinal nerve near the iliac crest. It then perforates the internal oblique muscle, and, piercing the aponeurosis of the external oblique, a little above the superficial inguinal opening, is distributed to the skin of the abdomen above the pubes.

The size of the iliac branch of this nerve varies inversely with that of the lateral cutaneous branch of the twelfth dorsal. The hypogastric branch is not unfrequently joined with the last dorsal nerve between the muscles, near the crest of the innominate bone.

The *ilio-inguinal nerve*, smaller than the preceding, supplies the integument of the groin. Descending obliquely outwards over the quadratus lumborum, it crosses the fibres of the iliacus muscle, being placed lower down than the ilio-hypogastric: it then perforates the transverse muscle further forwards than the ilio-hypogastric; communicating with that nerve between the abdominal muscles. Then piercing the internal oblique muscle, it descends in the inguinal canal, and emerging at the superficial inguinal ring, is distributed to the skin upon the groin, as well as to that upon the scrotum and penis in the male, or the labium pudendi in the female, communicating with the inferior pudendal nerve. In its progress this nerve furnishes branches to the internal oblique muscle.

The ilio-inguinal nerve occasionally arises from the loop connecting the first and second lumbar nerves. It is sometimes small, and ends near the iliac crest by joining the ilio-hypogastric nerve; in that case the last nerve gives off an inguinal branch having a similar course and distribution to the ilio-inguinal nerve, the place of which it supplies.

GENITO-CRURAL NERVE.

The genito-crural nerve belongs partly to the external genital organs and partly to the thigh. It is derived chiefly from the second lumbar nerve, but receives also a few fibres from the connecting cord between that and the first nerve. The nerve descends obliquely through the psoas muscle, and afterwards on its fore part, towards Poupart's ligament, dividing at a variable height into an internal or genital, and an external or crural branch.

It often bifurcates close to its origin from the plexus, in which case its two branches perforate the psoas muscle in different places.

(a) The *genital branch* (external spermatic, Schmidt), lies upon or near the external iliac artery, and sends filaments along that vessel ; then perforating the transversalis fascia, it passes through the inguinal canal with the spermatic cord, and is lost upon

Fig. 437.

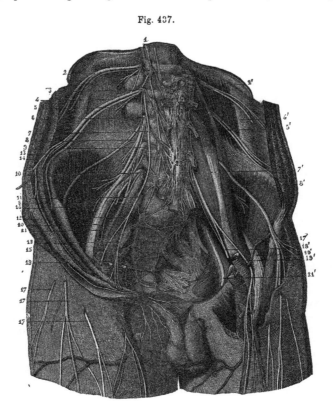

Fig. 437.—VIEW FROM BEFORE OF THE ANTERIOR BRANCHES OF THE LUMBAR AND SACRAL NERVES WITH THE PLEXUSES (from Sappey, after Hirschfeld and Leveillé). ¼

1, lumbar cord of the great sympathetic nerve ; 2, 2', anterior primary division of the twelfth dorsal nerve ; 3, first lumbar ; 4, 4', ilio-inguinal branch of this nerve ; 5, 5', ilio-hypogastric branch ; 6, second lumbar nerve ; 7, 7', genito-crural nerve rising from the first and second lumbar ; 8, 8', external cutaneous nerve of the thigh ; 9, third lumbar nerve ; 10, fourth ; 11, fifth ; 12, lumbo-sacral trunk ; 13, iliac branch of the ilio-inguinal ; 14, its abdominal branch ; 15, its genital branch ; 16, external cutaneous nerve of the right side passing out of the pelvis under Poupart's ligament ; 17, 17, 17, cutaneous ramifications of this nerve ; 17', the same nerve exposed on the left side ; 18, 18', genital branch of the genito-crural ; 19, its crural branch on the right side becoming cutaneous ; 19', the same on the left side exposed as it descends in front of the femoral artery ; 20, anterior crural nerve ; 21, 21', obturator nerve ; 22, left sciatic plexus ; 23, aortic plexus of the sympathetic nerve connected superiorly with the other pre-aortic plexuses and the lumbar ganglia, and inferiorly with the hypogastric plexus.

the cremaster muscle. In the female it accompanies the round ligament of the uterus.

(b) The *crural branch* (lumbo-inguinal nerve, Schmidt), descends upon the psoas muscle beneath Poupart's ligament into the thigh. Immediately below that ligament, and at the outer side of the femoral artery, it pierces the fascia lata, and supplies the skin on the upper part of the thigh, communicating with the middle cutaneous branch of the anterior crural nerve. Whilst it is passing beneath Poupart's ligament, some filaments are prolonged from this nerve on the femoral artery. It is stated by Schmidt, that when the crural branch of the genito-crural nerve is large, and commences near the plexus, he has observed it to give a muscular branch to the lower border of the internal oblique and transversalis muscles.

EXTERNAL CUTANEOUS NERVES.

This nerve, commencing from the loop formed between the second and third lumbar nerves, on emerging from the outer border of the psoas muscle, crosses the iliacus muscle below the ilio-inguinal nerve, and passing beneath Poupart's ligament, reaches the thigh beneath the anterior superior iliac spine, where it divides into an anterior and a posterior branch distributed to the integument of the outer side of the hip and thigh.

(a) The *posterior branch* perforates the fascia lata, and subdivides into two or three others, which turn backwards and supply the skin upon the outer surface of the limb, from the upper border of the hip-bone nearly to the middle of the thigh. The highest among them are crossed by the cutaneous branches from the last dorsal nerve.

(b) An *anterior branch*, the continuation of the nerve, is at first contained in a sheath or canal formed in the substance of the fascia lata; but, about four inches below Poupart's ligament, it enters the subcutaneous fatty tissue, and is distributed along the outer part of the front of the thigh, ending near the knee. The principal offsets spring from its outer side. In some cases, this branch reaches quite down to the knee, and communicates there with the internal saphenous nerve.

OBTURATOR NERVE.

The obturator nerve (internal crural) is distributed to the adductor muscles of the thigh, and to the hip and knee-joints. It arises from the lumbar plexus by two roots, one from the third and the other from the fourth lumbar nerve. Having emerged from the inner border of the psoas muscle, opposite to the brim of the pelvis, it runs along the side of the pelvic cavity, above the obturator vessels, as far as the opening in the upper part of the thyroid foramen, through which it escapes from the pelvis into the thigh. Here it immediately divides into an anterior and a posterior branch, which are separated from one another by the short adductor muscle.

A.—The *anterior portion* communicates with the accessory obturator nerve, when that nerve is present, and descends in front of the adductor brevis and behind the pectineus and adductor longus muscles. It gives branches as follows.

(a) An *articular branch* to the hip-joint arises near the thyroid membrane.

(b) *Muscular branches* are given to the gracilis and adductor longus muscles, and occasionally also others to the adductor brevis and pectineus.

(c) The *terminal twig* turns outwards upon the femoral artery, and surrounds that vessel with small filaments.

(d) An offset at the lower border of the adductor longus communicates beneath the fascia with the internal cutaneous branch of the anterior crural nerve, and with a branch of the internal saphenous nerve, forming a sort of plexus.

Occasional cutaneous nerve.—In some instances the communicating branch described is larger than usual, and descends along the posterior border of the sartorius to the inner side of the knee, where it perforates the fascia, communicates with the internal saphenous nerve, and extends down the inner side of the limb, supplying the skin as low as the middle of the leg.

Fig. 438.—THE LUMBAR PLEXUS FROM BEFORE, WITH THE DISTRIBUTION OF SOME OF ITS NERVES (slightly altered from Schmidt). ½

Fig. 438.

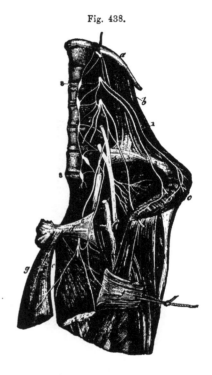

a, the last rib ; *b*, quadratus lumborum muscle ; *c*, oblique and transverse muscles cut near the crest of the ilium and turned down ; *d*, pubes ; *e*, adductor brevis muscle ; *f*, pectineus divided and turned outwards ; *g*, adductor longus; 1, ilio-hypogastric nerve ; 2, ilio-inguinal ; 3, external cutaneous; 4, anterior crural ; 5, accessory obturator ; 6, obturator united with the accessory by a loop round the pubes; 7, genito-crural in two branches cut short near their origin ; 8, 8, lumbar portion of the gangliated sympathetic cord.

When this cutaneous branch of the obturator nerve is present, the internal cutaneous branch of the anterior crural nerve is small, the size of the two nerves bearing an inverse proportion to each other.

B.—The *posterior* or *deep* part of the obturator nerve, having passed through some fibres of the external obturator muscle, crosses behind the short adductor to the fore part of the adductor magnus, where it divides into many branches, all of which enter those muscles, excepting one which is prolonged downwards to the knee-joint.

(*a*) The *muscular* branches supply the external obturator and the great adductor muscle, with the short adductor also when this muscle receives no branch from the anterior division of the nerve.

(*b*) The *articular* branch for the knee rests at first on the adductor magnus, but perforates the lower fibres of that muscle, and thus reaches the upper part of the popliteal space. Supported by the popliteal artery, and sending filaments around that vessel, the nerve then descends to the back of the knee-joint, and enters the articulation through the posterior ligament. (Thomson, "London Med. and Surg. Journal," No. xcv.)

ACCESSORY OBTURATOR NERVE.

The accessory obturator nerve, a small and inconstant nerve, arising from the obturator nerve near its upper end, or separately from the same nerves of the plexus, descends along the inner border of the psoas muscle, over the pubic bone, and, passing behind the pectineus muscle, ends by dividing into several branches. Of these one joins the anterior branch of the obturator nerve ; another penetrates the pectineus on the under surface ; whilst a third enters the hip-joint with the articular artery.

near the saphenous opening, and reaches down to the middle of the thigh. The others appear beneath the skin lower down by the side of the vein ; one, larger than the rest, passes through the fascia about the middle of the thigh, and extends to the knee. In some instances, these small branches spring directly from the anterior crural nerve, and they often communicate with each other.

Fig. 440.

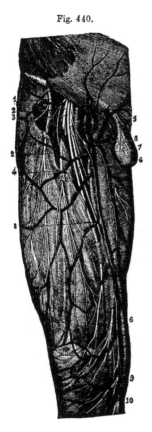

Fig. 440.—CUTANEOUS NERVES OF THE ANTERIOR AND INNER PART OF THE THIGH (from Sappey after Hirschfeld and Leveillé). ½

1, external cutaneous nerve ; 2, 2, middle cutaneous branch of the anterior crural passing through the sartorius muscle and the fascia ; 3, 3, anterior division of the internal cutaneous ; 4, filament to the sartorius ; 5, inner or posterior division of the internal cutaneous ; 6, its superficial branch to the inside of the knee after perforating the fascia ; 7, deep or communicating branch ; 8, superficial branch of the musculo-cutaneous of the crural ; 9, patellar branch of the internal saphenous nerve ; 10, continuation of the saphenous down the leg.

(b) The *anterior branch*, descending in a straight line to the knee, perforates the fascia lata in the lower part of the thigh ; it afterwards runs down near the intermuscular septum, giving off filaments on each side to the skin, and is finally directed over the patella to the outer side of the knee. It communicates above the joint with a branch of the long saphenous nerve ; and sometimes it takes the place of the branch usually given by the latter to the integument over the patella.

This branch of the internal cutaneous nerve sometimes lies above the fascia in its whole length. It occasionally gives off a cutaneous filament, which accompanies the long saphenous vein, and in some cases it communicates with the branch to be next described.

The *inner branch* of the internal cutaneous nerve, descending along the posterior border of the sartorius muscle, perforates the fascia lata at the inner side of the knee, and communicates by a small branch with the internal saphenous nerve, which here descends in front of it. It gives some cutaneous filaments to the lower part of the thigh on the inner side, and is distributed to the skin upon the inner side of the leg. Whilst beneath the fascia, this branch of the internal cutaneous nerve joins in an interlacement with offsets of the obturator nerve below the middle of the thigh, and with the branch of the saphenous nerve nearer the knee.

D.　INTERNAL SAPHENOUS NERVE.

The internal or long saphenous nerve is the largest of the cutaneous branches of the anterior crural nerve. In some cases it arises in connection with one of the deep or muscular branches.

This nerve is deeply placed as far as the knee, and is subcutaneous in the rest of its course. In the thigh it accompanies the femoral vessels, lying at first somewhat to their outer side, but lower down approaching close to them, and passing beneath the same aponeurosis.   When the vessels pass through the opening in the adductor muscle into the popliteal space, the saphenous nerve separates from them, and is continued downwards beneath the sartorius muscle to the inner side of the knee; where, having first given off, as it lies near the inner condyle of the femur, a branch which is distributed over the front of the patella, it becomes subcutaneous by piercing the fascia between the tendons of the sartorius and gracilis muscles.

The nerve then accompanies the saphenous vein along the inner side of the leg, and passing in front of the ankle is distributed to the inner side of the foot.   In the leg it is connected with the internal cutaneous nerve.

The distribution of the branches is as follows.

(*a*) A *communicating branch* is given off about the middle of the thigh to join in the interlacement formed beneath the fascia lata by this nerve and branches of the obturator and internal cutaneous nerves.   After it has left the aponeurotic covering of the femoral vessels, the internal saphenous nerve has, in some cases, a further connection with one or other of the nerves just referred to.

(*b*) The *branch to the integument in front of the patella* perforates the sartorius muscle and the fascia lata; and, having received a communicating offset from the internal cutaneous nerve, spreads out upon the fore part of the knee; and, by uniting with branches of the middle and external cutaneous nerves, forms a plexus—plexus patellæ.

(*c*) A branch to the inner ankle is given off in the lower third of the leg, and descends along the margin of the tibia.

(*d*) Filaments from this nerve enter the tarsal ligaments.

Summary.—The anterior crural nerve is distributed to the skin upon the fore part and inner side of the thigh, commencing below the termination of the ilio-inguinal and genito-crural nerves.   It furnishes also a cutaneous nerve to the inner side of the leg and foot.   All the muscles on the front and outer side of the thigh receive their nerves from the anterior crural, and the pec-tineus is also in part supplied by this nerve, and in part by the obturator. The tensor muscle of the fascia lata is supplied from a different source, viz., the superior gluteal nerve.   Lastly, two branches are given from the anterior crural nerve to the knee-joint.

FIFTH LUMBAR NERVE.

The anterior branch of the fifth lumbar nerve, having received a fasciculus from the nerve next above it, descends to join the first sacral nerve, and form part of the sacral plexus.   The cord resulting from the union of the fifth with a part of the fourth nerve, is named the *lumbo-sacral* nerve.

SUPERIOR GLUTEAL NERVE.

Before joining the first sacral nerve the lumbo-sacral cord gives off from behind the superior gluteal nerve; this offset leaves the pelvis through the large sacro-sciatic foramen, above the pyriformis muscle, and divides like the gluteal artery into two branches, which are distributed chiefly to the smaller gluteal muscles and tensor of the fascia lata.

(*a*) The *upper* branch runs with the gluteal artery along the origin of the gluteus minimus, and is lost in it and in the gluteus medius.

(*b*) The *lower* branch crosses over the middle of the gluteus minimus, between this and the gluteus medius, and supplying filaments to both those muscles, is con-tinued forwards, and terminates in the tensor muscle of the fascia lata.

x x 2

ANTERIOR PRIMARY DIVISIONS OF THE SACRAL AND COCCYGEAL NERVES.

THE SACRAL NERVES.

The anterior divisions of the first four sacral nerves emerge from the spinal canal by the anterior sacral foramina, and the fifth passes out between the sacrum and coccyx.

The first two sacral nerves are large, and of nearly equal size ; the others diminish rapidly, and the fifth is exceedingly slender. Like the anterior divisions of the other spinal nerves, those of the sacral nerves communicate with the sympathetic : the communicating cords are very short, as the sympathetic ganglia are close to the inner margin of the foramina of the sacrum.

The first three nerves and part of the fourth contribute to form the sacral plexus. The fifth has no share in the plexus,—it ends on the back of the coccyx. As the description of the fourth and fifth sacral nerves and of the coccygeal will occupy only a short space, these three nerves may be noticed first, before the other nerves and the numerous branches to which they give rise are described.

THE FOURTH SACRAL NERVE.

Only one part of the anterior division of this nerve joins the sacral plexus ; the remainder, which is nearly half the nerve, supplies branches to the viscera and muscles of the pelvis, and sends downwards a connecting filament to the fifth nerve.

(a) The *visceral branches* of the fourth sacral nerve are directed forwards to the lower part of the bladder, and communicate freely with branches from the sympathetic nerve. Offsets are distributed to the neighbouring viscera, according to the sex. They will be described with the pelvic portion of the sympathetic nerve. The foregoing branches are, in some instances, furnished by the third sacral nerve instead of the fourth, and not unfrequently from both of these nerves.

(b) Of the *muscular branches*, one supplies the *levator ani*, piercing that muscle on the pelvic surface ; another enters the *coccygeus*, whilst a third ends in the *external sphincter* muscle of the rectum. The last branch, after passing either through the coccygeus, or between it and the levator ani, reaches the perinæum, and is distributed likewise to the integuments between the anus and the coccyx.

THE FIFTH SACRAL NERVE.

The anterior branch of this, the lowest sacral nerve, comes forwards through the coccygeus muscle opposite the junction of the sacrum with the first coccygeal vertebra ; it then descends upon the coccygeus nearly to the tip of the coccyx, where it turns backwards through the fibres of that muscle, and ends in the integument upon the posterior and lateral aspect of the bone.

. As soon as this nerve appears in front of the bone (in the pelvis) it is joined by the descending filament from the fourth nerve, and lower down by the small anterior division of the coccygeal nerve. It supplies small filaments to the coccygeus muscle.

THE COCCYGEAL NERVE.

The anterior branch of the coccygeal, or, as it is sometimes named, the sixth sacral nerve, is a very small filament. It escapes from the spinal canal by the terminal opening, pierces the sacro-sciatic ligament and the coccygeus muscle, and, being joined upon the side of the coccyx with the fifth sacral nerve, partakes in the distribution of that nerve.

THE SACRAL PLEXUS.

The lumbo-sacral cord (resulting as before described from the junction of

Fig. 441. — Diagrammatic Outline of the Lumbar and Sacral Plexuses with the principal Nerves arising from them. ½

Fig. 441.

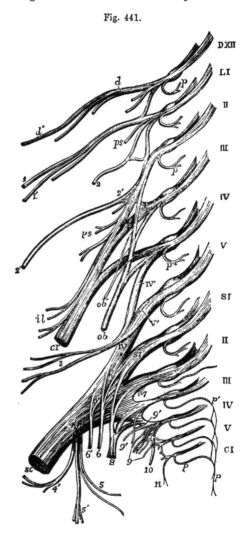

The references to the nerves of the lumbar plexus will be found at p. 659. DXII, roots of the last dorsal nerve ; LI to V, roots of the five lumbar nerves ; SI to V, and CI, roots of the five sacral and the coccygeal nerves ; IV', V', loop from the anterior primary branches of the fourth and fifth lumbar nerves, forming the lumbo-sacral cord ; 3, superior gluteal nerve ; SC, sacral plexus ending in the great sciatic nerve ; 4, lesser sciatic nerve, rising from the plexus posteriorly ; 4', inferior gluteal branches ; 5, inferior pudendal ; 5', posterior cutaneous of the thigh and leg ; 6, 6, branches to the obturator internus and gemellus superior ; 6', 6', branches to the gemellus inferior, quadratus and hip-joint ; 7, twigs to the pyriformis ; 8, 8, pudic from the first, second, third, and fourth sacral ; 9, visceral branches ; 9', twig to the levator ani ; 10, cutaneous from the fourth, which passes round the lower border of the gluteus maximus ; 11, coccygeal branches.

the fifth and part of the fourth lumbar nerves), the anterior divisions of the first three sacral nerves, and part of the fourth, unite to form this plexus. Its construction differs from that of the other spinal nervous plexuses in this respect, that the several constituent nerves entering into it

unite into one broad flat cord.   To the place of union the nerves proceed in
different directions, that of the upper ones being obliquely downwards, while
that of the lower is nearly horizontal ; and, as a consequence of this
difference, they diminish in length from the first to the last.   The sacral
plexus rests on the anterior surface of the pyriform muscle, opposite the side
of the sacrum, and escaping through the great sacro-sciatic foramen, ends in
the great sciatic nerve.

   Branches.—The sacral plexus gives rise to the great sciatic nerve, and to
various smaller branches ; viz., the pudic nerve, the small sciatic nerve, and
branches to the obturator internus, pyriformis, gemelli, and quadratus
femoris muscles.

MUSCULAR BRANCHES.

   a. To *the pyriformis muscle,* one or more branches are given, either from the plexus
or from the upper sacral nerves before they reach the plexus.
   b. The *nerve of the internal obturator muscle* arises from the part of the plexus
formed by the union of the lumbo-sacral and the first sacral nerves.   It turns over
the ischial spine of the hip-bone with the pudic vessels, and is then directed forwards
through the small sacro-sciatic foramen to reach the inner surface of the obturator
muscle.
   c. To the *levator ani* one or more twigs proceed from the lower part of the plexus.
   d. The *superior gemellus* receives a small branch, which arises from the lower part
of the plexus.
   e. The small nerve which supplies *the lower gemellus* and *quadratus femoris*
muscles springs from the lower part of the plexus.   Concealed at first by the great
sciatic nerve, it passes beneath the gemelli and the tendon of the internal obturator,
—between those muscles and the capsule of the hip-joint,—and reaches the deep
(anterior) surface of the quadratus.   It furnishes a small articular filament to the
back part of the hip-joint.

THE PUDIC NERVE.

   This nerve, arising from the lower part of the sacral plexus, turns over
the spine of the ischium, and then passes forwards through the small sacro-
sciatic foramen, where it usually gives off the inferior hæmorrhoidal branch.
It is next directed along the outer part of the ischio-rectal fossa, in a sheath
of the obturator fascia, along with the pudic vessels, and divides into two
terminal branches, the perinæal nerve and the dorsal nerve of the penis.

   A.—The *perinæal nerve,* the lower and much the larger of the two
divisions of the pudic nerve, lies below the pudic artery, and is expended in
superficial and muscular branches.

   a. The *superficial perinæal* branches are two in number, anterior and posterior.
The *posterior* branch, which first separates from the perinæal nerve, reaching the back
part of the ischio-rectal fossa, gives filaments inwards to the skin in front of the
anus, and turns forwards in company with the anterior branch to reach the scrotum.
The *anterior* branch descends to the fore part of the ischio-rectal fossa ; and, passing
forwards with the superficial perinæal artery, ramifies in the skin on the fore part of
the scrotum and on the penis.   This branch sends small twigs to the levator ani
muscle.   The superficial perinæal nerves are accompanied to the scrotum by the
inferior pudendal branch of the small sciatic nerve.   The three branches are sometimes named *long scrotal nerves.*
   In the female, both the superficial perinæal branches terminate in the external
labium pudendi.
   b. The *muscular branches* generally arise by a single trunk, which is directed
inwards under cover of the transversalis perinæi muscle, and divides into offsets which

are distributed to the tranversalis perinæi, erector penis, accelerator urinæ, and compressor urethræ.

c. Slender filaments are sent inwards to the corpus spongiosum urethræ; some of these, before penetrating the erectile tissue, run a considerable distance over its surface.

Fig. 442.—Right Side of the Interior of the Male Pelvis, with the Principal Nerves displayed (from Hirschfeld and Leveillé). ¼

The left wall has been removed as far as the sacrum behind and the symphysis pubis in front; the viscera and the lower part of the levator ani have been removed; a, the lower part of the aorta; a', placed on the fifth lumbar vertebra, between the two common iliac arteries, of which the left is cut short; b, the right external iliac artery and vein; c, the symphysis pubis; d, the divided pyriformis muscle, close to the left auricular surface of the sacrum; e, bulb of the urethra covered by the accelerator urinæ muscle; the membranous part of the urethra cut short is seen passing into it; 1, placed on the crest of the ilium, points to the external cutaneous nerve of the thigh passing over the iliacus muscle; 2, placed on the psoas muscle, points to the

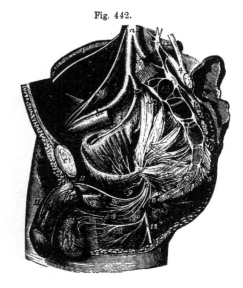

Fig. 442.

genito-crural nerve; 3, obturator nerve; 4, 4, placed on the lumbo-sacral cords; that of the right side points to the gluteal artery cut short; 4', the superior gluteal nerve; 5, placed on the inside of the right sacral plexus, points by four lines to the anterior divisions of the four upper sacral nerves, which, with the lumbo-sacral cord, unite in the plexus; 5', placed on the fifth piece of the sacrum, points to the fifth sacral nerves; 5", the visceral branches proceeding from the third and fourth sacral nerves; 6, placed on the lower part of the coccyx, below the coccygeal nerves; 7, placed on the line of division of the pelvic fascia, points to the nerve of the levator ani muscle; 8, placed at the lower border of the great sacro-sciatic ligament, points to the cutaneous nerves of the anus; 9, nerve of the obturator internus; 10, the pudic nerve; 10', is placed above the muscular branches of the perineal nerve; 10", the anterior and posterior superficial perineal nerves, and on the scrotum the distribution of these nerves and the inferior pudendal nerve; 11, the right dorsal nerve of the penis; 11', the nerve on the left crus penis which is cut short; 12, the continuation of the lesser sciatic nerve on the back of the thigh; 12', the inferior pudendal branch; 13, placed on the transverse process of the fifth lumbar vertebra, marks the lowest lumbar sympathetic ganglion; 14, placed on the body of the first piece of the sacrum, points to the upper sacral sympathetic ganglia; between 14 and 6, are seen the remaining ganglia and sympathetic nervous cords, as well as their union with the sacral and coccygeal nerves, and at 6, the lowest ganglion or ganglion impar.

B.—The *dorsal nerve of the penis*, the upper division of the pudic nerve, accompanies the pudic artery in its course between the layers of the deep perinæeal or subpubic fascia, and afterwards through the suspensory ligament, to reach the dorsum of the penis, along which it passes as far as the glands, where it divides into filaments for the supply of that part. On the penis, this nerve is joined by branches of the sympathetic system, and it sends outwards numerous offsets to the integument on the upper surface and sides of the organ, including the prepuce.   One large branch penetrates the corpus cavernosum penis.

In the female, the dorsal nerve of the clitoris is much smaller than the corresponding branch in the male; it is similarly distributed.

C.—The *inferior hæmorrhoidal* nerve arises from the pudic nerve at the back of the pelvis, or it may come directly from the sacral plexus, and be transmitted through the small sacro-sciatic foramen to its distribution in the lower end of the rectum.

Fig. 443.

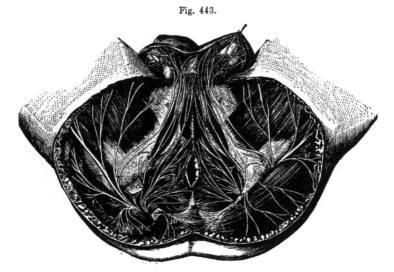

Fig. 443.—Dissection of the Perinæum of the Male to show the Distribution of the Pudic and other Nerves (from Hirschfeld and Leveillé). ¼

On the right side a part of the gluteus maximus muscle and the great sacro-sciatic liga-ment have been removed to show the descent of the nerves from the great sacro-sciatic foramen.   1, great sciatic nerve of the right side ; 2, lesser sciatic nerve ; 2′, its muscular branches to the gluteus maximus (right side) ; 2″, cutaneous branches to the buttock (left side) ; 3, continuation of the nerve as posterior middle cutaneous of the thigh ; 3, internal and external cutaneous branches ; 4, 4, inferior pudendal branch ; 4′, network of this and the perineal nerves on the scrotum ; 5, right pudic nerve ; 6, superior branch or nerve to the penis ; 7, the external superficial perineal branch ; 7′, the internal superficial perineal branch ; 8, musculo-bulbal branches ; 9, hemorrhoidal or cutaneous anal branches ; 10, cutaneous branch of the fourth sacral nerve.

Some of the branches of this nerve end in the external sphincter and in the adjacent skin of the anus ; others reach the skin in front of that part, and communicate with the inferior pudendal branch of the small sciatic nerve, and with the superficial perineal nerves.

Summary.—The pudic nerve supplies the perinæum, the penis, and part of the scrotum, also the urethral and anal muscles in the male ; and the clitoris, labia, and other corresponding parts in the female. It communicates with the inferior pudendal branch of the small sciatic nerve.

SMALL SCIATIC NERVE.

The small sciatic nerve (nervus ischiadicus minor) is chiefly a cutaneous nerve, supplying the integument of the lower part of the buttock, the back of the thigh, and upper part of the calf of the leg; it furnishes also branches to one muscle—the gluteus maximus.

This nerve is formed by the union of two or more nervous cords, derived from the lower and back part of the sacral plexus. Arising below the pyriform muscle, it descends beneath the gluteus maximus, and at the lower border of that muscle comes into contact with the fascia lata. Continuing its course downwards along the back of the limb, it perforates the fascia a little below the knee.

Fig. 444.—Deep Nerves in the Gluteal and Inferior Pudendal Regions (from Hirschfeld and Leveillé). ¼

Fig. 444.

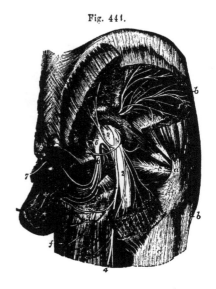

a, back part of the great trochanter; b, tensor vaginæ femoris muscle; c, tendon of the obturator internus muscle near its insertion; d, upper part of the vastus externus; e, coccyx; f, gracilis muscle; between f and d, the adductor magnus, semitendinosus, and biceps muscles; * placed at the meeting of the crura penis above the urethra; 1, placed upon the ilium close above the sacro-sciatic notch, marks the superior gluteal nerve, and on the divided parts of the gluteus medius muscle, the superior branch of the nerve; 1', on the surface of the gluteus minimus muscle, the inferior branch of the nerve; 1", branch of the nerve to the tensor vaginæ femoris; 2, sacral plexus and great sciatic nerve; 2', muscular twig from the plexus to the pyriformis; 2", muscular branches to the gemellus superior and obturator internus; 3, lesser sciatic nerve; 3', placed on the upper and lower parts of the divided gluteus maximus, the inferior gluteal muscular branches of the lesser sciatic nerve; 3", the cutaneous branches of the same nerve winding round the lower border of the gluteus maximus; 4, the continuation of the lesser sciatic nerve as posterior cutaneous nerve of the thigh; 4', inferior pudendal branch of the lesser sciatic; 5, placed on the lower part of the sacral plexus, points to the origin of the pudic nerve; 6, its perineal division with its muscular branches; 6', anterior or superior superficial perineal branch; 6", posterior or inferior superficial perineal; + +, distribution of these nerves and the inferior pudendal on the scrotum; 7, dorsal nerve of the penis.

The *branches* of the small sciatic nerve are as follows.

A. The *inferior gluteal* branches, given off under the gluteus maximus, supply the lower part of that muscle.—A distinct gluteal branch commonly proceeds from the sacral plexus to the upper part of the muscle.

B. The *cutaneous branches* of the nerve principally emerge from beneath the lower border of the gluteus maximus, arranged in an external and an internal set. Others appear lower down.

a. The *internal* are mostly distributed to the skin of the inner side of the thigh at the upper part. One branch, however, which is much larger than the rest, is distinguished as the inferior pudendal.

Fig. 445.   Fig. 446.

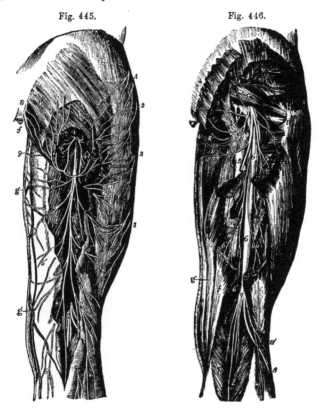

Fig. 445.—POSTERIOR CUTANEOUS NERVES OF THE HIP AND THIGH
(from Hirschfeld and Leveillé). ⅓

f *a*, gluteus maximus muscle partially uncovered by the removal of a part of the fascia lata, and divided at its inferior part to show the lesser sciatic nerve ; *b*, fascia lata over the glutei muscles and the outer part of the hip ; *c*, *d*, part of the semitendinosus, biceps, and semimembranosus muscles exposed by the removal of the fascia ; *e*, gastrocnemius ; *f*, coccyx ; *g*, internal branches of the saphena vein ; 1, iliac cutaneous branches of the ilio-inguinal and ilio-hypogastric nerves ; 2, cutaneous iliac branches of the last inter-costal ; 3, posterior twigs of the external cutaneous nerve of the thigh ; 4, lesser sciatic nerve issuing from below the gluteus maximus muscle ; 4′, its muscular branches ; 4″, its cutaneous gluteal branches ; 5, posterior middle cutaneous continued from the lesser sciatic ; 5′, 5′, its inner and outer branches spreading on the fascia of the thigh ; 6, 6, its terminal branches descending on the calf of the leg ; 7, posterior tibial and fibular nerves separating in the popliteal space ; 8, lower posterior divisions of the sacral and coccygeal nerves ; 9, inferior pudendal nerve.

Fig. 446.—DEEP POSTERIOR NERVES OF THE HIP AND THIGH
(from Hirschfeld and Leveillé). ⅓

a, gluteus medius muscle ; *b*, gluteus maximus ; *c*, pyriformis ; *d*, placed on the trochanter major, points to the tendon of the obturator internus; *e*, upper part of the femoral head of the biceps; *f*, semitendinosus; *g*, semimembranosus; *h*, gastrocnemius; *i*, popliteal artery; 1, placed on the gluteus minimus muscle, points to the superior gluteal nerves; 2, inferior gluteal branches of the lesser sciatic; 3, placed on the greater sacro-sciatic ligament, points to the pudic nerve; 3', its farther course; 4, inferior pudendal; 5, placed on the upper divided part of the semitendinosus and biceps, points to the posterior middle cutaneous nerve of the thigh; 6, great sciatic nerve, 6', 6', some of its muscular branches to the flexors; 7, internal popliteal or posterior tibial nerve; 7', its muscular or sural branches; 8, external popliteal or peroneal nerve; 8', its external cutaneous branch; 9, communicating tibial; 9', communicating peroneal branch to the external saphenous nerve.

The *inferior pudendal* branch turns forwards below the ischial tuberosity to reach the perinæum. Its filaments then extend forwards to the front and outer part of the scrotum, and communicate with one of the superficial perineal nerves. In the female, the inferior pudendal branch is distributed to the external labium pudendi.

b. The *external* cutaneous branches, two or three in number, turn upwards in a retrograde course to the skin over the lower and outer part of the great gluteal muscle. In some instances one takes a different course, descending and ramifying in the integuments on the outer side of the thigh nearly to the middle.

c. Of the *lower branches* some small cutaneous filaments pierce the fascia of the thigh above the popliteal space. One of these, arising somewhat above the knee-joint, is prolonged over the popliteal region to the upper part of the leg.

Of the *terminal twigs*, perforating the fascia lata opposite the lower part of the popliteal space, one accompanies the short saphenous vein beyond the middle of the leg, and others pass into the integument covering the inner and outer heads of the gastrocnemius muscle. Its terminal cutaneous branches communicate with the short saphenous nerve.

GREAT SCIATIC NERVE.

The great sciatic nerve (nervus ischiadicus), the largest nerve in the body, supplies the muscles at the back of the thigh, and by its branches of continuation gives nerves to all the muscles below the knee and to the greater part of the integument of the leg and foot. The several joints of the lower limb receive filaments from it and its branches.

This large nerve is continued from the lower end of the sacral plexus. It escapes from the pelvis through the great sacro-sciatic foramen, below the pyriformis muscle, and reaches down below the middle of the thigh, where it separates into two large divisions, named the *internal* and *external popliteal* nerves. At first it lies in the hollow between the great trochanter and the ischial tuberosity, covered by the gluteus maximus and resting on the gemelli, obturator internus and quadratus femoris muscles, in company with the small sciatic nerve and the sciatic artery, and receiving from that artery a branch which runs for some distance in its substance. Lower down it rests on the adductor magnus, and is covered behind by the long head of the biceps muscle.

The bifurcation of the sciatic nerve may take place at any point intermediate between the sacral plexus and the lower part of the thigh; and, occasionally, it is found to occur even within the pelvis, a portion of the pyriformis muscle being interposed between the two great divisions of the nerve.

Branches of the trunk.—In its course downwards, the great sciatic nerve supplies offsets to some contiguous parts, viz., to the hip-joint, and to the muscles at the back of the thigh.

a. The *articular branches* are derived from the upper end of the nerve, and enter the capsular ligament of the hip-joint, on the posterior aspect. They sometimes arise from the sacral plexus.

b. The *muscular branches* are given off under cover of the biceps muscle; they supply the flexors of the leg, viz., the biceps, semitendinosus, and semimembranosus. A branch is likewise given to the adductor magnus.

INTERNAL POPLITEAL NERVE.

The internal popliteal nerve, the larger of the two divisions of the great sciatic nerve, following the same direction as the parent trunk, continues along the middle of the popliteal space to the lower border of the popliteus muscle, beneath which point the continuation of the trunk receives the name of *posterior tibial.* The interior popliteal nerve lies at first at a considerable distance from the popliteal artery, at the outer side and nearer to the surface; but, from the knee-joint downwards, the nerve, continuing a straight course, is close behind the artery, and then crosses it rather to the inner side.

Fig. 447.

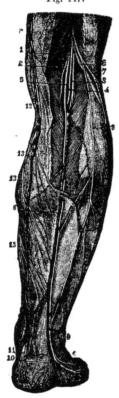

Fig. 447.—POSTERIOR CUTANEOUS NERVES OF THE LEG (from Sappey after Hirschfeld and Leveillé). ⅓

1, internal popliteal division of the great sciatic nerve; 2, branch to the internal part of the gastrocnemius muscle; 3, 4, branches to the external part and plantaris; 5, communicating branch to the external saphenous nerve; 6, external popliteal nerve; 7, cutaneous branch; 8, communicating branch descending to unite with that from the internal popliteal in, 9, the external saphenous nerve; 10, calcaneal branch from this nerve; 11, calcaneal and plantar cutaneous branches from the posterior tibial nerve; 12, internal saphenous nerve; 13, posterior branches of this nerve.

The inner division of the sciatic nerve, from its commencement to its partition at the foot, is often described in anatomical works under the same appellation throughout; the name varying, however, with different writers, as for example, " cruralis internus," or " popliteus internus,"—Winslow: " tibialis posterior,"—Haller : " tibialis vel tibieus,"—Fischer, &c.

Branches.—The internal popliteal nerve supplies branches to the knee-joint and to the muscles of the calf of the leg, and also part of a cutaneous branch, the external or short saphenous nerve.

ARTICULAR NERVES.—The *articular branches* are generally three in number ; two of these accompany the upper and lower articular arteries of the inner side of the knee-joint, the third follows the middle or azygos artery. These nerves pierce the ligamentous tissue of the joint.—The upper one is often wanting.

MUSCULAR BRANCHES.—The muscular branches of the internal popliteal nerve arise behind the knee-joint, while the nerve is between the heads of the gastrocnemius muscle.

a. The nerves to the *gastrocnemius* consist of two branches, which separate, one to supply each part of the muscle.

b. The small nerve of the *plantaris* muscle is derived from the outer of the branches just described, or directly from the main trunk (internal popliteal).

c. The *soleus* receives a branch of considerable size, which enters the muscle on the posterior aspect after descending to it in front of the gastrocnemius.

d. The *nerve* of the *popliteus* muscle lies deeper than the preceding branches, and arises somewhat below the joint; it descends along the outer side of the popliteal vessels, and, after turning beneath the lower border of the muscle, enters the deep or anterior surface.

EXTERNAL OR SHORT SAPHENOUS NERVE.

The cutaneous branch of the internal popliteal nerve (ramus communicans tibialis) descends along the leg beneath the fascia, resting on the gastro-cnemius, in the furrow between the heads of the muscle, to about midway between the knee and the foot. Here it perforates the fascia, and a little lower down is usually joined by a branch from the external popliteal nerve (communicans peronei). After receiving this communicating branch, the external saphenous nerve descends beneath the integument near the outer side of the tendo Achillis in company with the short saphenous vein, and turns forwards beneath the outer malleolus to end in the skin at the side of the foot and on the little toe. On the dorsum of the. foot this nerve communicates with the musculo-cutaneous nerve.

In many cases, the external saphenous nerve supplies the outer side of the fourth toe, as well as the little toe. The union between the saphenous nerve and the branch of the external popliteal nerve occurs in some cases higher than usual, occasionally even at or close to the popliteal space. It sometimes happens that the communication between the nerves is altogether wanting; in which case the cutaneous nerve to the foot is generally continued from the branch of the internal popliteal nerve.

POSTERIOR TIBIAL NERVE.

The internal popliteal nerve receives the name of posterior tibial at the lower margin of the popliteus muscle. It passes down the leg with the posterior tibial artery, lying for a short distance at the inner side of the vessel, and afterwards at the outer side ; the artery inclining inwards from its origin while the nerve continues its straight course. In the interval between the inner malleolus and the heel, it divides into the two plantar nerves (internal and external). The posterior tibial nerve, like the accompanying vessels, is covered at first by the muscles of the calf of the leg, afterwards only by the integument and fascia, and it rests upon the deep-seated muscles.

Lateral branches.—The deep muscles on the back of the leg and the integument of the sole of the foot receive branches from the posterior tibial nerve in its course along the leg.

a. The *muscular branches* emanate from the upper part of the nerve, either separately or by a common trunk ; and one is distributed to each of the deep muscles, viz., the tibialis posticus, the long 'flexor of the toes, and the long flexor of the great toe. The branch which supplies the last-named muscle runs along the peroneal artery before penetrating the muscle.

b. A *calcaneo-plantar cutaneous* branch is furnished from the posterior tibial nerve ; the plantar part perforates the internal annular ligament, and ramifies in the integument at the inner side of the sole of the foot, and beneath the heel.

INTERNAL PLANTAR NERVE.

The internal plantar, the larger of the two nerves to the sole of the foot, into which the posterior tibial divides, accompanies the internal or smaller plantar artery, and supplies nerves to both sides of the three inner toes, and to one side of the fourth. From the point at which it separates from the posterior tibial nerve, it is directed forwards under cover of the first part of the abductor of the great toe, and, passing between that muscle and the short flexor of the toes, it gives off the internal cutaneous branch for the great toe, and divides opposite the middle of the foot into three digital branches. The outermost of these branches communicates with the external plantar nerve.

Branches.—a. Small *muscular* branches are supplied to the abductor pollicis and flexor brevis digitorum.

b. Small *plantar cutaneous* branches perforate the plantar fascia to ramify in the integument of the sole of the foot.

c. The *digital branches* are named numerically from within outwards: the three outer pass from under cover of the plantar fascia near the clefts between the toes. The first or innermost branch continues single, but the other three bifurcate to supply the adjacent sides of two toes. These branches require separate notice.

The *first* digital branch is that destined for the inner side of the great toe; it becomes subcutaneous farther back than the others, and sends off a branch to the *flexor brevis pollicis.*

Fig. 448.

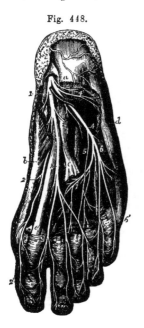

Fig. 448.—SUPERFICIAL AND DEEP DISTRIBUTION OF THE PLANTAR NERVES (from Hirschfeld and Leveillé, slightly modified). ½

The flexor communis brevis, the abductor pollicis and abductor minimi digiti, a part of the tendons of the flexor communis longus, together with the lumbricales muscles, have been removed so as to bring into view the transversus and interossei in the middle of the foot.

a, upon the posterior extremity of the flexor communis brevis, near which, descending over the heel, are seen ramifications of the calcaneal branch of the posterior tibial nerve; *b,* abductor pollicis; *c,* tendon of the flexor communis longus divided close to the place where it is joined by the flexor accessorius; *d,* abductor minimi digiti; *e,* tendon of the flexor longus pollicis between the two portions of the flexor brevis pollicis; 1, internal plantar nerve giving some twigs to the abductor pollicis, and 1', a branch to the flexor communis brevis, cut as it lies on the accessorius; 2, inner branch of the internal plantar nerve giving branches to the abductor pollicis, flexor brevis pollicis, and forming, 2', the internal cutaneous of the great toe; 3, continuation of the internal plantar nerve, dividing subsequently into three branches, which form, 3', 3', 3', the collateral plantar cutaneous nerves of the first and second, second and third, and third and fourth toes; 4, the external plantar nerve; 4', its branch to the abductor minimi digiti; 5, twig of union between the plantar nerves; 6, superficial branch of the external plantar nerve; subsequently dividing into 6', 6', the collateral cutaneous nerves of the fourth and fifth toes and the external nerve of the fifth; 7, deep branch of the external plantar nerve giving twigs to the adductor pollicis, the interossei, the transversalis, and to the third and fourth lumbricales muscles.

The *second* branch having reached the interval between the first and second metatarsal bones, furnishes a small twig to the *first lumbricalis* muscle, and bifurcates behind the cleft between the great toe and the second to supply their contiguous sides.

The *third* digital branch, corresponding with the second interosseous space, gives a slender filament to the *second lumbricalis* muscle, and divides in a manner similar to that of the second branch into two offsets for the sides of the second and third toes.

The *fourth* digital branch distributed to the adjacent sides of the third and fourth toes, receives a communicating branch from the external plantar nerve.

Along the sides of the toes, cutaneous and articular filaments are given from these digital nerves; and, opposite the ungual phalanx, each sends a dorsal branch to the pulp beneath the nail, and then runs on to the ball of the toe, where it is distributed like the nerves of the fingers.

EXTERNAL PLANTAR NERVE.

The external plantar nerve completes the supply of digital nerves to the toes, furnishing branches to the little toe and half the fourth: it also gives a deep branch of considerable size, which is distributed to several of the short muscles in the sole of the foot. There is thus a great resemblance between the distribution of this nerve in the foot and that of the ulnar nerve in the hand.

The external plantar nerve runs obliquely forwards towards the outer side of the foot, along with the external plantar artery, between the flexor brevis digitorum and the flexor accessorius, as far as the interval between the former muscle and the abductor of the little toe. Here it divides into a superficial and a deep branch, having previously furnished offsets to the *flexor accessorius* and the *abductor minimi digiti.*

a. The *superficial portion* separates into two digital branches, which have the same general arrangement as the digital branches of the internal plantar nerve. They are distributed thus.

Digital branches.—One of the digital branches continues undivided, and runs along the outer side of the little toe: it is smaller than the other, and pierces the plantar fascia further back. The *short flexor muscle of the little toe,* and occasionally *one or two interosseous* muscles of the fourth metatarsal space receive branches from this nerve.

The larger digital branch communicates with an offset from the internal plantar nerve, and bifurcates near the cleft between the fourth and fifth toes to supply one side of each.

b. The *deep* or *muscular* branch of the external plantar nerve dips into the sole of the foot with the external plantar artery, under cover of the tendons of the flexor muscles and the adductor pollicis, and terminates in numerous branches for the following muscles:—all the interossei (dorsal and plantar) except occasionally one or both of those in the fourth space, the two outer lumbricales, the adductor pollicis, and the transversalis pedis.

Summary of the internal popliteal nerve.—This nerve supplies all the muscles of the back of the leg and sole of the foot, and the integument of the plantar aspect of the toes, the sole of the foot, and in part that of the leg.

EXTERNAL POPLITEAL OR PERONEAL NERVE.

This nerve descends obliquely along the outer side of the popliteal space, lying close to the biceps muscle. Continuing downwards over the outer part of the gastrocnemius muscle (between it and the biceps) below the head of the fibula, the nerve turns round that bone, passing between it and the peroneus longus muscle, and then divides into the *anterior tibial* and the *musculo-cutaneous* nerves.

Lateral branches.—Some articular and cutaneous branches are derived from the external popliteal nerve before its final division.

ARTICULAR NERVES.—The articular branches are conducted to the outer side of the capsular ligament of the knee-joint by the upper and lower articular arteries of that side. They sometimes arise together, and the upper one occasionally springs from the great sciatic nerve before the bifurcation.

From the place of division of the external popliteal nerve, a *recurrent articular* nerve ascends through the tibialis anticus muscle with the recurrent artery to reach the fore part of the knee-joint.

Fig. 449.

Fig. 449.—CUTANEOUS NERVES OF THE OUTER SIDE OF THE LEG AND FOOT (from Sappey after Hirschfeld and Leveillé). ⅙

1, external popliteal nerve ; 2, its external cutaneous branch ; 3, communicating branch which unites with 4, that form the internal popliteal in 5, the external saphenous nerve ; 6, calcaneal branch of the external saphenous ; 7, external dorsal digital branch to the fifth toe ; 8, collateral dorsal digital branch of the fourth and fifth toes ; 9, musculo-cutaneous nerve ; 10, its cutaneous branches ; 11, loop of union with the external saphenous ; 12, union between its outer and inner branches ; 13, anterior tibial nerve, shown by the removal of a part of the muscles, and giving muscular branches superiorly ; 14, its terminal branch emerging in the space between the first and second toes, where it gives the collateral dorsal digital branches to their adjacent sides ; 15, branches to the peronei muscles.

CUTANEOUS NERVES.—The cutaneous branches, two or three in number, supply the skin on the back part and outer side of the leg.

The *peroneal communicating branch* (r. communicans fibularis), which joins the short saphenous nerve below the middle of the back of the leg, is the largest of these nerves. In some instances, it continues a separate branch and its cutaneous filaments reach down to the heel or on to the outside of the foot.

Another cutaneous branch extends along the outer side of the leg to the middle or lower part, sending offsets both backwards and forwards.

MUSCULO-CUTANEOUS NERVE.

The musculo-cutaneous (peroneal) nerve descends between the peronei muscles and the long extensor of the toes, and reaches the surface by perforating the fascia in the lower part of the leg on the anterior aspect. It then divides into two branches, distinguished as external and internal, which proceed to the toes. The two branches sometimes perforate the fascia at a different height.

(a) *Muscular branches* are given to the peroneus longus and peroneus brevis.

(b) *Cutaneous branches* given off before the final division are distributed to the lower part of the leg.

(c) The *internal branch* of the musculo-cutaneous nerve, passing forwards along the dorsum of the foot, furnishes one branch to the inner side of the great toe, and others to the contiguous sides of the second and third toes. It gives other offsets, which extend over the inner ankle and side of the foot. This nerve communicates with the long saphenous nerve on the inner side of the foot, and with the anterior tibial nerve between the first and second toes.

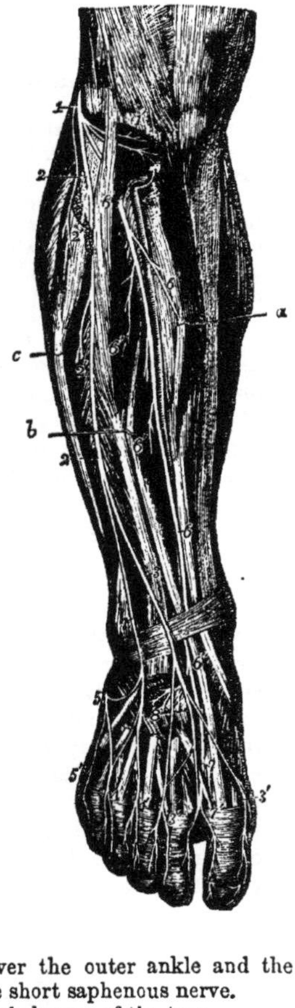

Fig. 450. — VIEW OF THE DISTRIBUTION OF THE BRANCHES OF THE EXTERNAL POPLITEAL NERVE IN THE FRONT OF THE LEG AND DORSUM OF THE FOOT (from Hirschfeld and Leveillé).

Fig. 450.

The upper part of the peroneus longus muscle has been removed, the tibialis anticus, the long extensor of the great toe and the peroneus longus have been drawn separate in the leg by hooks marked *a*, *b*, and *c*, and the tendons of the extensor muscles have been removed in the dorsum of the foot, to show the deeper seated nerves ; 1, the external popliteal or peroneal nerve winding round the other part of the fibula ; 1', its recurrent articular branches exposed by the dissection of the upper part of the tibialis anticus muscle ; 2, 2, the musculo-cutaneous nerve ; 2', 2', twigs to the long and short peroneal muscles; 3, internal branch of the musculo-cutaneous nerve ; 3', 3', its dorsal digital branches to the inside of the great toe, and to the adjacent sides of the second and third toes ; 4, the external branch ; 4', 4', its dorsal digital branches to the adjacent sides of the third and fourth toes, and in part to the space between the fourth and fifth toes ; 5, the external saphenous nerve descending on the outer border of the foot, and uniting at two places with the outer branch of the musculo-cutaneous ; 5', its branch to the outer side of the fifth toe ; 6, placed on the upper part of the extensor communis digitorum, marks the anterior tibial nerve passing beneath the muscles ; 6, placed farther down on the tendon of the tibialis anticus, points to the nerve as it crosses to the inside of the anterior tibial artery ; 6', its muscular branches in the leg ; 6", on the tendon of the extensor longus pollicis points to the anterior tibial nerve after it has passed into the foot behind that tendon ; 7, its inner branch uniting with a twig of the musculo-cutaneous, and giving the dorsal digital nerves to the adjacent sides of the first and second toes; 8, distribution of its outer branch to the extensor brevis digitorum and tarsal articulations.

(d) The *external branch*, larger than the internal one, descends over the foot towards the fourth toe, which, together with the contiguous borders of the third and fifth toes, it supplies with branches. Cutaneous nerves, derived from this branch, spread over the outer ankle and the outer side of the foot, where they are connected with the short saphenous nerve.

The dorsal digital nerves are continued on to the last phalanges of the toes.

The number of toes supplied by each of the two divisions of the musculo-cutaneous nerve is liable to vary ; together these nerves commonly supply all the toes on the dorsal aspect, excepting the outer side of the little toe, which receives a branch from the short saphenous nerve, and the adjacent sides of the great toe and the second toe, to which the anterior tibial nerve is distributed : with this latter branch, however, it generally communicates.

The anterior tibial (interosseous nerve), commencing between the fibula and the peroneus longus, inclines obliquely beneath the long extensor of the toes to the fore part of the interosseous membrane, and there comes into contact with the anterior tibial vessels, and with those vessels it descends to the front of the ankle-joint, where it divides into an external and an internal branch. The nerve first reaches the outer side of the anterior tibial artery, above the middle of the leg ; and, after crossing in front of that vessel once or oftener, lies to the inner side of it at the bend of the ankle.

(a) *Muscular branches.*—In its course along the leg, the anterior tibial nerve gives slender filaments to the muscles between which it is placed, namely, the tibialis anticus, the long extensor of the toes, and the special extensor of the great toe.

(b) The *external branch* of the anterior tibial nerve turns outwards over the tarsus beneath the short extensor of the toes ; and, having become enlarged (like the posterior interosseous nerve on the wrist) terminates in branches which supply the short extensor muscle, and likewise the articulations of the foot.

(c) The *internal branch*, continuing onwards in the direction of the anterior tibial nerve, accompanies the dorsal artery of the foot to the first interosseous space, and ends in two branches, which supply the integument on the neighbouring sides of the great toe and the second toe on their dorsal aspect. It communicates with the internal division of the musculo-cutaneous nerve.

Summary of the external popliteal nerve.—This nerve supplies, besides articular branches to the knee, ankle, and foot, the peronei muscles, extensor muscles of the foot, also the integument of the front of the leg and dorsum of the foot. It gives the ramus communicans fibularis to the short saphenous branch of the internal popliteal nerve, and communicates with the long saphenous nerve.

SYNOPSIS OF THE CUTANEOUS DISTRIBUTION OF THE CEREBRO-SPINAL NERVES.

HEAD.—The *face and head in front of the ear* are supplied with sensory nerves from the fifth cranial nerve. The ophthalmic division supplies branches to the forehead, upper eyelid, and dorsum of the nose. The superior maxillary division supplies the cheek, ala of the nose, upper lip, lower eyelid, and the region behind the eye, over the temporal fascia. The inferior maxillary division supplies the chin and lower lip, the pinna of the ear on its outer side, and the integument in front of the ear and upwards to the vertex of the head.

The *head, behind the ear,* is mainly supplied by the great occipital branch of the posterior division of the second spinal nerve, but above the occipital protuberance there is also distributed the branch from the posterior division of the third spinal nerve; and, in front of the area of the great occipital nerve, is a space supplied by anterior divisions of spinal nerves, viz., the back of the pinna of the ear, together with the integument behind and that in front over the parotid gland, which are supplied by the. great auricular nerve ; while between the area of that nerve and the great occipital the small occipital nerve intervenes. The auricular branch of the pneumo-gastric nerve also is distributed on the back of the ear.

TRUNK.—The *posterior divisions of the spinal nerves* supply an area, extending on the back from the vertex of the skull to the buttock. This area is narrow in the neck ; it is spread out over the back of the scapula ;

and on the buttock the distribution of the lumbar nerves extends to the trochanters.

The *area supplied by the cervical plexus*, besides extending upwards, as already mentioned, on the lateral part of the skull, stretches over the front and sides of the neck, and the upper part of the shoulder and breast.

The *area of the anterior divisions of the dorsal and first lumbar nerves* meets superiorly with that of the cervical plexus, and posteriorly with that of the posterior divisions of dorsal and lumbar nerves. It passes down over the haunch and along by the outer part of Poupart's ligament, and includes part of the scrotum and a small portion of the integument of the thigh internal to the saphenous opening.

The *perinæum* and *penis* are supplied by the pudic nerve ; the *scrotum* by branches of the pudic, inferior pudendal, and ilio-inguinal nerves.

UPPER LIMB.—The *shoulder*, supplied superiorly by the cervical plexus, receives its cutaneous nerves inferiorly as far as the insertion of the deltoid from the circumflex nerve.

The *arm* internally is supplied by the intercosto-humeral nerve and the nerve of Wrisberg. The inner and anterior part is supplied by the internal cutaneous nerve ; and the posterior and outer part by the internal and external branches of the musculo-spiral nerve.

The *forearm*, anteriorly and on the outer side, is supplied by the external cutaneous ; on its outer and posterior aspect, superiorly by the external cutaneous branches of the musculo-spiral, and inferiorly by the radial branch of the same nerve. On the inner side, both in front and behind, is the internal cutaneous nerve, and inferiorly are branches of the ulnar.

On the back of the hand are the radial and ulnar nerves, the radial supplying about three fingers and a half or less, and the ulnar one and a half or more.

On the front of the hand, the median nerve supplies three fingers and a half, and the ulnar one and a half. In the *palm* is a branch of the median given off above the wrist. On the *ball of the thumb* are branches of the musculo-cutaneous, median, and radial nerves.

LOWER LIMB.—The *buttock* is supplied from above by the cutaneous branches of the posterior divisions of the lumbar nerves, with the ilio-hypogastric and lateral branches of the last dorsal nerves ; internally by the posterior divisions of the sacral nerves ; externally by the posterior branch of the external cutaneous nerve proceeding from the front ; and inferiorly by branches of the small sciatic nerve proceeding from below.

The *thigh* is supplied externally by the external cutaneous nerve ; posteriorly, and in the upper half of its inner aspect, by the small sciatic ; anteriorly, and in the lower half of the inner aspect, by the middle and internal cutaneous.

The *leg* is supplied posteriorly by the small sciatic and short saphenous nerves ; internally by the long saphenous and branches of the internal cutaneous of the thigh ; and outside and in front by cutaneous branches of the external popliteal nerve, and by its musculo-cutaneous branch.

On the dorsum of the foot are the branches of the musculo-cutaneous, supplying all the toes with the exception of the adjacent sides of the first and second, which are supplied by the anterior tibial, and the outer side of the little toe, which, with the outer side of the foot, is supplied by the short saphenous nerve. The long saphenous is the cutaneous nerve on the inner side of the foot. .

The *sole of the foot* is supplied by the plantar nerves. The internal plantar nerve gives branches to three toes and a half; the external to the remaining one toe and a half.

SYNOPSIS OF THE MUSCULAR DISTRIBUTION OF THE CEREBRO-SPINAL NERVES.

MUSCLES OF THE HEAD AND FORE PART OF THE NECK.

The *muscles of the orbit* are mostly supplied by the third cranial nerve—the superior division of that nerve being distributed to the levator palpebræ and the superior rectus muscles; and the inferior division to the inferior and internal recti and the inferior oblique. The superior oblique muscle is supplied by the fourth nerve, the external rectus by the sixth; while the tensor tarsi has no special nerve apart from those of the orbicularis palpebrarum, which are derived from the facial.

The *superficial muscles of the face and scalp,* which are associated in their action as a group of muscles of expression, are supplied by the portio dura of the seventh cranial nerve; the retrahens auriculam and occipitalis muscles being supplied by the posterior auricular branch.

The *deep muscles of the face,* employed in mastication, viz., the temporal, masseter, buccinator, and two pterygoid muscles, are supplied by the inferior maxillary division of the fifth cranial nerve.

Muscles above the hyoid bone.—The mylo-hyoid muscle and anterior belly of the digastric are supplied by a special branch of the inferior maxillary division of the fifth cranial nerve; the posterior belly of the digastric muscle, and the stylo-hyoid, are supplied by branches of the portio dura. The genio-hyoid and the muscles of the tongue receive their nervous supply from the hypoglossal nerve.

The *muscles ascending to the hyoid bone and larynx,* viz., the sterno-hyoid, omo-hyoid, and sterno-thyroid, are supplied from the ramus descendens noni and its loop with the cervical plexus, while the thyro-hyoid muscle receives a separate twig from the ninth nerve.

The larynx, pharynx, and soft palate.—The crico-thyroid muscle is supplied by the external laryngeal branch of the pneumo-gastric nerve, and the other intrinsic muscles of the larynx by the recurrent laryngeal. The muscles of the pharynx are supplied principally by the pharyngeal branch of the pneumo-gastric; the stylo-pharyngeus, however, is supplied by the glosso-pharyngeal nerve. Of the muscles of the soft palate unconnected with the tongue or pharynx, the tensor palati receives its nerve from the otic ganglion (which also supplies the tensor tympani); the levator palati gets a twig (Meckel) from the posterior palatine branch of the spheno-palatine ganglion, and the azygos uvulæ is probably supplied from the same source.

MUSCLES BELONGING EXCLUSIVELY TO THE TRUNK, AND MUSCLES ASCENDING TO THE SKULL.

All those *muscles of the back* which are unconnected with the upper limb, viz., the posterior serrati, the splenius, complexus, erector spinæ, and the muscles more deeply placed, receive their supply from the posterior divisions of the spinal nerves.

The *sterno-mastoid* is supplied by the spinal accessory nerve and a twig of the cervical plexus coming from the second cervical nerve.

The *rectus capitis anticus major* and *minor* are supplied by twigs from the upper cervical nerves; the *longus colli* and *scaleni* muscles by twigs from the lower cervical nerves.

The *muscles of the chest*, viz , the intercostals, subcostals, levatores costarum, and triangularis sterni, are supplied by the intercostal nerves.

The *obliqui, transversus*, and *rectus* of the abdomen are supplied by the lower intercostal nerves; and the oblique and transverse muscles also get branches from the ilio-inguinal and ilio-hypogastric nerves. The *cremaster* muscle is supplied by the genital branch of the genito-crural nerve.

The *quadratus lumborum* (like the psoas) receives small branches from the lumbar nerves before they form the plexus.

The *diaphragm* receives the phrenic nerves from the fourth and fifth cervical nerves, and likewise sympathetic filaments from the plexuses round the phrenic arteries.

The *muscles of the urethra and penis* are supplied by the pudic nerve; the *levator and sphincter ani* by the pudic and by the fourth and fifth sacral and the coccygeal nerves; and the *coccygeus* muscle by the three last named nerves.

MUSCLES ATTACHING THE UPPER LIMB TO THE TRUNK.

The *trapezius* and the *sterno-cleido-mastoid* receive the distribution of the *spinal accessory* nerve, and, in union with it, filaments from the cervical plexus.

The *latissimus dorsi* receives the long subscapular nerve.

The *rhomboidei* are supplied by a special branch from the anterior division of the fifth cervical nerve.

The *levator anguli scapulæ* is supplied by branches from the anterior division of the third cervical nerve, and sometimes partly also by the branch to the rhomboid muscles.

The *serratus magnus* has a special nerve, the posterior thoracic, derived from the fifth and sixth cervical nerves.

The *subclavius* receives a special branch from the place of union of the fifth and sixth cervical nerves.

The *pectorales* are supplied by the anterior thoracic branches of the brachial plexus, the larger muscle receiving filaments from both these nerves, and the smaller from the inner only.

MUSCLES OF THE UPPER LIMB.

Muscles of the shoulder.—The supraspinatus and infraspinatus are supplied by the suprascapular nerve; the subscapularis by the two smaller subscapular nerves; the teres major by the second subscapular, and the deltoid and teres minor by the circumflex nerve.

Posterior muscles of the arm and forearm.—The triceps, anconeus, supinator longus, and extensor carpi radialis longior are supplied by direct branches of the musculo-spiral nerve; while the extensor carpi radialis brevior and the other extensor muscles in the forearm receive their branches from the posterior interosseous division of that nerve.

Anterior muscles of the arm and forearm.—The coraco-brachialis, biceps, and brachialis anticus are supplied by the musculo-cutaneous nerve : the brachialis anticus likewise generally receives a twig from the musculo-spiral nerve. The muscles in front of the forearm are supplied by the median nerve, with the exception of the flexor carpi ulnaris and the ulnar half of the flexor profundus digitorum, which are supplied by the ulnar nerve, and the supinator longus, which is supplied by the musculo-spiral.

Muscles of the hand.—The abductor and opponens pollicis, the outer half of the flexor brevis pollicis, and the two outer lumbricales muscles, are supplied by the median nerve : all the other muscles receive their nerves from the ulnar.

MUSCLES OF THE LOWER LIMB.

Posterior muscles of the hip and thigh.—The gluteus maximus is mainly supplied by the small sciatic nerve, and receives at its upper part a separate branch from the sacral plexus.   The gluteus medius and minimus, together with the tensor vaginæ femoris, are supplied by the gluteal nerve.   The pyriformis, gemelli, obturator internus, and quadratus femoris receive special branches from the sacral plexus.   The hamstring muscles are supplied by branches from the great sciatic nerve.

Anterior and internal muscles of the thigh.—The psoas muscle is supplied by separate twigs from the lumbar nerves.   The iliacus, quadriceps extensor femoris, and sartorius are supplied by the anterior crural nerve.   The adductor muscles, the obturator externus and the pectineus, are supplied by the obturator nerve, but the adductor magnus likewise receives a branch from the great sciatic, and the pectineus sometimes has a branch from the anterior crural.

Anterior muscles of the leg and foot.—The muscles in front of the leg, together with the extensor brevis digitorum, are supplied by the anterior tibial nerve.

The *peroneus longus* and *brevis* are supplied by the musculo-cutaneous nerve.

Posterior muscles of the leg.—The gastrocnemius, plantaris, soleus, and popliteus are supplied by branches from the internal popliteal nerve ; the deep muscles, viz., the flexor longus digitorum, flexor longus pollicis, and tibialis posticus, derive their nerves from the posterior tibial.

Plantar muscles.—The flexor brevis digitorum, the abductor and flexor brevis pollicis, and the two inner lumbricales, are supplied by the internal plantar nerve ; all the others, including the flexor accessorius and interossei, are supplied by the external plantar nerve.

III. SYMPATHETIC NERVES.

The nerves of the sympathetic system (nervus intercostalis ; nerves of organic life—Bichat) are distributed in general to all the internal viscera, but some organs receive their nerves also from the cerebro-spinal system, as the lungs, the heart, and the upper and lower parts of the alimentary canal.   It appears from physiological researches to be also the special province of the sympathetic system to supply nerves to the coats of the blood-vessels.

This division of the nervous system consists of a somewhat complicated collection of ganglia, cords and plexuses, the parts of which may, for convenience, be classified in three groups, viz., the principal gangliated cords, the great prevertebral plexuses with the nerves proceeding from them, and the ganglia of union with cranial nerves.

The *gangliated cords* consist of two series, in each of which the ganglia are connected by intervening cords.   These cords are placed symmetrically in

front of the vertebral column, and extend
from the base of the skull to the coccyx.
Superiorly they are connected with plexuses which enter the cranial cavity, while
inferiorly they converge on the sacrum,
and terminate in a single ganglion on the
coccyx. The several portions of the cords
are distinguished as cervical, dorsal, lumbar, and sacral, and in each of these parts
the ganglia are equal in number, or nearly
so, to the vertebræ on which they lie,
except in the neck, where there are only
three.

Fig. 451.—DIAGRAMMATIC OUTLINE OF THE SYM-
PATHETIC CORD OF ONE SIDE IN CONNECTION
WITH THE SPINAL NERVES.

The full description of this figure will be found
at p. 629.

On the right side the following letters indicate parts of the sympathetic nerves; viz.
a, the superior cervical ganglion, communicating with the upper cervical spinal nerves
and continued below into the great sympathetic
cord; *b*, the middle cervical ganglion; *c, d*, the
lower cervical ganglion united with the first
dorsal; *d'*, the eleventh dorsal ganglion; from
the fifth to the ninth dorsal ganglia the origins
of the great splanchnic nerve are shown; *l*, the
lowest dorsal or upper lumbar ganglion; *ss*, the
upper sacral ganglion. In the whole extent of the
sympathetic cord, the twigs of union with the
spinal nerves are shown.

*Connection of the gangliated cords with
the cerebro-spinal system.*—The ganglia are
severally connected with the spinal nerves
in their neighbourhood by means of short
cords; each connecting cord consisting of
a white and a grey portion, the former
of which may be considered as proceeding
from the spinal nerve to the ganglion,
the latter from the ganglion to the spinal
nerve. At its upper end the gangliated
cord communicates likewise with certain
cranial nerves. The main cords intervening between the ganglia, like the smaller
ones connecting the ganglia with the spinal
nerves, are composed of a grey and a white
part, the white being continuous with the
fibres of the spinal nerves prolonged to the
ganglia.

The *great prevertebral plexuses* comprise
three large aggregations of nerves, or
nerves and ganglia situated in front of
the spine, and occupying respectively the
thorax, the abdomen, and the pelvis.
They are single and median, and are

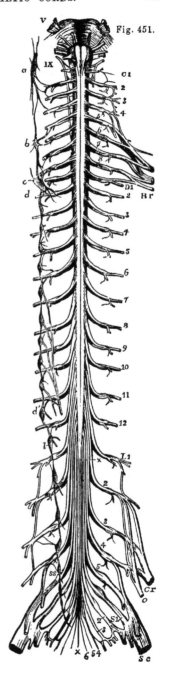

Fig. 451.

named respectively the cardiac, the solar, and the hypogastric plexus. These plexuses receive branches from both the gangliated cords above noticed, and they constitute centres from which the viscera are supplied with nerves.

The *cranial ganglia of the sympathetic* are the ophthalmic, spheno-palatine, submaxillary, and otic, which, being intimately united with the fifth cranial nerve, have already been described along with that nerve. They are also more or less directly connected with the upper end of the sympathetic gangliated cords; but it will be unnecessary to give any special description of them in this place.

A. THE GANGLIATED CORDS.

THE CERVICAL PART.

In the neck, each gangliated cord is deeply placed behind the sheath of the great cervical blood-vessels, and in contact with the muscles which immediately cover the fore part of the vertebral column. It comprises three ganglia, the first of which is placed near the base of the skull, the second in the lower part of the neck, and the third immediately above the head of the first rib.

THE UPPER CERVICAL GANGLION.

This is the largest ganglion of the great sympathetic cord. It is continued superiorly into an ascending branch, and tapers below into the connecting cord, so as to present usually a fusiform shape; but there is considerable variety in this respect in different cases, the ganglion being occasionally broader than usual, and sometimes constricted at intervals. It has the reddish-grey colour characteristic of the ganglia of the sympathetic system. It is placed on the larger rectus muscle, opposite the second and third cervical vertebræ, and behind the internal carotid artery.

Connection with spinal nerves.—At its outer side the superior cervical ganglion is connected with the first four spinal nerves, by means of slender cords, which have the structure pointed out in the general description as being common to the series.

The circumstance of this ganglion being connected with so many as four spinal nerves, together with its occasionally constricted appearance, is favourable to the view that it may be regarded as consisting of several ganglia which have coalesced.

Connection with cranial nerves.—Small twigs connect the ganglion or its cranial cord with the second ganglion of the pneumo-gastric, and with the ninth cranial nerve, near the base of the skull; and another branch, which is directed upwards from the ganglion, divides at the base of the skull into two filaments, one of which ends in the second (petrosal) ganglion of the glosso-pharyngeal nerve; while the other, entering the jugular foramen, joins the ganglion of the root of the pneumo-gastric.

Besides the branches connecting it with cranial and spinal nerves, the first cervical ganglion gives off also the ascending branch, the upper cardiac nerve, pharyngeal nerves, and branches to blood-vessels.

1. ASCENDING BRANCH AND CRANIAL PLEXUSES.

The ascending branch of the first cervical ganglion is soft in texture and of a reddish tint, seeming to be in some degree a prolongation of the ganglion itself. In its course to the skull, it is concealed by the internal carotid artery, with which it enters the carotid canal in the temporal bone, and it is then divided into two parts, which are placed one on the outer, the other on the inner side of the vessel.

Fig. 452

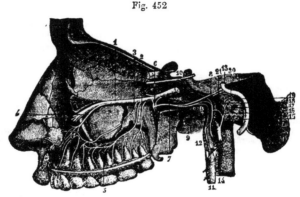

Fig. 452.—CONNECTIONS OF THE SYMPATHETIC NERVE THROUGH ITS CAROTID BRANCH WITH SOME OF THE CRANIAL NERVES.

The full description of this figure will be found at p. 602. The following numbers refer to sympathetic nerves and their connections :—6, spheno-palatine ganglion ; 7, Vidian nerve ; 9, its carotid branch ; 10, a part of the sixth nerve, receiving twigs from the carotid plexus of the sympathetic ; 11, superior cervical sympathetic ganglion ; 12, its prolongation in the carotid branch ; 15, anastomosing nerve of Jacobson ; 16, twig uniting it to the sympathetic.

The *external division* distributes filaments to the internal carotid artery, and, after communicating by means of other filaments with the internal division of the cord, forms the *carotid plexus*.

The *inner division*, rather the smaller of the two, supplies filaments to the carotid artery, and goes to form the *cavernous plexus*. The terminal parts of these divisions of the cranial cord are prolonged on the trunk of the internal carotid, and extend to the cerebral and ophthalmic arteries, around which they form secondary plexuses, those on the cerebral artery ascending to the pia mater. One minute plexus enters the eye-ball with the central artery of the retina.

It was stated by Ribes (Mem. de la Société Méd. d'Emulation, tom. viii. p. 606,) that the cranial prolongations of the sympathetic nerve from the two sides coalesce with one another on the anterior communicating artery,—a small ganglion or a plexus being formed at the point of junction ; but this connection has not been satisfactorily made out by other observers.

CAROTID PLEXUS.—The carotid plexus, situated on the outer side of the internal carotid artery at its second bend (reckoning from below), or between the second and third bends, joins the fifth and sixth cranial nerves, and gives many filaments to the vessel on which it lies.

Branches.—(a) The connection with the *sixth nerve* is established by means of one or two filaments of considerable size, which are supplied to that nerve where it lies by the side of the internal carotid artery.

(*b*) The filaments connected with the *Gasserian ganglion* of the fifth nerve proceed sometimes from the carotid plexus, at others from the cavernous.

(*c*) The *deep branch* of the Vidian nerve passes backwards to the carotid plexus, and after leaving the Vidian canal, lies in the cartilaginous substance which closes the foramen lacerum medium. Valentin describes nerves as furnished to the dura mater from the carotid plexus.

CAVERNOUS PLEXUS.—The cavernous plexus, named from its position in the sinus of the same name, is placed below and rather to the inner side of the highest turn of the internal carotid artery. Besides giving branches on the artery, it communicates with the third, the fourth and the ophthalmic of the fifth cranial nerves.

Branches.—(*a*) The filament which joins the *third nerve* comes into connection with it close to the point of division of that nerve.

(*b*) The branch to the *fourth nerve*, which may be derived from either the cavernous or the carotid plexus, joins the nerve where it lies in the wall of the cavernous sinus.

(*c*) The filaments connected with the *ophthalmic trunk* of the *fifth* nerve are supplied to its inner surface. One of them is continued forwards to the lenticular ganglion, either in connection with or distinct from the nasal nerve.

2. PHARYNGEAL NERVES AND PLEXUS.

These nerves arise from the inner part of the ganglion, and are directed obliquely inwards to the side of the pharynx. Opposite the middle constrictor muscle they unite with branches of the pneumo-gastric and glossopharyngeal nerves ; and by their union with those nerves the *pharyngeal plexus* is formed. Branches emanating from the plexus are distributed to the muscles and mucous membrane of the pharynx.

3. UPPER CARDIAC NERVE.

Each of the cervical ganglia of the sympathetic furnishes a cardiac branch, the three being named respectively the upper, middle and lower cardiac nerves.

These branches are continued singly, or in connection, to the large prevertebral centre (cardiac plexus) of the thorax. Their size varies considerably, and where one branch is smaller than common, another will be found to be increased in size, as if to compensate for the defect. There are some differences in the disposition of the nerves of the right and left sides.

The *upper cardiac nerve* (n. cardiacus superficialis) of the *right side* proceeds from two or more branches of the ganglion, with, in some instances, an offset from the cord connecting the first two ganglia. In its course down the neck the nerve lies behind the carotid sheath, in contact with the longus colli muscle ; and it is placed in front of the lower thyroid artery and the recurrent laryngeal nerve. Entering the thorax, it passes in some cases before, in others behind the subclavian artery, and is directed along the innominate artery to the back part of the arch of the aorta, where it ends in the deep cardiac plexus, a few small filaments continuing also to the front of the great vessel. Some branches distributed to the thyroid body accompany the inferior thyroid artery.

In its course downwards the cardiac nerve is repeatedly connected with other branches of the sympathetic, and with the pneumo-gastric nerve. Thus about the middle of the neck it is joined by some filaments from the external laryngeal nerve; and, rather lower down, by one or more filaments from the trunk of the pneumo-gastric nerve; lastly, on entering the chest, it joins with the recurrent laryngeal.

Fig. 453.

Fig. 453.—CONNECTIONS OF THE CERVICAL AND UPPER DORSAL SYMPATHETIC GANGLIA AND NERVES ON THE LEFT SIDE.

The full description of this figure will be found at p. 620. The following numbers refer to the sympathetic ganglia and nerves, and those immediately connected with them :—3, pharyngeal plexus; 8, laryngeal plexus; 13, pulmonary plexus; and to the reader's left, above the pulmonary artery, a part of the cardiac plexus; 24, superior cervical ganglion of the sympathetic; 25, middle cervical ganglion; 26, inferior cervical ganglion united with the first dorsal ganglion; 27, 28, 29, 30, second, third, fourth, and fifth dorsal ganglia.

Variety.—Instead of passing to the thorax in the manner described, the superior cardiac nerve may join the cardiac branch furnished from one of the other cervical ganglia. Scarpa describes this as the common disposition of the nerve; but Cruveilhier (Anat. Descript., t. iv.) states that he has not in any case found the cardiac nerves to correspond exactly with the figures of the "Tabulæ Neurologicæ."

The superficial cardiac nerve of the *left side* has, while in the neck, the same course and connections as that of the right side.   But within the chest it follows the left carotid artery to the arch of the aorta, and ends in some instances in the superficial cardiac plexus, while in others it joins the deep plexus ; and accordingly it passes either in front of or behind the arch of the aorta.

4. BRANCHES TO BLOOD-VESSELS.

The nerves which ramify on the arteries (nervi molles) spring from the front of the ganglion, and twine round the trunk of the carotid artery. They are prolonged on each branch of the external carotid, and form slender plexuses upon them.

Communications with other nerves.—From the plexus on the facial artery is derived the filament which joins the submaxillary ganglion ; and, from that on the middle meningeal artery, twigs have been described as extending to the otic ganglion, as well as to the gangliform enlargement of the facial nerve.   Lastly, a communication is established between the plexus on the carotid artery, and the digastric branch of the facial nerve.

Small *ganglia* are occasionally found on some of the vascular plexuses, close to the origin of the vessels with which they are associated.   Thus lingual, temporal, and pharyngeal ganglia have been described ; and besides these there is a larger body, the ganglion intercaroticum, placed· on the inner side of the angle of division of the common carotid artery.   This body, long known to anatomists as a ganglion, has been stated by Luschka to have a structure very different from the nervous ganglia in general, and has been named by him the "glandula intercarotica."

The *ganglion intercaroticum* was described by Luschka as presenting principally a follicular structure, and regarded by him as being of a nature similar to the glandula coccygea, which he had previously discovered.   It appears, however, from the researches of Julius Arnold, that the follicular appearances observed by Luschka, both in this instance and in the coccygeal gland, were produced by arterial glomeruli seen in section ; and that the ganglion intercaroticum consists of numbers of those glomeruli gathered into several larger masses, and of dense plexuses of nerves surrounding respectively the glomeruli, the masses, and the whole structure.   Within those plexuses nerve-cells are scattered, but not in very great number.   The ganglion is usually about one-fourth of an inch long ; but, according to Luschka, may be divided into small separate masses, and thus escape attention, or be supposed to be absent.— (Luschka, Anat. d. Menschen, vol. i. 1862 ; and Julius Arnold, in Virchow's Archiv., June, 1865.)

MIDDLE CERVICAL GANGLION.

The middle ganglion (ganglion thyroideum), much the smallest of the cervical ganglia, is placed on or near the inferior thyroid artery.   It is usually connected with the fifth and sixth spinal nerves, but in a somewhat variable manner.   It gives off thyroid branches and the middle cardiac nerve.

THYROID BRANCHES.—From the inner side of the ganglion some twigs proceed along the inferior thyroid artery to the thyroid body, where they join the recurrent laryngeal and the external laryngeal nerves.   Whilst on the artery, these branches communicate with the upper cardiac nerve.

THE MIDDLE CARDIAC NERVE (nervus cardiacus profundus v. magnus) of the *right side* is prolonged to the chest behind the sheath of the common carotid artery, and either in front of or behind the subclavian artery.   In the chest it lies on the trachea, where it is joined by filaments of the recurrent laryngeal nerve, and it ends in the right side of the deep cardiac plexus. While in the neck, the nerve communicates with the upper cardiac nerve and the recurrent branch of the pneumo-gastric.

On the *left side*, the middle cardiac nerve enters the chest between the left carotid and subclavian arteries, and joins the left side of the deep cardiac plexus.

When the middle cervical ganglion is small, the middle cardiac nerve may be found to be an offset of the inter-ganglionic cord.

LOWER CERVICAL GANGLION.

The lower or third cervical ganglion is irregular in shape, usually somewhat flattened and round or semilunar, and is frequently united in part to the first thoracic ganglion. Placed in a hollow between the transverse process of the last cervical vertebra and the neck of the first rib, it is concealed by the vertebral artery. It is connected by short communicating cords with the two lowest cervical nerves. Numerous branches are given off from it, among which the largest is the lower cardiac nerve.

THE LOWER CARDIAC NERVE, issuing from the third cervical ganglion or from the first thoracic, inclines inwards on the *right side*, behind the subclavian artery, and terminates in the cardiac plexus behind the arch of the aorta. It communicates with the middle cardiac and recurrent laryngeal nerves behind the subclavian artery.

On the *left side*, the lower cardiac often becomes blended with the middle cardiac nerve, and the cord resulting from their union terminates in the deep cardiac plexus.

BRANCHES TO BLOOD-VESSELS.—From the lowest cervical and first dorsal ganglia a few slender branches ascend along the vertebral artery in its osseous canal, forming a plexus round the vessel by their inter-communications, and supplying it with offsets. This plexus is connected with the cervical spinal nerves as far upwards as the fourth.

One or two branches frequently pass from the lower cervical ganglion to the first dorsal ganglion in front of the subclavian artery, forming loops round the vessel (ansæ Vieussenii), and supplying it with small offsets.

THORACIC PART OF THE GANGLIATED CORD.

In the thorax the gangliated cord is placed towards the side of the spinal column, in a line passing over the heads of the ribs. It is covered by the pleura, and crosses the intercostal blood-vessels.

Opposite the head of each rib the cord usually presents a ganglion, so that there are commonly twelve in all ; but, from the occasional coalescence of two, the number varies slightly. The first ganglion when distinct is larger than the rest, and is of an elongated form ; but it is often blended with the lower cervical ganglion. The rest are small, generally oval, but very various in form.

Connection with the spinal nerves.—The branches of connection between the spinal nerves and the ganglia of the sympathetic are usually two in number for each ganglion ; one of these generally resembling the spinal nerve in structure, the other more similar to the sympathetic nerve.

BRANCHES OF THE GANGLIA.

The branches furnished by the *first five or six ganglia* are small, and are

Fig. 454

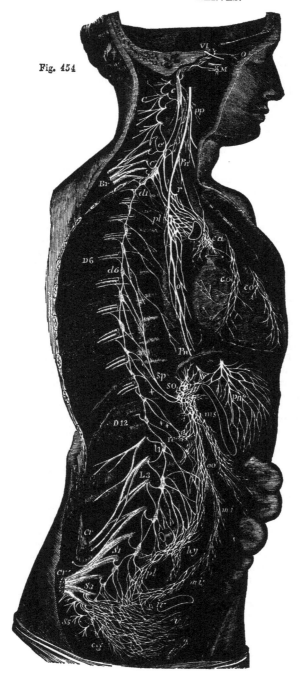

Fig. 454.—Diagrammatic View of the Sympathetic Cord of the Right Side, showing its Connections with the Principal Cerebro-Spinal Nerves and the Main Preaortic Plexuses. ¼

Cerebro-spinal Nerves.—VI, a portion of the sixth cranial nerve as it passes through the cavernous sinus, receiving two twigs from the carotid plexus of the sympathetic nerve; O, ophthalmic ganglion connected by a twig with the carotid plexus; M, connection of the spheno-palatine ganglion by the Vidian nerve with the carotid plexus; C, cervical plexus; Br, brachial plexus; D 6, sixth intercostal nerve; D 12, twelfth; L 3, third lumbar nerve; S 1, first sacral nerve; S 3, third; S 5, fifth; Cr, anterior crural nerve; Cr', great sciatic; *pn*, pneumo-gastric nerve in the lower part of the neck; *r*, recurrent nerve winding round the subclavian artery.

Sympathetic Cord.—*c*, superior cervical ganglion; *c'*, second or middle; *c"*, inferior; from each of these ganglia cardiac nerves (all deep on this side) are seen descending to the cardiac plexus; *d* 1, placed immediately below the first dorsal sympathetic ganglion: *d* 6, is opposite the sixth; *l* 1, first lumbar ganglion; *c g*, the terminal or coccygeal ganglion.

Preaortic and Visceral Plexuses.—*p p*, pharyngeal, and, lower down, laryngeal plexus; *p l*, posterior pulmonary plexus spreading from the pneumo-gastric on the back of the right bronchus; *c a*, on the aorta, the cardiac plexus, towards which, in addition to the cardiac nerves from the three cervical sympathetic ganglia, other branches are seen descending from the pneumo-gastric and recurrent nerves; *c o*, right or posterior, and *c o'*, left or anterior coronary plexus; *o*, œsophageal plexus in long meshes on the gullet; *s p*, great splanchnic nerve formed by branches from the fifth, sixth, seventh, eighth, and ninth dorsal ganglia; +, small splanchnic from the ninth and tenth; + +, smallest or third splanchnic from the eleventh: the first and second of these are shown joining the solar plexus, *s o ;* the third descending to the renal plexus, *r e ;* connecting branches between the solar plexus and the pneumo-gastric nerves are also represented; *p n'*, above the place where the right pneumo-gastric passes to the lower or posterior surface of the stomach; *p n"*, the left distributed on the anterior or upper surface of the cardiac portion of the organ: from the solar plexus large branches are seen surrounding the arteries of the cœliac axis, and descending to *m s*, the superior mesenteric plexus; opposite to this is an indication of the suprarenal plexus; below *r e* (the renal plexus), the spermatic plexus is also indicated; *a o*, on the front of the aorta, marks the aortic plexus, formed by nerves descending from the solar and superior mesenteric plexuses and from the lumbar ganglia; *m i*, the inferior mesenteric plexus surrounding the corresponding artery; *h y*, hypogastric plexus placed between the common iliac vessels, connected above with the aortic plexus, receiving nerves from the lower lumbar ganglia, and dividing below into the right and left pelvic or inferior hypogastric plexuses; *p l*, the right pelvic plexus; from this the nerves descending are joined by those from the plexus on the superior hemorrhoidal vessels, *m i'*, by sympathetic nerves from the sacral ganglia, and by numerous visceral nerves from the third and fourth sacral spinal nerves, and there are thus formed the rectal, vesical, and other plexuses, which ramify upon the viscera from behind forwards and from below upwards, as towards *i r*, and *v*, the rectum and bladder.

distributed in a great measure to the thoracic aorta, the vertebræ, and ligaments. Several of these branches enter the posterior pulmonary plexus.

The branches furnished by the *lower six or seven ganglia* unite into three cords on each side, which pass down to join plexuses in the abdomen, and are distinguished as the great, the small, and the smallest splanchnic nerve.

THE GREAT SPLANCHNIC NERVE.

This nerve is formed by the union of small cords (roots) given off by the thoracic ganglia from the fifth or sixth to the ninth or tenth inclusive. By careful examination of specimens after immersion in acetic or diluted nitric acid, small filaments may be traced from the splanchnic roots upwards as far as the third ganglion, or even as far as the first (Beck, in the "Philosophical Transactions," Part 2, for 1846).

Gradually augmented by the successive addition of the several roots, the cord descends obliquely inwards over the bodies of the dorsal vertebræ; and, after perforating the crus of the diaphragm at a variable point, terminates in the semilunar ganglion, frequently sending some filaments to the renal plexus and the suprarenal body.

The splanchnic nerve is remarkable from its white colour and firmness, which are owing to the preponderance of the spinal nerve-fibres in its composition.

In the chest the great splanchnic nerve is not unfrequently divided into parts, and forms a plexus with the small splanchnic nerve. Occasionally also a small ganglion (ganglion splanchnicum) is formed on it over the last dorsal vertebra, or the last but one; and when it presents a plexiform arrangement, several small ganglia have been observed on its divisions.

In eight instances out of a large number of bodies, Wrisberg observed a fourth splanchic nerve (nervus splanchnicus supremus). It is described as formed by offsets from the cardiac nerves, and from the lower cervical as well as some of the upper thoracic ganglia. ("Observ. Anatom. de Nerv. Viscerum particula prima," p. 25, sect. 3.)

SMALL SPLANCHNIC NERVE.

The small or second splanchnic nerve springs from the tenth or eleventh ganglia, or from the neighbouring part of the cord. It passes along with the preceding nerve, or separately through the diaphragm, and ends in the cœliac plexus. In the chest this nerve often communicates with the large splanchnic nerve; and in some instances it furnishes filaments to the renal plexus, especially if the lowest splanchnic nerve is very small or wanting.

SMALLEST SPLANCHNIC NERVE.

This nerve (nerv. renalis posterior—Walter) arises from one of the lowest thoracic ganglia, and communicates sometimes with the nerve last described. After piercing the diaphragm, it ends in the renal plexus, and in the inferior part of the cœliac plexus.

LUMBAR PART OF THE GANGLIATED CORD.

In the lumbar region the two gangliated cords approach one another more nearly than in the thorax. They are placed before the bodies of the vertebræ, each lying along the inner margin of the psoas muscle; and that of the right side is partly covered by the vena cava.

The ganglia are small, and of an oval shape. They are commonly four in number, but occasionally, when their number is diminished, they are of larger size.

Connection with spinal nerves.—In consequence of the greater distance at which the lumbar ganglia are placed from the intervertebral foramina, the branches of connection with the spinal nerves are longer than in other parts of the gangliated cord. There are generally two connecting branches for each ganglion, but the number is not so uniform as it is in the chest; nor are those belonging to any one ganglion connected always with the same spinal nerve. The connecting branches accompany the lumbar arteries, and, as they cross the bodies of the vertebræ, are covered by the fibrous bands which give origin to the larger psoas muscle.

BRANCHES.—The branches of these ganglia are uncertain in their number. Some join a plexus on the aorta; others descending go to form the hypogastric plexus. Several filaments are distributed to the vertebræ and the ligaments connecting them.

SACRAL PART OF THE GANGLIATED CORD.

Over the sacrum the gangliated cord of the sympathetic nerve is much diminished in size, and gives but few branches to the viscera. Its position on the front of the sacrum is along the inner side of the anterior sacral foramina ; and, like the two series of those foramina, the two cords approach one another in their progress downwards. The upper end of each is connected with the last lumbar ganglion by a single or a double interganglionic cord ; and at the lower end, they are connected by means of a loop with a single median ganglion, *ganglion impar*, placed on the fore part of the coccyx. The sacral ganglia are usually five in number ; but the variation both in size and number is more marked in these than in the thoracic or lumbar ganglia.

Connection with spinal nerves.—From the proximity of the sacral ganglia to the spinal nerves at their emergence from the foramina, the communicating branches are very short : there are usually two for each ganglion, and these are in some cases connected with different sacral nerves. The coccygeal nerve communicates with the last sacral, or the coccygeal ganglion.

Branches.—The branches proceeding from the sacral ganglia are much smaller than those from other ganglia of the cord. They are for the most part expended on the front of the sacrum, and join the corresponding branches from the opposite side. Some filaments from one or two of the first ganglia enter the hypogastric plexus, while others go to form a plexus on the middle sacral artery. From the loop connecting the two cords on which the coccygeal ganglion is formed, filaments are given to the coccyx and the ligaments about it, and to the coccygeal gland.

COCCYGEAL GLAND.

Under this name has been described by Luschka a minute structure, which has since received the attention of a number of writers. It is usually, according to Luschka, of the size of a lentil, and sometimes as large as a small pea ; its colour is reddish grey ; its surface lobulated ; and it occupies a hollow at the tip of the coccyx, between the tendons attached to that part. It receives terminal twigs of the middle sacral artery and minute filaments from the ganglion impar. It consists of an aggregation of grains or lobules, which in some instances remain separate one from another. These lobules are principally composed of thick-walled cavities of vesicular and tubular appearance, described by Luschka and subsequent writers as closed follicles filled with cellular contents, but recently demonstrated by Julius Arnold to be clumps of dilated and tortuous small arteries, with thickened muscular and epithelial coats. Nerve-cells are found scattered in the stroma of the organ.

The coccygeal gland is a structure evidently of a similar nature to the ganglion intercaroticum, the principal differences apparently being, that the glomeruli of the ganglion intercaroticum are produced principally by the convolution and ramification of arterial twigs, while in the coccygeal gland there is dilation of the branches and thickening of their walls ; and that the nervous element 'is more developed in the intercarotid ganglion than in the coccygeal gland. Arnold, with Luschka, appears inclined to consider both structures as allied in nature to the suprarenal capsules. According to Arnold, there is always a number of small grape-like appendages on the coccygeal part of the middle sacral artery, microscopic in size, but similar in nature to the lobules of which the coccygeal gland is composed. (Luschka, " Der Hirnanhang und die Steissdruse des Menschen." Berlin, 1860. Also " Anat. d. Mensch.," vol. ii. part 2, p. 187. Julius Arnold in Virchow's " Archiv," March, 1865.)

B. THE GREAT PLEXUSES OF THE SYMPATHETIC.

Under this head may be included certain large plexuses of nerves placed further forwards in the visceral cavity than the gangliated cords, and furnishing branches to the viscera. The principal of these plexuses are the cardiac, the solar, and the hypogastric with the pelvic plexuses prolonged from it. They are composed of assemblages of nerves, or of nerves and ganglia, and from them smaller plexuses are derived.

CARDIAC PLEXUS.

This plexus receives the cardiac branches of the cervical ganglia and those of the pneumo-gastric nerves, and from it proceed the nerves which supply the heart, besides some offsets which contribute to the nervous supply of the lungs. It lies upon the aorta and pulmonary artery, where these vessels are in contact, and in its network are distinguished two parts, the superficial and the deep cardiac plexuses, the deep plexus being seen behind the vessels, and the superficial more in front, but both being closely connected. The branches pass from these plexuses chiefly forward in two bundles, accompanying the coronary arteries.

SUPERFICIAL CARDIAC PLEXUS.

The superficial cardiac plexus lies in the concavity of the arch of the aorta, in front of the right branch of the pulmonary artery. In it the superficial or first cardiac nerve of the sympathetic of the left side terminates, either wholly or in part, together with the lower cardiac branch of the left pneumo-gastric nerve, and in some cases also that of the right side. In the superficial plexus a small ganglion, the *ganglion of Wrisberg*, is frequently found at the point of union of the nerves. Besides ending in the anterior coronary plexus, the superficial cardiac furnishes laterally filaments along the pulmonary artery to the anterior pulmonary plexus of the left side.

DEEP CARDIAC PLEXUS.

The deep cardiac plexus, much larger than the superficial one, is placed behind the arch of the aorta, between it and the end of the trachea, and above the point of division of the pulmonary artery.

This plexus receives all the cardiac branches of the cervical ganglia of the sympathetic nerve, except the first or superficial cardiac nerve of the left side. It likewise receives the cardiac nerves furnished by the vagus and by the recurrent laryngeal branch of that nerve, with the exception of the left lower cardiac nerve.

Of the branches from the *right side* of the plexus, the greater number descend in front of the right pulmonary artery, and join branches from the superficial part in the formation of the anterior coronary plexus, while the rest, passing behind the right pulmonary artery, are distributed to the right auricle of the heart, and a few filaments are continued into the posterior coronary plexus.

On the *left side*, a few branches pass forwards by the side of the ductus arteriosus to join the superficial cardiac plexus; but the great majority end in the posterior coronary plexus.

The deep cardiac plexus sends filaments to the anterior pulmonary plexus on each side.

Coronary Plexuses.—The *anterior coronary plexus*, formed at first from the fibres of the superficial cardiac plexus, passes forwards between the aorta and pulmonary artery, and, having received an accession of fibres from the deep cardiac plexus, follows the course of the left or anterior coronary artery.

The *posterior coronary plexus*, derived chiefly from the left part of the deep cardiac plexus, but joined by nerves from the right portion of that plexus, surrounding the branches of the right coronary artery accompany them to the back of the heart.

Nervous filaments ramify in great number under the lining membrane of the heart. They are not so easily distinguished in man as in some animals. In the heart of the calf or lamb they are distinctly seen without dissection, running in lines which cross obliquely the muscular fibres. Remak was the first to observe that these branches are furnished with small ganglia, both on the surface and in the muscular substance. (Müller's "Archiv," 1844.)

SOLAR OR EPIGASTRIC PLEXUS.

The solar or epigastric plexus, which is the largest of the prevertebral centres, is placed at the upper part of the abdomen, behind the stomach, and in front of the aorta and the pillars of the diaphragm. Surrounding the origin of the cœliac axis and the upper mesenteric artery, it occupies the interval between the suprarenal bodies, and extends downwards as far as the pancreas. The plexus consists of nervous cords, with several ganglia of various sizes connected with them. The large splanchnic nerves on both sides, and some branches of the pneumo-gastric, terminate in it. The branches given off from it are very numerous, and accompany the arteries to the principal viscera of the abdomen, constituting so many secondary plexuses on the vessels. Thus diaphragmatic, cœliac, renal, mesenteric, and other plexuses are recognised, which follow the corresponding arteries.

Semilunar ganglia.—The solar plexus contains, as already mentioned, several ganglia ; and by the presence of these bodies, and their size, it is distinguished from the other prevertebral plexuses. The two principal ganglionic masses, named *semilunar*, though they have often little of the form the name implies, occupy the upper and outer part of the plexus, one on each side, and are placed close to the suprarenal bodies by the side of the cœliac and the superior mesenteric arteries. At the upper end, which is expanded, each ganglion receives the great splanchnic nerve.

Diaphragmatic Plexus.—The nerves (inferior diaphragmatic) composing this plexus are derived from the upper part of the semilunar ganglion, and are larger on the right than on the left side. Accompanying the arteries along the lower surface of the diaphragm, the nerves sink into the substance of the muscle. They furnish some filaments to the suprarenal body, and join with the spinal phrenic nerves.

At the right side, on the under surface of the diaphragm, and near the suprarenal body, there is a small ganglion, *ganglion diaphragmaticum*, which marks the junction between the phrenic nerves of the spinal and sympathetic systems. From this small ganglion filaments are distributed to the vena cava, the suprarenal body, and the hepatic plexus. On the left side the ganglion is wanting, but some filaments are prolonged to the hepatic plexus.

Suprarenal Plexus.—The suprarenal nerves issue from the solar plexus and the outer part of the semilunar ganglion, a few filaments being added from the diaphragmatic nerve. They are short, but numerous in comparison with the size of the body which they supply : they enter the upper and inner parts of the suprarenal capsule. These nerves are continuous below with the renal plexus. The plexus is joined by branches from one of the splanchnic nerves, and presents a ganglion (*gangl. splanchnico-suprarenale*), where it is connected with those branches. The plexus and ganglion are smaller on the left than on the right side.

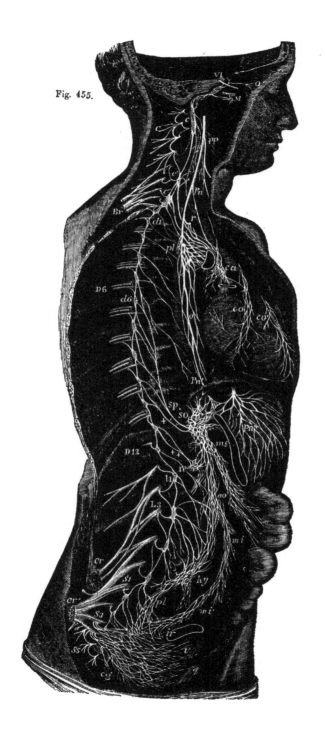

Fig. 455.

Fig. 455.—DIAGRAMMATIC VIEW OF THE SYMPATHETIC CORD OF THE RIGHT SIDE, WITH ITS PRINCIPAL GANGLIA, PLEXUSES, AND NERVES.

This figure is repeated in illustration of the sympathetic nerves in the lower half of the body.

c, superior cervical ganglion ; c', second or middle ; c'', inferior : from each of these ganglia cardiac nerves (deep on this side) are seen descending to the cardiac plexus ; d 1, placed immediately below the first dorsal sympathetic ganglion ; d 6, is opposite the sixth ; l 1, first lumbar ganglion ; cg, the terminal or coccygeal ganglion ; pp, pharyngeal, and, lower down, laryngeal plexus ; pl, posterior pulmonary plexus spreading from the pneumo-gastric on the back of the right bronchus ; ca, on the aorta, the cardiac plexus, towards which, in addition to the cardiac nerves from the three cervical sympathetic ganglia, other branches are seen descending from the pneumogastric and recurrent nerves ; co, right or posterior, and co', left or anterior coronary plexus ; o, œsophageal plexus in long meshes on the gullet ; sp, great splanchnic nerve formed by branches from the fifth, sixth, seventh, eighth, and ninth dorsal ganglia ; $+$, small splanchnic from the ninth and tenth ; $++$, smallest or third splanchnic from the eleventh : the first and second of these are shown joining the solar plexus, so ; the third descending to the renal plexus, re ; connecting branches between the solar plexus and the pneumo-gastric nerves are also represented ; pn', above the place where the right pneumo-gastric passes to the lower or posterior surface of the stomach ; pn'', the left distributed on the anterior or upper surface of the cardiac portion of the organ : from the solar plexus large branches are seen surrounding the arteries of the cœliac axis, and descending to ms, the superior mesenteric plexus ; opposite to this is an indication of the suprarenal plexus ; below re (the renal plexus), the spermatic plexus is also indicated ; ao, on the front of the aorta, marks the aortic plexus, formed by nerves descending from the solar and superior mesenteric plexuses and from the lumbar ganglia ; mi, the inferior mesenteric plexus surrounding the corresponding artery ; hy, hypogastric plexus placed between the common iliac vessels, connected above with the aortic plexus, receiving nerves from the lower lumbar ganglia, and dividing below into the right and left pelvic or inferior hypogastric plexuses ; pl, the right pelvic plexus ; from this the nerves descending are joined by those from the plexus on the superior hemorrhoidal vessel, mi', by sympathetic nerves from the sacral ganglia, and by numerous visceral nerves from the third and fourth sacral spinal nerves, and there are thus formed the rectal, vesical, and other plexuses, which ramify upon the viscera from behind forwards and from below upwards, as towards ir, and v, the rectum and bladder.

RENAL PLEXUS.— The nerves forming the renal plexus, fifteen or twenty in number, emanate for the most part from the outer part of the semilunar ganglion ; but some are added from the solar plexus and the aortic plexus. Moreover, filaments from the smallest splanchnic nerve, and occasionally from the other splanchnic nerves, terminate in the renal plexus. In their course along the renal artery, ganglia of different sizes are formed on these nerves. Lastly, dividing with the branching of the vessel, the nerves follow the renal arteries into the substance of the kidney. On the right side some filaments are furnished to the vena cava, behind which the plexus passes with the renal artery ; and others go to form the spermatic plexus.

SPERMATIC PLEXUS.—This small plexus commences in the renal, but receives in its course along the spermatic artery an accession from the aortic plexus. Continuing downwards to the testis, the spermatic nerves are connected with others which accompany the vas deferens and its artery from the pelvis.

In the female, the plexus, like the artery, is distributed to the ovary and the uterus.

CŒLIAC PLEXUS.—This plexus is of large size, and is derived from the fore part of the great epigastric plexus. It surrounds the cœliac axis in a kind of membranous sheath, and subdivides, with the artery, into coronary, hepatic, and splenic plexuses, the branches of which form communications corresponding with the arches of arterial anastomosis. The plexus receives offsets from one or more of the splanchnic nerves, and on the left side a branch from the pneumo-gastric nerve is continued into it. (Swan.)

The *coronary plexus* is placed with its artery along the small curvature of the stomach, and unites with the nerves which accompany the pyloric artery, as well as with branches of the pneumo-gastric nerves. The nerves of this plexus enter the coats of the stomach, after running a short distance beneath the peritoneum.

The *hepatic plexus*, the largest of the three divisions of the cœliac plexus, ascends with the hepatic vessels and the bile-duct, and, entering the substance of the liver, ramifies on the branches of the vena portæ and the hepatic artery. Offsets from the left pneumo-gastric and diaphragmatic nerves join the hepatic plexus at the left side of the vessels. From this plexus filaments are furnished to the right supra-renal plexus, as well as other secondary plexuses which follow the branches of the hepatic artery. Thus there is a *cystic* plexus to the gall-bladder; and there are *pyloric*, *gastro-epiploic*, and *gastro duodenal* plexuses, which unite with coronary, splenic, and mesenteric nerves.

The *splenic plexus*, continued on the splenic artery and its branches into the substance of the spleen, is reinforced at its beginning by branches from the left semi-lunar ganglion, and by a filament from the right vagus nerve. It furnishes the *left gastro-epiploic* and *pancreatic* plexuses, which course along the corresponding branches of the splenic artery, and, like the vessels, are distributed to the stomach and pancreas.

SUPERIOR MESENTERIC PLEXUS.—The plexus accompanying the superior mesenteric artery, whiter in colour and firmer than either of the preceding offsets of the solar plexus, envelopes the artery in a membraniform sheath, and receives a prolongation from the junction of the right pneumo-gastric nerve with the cœliac plexus. Near the root of the artery, ganglionic masses (gangl. meseraica) occur in connection with the nerves of this plexus.

The offsets of the plexus are in name and distribution the same as the vessels. In their progress to the intestine some of the nerves quit the arteries which first sup-ported them, and are directed forwards in the intervals between the vessels. As they proceed, they divide, and unite with lateral branches, like the arteries, but without the same regularity: they finally pass upon the intestine along the line of attachment of the mesentery.

THE AORTIC PLEXUS.

The aortic or intermesenteric plexus, placed along the abdominal aorta, and occupying the interval between the origin of the superior and inferior mesenteric arteries, consists, for the most part, of two lateral portions, connected with the semilunar ganglia and renal plexuses, which are extended on the sides of the aorta, and which meet in several larger communicating branches over the middle of that vessel. It is joined by branches from some of the lumbar ganglia, and presents not unfrequently one or more distinct ganglionic enlargements towards its centre.

The aortic plexus furnishes the inferior mesenteric plexus and part of the spermatic, gives some filaments to the lower vena cava, and ends below in the hypogastric plexus.

INFERIOR MESENTERIC PLEXUS.—This plexus is derived principally from the left lateral part of the aortic plexus, and closely surrounds with a network the inferior mesenteric artery. It distributes nerves to the left or descending part and the sigmoid flexure of the colon, and assists in supplying the rectum. The nerves of this plexus, like those of the superior mesenteric plexus, are firm in texture, and of a whitish colour.

The highest branches (those on the left colic artery) are connected with the last branches (middle colic) of the superior mesenteric plexus, while others in the pelvis unite with offsets derived from the pelvic plexus.

HYPOGASTRIC PLEXUS.

The hypogastric plexus, the assemblage of nerves destined for the supply of the viscera of the pelvis, lies invested in a sheath of dense connective tissue, in the interval between the two common iliac arteries. It is formed by eight or ten nerves on each side, which descend from the aortic plexus, receiving considerable branches from the lumbar ganglia, and, after crossing the common iliac artery, interlace in the form of a flat plexiform mass placed in front of the lowest lumbar vertebra. The plexus contains no distinct ganglia. At the lower end it divides into two parts, which are directed forwards, one to each side of the pelvic viscera, and form the pelvic plexuses.

PELVIC PLEXUS.

The pelvic or inferior hypogastric plexuses, one on each side, are placed in the lower part of the pelvic cavity by the side of the rectum, and of the vagina in the female. The nerves, prolonged from the hypogastric plexus, enter into repeated communications as they descend, and form at the points of connection small knots, which contain a little ganglionic matter. After descending some way, they become united with branches of the spinal nerves, as well as with a few offsets of the sacral ganglia, and the union of all constitutes the pelvic plexus. The spinal branches, which enter into the plexus, are furnished from the third and fourth sacral nerves, especially the third ; and filaments are likewise added from the first and second sacral nerves. Small ganglia are formed at the places of union of the spinal nerves, as well as elsewhere in the plexus (plexus gangliosus—Tiedemann).

From the plexus so constituted numerous nerves are distributed to the pelvic viscera. They correspond with the branches of the internal iliac artery, and vary with the sex ; thus, besides hæmorrhoidal and vesical nerves, which are common to both sexes, there are nerves special to each :—namely in the male, for the prostate, vesicula seminalis, and vas deferens ; in the female, for the vagina, uterus, ovary, and Fallopian tube.

The nerves distributed to the urinary bladder and the vagina contain a larger proportion of spinal nerves than those furnished to the other pelvic viscera.

INFERIOR HÆMORRHOIDAL NERVES.—These slender nerves proceed from the back part of the pelvic plexus. They join with the nerves (superior hæmorrhoidal) which descend with the inferior mesenteric artery, and penetrate the coats of the rectum.

VESICAL PLEXUS.—The nerves of the urinary bladder are very numerous. They are directed from the anterior part of the pelvic plexus to the side and lower part of the bladder. At first, these nerves accompany the vesical blood-vessels, but afterwards they leave the vessels, and subdivide into minute branches before perforating the muscular coat of the organ. Secondary plexuses are given in the male to the vas deferens and the vesicula seminalis.

The *nerves of the vas deferens* ramify round that tube, and communicate in the spermatic cord with the nerves of the spermatic plexus. Those furnished to the *vesicula seminalis* form an interlacement on the vesicula, and some branches penetrate its substance. Other filaments from the prostatic nerves reach the same structure.

PROSTATIC PLEXUS.—The nerves of this plexus are of considerable size, and pass onwards between the prostate gland and the levator ani. Some are furnished to the prostate and to the vesicula seminalis ; and the plexus is then continued forwards to supply the erectile substance of the penis, where its nerves are named " cavernous."

Cavernous nerves of the penis.— These are very slender, and difficult to dissect. Continuing from the prostatic plexus they pass onwards, beneath the arch of the pubes, and through the muscular structure connected with the membranous part of the urethra, to the dorsum of the penis. At the anterior margin of the levator ani muscle the cavernous nerves are joined by some short filaments from the pudic nerve.   After distributing twigs to the forepart of the prostate, these nerves divide into branches for the erectile substance of the penis, as follows :—

Small cavernous nerves (Müller), which perforate the fibrous covering of the corpus cavernosum near the root of the penis, and end in the erectile substance.

The *large cavernous nerve*, which extends forward on the dorsum of the penis, and dividing, gives filaments that penetrate the corpus cavernosum, and pass with or near the cavernous artery (art. profunda penis).   As it continues onwards, this nerve joins with the dorsal branch of the pudic nerve about the middle of the penis, and is distributed to the corpus cavernosum.   Branches from the foregoing nerves reach the corpus spongiosum urethræ.   (Müller, " Ueber die organischen Nerven der erectilen männlichen Geschlechtsorgane," &c. Berlin, 1836.)

NERVES OF THE OVARY.—The ovary is supplied chiefly from the plexus prolonged on the ovarian artery from the abdomen ; but it receives another offset from the uterine nerves.

VAGINAL PLEXUS.—The nerves furnished to the vagina leave the lower part of the pelvic plexus—that part with which the spinal nerves are more particularly combined.   They are distributed to the vagina without previously entering into a plexiform arrangement ; and they end in the erectile tissue on the lower and anterior part, and in the mucous membrane.

NERVES OF THE UTERUS.—These nerves are given more immediately from the lateral fasciculus prolonged to the pelvic plexus from the hypogastric plexus, above the point of connection with the sacral nerves. Separating opposite the neck of the uterus, they are directed upwards with the blood-vessels along the side of this organ, between the layers of its broad ligament.   Some very slender filaments form round the arteries a plexus, in which minute ganglia are found scattered at intervals, and these nerves continue their course in the substance of the organ in connection with the blood-vessels.   But the larger part of the nerves soon leave the vessels ; and after dividing repeatedly, without communicating with each other and without forming any gangliform enlargements, sink into the substance of the uterus, penetrating for the most part its neck and the lower part of its body.   One branch, continued directly from the common hypogastric plexus, reaches the body of the uterus above the rest ; and a nerve from the same source ascends to the Fallopian tube.   Lastly, the fundus of the uterus often receives a branch from the ovarian nerve. (Fr. Tiedemann, Tab. Nerv. Uteri, Heidelberg, 1822 ; Robert Lee, in Phil. Trans., 1841, 1842, 1846, and 1849 ; and Snow Beck, in Phil. Trans., 1846, part ii.)

The *nerves of the gravid uterus* have been frequently investigated, with a view to discover if they become enlarged along with the increase in size of the organ.   It is ascertained that the increase which takes place is confined, for the most part, to thickening of the fibrous envelopes of the nerves ; but it appears also, from the researches of Kilian, that fibres furnished with a medullary sheath, which in the unimpregnated state of the uterus lose that sheath as they proceed to their distribution, in the impregnated condition of the uterus continue to be surrounded with it as they run between the muscular fibres. (Farre, in Supplement of Cyclopædia of Anat. and Phys., " Uterus and Appendages.")

IV. ORGANS OF THE SENSES.

In this place it is intended to describe the organs of sight, hearing and smelling, which, considered with reference to their anatomy and development, are regarded as the higher organs of special sense. The description of the organ of touch is given along with the skin in the histological part of the work, and that of the organ of taste along with the descriptive anatomy of the digestive system.

THE EYE.

The organ of vision, strictly speaking, consists only of the ball or globe of the eye, a spheroidal structure enclosed by strong membranous coverings, receiving the optic nerve posteriorly, and containing the sensitive terminations of that nerve, together with a series of transparent media, which constitute an optical instrument of variable focus, through which the rays of light are transmitted to the sensitive part, and so brought into focus as to form upon it a distinct inverted image of the objects from which they proceed. But there are likewise various structures external to the eyeball which contribute to the production of perfect vision, such as the straight and oblique muscles by which the eyeball is moved in different directions, and the various supporting and protective structures known as *appendages of the eye* (tutamina oculi), including the eyebrows, eyelids, and conjunctiva, and the lachrymal apparatus.

APPENDAGES OF THE EYE.

THE EYELIDS AND CONJUNCTIVA.

The eyelids (palpebræ) are moveable folds of integument, strengthened toward their margins by a thin lamina of cartilage. The mucous membrane, which lines their inner surface, and which is reflected thence in the form of a pellucid covering on the surface of the eyeball, is named membrana conjunctiva.

The upper lid is larger and more moveable than the lower: the transparent part of the globe is covered by it when the eye is closed; and the eye is opened chiefly by the elevation of this lid by a muscle (levator palpebræ) devoted exclusively to this purpose. The eyelids are joined at the outer and inner angles (canthi) of the eye. The interval between the angles, *fissura palpebrarum*, varies in length in different persons, and, according to its extent, the size of the globe being nearly the same, gives the appearance of a larger or a smaller eye. The greater part of the edge of each eyelid is flattened, but towards the inner angle it is rounded off for a short space, at the same time that it somewhat changes its direction; and, where the two differently formed parts join, there exists on each lid a slight conical elevation— *papilla lachrymalis*, the apex of which is pierced by the aperture or punctum of the corresponding lachrymal canalicule.

In the greater part of their extent the lids are applied to the surface of the eyeball; but at the inner canthus, opposite the puncta lachrymalia, there intervenes a vertical fold of conjunctiva, the *plica semilunaris*, resting on the eyeball; while, occupying the recess of the angle internal to the border of this fold, is a spongy-looking reddish elevation, formed by a group of

3 A

glandular follicles, and named the *caruncula lachrymalis*. The plica semilunaris is the rudiment of the third eyelid (membrana nictitans) found in some animals.

Structure of the lids.—The skin covering the eyelids is thin and delicate; and at the line of the eyelashes, altered in its character, it joins the conjunctival mucous membrane which lines the inner surface of the lids. Beneath the skin, and between it and the conjunctiva, the following structures are successively met with, viz. :—The fibres of the orbicularis muscle; loose connective tissue; the tarsal cartilages, together with a thin fibrous membrane, the palpebral ligament, which attaches them to the margin of the orbit; and, finally, the Meibomian glands. In the upper eyelid there is, in addition, the insertion of the levator palpebræ superioris, in the form of a fibrous expansion fixed to the anterior surface of the tarsal cartilage.

Fig. 456.

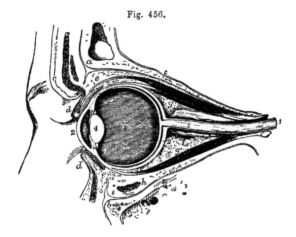

Fig. 456.—VERTICAL SECTION OF THE LEFT ORBIT AND ITS CONTENTS.

The section has been carried through the middle of the optic foramen and optic nerve obliquely as far as the back of the eyeball, and thence forward through the eyeball, eyelids, &c., in an antero-posterior direction. *a*, the frontal bone; *b*, the superior maxillary bone; *c*, the eyebrow with the orbicularis palpebrarum, integument, &c., divided; *d*, the upper, and *d'*, the lower eyelid, partially open, showing the section of the tarsal cartilages and other component parts, the eyelashes, &c.; *e, e*, the reflection of the conjunctiva from the upper and lower eyelids to the surface of the eyeball; *f*, the levator palpebræ superioris muscle; *g*, the upper, *g'*, the lower rectus muscle; *h*, the inferior oblique muscle divided; 1, 1, the optic nerve divided in its sheath; 2, the cornea; 2', the sclerotic; 3, the aqueous chamber; 4, the crystalline lens; 5, the centre of the vitreous humour.

The *fibres of the orbicularis muscle* are closely adherent to the skin by fine connective tissue, entirely devoid of fat. A marginal fasciculus of its fibres has been found within the line of the eyelashes, separated by the bulbs of the lashes from the other fibres, and constituting the *ciliary muscle* of Riolan. The fibres of the orbicular muscle, while adherent to the skin, glide loosely on the tarsal cartilages.

The *tarsal cartilages* (tarsi) are two thin elongated plates of cartilages of the yellow kind placed one in each lid, and serving to give shape and firmness to those parts. The upper cartilage, the larger, is half oval in form, being broader near the centre and narrowing towards the angles of the lids. The lower is thinner, much narrower, and more nearly of an uniform breadth throughout. The free or ciliary edge of the cartilages, which is straight, is thicker than any other part. At the inner canthus the cartilages are fixed by the fibrous slips of the tendo palpebrarum (p. 172); and at the outer angle they are attached to the malar bone by a fibrous band belonging to the palpebral ligament, and named the external *tarsal ligament*.

The *palpebral ligament* is a fibrous membrane placed beneath the orbicularis muscle, attached peripherally to the margin of the orbit, and internally to the tarsal cartilages near the inner free edge. The membrane is thickest at the outer part of the orbit.

Meibomian glands.—On the ocular surface of each lid are seen from twenty to thirty parallel vertical lines of yellow granules, lying immediately under the conjunctival mucous membrane. They are compound sebacious follicles,

Fig. 457.—MEIBOMIAN GLANDS, LACHRYMAL GLAND, &c., AS SEEN FROM THE DEEP SURFACE OF THE EYELIDS OF THE LEFT SIDE.

Fig. 457.

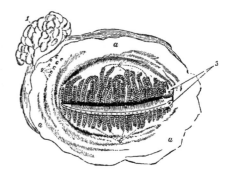

a, palpebral conjunctiva; 1, lachrymal gland; 2, openings of seven or eight glandular ducts; 3, upper and lower puncta lachrymalia; 6, 6, shut ends of the upper and lower Meibomian glands, of which the openings are indicated along the margins of the eyelids.

embedded in grooves at the back of the tarsal cartilages; and they open on the free margin of the lids by minute orifices, generally as many in number as the lines of follicles themselves. These glands consist of nearly straight excretory tubes, each of which is closed at the end, and has numerous small cæcal appendages projecting from its sides. The tubes are lined by mucous membrane, on the surface of which is a layer of scaly or pavement epithelium cells.

According to Heinrich Müller there is likewise a layer of unstriped muscular fibre contained in each eyelid; that of the upper lid arising from the under surface of the levator palpebræ, that of the lower lid arising from the neighbourhood of the inferior oblique muscle, and each being inserted near the margin of the tarsal cartilage. It may also be mentioned in this place that the same writer describes a layer of unstriped muscle crossing the spheno-maxillary fissure, corresponding to a more largely developed layer found in the extensive aponeurotic part of the orbital wall of various mammalia. This set of fibres has been more particularly described by Turner (H. Müller, in Zeitschr. f. Wiss. Zool. 1858, p. 541; W. Turner, in Nat. Hist. Rev. 1862, p. 106).

The *eyelashes* (cilia) are strong short curved hairs, arranged in two or more rows along the margin of the lids, at the line of union between the skin and the conjunctival mucous membrane. The lashes of the upper lid, more numerous and longer than the lower, have the convexity of their curve directed downwards and forwards; whilst those of the lower lid are arched in the opposite direction. Near the inner canthus these hairs are weaker and more scattered.

Structure of the conjunctiva.—The conjunctiva consists of the palpebral part, along with which may be grouped the plica semilunaris and caruncula lachrymalis, and of the ocular part or conjunctiva bulbi, in which may be distinguished the sclerotic and corneal portions : each of these several parts presents peculiar and distinctive characters. The epithelium is stratified and thick ; the cells of the superficial strata scaly, delicate, and each with a distinct nucleus.

The *palpebral portion* of the conjunctiva is opaque and red, is thicker and more vascular than any other part of the membrane, and presents numerous fine papillæ freely supplied with nerves. At the margins of the lids the palpebral conjunctiva enters the ducts of the Meibomian glands ; through the puncta lachrymalia it passes into the canaliculi, and is continuous with the lining membrane of the lachrymal sac ; and it is prolonged into the orifices of the ducts of the lachrymal gland.

The *sclerotic portion* of the conjunctiva, changing its character at the line of reflection from the eyelids, becomes thinner, and loses its papillary structure : it is loosely connected to the eyeball by submucous tissue. It is also transparent and nearly colourless, but a few scattered branches of blood-vessels are generally visible on it in the healthy condition, and under the influence of inflammatory congestion a copious network of vessels very irregularly disposed comes into view. This network is derived from the palpebral and lachrymal arteries. It may be easily made to glide loosely on the surface of the eyeball by pressing the eyelid against it. But another set of vessels likewise exists on the surface of the sclerotic, and may be brought into view by congestion. The position of this set is entirely sub-conjunctival, adherent to the sclerotic coat ; they are less tortuous than the conjunctival set, and are derived from the muscular and anterior ciliary branches of the ophthalmic artery : they remain immoveable on pressure of the eyelid. They dip into the sclerotic near the cornea, and appear to unite with a more deeply connected minute network disposed in closely set straight lines, radiating from the margin of the cornea, and the gorged condition of which is well known to ophthalmic surgeons as characteristic of sclerotitis.

The *corneal conjunctiva* consists almost entirely of epithelium, any underlying membrane being extremely thin, transparent, and adherent to the anterior elastic layer of the cornea, in connection with which it will be again referred to. Vessels lie between it and the cornea, and form a circle of anastomotic capillary loops around the circumference. This plexus of vessels extends farther inwards in the fœtus.

A well developed network of *lymphatics* exists throughout the sclerotic and palpebral portions of the conjunctiva ; but at the margin of the cornea a sudden diminution takes place in the size of the meshes and diameter of the vessels. Of the network referred to, only a narrow circle $\frac{1}{25}$th of an inch in diameter exists on the corneal conjunctiva, and this circle has a well defined inner margin within which no lymphatics exist (Teichmann).

The *nerves* in the membrane, as far as the cornea, seem to have the same arrangement as in the skin in general.

In the submucous tissue of the eyelids there are small follicular glands spread over the whole surface of the conjunctiva palpebrarum, and in the vicinity of the reflection of the conjunctiva upon the eyeball a set of larger more complex glands of a racemose structure, somewhat similar to that of the lachrymal gland (Sappey, C. and W. Krause).

Closed follicles have also been observed in the conjunctiva by Bruch, and, after him, by other observers.

THE LACHRYMAL APPARATUS.

The parts which constitute the lachrymal apparatus are the following, viz. :—The gland by which the tears are secreted, situated at the upper and outer side of the orbit, together with its excretory ducts ; the two canals into which the fluid is received near the inner angle ; and the sac with the nasal duct continued from it, through which the tears pass into the inferior meatus of the nose.

The *lachrymal gland*, an oblong flattened body, about the size of a small almond, is placed in the upper and outer part of the orbit, a little behind the anterior margin. The upper surface of the gland, convex, is lodged in a slight depression in the orbital plate of the frontal bone, to the periosteum of which it adheres by fibrous bands ; the lower surface is adapted to the convexity of the eyeball, and is in contact with the upper and the outer recti muscles. The fore part of the gland, separated from the rest by a slight depression, and sometimes described as a second lobe, or as a distinct gland, is closely adherent to the back of the upper eyelid, and is covered on the ocular surface only by a reflection of the conjunctiva. The glandular ducts, usually from six to eight in number, are very small, and emerge from the thinner portion of the gland. After running obliquely under the mucous membrane, and separating at the same time from each other, they open in a row by separate orifices, the greater number in the fold above the outer canthus, and two of them (Hyrtl) in the fold below.

Fig. 458.—FRONT OF THE LEFT EYELIDS, WITH THE LACHRYMAL CANALS AND NA-SAL DUCT EXPOSED.

Fig. 458.

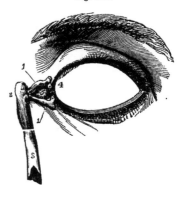

1, 1, upper and lower lachrymal canals, showing towards the eyelids the narrow bent portions and the puncta lachrymalia ; 2, lachrymal sac ; 3, the lower part of the nasal duct ; 4, plica semilunaris ; 5, caruncula lachrymalis.

Lachrymal canals.—On the margin of each lid, near the inner angle, and in front of the fold of membrane called plica semilunaris, is a small elevation (papilla lachrymalis), already described. Each papilla is perforated by a small aperture, *punctum lachrymale ;* and at these apertures commence two small canals, *canaliculi*, which convey the tears from the eye to the lachrymal sac. The upper canal is rather the smaller and longer of the two : it first ascends from the punctum ; then makes a sudden bend, and is directed inwards and downwards to join the lachrymal sac. The lower canal descends from the corresponding punctum ; and soon changing its direction like the upper one, takes a nearly horizontal course inwards. Both canals are dilated where they are bent. In some cases they unite near the end to form a short common trunk ; more commonly they open separately, but close together, into the sac.

The *lachrymal sac* and *nasal duct* constitute together the passage by which the tears are conveyed from the lachrymal canals to the cavity of the nose.

The lachrymal sac, the upper dilated portion of the passage, is situated at the side of the nose, near the inner canthus of the eye, and lies embedded in a deep groove in the lachrymal and upper maxillary bones. It is of an oval form ; the upper end closed and rounded, and the lower end gradually narrowing somewhat into the nasal duct. On the outer side, and a little in front, it receives the lachrymal canals ; and here it is covered by the tendo palpebrarum, and by some of the inner fibres of the orbicular muscle of the lids ; while on its inner or posterior surface the tensor tarsi muscle is placed. The sac is composed of fibrous and elastic tissues, adhering closely to the bones above mentioned, and strengthened by fibrous processes sent from the tendo palpebrarum, which crosses a little above its middle. The inner surface is lined by a reddish mucous membrane, which is continuous through the canaliculi with the conjunctiva, and through the nasal duct with the mucous membrane of the nose.

The nasal duct (ductus ad nasum), about six or seven lines in length, grooving the upper maxillary bone, descends to the fore part of the lower meatus of the nose, the osseous canal being completed by the ungual and lower turbinated bones. A tube of fibrous membrane, continuous with the lachrymal sac, adheres to the parietes of this canal, and is lined by mucous membrane, which, at the opening into the nose, is often arranged in the form of an imperfect valve. The nasal duct is rather narrower in the middle than at either end ; its direction is not quite vertical, but inclined slightly outwards and backwards.

The mucous membrane in the canaliculi possesses a laminar epithelium, but in the nasal sac and duct a ciliated epithelium as in the nose.

Various valves have been described in connection with the lachrymal sac and canals. One, the valve of Hasner, is formed by the mucous membrane of the nose overhanging the inferior orifice of the nasal duct, and has had imputed to it the function of preventing entrance of foreign matters in violent expiratory movements; but the disposition of the mucous membrane at this orifice appears to be subject to some variation. Another fold, the valve of Huschke, placed at the deep orifice of the canaliculi, is supposed by some to prevent the return of the tears from the sac into those tubes, but by others, it is declared to be inconstant, and insufficient, even when found, to close the orifice. A third fold, the valve of Foltz, is described as forming a projection inwards on one side of the vertical part of each canaliculus, near the punctum lachrymale, and as being sufficient to close the tube when it is flattened by the pressure of the fibres of the orbicularis and tensor tarsi muscles as in winking. The experiments of Foltz on rabbits go to prove that the punctum lachrymale having been turned backwards towards the eye in winking, and the canaliculus being compressed by the muscles, as soon as the pressure is removed the canaliculus resumes its open form, and so sucks in tears which by the next compression in winking are forced onwards into the lachrymal sac ; and also, that when the muscles are paralysed, the canaliculi cease to carry away the tears. See review of Foltz's paper in Dublin Quarterly Journal, Feby. 1863 ; also, Hyrtl, Topogr. Anatomie.

THE GLOBE OF THE EYE.

The globe or ball of the eye is a composite structure of an irregularly spheroidal form, placed in the fore part of the orbital cavity, and receiving the thick stem of the optic nerve behind. The recti and obliqui muscles closely surround the greater part of the eyeball, and are capable of changing its position within certain limits : the lids, with the plica semilunaris and caruncle, are in contact with its covering of conjunctiva in front ; and behind it is supported by a quantity of loose fat and connective tissue.

The eyeball, when viewed in profile, is found to be composed of segments of two spheres, of which the anterior is the smaller and more prominent : the segment of the larger posterior opaque sphere corresponds with the limit of the sclerotic coat, and the translucent portion of the smaller sphere with that of the cornea.

From before backwards the ball measures about nine-tenths of an inch, and its transverse diameter exceeds this measurement by about a line.

Except when directed towards near objects, the axes of the eyes are nearly parallel ; the optic nerves, on the contrary, diverge considerably from one another, and each nerve enters the corresponding eye about a tenth of an inch to the inner or nasal side of the axis of the globe.

The eyeball is composed of several investing membranes, concentrically arranged, and of certain fluid and solid parts contained within them. The membranes are three in number, with the following designations and general structure :—An external fibrous covering, named *sclerotic* and *cornea ;* a middle vascular, pigmentary, and in part also muscular membrane, the *choroid* and the *iris ;* and an internal nervous stratum, the *retina.* The enclosed refracting media, three in number, are the *aqueous humour,* the *vitreous body,* and the *lens* with its *capsule.*

Around the eyeball there is an adventitious tunic of fascia, *tunica vaginalis oculi,* or capsule of Tenon, which is perforated by the tendons of the recti and obliqui muscles, and connected with the sclerotic by merely the most delicate connective tissue. This capsule separates the eye-ball from the orbital fat, and enables it to glide freely in its movements. (See, for details, Richet, Traité d'Anatomie Médico-Chirurgicale ; and O'Ferrall, in Dublin Quart. Journ. Med. Science, July, 1841.)

EXTERNAL COAT OF THE EYEBALL.

The external investing membrane, which forms a complete covering for the ball, consists of two parts of different appearance and structure. Of these the hinder part, much the largest, is opaque and densely fibrous, and is named the sclerotic coat, while the anterior smaller segment is transparent, and is named the cornea.

THE SCLEROTIC COAT.

The sclerotic (cornea opaca), the tunic of the eye on which the maintenance of the form of the greater part of the organ chiefly depends, is a strong, opaque, unyielding, fibrous structure. The membrane covers about five-sixths of the eye-ball, and is pierced behind by the optic nerve. The outer surface is white and smooth, except where the tendons of the recti and obliqui muscles are inserted into it. The inner surface is of a light brown colour, and rough from the presence of a delicate connective tissue (*membrana fusca*), through which branches of the ciliary vessels and nerves cross obliquely. The sclerotic is thickest at the back part of the eye, and thinnest at about a quarter of an inch from the cornea : at the junction with the cornea, it is again somewhat thickened. The optic nerve pierces this coat about one-tenth of an inch internal to the axis of the ball, and the opening is somewhat smaller at the inner than at the outer surface of the coat. The fibrous sheath of the nerve, together with the membranous processes which separate the funiculi of its fibres, blend with the sclerotic at the margin of the aperture : in consequence of this arrangement, when the nerve is cut off close to the eye-ball, the funiculi are seen to enter by a group of pores ; and to the part of the sclerotic thus perforated the name of *lamina cribrosa* is sometimes given. Around this cribrous opening are smaller apertures for vessels and nerves.

Fig. 459.

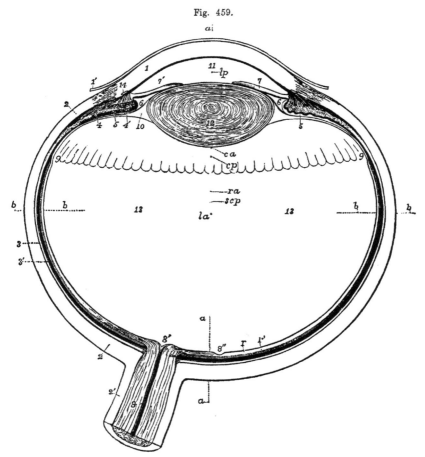

Fig. 459.—VIEW OF THE LOWER HALF OF THE RIGHT ADULT HUMAN EYE, DIVIDED HORIZONTALLY THROUGH THE MIDDLE. †

The specimen from which this outline is taken was obtained by dividing the eye of a man of about forty years of age in the frozen state. It was carefully compared with other specimens obtained in a similar manner ; and in the drawing averages have been given in any particulars in which differences among them presented themselves.

1, the cornea ; 1', its conjunctival layer ; 2, the sclerotic ; 2', sheath of the optic nerve passing into the sclerotic ; 3, external or vascular layer of the choroid ; 3', its internal pigmental layer ; 4, ciliary muscle, its radiating portion ; 4', cut fibres of the circular portion ; 5, ciliary fold or process ; 6, placed in the posterior division of the aqueous chamber, in front of the suspensory ligament of the lens ; 7, the iris (outer side) ; 7', the smaller inner side ; 8, placed on the divided optic nerve, points to the arteria centralis retinæ ; 8', colliculus or eminence at the passage of the optic nerve into the retina ; 8″, fovea centralis retinæ ; r, the nervous layer of the retina ; r′, the bacillar layer ; 9, ora serrata at the commencement of the ciliary part of the retina ; 10, canal of Petit ; 11, anterior division of the aqueous chamber in front of the pupil ; 12, the crystalline lens, within its capsule ; 13, the vitreous humour ; a, a, a, parts of a dotted line in the axis of the eye ; b, b, b, b, a line in the transverse diameter. It will be observed that from the pupil being placed nearer the inner side the axis of the eye-ball a, a, does not pass exactly through the centre of the pupil, and that this line falls a little to the inner side of the fovea centralis. The following letters indicate the centres of the curvatures of the different surfaces ; assuming them to be nearly spherical, viz. : c a, anterior surface of the cornea ; c p, posterior surface ; l a, anterior surface of the lens ; l p, posterior surface ; s c p, posterior surface of the sclerotic ; r a, anterior surface of the retina.

In connection with this figure the following average dimensions of the parts of the adult eye in fractions of an English inch may be stated :—

Transverse diameter of the eyeball	1·
Vertical diameter (Krause)	0·96
Antero-posterior diameter	0·96
Diameter of the optic nerve with its sheath	0·16
Diameter of the nervous part at its passage through the choroid membrane	0·09
Greatest thickness of the sclerotic, choroid, and retina together	0·08
Greatest thickness of the sclerotic posteriorly	0·05
Smallest thickness at the sides and in front	0·025
Greatest thickness of the cornea	0·055
Distance from the middle of the posterior surface of the cornea to the front of the lens	0·07
Antero-posterior diameter of the lens	0·19
Transverse ditto	0·35
Greatest thickness of the ciliary muscle and ciliary processes together	0·06
Greatest thickness of the ciliary muscle	0·035
Thickness of the iris	0·015
Length of the radius of curvature of the anterior surface of the cornea (regarding it approximately as spherical)	0·305
Radius of the posterior surface	0·275
Radius of curvature of the anterior surface of the lens	0·36
Radius of the posterior surface	0·21
Approximate length of the radius of curvature of the outer surface in the posterior half of the retina	0·485
Approximate radius of curvature of the external surface of the posterior part of the sclerotic coat	0·5
Distance of the middle of the posterior surface of the lens from the middle of the retina	0·575
Distance between the centre of the spot of entrance of the optic nerve and the middle of the fovea centralis retinæ	0·14
Diameter of the base of the cornea	0·48
Diameter of the base of the iris transversely	0·45
Diameter of the base of the iris vertically	0·43
Diameter of the pupil	0·14

STRUCTURE.—The sclerotic coat is formed of connective tissue, and yields gelatine on boiling. Its fibres are combined with fine elastic tissue, and with fusiform and stellate nucleated cells, and are aggregated into bundles, which are disposed in layers both longitudinally and transversely, the longitudinal arrangement being most marked at the surfaces. These layers communicate at intervals, and the sclerotic presents a ramified and laminar appearance on a vertical section.

A few blood-vessels permeate the fibrous texture in the form of a net-work of the smallest capillaries with very wide meshes ; and in the neighbourhood of the cornea a ring of greater vascularity exists, which has been already noticed in the description of the sclerotic conjunctiva. The existence of nerves in the sclerotic has not yet been allowed by all anatomists.

The cornea (cornea pellucida), the transparent fore part of the external coat, admits light into the interior of the ball. It is nearly circular in shape, and its arc extends to about one-sixth of the circumference of the whole globe; it is occasionally widest in the transverse direction. Being of a curvature of a smaller radius than the sclerotic, it projects forwards beyond the general surface of curvature of that membrane, somewhat like the glass of a watch: the degree of its curve varies, however, in different persons, and at different periods of life in the same person, being more prominent in youth and flattened in advanced age. Its thickness is in general nearly the same throughout, viz., from $\frac{1}{22}$ to $\frac{1}{32}$ of an inch, excepting towards the outer margin, where it becomes somewhat thinner. The posterior concave surface exceeds slightly in extent the anterior or convex, in consequence of the latter being encroached on by the opacity of the sclerotic.

Fig. 460.　　　　　　　　　　Fig. 461.

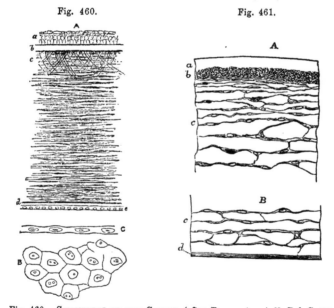

Fig. 460.—STRUCTURES OF THE CORNEA (after Bowman). A $\frac{40}{1}$, B & C, $\frac{300}{1}$

A, small portion of a vertical section of the cornea in the adult; *a*, conjunctival epithelium; *b*, anterior elastic lamina; *c* to *d*, fibrous laminæ with nuclear bodies interspersed between them; *c*, fibres shooting through some of these layers from the external elastic lamina; *d*, posterior elastic lamina; *e*, internal epithelium. B, epithelium of the membrane of Demours, as seen looking towards its surface. C, the same seen in section.

Fig. 461.—SMALL PORTIONS OF A VERTICAL SECTION OF THE CORNEA AT BIRTH (from Kölliker). $\frac{350}{1}$

The preparation has been treated with acetic acid. A, the anterior part; *a*, anterior elastic lamina; *b*, layer of closely set granules (probably small cells) placed under the anterior elastic layer, with little fibrous structure; *c*, developed fibrous tissue, with united connective-tissue corpuscles; B, posterior part of the cornea; *c*, as before; *d*, posterior elastic layer.

At its circumference the cornea joins the sclerotic part by continuity of tissue, but always so as to be overlapped by the opacity of that structure like a watch glass by the edge of the groove into which it is received.

STRUCTURE.—The cornea consists of a central thick fibrous part, the cornea proper, covered in front by the conjunctival epithelium and the anterior elastic lamina, and behind by the posterior elastic lamina or membrane of Demours.

The *cornea proper* is a stratified structure, the constituent fibres of which, continuous externally with those of the opaque sclerotic, are soft and comparatively indistinct, and between the strata of which are numerous delicate anastomosing nucleated cells, of fusiform appearance as seen in vertical sections, but expanded in the direction of the laminæ, and presenting in sections parallel to the surface a stellate appearance. The strata, about sixty in number, at a given spot (Bowman),* maintain frequent communications with contiguous layers, so that they can be detached only for a very short distance : in consequence of this stratified composition the cornea may be penetrated or torn most readily in the direction of the supposed laminæ. The transparency of the cornea is impaired by derangement of the relative position, or by approximation of the strata to each other. The cornea proper is permeable to fluid, and affords chondrin, not gelatine, on boiling (J. Müller).

There have been observed by v. Recklinghausen in the cornea of the frog, when examined in a chamber of liquid connected with the microscope, not only a rich network of anastomosing cells, but other cells also which change both their form and position by means of processes thrown out from and disappearing again into their substance, like the pseudopods of amœbæ. (Virchow's Archiv, Vol. 28, p. 157).

According to Henle, the anastomosing cells of the cornea are mere spaces devoid of any walls distinct from the surrounding matrix, and are the only interlaminar spaces naturally existing. (Systematische Anatomie, Vol. ii. p. 599).

The membranes investing the fibrous part of the cornea before and behind are both of them structureless, with epithelium on their free surface.

The anterior elastic lamina (Bowman) is a transparent glassy stratum without recognised texture, from $\frac{1}{2000}$th to $\frac{1}{1200}$th of an inch thick, and not rendered opaque by acids. From the surface resting on the fibrous strata of the cornea, a few fine threads are prolonged in a slanting direction, and are lost among the more superficial of those strata : their action is supposed to be to keep the membrane tied down smoothly to the cornea. The *epithelium* on the front of this lamina is stratified, the superficial cells being flat, and the main thickness formed of three or four layers of rounded cells, the deepest of which are vertically elongated, so as to be nearly twice as long as broad.

It is right to mention that this epithelium in the horse, the ox, and the sheep, has a much more remarkable appearance than in man, and one not to be accounted for by the ordinarily presumed mode of growth of stratified epithelia ; for the deepest cells are greatly elongated and larger than those which are immediately superimposed, and have precisely the appearance of true columnar epithelium, the flat ends resting on the subjacent elastic lamina, and the pointed extremities directed forwards.

The *membrane* of *Demours* or *Descemet* (posterior elastic lamina, Bowman), not very closely united with the fibrous part of the cornea, is transparent and glassy in appearance, firm and structureless, but very brittle and elastic ; and

* Lectures on the parts concerned in the operations on the eye, and on the structure of the retina. London, 1849.

when shreds are removed they curl up always with the attached surface innermost. Its transparency is not impaired by acids, by boiling in water, or by maceration in alkalies. In thickness it varies between $\frac{1}{3000}$th and $\frac{1}{2000}$th of an inch. At its circumference the membrane breaks up into bundles of fine threads, which are partly continued into the front of the iris, forming the "pillars of the iris," and partly into the fore part of the choroid and sclerotic coats. It is lined with an epithelial covering, which resembles that on serous membranes, consisting of a single layer of flat poly-gonal transparent cells with distinct nuclei.

Blood-vessels and nerves.—In a state of health the cornea is not provided with blood-vessels, except at the circumference, where they form very fine capillary loops and accompany the nerves. The existence of *lymphatics* has not been satisfactorily ascertained. The *nerves* of the cornea are very numerous, according to Schlemm.[*] Derived from the ciliary nerves they enter the fore part of the sclerotic, and are from twenty-four to thirty-six in number. Continued into the fibrous part of the cornea, they retain their dark outline for $\frac{1}{20}$th to $\frac{1}{10}$th of an inch, and then becoming trans-parent, ramify and form a network through the laminated structure.

MIDDLE TUNIC OF THE EYEBALL.

This coat consists of two parts, one a large posterior segment—the choroid, reaching as far as the cornea, and formed chiefly of blood-vessels and pig-mentary material; the other, a small anterior muscular part—the iris. Between these and connected with both is situated the white ring of the ciliary muscle.

Fig. 462.

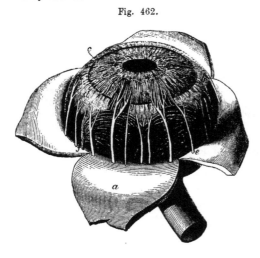

Fig. 462.—CHOROID MEM-BRANE AND IRIS EXPOSED BY THE REMOVAL OF THE SCLEROTIC AND CORNEA (after Zinn). $\frac{2}{1}$

a, one of the segments of the sclerotic thrown back; *b*, ciliary muscle and liga-ment; *c*, iris; *e*, one of the ciliary nerves; *f*, one of the vasa vorticosa or choroidal veins.

THE CHOROID COAT.

The choroid coat of the eye (tunica choro-idea s. vasculosa) is a dark brown membrane lying between the scle-rotic and the retina. It reaches forwards to the ciliary ligament, or nearly to the cornea, where it ends by a series of plaits or folds named ciliary processes, disposed in a circle projecting inwards at the back of the circumferential portion of the iris. At the hinder part, where the tunic is thickest, the optic nerve is transmitted through a circular opening. The outer surface is rough, and is connected to the sclerotic by loose connective tissue (lamina fusca of

[*] Berl. Encycl. Wört. art. Augapfel, Vol. iv. p. 22.

authors), and by vessels and nerves. The inner surface, which is smooth, is lined by a continuous layer of pigmentary cells.

The *ciliary processes*, about eighty-five in number, are arranged radiately in a circle. They consist of larger and smaller folds, without regular alternation, and the small folds number about one-third of the large. Each of the larger folds, measuring about $\frac{1}{10}$th of an inch in length and $\frac{1}{40}$th in depth, forms a rounded projection at its inner end, which is free from the

Fig. 463. — CILIARY PROCESSES AS SEEN
FROM BEHIND. $\frac{2}{1}$

Fig. 463.

1, posterior surface of the iris, with the sphincter muscle of the pupil ; 2, anterior part of the choroid coat; 3, one of the ciliary processes, of which about seventy are represented.

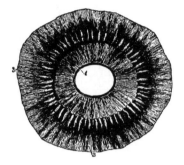

pigment which invests the rest of the structure; but externally they become gradually narrower, and disappear in the choroid coat : the smaller processes are only half as deep as the others. At and near their internal or anterior extremities the processes are connected by lateral loop-like projections, and are separated from the iris by pigment. The plications of the ciliary processes fit into corresponding plications of the suspensory ligament of the lens.

STRUCTURE.—From a difference in the fineness of its constituent blood-vessels, the choroidal coat resolves itself into two strata, inner and outer ;—the latter containing the larger branches, and the former the capillary ramifications.

In the *outer part* of the coat are situated the branches of the vessels. The arteries are large, and are directed forwards before they bend downwards to end on the inner surface ; whilst the veins (vasa vorticosa) are disposed in curves as they converge to four of five principal trunks issuing from the eyeball. In the intervals between those vessels are lodged elongated and star-shaped pigment cells with very fine offsets, which intercommunicate and form a network or stroma. Towards the inner part of the tunic, this network passes gradually into a web without pigment : it resembles elastic tissue in its chemical and physical properties.

The *inner part* of the choroid coat (tunica Ruyschiana s. chorio-capillaris) is formed by the capillaries of the choroidal vessels. From the ends of the large arteries the capillaries radiate in a star-like manner, and form meshes which are more delicate and smaller than in any other texture, and are finer at the back than the front of the ball. This fine network reaches as far forwards as about $\frac{1}{8}$th of an inch from the cornea, or opposite to the ending of the expansion of the optic nerve, where its meshes become larger, and join those of the ciliary processes.

On the inner surface of the tunica Ruyschiana may be detected, according to various authors, a structureless transparent membrane, the *membrane of Bruch*, underlying the pigmentary layer.

The *ciliary processes* have the same structure as the choroid, of which they are a part ; but the capillary plexus of the vessels, less fine, has meshes with chiefly a longitudinal direction; and the ramified cells, fewer in number, are devoid of pigment towards the free extremities of the folds.

The *pigmentary layer* (choroidal epithelium, membrane of the black pigment) forms a thin dark lining to the whole inner surface of the choroid and

Fig. 464.

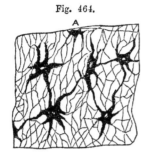

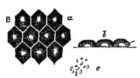

Fig. 464.—PIGMENT CELLS OF THE MIDDLE COAT (after Kölliker).

A, small portion of the choroid with the stellate or ramified cells which form its stroma. B, pigment cells, which cover the inner surface of the choroid; *a*, these cells seen from the surface, of hexagonal form, and showing nuclei in their interior; *b*, three of the same cells viewed edgeways; *c*, molecular pigment, which fills the cells.

the iris. As far forwards as the ciliary processes it consists of only a single layer of flat six-sided cells, applied edge to edge like mosaic work. Each cell contains a nucleus and more or less dense molecular contents, accumulated in greatest abundance towards the circumference of the cell, and partly obscuring the nucleus. On the ciliary processes and the iris the pigment is several layers deep, and the cells, smaller and rounded, are so filled with dark pigment as to cover up the nucleus. In the eye of the albino, pigment is absent both from the hexagonal cells and the ramified corpuscles of the choroidal tunic.

It may be mentioned that in fishes, and in many mammals, including the ox and the sheep, the eyes of which are often selected for dissection, the choroid, instead of being uniformly lined with dark pigment, presents on a greater or less extent of its back part a silvery layer named *tapetum*. The tapetum in ruminants consists of tendinous fibres, and in carnivora and fishes of cells, filled, in the carnivora, with granular matter (Leidig), in fishes with slender rods. On its inner surface is the tunic of Ruysch, as well as the layer of hexagonal cells, which, however, is here destitute of pigment.

THE IRIS.

The iris is the contractile and coloured membrane which is seen behind the transparent cornea, and gives the tint to the eye. In its centre it is perforated by an aperture—the pupil.

By its circumferential border, which is nearly circular, the iris is connected with the choroid, the cornea, and the ciliary ligament and muscles: the free inner edge is the boundary of the pupil, and is constantly altering its dimensions during life. The iris measures $\frac{1}{2}$ an inch across, and, in a state of rest, from the circumference to the pupil about $\frac{1}{5}$th of an inch. Its surfaces look forwards and backwards. The anterior, variously coloured in different eyes, is marked by waving lines converging towards the pupil, near which they join in a series of irregular elevations; and, internal to these, other finer lines pass to the pupil. The posterior surface is covered with dark pigment; and this being removed, there is seen at the margin of the pupil a narrow circular band of fibres (sphincter muscle of the pupil), with which lines radiating inwards are blended.

The *pupil* is nearly circular in form, and is placed a little to the inner side of the centre of the iris. It varies in size according to the contraction or relaxation of the muscular fibres, and this variation ranges from $\frac{1}{20}$th to $\frac{1}{3}$rd of an inch. The movements of the iris regulate the quantity of light admitted to the eyeball, and are associated with convergence of the optic axes, and with the focal adjustment of the eye.

STRUCTURE.—Fibrous and muscular tissues form the framework of the iris, and pigment is scattered through the texture. In front and behind is placed a distinct layer of pigment cells. It is still matter of discussion whether or not in the adult a delicate epithelium is continued from the margin of the cornea over the front of the iris : it is admitted to exist in childhood.

The *fibrous* stroma consists of fibres of connective tissue directed radiatingly towards the pupil, and circularly at the circumference ; these, interweaving with one another, form a net-like web which is less open towards the surfaces.

The *muscular fibre* is of the non-striated kind, and is disposed as a ring (sphincter) around the pupil, and as rays (dilatator) from the centre to the circumference.

Fig. 465.—A SMALL PART OF THE IRIS, SHOWING THE MUSCULAR STRUCTURE (from Kölliker). $\frac{350}{1}$

The specimen is from the albino-rabbit, and has been treated with acetic acid : *a*, the sphincter muscle at the margin of the pupil ; *b*, fasciculi of the dilatator muscle ; *c*, connective tissue with nuclear cells rendered clear by the acid.

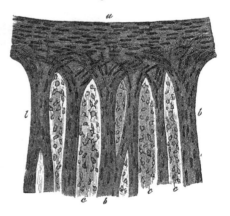

Fig. 465.

The *sphincter* is the flat narrow band on the posterior surface of the iris, close to the pupil, and is about $\frac{1}{40}$th of an inch wide. At the edge of the pupil the fibres are close together, but at the peripheral border they are separated, and form less complete rings.

The *dilatator*, less apparent than the sphincter, begins at the ciliary or outer margin of the iris, and its fibres, collected into bundles, are directed inwards between the vessels and nerves, converging towards the pupil, and forming a net-work by their intercommunications. At the pupil they blend with the sphincter, some reaching near to its inner margin.

Pigmentary elements.—In the substance of the iris anteriorly and throughout its thickness are variously-shaped and ramified pigment cells like those in the choroid membrane. The pigment contained in them is yellow, or of lighter or darker shades of brown, according to the colour of the eye. On the fore part of the iris is a thin stratum of rather oval or rounded cells with granular ramified offsets (an epithelial layer—Kölliker). At the posterior surface is a covering of dark pigment—the *uvea* of authors ; this is continuous with the pigmentary layer lining the choroid and the ciliary processes, and consists of several strata of small roundish cells filled with dark pigment. The colour of the iris depends on the pigment; in the different shades of blue eye it arises from the black pigment of the posterior surface appearing more or less through the texture, which is only slightly coloured or is colourless; and in the black, brown, and grey eye, the colour is due to the pigment scattered through the iris substance.

Fig. 466.

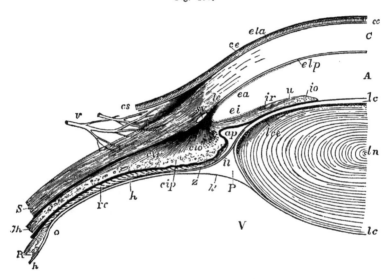

Fig. 466.—Sectional view of the Connections of the Cornea, Sclerotic, Iris, Ciliary Muscle, Ciliary Processes, Hyaloid Membrane and Lens. ⁸⁄₁

The specimen extends from the middle of the lens to the ora serrata on the inner side of the right eye. C, the laminated cornea; cc, conjunctiva corneæ; cs, conjunctiva scleroticæ; ce, epithelium of the conjunctiva; ela, anterior elastic layer of the cornea passing outwards in part into the conjunctiva; elp, posterior elastic layer; le, ligamentum pectinatum iridis, elastic ligament, spreading into the base of the iris, the sclerotic, and the attachment of the radiated ciliary muscle; S, the sclerotic at its thinnest part; A, the anterior aqueous chamber; ap, the recess forming the posterior division of the aqueous chamber; sv, placed at the junction of the cornea and sclerotic, points to the circular venous sinus or canal of Schlemm; ea, epithelium behind the cornea indicated by a dotted line; ei, epithelium in front of the iris similarly indicated; ir, radiating muscle of the iris; io, divided fibres of the orbicular muscle; u, pigment layer or uvea; ln, centre of the crystalline lens; lc, capsule of the lens; lce, layer of cells in front of the lens; cir, radiating ciliary muscle or tensor choroideæ; cio, divided orbicular fibres; cip, ciliary process, along the inner border of which a layer of pigment is continued from the choroid to the uvea, excepting at the end of the process; Ch, choroid membrane; R, the retina close to the ora serrata; rc, the ciliary part of the retina, the structure of which is imperfectly represented; V, the vitreous humour; h, the hyaloid membrane; P, canal of Petit; h', the hyaloid membrane continued behind the canal to the capsule of the lens; Z, zonule of Zinn, and ll, suspensory ligament of the lens proceeding from the hyaloid covering the ciliary process to the front of the capsule of the lens.

The *vessels* and *nerves* have a radiating arrangement through the stroma; the former giving rise to rings, one at the circumference, the other near the pupil; and the latter forming a network. (See the description of the vessels and nerves of the vascular coat.)

Pupillary membrane (membrana pupillaris).—In fœtal life a delicate transparent membrane thus named closes the pupil, and completes the curtain of the iris. The pupillary membrane contains minute vessels, continuous with those of the iris and of the capsule of the crystalline lens; they are arranged in loops, which converge towards each other, but do not quite meet at the centre of the pupil. At about the seventh or eighth month of fœtal life these vessels gradually disappear; and, in proportion as the vascularity diminishes, the membrane itself is absorbed from near the centre towards the circumference. At the period of birth, often a few shreds, sometimes a larger portion, and occasionally the whole membrane is found persistent. (See also the account of the development of the eye.)

CILIARY MUSCLE, LIGAMENTUM PECTINATUM, AND CIRCULAR SINUS.

When the outer coat of the eyeball is separated from the choroid, a circular groove is seen passing round on the inner surface of the sclerotic, at its corneal margin. This groove is the outer wall of a venous canal, the *sinus circularis iridis* or *canal of Schlemm*. On the middle coat a corresponding groove, which completes the canal, is seen,—and this is bounded in front by a torn membranous edge bounding the anterior surface of the iris, the ligamentum pectinatum, while the thickest part of the white ring of the ciliary muscle is behind it. This canal communicates with other venous spaces which give an erectile appearance to the tissue at the base of the ciliary processes.

The *ligamentum pectinatum* consists of slight festoon-like processes of the fibres of the iris, lying in a transparent elastic fibrous tissue continuous with the posterior elastic layer of the cornea. It is a more developed structure in the eyes of the sheep and ox than in the human eye, and in them the festooned processes are prominent, giving a milled appearance like that of the edge of a coin.

The *ciliary muscle* (Bowman) forms a ring of unstriped muscular tissue about $\frac{1}{10}$th of an inch broad on the fore part of the choroid. Its fibres, yellowish-white in colour, and longitudinal in direction, are attached in front to the inner surface of the sclerotic coat ; and are also connected with the terminal fibres of the posterior elastic layer of the cornea. From that origin the fibres are directed inwards and backwards in a manner which in a section appears radiated, and end by joining the choroid coat opposite and beyond the ciliary processes. The muscle is soft, and ramified pigment-cells are scattered through its substance.

Concealed by the longitudinal or radiated fibres is a ring of fibres taking a circular direction, and which were still described as the ciliary ligament after the radiated fibres had been admitted to be muscular. This set constitutes the circular muscle of H. Müller.

The ciliary muscle appears to be in some way effective in producing the change in the form of the lens which takes place in accommodation of the eye to near vision (see Allen Thomson in "Glasgow Medical Journal" for 1857).

VESSELS AND NERVES OF THE MIDDLE TUNIC OF THE EYE.

The *arteries of the choroid* and the *ciliary processes* are derived from the posterior and anterior ciliary vessels. The posterior consist of two sets, distinguished as the short and the long. The *short* (posterior) *ciliary branches* of the ophthalmic artery pierce the sclerotic close to the optic nerve, and divide into branches which pass forward in meridional directions in the choroid membrane. Communicating freely they diminish in size, and entering the choroid form a close network of fine capillaries (*tunica Ruyschiana*) already described.

Fig. 467.

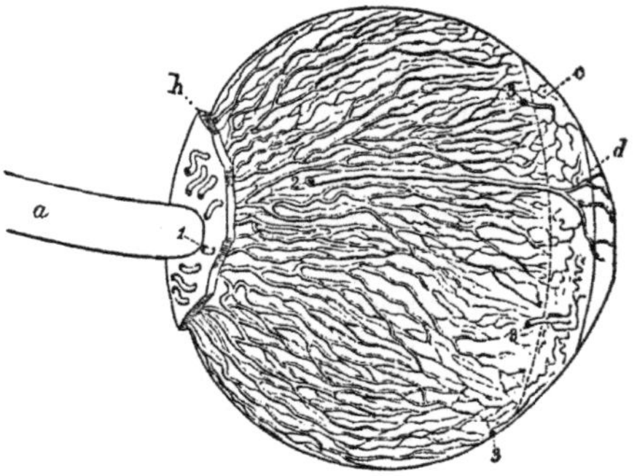

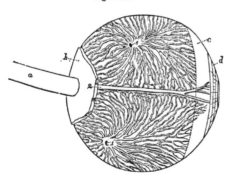

Fig. 468.

Fig. 467.—LATERAL VIEW OF THE ARTERIES OF THE CHOROID AND IRIS (from Arnold). ⅔

a, optic nerve ; *b*, part of the sclerotic left behind, the greater part and the cornea having been removed anteriorly ; *c*, ciliary muscle ; *d*, iris ; 1, posterior ciliary arteries piercing the sclerotic and passing along the choroid ; 2, one of the long posterior ciliary arteries ; 3, several of the short or anterior ciliary arteries.

The *veins of the choroid coat* constitute an outer layer, partially separable from the arterial network, and easily recognised by the direction of the larger vessels. These converge to four nearly equidistant trunks, which pass through the sclerotic about half way between the margin

Fig. 468.—LATERAL VIEW OF THE VEINS OF THE CHOROID (from Arnold).

The preparation is similar to that represented in the previous figure. 1, 1, two trunks of the venæ vorticosæ at the place where they leave the choroid and pierce the sclerotic coat.

of the cornea and the entrance of the optic nerve, and pour their contents into the ophthalmic vein. From their whorl-like arrangement they are known as the *vasa vorticosa*.

The *blood-vessels of the ciliary processes* are very numerous, and are derived from the anterior ciliary, and from those of the fore part of the choroidal membrane. Several small arterial branches enter the outer part of each ciliary process, at first running parallel to each other and communicating sparingly. As they enter the prominent folded portion, the vessels become tortuous, subdivide minutely, and inosculate frequently by cross branches. Finally they form short arches or loops, and turn backwards to pour their contents into the radicles of the veins.

On the free border of the fold, one artery, larger than the rest, extends

the whole length of each ciliary process, and communicates through intervening vessels with a long venous trunk which runs a similar course on the
attached surface.

Fig. 469.

Fig. 469.—INJECTED BLOOD-VESSELS OF THE CHOROID COAT (from Sappey). $\frac{20}{1}$

1, one of the larger veins; 2, small communicating vessels; 3, branches dividing into
the smallest vorticose vessels.

Arteries of the iris.—The special arteries of the iris are the long ciliary
and the anterior ciliary.

The *long* (posterior) *ciliary arteries,* two in number, and derived from the ophthalmic, pierce the sclerotic a little before, and one on each side of, the optic nerve.
Having gained the interval between the sclerotic and choroid coats, they extend
horizontally forwards through the loose connective tissue (*membrana fusca*) to the
ciliary muscle. In this course they lie nearly in the horizontal plane of the axis of the
eye-ball, the outer vessel being however a little above, and the inner one a little below
the level of that line. A short space behind the fixed margin of the iris each vessel
divides into an upper and a lower branch, and these, anastomosing with the corresponding vessels on the opposite side and with the anterior ciliary, form a vascular
ring (*circulus major*) in the ciliary muscle. From this circle smaller branches arise
to supply the muscle; whilst others converge towards the pupil, and there, freely communicating by transverse offsets from one to another, form a second circle of anastomosis (*circulus minor*), and end in small veins.

The *anterior ciliary arteries,* five or six in number, but smaller than the vessels
just described, are supplied from the muscular and lachrymal branches of the ophthalmic artery, and pierce the sclerotic about a line behind the margin of the cornea;
finally, they divide into branches which supply the ciliary processes, and join the
circulus major.

Besides these special arteries, numerous minute vessels enter the iris from the ciliary
processes.

The *veins of the iris* follow closely the arrangement of the arteries just
described. The circular sinus communicates with this system of vessels.

3 B 2

The *nerves* for the supply of the iris are named ciliary : they are numerous and large ; and, before entering the iris, divide in the substance of the
ciliary muscle.

Fig. 470.   Fig. 471.

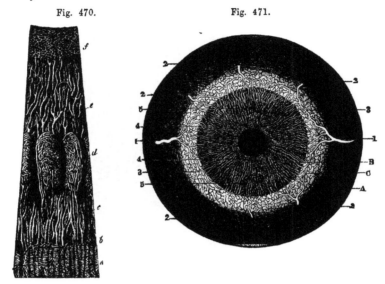

Fig. 470.—VESSELS OF THE CHOROID, CILIARY PROCESSES AND IRIS OF A CHILD (from
Kólliker after Arnold). $\frac{19}{1}$

a, capillary network of the posterior segment of the choroid ending at *b*, the ora
serrata ; *c*, arteries of the corona ciliaris, supplying the ciliary processes *d*, and passing
into the iris *e ; f*, the capillary network close to the pupillary margin of the iris.

Fig. 471.—FRONT VIEW OF THE BLOODVESSELS OF THE CHOROID COAT AND IRIS FROM
BEFORE (from Arnold). $\frac{2\frac{1}{2}}{1}$

A, interior part of the choroid : B, iris; C, ciliary muscle, &c. ; 1, 1, long posterior
ciliary arteries ; 2, five of the anterior ciliary arteries ramifying towards the outer margin
of the iris ; 3, loop of communication between one of the anterior and one of the long
posterior ciliary arteries ; 4, internal circle and network of the vessels of the iris ; 5,
external radial network of vessels.

Fig. 472.   Fig. 472.—LATERAL VIEW OF THE CILIARY
                                 NERVES (from Arnold).

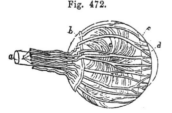

a, optic nerve ; *b*, back part of the sclerotic ;
c, ciliary muscle, &c. ; *d*, iris ; *e*, outer surface
of the choroid coat; 1, five of the ciliary nerves
passing along the sheath of the optic nerve,
piercing the sclerotic posteriorly, and thence
passing forward on the choroid membrane to
the ciliary muscle and iris. The nerves are
represented too large.

The *ciliary nerves*, about fifteen in number, and derived from the lenticular ganglion
and the nasal branch of the ophthalmic division of the fifth nerve, pierce the
sclerotic near the entrance of the optic nerve, and come immediately into contact with

the choroid. They are somewhat flattened in form, are partly embedded in grooves on the inner surface of the sclerotic, and communicate occasionally with each other before supplying the cornea and entering the ciliary muscle. When the sclerotic is

Fig. 473. — DISTRIBUTION OF NERVES IN THE IRIS (from Kölliker). $\frac{50}{1}$

The preparation was taken from the eye of an albino rabbit, and was treated with soda. *a*, smaller branches of the ciliary nerves advancing from the choroid ; *b*, loops of union between them at the margin of the iris ; *c*, arches of union in the iris ; *c′*, finer network in the inner part ; *d*, some of the terminations of single nerve-filaments in the outer part of the iris ; *e*, sphincter pupillæ muscle.

Fig. 473.

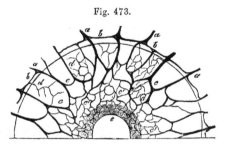

carefully stripped from the subjacent structures, these nerves are seen lying on the surface of the choroid. Within the ciliary muscle the nerves subdivide minutely, a few being lost in its substance, but the greater number pass on to the iris. In the iris the nerves follow the course of the blood-vessels, dividing into branches, which communicate with one another as far as the pupil. In the iris they soon lose their dark outline, and their mode of termination is not ascertained.

RETINA OR NERVOUS TUNIC.

The retina is a delicate almost pulpy membrane, which contains the terminal part of the optic nerve. It lies within the choroid coat, and rests on the hyaloid membrane of the vitreous humour. It extends forwards nearly to the outer edge of the ciliary processes of the choroid, where it ends in a finely indented border—*ora serrata*. From this border there is continued

Fig. 474.—THE POSTERIOR HALF OF THE RETINA OF THE LEFT EYE VIEWED FROM BEFORE (after Henle). $\frac{2}{1}$

s, the cut edge of the sclerotic coat ; *ch*, the choroid ; *r*, the retina : in the interior at the middle the macula lutea with the depression of the fovea centralis is represented by a slight oval shade ; towards the left side the light spot indicates the colliculus or eminence at the entrance of the optic nerve, from the centre of which the arteria centralis is seen spreading its branches into the retina, leaving the part occupied by the macula comparatively free.

onwards a thin layer of transparent nucleated cells (not nerve-elements) of an elongated or cylindrical form, constituting the pars ciliaris retinæ,

Fig. 474.

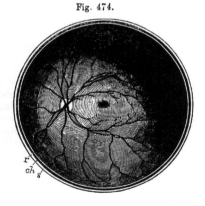

which reaches as far as the tips of the ciliary processes, and there gradually disappears. The thickness of the retina diminishes from behind forwards. In the fresh eye it is translucent and of a light pink colour ; but after death it soon becomes opaque, and this change is most marked under the action of water, alcohol, and other fluids. The outer surface is rough or slightly flocculent when the choroid is detached, and is in contact with the pigmentary layer; and from it a more or less complete stratum may be raised with care in a perfectly fresh eye. This layer, at first called membrane of Jacob from its discoverer, is now generally recognised as the columnar layer. The inner surface of the retina is smooth, and is merely applied to the vitreous body within it: on it the following objects may be seen. In the axis of the ball is a *yellow spot*—macula lutea (limbus luteus, Sömmerring), which is somewhat elliptical in shape, and about $\frac{1}{20}$th of an inch in diameter: in its centre is a slight hollow, *fovea centralis*, and, as the retina is thinner

Fig. 475.

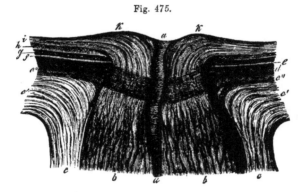

Fig. 475.—Section through the Middle of the Optic Nerve and the Tunics of the Eye at the Place of its Passage through them (from Kölliker after Ecker). $\frac{6}{1}$

The drawing was taken from a chromic acid preparation: *a*, arteria centralis retinæ; *b*, fasciculi of optic nerve fibres with neurilemma; *c*, sheath of the optic nerve, passing into *c'*, the sclerotic coat; *c''*, outermost pigmental layer of the choroid or membrana fusca; *d*, choroid and inner pigment-layer; *e, f*, columnar layer of the retina; *g*, the two granular layers; *h*, layer of nerve-cells; *i*, layer of nerve-fibres; *k*, colliculus or eminence at the entrance of the optic nerve; *l*, lamina cribrosa.

here than elsewhere, the pigmentary layer of the choroid is visible through it, giving rise to the appearance of a hole through the tunic. About $\frac{1}{10}$th of an inch inside the yellow spot is the round disc, *porus opticus*, where the optic nerve expands, and in its centre the point from which the vessels of the retina branch. At this place the nervous substance is slightly elevated so as to form an eminence (colliculus nervi optici).

STRUCTURE.—The retina, when examined microscopically in vertical sections, exhibits a series of dissimilar strata, together with structures not confined to one stratum. (1st) Externally is the columnar layer; (2nd), in the middle is the granular layer, comprising the external nuclear, the internuclear, the internal nuclear, and the molecular layers; and (3rd) internally is the nervous layer, consisting of three strata, one of nerve-cells, another of nerve-fibres—the ramifications of the optic nerve, and, on the inner surface of this last, a limiting membrane. (4th) Traversing the strata from the columnar layer to the limiting membrane, are placed vertical fibrils of varying kinds at different depths, and not fully ascertained to be continuous,—the radiating fibres of Müller. (5th) Blood-vessels distributed in the retina, are placed chiefly towards the inner surface.

1. The *columnar layer* (stratum bacillorum), consists of innumerable thin *rods*, placed vertically side by side like palisades, and of other larger bodies, more or less thickly interspersed among these, and named *cones*. These

Fig. 476.—VERTICAL SECTION OF A SMALL PART OF THE RETINA (after Kölliker). $\frac{350}{1}$

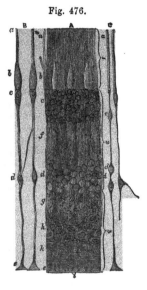

Fig. 476.

A, entire section of a small part of the retina; B, two cones represented separately in their connection with the fibres of Müller and other structures; C, two rods represented separately in their connection with the granules, fibres of, Müller, and the nerve-cells; 1, columnar layer; *a*, in A and C, the rods, in B, the terminal part of the cone; *b*, cones; 2, granular layer; *c*, outer layer of nuclei (striated corpuscles of Henle); *d*, inner layer of nuclei : *f*, internuclear layer; 3, nervous layer; *g*, fine molecular substance outside *h*, the nerve-cells; *k*, nerve-fibres; *l*, membrana limitans ; *e*, inner ends of the fibres of Müller resting on the limiting membrane.

structures are glistening, soft, easily destroyed, and lose their characters quickly in fluids. The rods are of uniform diameter, and are abruptly truncated externally. The cones are flask-shaped in the inner part of their extent, and taper to a rod-like extremity externally. Each cone rests on a pyriform cell continuous with it, and forming the extremity of a fibre of Müller ; while the rods end internally in pointed extremities ranging with these pyriform cells, and represented as formed by similar bodies (Kölliker); but this continuity with Müllerian fibres is still disputed. The dilated portions of the cones present granular contents, and a similar appearance is described in the inner halves of the rods. At the outer ends the rods project somewhat farther than the cones. When the outer surface of the retina is viewed about midway between its centre and margin with a strong enough magnifying power, a number of minute globular-looking bodies, the ends of the rods, appear ; and between them, at a deeper level, other

Fig. 477.—OUTER SURFACE OF THE COLUMNAR LAYER OF THE RETINA (from Kölliker). $\frac{350}{1}$

Fig. 477.

a, part of the columnar layer within the macula lutea, where only cones are present ; *b*, part near the macula, where a single row of rods intervenes between the cones; *c*, from a part of the retina midway between the macula and the ora serrata, showing a preponderance of the rods.

transparent larger bodies, the swellings of the cones, are seen, with a smaller circle within each—the end of its narrower part. Towards the margin the rods become more numerous ; near the centre the cones predominate ; and in the macula lutea the cones alone are seen.

2. The *external and internal nuclear* divisions of the *granular layer* are two collections of rounded and oval corpuscles, refracting light pretty strongly. The corpuscles of the internal nuclear layer are small cells with large nuclei,

as are also some, at least,' of those of the outer layer, namely, the pyriform
bodies supporting the cones.   But, according to recent investigations of
Henle, whose statements have been corroborated by Ritter as holding good
in the mammals generally, the bodies which constitute the bulk of the outer
nuclear layer, are elliptical corpuscles, which, when perfectly fresh, exhibit
transverse striation similar to muscular fibre, to the extent of three dark
lines alternating with clear substance in each, but which soon break up into
globules.

Fig. 478.   Fig. 478.—Striped Elliptical Corpuscles of the External Nuclear
                         Layer of the Retina (from Henle). $\frac{8\,0\,0}{1}$

The *internuclear layer*, which lies between the layers now
referred to, is a clear space of unequal depth, vertically striated,
and having likewise a molecular appearance.   The molecular
basis is more marked in a thin stratum which intervenes between the internal nuclear layer and the nerve-cells, and which,
therefore, has been distinguished as the *molecular layer*.

3. *Nervous layer.*—*a.* The *cellular layer* consists of nerve-cells with a
fine molecular material interspersed among them.   At the bottom of the
eye over the yellow spot they are thickest (from 8 to 10 cells deep), and
decrease in quantity in front ; so that at a fifth of an inch from the ora
serrata they are only scattered in clusters.   Around the entrance of the
optic nerve there is only a single stratum of these elements.   The cells when
fresh are transparent and nucleated, being roundish or pear-shaped in outline, and are furnished with from two to six ramified offsets.   By their
internal offsets the cells are continuous with the nerve-fibres beneath ; by
horizontal offsets they are united one with another ; and by those which
pass outwards they are connected with the corpuscles of the internal nuclear
layer.

b. The *nerve-fibre layer* consists of nerve-fibres directed forwards from the
optic nerve, and collected into small bundles, which, compressed laterally,
intercommunicate and form a delicate web with narrow elongated meshes.
This stratum diminishes in thickness forwards, and ends at the ora serrata :
it forms a continuous membrane, except at the yellow spot, where the nerve
fibres are wanting.   According to Bowman, the fibres, which lose their dark
outline on reaching the retina, consist there of an axis-cylinder only.   It is
now well established that they terminate in the nerve-cells on which they
lie, and this is the only mode of their termination which has been fully
ascertained.

c. *Membrana limitans and connective tissue.*—The limiting membrane lines
the inner surface of the retina, separating it from the vitreous body.   It is
an extremely thin and delicate membrane, which can be detached in shreds ;
and it agrees with the other glassy membranes of the eye-ball in not being
affected by alkalies, maceration, or boiling.   On its retinal surface it is
studded with the broadened insertions of vertical threads of connective tissue,
which separate the nerve-fibres into bundles, and form the inner parts of the
Müllerian fibres.   Nuclei apparently exist both in these and in the membrana limitans itself.   Delicate homogeneous connective tissue, likewise,
enters into the composition of the layers of the retina as far outwards as the
bases of the rods and cones, and gives there the appearance of a horizontal
line, the external limitary membrane of Schultze.

4. *Radiating fibres of Müller, and connections of the different elements of the
retina.*—From the foregoing description it will be gathered that the history of

.the Müllerian fibres is still incomplete. Indeed, the minuteness and delicacy of their structure renders their investigation one of the most difficult subjects of anatomical inquiry.   Heinrich Müller, to whom science chiefly owes the advance which has of late years been made towards the elucidation of the minute structure of the retina, described radiating fibres, extending vertically from the rods and cones to the membrana limitans, interrupted in their course by the corpuscles of the outer and inner nuclear layers, and connected with the nerve-cells.   He subsequently recognised the vertical fibres in the internal layers as connective tissue,—a view now universally adopted.   It appears to be clearly established, that from the pyriform corpuscles at the base of each cone a thread passes inwards to a corpuscle of the internal nuclear layer.   It is also stated that more slender threads unite the rods with the deep layers ; and Kölliker represents a thread passing out from a corpuscle of the internal nuclear layer as afterwards dividing into branches, on which are placed corpuscles of the external nuclear layer, and which terminate in rods.   This account of the structure seems best to accord with the physiological view now very generally held, that the columnar layer is the more immediate seat of the formation of a distinct image in vision, and of the reception of visual impressions from rays of light impinging upon the retina. It is right to state, however, that a different view is taken by Henle, who believes that the rods are free, and that the fibres observed by H. Müller and Kölliker are artificial products, the result of coagulation by re-agents. Henle regards the retina as composed of an outer part, which he terms the mosaic layer, and which comprises the columnar structures of Jacob's membrane, and the external nuclear layer, and is destitute of blood-vessels ; and an inner nervous part comparable to the structure found in the cerebral convolutions, and consisting of a stratum of nerve fibres and of two strata of nerve cells alternating with granular strata ; the corpuscles of the internal nuclear layer being considered by him as nerve cells of a smaller order than those of the cellular layer.

5. *Vessels of the retina.*—An artery enters and a vein leaves the retina between the bundles of fibres of the optic nerve.

The artery (arteria centralis retinæ) is an offset of the ophthalmic, and divides into four or five primary branches as soon as it enters the eye-ball. These larger offsets are situated at first on the inner surface of the nerve fibres, but they soon pass between these into the stratum of nerve-cells, where they form a network of very fine capillaries with rather wide meshes, which reaches in front to the ora serrata.

The vein corresponding to this artery has a similar distribution : it terminates in the ophthalmic vein.   In animals there is a circular vessel (circulus venosus retinæ) following the line of the ora serrata.

Constituents of the retina in the yellow spot.—In this part of the retina the several layers above described undergo some modification : the following are the alterations in the strata from without inwards.   In the columnar layer, only the cones are present, but they are set close together, and are smaller than elsewhere.   The granular layer is absent opposite the fovea centralis.   The nervous layer is thus modified : the nerve-cells cover the whole spot, like laminated epithelium, and rest internally on the membrana limitans ; but the molecular substance outside them is absent over the fovea centralis ; the nerve-fibres extend only into the circumference of the spot amongst the cells, without forming a layer over it.   The fibres of Müller are found at the circumference but not over the fovea centralis; they have an oblique, almost horizontal direction, and present a specially nerve-like appearance. Only capillary vessels occupy the yellow spot, the larger branches passing round it.

Fig. 479.

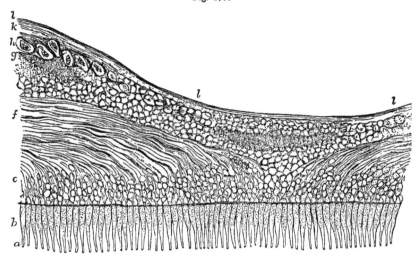

Fig. 479.—Vertical Section of the Retina through the Middle of the Fovea Centralis (from Henle.) $\frac{300}{1}$

This figure is taken from a preparation of the human retina hardened in alcohol, and is designed to show the peculiarities of this part as compared with other regions of the retina, viz., the obliquity of the Müllerian fibres, the thinness of the layer of nerve-fibres, and the absence of the granular layer in the centre. *a, b,* cones of the columnar layer; *c,* external nuclear layer; *d,* internal nuclear layer; *f,* external fibrous layer; *g,* molecular substance next to *h,* the ganglionic layer; *k,* the layer of nerve-fibres; *l,* the internal limiting membrane.

The *yellow colour* of the macula lutea is deepest towards the centre, and is due to a pigment which imbues all the layers except the columnar: it does not appear to be contained in cells, and is soon removed after death by the action of water.

Ciliary part of the retina.—The structure which has been named the ciliary part of the retina is situated in front of the ora serrata, and extends thence over the inner ends of the ciliary processes to the base of the iris (therefore, over the whole corona ciliaris). Though entirely destitute of the nervous parts of the retinal structure, it is still in continuity with the substance of the retina, and in the form of a grey membrane, adheres to the ciliary processes and zonule of Zinn, and is usually in great part detached from the neighbouring parts along with the latter. According to Kölliker, this layer consists of elongated nucleated cells, which in the human subject are broad externally, and with flat or forked bases set upon the internal limiting membrane. He regards these cells as probably corresponding to the Müllerian fibres, and as constituting in this place the only representative of the retinal structure.

On the structure of the retina may be consulted Heinrich Müller, in Siebold und Kölliker's Zeitschrift, 1851 and 1856; M. Schultze, "Obs. de retinæ Struct. penit.," 1859; Goodsir, in Edin. Med. Journal, 1855; Kölliker, Handbuch d. Gewebelehre, 4th ed., 1863; and Henle, Handbuch d. System. Anatomie, vol. ii., 1866.

Fig. 480.   Fig. 481.

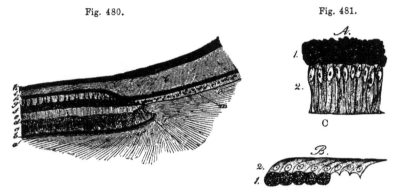

Fig. 480.—VERTICAL SECTION THROUGH THE CHOROID AND RETINA NEAR THE ORA SERRATA (from Kölliker). ♀

a, hyaloid membrane; *a'*, indications of fibres which radiate from the anterior margin of the retina into the vitreous body; *b*, limiting membrane and fibrous layer of the retina; *c*, ganglionic layer with a few cells shown; *d*, inner nuclear layer; *e*, inter-nuclear substance; *f*, outer nuclear layer; *g*, columnar layer; *h*, dark pigment; *i*, middle layer of the choroid; *l*, beginning of one of the ciliary processes; *m*, ciliary part of the retina. (The recess shown at *a'* is not constant.)

Fig. 481.—A SMALL PORTION OF THE CILIARY PART OF THE RETINA (from Kölliker). ²⁴⁰⁄₁

A, human; B, from the ox; 1, pigment-cells; 2, cells forming the ciliary part.

THE VITREOUS BODY.

The vitreous body is the largest of the transparent parts occupying the centre of the eye-ball. Globular in form, it occupies about four-fifths of the ball, and supports the delicate retina, being in contact with the membrana limitans. On the fore part it is hollowed out for the reception of the lens and its capsule, and behind it is more closely connected with the retina than at the sides, having received at that part offsets of the retinal vessels in fœtal life. It is quite transparent, and like a thin jelly in its interior. Its surface is formed by a thin enveloping glassy membrane, named hyaloid, and as long as this membrane is entire, it retains its form in water. No vessels enter it, and its nutrition must be therefore dependent upon the surrounding vascular textures—viz., the retina, and the ciliary processes.

The *hyaloid* is an extremely thin and clear membrane. When traced forwards it is found to be connected, opposite the outer part of the ciliary processes, with a firm membrane passing in front of the marginal part of the lens (suspensory ligament), while a thinner layer, proceeding inwards from this, becomes united with the posterior layer of the capsule of the lens, so that it is doubtful whether or not the membrane is prolonged between the capsule and the vitreous body. On the inner surface of the hyaloid are a few delicate nuclei. Fibres have been supposed to be prolonged inwards from it, to form cells for the contained fluid, but observations with the microscope do not show any in the adult, though in the fœtus there are fibres in the interior of the vitreous mass, with "minute

Fig. 482.

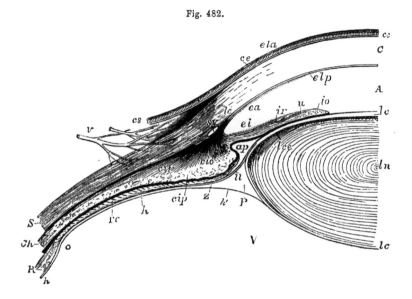

Fig. 482.—Vertical Section of a Part of the Eyeball, showing the Connections of the Cornea, Sclerotic, Iris, Ciliary Muscle, Hyaloid, and Lens. ⁹⁄₂

The full description of this figure will be found at p. 720; the following references apply to the lens and parts connected with it. A, the anterior aqueous chamber in front of the pupil; ap, the recess forming the posterior division of the aqueous chamber, the iris resting between this and the pupil on the surface of the lens; ir, radiating fibres of the iris or dilatator pupillæ muscle; io, orbicular fibres or sphincter muscle; u, pigment layer of the iris or uvea; ln, the lens at its centre; lc, its capsule; lce, granular or cellular layer in front of the lens; this layer is seen to terminate abruptly at the margin of the lens, where the new fibres of the lens are developed, and from whence the nuclei of the fibres extend for a certain depth inwards in an irregular plane in the growing lens; h, the hyaloid membrane; Z, the zonule of Zinn; P, the canal of Petit; ll, in front of it the suspensory ligament of the lens; h', the part of the hyaloid which closes the canal of Petit behind and extends to the posterior surface of the lens; V, the vitreous humour.

nuclear granules" at their point of junction. (Bowman.) It is still doubtful how far the appearances of lamination produced by the action of chromic acid, or of radiated fibrillation resulting from congelation, are true indications of any actually existing structure in the interior of the vitreous humour.

The *fluid* collected from the vitreous body by puncturing it resembles water: it contains, however, some salts with a little albumen.

THE LENS AND ITS CAPSULE.

The lens, enclosed in a capsule, is situated behind the pupil, and in front of the vitreous body.

The *capsule* of the lens, a transparent glass-like membrane closely surrounding the contained body, is hard and brittle, especially in front, but very elastic and permeable to fluid. The anterior surface is in contact with the iris towards the pupil, and recedes from it slightly at the circumference ; the posterior rests closely on the vitreous body. Around the circumference is a space to be afterwards noticed, the canal of Petit. The fore part of the capsule is several times, thicker than the back, as far out as to $\frac{1}{15}$th of an inch from the circumference, where the suspensory ligament joins it ; but beyond that spot it becomes thinner, and it is thinnest behind. In its nature the capsule of the lens resembles the glassy membrane at the back of the cornea, for it is structureless, and remains transparent under the action of acids, alcohol, and boiling water ; and when ruptured, the edges roll up with the outer surface innermost. (Bowman.)

Connecting the anterior wall of the capsule closely to the lens is a single layer of granular and nucleated polygonal cells, which ends abruptly where the capsule comes in contact with the hyaloid membrane. The place of termination of this cellular layer round the margin of the lens corresponds to the line from which the fibres of the lens are developed. There is no such layer of cells on the posterior wall of the capsule, but in hardened specimens various reticulated appearances may be detected, which probably arise, as supposed by Henle, from the pressure one on another of globules of a fluid separated from the lens after death, and known as liquor Morgagni.

No vessels enter the capsule of the lens in the adult. In the fœtus it receives an artery behind, which is named the *capsular* artery. This vessel leaves the arteria centralis retinæ at the centre of the optic nerve, and passing through the substance of the corpus vitreum, enters the posterior portion of the capsule of the lens, where it divides into radiating branches. These form a fine network, turn round the margin of the lens, and extend forwards to become continuous with the vessels in the pupillary membrane and the iris.

Some authors (Albinus, Zinn, &c.) state that they have traced vessels from the capsule into the substance of the lens itself.

THE LENS.

The lens (lens crystallina) is a doubly convex transparent solid body, with a rounded circumference. Its convexity is not alike on the two surfaces,

Fig. 483.—LAMINATED STRUCTURE OF THE CRYSTAL-LINE LENS (from Arnold). ‡

Fig. 483.

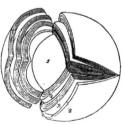

The laminæ are split up after hardening in alcohol. 1, the denser central part or nucleus; 2, the successive external layers.

being greatest behind, and the curvature is less at the centre than towards the margin. It measures about $\frac{1}{3}$rd of an inch across, and $\frac{1}{5}$th from before backwards. In a fresh lens the outer portion is soft and easily detached ; the succeeding layers are of a firmer consistence ; and in the centre the substance becomes much harder, constituting the nucleus. On the anterior and· posterior surfaces are faint white lines directed from the poles towards the circumference ; these in the adult are somewhat variable and numerous on the surface, but in the fœtal lens throughout, and towards the centre of the lens in the adult, they are three in number, diverging from each other like rays at equal angles of 120°. The lines at opposite poles have an intermediate position (not being over one another) : they are the edges of planes or septa projecting vertically inwards to the centre of the lens, and receiving the ends of the lens-fibres which are collected upon them.

Fig. 484.

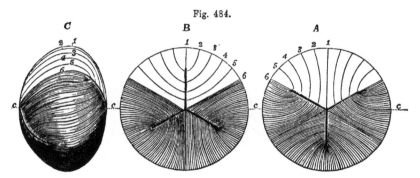

Fig. 484.—Outlines illustrating the Course of the Fibres in the Fœtal Crystalline Lens. ⅞

This diagram represents the typical or more simple state of the fibres in the full-grown fœtal or infantile condition ; the three dotted lines radiating at equal angles of 120° from the centre indicate the position of the intersecting planes, where they reach the surface ; the figures 1, 2, 3, 4, 5, and 6, indicate certain fibres selected arbitrarily at equal distances in one-sixth part of the lens to show their course from the front to the back ; A, the anterior surface ; B, the posterior surface ; C, the lateral aspect : in these several figures, for the sake of clearness, a few lines only are introduced into the upper third, while in the lower two-thirds a greater number are marked ; but no attempt is made to represent the number existing in nature ; the parts of the dotted line marked c, are on a level with the centre of the several lenses.

Fig. 485.

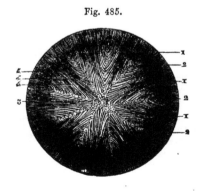

Fig. 485.—Front View of the Fibrous Structure of the Adult Lens (from Sappey after Arnold). ⅚

In this figure more numerous planes of intersection of the fibres are shown than in fig. 484.

STRUCTURE.—When the lens has been dried, or hardened by immersion in spirit, boiling water, or other fluid capable of rendering it firm and white, concentric laminæ, narrowing to a point at the poles, may be detached from it. The laminæ are further composed of microscopic fibres, which adhere together by wavy or slightly serrated margins. The lens is albuminous in its composition, and is devoid of blood-vessels ; and at the planes of intersection a finely granular homogeneous material takes the place of the fibres.

The *fibres of the lens* are somewhat flattened threads, about $\frac{1}{5000}$th of an inch wide, and are directed over the edge of the lens from the planes on one surface to those on the other. In their course between the opposite surfaces, no fibre passes from pole to pole, but the fibres beginning in the pole or centre of one surface terminate in the end of a plane on the opposite surface, and vice versâ; the intervening fibres passing to their corresponding places between. Some of the superficial fibres possess transparent nuclei, at nearly regular intervals. In the more superficial fibres of the growing lens the nuclei occupy very regularly the equatorial part. At their ends, where the fibres meet the planes, they are soft and indistinct; and at the

Fig. 486.—Magnified View of the Fibres of the Crystalline Lens.

Fig. 486.

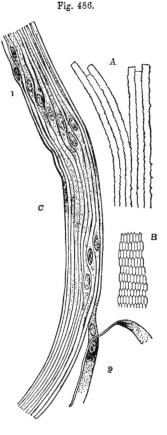

A, longitudinal view of the fibres of the lens from the ox, showing the serrated edges. B, transverse section of the fibres of the lens from the human eye (from Kölliker). C, longitudinal view of a few of the fibres from the equatorial region of the human lens (from Henle). $\frac{250}{1}$ The most of the fibres are seen edgeways, and, towards 1, present the swellings and nuclei of the "nuclear zone;" at 2, the flattened sides of two fibres are seen.

middle part, where they are placed on the margin of the lens, they are widest and best marked. The fibres are six-sided prisms, flattened in the plane of the lamina in which they lie. The edges are bevelled and sinuous; they are very regularly toothed at the edges in fishes and some other animals

Fig. 487.

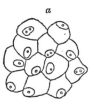

Fig. 487.—Cells connecting the Lens with its Capsule (from Bowman). $\frac{250}{1}$

for more perfect junction with those in the same plane; but in man and mammals, the edge is only slightly jagged or irregularly serrated.

Changes in the lens by age.—In the *fœtus*, the lens is nearly spherical: it has a slightly reddish colour, is not perfectly transparent, and is softer, and more readily broken down than at a more advanced age.

· In the *adult*, the anterior surface of the lens becomes more obviously less convex than the posterior ; and the substance of the lens is firmer, colourless, and transparent.

Fig. 488.

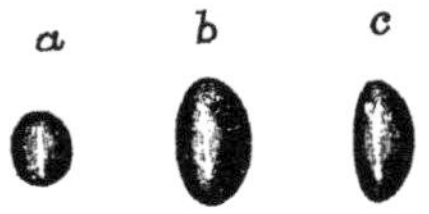

Fig. 488.—Side Views of the Lens at Different Ages.

a　*b*　*c*

a, at birth with the deepest convexity; *b*, in adult life with medium convexity; *c*, in old age with considerable flattening of the curvatures.

In *old age*, it is more flattened on both surfaces ; it assumes a yellowish or amber tinge, and is apt to lose its transparency as it gradually increases in toughness and specific gravity.

SUSPENSORY LIGAMENT OF THE LENS AND CANAL OF PETIT.

The *suspensory ligament* of the lens—Retzius—(Zonula of Zinn) is a slender but tolerably firm transparent membrane, which, attached to the fore part of the capsule of the lens close to its circumference, passes outwards to join the hyaloid membrane of the vitreous humour at its most anterior convex part, opposite the ora serrata of the retina, and assists in retaining the lens in its place. On the anterior surface small streaks of pigment are observable after its separation from the other membranes, and when this pigment is removed by washing, small but regular folds—processus ciliares zonulæ—come into view near the lens ; these are plaits in the membrane, and are received into the intervals between the ciliary processes of the choroid coat, into which they fit. Between the folds and the lens-capsule is a slight interval free from plaits, which forms part of the boundary of the posterior aqueous chamber. The posterior surface is turned towards the hyaloid membrane, from which it is separated near the lens by a space named the canal of Petit.

The suspensory ligament has chemical properties similar to those of the capsule of the lens, but in it parallel or slightly radiating longitudinal fibres may be recognised, which are stiff, elastic, and pale, resembling those of elastic tissue, being less pliable, and less acted on by acetic acid than those of connective tissue.

Fig. 489.

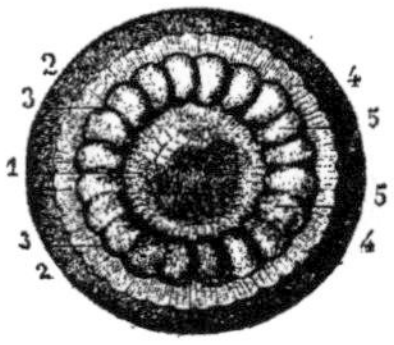

Fig. 489.—View from before of the Canal of Petit inflated (from Sappey).

The anterior parts of the sclerotic, choroid, iris and cornea having been removed, the remaining parts are viewed from before, and the canal of Petit has been inflated with air through an artificial opening. 1, front of the lens ; 2, vitreous body ; 3, outer border of the canal of Petit ; 4, outer part of the zonule of Zinn ; 5, appearance of sacculated dilatations of the canal of Petit.

The *canal of Petit* is the interval surrounding the edge of the lens-capsule, bounded in front by the suspensory ligament, and behind by the hyaloid membrane. Its width is about $\frac{1}{10}$th of an inch. On blowing air into it through an opening in the anterior boundary, the plaits of the suspensory ligament on its front are distended, and the canal presents a sacculated appearance.

AQUEOUS HUMOUR AND ITS CHAMBER.

The aqueous humour fills the space in the fore part of the eyeball, between the cornea and the capsule of the lens with its suspensory ligament. The iris, resting in part upon the lens, divides the aqueous chamber partially into two. The aqueous humour differs little from water in its physical characters; but it contains a small quantity of some solid matter, chiefly chloride of sodium, dissolved in it.

The chambers, into which the space containing the aqueous humour is divided by the iris, are named respectively the anterior and posterior. This subdivision is incomplete in the adult, but in the fœtus before the seventh month it is completed by means of the membrana pupillaris, which by its union with the margin of the pupil closes the aperture of communication between the two chambers.

The *anterior chamber* is limited in front by the cornea and behind by the iris, while opposite the pupil it is bounded by the capsule of the lens.

The *posterior chamber* was originally so named in the belief that a free space intervened between the iris and the capsule of the lens. It is now, however, well ascertained by observations on the living eye, and by sections made in the frozen state, that the pupillary margin and part of the posterior surface of the iris are in contact with the capsule of the lens; and the term posterior chamber can therefore be employed only to indicate the want of continuity between those opposed structures, where no space actually intervenes, and to the angular interval existing at the circumference between the ciliary processes, the iris, and the suspensory ligament.

DEVELOPMENT OF THE EYE.

The eyes begin to be developed at a very early period, in the form of two hollow processes projecting one from each side of the first primary cerebral vesicle. Each process becomes converted into a flask-shaped vesicle, called the *primary optic vesicle*, which communicates by a hollow pedicle with the base of the posterior division of the first primary cerebral vesicle. (See p. 578, and fig. 386 B.) According to the observations of Remak on the chick, the pedicles, originally separate, come together, and their cavities temporarily communicate,—a condition which may explain the formation of the optic commissure. The primary optic vesicle comes into contact at its extremity with the cuticle, which somewhat later becomes invaginated at this point, and forms a small pouch pressing inwards on the optic vesicle; the aperture of this

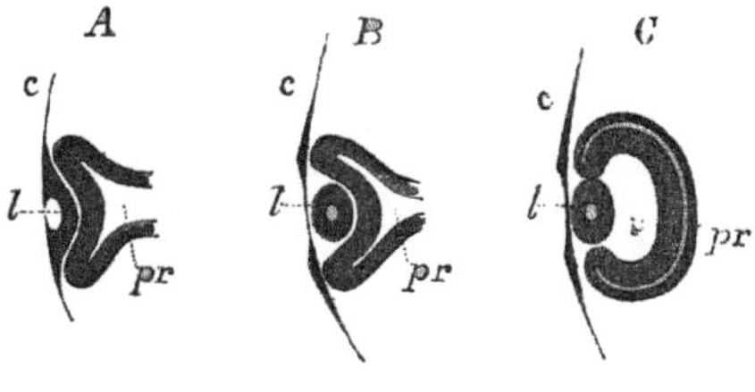

Fig. 490.—LONGITUDINAL SECTION OF THE PRIMARY OPTIC VESICLE IN THE CHICK MAGNIFIED (from Remak).

A, from an embryo of sixty-five hours; B, a few hours later; C, of the fourth day; c, the corneous layer or epidermis, presenting in A, the open depression for the lens, which is closed in B and C; l, the lens-follicle and lens; pr, the primary optic vesicle; in A and B, the pedicle is shown; in C, the section being to the side of the pedicle, the latter is not shown; v, the secondary ocular vesicle and vitreous humour.

pouch becomes constricted and closed, and the pouch is soon converted into a shut sac, within which the contents subsequently becoming solid form the lens and its capsule. After the lens has been separated from the cuticle, the deeper tissue sends a

3 c

projection from below upwards between the lens and the optic vesicle, in such a
manner as to invaginate the superficial and lower walls of the vesicle, pressing them
upwards and inwards on the superior and deep walls, and giving them the form of a
cup imperfect below, the *secondary optic vesicle.* The involution gives rise to the

Fig. 491.

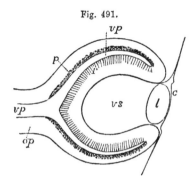

(Fig. 491.—DIAGRAMMATIC SKETCH OF A
VERTICAL LONGITUDINAL SECTION
THROUGH THE EYEBALL OF A HUMAN
FŒTUS OF FOUR WEEKS (after Kölli-
ker). $\frac{100}{1}$

The section is a little to the side so as to
avoid passing through the ocular cleft. *c,*
the cuticle, where it becomes later the
cornea ; *l,* the lens; *o p,* optic nerve
formed by the pedicle of the primary optic
vesicle ; *vp,* primary medullary cavity or
optic vesicle ; *p,* the pigment-layer of the
choroid coat of the outer wall ; *r,* the
inner wall forming the retina ; *v s,* secondary optic vesicle containing the rudiment
of the vitreous humour.

cavity in which the vitreous humour is formed ; and, the forepart of the optic nerve
participating in the invagination, it is by this means that the central artery of the
retina is introduced into the nerve and the eyeball, being, as it were, folded within
them. The deficiency in the wall of the cup of the secondary vesicle inferiorly is

Fig. 492.

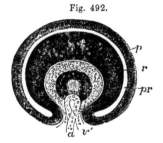

Fig. 492.—TRANSVERSE VERTICAL SECTION OF THE
EYEBALL OF A HUMAN EMBRYO OF FOUR WEEKS
(from Kölliker). $\frac{100}{1}$

The anterior half of the section is represented.
pr, the remains of the cavity of the primary optic
vesicle ; *p,* the inner part of the outer layer, forming the choroidal pigment; *r,* the thickened inner
part giving rise to the columnar and other structures of the retina ; *v,* the commencing vitreous
humour within the secondary optic vesicle ; *v',* the
ocular cleft through which the loop of the central
blood-vessel, *a,* projects from below ; *l,* the lens
with a central cavity.

gradually filled up by the growing downwards of the edges, until only a cleft is left,
which is discernible for a considerable time, and has been named the ocular cleft. The
history of this cleft is of interest in connection with congenital fissure of the iris
(coloboma iridis) and the accompanying condition of the choroid membrane. Some
difference of opinion exists with regard to the subsequent history of the walls of the
secondary vesicle, but the opinion of Kölliker appears to be well founded, that the
invaginated layer forms the retina, and the outer part the pigmentary epithelium
of the choroid. Thus the elements of Jacob's membrane and the hexagonal cells of
the choroid may be regarded as originally continuous, forming together the epithelial
lining of the cavity of the primary vesicle ; and the development of nervous tissue
underneath Jacob's membrane, while none exist beneath the choroidal layer, is a
circumstance which may be looked upon as analogous to the absence of nervous tissue
from various parts of the walls of the cerebral vesicles. The sclerotic coat and
cornea are formed from the surrounding tissue external to the parts of the eye which
they enclose ; and, according to Kölliker, the vascular part of the choroid is of later
formation. Still later, in the second month of fœtal life, the iris begins to be formed

as a septum projecting inwards from the forepart of the choroid coat, between the lens and the cornea.

Fig. 493.—VERTICAL LONGITUDINAL SECTION OF THE EYE OF AN EMBRYO CALF (from Kölliker). ♀

c, the cornea ; *cc*, conjunctiva of the cornea ; *l*, the lens ; *v*, vitreous humour ; *r*, retina ; *p*, pigment-layer of the choroid ; *s c*, commencement of the sclerotic and choroid coats ; *m*, superior and inferior recti muscles ; *pa*, folds of integument forming the commencement of the upper and lower eyelids.

Fig. 493.

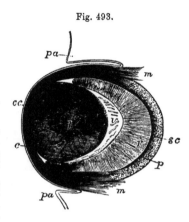

The crystalline lens in the fœtus is surrounded by a highly vascular tunic, supplied by a branch of the central artery of the retina, which passes forwards in the axis of the globe, and breaks up at the back of the lens into a brush of rapidly subdividing branches. The forepart of this tunic, adherent to the pupillary margin of the iris, forms the *pupillary membrane* by which the aperture of the pupil is closed. The whole tunic, however, together with the artery which supplies it, becomes atrophied, and is lost sight of before birth in the human subject, although in some animals it remains for a few days after. According to Kölliker, the anterior chamber is formed

Fig. 494.—BLOODVESSELS OF THE CAPSULO-PUPILLARY MEMBRANE OF A NEW-BORN KITTEN, MAGNIFIED (from Kölliker).

The drawing is taken from a preparation injected by Tiersch, and shows in the central part the convergence of the network of vessels in the pupillary membrane.

Fig. 494.

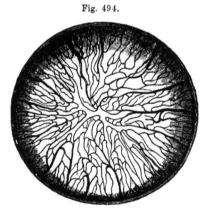

only a short time before birth by the intervention of the aqueous humour between the iris and cornea.

The *eyelids* make their appearance as folds of integument, subsequently to the formation of the globe. When they have met together in front of the eye, their edges become closely glued together ; and they again open before birth.

The lachrymal canal may be regarded as a persistently open part of the fissure between the lateral frontal process and maxillary lobe of the embryo. (See p. 65, and fig. 56 B, 4, 6.)

THE organ of hearing is divisible into three parts : the external ear, the tympanum or middle ear, and the labyrinth or internal ear. The first two of these are to be considered as accessories or appendages to the third, which is the sentient portion of the organ.

Fig. 495.

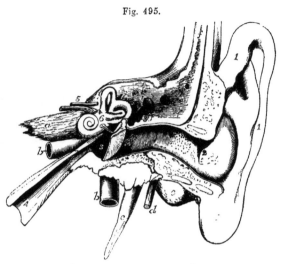

Fig. 495.—DIAGRAMMATIC VIEW FROM BEFORE OF THE PARTS COMPOSING THE ORGAN OF HEARING OF THE LEFT SIDE (after Arnold, and from nature).

The temporal bone of the left side, with the accompanying soft parts, has been detached from the head, and a section has been carried through it transversely so as to remove the front of the meatus externus, half the tympanic membrane, the upper and anterior wall of the tympanum and Eustachian tube. The meatus internus has also been opened, and the bony labyrinth exposed by the removal of the surrounding parts of the petrous bone. 1, the pinna and lobe ; 2, 2', meatus externus ; 2', membrana tympani ; 3, cavity of the tympanum ; 3', its opening backwards into the mastoid cells ; between 3 and 3', the chain of small bones ; 4, Eustachian tube ; 5, meatus internus containing the facial (uppermost) and the auditory nerves ; 6, placed on the vestibule of the labyrinth above the fenestra ovalis ; a, apex of the petrous bone ; b, internal carotid artery ; c, styloid process ; d, facial nerve issuing from the stylo-mastoid foramen ; e, mastoid process ; f, squamous part of the bone covered by integument, &c.

THE EXTERNAL EAR.

In the external ear are included the pinna,—the part of the outer ear which projects from the side of the head ; together with the meatus or passage which leads thence to the tympanum, and is closed at its inner extremity by a membrane (membrana tympani) interposed between it and the middle ear.

THE PINNA.

Superficial configuration.—The general form of the pinna or auricle is concave, as seen from the outside, to fit it for collecting and concentrating the undulations of sound ; it is thrown into various elevations and hollows, to which distinct names have been given. The largest and deepest concavity, a little below the centre of the organ, is called the *concha ;* it surrounds the entrance to the external auditory meatus, and is unequally divided at its upper part by a ridge, which is the beginning of the helix. In front of the concha, and projecting backwards over the meatus auditorius, is a conical prominence, the *tragus,* covered usually with hairs. Behind this, and separated from it by a deep notch (incisura intertragica), is another smaller elevation, the *antitragus.* Beneath the antitragus, and forming the lower end of the auricle, is the *lobule,* which is devoid of the firmness and elasticity that characterise the rest of the pinna. The thinner and larger portion of the pinna is bounded by a prominent and incurved margin, the *helix,* which, springing above and rather within the tragus, from the hollow

Fig. 496.—OUTER SURFACE OF THE PINNA OF THE RIGHT AURICLE. ⅔

Fig. 497.

1, helix ; 2, fossa of the helix ; 3, antihelix ; 4, fossa of the antihelix ; 5, antitragus ; 6, tragus ; 7, concha ; 8, lobule.

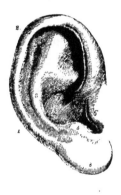

of the concha, surrounds the upper and posterior margin of the auricle, and gradually loses itself in the back part of the lobule. Within the helix is another curved ridge, the *antihelix,* which, beginning below at the antitragus, sweeps round the hollow of the concha, forming the posterior boundary of that concavity, and is divided superiorly into two diverging ridges. Between the helix and the antihelix is a narrow curved groove, the *fossa of the helix* (fossa innominata, scaphoidea) ; and in the fork of the antihelix is a somewhat triangular depression, the *fossa of the antihelix* (fossa triangularis vel ovalis).

Structure.—The pinna consists of a thin plate of cartilage and of integument, with a certain amount of adipose tissue. It presents also several ligaments and small muscles of minor importance.

The *skin of the pinna* is thin, closely adherent to the cartilage, and contains sebaceous follicles, which are most abundant in the hollows of the concha and scaphoid fossa.

The *cartilage* presents all the inequalities of surface already described as apparent on the outer surface of the pinna ; and on its cranial surface exhibits prominences the reverse of the concha and the fossa of the helix, while between these is a depression in the situation of the antihelix. This cartilage is not confined to the pinna, but enters likewise into the construction of the outer part of the external auditory canal. When dissected separate from other structures, it is seen to be attached by fibrous tissue to the rough and prominent margin of the external auditory meatus of the temporal bone. The tubular part is cleft in front from between the tragus and fore part of the helix inwards to the bone, the deficiency being filled with fibrous membrane ; thus the cartilage may be said to be a plate, a part of which assumes the tubular form by being folded so as to bring the upper margin, which lies in front of the tube of the ear, nearly into contact with the lower part, which being coiled inwards upon itself forms the upper border of the tragus. Following the free border of the plate backwards beneath the meatus, it is seen to pass round the lower margin of the concha, and to form the prominences of the tragus and antitragus, while the cartilage is absent altogether from the lobule, which contains only fat and tough connective tissue. Behind the antitragus is a deep notch, separating it from the cartilage of the helix, which here forms a tail-like process descending towards the lobule. At the fore part of the pinna, opposite the first bend of the helix, is a small conical projection of the cartilage, called the *process of the helix*, to which the anterior ligament is attached. Behind this process is a short vertical slit in the helix; and on the surface of the tragus is a similar but somewhat longer fissure. A deep fissure passes back between the commencement of the helix and the tube of the ear, and another passing outwards and backwards from the deep end of the longitudinal cleft separates the part forming the tragus from the rest of the tube, so that the tube is continuous with the pinna only by means of a narrow isthmus. One or two other irregular gaps or fissures partially divide the cartilaginous tube transversely, and the whole of these deficiencies are termed *fissures of Santorini*. The substance of the cartilage is very pliable, and is covered by a firm fibrous perichondrium.

Of the *ligaments of the pinna*, the most important are two, which assist in attaching it to the side of the head. The *anterior* ligament, broad and strong, extends from the process of the helix to the root of the zygoma. The *posterior* ligament fixes the back of the auricle (opposite the concha) to the outer surface of the mastoid process of the temporal bone. A few fibres attach the tragus also to the root of the zygoma. Ligamentous fibres are likewise placed across the fissures and intervals left in the cartilage.

Of the *muscles of the pinna*, those which are attached by one end to the side of the head, and move the pinna as a whole, have been already described (p. 170): there remain to be examined several smaller muscles, composed of thin layers of pale fibres, which extend from one part of the pinna to another, and may be named the special muscles of the organ. Six small muscles are distinguished; four being placed on the outer and two on the inner or deep surface of the pinna.

The *smaller muscle of the helix* (m. minor helicis) is a small bundle of oblique fibres, lying over, and firmly attached to, that portion of the helix which springs from the bottom of the concha.

Fig. 497   Fig. 498.

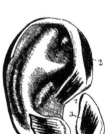 

Fig. 497.—CARTILAGE OF THE PINNA EXPOSED, WITH THE MUSCLES ON ITS OUTER SURFACE.

1, musculus helicis minor; 2, m. helicis major; 3, tragicus; 4, antitragicus.

Fig. 498.—INNER SURFACE OF THE CARTILAGE OF THE PINNA WITH THE SMALL MUSCLES ATTACHED.

5, transversus auriculæ muscle; 6, obliquus auriculæ muscle.

The *greater muscle of the helix* (m. major helicis) lies vertically along the anterior margin of the pinna. By its lower end it is attached to the process of the helix; and above, its fibres terminate opposite the point at which the ridge of the helix turns backwards.

The *muscle of the tragus* (m. tragicus) is a flat bundle of short fibres covering the outer surface of the tragus : its direction is nearly vertical.

The *muscle of the antitragus* (m. antitragicus) is placed obliquely over the antitragus and behind the lower part of the antihelix. It is fixed at one end to the antitragus, from which point its fibres ascend to be inserted into the tail-like extremity of the helix, above and behind the lobule.

The *transverse muscle* (m. transversus auriculæ) lies on the inner or cranial surface of the pinna, and consists of radiating fibres which extend from the back of the concha to the prominence which corresponds with the groove of the helix.

The *oblique muscle* (Tód) consists of a few fibres stretching from the back of the concha to the convexity directly above it, across the back of the inferior branch of the antihelix, and near the fibres of the transverse muscle.

Arteries of the pinna.—The *posterior auricular* artery, a branch from the external carotid, is distributed chiefly on the posterior or inner surface, but sends small branches · round and through the cartilage to ramify on the outer surface of the pinna. Besides this artery, the auricle receives others, the *anterior auricular* from the temporal in front, and a small artery from the occipital behind.

The *veins* correspond much in their course with the arteries. They join the temporal vein, and their blood is returned therefore through the external jugular.

Nerves of the pinna.—The *great auricular* nerve (p. 638), from the cervical plexus, supplies the greater part of the back of the auricle, and sends small filaments with the posterior auricular artery to the outer surface of the lobule and the part of the ear above it. The *posterior auricular* nerve, derived from the facial (p. 612), after communicating with the *auricular branch of the pneumogastric*, ramifies on the back of the ear and supplies the retrahent muscle. The upper muscles of the auricle receive their supply from the *temporal branches* of the same nerve. The *auriculo-temporal* branch of the third division of the fifth nerve (p. 606) gives filaments chiefly to the outer and anterior surface of the pinna.

THE EXTERNAL AUDITORY CANAL.

The external auditory canal (meatus auditorius externus) extends from the bottom of the concha to the membrane of the tympanum, and serves to convey to the middle chamber of the ear the vibrations of sound collected by the auricle. The canal is about one inch and a quarter in length. In

Fig. 499.—View of the Lower Half of the Auricle and Meatus in the Left Ear divided by a Horizontal Section (after Sömmerring).

1 and 2, cut surfaces of the bony part of the meatus ; 3, cut surface of the cartilage of the pinna ; 4, external meatus with the openings of numerous ceruminous glands indicated ; 5, lobule ; 6, membrane of the tympanum ; 7, dura mater lining the skull.

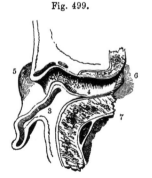

Fig. 499.

its inward course it is inclined somewhat forwards ; and it presents likewise a distinct vertical curve, being directed at first somewhat upwards, and afterwards turning somewhat abruptly over a convexity of the osseous part of its floor, and dipping downwards to its termination,—a change of direction which must be borne in mind by the surgeon in introducing specula into the ear. The calibre of the passage is smallest about the middle. The outer opening is largest from above downwards, but the tympanic end of the tube is slightly widest in the transverse direction. At the inner extremity the tube is terminated by the membrana tympani, which is placed obliquely, with the inferior margin inclined towards the mesial plane, and thus the floor of the meatus is longer than its roof.

The meatus is composed of a tube partly cartilaginous and partly osseous, and is lined by a prolongation of the skin of the pinna.

The *cartilaginous* part of the meatus forms somewhat less than half the length of the passage. It is formed by the deep part of the cartilage of the pinna, which has been already described.

The *osseous portion* of the meatus is a little longer and rather narrower than the cartilaginous part. At its inner end it presents a narrow groove, which extends round the sides and floor of the meatus, but is deficient above ; into this the margin of the membrana tympani is inserted.

The *skin* of the meatus is continuous with that covering the pinna, but is very thin, and becomes gradually thinner towards the bottom of the passage. In the osseous part of the canal it adheres very closely to the periosteum ; and at the bottom of the tube this lining is stretched over the surface of the membrana tympani, forming the outer layer of that structure. After maceration in water, or when decomposition is advanced, the epidermic lining of the passage may be separated and drawn out entire, and then it appears as a small tube closed at one end somewhat like the finger of a glove. Towards the outer part the skin possesses fine hairs and sebaceous glands ; and in the thick subdermic tissue over the cartilage are many small oval glands of a brownish-yellow colour, agreeing in form and structure with the sweat glands. The cerumen or ear-wax is secreted by these glands, *glandulæ ceruminosæ*, and their numerous openings may be seen to perforate the skin of the meatus. These accessory parts are absent over the bony part of the tube.

Vessels and nerves.—The external auditory meatus is supplied with arteries from the posterior auricular, internal maxillary and temporal arteries; and with nerves chiefly from the temporo-auricular branch of the fifth nerve. ·

State in the infant.—The auditory passage is in a very rudimentary state in the infant, for the osseous part begins to grow out of the tympanic bone only at the period of birth (p. 68), and thus the internal and middle parts of the ear are brought much closer to the surface than in the adult.

THE MIDDLE EAR OR TYMPANUM.

The tympanum or drum, the middle chamber of the ear, is a narrow irregular cavity in the substance of the temporal bone, placed between the inner end of the external auditory canal and the labyrinth. It receives the atmospheric air from the pharynx through the Eustachian tube, and contains a chain of small bones, by means of which the vibrations communicated from without to the membrana tympani are in part conveyed across the cavity to the sentient part of the internal ear, and by which also pressure is maintained on the contents of the internal ear, varying in amount according to the tension of the membrana tympani. The tympanum contains likewise minute muscles and ligaments, which belong to the bones referred to, as well as some nerves which end within this cavity, or pass through it to other parts.

The cavity of the tympanum may be considered as presenting for consideration a roof and a floor, an outer and an inner wall, and an anterior and a posterior boundary.

The *roof* of the tympanum is formed by a thin plate of bone, which may be easily broken through so as to obtain a view of the tympanic cavity from above ; it is situated on the upper surface of the petrous portion of the temporal bone, near the angle of union with the squamous portion, from which in its development it is derived.

The *floor* is narrow, in consequence of the outer and inner boundaries being inclined towards each other.

The *outer wall* is mainly formed by a thin semitransparent membrane—membrana tympani, which closes the inner end of the external auditory meatus ; and, to a small extent, by bone.   Immediately in front of the ring of bone into which the membrana tympani is inserted, is the inner extremity of the fissure of Glaser, which gives passage to the laxator tympani muscle, and attachment to the processus gracilis of the malleus.   Close to the back of this fissure is the opening of a small canal (named by Cruveilhier the canal of Huguier), through which the chorda tympani nerve usually escapes from the cavity of the tympanum and the skull.

Fig. 500.—Membrana Tympani as seen from the outer and inner side.

A, the outer surface ; B, the inner ; in the latter the small bones are seen adherent to the membrane and adjacent parts of the temporal bone; in A, the shaded part indicates the small bones as partially seen through the membrane ; 1, membrana tympani ; 2, malleus ; 3, stapes ; 4, incus.

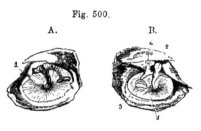

Fig. 500.

A.   B.

The *membrana tympani* is a nearly circular disc, slightly concave on its outer surface.   It is inserted into the groove already noticed at the end of the meatus externus, and so obliquely that the membrane inclines towards the anterior and lower part of the canal at an angle of about 45°.   The handle of the malleus, one of the small bones of the tympanum, descends between the middle and inner layers of the membrana tympani to a little below the centre, where it is firmly fixed ; and, as the direction of this process of the bone is slightly inwards, the outer surface of the membrane is thereby rendered concave, being held inwards in the shape of a shallow cone.

Though very thin, the membrana tympani is composed of three distinct structures.   A prolongation of the skin of the external meatus forms the outer layer ; the mucous membrane lining the cavity of the tympanum furnishes an inner layer ; and between those two is the proper substance of the membrane, made up of fine fibrous and elastic tissues with vessels and nerves.   The greater number of the fibres radiate from near the centre at the attachment of the handle of the malleus ; but close to the circumference are some circular fibres, which form a dense, almost ligamentous ring.

The *inner wall* of the tympanum, which separates it from the internal ear, is very uneven, presenting several elevations and foramina.   Near its upper part is an ovoid, or nearly kidney-shaped opening—*fenestra ovalis*, which leads into the cavity of the vestibule.   This opening, the long diameter of which is from before backwards, with a slight inclination downwards in front, is occupied in the recent state by the base of the stapes, and the annular ligament connected with that process of bone. Above the fenestra ovalis, and between it and the roof of the tympanum, a ridge indicates the position of the aqueduct of Fallopius, as it passes backwards, containing the portio dura of the seventh nerve. Below it is a larger and more rounded elevation, caused by the projection outwards of the first turn of the cochlea, and named the *promontory*, or *tuber cochleæ* ; it is marked by grooves, in which lie the nerves of the tympanic plexus.

Fig. 501.

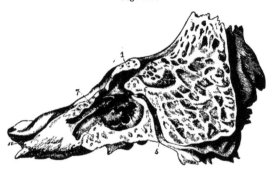

Fig. 501.—Inner Wall of the Osseous Tympanum as exposed by a Longitudinal Section of the Petrous and Mastoid Bone (from Gordon).

1, opening of the tympanum into the mastoid cells ; 2, fenestra ovalis ; 3, fenestra rotunda ; 4, promontory ; 5, aqueduct of Fallopius, or canal of the facial nerve ; 6, junction of the canal for the chorda tympani with the aqueduct ; 7, processus cochleariformis ; 8, groove above it for the tensor tympani muscle ; 9, Eustachian tube ; 10, anterior orifice of the carotid canal.

Below and behind the promontory, and somewhat hidden by it, is a slightly oval aperture named *fenestra rotunda*, which lies within a funnel-shaped depression. In the macerated and dried bone the fenestra rotunda opens into the scala tympani of the cochlea ; but, in the recent state it is closed by a thin membrane.

The membrane closing the fenestra rotunda—the *secondary membrane of the tympanum* (Scarpa)—is rather concave towards the tympanic cavity, and is composed of three strata like the membrana tympani ; the middle layer being fibrous, and the outer and inner derived from the membranes lining the cavities between which it is interposed, viz., the tympanum and the cochlea.

The *posterior wall* of the tympanum presents at its upper part one larger, and several smaller openings, which lead into irregular cavities, the *mastoid cells*, in the substance of the mastoid process of the temporal bone. These cells communicate freely with one another, and are lined by mucous membrane continuous with that which clothes the tympanum. Behind the fenestra ovalis, and directed forwards, is a small conical eminence, called the *pyramid*, or eminentia papillaris. Its apex is pierced by a foramen, through which the tendon of the stapedius muscle emerges from a canal which turns downwards in the posterior wall of the tympanum, and joins obliquely the descending part of the aqueduct of Fallopius.

The *anterior extremity* of the tympanum is narrowed by the gradual descent of the roof, and is continued into the Eustachian orifice. The lower compartment of this orifice, lined with mucous membrane, forms the commencement of the Eustachian tube; the upper compartment, about half an inch long, lodges the tensor tympani muscle, and opens into the tympanum immediately in front of the fenestra ovalis, surrounded by the expanded and everted end of the cochleariform process, which separates it from the lower compartment.

Fig. 502.

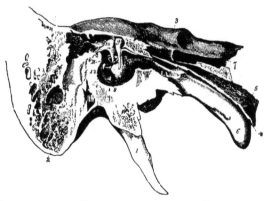

Fig. 502.—ANTERO-POSTERIOR SECTION OF THE TEMPORAL BONE, SHOWING THE INNER WALL OF THE TYMPANUM, WITH THE EUSTACHIAN TUBE AND SMALL BONES IN THE RECENT STATE (from Arnold).

1, styloid process ; 2, mastoid process ; 3, upper part of the petrous bone ; 4, pharyngeal end of the Eustachian tube ; 5, its cartilage ; 6, its mucous surface ; 7, carotid canal ; 8, fenestra rotunda ; 9, malleus ; 10, incus ; 11, stapes ; 12, pyramid and stapedius muscle ; above 9, and behind 10, the suspensory ligaments of the malleus and incus are also seen.

The *Eustachian tube* is a canal, formed partly of bone, partly of cartilage and membrane, which leads from the cavity of the tympanum to the upper part of the pharynx. From the tympanum it is directed forwards and inwards, with a little inclination downwards ; and its entire length is about an inch and a half. The *osseous* division of the Eustachian tube, already described in the Osteology, is placed in the angle of junction of the petrous portion of the temporal bone with the squamous portion. The *anterior* part of the tube is formed of a triangular piece of cartilage, the edges of which are slightly curled round towards each other, leaving an interval at the under side, in which the canal is completed by dense but pliable fibrous membrane. Narrow behind, the tube gradually expands till it becomes wide and trumpet-shaped in front ; and the anterior part is compressed from side to side, and is fixed to the inner pterygoid process of the sphenoid bone. The anterior opening is oval in form, and is placed obliquely at the side and upper part of the pharynx, into which its prominent margin projects behind the lower meatus of the nose, and above the level of the hard palate. Through this aperture the mucous membrane of the pharynx is continuous with that which lines the tympanum, and under certain conditions air passes into and out of that cavity.

Three small bones (ossicula auditûs) are contained in the upper part of the tympanum : of these, the outermost (malleus) is attached to the membrana tympani ; the innermost (stapes) is fixed in the fenestra ovalis ; and

Fig. 503.

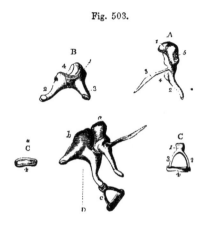

Fig. 503.—Bones of the Tympanum of the Right Side (from Arnold). ²⁄₁

A, malleus ; 1, its head ; 2, the handle ; 3, long or slender process ; 4, short process ; B, incus ; 1, its body ; 2, the long process with the orbicular process ; 3, short or posterior process ; 4, articular surface receiving the head of the malleus ; C, stapes ; 1, head ; 2, posterior crus ; 3, anterior crus ; 4, base ; C*, base of the stapes ; D, the three bones in their natural connection as seen from the outside ; a, malleus ; b, incus ; c, stapes.

the third (incus), placed between the other two, is connected to both by articular surfaces. The malleus and incus are placed in nearly a vertical, the stapes in a horizontal direction. They form together an angular and jointed connecting rod between the membrana tympani and the membrane which closes the fenestra ovalis.

The *malleus*, or hammer bone, consists of a central thicker portion, with processes of different lengths. At the upper end of the bone is a rounded *head* (capitulum), which presents internally and posteriorly an irregularly oval surface covered with cartilage, for articulation with the incus. Below the head is a constricted *neck* (cervix); and beneath this another slight enlargement of the bone, to which the processes are attached. The *handle* (manubrium) of the malleus is a tapering and slightly twisted process, com · pressed from before backwards to near its point, where it is flattened in the opposite direction : it descends with a slight inclination forwards and inwards, and is received between the middle and inner layers of the membrana tympani, to which it is closely attached. The *long process* (processus gracilis) is a very slender spiculum of bone, which in the adult is usually broken off in its removal from the tympanum, in consequence of its union with the temporal bone ; it projects at nearly a right angle from the front of the neck of the malleus, and extends thence obliquely downwards and forwards to the Glasserian fissure. Its end is flattened and expanded, and is connected by ligamentous fibres and by bone to the sides of the fissure. The *short process* (processus brevis vel obtusus) is a low conical eminence springing from the root of the manubrium, beneath the cervix, and projecting outwards towards the upper part of the membrana tympani.

The *incus* has been compared to an anvil in form ; but it resembles perhaps more nearly a tooth with two fangs widely separated. It consists of a body and two processes. The *body* presents in front a concavo-convex articular surface, which is directed upwards and forwards, and receives the head of the malleus. The surfaces of the joint thus formed are tipped with articular cartilage and enclosed by a synovial membrane. The *shorter* of the two processes (crus breve) of the incus projects nearly horizontally backwards from the upper part of the body of the bone, and is connected by ligamentous fibres with the posterior wall of the tympanum near the entrance of the mastoid cells. The *long process* (crus longum) tapers rather more gradually, and descends nearly vertically behind the handle of the malleus : at its extremity it is bent inwards, and is suddenly narrowed into a short neck ; and upon this is set a flattened rounded tubercle (processus lenticularis), tipped with cartilage. This tubercle, which articulates with the head of the stapes, was formerly, under the name of *os orbiculare seu lenticulare,* described as a separate bone, which indeed it originally is in childhood.

The *stapes,* the third and innermost bone of the ear, is in shape remarkably like a stirrup, and is composed of a head, a base, and two crura. The *head* is directed outwards, and has on its end a slight depression, covered with cartilage, which articulates with the lenticular process of the incus. The *base* is a plate of bone placed in the fenestra ovalis, to the margin of which it is fixed by ligamentous fibres. The form of the base is irregularly oval, the upper margin being curved, while the lower is nearly straight. The crura of the stapes diverge from a constricted part (*neck*) of the bone, situated close to the head, and are attached to the outer surface of the base near its extremities. The anterior crus is the shorter and straighter of the two. The crura, with the base of the stapes, enclose a small triangular or arched space, which in the recent state is occupied by a thin membrane stretched across. A shallow groove runs round the opposed surfaces of the bone, and into this the membrane is received.

LIGAMENTS AND MUSCLES OF THE TYMPANUM.

Ligaments.—In the articulations of the small bones of the ear with each other, the connection is strengthened by ligamentous fibres which cover the synovial membranes.

The attachment of the bones of the ear to the walls of the tympanum is effected partly by the reflections of the mucous membrane lining that cavity, but chiefly by muscles and by the following ligaments.

The *suspensory ligament of the malleus* consists of a small bundle of fibres, which descends perpendicularly from the roof of the tympanum to the head of the malleus.

The incus is likewise suspended by a small ligament (*the posterior ligament of the incus*), which extends from near the point of the short crus directly backwards towards the posterior wall of the tympanum, where it is attached near the entrance to the mastoid cells. ·

Arnold describes an upper ligament which attaches the incus, near its articulation with the malleus, to the roof of the tympanum. It lies close behind the suspensory ligament of the malleus.

The *annular* or *orbicular ligament of the stapes* connects the base of the bone to the margin of the fenestra ovalis, in which it is lodged.

Muscles.—There are three well-determined muscles of the tympanum. Sömmerring describes four, and some authors a larger number ; but the descriptions of these last muscles are not confirmed by later research. Of the three muscles generally recognised, two are attached to the malleus, and one to the stapes.

The *tensor tympani* (musculus internus mallei) is the largest of these muscles. It consists of a tapering fleshy part, about half an inch in length, and a slender tendon. The muscular fibres arise from the cartilaginous end of the Eustachian tube and the adjoining surface of the sphenoid bone, and from the sides of the upper compartment of the Eustachian orifice. In

Fig. 504.

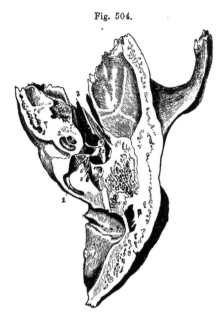

Fig. 504.—View of the Cavity of the Right Tympanum from above.

The cavity of the tympanum and some parts of the labyrinth have been exposed by a horizontal section removing the upper part of the temporal bone. 1, posterior semicircular canal opened ; 2, the cavity of the cochlea opened ; 3, osseous part of the Eustachian tube ; 4, head of the malleus ; 5, incus ; 6, stapes, with its base set in the fenestra ovalis ; 7, tensor tympani muscle ; 8, stapedius.

this canal the muscle is conducted nearly horizontally backwards to the cavity of the tympanum. Immediately in front of the fenestra ovalis the tendon of the muscle bends at nearly a right angle over the end of the processus cochleariformis as through a pulley, and, contained in a fibrous sheath, passes outwards to be inserted into the inner part of the handle of the malleus, near its root.

The *laxator tympani* (laxator tympani major of Sömmerring) is generally believed to be distinctly muscular, but being partly concealed by a band of fibrous tissue, doubts are still entertained by some observers as to whether the structure known under this name is of a muscular or ligamentous nature. Arising from the spinous process of the sphenoid bone, and slightly from the cartilaginous part of the Eustachian tube, it is directed backwards, passes through the Glaserian fissure, and is inserted into the neck of the malleus, just above the root of the processus gracilis.

The *laxator tympani minor* of Sömmerring (*posterior ligament of the malleus*, Lincke) is made up of reddish fibres, which are fixed at one end to the upper and back part of the external auditory meatus, pass forwards and inwards between the middle and inner layers of the membrana tympani, and are inserted into the outer border of the handle of the malleus, and the short process near it. Sömmerring. Icones Organi Auditûs Humani, 1801.

The *stapedius* is a very distinct muscle, but is hid within the bone, being lodged in the descending part of the aqueductus Fallopii and in the hollow of the pyramid. The tendon issues from the aperture at the apex of that little elevation, and passing forwards, surrounded by a fibrous sheath, is inserted into the neck of the stapes posteriorly, close to the articulation of that bone with the lenticular process of the incus.

A very slender spine of bone has been found occasionally in the tendon of the stapedius in man : and a similar piece of bone, though of a rounder shape, exists con-

stantly in the horse, the ox, and other animals. This circumstance is the more interesting when it is remembered that cartilage occupies the position of the stapedius
before the muscle is developed. (P. 66 and fig. 528.)

Actions.—The malleus and incus move together round an axis extending backwards
from the attachment of the processus gracilis of the malleus in the Glasserian fissure
to the attachment of the short process of the incus posteriorly. The tendon of the
tensor tympani muscle passing from within to be inserted below that line, pulls the
handle of the malleus inwards, while the laxator tympani inserted above that line, by
pulling the head of the bone inwards, moves the handle outwards. The incus, moving
along with the malleus, pushes the stapes inwards towards the internal ear when the
membrana tympani is made tight, and withdraws that bone from the fenestra ovalis,
when the membrana tympani is relaxed. But the cavity of the inner ear is full of

Fig. 505.—Outline of the Three Small Bones of
the Left Ear as seen from before. ²⁄₁

Fig. 505.

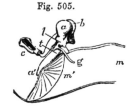

This figure is designed to illustrate the effect of
the action of the tensor and laxator muscles of the
tympanic membrane in connection with their relation
to the axis of rotation of the malleus. *a, a'*, the
malleus ; *b*, the incus seen behind it ; *c*, the stapes ;
m, m', the inner part of the meatus externus closed
by the tympanic membrane, of which the posterior
half is represented ; the axis of rotation of the
malleus being supposed to pass through a point at
the root of the processus gracilis, *g* ; the line *t*, indi-
cates the direction and position of the tendon of the tensor tympani pulling the lower
part of the malleus inwards, the line *l*, that of the laxator tympani pulling inwards the
upper half of the malleus.

liquid ; and its walls are unyielding, except at the fenestra rotunda ; when, therefore, the stapes is pushed inwards the secondary membrane of the tympanum, which
blocks up the fenestra rotunda, must be made tense by pressure from within. The
attachment of the handle of the malleus, however, to the membrana tympani allows
greater freedom of movement to that process than is allowed to the stapes by the
ligament of its base, and when the movement of the stapes ceases, it is plain that the
malleus in any movement must rotate on the head of the incus ; and hence, probably,
the necessity of a moveable articulation between those bones. The action of the
stapedius muscle is obviously to draw the head of the stapes backwards, in doing which
the hinder end of the base of that bone will be pressed against the margin of the
fenestra ovalis, while the fore part will be withdrawn from the fenestra. The object
gained by this movement of the stapes is not sufficiently ascertained ; but it is at least
evident that, if the stapes be pressed inwards by the incus in the action of the tensor
tympani, the stapedius muscle, if then contracted, will modify the pressure on the
internal ear. It is conceivable that the stapedius may thus protect the sensitive part
of the ear to a certain extent from excessive stimulation of the auditory nerve.

THE LINING MEMBRANE OF THE TYMPANUM.

The mucous membrane of the tympanum is continuous with that of the
pharynx through the Eustachian tube, and is further prolonged from the
tympanum backwards into the mastoid cells. Two folds which cross the
breadth of the cavity descend from the part of the membrane which lines
the roof. The anterior fold descends to turn round the tendon of the tensor
tympani muscle ; the posterior fold passes round the stapes. The malleus
and incus are invested by the lining of the outer wall of the cavity. The
mucous membrane which lines the cartilaginous part of the Eustachian tube
resembles much the membrane of the pharynx, with which it is immediately
continuous ; it is thick and vascular, and is covered by several layers of
laminar epithelium with vibratile cilia, and is provided with many simple
mucous glands which pour out a thick secretion : in the osseous part of the

tube, however, this membrane becomes gradually thinner.   In the tympanum and the mastoid cells it is paler, thinner and less vascular, and secretes
a less viscid, but yellowish fluid.   The epithelium in the tympanic cavity is
also ciliated.   The cilia, however, are usually absent from the part which
lines the membrana tympani (Kölliker, Handbuch, p. 691).

THE VESSELS AND NERVES OF THE TYMPANUM.

The *arteries* of the tympanum, though very small, are numerous, and
are derived from several branches of the external, and from the internal
carotid.

The fore part of the cavity is supplied chiefly by the *tympanic branch* of the internal
maxillary (p. 356), which enters by the fissure of Glaser.   The back part of the
cavity, including the mastoid cells, receives its arteries from the *stylo-mastoid branch*
of the posterior auricular artery (p. 353), which is conducted to the tympanum by the
aqueduct of Fallopius.   These two arteries form by their anastomosis a vascular circle
round the margin of the membrana tympani.   The smaller arteries of the tympanum
are, the *petrosal branch* of the middle meningeal, which enters through the *hiatus
Fallopii;* branches through the bone from the *internal carotid* artery, furnished from
that vessel whilst in the carotid canal; and occasionally a twig along the Eustachian
tube from the *ascending pharyngeal* artery.

The *veins* of the tympanum pour their contents through the middle meningeal and
pharyngeal veins, and through a plexus near the articulation of the lower jaw, into the
internal jugular vein.

Nerves.—The tympanum contains numerous nerves; for, besides those
which supply the parts of the middle ear, there are several which serve
merely to connect nerves of different origin.

The lining membrane of the tympanum is supplied by filaments from the
plexus (tympanic plexus), which occupies the shallow grooves on the inner
wall of the cavity, particularly on the surface of the promontory.

The *tympanic plexus* is formed by the communications between, 1st, the
tympanic branch (nerve of Jacobson) from the petrous ganglion of the glosso-
pharyngeal; 2nd, a *filament from the carotid plexus* of the sympathetic; 3rd,
a branch which joins the *great superficial petrosal nerve*, from the Vidian;
4th and lastly, the *small superficial petrosal nerve*, from the otic ganglion.

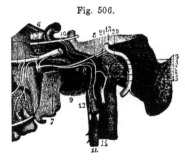

Fig. 506.

Fig. 506.—VIEW OF THE TYMPANIC
PLEXUS OF NERVES (after Hirschfeld
and Leveillé).

6, spheno-palatine ganglion; 7, Vidian
nerve; 8, great superficial petrosal
nerve; 9, carotid branch of the Vidian
nerve; 10, part of the sixth nerve connected by twigs with the sympathetic;
11, superior cervical ganglion of the
sympathetic; 12, carotid branch; 13,
facial nerve; 14, glosso-pharyngeal
nerve; 15, nerve of Jacobson; 16, its
twig to the sympathetic; 17, filament
to the fenestra rotunda; 18, filament to
the Eustachian tube; 19, filament to the
fenestra ovalis; 20, union of external deep petrosal nerve with the lesser superficial
petrosal; 21, internal deep petrosal twig uniting with the great superficial petrosal.

The nerve of Jacobson enters the tympanum by a small foramen near its floor,
which forms the upper end of a short canal in the petrous portion of the temporal
bone, beginning at the base of the skull between the carotid foramen and the jugular

fossa. The nerve from the carotid plexus is above and in front of this, and passes through the bone directly from the carotid canal. The branch to the great superficial petrosal nerve is lodged in a canal which opens on the inner wall of the tympanum in front of the fenestra ovalis. The small superficial petrosal nerve also enters at the fore part of the cavity beneath the canal for the tensor tympani.

Nerves to Muscles.—The tensor tympani muscle obtains its nerve from the otic ganglion (see fig. 410); the laxator tympani is said to be supplied by the chorda tympani: and the stapedius is figured by Sömmerring as receiving a filament from the facial nerve.

The *chorda tympani* is invested by a tubular reflection of the lining membrane of the tympanum; its course across the cavity has already been described (p. 611).

THE INTERNAL EAR, OR LABYRINTH.

The inner, or sensory part of the organ of hearing, is contained in the petrous portion of the temporal bone. It consists of a cavity—the osseous labyrinth—hollowed out of the bone, and of the membranous labyrinth contained within the osseous walls.

Fig. 507.—Right Bony Labyrinth, viewed from the Outer Side (after Sömmerring). $\frac{2\frac{1}{2}}{1}$

Fig. 507.

The specimen here represented is prepared by separating piecemeal the looser substance of the petrous bone from the dense walls which immediately enclose the labyrinth. 1, the vestibule; 2, fenestra ovalis; 3, superior semicircular canal; 4, horizontal or external canal; 5, posterior canal; *, ampullæ of the semicircular canals; 6, first turn of the cochlea; 7, second turn; 8, apex; 9, fenestra rotunda. The smaller figure in outline below shows the natural size.

The *osseous labyrinth* is incompletely divided into three parts, named the vestibule, the semicircular canals, and the cochlea. They are lined throughout by a thin membrane, within which there is a clear fluid named perilymph.

The *membranous labyrinth* is contained within the bony labyrinth, and, being smaller than it, leaves a space between the two, occupied by the peri-lymph just referred to. The membranous structure supports numerous minute ramifications of the auditory nerve, and encloses a fluid named the endolymph.

THE OSSEOUS LABYRINTH.

The *vestibule* forms a central chamber of the labyrinth, which communi-cates in front with the cochlea, behind with the semicircular canals, on the outer side with the cavity of the tympanum, and on the inner side with the meatus auditorius internus. The vestibule is irregularly ovoidal in shape from before backwards, and is slightly flattened or compressed from without

3 D

inwards : except in the last-mentioned direction, in which it is somewhat smaller, it measures about $\frac{1}{5}$th of an inch in diameter.

The outer wall which separates it from the cavity of the tympanum, is perforated by the fenestra ovalis, which in the recent state is closed by the base of the stapes and its annular ligament.

At the fore part of the inner wall is a small round pit, the *fovea hemispherica*, pierced with many small holes, which serve to transmit branches of the auditory nerve from the internal auditory meatus. This fossa is limited behind by a vertical ridge named *crista vestibuli* or eminentia pyramidalis. Behind the crest is the small oblique opening of a canal, the *aqueduct of the vestibule*, which extends to the posterior surface of the bone, and transmits a small vein in a tubular prolongation of membrane.

In the roof is an oval depression, placed somewhat transversely, *fovea hemi-elliptica*, whose inner part is separated by the crest from the hemispherical fossa.

At the back part of the vestibule are five round apertures, leading into the semicircular canals : and at the lower and fore part of the cavity is a larger opening, which communicates with the scala vestibuli of the cochlea— *apertura scalæ vestibuli.*

The *semicircular canals* are three bony tubes, situate above and behind the vestibule, into which they open by five apertures, the contiguous ends of

Fig. 508.

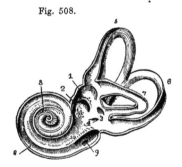

Fig. 508.—View of the Interior of the Left Labyrinth (from Sömmerring). $\frac{2\frac{1}{2}}{1}$

The bony wall of the labyrinth is removed superiorly and externally. 1, fovea hemi-elliptica; 2, fovea hemispherica; 3, common opening of the superior and posterior semicircular canals; 4, opening of the aqueduct of the vestibule; 5, the superior, 6, the posterior, and 7, the external semicircular canals; 8, spiral tube of the cochlea (scala tympani); 9, opening of the aqueduct of the cochlea; 10, placed on the lamina spiralis in the scala vestibuli.

two of the canals being joined. They are unequal in length, but each tube is bent so as to form about two-thirds of a circle ; and each presents, at one end, a slightly dilated part, called the *ampulla*. The canals are compressed laterally, and measure across about $\frac{1}{20}$th of an inch ; but in the ampulla each has a diameter of $\frac{1}{10}$th of an inch.

The canals differ from one another in position with regard to the vestibule, in direction, and in length. The *superior* semicircular canal is *vertical* and transverse ; and, rising above any other part of the labyrinth, its place is indicated by a smooth arched projection on the upper surface of the bone. The ampullary end of this canal is the anterior, and opens by a distinct orifice into the upper part of the vestibule ; whilst the opposite extremity joins the non-dilated end of the posterior semicircular canal, and opens by a common aperture with it into the back part of the vestibule. The *posterior* semicircular canal, *vertical* and longitudinal in direction, is the longest of the three tubes: its ampullary end is placed at the lower and back part of the vestibule ; and the opposite end joins in the common canal above described. The *external* semicircular canal arches *horizontally* outwards, and opens by two distinct orifices into the upper and back part of the vestibule. This canal is shorter than either of the other two : its ampulla is at the outer end, just above the fenestra ovalis.

Fig. 509.

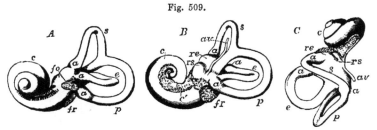

Fig. 509.—VIEWS OF A CAST OF THE INTERIOR OF THE LABYRINTH (from Henle). ⅔

Such casts may easily be made in fusible metal, and give a very correct view of the form of the different parts of the labyrinthic cavity. A, view of the left labyrinth from the outer side ; B, the right labyrinth from the inner side ; C, the left labyrinth from above ; *s,* the superior, *p,* the posterior, and *e,* the external semicircular canals ; *a,* their several ampullæ ; *r e,* fovea hemi-elliptica of the vestibule ; *rs,* fovea hemispherica ; *av,* aqueduct of the vestibule ; *fo,* fenestra ovalis ; *fr,* fenestra rotunda ; *c,* the coiled tube of the cochlea ; *c',* the first part of the tube towards the base with the tractus foraminosus spiralis.

The *cochlea* is the most anterior division of the internal ear. When the dense bony substance, in which it lies embedded, is picked away, the cochlea presents the form of a blunt cone, the base of which is turned towards the internal auditory meatus, whilst the apex is directed outwards, with an inclination forwards and downwards, and is close to the canal for the tensor tympani muscle. It measures about a quarter of an inch in length, and the same in breadth at the base. The osseous part of the cochlea consists of a gradually tapering spiral tube, the inner wall of which is formed by the central column, or *modiolus,* round which it winds, and which is partially divided along its whole extent by a spiral lamina, projecting into it from the modiolus. From this osseous spiral lamina membranous structures are stretched across to the outer wall of the tube, and thus are completely separated two passages or scalæ, one on each side of the spiral lamina, which communicate one with the other by only a small opening, named *helicotrema,* placed at the apex of the cochlea.

Fig. 510.

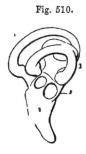

Fig. 510.—OSSEOUS LABYRINTH OF THE BARN-OWL (STRIX FLAMMEA) (from Breschet). ¼

1, semicircular canals ; 2, vestibule ; 3, cochlea in the form of a short straight tube.

That the cochlea is justly to be considered as an elongated tube, coiled spirally on the modiolus, is illustrated by the simple pouch-like form of the rudimentary cochlea of birds.

The *spiral canal of the cochlea* is about an inch and a half long, and about the tenth of an inch in diameter in its widest part at the commencement. ˙˙ From this point the canal makes two turns and ˙a half round the central pillar (from left to right in the right ear, and in the opposite direction in the left ear), and ends by an arched and closed extremity called the *cupola*, which forms the apex of the cochlea. The first coil, being much the widest in its curve and composed of the largest portion of the tube, nearly hides the second turn from view ; and bulging somewhat into the tympanum, forms the round elevation on the inner wall of that cavity called the promontory.

Fig. 511.   Fig. 512.

 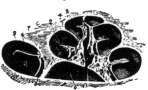

Fig. 511.—Diagrammatic View of the Canal of the Cochlea laid open. ⅚

1, modiolus or central pillar ; 2, placed on three turns of the lamina spiralis ; 3, scala tympani ; 4, scala vestibuli.

Fig. 512.—View of the Osseous Cochlea divided through the Middle (from Arnold). ⅚

1, central canal of the modiolus ; 2, lamina spiralis ossea ; 3, scala tympani ; 4, scala vestibuli ; 5, porous substance of the modiolus near one of the sections of the canalis spiralis modioli.

The *modiolus* (columella cochleæ) forms the central pillar or axis round which turn the spiral tube and the spiral lamina. It is much thickest within the first turn of the cochlea, and rapidly diminishes in size in the succeeding parts. The outer surface is dense, being, in fact, composed of the walls of the spiral tube ; but the centre is soft and spongy as far as the last half coil, and is pierced by many small canals, for the passage of the nerves and vessels to the lamina spiralis: one of these canals, larger than the rest (*canalis centralis modioli*), runs from the base through the centre of the modiolus.

The *lamina spiralis ossea* is a thin, flat plate, growing from and winding round the modiolus, and projecting into the spiral tube, so as to divide it partly into two. Its free margin, which gives attachment in the recent state to the membranous septum, or zone, does not reach farther than about half of the distance between the modiolus and the outer wall of the spiral tube. The osseous lamina terminates close to the apex of the cochlea in a hook-like process (hamulus), which partly bounds the helicotrema.

:The lamina is thin and dense towards its free margin ; but near the modiolus it is composed of two dense outer plates enclosing a more open and spongy structure, in which are numerous small canals, continuous, but running at right angles with the canals in the centre of the modiolus. In these the nerves and vessels are lodged : they terminate on the inferior or tympanic aspect of the lamina, and the line of their orifices forms the *tractus foraminosus spiralis*. Winding round the modiolus, close to the lamina spiralis, is a small canal, named by Rosenthal the *canalis spiralis modioli*.

The *scalæ* in the osseous cochlea are two in number, distinguished as the scala tympani and scala vestibuli.

The *scala tympani*, the portion of the tube on the basal side of the lamina spiralis, commences at the fenestra rotunda, where in the recent state it is separated from the tympanum by the secondary membrana tympani. Near its commencement is the orifice of a small canal *aqueductus cochleæ*, which extends downwards and inwards through the substance of the petrous part of the temporal bone to near the jugular fossa, and transmits a small vein. The surface of the spiral lamina which looks towards this scala is marked with numerous transverse striæ. The *scala vestibuli* is rather narrower than the scala tympani in the first turn of the cochlea; it commences from the cavity of the vestibule, and communicates, as already described, with the scala tympani at the apex of the modiolus.

The lining membrane of the osseous labyrinth.—This is a thin membrane (periosteum ?), which closely adheres to the whole inner surface of the several parts of ·the labyrinthic cavity just described. It has no continuity with the lining membrane of the tympanum, being stretched across the openings of the round and oval fenestræ. It is composed of fibres of connective tissue. Its outer surface is rough, and adheres closely, like periosteum, to the bone : the inner surface is pale and smooth, is covered with a single layer of epithelium, like that of the arachnoid, and secretes a thin, slightly albuminous or serous fluid. This secretion, known as the *liquor Cotunnii*, or *perilymph*, separates the membranous from the osseous labyrinth in the vestibule and semicircular canals, occupies the cavities of the scala tympani and scala vestibuli in the cochlea, and is continued into the aqueducts as far as the membrane lining these passages remains pervious.

THE MEMBRANOUS LABYRINTH.

Within the osseous labyrinth, and separated from its lining membrane by the perilymph, membranous structures exist in which the ultimate ramifications of the auditory nerve are spread. In the vestibule and semicircular canals these structures have a general resemblance in form to the complicated cavity in which they are contained. In the cochlea they complete the septum between the scalæ already mentioned, and enclose a third spiral passage, the canalis membranacea, the existence of which has only been discovered of late years. The liquid contained within the membranous labyrinth is distinguished as *endolymph*.

VESTIBULE.—The *membranous vestibule* consists of two closely connected sacs, and the parts by which they are united to the membranous semicircular canals and canal of the cochlea.

The larger of the two sacs, the *common sinus* or *utricle*, is of an oblong form and slightly flattened from without inwards. It is lodged in the upper and back part of the osseous vestibule, occupying the fovea hemielliptica. Opposite the crista vestibuli several small branches of the auditory nerve enter from the foramina in the bone ; and here the walls of the common sinus are thicker and more opaque than elsewhere. The extremities of the membranous semicircular canals terminate in the cavity of the common sinus. A small mass of calcareous particles, *otoliths* or *otoconia*, is lodged in the wall of the sac. These otoliths are crystals of carbonate of lime, and are described as six-sided, and pointed at their extremities. They are connected with the wall of the sac in a way not yet clearly determined.

The smaller vestibular vesicle, the *saccule*, is more nearly spherical than the common sinus, but, like it, is somewhat flattened. The saccule is situated in the lower and fore part of the cavity of the osseous vestibule, close to the opening from the scala vestibuli of the cochlea, and is received into the hollow of the fovea hemispherica, from the bottom of which many branches of nerve enter.   The sacculus appears to have a cavity distinct from that of the utricle, but is filled with the like thin and clear fluid, *endolymph,* and contains similar otoconia in its wall.   It is prolonged below into a short narrow duct, canalis reuniens, which opens abruptly into the membranous canal of the cochlea.

Fig. 513.

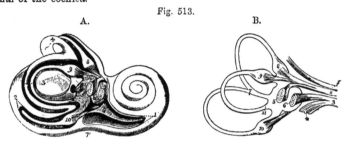

Fig. 513.—VIEWS OF THE INTERIOR OF THE RIGHT LABYRINTH WITH ITS MEMBRANOUS PARTS AND NERVES (from Breschet). ³⁄₁

A, the outer wall of the osseous labyrinth in part removed so as to display the membranous parts within. 1, commencement of the spiral tube of the cochlea ; 2, posterior semicircular canal partially opened, showing its membranous canal and ampulla; 3, external or horizontal canal entirely opened ; 4, superior canal ; 5, utriculus or common sinus with its group of otoliths ; 6, saccule with its otoliths ; 7, placed on the lamina spiralis in the commencement of the scala vestibuli ; 7', scala tympani ; 8, membranous ampulla of the superior semicircular canal ; 9, ampulla of the horizontal, and 10, that of the posterior semicircular canal.

B, membranous labyrinth and nervous twigs detached ; 1, facial nerve in the meatus auditorius internus ; 2, anterior division of the auditory nerve giving branches to 5, 8, and 9, the utricle and the ampullæ of the superior and external canals ; 3, posterior division of the auditory nerve, giving branches to the saccule ; 6, posterior ampulla, 10, and cochlea, 4 ; 7, the united part of the superior and posterior canals ; 11, the posterior extremity of the external canal.

SEMICIRCULAR CANALS.—The *membranous semicircular canals* are about one third the diameter of the osseous tubes in which they are lodged, and are dilated into ampullæ within the ampullary enlargements of those tubes. At the ampullæ they are thicker and less translucent than in the rest of their extent, and nearly fill their bony cases.   That part of each ampulla which is towards the concavity of the semicircle of the canal is free ; whilst the opposite portion is flattened, receives branches of nerves and bloodvessels, and presents on its inner surface a transverse projection, *septum transversum,* which partly divides the cavity into two.   The ampullæ likewise contain otoliths in their epithelial lining.

Auditory nerve: vestibular division.—At the bottom of the meatus auditorius internus the auditory nerve divides into an anterior and a posterior branch, which, broken up into minute filaments, pass through the perforations of the cribriform plate which separates the meatus from the internal ear, and are distributed respectively to the cochlea and vestibule.   In both branches, as well as in the trunk, there are numerous nerve-cells, apparently both with and without poles.   The *vestibular nerve* divides into five branches, which proceed respectively to the utricle, the saccule, and the three ampullæ of the semicircular canals : those for the utricle and the superior and external semicircular canals enter the cavity in a group along the crista vestibuli ; the fibrils for the sacculus enter the vestibule by a smaller group of foramina, which are situated below those just described, and open at the bottom of the fovea hemispherica ; the branch for the posterior semicircular canal is long and slender, and traverses a small passage

Fig. 515.

Fig. 514.

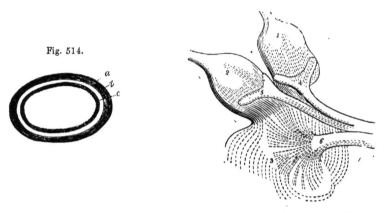

Fig. 514.—Transverse Section of one of the Membranous Semicircular Canals (from Kölliker). $\frac{250}{1}$

This specimen is from the ear of the calf: *a*, external fibrous layer with interspersed nuclei ; *b*, homogeneous layer ; *c*, epithelial lining.

Fig. 515.—Ampullæ of the Superior and External Semicircular Canals and Part of the Common Sinus showing the Arrangement of the Nerves (from Steifensand). $\frac{20}{1}$

1, membranous ampulla of the superior canal ; 2, that of the external canal ; 3, part of the common sinus ; 4 and 5, fork-like swellings of the nerves at their ampullar distribution ; 6, twig of the auditory nerve spreading in the common sinus.

in the bone behind the foramina for the nerve of the sacculus. The nerves of the ampullæ enter the flattened or least prominent side of the ampullæ, where they each form a forked swelling, which corresponds with the transverse septum already described, in the interior of the dilatation. No filaments have been found extending to any other parts of the semicircular canals.

Microscopic structure.—The walls of the common sinus, sacculus and membranous semicircular canals are in general semitransparent; but they are thicker and more opaque where nerves and vessels enter. On the outer surface is a layer of minutely ramified blood-vessels and loose tissue, which contains irregular pigment-cells: within this is a transparent layer, faintly fibrillated, and presenting elongated nuclei when acetic acid is added ; lining the interior is an epithelial layer of polygonal nucleated cells. The mode of ending of the nerves in the membranous substance of the vestibule and semicircular canals is difficult to investigate, on account of the minuteness and delicacy of the parts ; for this reason also observers have had recourse in great measure to the examination of the vestibule

and semicircular canals in fishes, in which they are of large size. The
subject still requires further research, but it appears to be pretty certain
from the observations of Reich, the successive papers of M. and F. E.
Schultze, and the corroborating observations of Kölliker, that the nerve
fibres break up in the transparent layer into minute ramifications, which
enter the epithelium and form between the epithelial cells spindle-shaped
nucleated bodies with elongated extremities. There have also been observed long hair-like processes, *fila acustica*, projecting into the cavity,
beyond the epithelial surface of the ridge of the ampullæ, and likewise
in the sacs ; and the actual continuity of these hairs with the nerve-
terminations has been in one instance observed by F. E. Schultze.
According to Lang the hairs are only the altered remains of a delicate cap
of tissue on the surface of the epithelium.—(Kölliker's Gewebelehre, 4th
ed., p. 694.)

Fig. 516.   Fig. 517.

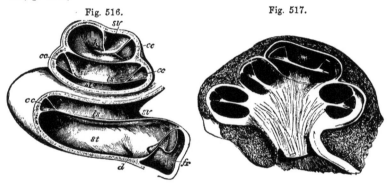

Fig. 516.—Left Cochlea of a Child some Weeks Old, opened (from Reichert). ♀

The drawing was taken from a specimen which had been preserved in alcohol, and was
afterwards dried ; the section is made so as to show the lamina spiralis, scalæ, and
cochlear canal in each of the three coils : the membranous spiral lamina is preserved,
but the appearances connected with the organ of Corti, &c., have been lost from drying.
f r, fenestra rotunda with its membrane ; *s t,* scala tympani ; *s v,* scala vestibuli ; *l s,*
lamina spiralis ; *h,* hamulus ; *c c,* canalis cochleæ ; *d,* opening of the aqueductus
cochleæ.

Fig. 517.—Vertical Section of the Cochlea of a Fœtal Calf (from Kölliker). ♀

In this specimen the external wall was ossified, but the modiolus and spiral lamina
were still cartilaginous ; the section shows in each part of the cochlear tube the two
scalæ with the intermediate canalis cochleæ and lamina spiralis ; the radiating lines in
the modiolus indicate the passage of the auditory nerves towards the spiral lamina.

Cochlea.—The *membranous cochlea* has the form of a three-sided
tube, the canalis membranacea, interposed between the scala vestibuli
and the scala tympani. The peripheral wall of this canal is formed by
part of the osseous cochlea, and on its other sides it is bounded by the
basilar membrane and membrane of Reissner respectively, while at its
inner angle is a structure named limbus laminæ spiralis, and in its interior
resting on the basilar membrane, is the organ of Corti with the membrana
tectoria covering it. Each of these parts requires description.

The *membrana basilaris,* or *lamina spiralis membranacea,* is stretched
across from the free margin (labium tympanicum) of the osseous lamina to

the outer part of the spiral canal, lying in the same plane as the osseous lamina, and attached peripherally through the medium of a thick structure, the *spiral ligament*. It increases in breadth from the base to the apex of the cochlea, while the osseous spiral lamina diminishes in breadth. Thus in the first turn of the cochlea this membrane forms about half of the breadth of the septum made by it and the osseous lamina; but towards the apex of the cochlea the proportion between the two parts is gradually reversed, until, near the helicotrema, the membranous part is left almost unsupported by any plate of bone.

Fig. 518.—SECTION THROUGH ONE OF THE COILS OF THE COCHLEA (altered from Henle). $\frac{30}{1}$

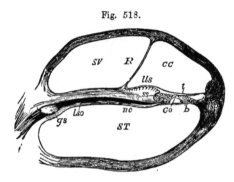

Fig. 518.

A, the section is made in a specimen softened by immersion in hydrochloric acid ; S T, scala tympani ; S V, scala vestibuli ; C C, canalis cochleæ ; R, membrane of Reissner forming its vestibular wall ; *l s o*, lamina spiralis ossea ; *l l s* to *l s p*, lamina spiralis membranacea ; *l l s*, limbus laminæ spiralis ; *s s*, sulcus spiralis ; *n c*, twigs of cochlear nerve ; *g s*, ganglion spirale ; *t*, membrana tectoria; *b*, membrana basilaris ; *C o*, organ of Corti ; *l s p*, ligamentum spirale.

The *limbus laminæ spiralis* (denticulate lamina of Todd and Bowman) is a thick periosteal development near the edge of the osseous spiral lamina on the side which looks towards the vestibular scala. It makes a somewhat convex elevation, presenting externally a sharp margin which overhangs that to which the basilar membrane is attached, being separated from it by a groove. The groove is termed *sulcus spiralis*, and the margins *labium vestibulare* and *labium tympanicum* respectively. The *membrane of Reissner* (membrana vestibularis) arises from the inner part of the limbus, and extends outwards at a considerable angle with the osseous spiral lamina.

The *membrana tectoria* (Claudius), or membrane of Corti (Kölliker), has been variously described, but, according to the most recent researches, is an elastic membrane attached on its one border close to the membrane of Reissner, and on the other by an extremely delicate portion to the peripheral wall of the cochlea, a little above the membrana basilaris (Claudius and Henle). It thus divides the canalis membranacea into two parts : the large part placed between it and the membrane of Reissner, and containing endolymph ; the other, a narrow interval dividing it from the membrana basilaris, and occupied by various cellular and rod-like structures of a highly complicated description, which together are designated as the *organ of Corti*.

The *canalis membranacea*, or *ductus cochlearis*, bounded in the manner already described, presents a blind pointed extremity at the apex and another at the base. That at the apex extends beyond the hamulus, fixed to the wall of the cupola, and partly bounding the helicotrema; that at the base fits into the angle at the commencement of the osseous spiral lamina in front of the floor of the vestibule. Near to this blind extremity the canalis membranacea receives a small duct, *canalis reuniens* (Hensen), which is continued downwards from the saccule of the vestibule like the neck of a flask, and enters the membranous canal abruptly nearly at a right angle to it. Thus the cavity of the canalis membranacea is rendered continuous with that of the saccule.

Fig. 519.

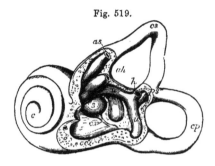

Fig. 519.—THE LEFT LABYRINTH OF A CHILD AT BIRTH, PARTIALLY OPENED ON ITS OUTER SIDE TO SHOW THE COMMENCEMENT OF THE MEMBRANOUS CANAL OF THE COCHLEA (slightly altered after Reichert). $\frac{3}{1}$

The external or horizontal canal has been removed; *c s*, superior canal; *c p*, posterior canal; *a s*, membranous ampulla and tube of the superior canal cut short; *a h*, that of the external or horizontal canal; *h*, undilated end of the horizontal canal in front of the common opening of the superior and posterior canals; *p s*, united superior and posterior canals; *u*, utriculus; *s*, sacculus; *c c*, vestibular part or commencement of the membranous canal of the cochlea; *c r* canalis reuniens connecting it with the sacculus; *c*, cochlea.

It is necessary to explain that, although the canalis membranacea was described by Reissner so long ago as in 1851, yet, owing to some confusion having arisen between the membrane of Reissner and the membrana tectoria described by Corti, whose work appeared at the same time, the nature of this canal has until comparatively

Fig. 520.

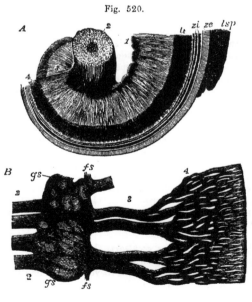

Fig. 520.—DISTRIBUTION OF THE COCHLEAR NERVES IN THE LAMINA SPIRALIS (after Henle).

A, part of the modiolus and spiral lamina showing the cochlear nerves forming a network, viewed from the base; 1, the twigs of the nerve issuing from the tractus spiralis foraminosus; 2, the branches of the nerve entering by the central canal of the modiolus; 3, wide plexus in the bony lamina spiralis; 4, close plexus at its border; *l t*, labium tympanicum; *z i*, zona interna; *z e*, zona externa; *l s p*, ligamentum spirale. B, part of the nerves extracted and more highly magnified; 2, twigs of the nerve from the modiolus close to the lamina spiralis ossea; *g s*, spiral gangliform enlargement of the nerve (habenaria ganglionaris); *f s*, nerve-fibres running spirally along the gangliform swelling (Henle); 3, wide plexus; 4, close plexus of nerve-fibres as in A.

recently been generally misconceived. The history of the discovery and subsequent appreciation of the nature of the canalis membranacea is fully given by Reichert. (Abhandl. d. Königl. Akad. d. Wissensch., Berlin, 1864.)

Cochlear division of the auditory nerve.—The nerve of the *cochlea* is shorter, flatter, and broader than any of the other nerves of the internal ear, and perforates the bone by a number of foramina at the bottom of the internal meatus, below the opening of the Fallopian aqueduct. These foramina are arranged in a shallow spiral groove (tractus spiralis foraminulentus) in the centre of the base of the cochlea ; and they lead into small bony canals, which follow first the direction of the axis of the cochlea, through the modiolus, and then radiate outwards, between the plates of the bony lamina spiralis. In the centre of the spiral groove is a larger foramen which leads to the canalis centralis modioli. Through the central foramen and straight canal the filaments for the last half-turn of the lamina spiralis are conducted ; whilst the first two turns are supplied by filaments which occupy the smaller foramina and bent canals. In the bone the nerves have dark outlines, and near the edge of the spiral lamina they form a plexus which contains ganglion-cells, and may be considered as a spiral ganglion contained in an osseous canal, *canalis spiralis modioli*, already mentioned. From the outer side of this ganglion the fibres, still possessing the dark outline, pass onwards with a plexiform arrangement, and, emerging from the bone beneath the labium tympanicum of the limbus, are collected into bundles, which, opposite a line of perforations situated at the junction with the membrana basilaris and named habenula perforata, present the appearance of conical extremities entering those perforations. Beyond this they have not yet been traced with certainty, although it seems probable, as suggested by Kölliker, that the nerves are in continuity with spindle-shaped cells in the organ of Corti.

Microscopic structure.—The *limbus lamina spiralis* is a thick structure continuous with the periosteum of the vestibular surface of the osseous lamina. Its free surface is thrown into a number of fungiform elevations narrower at the base than at their extremities. Towards the inner part of the limbus these elevations are short and vertical, but those which are placed further out are more and more oblique and longer, and the labium vestibulare is formed by the outermost of them, which are lengthened into rib-like processes with flat extremities placed edge to edge, overhanging the sulcus spiralis like teeth. In the spaces between the elevations numerous small bodies like nuclei are disposed. In the floor of the sulcus spiralis where the labium tympanicum is continued into the membrana basilaris a series of elevations (apparent teeth of Corti) are directed into the membrane, and between their outer extremities are the oblique perforations occupied by the conical extremities of the nerve-bundles. This part is the *habenula perforata* of Kölliker : it is described by him along with the membrana basilaris, and by Henle along with the limbus. Henle considers the appearance of elevations as caused merely by the nerve-bundles grooving the under surface and leaving thicker structure between.

The *membrana basilaris*.—The *membrana basilaris* is divisible into an inner and an outer zone. The inner zone (habenula tecta vel arcuata) is covered over by the rods of Corti ; the outer zone (zona pectinata) is attached peripherally to the walls of the canal through the medium of the cochlear ligament. The inner zone, together with the apparatus on its surface, continues, according to Henle, of an uniform breadth of about $\frac{1}{250}$th of an inch, both in the different parts of the same cochlea, and likewise in different animals : so that

Fig. 521.

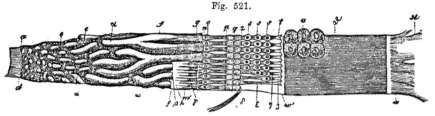

Fig. 521.—UPPER OR VESTIBULAR SURFACE OF A NARROW STRIP OF THE LAMINA
SPIRALIS MEMBRANACEA (from Kölliker after Corti). $\frac{205}{1}$

The drawing is defective as regards the organ of Corti, but explains the nomenclature
of the parts introduced by that author, and more or less adhered to by subsequent writers,
although variously departed from in some of its details. The nomenclature adopted in the
text has been selected from various writers, and it will be observed differs considerably
from the following : a, periosteum of the zona spiralis ossea; $d\,w$, lamina spiralis membranacea ; $d\,w'$, zona denticulata; $d\,f$, habenula sulcata ; d, place where the periosteum thickens ; e, granules in the areolæ of the habenula sulcata ; $f\,g$, teeth of the first
series ; $g\,f\,h$, sulcus vel semicanalis spiralis: h, its lower wall ; $h\,w$, habenula denticulata ; $h\,m$, apparent teeth ; $n\,t$, teeth of the second series ; $n\,p$, inner segments of the
same ; o, swellings with nuclei ; $p\,q$ and $q\,z$, articulating pieces of the same : t, anterior
segments of the second series ; $s\,s\,s$, three cylindrical cells placed on them ; u, epithelial
cells placed under the membrane of Corti ; $w'\,w$, zona pectinata ; $a\,a$, band-like
elevations of the habenula sulcata ; β, placed where a tooth of the first series takes
its origin ; γ, holes between the apparent teeth ; δ, fore part of one of the teeth of the
second series thrown back ; ϵ, one of them in its place without its epithelial cells ; ζ, one
with only the lowest epithelial cell ; η, one with the two lowest cells ; δ, striæ or slight
elevations of the zona pectinata ; κ, periosteum attaching the lamina spiralis, with λ,
apertures between the bundles.

the increasing breadth of the membrane from base to apex of the cochlea
is due to broadening of the zona pectinata. According to the same
observer the membrane is mainly homogeneous, and in the outer zone is
thicker than in the inner, and somewhat tuberculated ; but on the surface
towards the membranous canal it is transversely striated by a layer of
extremely delicate fibres ; and on the other surface is a less perfect layer
of fibres with spindle-shaped corpuscles, which are placed longitudinally,
and in young subjects are arranged so as to cover the inner zone and the
attachment to the spiral ligament, leaving the outer zone free. A single
layer of epithelium lies on the surface.

The *ligamentum spirale* (*musculus cochlearis* of Todd and Bowman) is
triangular in section, receiving at its inner angle the basilar membrane, and
spreading out rapidly to be attached by a broad base to the wall of the
cochlea. Its fibres are directed outwards from the membrane to the bone,
and it exhibits nuclei, like the ciliary muscle, whence Todd and Bowman
conceived it to be muscular. Hensen represents it as composed of branching nucleated cells.

The organ of Corti.—Under this name may be comprised the whole of
the structures intervening between the membrana basilaris and membrana
tectoria. The most prominent part of it is formed by an outer and an inner
series of rods, which, attached respectively to the inner and outer margins of
the inner zone of the basilar membrane, meet together like the beams of a
roof, and cover in a three-sided space, of which the inner zone of the basilar
membrane is the floor. These structures, the *fibres* or *rods of Corti*, are
closely adherent by their lower extremities to the basilar membrane. They
are placed with the regularity of piano keys, and have been likened in con-

Fig. 522.

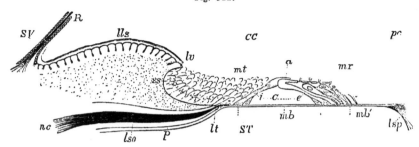

Fig. 522.—Diagrammatic Outline of a Radial Section through the Lamina Spiralis Membranacea, Organ of Corti, &c. (after Kölliker, Henle, and others). $\frac{250}{1}$

This figure may be regarded as a more enlarged and explanatory view of the part of fig. 518 representing the organ of Corti : S V, part of the scala vestibuli ; C C, canalis cochleæ ; S T, scala tympani ; R, membrane of Reissner, forming the partition between the scala vestibuli and the canalis cochleæ ; *l s o*, a small part of the lamina spiralis ossea cut in the direction of one of the canals transmitting the cochlear nerves, *n c* ; *p*, periosteum lining the scala tympani ; *l l s*, limbus laminæ spiralis, presenting a great thickening of the periosteum, in which over the extremity of the osseous spiral lamina is found the sulcus spiralis *s s*, and upon the upper surface of which are the toothed projections ; *l v*, labium vestibulare ; *l t*, labium tympanicum of the sulcus spiralis ; *l v* to *l s p*, the lamina spiralis membranacea with its contained parts ; *m t*, membrana tectoria passing from the limbus laminæ spiralis to the outer wall of the cochlear tube ; *m b*, membrana basilaris, stretched from the labium tympanicum to the outer wall of the cochlear tube, where it expands in the ligamentum spirale, *l s p* ; the part marked by the letters *m b*, between two short dotted lines, forms the zona tecta or z. arcuata ; the part indicated by *m b'* and between the adjacent dotted lines is the zona pectinata ; C, the organ of Corti ; *i*, the internal rods ; *e*, the external rods ; these are set by their lower flattened ends on the basilar membrane, and are articulated together at their upper parts, *a*, the inner overlapping the outer ; a nucleus is seen close to the base of each of the rods close on the basilar membrane ; *m r*, membrana reticularis, stretched to the outer wall of the cochlea, *p c* ; below *m r*, the cells of Corti lying obliquely on the outer rods, and between them the cells of Deiters, and between these and the outer wall of the cochlea epithelial cells ; between *a* and *m r*, are indicated the perforations through which the hair-like terminations of the cells of Corti project : the sulcus spiralis is seen filled with cylindrical and other epithelium.

sistency to cartilage. The inner rods are more closely set and more numerous than the outer, and appear generally to be of a uniform breadth, flattened, and with a nucleiform body placed subjacent to the lower extremity. The outer rods are narrow and cylindrical in their shafts, and expanded at the lower extremity, which has a nucleiform body subjacent to it, as in the case of the inner rods. At their upper ends where they meet together, both sets of rods are thickened, and the parts which are in contact (*coins articulaires externes et internes* of Corti) have the appearance of quadrilateral plates directed outwards so that those of the inner row lie over those of the outer row, and those of the outer row are bent backwards from the direction in which the rods to which they belong are placed. From the junction line of the rods there extends outwards an extremely delicate network, the *lamina reticularis* of Kölliker (i. velamentosa, Deiters), which, it may be gathered from different accounts, is mainly constructed of a layer of squamous cells so disposed as to leave at least three rows of large perforations between them, and which are cemented together by a network of intervening substance which is sometimes detected when the cells are not. At its inner margin this lamina is united by flat plates to the inner series of rods, and by narrow bodies with flattened extremities to the outer series : at its outer margin it has not yet been demonstrated that it is attached to the wall of the cochlea, although it has been supposed that its function might be to give fixity to the rods of Corti. Besides the rods and the lamina reticularis the organ of Corti presents various cellular elements. Of these the most important are an outer and inner series of cells with stiff

Fig. 523.

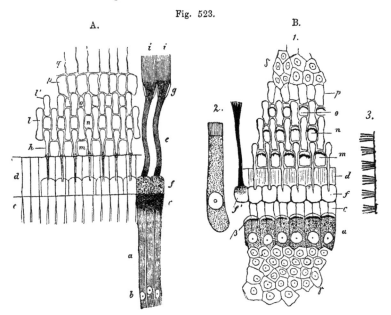

Fig. 523 A.—View from above of the Organ of Corti and Lamina Reticularis in the Ox (from Kölliker). $\frac{540}{1}$

a, inner rods or fibres of Corti ; b, inner ends of the same with the deeper attached nuclei ; c, articulating part of the same ; d, clear plates appended, which with others from the outer rods form the commencement of the membrana reticularis ; e, outer rods or fibres of Corti ; f, their articulating portions ; g, their terminations at the membrana basilaris ; h, plates of the outer rods belonging to the membrana reticularis ; i i, apparent extension of the ends of the fibres of Corti in the striæ of the zona pectinata of the basilar membrane ; l, their inner connecting plates ; l', their outer connecting plates ; m, n, o, first, second, and third series of perforations ; p, rectangular terminal part of the lamina ; q, prolongation of this in the form of fibres upon the large epithelial cells of the organ of Corti.

Fig. 523 B.—The Organ of Corti of the Cat (from Kölliker). $\frac{540}{1}$

1, the organ of Corti from above ; c, the articulated part of the inner fibres or rods ; d, connected plates which form the commencement of the membrana reticularis ; f, articulating portions of the outer rods ; f', one of these connected with a filamentous process, and presenting granular or punctated contents; m, n, o, first, second, and third row of perforations, in which the cilia of Corti's cells are represented as dark arched lines; a, inner ciliated cells with (β) their cilia, forming the outermost part of the thick epithelium of the sulcus spiralis (γ), and which covers the inner fibres (rods) of Corti as far as their articulating parts ; δ, outer part of the network of the lamina reticularis ; 2, a cell of Corti with its hairs, but no visible filamentous appendage ; 3, lateral view of the lamina reticularis with the bundles of cilia of the cells of Corti.

cilia projecting from their upper extremities. The *inner ciliated cells* form a single row resting on the articulating ends of the inner rods : the *outer ciliated cells* (pedunculated cells of Corti) are placed in three rows external to the outer rods, and are described as attached by pointed extremities to the membrana basilaris, and with their ciliated ends opposite the three rows of openings in the lamina reticularis ; so that sometimes when the lamina is detached the ends of the cells are detached with it. Alternating with the outer ciliated cells are the *cells of Deiters,* which are fusiform and prolonged into a thread at each extremity, one passing up to the lamina reticularis, and the other down to the outer zone of the membrana basilaris. The upper surface of the remaining part of the basilar membrane is covered with hexagonal epithelium-cells. The sulcus spiralis is likewise filled with large epithelial cells, which, according to Kölliker, project in a swelling distinct from the proper organ of Corti.

The mode of termination of the nerves, as has been already said, is uncertain, but minute fibres, consisting of axis-cylinders only, have been traced by Deiters into the organ of Corti, and his statements receive some support from Kölliker and Henle. These fibres are said to divide into a radiating set distributed both above and beneath the rods, and into a spiral set which are continued in the longitudinal direction of the canal.

The *membrana tectoria* is described by Henle as presenting three zones. The inner of these is delicate and presents large openings corresponding to elevations of the limbus ; the middle or generally recognised part is formed of layers of fibres directed outwards, but yet crossing each other ; and the outer part, unrecognised by most observers, is extremely delicate, forming a network, the openings in which are elongated in the direction of the canal.

The *membrane of Reissner* is an extremely easily torn membrane, on both sides of which epithelium has been described.

On the microscopic anatomy of the cochlea may be consulted Henle's Systematische Anatomie ; Kölliker's Gewebelehre, 4th edition ; also the papers of Corti, Claudius, Deiters, and Hensen, in Vols., III. VII., X., and XIII. of Siebold and Kölliker's Zeitsch. f. Wissensch. Zoologie ; and Deiters Untersuchungen über die Lamina Spiralis Membranacea.

BLOOD-VESSELS OF THE LABYRINTH.

Arteries.—The *internal auditory artery,* a branch from the basilar, enters the internal meatus of the ear with the auditory and facial nerves, and at the bottom of that shallow canal divides into vestibular and cochlear branches.

The *vestibular* branches are distributed to the common sinus, sacculus, and semicircular canals, with the branches of nerve which they accompany through the bony foramina. At first they ramify on the exterior of the membranous labyrinth, and end in capillaries both on the outer surface and in the substance of the special glassy layer. The plexus is best marked internally near the ending of the nerves.

The *cochlear* branches, twelve or fourteen in number, traverse the many small canals in the modiolus and bony lamina spiralis, and form in the latter a capillary plexus that joins at intervals the vas spirale, to be presently described. From this plexus offsets are distributed in the form of a fine network on the periosteum, but the vessels do not anastomose across the membrana basilaris. The *vas spirale* is a single, sometimes branched vessel which runs along the under surface of the membranous zone, near the bone : it is like a capillary in texture, but larger in size, and is probably venous. On the outer wall of the membranous canal there is a specially vascular strip which has received the name of *stria vascularis.*

Besides the foregoing vessel, which is the chief artery of the internal ear, the *stylo-mastoid* branch of the posterior auricular, and occasionally the occipital artery (Jones), send twigs to the vestibule and the posterior semicircular canal.

Veins.—The veins of the *cochlea* issue from the grooves of the cochlear axis and join the veins of the *vestibule* and semicircular canals: these accompany the arterial branches, and, uniting with those of the cochlea at the base of the modiolus, pour their contents into the superior petrosal sinus.

DEVELOPMENT OF THE EAR.

In the very young embryo the first rudiment of the ear is seen in the form of a small vesicle—the *primary auditory vesicle* lying at the side of the third primary cerebral vesicle. It has to a certain extent an appearance similar to that of the primary optic vesicle situated further forwards, and was long very naturally supposed to be formed like it by a protrusion of the wall of the primary medullary cavity of the brain; but it has latterly been established by various observers that it is produced solely by invagination of the integument, and has no original connection with the brain. During the third day of incubation it can be seen in the chick, still open to the outside, above and behind the second branchial lappet. It soon becomes completely closed, and is afterwards developed into the membranous labyrinth. The first complication which the vesicle exhibits is by the extension of a process upwards and backwards, which remains permanent in the lower vertebrata, but in mammals is obliterated, its vestiges remaining in the aqueduct of the vestibule. The semicircular canals next appear as elongated elevations of the surface of the primary vesicle: the middle portion of each elevation becomes separated from the rest of the

Fig. 524.

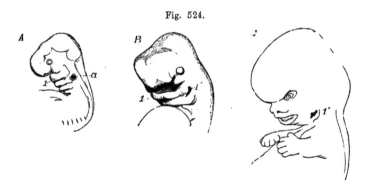

Fig. 524.—Outlines showing the Formation of the External Ear in the Fœtus.

A, head and upper part of the body of a human fœtus of about four weeks (from nature). ♀ Four branchial plates (the first, forming the lower jaw, is marked 1), and four clefts are shown; the auditory vesicle (*a*), though closed, is visible from the transparency of the parts, and is placed behind the second branchial plate.

B, the same parts in a human fœtus of about six weeks (from Ecker). ⁵⁄₁ The third and fourth plates have nearly disappeared, and the third and fourth clefts are closed; the second is nearly closed; but the first (1′) is somewhat widened posteriorly in connection with the formation of the meatus externus.

C, human fœtus of about nine weeks (from nature). ³⁄₁ The first branchial cleft is more dilated, and has altered its form along with the integument behind it in connection with the formation of the meatus externus and the auricle.

vesicle by bending in of its walls under it, and thus the elevation is converted into a tube open at each end, which subsequently becomes elongated and presents an ampullar dilatation. The cartilage which forms the osseous labyrinth is continuous with that of the rest of the primordial cranium. The cartilaginous walls of the cavity are united by connective tissue to the vesicle : this connective tissue, according to Kölliker, becomes divided into three layers, of which the outer forms the lining periosteum, the inner forms the external walls of the membranous labyrinth, while the intervening layer swells up into gelatinous tissue, the meshes of which become wider and wider, till at last the space is left which ultimately is found containing perilymph.

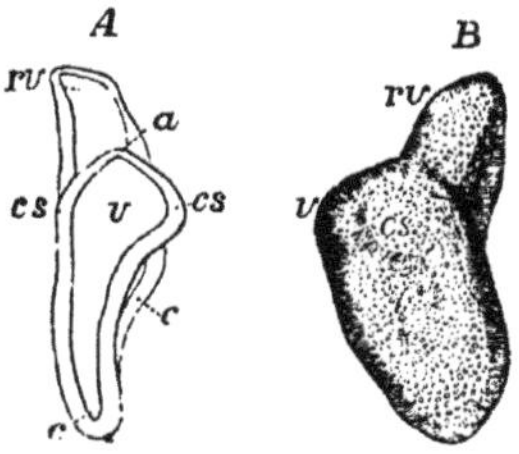

Fig. 525.—LABYRINTH OF THE HUMAN FŒTUS OF FOUR WEEKS, MAGNIFIED (from Kölliker).

Fig. 525.

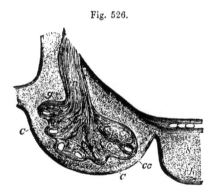

A, from behind ; B, from before; v, the vestibule ; r v, recessus vestibuli, giving rise later to the aqueduct ; c s, commencement of the semicircular canals ; a, upper dilatation, belonging perhaps to another semi-circular canal ; c, cochlea.

The cochlea appears at first as a prolongation downwards from the auditory vesicle, but afterwards become tilted forwards. This prolongation of the auditory vesicle is the rudimentary canalis membranacea. Close to it is placed the cochlear nerve, with a gangliform extremity. The canal becomes elongated in a spiral direction, and the ganglion, which is elongated with it, becomes the ganglion spirale. Between the canal and the cartilaginous wall which afterwards surrounds it a large amount of connective tissue intervenes, and in this the cavities of the scala vestibuli and scala tympani appear at a later period, precisely as does the space for the perilymph in the

Fig. 526.—TRANSVERSE SECTION OF THE COCHLEA IN A FŒTAL CALF, MAGNIFIED (from Kölliker).

Fig. 526.

C, the wall of the cochlea, still carti-laginous ; c c, canalis cochleæ ; l s, placed in the tissue occupying the place of the scala vestibuli indicates the lamina spiralis ; n, the central cochlear nerve ; g, the place of the spiral ganglion ; S, the body of the sphenoid ; c h, chorda dorsalis.

vestibule. The modiolus and spiral lamina, according to Kölliker, are ossified without intervention of cartilage. Within the canalis membranacea Kölliker finds in the embryo a continuous epithelial lining, thin on the membrane of Reissner and on the outer wall, but forming a thick eleva-tion in the position of the rods of Corti, and a larger elevation more internally, filling up the sulcus spiralis. On the surface of this latter elevation he observes a transparent body, the membrane of Corti.

With regard to the middle and external ear, it has been already explained at pages 65 and 66 that the external aperture, the tympanic cavity, and the Eustachian tube, are formed in the posterior or upper part of the first branchial cleft, which remains open except at the place where the passage is interrupted by the formation of the membrana tympani ; and also that the incus and malleus are formed in the first branchial lappet from the proximal part of Meckel's cartilage, and the stapes and stapedius muscle and the styloid process in the second lappet. It is pointed out by

Kölliker that during the whole period of fœtal life the tympanic cavity is occupied by connective tissue, in which the ossicles are imbedded; and that only after the breathing process is commenced this tissue recedes before an expansion of the mucous membrane. The pinna is gradually developed on the posterior margin of the first branchial cleft. It is deserving of notice that congenital malformation of the external ear, with occlusion of the meatus and greater or less imperfection of the tympanic

Fig. 527.

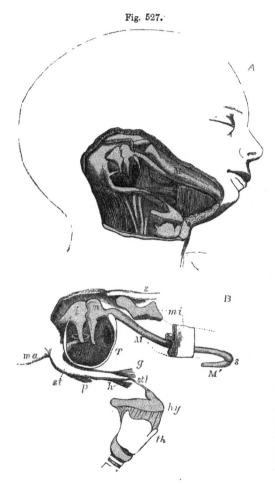

Fig. 527.—VIEWS OF THE CARTILAGE OF MECKEL AND PARTS CONNECTED WITH THE FIRST AND SECOND BRANCHIAL PLATES.

A (after Kölliker), head of a fœtus of about eighteen weeks, showing the cartilage of Meckel in connection with the malleus, &c. M, the cartilage of Meckel of the right side. B (from nature). An enlarged sketch explanatory of the above view; z, the zygomatic arch; ma, the mastoid process; mi, portions of the lower jaw of which the parts near the angle and the symphysis have been removed; M, the cartilage of Meckel of the right side; M', a small part of that of the left side, joining the left cartilage at s, the symphysis; T, the tympanic ring; m, the malleus; i, the incus; s, the stapes; sta, the stapedius muscle; st, the styloid process; p, h, g, the stylo-pharyngeus, stylohyoid and stylo-glossus muscles, stl, stylo-hyoid ligament attached to the lesser cornu of the hyoid bone; hy, the hyoid bone; th, thyroid cartilage. In A, the head being turned somewhat upwards, the same parts are shown, together with the surrounding muscles, the carotid artery, jugular vein, &c.

apparatus, are observed in connection with abnormal development of the deeper parts of the first and second branchial lappets and the intermediate cleft; while cases have been observed of the persistence in the neck of the adult of one or more of the branchial clefts situated behind the first. (Allen Thomson, Proceed. Roy. Soc. of Edin. 1844, and Edin. Journ. of Med. Sc. 1847.)

THE NOSE.

The nose is the special organ of the sense of smell. It has also other functions to fulfil ;— for, communicating freely with the cavities of the mouth and lungs, it is concerned in respiration, voice, and taste ; and by means of muscles on its exterior, which are closely connected with the muscles of the face, it assists in the expression of the different passions and feelings of the mind.

Fig. 528.—LATERAL VIEW OF THE CARTILAGES OF THE NOSE (from Arnold). ¾

Fig. 528.

a, right nasal bone; *b*, nasal process of the superior maxillary bone ; 1, upper lateral cartilage or wing-like expansion of the septal cartilage ; 2, lower lateral cartilage (outer part) ; 2*, inner part of the same ; 3, sesamoid cartilages.

This organ consists of, first, the anterior prominent part, composed of bone and cartilages, with muscles already described, which slightly move the cartilages, and two orifices, *anterior nares*, opening downwards ; and, secondly, of the two nasal fossæ, in which the olfactory nerves are expanded. The nasal fossæ are separated from each other by a partition, *septum nasi*, formed of bone and cartilage : they communicate at the outer side with hollows in the neighbouring bones (ethmoid, sphenoid, frontal, and superior maxillary); and they open backwards into the pharynx through the posterior nares. The skin of the nose is studded, particularly in the grooves of the alæ or outer walls of the nostrils, with numerous small openings, which lead to sebaceous follicles. Within the margin of the nostrils there is a number of short, stiff, and slightly curved hairs—*vibrissæ*, which grow from the inner surface of the alæ and septum nasi, as far as the place where the skin is continuous with the mucous membrane lining the cavity of the nose.

CARTILAGES OF THE NOSE.

These are the chief support of the outer part of the organ. They occupy the triangular opening seen in front of the nasal cavity in the dried skull, and assist in forming the septum between the nasal fossæ. There are usually reckoned two larger and three smaller cartilages on each side, and one central piece or cartilage of the septum.

The *upper lateral cartilages* (cartilagines laterales nasi) are situated in the upper part of the projecting portion of the nose, immediately below the free margin of the nasal bones. Each cartilage is flattened and triangular in shape, and presents one surface outwards, and the other inwards towards

3 E 2

the nasal cavity. The anterior margin, thicker than the posterior one, meets the lateral cartilage of the opposite side above, but is closely united with the edge of the cartilage of the septum below ; so closely indeed, that by some, as Henle, the upper lateral are regarded as reflected wings of the median cartilage. The inferior margin is connected by fibrous membrane with the lower lateral cartilage ; and the posterior edge is inserted into the ascending process of the upper maxilla and the free margin of the nasal bone.

Fig. 529.   Fig. 530.

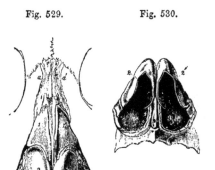

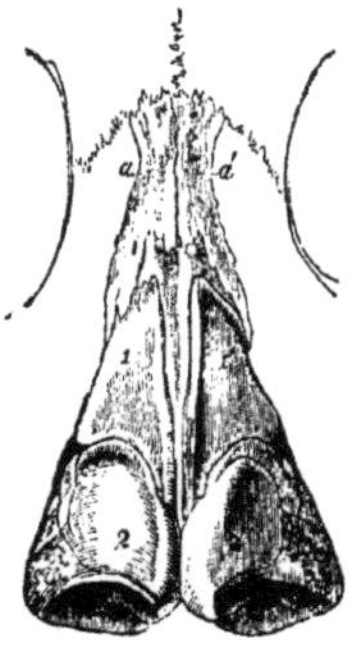

Fig. 529.—FRONT VIEW OF THE CARTILAGES OF THE NOSE (from Arnold). ¾

a, a', nasal bones ; 1, 1', upper lateral cartilages or wing-like expansions of the septal cartilage ; 2, 2', lower lateral cartilages.

Fig. 530.—VIEW OF THE CARTILAGES OF THE NOSE FROM BELOW (from Arnold). ¾

2, 2', outer part of the lower lateral cartilages ; 2*, 2*, inner part of the same ; 4, lower edge of the cartilage of the septum.

The *lower lateral cartilages* (cartilagines alarum nasi) are thinner than the preceding, below which they are placed, and are chiefly characterised by their peculiar curved form. Each cartilage consists of an elongated plate, so bent upon itself as to pass in front and on each side of the nostril to which it belongs, and by this arrangement serves to keep it open. The outer portion is somewhat oval and flattened, or irregularly convex externally. Behind, it is attached to the margin of the ascending process of the upper maxilla, by tough fibrous membrane, in which are two or three cartilaginous nodules (cartilag. minores vel sesamoideæ); above, it is fixed, also by fibrous membrane, to the upper lateral cartilage, and to the lower and fore part of the cartilage of the septum. Towards the middle line it is curved backwards, bounding a deep mesial groove, at the bottom of which it meets with its fellow of the opposite side, and continues to pass backwards, forming a small part of the columna nasi, below the level of the cartilage of the septum. This inner part of the cartilage of the ala is thick and narrow, curls outwards, and ends in a free rounded margin which projects outwards towards the nostril. The lower and most prominent portion of the ala of the nose, like the lobule of the ear, is formed of thickened skin with subjacent tissue, and is unsupported by cartilage.

The *cartilage of the septum* has a somewhat triangular outline, and is thicker at the edges than near the centre. It is placed nearly vertically in the middle line of the nose, and completes, at the fore part, the separation between the nasal fossæ. The anterior margin of the cartilage, thickest above, is firmly attached to the back of the nasal bones near their line of junction ; and below this it lies successively between the upper and the lower lateral cartilages, united firmly with the former and loosely with the latter. The posterior margin is fixed to the lower and fore part of the central plate of the ethmoid bone ; and the lower margin is received into the groove of the vomer, as well as into the median ridge between the superior maxillæ.

Fig. 531.—Osseous and Cartilaginous Septum of the Nose, seen from the Left Side (from Arnold). ⅔

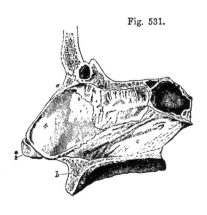

Fig. 531.

a, right nasal bone; *b*, superior maxillary bone; *c*, sphenoidal sinus; *d*, perpendicular plate of the ethmoid bone; *e*, vomer; 2*, inner part of the right lower lateral cartilage; 4, cartilage of the septum.

This cartilage is the persistent anterior extremity of the primordial cranium. In young subjects it is prolonged back to the body of the presphenoid bone; and in many adults an irregular thin band remains between the vomer and the central plate of the ethmoid.

NASAL FOSSÆ.

The nasal fossæ, and the various openings into them, with the posterior nares, have been previously described as they exist in the skeleton, and the

Fig. 532.

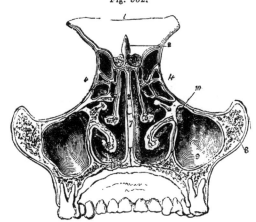

Fig. 532.—Transverse Vertical Section of the Nasal Fossæ seen from behind (from Arnold). ¾

1, part of the frontal bone; 2, crista galli; 3, perpendicular plate of the ethmoid; between 4 and 4, the ethmoid cells; 5, right middle spongy bone; 6, left lower spongy bone; 7, vomer; 8, malar bone; 9, maxillary sinus; 10, its opening into the middle meatus.

greater part of that description is also applicable generally to the nose in a recent state; but it is proper to mention certain differences in the form and dimension of parts, which depend on the arrangement of the lining mem-brane, viz.—

Throughout the whole of the nasal fossæ it is to be observed that—

First, owing to the thickness of the membrane in question, (which not only lines the walls of the fossæ, but covers the spongy bones on both sides,) the nasal cavity is much narrower in the recent state. Second, in consequence of the prolongations of membrane on their free margins, the turbinate bones, and more particularly the lower pair, appear in the recent state to be both more prominent, and longer in the direction from before backwards, than in the dried skull. Third, by the arrange-ment of the mucous membrane round and over the orifices which open into the nasal fossæ, some of the foramina in the bones are narrowed, and others completely closed.

In the individual parts of the nasal fossæ the following particulars are to be noticed.

In the *upper meatus,* the small orifice which leads into the posterior ethmoidal cells is lined by a prolongation of the thin mucous membrane which continues into those cavities; but the spheno-palatine foramen is covered over by the Schneiderian membrane, so that no such opening exists in the recent nasal fossa.

In the *middle meatus* the aperture of the infundibulum is nearly hidden by an overhanging fold of membrane ; it leads directly into the anterior ethmoidal cells, and through them into the frontal sinus. Below and behind this, the passage into the antrum of Highmore is surrounded by a circular fold of the pituitary membrane, (sometimes prominent and even slightly valvular,) which leaves a circular aperture much smaller than the foramen in the bony meatus.

In the *lower meatus,* the inferior orifice of the nasal duct is defended by one or two folds of membrane ; and when there are two, the folds are often adapted so accurately together as to prevent even air from passing back from the cavity of the nose to the lachrymal sac.

In the *roof* the apertures in the cribriform plate of the ethmoid bone are closed by the membrane, but the openings into the sphenoidal sinuses receive a prolonga-tion from it.

In the *floor* the incisor foramen is in the recent state generally closed. Some-times, however, a narrow funnel-shaped tube of the mucous membrane descends for a short distance into the canal, but is closed before it reaches the roof of the palate. Vesalius, Stenson, and Santorini, believed that this tube of membrane opened generally into the roof of the mouth by a small aperture close behind the interval between the central incisor teeth. Haller, Scarpa, and, more recently, Jacob-son, find that in man it is usually closed, and often difficult of detection. (See Cuvier's Report on a paper by Jacobson, " Annales du Museum d'Hist. Naturelle ; " Paris, 1811 ; vol. xviii. p. 412.)

MUCOUS MEMBRANE.

The pituitary or Schneiderian membrane, which lines the cavities of the nose, is a highly vascular mucous membrane, inseparably united, like that investing the cavity of the tympanum, with the periosteum and peri-chondrium over which it lies. It is continuous with the skin through the nostrils ; with the mucous membrane of the pharynx through the pos-terior apertures of the nasal fossæ ; with the conjunctiva through the nasal duct and lachrymal canaliculi ; and with the lining membrane of the several sinuses which communicate with the nasal fossæ. The pituitary membrane, however, varies much in thickness, vascularity, and general appearance in these different parts. It is thickest and most vascular over the turbinate bones (particularly the inferior), from the most dependent parts of which it forms projections in front and behind, thereby increasing the surface to some extent. On the septum nasi the pituitary membrane is still very thick and spongy ; but in the intervals between the turbinate bones, and over the floor of the nasal fossæ, it is considerably thinner. In the maxillary, frontal, and sphenoidal sinuses, and in the ethmoidal cells, the mucous lining membrane, being very thin and pale, contrasts strongly with that which occupies the nasal fossæ.

Fig. 533.

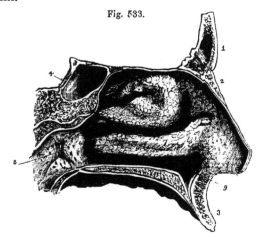

Fig. 533.—OUTER WALL OF THE LEFT NASAL FOSSA, COVERED BY THE PITUITARY MEMBRANE (from Arnold). ¾

1, frontal bone; 2, left nasal bone; 3, superior maxillary; 4, body of the sphenoid with the sphenoidal sinus; 5, projection of the membrane covering the upper spongy bone; 6, that of the middle; 7, that of the lower; the upper, middle, and lower meatuses are seen below the corresponding spongy bones; 8, opening of the Eustachian tube; 9, depression of the lining membrane of the nose in the anterior palatine canal.

In respect of the characters of the mucous membrane, three regions of the nasal fossæ may be distinguished. Thus, the region of the nostrils, including all the part which is roofed by the nasal cartilages, is lined with stratified squamous epithelium; the remainder of the fossæ is divisible into two parts, viz., the olfactory region in which the epithelium is non-ci-

Fig. 534.—VERTICAL SECTION OF A SMALL PORTION OF THE MEMBRANE OF THE NOSE FROM THE OLFACTORY REGION (from Ecker). ⁵⁰⁄₁

Fig. 534.

a, coloured part of the epithelium; *a'*, nuclei; *b*, deeper part containing the olfactory cells and filaments; *c*, connective tissue of the mucous membrane; *d*, one of the mucous glands; *d'*, its duct; *e*, twig of the olfactory nerve; *e'*, small twig passing to the surface.

liated and columnar, and the respiratory region in which it is ciliated and columnar. The membrane in the respiratory part, consisting of the inferior turbinated and all the lower portions of the fossæ, is studded with numerous mucous glands, which are of branched acinated appearance, and open by apparent orifices on the surface. These are most numerous about the middle and hinder parts of the nasal fossæ, and are largest at the back of the septum near the floor of the nasal cavity. They are much smaller and less numerous in the membrane lining the several cavities which communicate with the nasal fossæ.

The olfactory region or that in which the olfactory nerve is distributed, includes the upper and middle turbinated parts, and the upper portion of the septum. Its mucous membrane is thicker and more delicate in consistence than that of the ciliated region, being soft and pulpy. The columnar cells on its surface are prolonged at their deep extremities into threads, which have been observed to communicate with stellate cells of the connective tissue. Beneath the columnar cells is a considerable thickness of densely nucleated tissue, compared by Henle to the cortical brain-substance. The glands of this region are numerous ; but are of a more simple structure than those in the lower part of the fossæ.

Fig. 535.

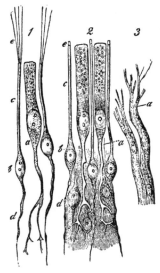

Fig. 535.—CELLS AND TERMINAL NERVE-FIBRES OF THE OLFACTORY REGION (from Frey after Schultze).

1, from the frog ; 2, from man ; *a*, epithelial cell, extending deeply into a ramified process; *b*, olfactory cells ; *c*, their peripheral rods; *e*, their extremities, seen in 1 to be prolonged into ciliary hairs ; *d*, their central filaments ; 3, olfactory nerve-fibres from the dog ; *a*, the division into fine fibrillæ.

Olfactory Cells.—Intermixed with the columnar epithelial cells of the olfactory region, and so numerous as to surround each of them, are certain peculiar bodies, each consisting of a spindle-shaped nucleated cell, from which proceed a superficial and a deep process. The superficial process is a cylindrical or slightly tapering thread passing directly to the surface, and terminating abruptly at the same level as the epithelial cells between which it lies : the deep process is more slender and passes vertically inwards. Both processes frequently present a beaded appearance similar to that observed in fine nerve-filaments, and considered to be of a similar accidental origin. It was suggested by Max Schultze, the discoverer of the olfactory cells, and is highly probable, that the deep processes are directly continuous with the filaments of the olfactory nerve, but the continuity does not appear to have been actually observed.

The superficial process of the olfactory cell was observed by Schultze to be surmounted by a short stiff hair-like process, and has been so described by others ; but both the discoverer and others are now agreed that this appearance results from the coagulation of albumen escaped from the interior of the process. Long and fine hair-like processes do, however, exist on the olfactory membranes of amphibia, reptiles, and birds, and had been previously pointed out by Schultze.

Olfactory Nerve.—The filaments of this nerve, lodged at first in grooves on the surface of the bone, enter obliquely the substance of the Schneiderian membrane, and pass to their distribution between its mucous and fibrous layers. The nerves of the septum are rather larger than those of the outer wall of the nasal fossæ ; they extend over the upper third of the septum, and as they descend become very indistinct. The nerves of the outer wall are divided into two groups—the posterior branches being distributed over the surface of the upper spongy bone, and the anterior branches descending ever the plain surface of the ethmoid and the middle spongy bone.

Fig. 536.

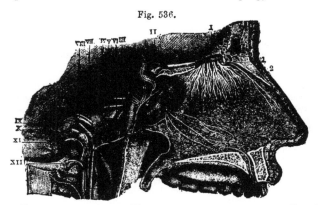

Fig. 536.—Nerves of the Septum Nasi, seen from the right side (from Sappey after Hirschfeld and Leveillé). ⅔

I, the olfactory bulb; 1, the olfactory nerves passing through the foramina of the cribriform plate, and descending to be distributed on the septum; 2, the internal or septal twig of the nasal branch of the ophthalmic nerve; 3, naso-palatine nerves.

The olfactory nerves as they descend ramify and unite in a plexiform manner, and the filaments join in brush-like and flattened tufts, which, spreading out laterally and communicating freely with similar offsets on

Fig. 537.—Nerves of the Outer Wall of the Nasal Fossæ (from Sappey after Hirschfeld and Leveillé). ⅔

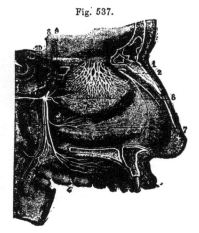

Fig. 537.

1, network of the branches of the olfactory nerve, descending upon the region of the superior and middle turbinated bones; 2, external twig of the ethmoidal branch of the nasal nerve; 3, spheno-palatine ganglion; 4, ramification of the anterior palatine nerves; 5, posterior, and 6, middle divisions of the palatine nerves; 7, branch to the region of the inferior turbinated bone; 8, branch to the region of the superior and middle turbinated bones; 9, naso-palatine branch to the septum cut short.

each side, form a fine net-work with elongated and narrow intervals between the points of junction; but it is impossible to trace by dissection the termination of the nerves in the membrane, in consequence of the difficulty of recognising the filaments, destitute of dark outline, as they lie among the other nucleated tissues.

In their nature the olfactory filaments differ much from the fibres of the cerebral and spinal nerves: they contain no white substance of Schwann, are pale, and finely granular in texture, firmly adherent one to another, and have oval corpuscles on their surface.

The greater part of the mucous membrane of the nasal fossæ is provided with nerves of common sensibility, derived from branches of the fifth pair : these have already been described at pp. 599, 603 and 604.

Blood-vessels.—The arteries and veins of the nose are derived from numerous sources : those of the interior form rich plexuses of capillaries in the lining membrane. The description of the arteries will be found at pp. 350, 356, 361 and 362 ; that of the veins at pp. 456 and 464.

DEVELOPMENT OF THE NOSE.

The organ of smell, as was first pointed out by V. Baer, owes its origin, like the primary auditory vesicle and the crystalline lens of the eye, to a depression of the integument. This depression, the primary olfactory groove, is at first encircled by a uniform wall, and is unconnected with the mouth. This stage has been observed by Kölliker in the human embryo of four weeks. Soon, however, by the unequal growth of the surrounding parts, a groove is formed, descending from the pit and passing into the mouth. Thus the middle frontal process is isolated between the grooves of opposite sides, while the lateral frontal process separates the nostril from the eye (p. 65). The maxillary lobes, growing forwards from behind the eyes, complete the boundaries of the nostrils, which then open into the fore part of the mouth. Kölliker observes this stage in the latter half of the second month. The palate subsequently grows inwards to the middle line, as has been elsewhere stated, and separates the nasal from the buccal cavity; leaving only the extremely minute communication of the incisor foramen. Meanwhile, with the growth of the face, the nasal fossæ deepen, and the turbinated bones make their appearance as processes from their walls. Observations are still wanting to determine whether the olfactory nerves are developed from the bulbs, and have thus a cerebral origin, or are separately formed from peripheral blastema like all other nerves, with the exception of the optic.

Fig. 537.*

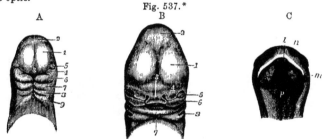

A   B   C

Fig. 537.*—VIEWS OF THE HEAD OF HUMAN EMBRYOES, ILLUSTRATING THE DEVELOPMENT
OF THE NOSE.

A, Head of an embryo of three weeks (from Ecker). ♀ 1, anterior cerebral vesicle ; 2, middle vesicle ; 3, nasal or middle frontal process ; 4, superior maxillary process ; 5, eye ; 6, inferior maxillary process or first visceral plate, and below it the first cleft ; 7, 8, and 9, second, third, and fourth plates and clefts.

B, Head of an embryo of about five weeks (from Ecker). ♀ 1, 2, 3, and 5, the same parts as in A ; 4, the external nasal or lateral frontal process, inside which is the nasal groove ; 6, the superior maxillary process ; 7, the inferior maxilla ; ×, the tongue seen within the mouth ; 8, the first branchial cleft which becomes the outer part of the meatus auditorius externus.

C, View of the head of an embryo of eight weeks seen from below, the lower jaw having been removed (from Kölliker). ‡ *n*, the external nasal apertures ; *i*, intermaxillary or incisor process, and to the outer side of this the internal nasal aperture ; *m*, one of the palatal processes of the upper jaw, which advancing inwards from the sides form the partition between the mouth and nose ; *p*, common cavity of the nose, mouth, and pharynx.

SECTION VI.—SPLANCHNOLOGY.

UNDER the division Splanchnology will be described those organs of the body which have not found a place in any of the foregoing parts of the work. These consist of the organs of digestion, the organs of respiration, the urinary organs, and the organs of generation.

ORGANS OF DIGESTION.

The *digestive apparatus* includes that portion of the organs of assimilation within which the food is received and partially converted into chyle, and from which, after the chyle has been absorbed, the residue or excrement is expelled. It consists mainly of a tubular part,—the *alimentary canal*, together with various glands of which it receives the secretions.

The alimentary canal is a long membranous tube commencing at the mouth and terminating at the anus, composed of certain tunics or coats, and lined by a continuous mucous membrane from one end to the other. Its average length is about thirty feet, being about five or six times the length of the body. Its upper extremity is placed beneath the base of the skull, the succeeding portion traverses the thorax, and by far the greater part is contained within the cavities of the abdomen and pelvis.

The part situated above the diaphragm consists of the organs of mastication, insalivation, and deglutition, and comprises the *mouth*, with the teeth and salivary glands, the *pharynx*, and the *œsophagus* or gullet. The remainder includes that part of the canal which is more immediately engaged in the digestive process, in absorption, and in defecation, as the stomach and the small and large intestine. The glands which are most intimately connected with digestion consist of those very numerous smaller glandular organs which are situated in the mucous membrane of the alimentary canal, and the larger glands, such as the pancreas and liver, whose ducts open within the canal.

THE MOUTH.

The *mouth*, or, more definitely, the *buccal cavity*, is the space included between the lips and the throat. Bounded by the lips, cheeks, tongue, and the hard and soft palate, it communicates behind with the pharynx through an opening called the *fauces* (isthmus faucium). The cavity of the mouth is lined throughout by a mucous membrane, which is of a pink rosy hue during life, but pale grey after death, and which presents peculiarities of surface and structure to be noticed hereafter.

The *lips* and *cheeks* are composed of an external layer of skin, and of an internal layer of mucous membrane, together with muscles, vessels, and nerves already fully described in other parts of this work, some areolar tissue, fat, and numerous small glands. The free border of the lips is protected by a dry mucous membrane, which becomes continuous with the skin, is covered with numerous minute papillæ, and is highly sensitive. On the inner surface of each lip, the mucous membrane forms a fold in the middle line, connecting the lip with the gums of the corresponding jaw. These are the *fræna* or *frænula* of the lips : that of the upper lip is much the larger.

Numerous small glands,. called *labial glands*, are found beneath the mucous membrane of the lips, around the opening of the mouth. They are situated between the mucous membrane and the orbicularis oris muscle. They are compound glands of a rounded form, the largest of them not exceeding the size of a split pea ; and they open into the mouth by distinct orifices.

Between the buccinator muscle and the mucous membrane of the cheek, by which it is lined in its whole extent, are the *buccal* glands, similar to the labial glands, but smaller. Two or three glands, larger than the rest, found between the masseter and buccinator muscles, and opening by separate ducts near the last molar tooth, are called the *molar* glands. The duct of the parotid gland also opens upon the inner surface of the cheek, opposite to the second upper molar tooth.

Immediately within the lips and cheek, are the *dental arches*, consisting of the teeth, gums, and maxillæ. The jaw-bones, the articulation and movements of the lower maxilla, and the muscles used in mastication, are elsewhere described. The *gums* (gingivæ) are composed of a dense fibrous tissue, connected very closely with the periosteum of the alveolar processes, and covered by a red and highly vascular but not very sensitive mucous membrane, which is smooth in its general surface, but is beset with fine papillæ in the immediate vicinity of the teeth.

THE TEETH.

In the human subject, as in mammalia generally, two sets of teeth make their appearance in the course of life, of which the first constitutes the *temporary, deciduous,* or *milk* teeth, whilst the second is named the *permanent* set. The temporary teeth are twenty in number, ten in each jaw, and the permanent set consists of thirty-two, sixteen above and sixteen below.

Deficiencies in the number of the teeth sometimes occur, and the number is frequently increased by one or more supernumerary teeth. These are usually small, and provided with only a single fang ; and, though generally distinct, they are sometimes attached to other teeth : they occur more frequently near the front than the hinder teeth, and are more often met with in the upper than in the lower jaw.

General Characters of the Teeth. —Every tooth consists of three portions, viz., one which projects above the gums and is named the body or *crown,*— another which is lodged in the alveolus or socket, and constitutes the root or *fang,*—and a third, intermediate between the other two, and, from being more or less constricted, named the cervix or *neck.* The size and form of each of these parts vary in the different kinds of teeth.

The roots of all the teeth are accurately fitted to the alveoli of the jaws, in which they are implanted. Each alveolus is lined by the periosteum, which also invests the contained fang as high as the cervix. This dental periosteum, sometimes named the periodontal membrane, is blended with the dense and slightly sensitive tissue of the gums, which closely surrounds the neck of the tooth. The roots of all the teeth taper from the cervix to the point, and this form, together with the accurate adjustment to the alveolus, has the effect of distributing the pressure during use over the whole socket, and of preventing its undue action on the apex of the fang through which the blood-vessels and nerves enter.

The thirty-two permanent teeth consist of four incisors, two canines, four bicuspids, and six molars in each jaw. The twenty temporary teeth are four incisors, two canines, and four molars above and below. There are no bicuspids among the temporary teeth, but the eight deciduous molars are succeeded by the eight bicuspids of the permanent set. The relative position and arrangement of the different kinds of teeth in the jaws may be expressed by the following formula, which also exhibits the relation between the two sets in these respects :—

Temporary teeth
$$\left\{\begin{array}{l}\text{Upper} \\ \text{Lower}\end{array}\right.$$

	MO.	CA.	IN.	CA.	MO.	
Upper	2	1	4	1	2	=10
Lower	2	1	4	1	2	=10

=20

Permanent teeth
$$\left\{\begin{array}{l}\text{Upper} \\ \text{Lower}\end{array}\right.$$

	MO.	BI.	CA.	IN.	CA.	BI.	MO.	
Upper	3	2	1	4	1	2	3	=16
Lower	3	2	1	4	1	2	3	=16

=32

Special Characters of the Permanent Teeth.—The *incisors*, eight in number, are the four front teeth in each jaw, and are so named from being adapted for cutting or dividing the soft substances used as food. Their *crowns* are chisel-shaped, and have a sharp horizontal cutting edge, which by continued use is bevelled off behind in the upper teeth, but in the lower teeth is worn down in front, where it comes into contact with the overlapping edges of the upper teeth. Before being subjected to wear, the horizontal edge of each incisor tooth is serrated or marked by three small prominent points. The

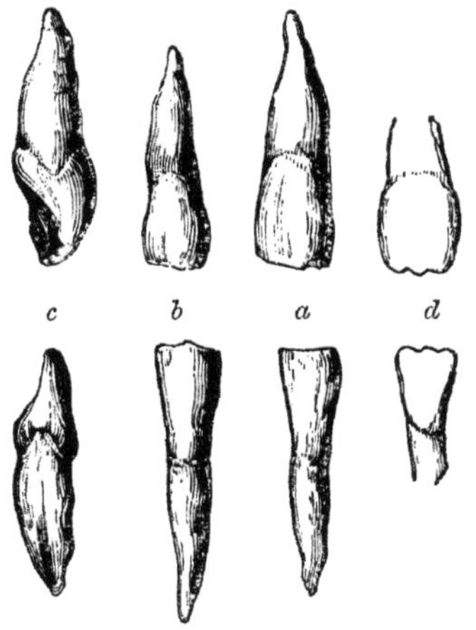

Fig. 538.—Incisor Teeth of the Upper and Lower Jaws.

Fig. 538.

a, front view of the upper and lower middle incisors ; *b*, front view of the upper and lower lateral incisors ; *c*, lateral view of the upper and lower middle incisors, showing the chisel shape of the crown ; a groove is seen marking slightly the fang of the lower tooth ; *d*, the upper and lower middle incisor teeth before they have been worn, showing the three pointed projections of the cutting edge.

c   *b*   *a*   *d*

anterior surface of the crown is slightly convex, and the posterior concave. The *fang* is long, single, conical, and compressed at the sides, where it sometimes though rarely presents a slight longitudinal furrow.

The lower incisor teeth are placed vertically in the jaw, but the corresponding upper teeth are directed obliquely forwards. The upper incisors are, on the whole, larger than the lower ones.

In the upper jaw the central incisors are larger than the lateral ; the reverse is the case in the lower jaw, the central incisors being there the smaller, and being, moreover, the smallest of all the incisor teeth.

The *canine* teeth (cuspidati), four in number, are placed one on each side, above and below, next to the lateral incisors. They are larger and stronger than the incisor teeth. The *crown* is thick and conical, convex in front and hollowed behind, and may be compared to that of a large incisor tooth the angles of which have been removed, so as to leave a single central point or *cusp*, whence the name *cuspidate* applied to these teeth.

Fig. 539.

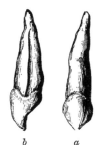

b   *a*

Fig. 539.—CANINE TOOTH OF THE UPPER JAW.

a, front view; *b*, lateral view, showing the long fang grooved on the side.

The point always becomes worn down by use. The *fang* of the canine teeth is single conical, and compressed at the sides : it is longer than the fangs of any of the other teeth, and is so thick as to cause a corresponding prominence of the alveolar arch : on the sides it is marked by a groove, an indication, as it were, of the cleft or division which appears in the teeth next following.

The upper canines, popularly called the *eye-teeth,* are larger than the lower, and in consequence of this, as well as of the greater width of the upper range of incisors, they are thrown a little farther outwards than the lower canine teeth. In the dog-

Fig. 540.   Fig. 541.

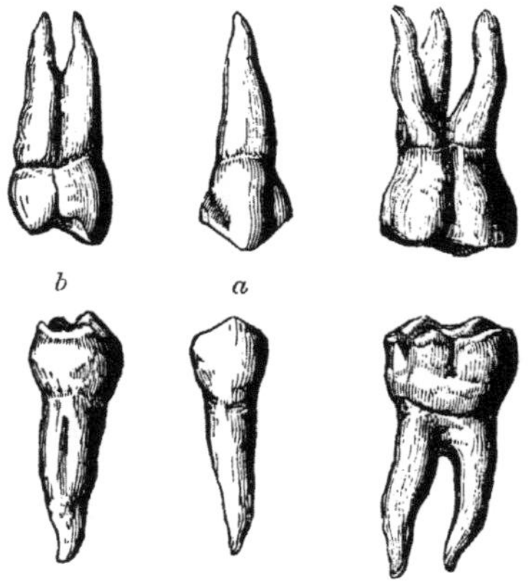

Fig. 540.—FIRST BICUSPID TOOTH OF THE UPPER AND LOWER JAWS.

a, front view; *b*, lateral view, showing the lateral groove of the fang, and the tendency in the upper to division.

Fig. 541.—FIRST MOLAR TOOTH OF THE UPPER AND LOWER JAWS.

They are viewed from the outer aspect.

b   *a*

tribe, and in the carnivora generally, these teeth acquire a great size, and are fitted for seizing and killing prey, and for gnawing and tearing it when used as food.

The *bicuspids* (bicuspidati), also called premolars, are four in each jaw ; they are shorter and smaller than the canines, next to which they are placed, two on each side.

The *crown* is compressed before and behind, its greater diameter being across the jaw. It is convex, not only on its outer or labial surface, like the preceding teeth, but on its inner surface also, which rises vertically from the gum : its free extremity is broader than that of an incisor or canine tooth, and is surmounted by two pointed tubercles or cusps, of which the external one is larger and higher than the other. The *fang* is also flattened, and is deeply grooved in all cases, showing a tendency to become double. The apex of the fang is generally bifid, and in the first upper bicuspid the root is often cleft for a considerable distance; but the bicuspid teeth are very variable in this respect, and may be, all four, free from any trace of bifidity of the root. The upper bicuspids are larger than the lower ones, and their cups are more deeply divided. Sometimes the first lower bicuspid has only óne tubercle distinctly marked, *i. e.*, the external, and in that case approaches in figure to a canine tooth.

The *molar* teeth, true or large molars, or multicuspid teeth, are twelve in number, and are arranged behind the bicuspid teeth, three on each side, above and below. They are distinguished by the large size of the crown, and by the great width of its grinding surface. The first molar is the largest, and the third is the smallest, in each range, so as to produce a gradation of size in these teeth. The last molar in each range, owing to its late appearance through the gums, is called the *wisdom-tooth*, dens sapientiæ. The *crowns* of the molar teeth are low and cuboid in their general form. Their outer and inner surfaces are convex, but the crowns are rather flattened before and behind. The grinding surface it nearly square in the lower teeth, and rhomboidal in the upper, the corners being rounded off : it is not smooth, but is provided with four or five trihedral tubercles or cusps (whence the name of multicuspidati), separated from each other by a crucial depression. The upper molars have four cusps situated at the angles of the masticating surface ; of these the internal and anterior cusp is the largest, and is frequently connected with the posterior external cusp by a low oblique ridge. In the upper wisdom-teeth, the two internal tubercles are usually blended together. The crowns of the lower molars, which are larger than those of the upper, have five cusps, the additional one being placed between the two posterior cusps, and rather to the outer side : this is especially evident in the lower wisdom-teeth, in which the crown is smaller and rounder than in the others. The *fangs* of all the molar teeth are multiple. In the two anterior molars of the upper jaw, they are three in number, viz. two placed externally, which are short, divergent, and turned towards the antrum of the superior maxilla ; and a third or internal fang, which is larger and longer, and is directed towards the palate, the posterior border of which extends as far back as that of the posterior external fang. This third fang is often slightly grooved, especially when the two internal cusps are very distinct, and sometimes it is divided into two smaller fangs. The two anterior molars of the lower jaw have each two fangs, one anterior, the other posterior, which are broad, compressed, and grooved on the faces that are turned towards each other, as if each consisted of two fangs fused together : they have an inclination or curve backwards in the jaw, and are slightly divergent, or sometimes parallel, or even nearly in contact with each other : more rarely one or both of them is divided into two smaller fangs. In the wisdom-teeth of both jaws the fangs are often collected into a single irregular conical mass, which is either directed backwards in the substance of the jaw, or curved irregularly : this composite fang sometimes shows traces of subdivision, and there are occasionally two fangs in the lower tooth and three in the upper.

The bicuspid and the molar teeth, from the breadth and uneven form of their crowns, are fitted for bruising, crushing, and grinding the food in mastication.

The range of teeth in each jaw forms a nearly uniform curve, which is not broken by any intervals, as is the case in the dental apparatus of many animals, even in the Quadrumana. The upper dental arch is rather wider than the lower one, so that the teeth of the upper jaw slightly overhang those of the lower. This is owing principally to the fact that the lower teeth are placed either vertically, as in front, or are inclined somewhat inwards, as is seen behind and at the sides, while the corresponding teeth of the upper jaw havé an inclination forwards in front, and outwards behind. While there is a slight diminution in the height of the exposed parts of the teeth from the incisors backwards to the wisdom-teeth, there is in man a general uniformity in the amount of projection of the crowns throughout the whole series. In consequence of the large proportionate breadth of the upper central incisors, the other teeth of the upper jaw are thrown somewhat outwards, so that in closure of the jaws the canine and bicuspid teeth come into contact partly with the corresponding lower teeth and partly with those next following; and in the case of the molar teeth, each cusp of the upper lies behind the corresponding cusp of the lower teeth. Since, however, the upper wisdom-teeth are smaller than those below, the dental ranges terminate behind nearly at the same point in both jaws.

The Milk-teeth.—The temporary incisor and canine teeth resemble those of the permanent set in their general form; but they are of smaller dimensions. The temporary molar teeth present some peculiarities. The hinder of the two is much the larger; it is the largest of all the milk-teeth, and is larger even than the second permanent bicuspid, by which it is afterwards replaced. The *crown* of the first upper milk molar has only three cusps,

Fig. 542.

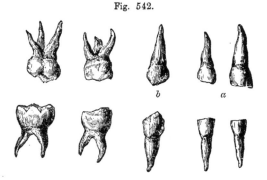

Fig. 542.—Milk Teeth of the Right Side of the Upper and Lower Jaws.

a, the incisors; *b*, the canines; *c*, the molar teeth.

two external and one internal; that of the second has four distinct cusps. The first lower temporary molar has four cusps, and the second five, of which in the latter case three are external. The *fangs* of the temporary molars resemble those of the permanent set, but they are smaller, and are more divergent from the neck of the tooth.

Structure.—On making a section of a tooth, the hard substance of which it is composed is found to be hollow in the centre. The form of the cavity bears a general resemblance to that of the tooth itself: it occupies the interior of the crown, is widest opposite to or a little above the neck, and extends down each fang, at the point of which it opens by a small orifice. In the crown of the incisor teeth the cavity is prolonged into two fine linear canals, which proceed one to each corner of the crown; in the bicuspid and molar teeth it advances a short distance into each cusp. In the case of a root formed by the blending of two or more fangs, as occurs occasionally in the wisdom-teeth, each division has a separate canal prolonged down to its apex.

The central cavity of a tooth is called the *pulp-cavity*, because it is occupied and accurately filled by a soft, highly vascular, and sensitive substance,

Fig. 543.—SECTIONS OF AN INCISOR AND MOLAR TOOTH.

Fig. 543.

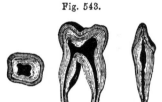

The longitudinal sections show the whole of the pulp-cavity in the incisor and molar teeth, its extension upwards within the crown and its prolongation downwards into the fangs with the small aperture at the point of each; these and the cross section show the relation of the dentine and enamel.

called the *dental pulp.* This pulp consists of areolar filaments, amongst which numerous nuclei and cells are rendered visible by the action of acetic acid. It is well supplied with vessels and nerves, which are derived from

Fig. 544.—MAGNIFIED LONGITUDINAL SECTION OF A BICUSPID TOOTH (after Retzius).

Fig. 544.

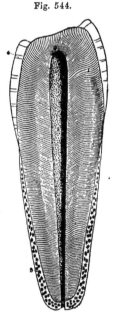

1, the ivory or dentine, showing the direction and primary curves of the dental tubuli; 2, the pulp-cavity with the small apertures of the tubuli into it; 3, the cement or crusta petrosa covering the fang as high as the border of the enamel at the neck, exhibiting lacunæ; 4, the enamel resting on the dentine; this has been worn away by use from the upper part.

the internal maxillary artery and the fifth pair, and which enter the cavity through the small aperture at the point of each fang.

The solid portion of the tooth is composed of three distinct substances, viz. the proper dental substance, *ivory* or *dentine,* the *enamel,* and the *cement* or *crusta petrosa.* The dentine constitutes by far the larger part of the hard substance of a tooth; the enamel is found only upon the exposed part or crown; and the cement covers with a thin layer the surface of the implanted portion or fang. A fourth variety of tissue, *osteodentine,* is formed within the dentine, at the expense of the pulp, as age advances.

A. The *dentine,* (Owen,) forming the principal mass or foundation of the body and root of a tooth, gives to both of these parts their general form, and immediately encloses the central cavity. It resembles very compact bone in its general aspect and chemical relations, but is not identical with it in structure, or in the exact proportions of its earthy and animal constituents.

According to the analyses of Berzelius and Bibra, the dentine of human teeth consists of 28 parts per cent. of animal, and 72 of earthy matter. The former is resolvable into gelatin by boiling. The composition of the latter, according to Bibra, is as follows, viz., phosphate of lime 66·7 per cent., carbonate of lime 3·3, phosphate of magnesia and other salts, including a trace of fluoride of calcium, 1·8. Berzelius found 5·3 carbonate of lime.

3 F

Fig. 545.

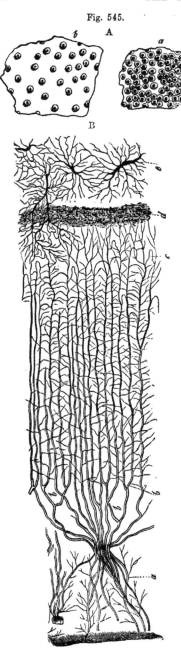

Examined under the micro-
scope, dentine is seen to consist
of an immense number of very
fine tubes, imbedded closely
together in a hard intertubular
matrix, and having the appearance of possessing distinct parietes. These *dental tubules* open
at their inner ends into the pulp-
cavity, the wall of which presents very numerous minute orifices over the whole of its inner
surface. From thence they
pass in a radiated manner

Fig. 545.—Sections of Dentine
(from Kölliker).

A, highly magnified cross sections of
the tubuli of dentine. ⁴⁵⁰⁄₁. *a*, from
a part in which the tubuli are very
closely set; *b*, from a part where
they are widely set.

B, longitudinal section of the root.
³⁵⁰⁄₁ *a*, the dental tubes near the inner
surface of the dentine with few tubuli;
b, subdivision of tubuli; *c*, looped
disposition of the tubuli; *d*, granular
layer consisting of small dental globules at the margin of the dentine;
e, lacunæ of the cement, one of them
connected by tubuli with those of the
dentine.

through every part of the ivory
towards its periphery. In the
upper portion of the crown they
have a vertical direction; but
towards the sides, and in the
neck and root, they become
gradually oblique, then horizontal, and are finally even inclined downwards towards the
point of the fang. The course
of the tubules is not straight,
but each describes, in passing
from the central to the peripheral part of the dentine, two
or three gentle curves *(primary
curvatures, Owen), and is besides bent throughout its whole
length into numerous fine undulations, which follow closely one
upon another; these are the
secondary curvatures. The curvatures of adjacent tubules so

far correspond, that the tubes are on the whole nearly parallel, being only slightly divergent as they pass towards the surface ; and as they divide several times dichotomously, and at first without being much diminished in size, they continue to occupy the substance of the dentine with nearly equidistant tubes, and thus produce, when seen in fine sections of the tooth made parallel to their course, a striated appearance, as if the dentine were made up of fine parallel fibres. The concurrence of many of these parallel curvatures of the dental tubuli produces, by the manner in which they reflect the light, an appearance of concentric undulations in the dentine, which may be well seen with a low magnifying power. This, however, is not to be confounded with another set of curved marks called contour lines, which depend on conditions of the matrix, and will be afterwards described. The average diameter of each tubule near its inner and larger end is $\frac{1}{4500}$th of an inch, and the distance between adjacent tubules is about two or three times their width. (Retzius.) From their sides numerous immeasurably fine branches are given off, which penetrate the hard intertubular substance, where they either anastomose or terminate blindly. These lateral ramuscles are said to be more abundant in the fang. Near the periphery of the ivory they are very numerous, and, together with the main tubules themselves, which there, by rapid division and subdivision, also become very fine, terminate by joining together in loops, or end in little dilatations, or in the cells of the granular layer to be described.

The dental tubules, when highly magnified, appear like dark lines against transmitted light, but are white when seen upon a black ground. Their tubular character is proved by the fact that ink, or other coloured fluids, together with minute bells of air, can be made to pass along them, in sections of dry teeth. Their walls, in transverse sections, may often appear thicker than they are in reality, owing to a certain length of the tubes being seen in the section : but if the orifice of the canal be brought exactly into focus, the wall appears as only a very thin, yellowish border ; and, indeed, Kölliker denies the existence of any wall distinct from the matrix. From the researches of Nasmyth, Tomes, and Kölliker, it appears that in the recent state the tubules are filled with substance (dental fibres), continuous with the pulp of the tooth : and it is suggested by Tomes that this is not only subservient to the nutrition of the dentine, but probably also confers on it a certain degree of sensibility. It has been noticed, indeed, that the dentine is more sensitive near the surface than deeper in its substance,—a fact not easily intelligible on the supposition that the sentient tissue is confined to the pulp-cavity.

In the temporary, and sometimes even in the permanent teeth, the tubules are constricted at short intervals, so as to present a moniliform character. The terminal branches of tubules are occasionally seen to pass on into the cement which covers the fang, and to communicate with the small ramified canals of the characteristic lacunæ found in that osseous layer. Tubules have likewise been observed by Tomes passing on into the enamel, more especially in the teeth of marsupial animals, but in a less marked degree in human teeth.

The *intertubular substance* is translucent. The animal matter which remains in it, after the earthy matter has been removed by an acid, exhibits a tendency to tear in the direction of the tubules, but is in reality a homogeneous substance, deposited in a laminated manner. This was shown by Sharpey, who observed that in the softened teeth of the cachalot or sperm-whale the animal substance was readily torn into fine lamellæ, disposed parallelly with the internal surface of the pulp-cavity, and therefore across the direction of the tubules. In these lamellæ the sections of the tubules appeared as round or oval apertures, the lamellæ having the same relation to the tubules as those of true bone to the canaliculi. The same tendency to lamination may be exhibited by boiling a longitudinal section of tooth with caustic potash, after which it presents closely set, short, and regular fissures, lying at right angles to the tubules, throughout the extent of the dentine. (Cleland.)

Fig. 546.

Fig. 547.

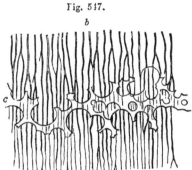

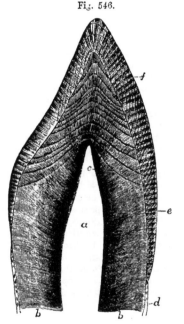

Fig. 546.—Vertical Section of the Upper Part of an Incisor Tooth (from Kölliker). $\frac{7}{1}$

a, the pulp-cavity; b, dentine or tubular substance; c, arched contour lines with interglobular spaces; d, cement; e, enamel with an indication of the direction of the columns; f, coloured lines of the enamel.

Fig. 547.—A Small Portion of the Dentine with Interglobular Spaces (from Kölliker). $\frac{350}{1}$

b, the tubules; c, the interglobular spaces filled with air.

A laminated structure of a more distinct description has been observed in the dentine of the crown, giving rise to the appearances in longitudinal sections termed *contour lines*. Czermak states that transverse sections of the tooth present concentric lines resembling the year-rings of wood : and Salter has shown that decalcified specimens readily break up in these lines ; the crowns of the teeth consisting of a series of superimposed hollow cones: the intervals between their strata, in longitudinal sections, appearing as contour markings, in transverse sections as annular lines ; in both cases corresponding with the surface of the pulp, as it existed during the formation of the tooth. The contour markings, when examined with the microscope, are seen to be caused by irregularities of the intertubular tissue, which, opposite these marks, presents the appearance of spaces or clefts bounded by globular masses of the ordinary tubular and dense substance. These globules vary in size from $\frac{1}{250}$th to $\frac{1}{10000}$th of an inch ; the largest being in the crown, the smallest in the fang. The tubuli pass through these globules, and appear to be continuous in direction across the interspaces from one globule to another.

Another kind of irregularity in the structure of the dentine gives rise to the *granular layer* of Purkinje ; the peculiarity of which consists in the presence of a number of minute cell-like cavities, which break up the uniformity of the matrix, and by branches anastomose one with another and receive terminations of dental tubuli. They are found principally in a layer beneath the cement, and also beneath the enamel. The circumstance of their forming connections with the tubules points to a difference in nature between these cavities and the much larger interglobular spaces.

The surface of the dentine where it is in contact with the enamel is marked by undulating grooves and ridges, and also by numerous minute hexagonal depressions, to which the microscopic fibres of the enamel are accurately adapted.

Fig. 548.

B. The *enamel* is that hard white covering which encrusts and protects the exposed portion or crown of a tooth. It is the hardest of all the dental tissues, but it is gradually worn down by protracted use. It is thickest on the grinding surface and cutting edges of the teeth, and becomes gradually thinner towards the neck, where it ceases. Its extent and thickness are readily

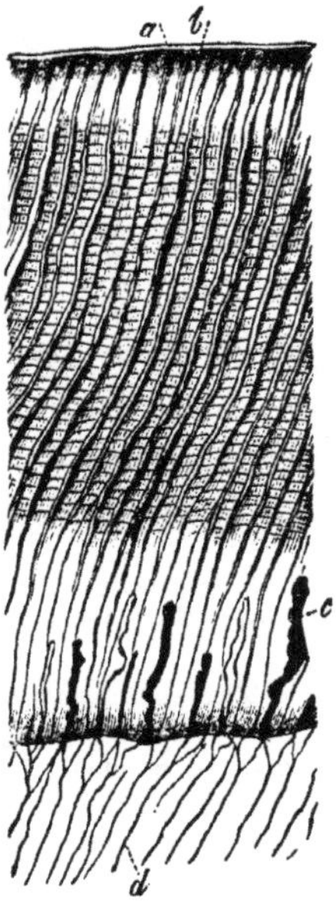

Fig. 548.—Thin Section of the Enamel and a Part of the Dentine (from Kölliker). $\frac{250}{1}$

a, cuticular pellicle of the enamel ; *b*, enamel-fibres or columns with fissures between them and cross striæ ; *c*, larger cavities in the enamel communicating with the extremities of some of the tubuli (*d*).

seen on charring the tooth, by which the dentine becomes blackened, whilst the enamel, owing to the very small quantity of animal matter in its composition, remains white. According to Bibra it contains of earthy constituents 96·5 per cent., viz. phosphate of lime with traces of fluoride of calcium 89·8, carbonate of lime 4·4, phosphate of magnesia and other salts 1·3 ; and has only 3·5 per cent. of animal matter. Berzelius, however, gives the proportion of carbonate of lime as 8, and of animal matter as only 2 per cent.

The enamel is made up entirely of very hard and dense microscopic fibres or prisms, composed almost wholly of earthy matter, arranged closely together, side by side, and set by one extremity upon the subjacent surface of the dentine. On the summit of the coronal portion of the tooth these enamel fibres are directed vertically, but on the sides they are nearly horizontal. As seen in a section they are disposed in gently waving lines, parallel with each other, but not so regular as the curvatures of the tubuli of the dentine, with which they have no agreement. The concurrence of these parallel curvatures produces, as in the case of the dentine, an appearance of concentric undulations in the enamel, which may be seen with a lens of low power. A series of concentric lines is likewise to be seen crossing the enamel fibres, as the contour lines cross the dentine : these are termed *coloured lines* from their brown appearance, but they seem rather to depend on lamination than on pigmentary deposit. Minute fissures not unfrequently exist in the deep part of the enamel, which run between clusters of the fibres down to the surface of the dentine ; and other much larger and more evident fissures are often observed leading down from the depressions or crevices between the cusps of the molar and premolar teeth. The surface of the enamel, especially in the milk-teeth, is marked by transverse ridges, which may be distinguished with a common magnifying glass.

Fig. 549.

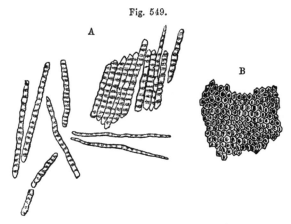

Fig. 549.—ENAMEL FIBRES (from Kölliker). $\frac{350}{1}$

A, fragments and single fibres of the enamel, isolated by the action of hydrochloric acid. B, surface of a small fragment of enamel, showing the hexagonal ends of the fibres.

The enamel-fibres have the form of solid hexagonal prisms.   Their diameter varies slightly, and is ordinarily about $\frac{1}{5000}$th of an inch.   They are marked at small intervals by dark transverse lines.   According to Tomes, the fibre is not in all cases solid, but has occasionally an extremely minute cavity in part or in the whole of its length, which is best seen in newly-developed enamel, but is also visible in adult teeth.   The inner ends of the prisms are implanted, as it were, into the minute hexagonal depressions found on the surface of the dentine, whilst the outer ends, somewhat larger in diameter, are free, and present, when examined with a high magnifying power, a tesselated appearance.

When submitted to the action of dilute acids, the enamel is almost entirely dissolved, and leaves scarcely any discernible traces of animal matter.   Near the deep surface this is rather more abundant, according to the observations of Retzius, who conceived that it there aided in fixing the enamel fibres.   By the action of an acid, the enamel of newly formed or still growing teeth may be broken up, and its structural elements more easily distinguished.   The prisms are then found to have interposed between them a delicate membranous structure, forming sheaths in which the calcarcous matter is deposited.   As this latter accumulates, the membranous structure becomes almost or entirely obliterated, and the now earthy prisms are inseparably consolidated.   Each membranous sheath, according to Tomes, contains a line of granular cells or masses, arranged in single series like the sarcous elements in muscular fibres, and thus occasioning the transverse markings.

It is also found, on treatment with acid, that a very thin membrane called by Kölliker "cuticle of the enamel,"—and by Busk and Huxley "*Nasmyth's membrane*" (after its discoverer), entirely covers the enamel upon its outer surface. This membrane, which is calcified in the natural state, forms a protective covering to the enamel. Berzelius and Retzius say that a similar membrane also exists between the enamel and the dentine, but Kölliker has been unable to find any in that situation.

c. The *crusta petrosa* or *cement* is the third substance which enters into the formation of the teeth. This is a layer of true bone, slightly modified in structure, and investing that part of the dentine which is not protected by the enamel. It covers the whole fang, towards the lower end of which it becomes gradually thicker, and is especially developed at the apex, and along the grooves of the compound fangs. Besides this, the calcified membrane or cuticle on the surface of the enamel has been regarded by various writers as a coating of cement in that situation, the representative of the coronal cement on the compound teeth of many herbivorous animals. As life advances, the cement generally becomes thicker, especially near the apex of the fang, where it sometimes blocks up the orifice leading into the pulp-cavity.

The crusta petrosa contains cells and canaliculi resembling those of bone ; they are placed lengthwise around the fang, and give off minute radiated ramifications, which are often found to proceed from one side only of a cell, towards the *periodontal* surface (Tomes). In the deeper layers of the cement the fine canaliculi sometimes anastomose with some of the terminal tubules of the subjacent dentine. Where the cement is very thick it may contain vascular canals, analogous to the Haversian canals of bone. On the deciduous teeth the cement is thinner, and contains fewer cells. It has been shown by Sharpey that perforating fibres, similar to those of ordinary bone, run abundantly through the cement. In chemical composition it resembles bone, and contains 30 per cent. of animal matter. The cement is, according to some, extremely sensitive at the neck of the tooth, if it be exposed by

Fig. 550.

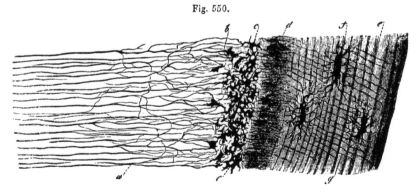

Fig. 550.--Section of a Portion of the Dentine and Cement from the Middle of the Root of an Incisor Tooth (from Kölliker). $\frac{350}{1}$

a, dental tubuli ramifying and terminating, some of them in the interglobular spaces (b and c), which resemble somewhat bone-lacunæ ; d, inner layer of the cement with numerous closely set canaliculi ; e, outer layer of cement; f, lacunæ ; g, canaliculi.

retraction of the gum. By its connection with the surrounding membranous structures it contributes to fix the tooth in the socket. It is the seat of the bony growths or exostoses sometimes found upon the teeth.

D. Osteodentine (Owen), *secondary dentine* (Tomes), or the *horny substance* of Blumenbach, is a hard substance which begins to be deposited on the inner surface of the dentine after the age of twenty years or later, so that the central cavity of a tooth becomes gradually diminished in size, whilst the pulp slowly shrinks or disappears. This additional substance, formerly regarded as an extension of the cement into the interior of the tooth, has been shown to have a distinct structure, in part resembling dentine, and in part bone. It is traversed by canals, which contain blood-vessels, and are surrounded by concentric lamellæ, like the Haversian canals of bone. From these canals, numerous tubules radiate in all directions, larger than the canaliculi of bone, resembling, in this respect, and also in their mode of ramification, the tubes of the dentine. This newly added structure may or may not coalesce with the previously formed dentine ; it appears to be produced by a slow conversion of the dental pulp.

Among special works on the teeth may be noticed, Retzius, in Müller's Archiv, 1837 ; Nasmyth, Researches on the Teeth, 1839 ; Owen, Odontography, 1840-45 ; Tomes, Lectures on Dental Physiology and Surgery, 1848, also in the Phil. Transactions, 1849 and 1850, and in Quart. Journ. of Micr. Science, 1856 ; Salter, in Quarterly Journal of Microscopic Science, 1853, in Guy's Hospital Reports, third series, vol. i. ; and in Trans. Path. Soc., 1854 and 1855 ; Czermak in Zeitschrift für wissensch. Zoologie, 1850 ; Huxley in Quarterly Journal of Microscopic Science, 1853.

DEVELOPMENT OF THE TEETH.

Although the general phenomena of the growth and succession of the teeth had received considerable attention from various anatomists, the observations of Arnold and Goodsir, made independently of each other, more especially the latter, were the first to give precision to our knowledge concerning their origin and the earlier stages of their formation. More recent researches have, it is true, shown that their account of the primordial condition of the dental germs may require some modification ; but nevertheless these authors were the first to establish the primordial connection of the teeth with the mucous membrane covering the edges of the maxillary arches, and Goodsir was the first to give a consistent view of the earlier steps of the formative process in the temporary and permanent series of teeth. (Arnold in Salzbürger Med. Zeitung, 1831 ; Goodsir in Edin. Med. and Surg. Journal, 1839.) The changes which take place in the bones of the jaws relate only to the formation of the sockets of the teeth. In their earliest condition these bones present no appearance of alveoli, but, concurrently with certain changes in the mucous membrane, to be immediately described, a wide groove is developed along the edge of the jaw, which gradually becomes deeper, and is at length divided across by thin bony partitions, so as to form a series of four-sided cells. These bony septa are not distinctly formed until near the fifth month of fœtal life. By the subsequent growth of the bone, these cavities or loculi are gradually closed round, except where they remain open at the edge of the jaw. By the end of the sixth month they are distinctly formed, but continue afterwards, in proportion to the growth of the teeth, to increase in size and depth, by the addition of new matter which widens and deepens the jaw.

The first stages in the development of the teeth, as observed by Arnold and Goodsir, consist of certain changes in the mucous membrane covering the borders of the maxillæ. About the sixth week of embryonic life, a depression or groove, having the form of a horse-shoe, appears along the edge of the jaw, in the mucous membrane of the gum ; this is the *primitive dental groove* (Goodsir). From the floor of this groove (supposed to be represented in a transverse section, in the diagrammatic figure 551,1)

Fig. 551.

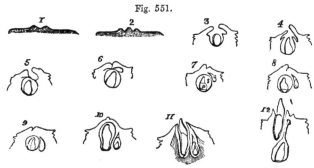

Fig. 551.—Diagrammatic Outlines of Sections through the Dental Germs and Sacs, at Different Stages of Development (from Goodsir).

1, the primitive dental groove of the gum cut across in a fœtus of about six weeks; 2, a papilla rising within the dental groove ; 3, 4, and 5, represent the follicular stage in which the papilla (or future tooth-pulp) is seen sunk within the follicle, and the lips of the follicle or opercula advancing towards each other gradually meet and close in the follicle ; 5, may be looked upon as representing the section indicated by the line *a b*, in fig. 559, through the sac of an incisor tooth, in which a lunated depression (*c*) is left behind ; in 6, the lips of the groove are seen to come together ; in 7, the union of the lips being complete, the follicle becomes a closed sac *s*, containing the dental pulp *p*, and having behind it the lunated depression *c*, now also enclosed, and forming the cavity of reserve for the germ of the corresponding permanent tooth ; in the remaining outlines, 8 to 12, are shown the commencement of the cap of dentine on the pulp, the subsequent steps in the formation of the milk tooth, and its eruption through the gum (11); also the gradual changes in the cavity of reserve, the appearance of its laminæ and papilla, its closure to form the sac of the permanent tooth, its descent into the jaw, behind and below the milk tooth, and the long pedicle (12) formed by its upper obliterated portion.

a series of ten papillæ, as at 2, arise in succession in each jaw, and constitute the germs or rudimentary pulps of the milk-teeth. These pulps or papillæ are processes of the mucous membrane itself, and not mere elevations of its epithelium. The order in which they appear is very regular. The earliest is that for the first milk molar tooth : it is seen at the seventh week, as soon as the dental groove is formed ; at the eighth week that for the canine tooth appears ; the two incisor papillæ follow next, at about the ninth week, the central one before the lateral ; lastly, the second molar papilla is visible at the tenth week, at which period this, the *papillary stage* of the rudiments of the teeth is completed. The papillæ in the upper jaw appear a little earlier than those in the lower jaw.—In the next place, the margins of the dental groove become thickened and prominent, especially the inner one ; and membranous septa or prolongations of the mucous membrane pass across between the papillæ from one margin to the other, so as to convert the bottom of the groove into a series of follicles, each containing one of the papillæ. These changes constitute the *follicular stage ;* they take place in the same order as that in which the papillæ make their appearance, and are completed about the fourteenth week, During the early part of this period the papillæ grow rapidly, they begin to show peculiarities of form, and project from the mouths of the follicles. Soon, however, the follicles become deeper, so as to hide the papillæ, which now assume a shape corresponding with that of the crowns of the future teeth. Small laminæ or opercula of membrane are then developed from the sides of each follicle, their number and position being regulated, it is said, by the form of the cutting edges and tubercles of the coming teeth : the incisor follicles having two laminæ, one external and one internal ; the canine, three, of which two are internal ; and the molars, four or five each.—The lips of the dental groove, as well as the opercula, now begin to cohere over the follicles from behind forwards, the posterior lip being very much thickened ; the groove itself is thus

Fig. 552.

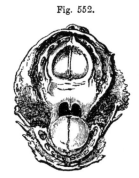

Fig. 552.—ENLARGED VIEW OF THE UPPER AND LOWER DENTAL ARCHES OF A FŒTUS OF ABOUT FOURTEEN WEEKS.

This specimen shows the follicular stage of development of all the milk teeth as described by Goodsir; in each follicle the papilla is seen projecting ; but this exposure of the papillæ and the cavity of the follicles probably arises from the accidental loss of the epithelial covering.

gradually obliterated, the follicles are converted into close sacs, and the *saccular stage* of the milk-teeth is thus completed about the end of the fifteenth week. Certain lunated depressions, which are formed one behind each of the milk-follicles about the fourteenth week, escape the general adhesion of the lips of the groove. From these depressions, as will be afterwards described, the sacs of the ten anterior permanent teeth are subsequently developed.

The first stages in the development of the teeth here described, the superficial origin and open condition of the dental sacs, and the free papillary commencement of the pulps, have been denied in recent years by Guillot, and by Robin and Magitot, who assert that the sacs with their contents make their first appearance in the submucous tissue, and are from the first closed sacs (Guillot in Annales des Sciences Naturelles, vol. ix., 1859 ; Robin and Magitot in Journal de la Physiologie, 1860, vol. iii., pp. 130 and 663). The observations of Kölliker, however, seem to furnish a clue to the explanation of what has been seen by these authors, at the same time that they confirm, in its most important features, Goodsir's mode of viewing the phenomena. In the fœtal lamb and calf, the first step in the formation of the tooth-germ, observed by Kölliker, consists in a depression of a part of the deepest layer of the epithelium into the subjacent mucous membrane. This depression, which, in common with Huxley, he regards as the commencement of the fœtal structure known as the enamel organ, to be afterwards described, widens subsequently, so as to become flask-shaped, remaining connected with the deep surface of the epithelium by a narrow neck. In the next stage the dental papilla rising from the surface of the mucous membrane, projects into, or indents the deepest side of the epithelial process or future enamel organ, and the dental sac is formed at a somewhat later period in the surrounding substance of the mucous membrane. In these animals, therefore, the epithelium of the edge of the jaw covers in completely the enamel-germ or primary tooth-follicle.

In man, Kölliker was unable to discover a similar arrangement, but found matters very much in the disposition described by Goodsir; that is, the follicles open, situated in a dental groove of the jaw, and containing at their deepest part the dental papillæ developed from the mucous membrane. But he conceives it not improbable that in Goodsir's specimens, as well as in his own, the whole of the epithelium had been abraded, and that the follicles and papillæ were thus unnaturally opened to the surface.

Fig. 553.—Diagrams of the Mode of Origin of the Dental Germ in the Ruminant (after Kölliker).

The three figures represent transverse sections of the gum and a part of the jaw at or shortly after the period of the formation of the germ, and are designed chiefly to show the relation of the germ to the epithelium.

A, represents the state in a very early condition, when the primitive dental follicle of a milk or temporary tooth has been formed by a depression from the deep layer of the epithelium.

B, represents a later stage, when the tooth-papilla has risen from the surface of the mucous membrane, and has inflected the primitive dental follicle.

C, represents a more advanced stage in which the dental sac has begun to be formed.

c, the superficial thick epithelium of the gum only sketched in outline; c', the deep layer of cylindrical cells; f, the primitive tooth-follicle; f', its cellular or granular contents and cavity; p, the dental papilla, and afterwards tooth-pulp; e, the inner inflected layer of the wall of the primitive follicle forming the inner part of the enamel organ; e', the outer wall of the same with the epithelial sprouts shooting into the tissue above; s, the commencement of the dental sac; fp, the follicle of the corresponding permanent tooth.

Fig. 553.

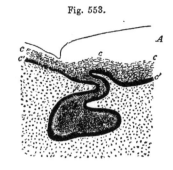

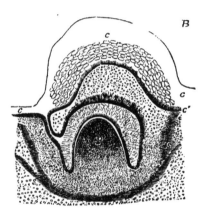

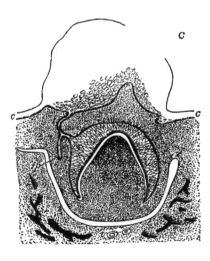

Waldeyer has shown by more recent observations, that in the human embryo the teeth arise in a manner essentially the same as that described by Kölliker in the ruminant. (Waldeyer, üb. die Entwick. der Zähne, Zeitsch. fur. ration. Medicin, 1865, and Henle's Bericht, &c. for 1864, p. 81.)

The *dental sacs*, after the closure of the follicles, continue to enlarge, as do also their contained papillæ. The walls of the sacs, which soon begin to thicken, consist of an outer fibroareolar membrane, and an internal highly vascular layer, lined by epithelium; their bloodvessels are derived partly from the dental arteries which course along the base of the sacs, and partly from those of the gums.

The papillæ, now the dental pulps, acquire a perfect resemblance to the crowns of the future teeth, and then the formation of the hard substance commences in them. This process begins very early, and by the end of the fourth month of fœtal life thin shells or caps of dentine are found on all the pulps of the milk-teeth, and a little later on that of the first permanent molar, while at the same time the coating of enamel begins to be deposited on each. The cap of dentine increases in extent by a growth around its edges, and in thickness by additions in its interior, at the expense of the substance of the pulp itself, which decreases in proportion. This growth of the tooth continues until the crown is completed of its proper width, and then the pulp undergoes a constriction at its base to form the cervix of the tooth, and afterwards elongates and becomes narrower, so as to serve as the basis of the fang. Sooner or later, after the completion of the crown, this part of the tooth appears through the gum, whilst the growth of dentine to complete the fang is continued at the surface of the elongating pulp, which gradually becomes encroached upon by successive formations of hard substance, until only a small cavity is left in the centre of the tooth, containing nothing but the reduced pulp, supplied by slender threads of vessels and nerves, which enter by a small aperture left at the point of the fang after the dentine is completed. In the case of teeth having complex crowns and more than a single fang, the process is somewhat modified. On the surface of the dental pulp of such a tooth, as many separate caps or shells of dental substance are formed as there are eminences or points ; these soon coalesce, and the formation of the tooth proceeds as before as far as the cervix. The pulp then becomes divided into two or more portions, corresponding with the future fangs, and the ossification advances in each as it does in a single fang. A horizontal projection or bridge of dentine shoots across the base of the pulp, between the commencing fangs, so that if the tooth be removed at this stage and examined on its under surface, its shell presents as many apertures as there are separate fangs. In all teeth, the pulp originally adheres by its entire base to the bottom of the sac ; but, when more than one fang is to be developed, the pulp is, as it were, separated from the sac in certain parts, so that it comes to adhere at two or three insulated points only, whilst the dentine continues to be formed along the intermediate and surrounding free surface of the pulp.

Formation of the hard tissues of the teeth.—Previously to the commencement of ossification, the primitive pulp is found to consist of microscopic nucleated cells (pulp-granules, Purkinje), more or less rounded in form, and imbedded in a clear

Fig. 554.

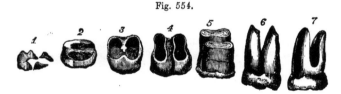

Fig. 554.—Different Stages in the formation of a Molar Tooth with Two Fangs (from Blake).

1, the distinct caps of dentine for five crowns in the earliest stage of formation ; in 2, and the remaining figures, the crown is downwards; in 2 and 3, the formation of the crown having proceeded as far as the neck, a bridge of dentine stretches across the base of the tooth-pulp ; and in 4, the division of the fangs is thus completed ; in 5, 6, and 7, the extension takes place in the fangs.

matrix containing a few very fine molecules, thinly disseminated in it. At the exterior of the pulp, the cells become elongated, and arranged perpendicularly to the

Fig. 555.—VERTICAL TRANSVERSE SEC-
TION OF THE DENTAL SAC, PULP, &C.,
OF A KITTEN (from Kölliker after a
preparation by Tiersch). ¼

Fig. 555.

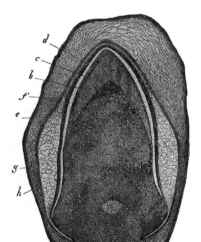

a, dental papilla or pulp, the outer darker part consisting of the dentine cells ; *b*, the cap of dentine formed upon the summit ; *c*, its covering of enamel ; *d*, inner layer of epithelium of the enamel organ ; *e*, gelatinous tissue ; *f*, outer epithelial layer of the enamel organ ; *g*, inner layer, and *h*, outer layer of the dental sac.

surface, so as to form a tolerably regular layer, resembling a columnar epithelium. The pulp contains white areolar fibres, without any elastic or yellow tissue, and it is highly vascular. The capillary vessels are most abundant at the points where ossification is to commence ; they form a series of loops between rows of cells arranged in a radiate manner, but they do not reach the surface. Besides this, the entire pulp is covered by a fine pellucid homogeneous membrane, named the *preformative membrane* (Purkinje, Raschkow), or *basement membrane*. The space between the pulp and the wall of the sac is occupied by a delicate substance accurately applied to its surface. This is the *outer pulp* of Hunter, termed also the *enamel-organ* (*organon adamantinæ*, Purkinje), being generally considered to be connected with the formation of the enamel. It presents three layers ; viz., externally, an epithelial layer with prominences which fit in between vascular processes of the surrounding mucous membrane ; internally, a layer of cylindrical nucleated cells, named the *enamel-membrane*, resting on the preformative membrane ; and between these, a bulky substance, consisting of small stellate cells anastomosing by long processes, and having the large meshes between them filled with clear fluid. This structure was formerly supposed to be similar to the primitive pulp ; but, as was first stated by Huxley and since confirmed by Kölliker, the whole enamel-organ is epithelial in nature, being derived by invagination from the cuticle.

The *dentine* is formed at the surface of the pulp, beneath the preformative membrane, but the precise manner in which it is derived from the soft tissues is still a matter for investigation. According to Purkinje, Retzius, and Raschkow, the preformative membrane is the part which first undergoes calcification, and afterwards the tissue of the pulp immediately beneath it. On gently separating the newly formed cap of dentine from the formative pulp, in the growing teeth of the human subject or of animals, and examining it under the microscope, the elongated cells of the pulp are found adhering in numbers to the inner surface of the newly-formed dentine. Owen states that the nuclei of the elongated cells, having themselves become lengthened, divide both longitudinally and transversely to develop secondary cells which continue included within the primary cells. The secondary cells then elongate, and together with their nuclei join end to end. Calcification proceeds in all parts, except in the nuclei of the secondary cells which remain as the cavities or lumina of the tubes ; the walls of the secondary cells are supposed to form the parietes of the tubes, and the material between the secondary cells together with the walls of the primary cells to be converted into the intertubular substance. The bifurcation of the tubuli is said to result from the junction of two secondary cells with a single one in a deeper layer of the pulp; and the constricted or moniliform appearance of the tubuli already mentioned as having been seen by some observers in growing or even in mature teeth, is thought to depend on an imperfect

Fig. 556.

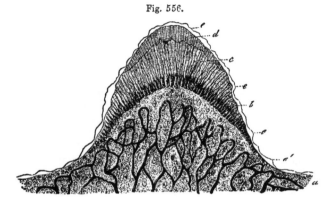

Fig. 556.—Vertical Section through the Point of a Human Fœtal Milk Tooth, in which the formation of the Dentine and Enamel has recently commenced (from Kölliker after Lent). $\frac{140}{1}$

a, dental pulp with blood-vessels; *b*, the dentine-cells upon its surface; *c*, the cap of dentine which has been formed on the summit, the tubuli being shown as prolongations from the tapering extremities of the dentine-cells; *d*, the enamel begun to be deposited; *e*, membranous layer, membrana præformativa of Huxley.

coalescence of the nuclei. In the teeth of young animals, Tomes has noticed the division of the cells and their subsequent coalescence to form the tubes, but he has failed to recognise the existence of primary cells including secondary ones. Lent finds that the superficial elongated cells of the dentinal pulp send off from their free ends long slender processes which form the tubes of the dentine, and which divide into branches, and anastomose together in the same manner as the tubes. Kölliker, who confirms Lent's observations, thinks it probable that a single cell may generate a tube in its whole length; at the same time a cell is sometimes constricted or incompletely divided into two, the more superficial of which becomes narrowed and lengthened into the dentinal tube.

With respect to the actual formation of the hard substance of the tooth, two views have been entertained; Kölliker conceiving it to proceed from the calcification of a soft matrix excreted from the dentinal cells and their thin prolongations already referred to; whereas Waldeyer, who denies the existence of a preformative membrane, maintains that the formation of the dentine consists in the conversion of a part of the protoplasm of the dentinal cells into a collagenous substance, which is subsequently calcified, while the remaining part of the cell-protoplasm continues in the form of soft fibres to occupy the interior of the tube surrounded by the calcified substance. (Op. cit. p. 189.) When the cap of dentine is examined in the newly formed state, besides the ordinary dentine, globules are commonly observed; but, if diluted hydrochloric acid be added, the globules disappear. Hence Czermak concludes that earthy impregnation proceeds for a time in a globular form, and that the after-presence of globular dentine is the result of arrested development; perfect development leading to the filling up of the spaces between the globules, and to the production of an uniformly compact tissue.

The *enamel* appears in the form of prismatic fibres which, until the point was contested by Huxley, have been generally supposed to be produced by calcification of the cells of the enamel-membrane, with which they correspond in figure. An enamel fibre may be formed by a single cell growing in length, while its previously formed portion becomes calcified, or by the union of a series of successively formed cells arranged vertically to the surface. During its formation the enamel is soft and chalky, and can easily be separated into its component prisms. Afterwards the membranous portion of it is nearly all obliterated, and the nuclei entirely disappear, or, according to Tomes, elongate

Fig. 557.

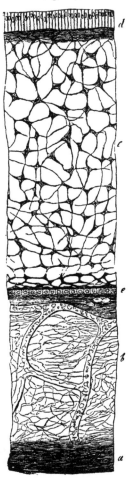

Fig. 557.—A Section through the Enamel Organ and Dental Sac from the Tooth of a Child at Birth (from Kölliker). $\frac{250}{1}$

a, outer dense layer of the dental sac; *b,* inner looser texture of the same with capillary blood-vessels and a somewhat denser layer towards the enamel organ; *c,* spongy substance; *d,* inner cells; and *e,* outer cellular layer of the enamel-organ.
B, four cells of the enamel-membrane. $\frac{350}{1}$

into a very fine central canal in each fibre. It is observed by Huxley that, if the pulp be treated with acetic acid, a voluminous, transparent membrane is raised from the whole surface in large folds, and that the ends of the enamel fibres are to be seen beneath it. The membrane is from $\frac{1}{7500}$th to $\frac{1}{100}$th of an inch in thickness; is clear, transparent, and exhibits little ridges bounding oval or quadrangular spaces; and is, according to him, continuous with the membrana præformativa. Huxley, therefore, considers that the enamel appears between the dentine and the preformative membrane, and that the enamel-organ takes no part in its formation. Tomes confirms the observation of Huxley with regard to the separability of this apparent membrane by acetic acid; but, upon closer examination, finds that it may be split into columns, which are, in conformity with his view of the structure of enamel, sheaths containing nuclei. Tomes, further, believes that these sheaths may be seen to pass through the membrane, which Huxley describes as limiting them superficially; and that, consequently, it is not, as Huxley imagines, the membrana præformativa. Waldeyer holds that the membrane described by Huxley between the enamel and the enamel-organ is only a layer of the most recently formed enamel, as he finds it possible always to detect enamel-cells with the ends partially calcified. He returns, therefore, to Schwann's original view, that the formation of the enamel-columns is due to the direct calcification of the enamel-cells. (Henle's Bericht, &c. for 1864, p. 81, and op. cit.)

The *Cement* appears to be formed simultaneously with the dentine of the fang by the periodontal membrane.

Eruption of the temporary teeth.—At the time of birth the crowns of the anterior milk-teeth, still enclosed in their sacs, are completed within the jaw, and their fangs begin to be formed. Their appearance through the gums follows a regular order, but the period at which each pair of teeth is cut varies within certain

B

limits. The eruption commences at the age of seven months, and is completed about the end of the second year. It begins with the central incisors

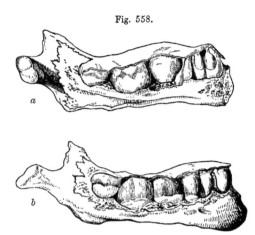

Fig. 558.

Fig. 558. — THE DENTAL SACS EXPOSED IN THE JAW OF A CHILD AT BIRTH.

a, the left half seen from the inner side; *b*, the right half seen from the outer side; part of the bone has been removed so as to expose the dental sacs as they lie below the gum; the lower figure shows the sacs of the milk-teeth and the first permanent molar, exposed by removing the bone from the outside; the upper figure shows the same from the inside, together with the pediculated sacs of the permanent incisor and canine teeth adhering to the gum.

of the lower jaw, which are immediately followed by those of the upper jaw; and, as a general rule, each of the lower range of teeth rises through the gum before the corresponding tooth of the upper set. The following scheme indicates, in months, the order and time of eruption of the milk-teeth.

MOLARS.	CANINES.	INCISORS.	CANINES.	MOLARS.
24 12	18	9 7 7 9	18	12 24

Before the teeth protrude through the gum, this undergoes some peculiar changes: its edge at first becomes dense and sharp, but, as the tooth approaches it, the sharp edge disappears, the gum becomes rounded or tumid, and is of a purplish hue; the summit of the tooth is seen like a white spot or line through the vascular gum, and soon afterwards rises through it. As the crown of the tooth advances to its ultimate position, the elongated fang becomes surrounded by a bony socket or alveolus. Before the eruption, the mucous membrane is studded with a number of small white bodies, which were described by Serres as glands (*dental glands*), and were supposed by him to secrete the tartar of the teeth. Meckel thought they were small abscesses, because no aperture could be detected in them. In a fœtus of six months, they were found by Sharpey to be small round pearl-like bodies situated in the corium of the mucous membrane, and having no aperture: they consist of small spherical capsules of various sizes, lined with a thick stratum of epithelium, the inner cells of which are flattened or scaly, like those lining the cheek, and are so numerous as almost to fill up the cavity. They are the prominences of the outer epithelial layer of the enamel organ, already referred to.

Development of the permanent teeth.—The preceding description of the structure of the dental sacs and pulps and of the mode of formation of the several parts of a tooth, applies to the permanent as well as to the milk-teeth.

The origin and progressive development of the sacs of the permanent teeth have still to be considered. There are six more permanent teeth in each jaw than there are milk-teeth, and it is found that the sacs of the ten anterior permanent teeth, which succeed the ten milk-teeth, have a different mode of origin from the six additional or superadded teeth, which are formed further back in the jaw.

Fig. 559.—Enlarged Diagram of the Dental Arch on the Left Side of the Lower Jaw of a Fœtus of about Fourteen Weeks (slightly altered from Goodsir).

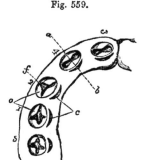

Fig. 559.

f, the follicles of the five milk-teeth, supposed to be open, showing the dental papillæ within them, and *o*, the opercula on their borders; they are numbered from 1 to 5 in the order of their first appearance; *c*, to the inside of each is the lunated depression forming the commencement of the germ of the corresponding permanent tooth; *a b*, line of the section shown in fig. 551, 5.

The sacs and pulps of the ten *anterior permanent* teeth have their foundations laid before birth, behind those of the milk set. Recurring to the follicular stage of the temporary teeth, which is completed about the fourteenth week, it will be remembered that behind each milk-follicle there is formed a small lunated recess, similar in form to an impression made by the nail. As already stated, the mucous membrane lining these recesses escapes the general adhesion of the lips and sides of the dental groove, so that when the latter closes they are converted into so many cavities, which are called by Goodsir, "*cavities of reserve*." They are ten in number in each jaw, and are formed successively from before backwards. They ultimately form the sacs for the permanent incisor, canine, and bicuspid teeth. These cavities soon elongate and recede into the substance of the gum behind the milk follicles, above and behind in the upper jaw, below and behind in the lower. In the meantime, a papilla appears in the bottom of each, (that for the central incisor appearing first, at about the sixth month,) and one or more folds or opercula, as in the case of the temporary teeth, are developed from the sides of the cavity, and, by their subsequent union, divide it into two portions, the lower portion containing the papilla, and now forming the dental sac and pulp of the permanent tooth, and the upper and narrower portion being gradually obliterated in the same manner as the primitive groove was closed over the milk-sacs. When these changes have taken place, the permanent sac adheres to the back of that for the temporary tooth. Both of them continue then to grow rapidly, and after a time it is found that the bony socket not only forms a cell for the reception of the milk-sac, but also a small posterior recess or niche for the permanent sac, with which the recess keeps pace in its growth. Confining our description now, for convenience, to the lower jaw only, it is found that at length the permanent sac so far recedes in the bone as to be lodged in a special osseous cavity at some distance below and behind the milk-tooth, the two being completely separated from each other by a bony partition. In descending into

3 G

the jaw, the permanent sac acquires at first a pear-shape, and is then connected with the gum by a solid membranous pedicle. The recess in the jaw
has a similar form, drawn out into a long canal for the pedicle, which opens

Fig. 560.

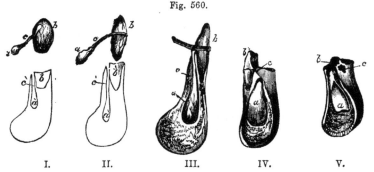

I.   II.   III.   IV.   V.

Fig. 560.—Sketches showing the Relations of the Temporary and Permanent
Dental Sacs and Teeth (after Blake, with some additions).

The lower parts of the three first figures, which are somewhat enlarged, represent
sections of the lower jaw through the alveolus of a temporary incisor tooth : *a*, indicates
the sac of the permanent tooth; *c*, its pedicle; *b*, the sac of the milk tooth or the milk
tooth itself; *a'*, *b'*, indicate the alveolar recesses in which the permanent and temporary
teeth are lodged, and *c*, the canal by which that of the former leads to the surface of
the bone behind the alveolus of the temporary tooth. The fourth and fifth figures, which
are nearly of the natural size, show the same relations in a more advanced stage, in IV,
previous to the change of teeth, in V, when the milk-tooth has fallen out and the permanent tooth begins to rise in the jaw ; *c*, the orifice of the bony canal leading to the
place of the permanent tooth.

on the edge of the jaw, by an aperture behind the corresponding milk-tooth.
The permanent tooth is thus separated from the socket of the milk-tooth by
a bony partition, against which, as well as against the root of the milk-tooth
just above it, it presses in its rise through the gum, so that these parts are in
a greater or less extent absorbed. When this has proceeded far enough, the
milk-tooth becomes loosened, falls out or is removed, and the permanent
tooth takes its place. The absorption of the dental substance commences
at or near the ends of the fangs, and proceeds upwards until nothing but
the crown remains. The cement is first attacked, and then the dentine :
but the process is similar in the two tissues. The change is not produced
merely by pressure, but through the agency of a special cellular structure
developed at the time, and applied to the surface of the tooth. Hollows or
indentations occur upon the latter, giving it a festooned appearance : and it
frequently happens that the dental tissues are deposited, absorbed, and
redeposited alternately in the same tooth (Tomes). The milk-teeth and
the permanent teeth are said by Serres to be supplied by two different
arteries, the obliteration of the one belonging to the temporary teeth being
regarded by him as the cause of their destruction ; but of this there is no
sufficient proof.

The six *posterior* (or "*superadded*") *permanent* teeth, that is, the three
permanent molars on each side, do not come in the place of other teeth.
They arise from successive extensions of the dental groove carried backwards
in the jaw, posterior to the milk-teeth, and named by Goodsir "*posterior
cavities of reserve.*"

During the general adhesion of the dental groove occurring at the fifteenth week, the part posterior to the last temporary molar follicle continues unobliterated, and thus forms a cavity of reserve, in the fundus of which a papilla ultimately appears, and forms the rudiment of the first permanent molar tooth : this takes place very early, viz., at the sixteenth week. The deepest part of this cavity is next converted by adhesion into a sac, which encloses the papilla, whilst its upper portion elongates backwards so as to form another cavity of reserve, in which, at the seventh month after birth, the papilla for the second molar tooth appears. After a long interval, during which the sac of the first permanent molar and its contained tooth have acquired great size, and that of the second molar has also advanced considerably in development, the same changes once more occur, and give rise to the sac and papilla of the wisdom tooth, the rudiments of which are visible at the sixth year. The subsequent development of the permanent molar teeth takes place from these sacs just like that of the other teeth.

Calcification begins first in the anterior permanent molar teeth. Its order and periods may be thus stated for the upper jaw, the lower being a little earlier : First molar, five or six months after birth ; central incisor, a little later ; lateral incisor and canine, eight or nine months ; two bicuspids, two years or more ; second molar, five or six years ; third molar, or wisdom tooth, about twelve years.

Fig. 561.

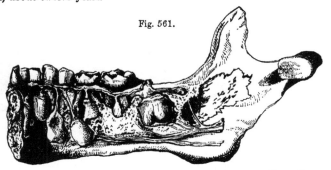

Fig. 561.—PART OF THE LOWER JAW OF A CHILD OF THREE OR FOUR YEARS OLD, SHOWING THE RELATIONS OF THE TEMPORARY AND PERMANENT TEETH.

The specimen contains all the milk-teeth of the right side, together with the incisors of the left; the inner plate of the jaw has been removed, so as to expose the sacs of all the permanent teeth of the right side, except the eighth or wisdom tooth, which is not yet formed. The large sac near the ramus of the jaw is that of the first permanent molar, and above and behind it is the commencing rudiment of the second molar.

Eruption of the permanent teeth.—The time at which this occurs in regard to each pair of teeth in the lower jaw is exhibited in the subjoined table. The corresponding teeth of the upper jaw appear somewhat later.

Molar, first .	6 years.
Incisors, central	7 „
„ lateral	8 „
Bicuspids, anterior	9 „
„ posterior	10 „
Canines .	11 to 12 „
Molars, second	12 to 13 „
„ third (or wisdom) .	17 to 25 „

3 G 2

It is just before the shedding of the temporary incisors, *i. e.*, about the sixth year, that there is the greatest number of teeth in the jaws. At that period there are all the milk-teeth, and all the permanent set except the wisdom teeth, making forty-eight.

Fig. 562.

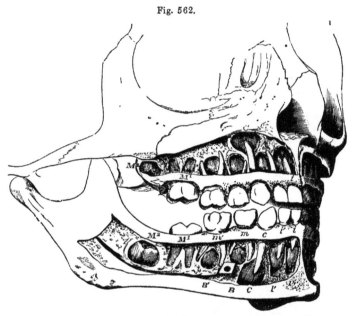

Fig. 562.—The Teeth of a Child of Six Years, with the Calcified Parts of the Permanent Teeth exposed (after Henle and from nature).

The whole of the teeth of the right side are shown, together with the three front teeth of the left side : in the upper and lower jaws the teeth are indicated as follows, viz. :— 1, *milk-teeth*—*i*, inner or first incisor ; *i'*, outer or second incisor ; *c*, canine ; *m*, first molar ; *m'* second molar. 2, *permanent teeth*—I, inner or first incisor ; I', outer or second incisor ; C, canine ; B, first bicuspid ; B' second bicuspid ; M¹, the first molar, which has passed through the gums ; M², the second molar, which has not yet risen above the gums : the third molar is not yet formed.

During the growth of the teeth the jaw increases in depth and length, and undergoes changes in form. In the child it is shallow, but it becomes much deeper in the adult. In the young subject the alveolar arch describes almost the segment of a circle ; but in the adult the curve is semi-elliptical. The increase which takes place in the length of the jaw arises from a growth behind the position of the milk-teeth, so as to provide room for the three additional teeth on each side belonging to the permanent set. At certain periods in the growth of the jaws there is not sufficient room in the alveolar arch for the growing sacs of the permanent molars ; and hence those parts are found at certain stages of their development to be enclosed in the base of the coronoid process of the lower jaw, and in the maxillary tuberosity in the upper jaw, but they afterwards successively assume their ultimate position as the bone increases in length. The space taken up by the ten anterior permanent teeth very nearly corresponds with that which had been occupied by the ten milk-teeth; the difference in width between the incisors of the two sets being compensated for by the smallness of the bicuspids in comparison with the milk-molars to which they succeed. Lastly, the angle formed by the ramus and body of the lower jaw differs at different ages ; thus it is obtuse in the infant, approaches nearer to a right angle in the adult, and again becomes somewhat obtuse in old age. (See p. 52.)

Relation of the blood-vessels and nerves to the tooth.—There is no evidence that the blood-vessels send branches into the hard substance. The red stain sometimes observed in the teeth, after death by asphyxia, and the red spots occasionally found in the dentine, are due to the imbibition of blood effused on the surface of the pulp. The dentine formed in young animals fed upon madder is tinged with that colouring matter, but this does not appear to take place when the growth of the tooth is completed. Nevertheless the tubules of the dentine may serve to convey through its substance nutrient fluid poured out by the blood-vessels of the pulp. The teeth are sometimes stained yellow in jaundice.

According to Czermak the primitive nerve-tubules run into the tooth-pulp in bundles, which are large towards the centre, and small at the periphery. They lose themselves in a plexus at the surface of the pulp. Czermak states that the fibres often divide, but that he has not seen loops frequently, and he is doubtful as to the precise mode of their termination.

THE TONGUE.

The *tongue* is a muscular organ covered with mucous membrane. By its muscular structure it takes part in the processes of mastication and deglutition, and in the articulation of speech, while its mucous membrane is endowed with common sensibility and is the seat of the sense of taste. The tongue occupies the concavity of the arch of the lower jaw : posteriorly it is connected with the hyoid bone, and the back part of its dorsum forms the floor of the arch of the fauces ; inferiorly it receives from base to apex the fibres of the genio-glossus muscle, and through the medium of that muscle is attached to the lower jaw.

A.—MUCOUS MEMBRANE.—On the *under surface* of the tongue the mucous membrane is smooth and thin. It forms a fold in the middle line, called the *frænum linguæ*, placed in front of the anterior border of the genio-glossi muscles. On each side below, as the mucous membrane passes from the tongue to the inner surface of the gums, it is reflected over the sublingual gland. Not far from the line continued forwards from the frænum, the ranine vein may be distinctly seen through the mucous membrane, and close to it lies the ranine artery. Further outwards is an elevated line with a fimbriated margin directed outwards, which extends to the tip. The ducts of the right and left submaxillary glands end by papillary orifices placed close together, one on each side of the frænum ; and further back, in the groove between the sides of the tongue and the lower jaw, are found the orifices of the several ducts belonging to the sublingual glands.

The *upper surface* or *dorsum* of the tongue is convex in its general outline, and is marked along the middle in its whole length by a slight furrow called the *raphe*, which indicates its bilateral symmetry. About half an inch from the base of the tongue, the raphe often terminates in a depression, closed at the bottom, which is called the *foramen cæcum* (Morgagui), and in which several mucous *glands* and follicles open. Three folds, named the glosso-epiglottic folds or frænula, of which the middle one is the largest (frænum epiglottidis), pass backwards from the base of the tongue to the epiglottis. The upper surface of the tongue is completely covered with numerous projections or eminences named *papillæ*. They are found also upon the tip and free borders, where, however, they gradually become smaller, and disappear towards its under surface. These papillæ are distinguished into three orders, varying both in size and form.

Fig. 563.

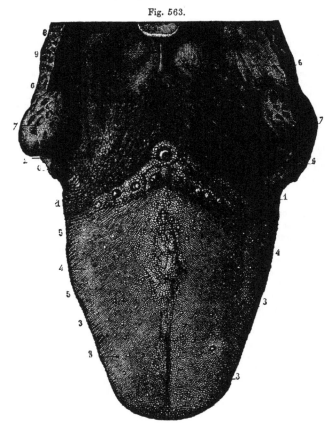

Fig. 563. —Papillar Surface of the Tongue, with the Fauces and Tonsils (from Sappey).

1, 2, circumvallate papillæ ; in front of 2, the foramen cæcum ; 3, fungiform papillæ ; 4, filiform and conical papillæ ; 5, transverse and oblique rugae ; 6, mucous glands at the base of the tongue and in the fauces ; 7, tonsils ; 8, part of the epiglottis ; 9, median glosso-epiglottidean fold or frænum epiglottidis.

The *large* or *circumvallate* papillæ, from seven to twelve in number, are found on the back part of the tongue, arranged in two rows, which run obliquely backwards and inwards, and meet towards the foramen cæcum, like the arms of the letter V. They are situated in cup-like cavities or depressions of the mucous membrane, and have the shape of an inverted cone, of which the apex is attached to the bottom of the cavity, and the broad flattened base appears on the surface. They are therefore surrounded by a circular furrow or trench, around which again is an annular elevation of the mucous membrane, covered with the smaller papillæ. The exposed surface of the papillæ vallatæ is beset with numerous smaller papillæ or filaments;

Fig. 564.—Vertical Section of the Circumvallate Papillæ (from Kölliker). $\frac{10}{1}$

A, the papilla; B, the surrounding wall; a, the epithelial covering; b, the nerves of the papilla and wall spreading towards the surface; c, the secondary papillæ.

Fig. 564.

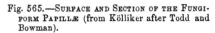

and in some of them there is found a central depression, into which mucous follicles open.

The *middle-sized* or *fungiform* papillæ, more numerous than the last, are small rounded eminences scattered over the middle and fore part of the dorsum of the tongue; but they are found in great numbers and closer

Fig. 565.—Surface and Section of the Fungiform Papillæ (from Kölliker after Todd and Bowman).

A, the surface of a fungiform papilla partially denuded of its epithelium, $\frac{25}{1}$; p, secondary papillæ; e, epithelium.
B, section of a fungiform papilla with the blood-vessels injected. a, artery; v, vein; c, capillary loops of simple papillæ in the neighbouring structure of the tongue; d, capillary loops of the secondary papillæ; e, epithelium.

Fig. 565.

together at the apex and upon the borders. They are easily distinguished in the living tongue by their deeper red colour. They are narrow at their point of attachment, but are gradually enlarged towards their free extremities, which are blunt and rounded, and are covered with smaller filamentous appendages or papillæ.

The *smallest papillæ, conical* and *filiform,* are the most numerous of all. They are minute, conical, tapering, or cylindrical processes, which are densely packed over the greater part of the dorsum of the tongue, but towards the base gradually disappear. They are arranged in lines, which correspond at first with the oblique direction of the two ridges of the papillæ vallatæ, but gradually become transverse towards the tip of the tongue. At the sides they are longer and more filiform, and arranged in parallel rows, perpendicular to the border of the tongue.

Considerable variety occurs in the appearance of the papillæ on the tongues of different persons. Thus occasionally instances occur in which the tongue has a quite smooth appearance, and others are seen in which numbers of the filiform papillæ are elongated into the appearance of short brown hairs, as shown in Fig. 566.

When examined microscopically in sections, all the kinds of papillæ now described are observed to be bearers of closely set secondary papillæ. The secondary papillæ are the structures which correspond with the papillæ of the general integument, and are occupied each by a long loop of capillary blood-vessel. Simple papillæ of the same description are likewise interspersed between the three large kinds, and are found on the back part of

Fig. 566.

Fig. 566.—Two Filiform Papillæ, one with Epithelium, the other without (from Kölliker, after Todd and Bowman). $\frac{35}{1}$

p, the substance of the papillæ dividing at their upper extremities into secondary papillæ ; *a*, artery, and *v*, vein, dividing into capillary loops ; *e*, epithelial covering, laminated between the papillæ, but extended into hair-like processes *f*, from the extremities of the secondary papillæ.

the tongue, behind the circumvallate range, as well as covering the under surface of the tongue and the rest of the mucous membrane of the mouth. The epithelium covering the tongue, like that of the mouth generally, is of the squamous kind. It is of considerable thickness, and the simple papillæ, together with the secondary papillæ surmounting those of the circumvallate and fungiform kinds, are concealed beneath it, or nearly so. But the secondary papillæ, borne by those of the filiform kind, are peculiar both in containing a number of elastic fibres, which give them greater firmness, and in the character of their epithelial covering, which is dense and imbricated, and which forms a separate process over each, greater in length than the papilla which it covers. Over some of the filiform papillæ these processes form a pencil of fine fibres ; and on others they approach closely in character and structure to hairs. The papillæ are undoubtedly the parts chiefly concerned in the special sense of taste ; but they also possess, in a very acute degree, common tactile sensibility ; and the filiform papillæ, armed with their denser epithelial covering, serve a mechanical use, in the action of the tongue upon the food, as is well illustrated by the more developed form which these papillæ attain in many carnivorous animals. The papillary surface of the tongue is supplied abundantly with nerves. It is difficult to trace the nerve-fibres in the papillæ filiformes, owing to the presence of elastic filaments. In the papillæ fungiformes the nerves are larger and more numerous, and form a plexus with brush-like branches : but they are still more abundant, and of greater size in the papillæ circumvallatæ.

Little that is satisfactory is known of the mode of termination of the nerve filaments in the human tongue. It is still a matter of doubt whether they enter the secondary papillæ surmounting the filiform set, the density of the tissue rendering the investigation peculiarly difficult in these. In the frog's tongue, Billroth and Axel Key believe that they have traced continuity of nerve-filaments with structures in the epithelium ; and, according to Axel Key, the arrangement is very similar to that of the olfactory cells —viz., rodlike bodies placed between the epithelial cells and continuous by their deep extremities with varicose fibres. (Billroth in Müller's Archiv, 1858, p. 159 ; Axel Key in Reichert's Archiv, 1861, p. 329).

Glands.—The mucous membrane of the tongue is provided with numerous follicles and glands. The follicles, simple and compound, are scattered over the surface ; but the rounded conglomerate glands, called *lingual glands,* are collected about the posterior part of the dorsum of

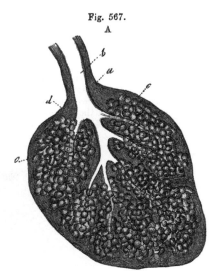

Fig. 567.

A

Fig. 567.—RACEMOSE MUCOUS GLAND, FROM THE FLOOR OF THE MOUTH (from Kölliker).

A, the entire gland as seen in section ; $\frac{50}{1}$ *a*, covering of connective tissue ; *b*, excretory duct; *c*, glandular vesicles ; *d*, duct of one of the lobules.

B, diagram of one of the lobules, more highly magnified ; *a*, excretory duct of the lobule ; *b*, secondary branch ; *c*, the glandular vesicles as they lie together in the gland ; *d*, the same separated, showing their connection as a glandular tube.

the tongue, near the papillæ vallatæ and foramen cæcum, into which last the ducts of several of these glands open. Other small glands are found also beneath the mucous membrane of the borders of the tongue.

B

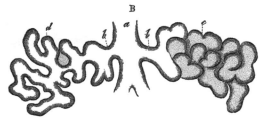

There is, in particular, a small group of these glands on the under surface of the tongue near the apex. They are there aggregated into a small oblong mass, out of which several ducts proceed and open separately on the mucous membrane. (Blandin, in Archives gén. de Médecine, 1823 ; Nuhn, Ueber eine noch nicht näher beschriebene Zungendrüse, Mannheim, 1845.)

B.—MUSCULAR SUBSTANCE.—The substance of the tongue is chiefly composed of muscular fibres, running in different but determinate directions ;— hence the variety and regularity of its movements, and its numerous changes of form. Many of the contractile fibres of the tongue belong to muscles which enter at its base and under surface, and attach it to other parts : these are called the *extrinsic muscles* of the tongue, and have been elsewhere described (pp. 185—186). Other bands of fibres which constitute the *intrinsic* or proper muscles, and are placed entirely within the substance of the organ, will be here more particularly noticed. They are as follow.

The *lingualis superficialis* (noto-glossus, Zaglas), consisting mainly of longitudinal fibres, is placed on the upper surface of the tongue, immediately beneath the mucous membrane, and is traceable from the apex of the organ backwards to the hyoid bone. The individual fibres do not run the whole of this distance, but are attached at intervals to the submucous and glandular tissues. The entire layer becomes thinner towards the base of the tongue, near which it is overlapped at the sides by a thin plane of oblique or nearly transverse fibres derived from the palato-glossus and hyo-glossus muscles. According to Zaglas, the fibres of this muscle are directed forwards and outwards.

The *lingualis inferior* (lingualis muscle of Douglas, Albinus, &c.) consists of a rounded muscular band, extending along the under surface of the tongue from base to apex, and lying outside the genio-hyo-glossus between that muscle and the hyo-glossus. Posteriorly, some of its fibres are lost in the substance of the tongue, and others reach the hyoid bone. In front,

Fig. 568.

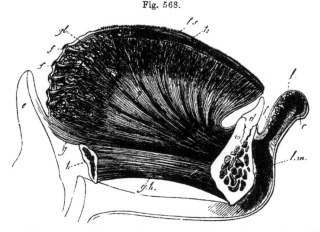

Fig. 568.—Longitudinal Vertical Section of the Tongue, Lip, &c. (from Kölliker and Arnold).

m, symphysis of the lower jaw; *d*, incisor tooth ; *h*, hyoid bone ; *g h*, genio-hyoid muscle; *g*, genio-hyo-glossus spreading into the whole extent of the tongue ; *t r*, transverse muscle ; *l s*, superior longitudinal muscle ; *g l*, lingual glands ; *f*, lingual follicles; *e*, epiglottis ; *l*, section of the lip and labial glands ; *o*, cut fibres of the orbicularis oris ; *l m*, levator menti.

having first been joined, at the anterior border of the hyo-glossus muscle, by fibres from the stylo-glossus, it is prolonged beneath the border of the tongue as far as its point.

The *transverse* muscular fibres of the tongue form together with the intermixed fat a considerable part of its substance. They are found in the interval between the upper and lower longitudinal muscles, and they are inter-woven extensively with the other muscular fibres. Passing outwards from the median plane, where they take origin from a fibrous septum, they reach the dorsum and borders of the tongue. In proceeding outwards, they separate, and the superior fibres incline upwards, forming a series of curves with the concavity turned upwards. The fibres of the palato-glossus muscle are found by Zaglas and Henle to be continuous with fibres of the transverse set.

Fig. 569. — TRANSVERSE VERTICAL SECTION OF THE TONGUE IN FRONT OF THE PAPILLÆ VALLATÆ, SEEN FROM BEFORE (from Kölliker).

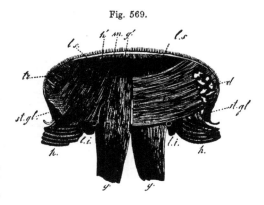

Fig. 569.

g, the genio-hyo-glossi muscles ; *g′,* the vertical fibres of the right side traced upwards to the surface ; *li,* inferior longitudinal muscle with the divided ranine artery ; *tr,* transverse muscle, entire on one side, but partially removed on the other, where the other muscles pass through it; *c,* septum linguæ ; *h,* hyo-glossus ; *h g l,* its fibres spreading upwards almost vertically outside the genio-hyo-glossus ; *h′,* vertical fibres reaching the surface ; *l s,* divided plates of the fibres of the superior longitudinal muscle between the vertical fibres ; *s t, g l,* stylo-glossus ; *d,* glands near the border of the tongue.

Vertical fibres (musculus perpendicularis externus of Zaglas), decussating with the transverse fibres and the insertions of the genio-glossus, form a set of curves in each half of the tongue with their concavity upwards, and ex-tending downwards and outwards from the dorsum to the under surface of the border, so that those which are outermost are shortest. (Zaglas, "On the Muscular Structure of the Tongue," in Goodsir's Annals, I. p. 1.)

Examined in transverse sections, the muscular fibres of the tongue are seen to be arranged so as to render the substance divisible into an outer part or *cortex* and a softer internal *medulla.* The fibres of the cortex are principally longitudinal, derived superiorly from the lingualis superior, further outwards from the hyo-glossus, on the side from the stylo-glossus, and beneath this from the lingualis inferior. They sheath the medullary part on all sides except inferiorly, where the genio-glossi muscles enter it between the infe-rior linguales. In the medullary part are found, imbedded in fat, the de-cussating fibres of the transverse muscle passing across, the genio-glossi radiating upwards and outwards, and the vertical muscles arching down-wards and outwards. In addition to the movements which may be given to the tongue by the extrinsic muscles, this organ is capable of being curved upwards, downwards, or laterally by its cortical fibres, it is flattened by the vertical fibres, and its margins are again drawn together by the transverse.

The septum of the tongue is a thin fibrous partition which extends forwards from the hyoid bone to the tip, and divides one half of the medullary part of the tongue from the other, but does not penetrate into the cortex. It cor-responds with the fusiform fibro-cartilage, found in the middle of the tongue of the dog, near its under surface.

The *arteries* of the tongue are derived from the lingualis, with some small branches from the facial and ascending pharyngeal. With these the veins for the most part correspond. (See pp. 348 and 456.)

Fig. 570.

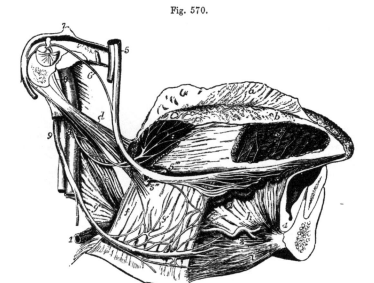

Fig. 570.—LATERAL VIEW OF THE NERVES AND BLOOD-VESSELS OF THE TONGUE (from Hirschfeld and Leveillé). ⅔

The lower jaw has been divided near the symphysis, and the right half removed ; the hyoid bone is entire, and the extrinsic muscles of the tongue are preserved on the right side. *a*, the epithelial covering of the tongue partially raised ; *b*, the papillar surface of the mucous membrane exposed ; *c*, the same near the papillæ vallatæ ; *d*, placed on the superior constrictor of the pharynx, points to the stylo-glossus muscle ; *e*, stylo-pharyngeus muscle, passing within the middle constrictor ; *f*, hyo-glossus ; *g*, middle constrictor of the pharynx ; *h*, genio-hyo-glossus ; *i*, genio-hyoideus ; 1, trunk of the lingual artery ; 2, ranine artery ; 3, sublingual branch ; 4, its terminal branches ; 5, trunk of the gustatory nerve ; 5′, distribution of its terminal twigs in the mucous membrane of the fore part of the tongue ; 5″, submaxillary ganglion ; 5‴, another small ganglion, connected with the gustatory nerve ; 6, chorda tympani nerve, passing from the facial nerve 7, to the trunk of the gustatory ; 8, trunk of the glosso-pharyngeal, receiving a twig of communication from the facial ; 8′, its distribution near the papillæ vallatæ ; 9, hypoglossal nerve ; 9′, its twigs to the hyo-glossus muscle and union with the gustatory ; further forward are seen its terminal branches to the muscular substance of the tongue.

The *nerves* of the tongue (exclusive of branches from the sympathetic nerves) are three : viz. the lingual or *gustatory* branch of the *fifth* pair, which supplies the papillæ and mucous membrane of the fore part and sides of the tongue to the extent of about two-thirds of its surface ; the lingual branch of the *glosso-pharyngeal*, which sends filaments to the mucous membrane at the base of the tongue, and especially to the papillæ vallatæ ; and, lastly, the *hypoglossal* nerve, which is distributed to the muscles. (See the description of these nerves.) Remak and Kölliker have discovered microscopic ganglia upon the expansion of the glosso-pharyngeal nerve, and in the sheep and calf upon the gustatory division of the fifth. Remak thought that they had some relation to the glands, but Kölliker finds them on branches not connected with those organs.

THE PALATE.

The roof of the mouth is formed by the palate, which consists of two portions ; the fore part being named the hard palate, and the back part, the soft palate.

The osseous framework of the *hard* palate, already described with the bones of the face, is covered by the periosteum, and by the lining membrane of the mouth, which adhere intimately together. The mucous membrane, which is continuous with that of the gums, is thick, dense, rather pale, and much corrugated, especially in front and at the sides ; but is smoother, thinner, and of a deeper colour behind. Along the middle line is a ridge or raphe, ending in front in a small eminence, which corresponds with the lower opening of the anterior palatine canal, and receives the terminal filaments of the naso-palatine and anterior palatine nerves. The membrane of the hard palate is provided with many muciparous glands, which form a continuous layer between the membrane and the bone, and it is covered with a squamous epithelium.

The *soft* palate (velum pendulum palati), is formed of a doubling of mucous membrane inclosing muscular fibres and numerous glands. It constitutes an incomplete and moveable partition between the mouth and the pharynx, continued from the posterior border of the hard palate, obliquely downwards and backwards. Its form and its inferior connections, bounding the isthmus of the fauces, have been already described, together with the muscles which enter into its composition, at p. 189.

The anterior or under surface of the velum, which is visible in the mouth, is concave. The mucous membrane, continuous with that of the hard palate, is thinner and darker than it, and is covered like it with scaly epithelium. The median ridge or raphe, which is continued backwards from the hard palate to the base of the uvula, indicates the original separation of the velum into two lateral halves.

The posterior surface of the soft palate, slightly convex or arched, is continuous above with the floor of the posterior nares. It is slightly elevated along the middle line, opposite to the uvula. The greater portion of its mucous membrane, as well as that of the free margin of the velum, is covered with a squamous epithelium ; but quite at its upper portion, near the orifice of the Eustachian tube, the epithelium is columnar and ciliated.

On both surfaces of the velum are found numerous small glands, called the *palatine* glands. They particularly abound on the upper surface, where they form quite a glandular layer ; they are also very abundant in the uvula.

THE TONSILS.

The *tonsils* (tonsillæ, amygdalæ) are two prominent bodies, which occupy the recesses formed, one on each side of the fauces, between the anterior and posterior palatine arches and the pillars of the fauces.

They are usually about six lines in length, and four in width and thickness ; but they vary much in size in different individuals.

The outer side of the tonsil is connected with the inner surface of the superior constrictor of the pharynx, and approaches very near to the internal carotid artery. Considered in relation to the surface of the neck, the tonsil corresponds to the angle of the lower jaw, where it may be felt beneath the skin when it is enlarged. Its inner surface, projecting into the fauces between the palatine arches, presents from twelve to fifteen orifices, which give it a perforated appearance. These orifices lead into recesses in the substance of the tonsil, from which other and smaller orifices conduct still

Fig. 571.

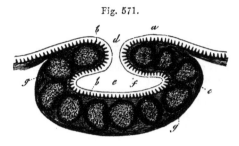

Fig. 571.—SECTION OF A FOL-
LICULAR GLAND FROM THE
ROOT OF THE TONGUE (from
Kölliker). $\frac{30}{1}$

a, epithelial lining; *b*, papillæ of the mucous membrane; *c*,
outer surface of the capsule,
formed of connective tissue: *d*,
outlet, and *e*, cavity of the capsule; *g*, follicles in the substance
of the capsular wall.

deeper into numerous fol-
licles. These follicles are
lined by the epithelium and
papillary mucous membrane of the throat, and have thick walls formed by a layer of closed capsules imbedded in the submucous tissue. The cap-sules, which may be compared to those of Peyer's glands of the intestine, besides having a mesh-work of capillary blood-vessels and delicate trabecular tissue within them, are filled with consistent greyish substance, containing cells and free nuclei, but without the characters of mucus. A substance having the same microscopic elements is found in the cavity of the follicle, but here it is liable to be mixed with mucus supplied by true mucous glands, the ducts of which pass into the follicle. The function of the tonsils is as little known as that of the other glands formed of closed capsules which are found in the mucous membrane of the alimentary canal.

Follicular recesses, surrounded by closed capsules, like the recesses and capsules of the tonsils, are also found at the root of the tongue, where they form a layer extending from the papillæ vallatæ to the epiglottis, and from one tonsil to the other, lying immediately beneath the mucous membrane and above the mucous glands, many of whose ducts they receive.

The tonsils receive a very large supply of blood from various sources, viz. from the tonsillar and palatine branches of the facial artery, and. from the descending palatine, the ascending pharyngeal and the dorsalis linguæ. From these arteries, fine branches and capillaries are distributed abundantly to the walls of the capsules and to the papillæ of the mucous membrane lining the follicles. The *veins* are numerous, and enter the tonsillar plexus on its outer side. The *nerves* come from the glosso-pharyngeal nerve, and from the fifth pair.

THE SALIVARY GLANDS.

The saliva, which is poured into the mouth, and there mixed with the food during mastication, is secreted by three pairs of glands, named from their respective situations, *parotid*, *submaxillary*, and *sublingual*. Agreeing in their general physical characters and minute structure, these glands differ in their size, form, and position.

The Parotid Gland.

The *parotid* is the largest of the three salivary glands. It lies on the side of the face, in front of the ear, and extends deeply into the space behind the ramus of the lower jaw. Its weight varies from five to eight drachms.

Its outer surface is convex and lobulated, and is covered by the skin and fascia, and partially by the platysma muscle. It is bounded above by the zygoma, below by a line drawn backwards from the lower border of the jaw to the sterno-mastoid muscle, and behind by the external meatus of the ear, the mastoid process, and sterno-mastoid muscle. Its anterior border, which lies over the ramus of the lower jaw, is less distinctly defined, and stretches forwards to a variable extent on the masseter muscle. It is from this anterior border of the gland that the excretory duct passes off; and there is

Fig. 572.

Fig. 572.—SKETCH OF A SUPERFICIAL DISSECTION OF THE FACE, SHOWING THE POSITION OF THE PAROTID AND SUBMAXILLARY GLANDS. ⅔

p, the larger part of the parotid gland; *p″*, the small part, which lies alongside the duct on the masseter muscle; *d*, the duct of Stenson before it perforates the buccinator muscle; *a*, transverse facial artery; *n, n*, branches of the facial nerve emerging from below the gland; *f*, the facial artery passing out of a groove in the submaxillary gland and ascending on the face; *s m*, superficial larger portion of the submaxillary gland lying over the posterior part of the mylo-hyoid muscle.

frequently found in connection with the duct, and lying upon the masseter muscle, a small process or a separated portion of the gland, which is called *glandula socia* (*socia parotidis*). On trying to raise the deeper part of the parotid gland from its position, it is found to extend far inwards, between the mastoid process and the ramus of the jaw, towards the base of the skull, and to be intimately connected with several deep-seated parts. Thus, above, it reaches into and occupies the posterior part of the glenoid cavity; behind and below, it touches the digastric muscle, and rests on the styloid process and styloid muscles ; and, in front, under cover of the ramus of the jaw, it advances a certain distance between the external and internal pterygoid muscles.

The internal carotid artery and internal jugular vein are close to the deep surface of the gland. The external carotid artery, accompanied by the temporal and internal maxillary veins, passes through the parotid gland ; and in that situation arise the temporal and internal maxillary arteries, as also the auricular and transverse facial branches of the temporal. The gland is also traversed by the facial nerve, which divides into branches within its substance, and it is pierced by branches of the great auricular nerve.

The *parotid duct*, named also *Stenson's duct* (d. Stenonianus), appears at the anterior border of the gland, about one finger's breadth below the zygoma, and runs forwards over the masseter muscle, accompanied by the socia parotidis, when that accessory portion of the gland exists, and receiving its ducts. At the anterior border of the masseter, the duct turns inwards through the fat of the cheek and pierces the buccinator muscle ; and then, after running for a short distance obliquely forwards beneath the mucous membrane, opens upon the inner surface of the cheek, by a small orifice opposite the crown of the second molar tooth of the upper jaw. Its direction across the face may be indicated by a line drawn from the lower margin of the concha of the ear to a point midway between the red margin of the lip and the ala of the nose. The length of the Stenonian duct is about two inches and a half, and its thickness about one line and a half. At the place where it perforates the buccinator, its canal is as large as a crow-quill, but at its orifice it is smaller than in any other part, and will only admit a very fine probe. The duct is surrounded by areolar tissue, and consists of an external, dense, and thick fibrous coat, in which contractile fibres are described, and of an internal mucous tunic, which is continuous with that of the mouth, but which is covered, from the orifice of the duct as far as to the smallest branches, with a columnar epithelium.

The parotid gland belongs to the class of compound racemose glands, and consists of numerous flattened lobes, held together by the ducts and vessels, and by a dense areolar web, which is continuous with the fascia upon its outer surface ; but the gland has no special or proper coat. The lobes are again divided into lobules, each of which consist of the branched terminations of the duct, and of vessels, nerves, and fine areolar tissue. The ducts terminate in closed vesicular extremities, about $\frac{1}{300}$th of an inch or more in diameter, which are lined with epithelium, and have capillary vessels ramifying upon them.

The vessels of the parotid gland enter and leave it at numerous points. The arteries are derived directly from the external carotid, and from those of its branches which pass through or near the gland. The veins correspond. The absorbents join the deep and superficial set in the neck ; and there are often one or more lymphatic glands imbedded in the substance of the parotid. The nerves come from the sympathetic (carotid plexus), and also, it is said, from the facial and the superficial temporal and great auricular nerves.

An instance is recorded by Gruber of a remarkable displacement of the parotid on one side ; the whole gland being situated on the masseter muscle as if it were an enlarged socia parotidis. (Virchow's Archiv, xxxii., p. 328.)

The Submaxillary Gland.

The *submaxillary gland*, the next in size to the parotid gland, is of a spheroidal form, and weighs about 2 or $2\frac{1}{2}$ drachms. It is situated immediately below the base and the inner surface of the inferior maxilla, and above the digastric muscle. In this position it is covered by the skin and the platysma myoides, and its inner surface rests on the mylo-hyoid, hyo-glossus, and stylo-glossus muscles ; above, it corresponds with a depression on the inner surface of the jaw-bone ; and it is separated behind from the parotid gland merely by the stylo-maxillary membrane. The facial artery, before it mounts over the jaw-bone, lies in a deep groove upon the back part and upper border of the gland.

The *duct* of the submaxillary gland, named *Wharton's duct*, which is about two inches in length, passes off from the gland, together with a thin process of the glandular substance, round the posterior border of the mylo-hyoid muscle, and then runs forwards and inwards above that muscle, between it and the hyo-glossus and genio-hyo-glossus, and beneath the sublingual gland, to reach the side of the frænum linguæ. Here it terminates, close to the duct of the opposite side, by a narrow orifice, which opens at the summit of a soft papilla seen beneath the tongue. The structure of this gland is like that of the parotid ; but its lobes are larger, its surrounding areolar web is finer, and its attachments are not so firm. Moreover, its duct has much thinner coats than the parotid duct.

The blood-vessels of the submaxillary gland are branches of the facial and lingual arteries and veins. The nerves include those derived from the small submaxillary ganglion, as well as branches from the mylo-hyoid division of the inferior dental nerve, and the sympathetic.

The Sublingual Gland.

The *sublingual gland*, the smallest of the salivary glands, is of a narrow oblong shape and weighs scarcely one drachm. It is situated along the

Fig. 573. — VIEW OF THE RIGHT SUB-MAXILLARY AND SUBLINGUAL GLANDS FROM THE INSIDE.

Fig. 573.

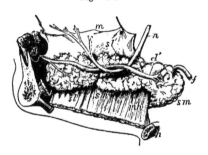

A part of the right side of the jaw, divided from the left at the symphysis, remains ; the tongue and its muscles have been removed ; but the mucous membrane of the right side is retained and is drawn upwards so as to expose the sublingual glands ; *s m*, the larger superficial part of the submaxillary gland ; *f*, the facial artery passing through it ; *s m'*, deep portion prolonged within the mylo-hyoid muscle *m h* ; *s l*, is placed below the anterior large part of the sublingual glands, with the duct of Bartholin partly shown ; *s l'*, placed above the hinder small end of the chain of glands, indicates the ducts of one or two perforating the mucous membrane ; *d*, the papilla, at which the duct of Wharton opens in front behind the incisor teeth; *d'*, the commencement of the duct ; *h*, the hyoid bone ; *n*, the gustatory nerve.

floor of the mouth, where it forms a ridge between the tongue and the gums of the lower jaw, covered only by the mucous membrane. It reaches from the frænum linguæ, in front, where it is in contact with the gland of the opposite side, obliquely backwards and outwards for rather more than an inch and a half. On its inner side it rests on the genio-hyo-glossus ; beneath, it is supported by the mylo-hyoid muscle, which is interposed

3 H

between it and the submaxillary gland ; but it is here in close contact with the Whartonian duct, with the accompanying deep portion of the last-named gland, and also with the lingual nerve.

The lobules of the sublingual gland are not so closely united together as those of the other salivary glands, and the ducts from many of them open separately into the mouth, along the ridge which indicates the posi-

A.   Fig. 574.   B.

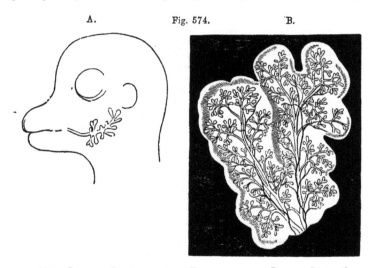

Fig. 574.—Sketches Illustrating the Formation of the Parotid Gland (from J. Müller).

A, head of a fœtal sheep magnified, showing the early simple condition of the parotid gland with the duct injected.
B, parotid gland of a fœtal sheep more advanced, the ducts and blood-vessels injected.

tion of the gland. These ducts, named *ducts of Rivini*, are from eight to twenty in number. Some of them open into the duct of Wharton. One, longer than the rest (which is occasionally derived in part also from the submaxillary gland), runs along the Whartonian duct, and opens either with it or very near it ; this has been named the duct of Bartholin.

The blood-vessels of this gland are supplied by the sublingual and submental arteries and veins. The nerves are numerous, and are derived from the lingual branch of the fifth pair.

Saliva.—The saliva is a clear limpid fluid, containing a few microscopic granular corpuscles. Its specific gravity is from 1·006 to 1·008, and it has only from 1 to 1½ parts of solid matter in 100. The saliva is always alkaline during the act of mastication ; but the fluid of the mouth becomes acid, and remains so until the next time of taking food : the reason being that the secretion of the mucous follicles of the mouth is acid, while that of the salivary glands is alkaline. Its chief ingredients, besides water and mucus, are a peculiar animal extractive substance, named salivine, with some alkaline and earthy salts. It is remarkable, besides, for containing a minute proportion of sulphocyanide of potassium.

Development.—In mammalia, according to Müller and Weber, the salivary glands, as shown in the case of the parotid gland in the embryo of the sheep, first appear in the form of a simple canal with bud-like processes lying in a blastema, and communicating with the cavity of the mouth. This canal becomes more and more ramified to form the ducts, whilst the blastema soon acquires a lobulated form, corresponding with that of the future gland, and at last wholly disappears, leaving the branched ducts, with their blood-vessels and connecting tissues. The submaxillary gland is said to be the first formed ; then the sublingual and the parotid.

THE PHARYNX.

The pharynx is that part of the alimentary canal which unites the cavities of the mouth and nose to the œsophagus. It extends from the base of the skull to the lower border of the cricoid cartilage, and forms a sac open at the lower end, and imperfect in front, where it presents apertures leading into the nose, mouth and larynx.

The velum pendulum palati projects backwards into the pharynx, and during the passage of the food completely separates an upper from a lower part by means of the contraction of the muscles connected with it which are placed in the posterior pillars of the fauces. Seven openings lead into the cavity of the pharynx ; viz., above the velum, the two posterior openings of the nares and, at the sides the apertures of the Eustachian tubes ; while, below the velum, there is first the passage leading from the mouth, then the superior opening of the larynx, and lastly the passage into the œsophagus.

The walls of the pharynx consist of a fascia or layer of fibrous tissues, named the pharyngeal aponeurosis, dense at its upper part but lax and weak below, surrounded by muscles, and lined by a mucous membrane. At its upper end this fibrous wall is attached to the posterior margin of the body of the sphenoid bone, and passes outwards to the petrous portion of the temporal. It is strengthened in the middle line by a strong band descending between the recti antici muscles from a part of the basilar process which often presents a marked tubercle.

The pharynx is usually described as attached superiorly to the basilar process of the occipital bone ; it is certain, however, from dissections in both young and old subjects, that the recti capitis antici muscles come quite forward to the anterior extremity of the basilar process ; that the posterior wall of the pharynx at its upper end forms a cul-de-sac on each side opposite the tip of the petrous bone, and lies in a curve, with its convexity forwards, in front of the recti muscles ; and that the only connection of the pharynx with the occipital bone is by means of the mesial band, which has been described, and which forms a cranio-pharyngeal ligament. The tubercle from which this band principally springs is sometimes named *tuberculum pharyngeum.*

Behind, the wall of the pharynx is loosely connected by areolar tissue to the prevertebral fascia covering the bodies of the cervical vertebræ and the muscles which rest upon them. At the sides, the walls have similar connections, by loose areolar tissue, with the styloid process and its muscles, and with the large vessels and nerves of the neck. In front, they are attached in succession to the sides of the posterior nares, the mouth and the larynx. Thus, commencing above by a tendinous structure only, at the petrous portion of the temporal bone and the Eustachian tube, the walls are connected by means of muscle and fibrous membrane, first, with the internal pterygoid plate, then with the pterygo-maxillary ligament, and next with the mylo-hyoid ridge of the lower jaw ; below this, they are attached

3 H 2

to the sides of the tongue, to the hyoid bone, and stylo-hyoid ligament ; and, still lower down, to the thyroid and cricoid cartilages.

The pharynx is about four inches and a half in length, and is considerably wider across than it is deep from before backwards. Its width above is moderate ; its widest part is opposite the cornua of the hyoid bone, and below this it rapidly contracts towards its termination, opposite the cricoid cartilage, where it is narrowest.

Structure. — The *muscles* of the pharynx are the superior, middle and inferior constrictors, the stylo-pharyngeus, and the palato-pharyngeus. They are described at page 187.

The *mucous* membrane lining the inner surface of the pharynx is continuous at the several apertures with that of the adjacent cavities. It varies somewhat in its character in different parts. Its upper portion is

Fig. 575.

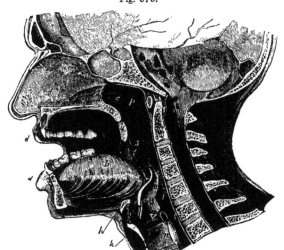

Fig. 575. —Antero-Posterior Vertical Section through the Head a little to the left of the Middle Line, showing the relations of the Nasal and Buccal Cavities, the Pharynx, Larynx, &c.

a, nasal septum, and below it the section of the hard palate ; *b*, the tongue ; *c*, soft palate ; *d*, the lips ; *u*, the uvula ; *r*, anterior pillar of the fauces ; *i*, posterior pillar ; *t*, the tonsil placed between the pillars ; *p*, upper part of the pharynx ; *h*, body of the hyoid bone ; *k*, thyroid cartilage ; *n*, cricoid cartilage ; *v*, on the upper vocal cords above the glottis ; *s*, epiglottis ; 1, posterior opening of the nares ; 3, behind the isthmus faucium ; 4, opposite the superior opening of the larynx ; 5, passage into the œsophagus ; 6, opening of the right Eustachian tube.

thick where it adheres to the periosteum of the basilar process, but is much thinner near the entrance of the Eustachian tube and the posterior nares : in this situation numerous *glands* are found collected in a layer beneath the mucous membrane. The glands are of two kinds, viz., racemose, which are especially numerous in the upper portion ; and simple or compound follicular, which exist throughout the whole of the pharynx. A chain of glands forming a glandular mass, exactly similar to that of the tonsils, stretches across the back of the fauces between the orifices of the two Eustachian tubes (Kölliker). In the part opposite the fauces, the mucous membrane exactly resembles that of the mouth. Lower down it becomes paler, and at the back of the larnyx it forms several longitudinal folds or plicæ. According to Henle, the epithelium upon the upper portion of the pharynx, as low down as a horizontal line level with the floor of the nares, is columnar and ciliated ; but, below that point, it is squamous and destitute of cilia.

THE ŒSOPHAGUS.

The *œsophagus* or gullet, the passage leading from the pharynx to the stomach, commences at the cricoid cartilage opposite the lower border of the fifth cervical vertebra, and, descending along the front of the spine, passes through the diaphragm opposite the ninth dorsal vertebra, and there ends by opening into the cardiac orifice of the stomach.

The length of the œsophagus is about nine or ten inches. It is of smaller diameter than any other division of the alimentary canal, its narrowest part being at the commencement behind the cricoid cartilage ; it is also slightly constricted in passing through the diaphragm, but, below that, gradually widens into the stomach. The œsophagus is not quite straight in its direction, but presents three slight curvatures. One of these is an antero-posterior flexure, corresponding with that of the vertebral column in the neck and thorax. The other two are slight lateral curves ; for the œsophagus, commencing in the median line, inclines to the left side as it descends to the root of the neck ; thence to the fifth dorsal vertebra it gradually resumes the mesial position ; and finally, it deviates again to the left, at the same time coming forward towards the œsophageal opening of the diaphragm. In the lower cervical and upper dorsal region the œsophagus is applied to the anterior surface of the spine, being connected with it and with the longus colli muscle by loose areolar tissue; but between it and the bodies of the upper dorsal vertebræ the thoracic duct ascends obliquely from right to left: its lower third is placed in front of the aorta. In the *neck*, the œsophagus lies close behind the trachea, and the recurrent laryngeal nerves ascend in the angles between them ; on each side is the common carotid artery, and also a part of the thyroid body, but, as the œsophagus inclines to the left side, it is in more immediate connection with the left carotid.

In the *thorax*, the œsophagus is successively covered in front by the lower part of the trachea, by the commencement of the left bronchus, and by the back of the pericardium. The aorta, except near the diaphragm, where the œsophagus is in front of the vessel, lies rather to the left, and the vena azygos to the right ; the pneumogastric nerves descend in close contact with its sides, and form a plexus around it, the left nerve proceeding gradually to the front, and the right nerve retiring behind it. Lastly, the œsophagus, which is here placed in the interval between the two pleuræ, comes partially in contact with both of those membranes.

Structure.—The walls of the œsophagus are composed of three coats ; viz., an external or muscular, a middle or areolar, and an internal or mucous coat. Outside the muscular strata, there is a layer of fibrous tissue, with well marked elastic fibres, which is sometimes spoken of as a distinct coat.

The *muscular* coat consists of an *external* longitudinal layer, and an *internal* circular layer. This twofold arrangement of the muscular fibres

prevails throughout the whole length of the alimentary canal ; but the two
layers are here much thicker, more uniformly disposed, and more evident
than in any other part except quite at the lower end of the rectum.   The
external or *longitudinal* fibres are disposed at the commencement of the
tube in three fasciculi, one in front, and one on each side of the œsophagus.   The lateral fasciculi are blended above with the inferior constrictor
of the pharynx ; the anterior fasciculus arises from the back of the cricoid
cartilage at the prominent ridge between the posterior crico-arytenoid
muscles, and its fibres spreading out obliquely on each side of the gullet as
they descend, soon blend with those of the lateral bundles to form a continuous layer around the tube.   The internal or *circular* fibres are separated

Fig 576.

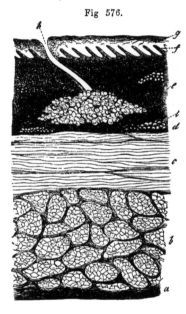

Fig. 576. — Section of the Coats of the
Human Œsophagus (from Kölliker). 41

The section is transverse, and from near
the middle of the gullet. *a*, fibrous covering ;
b, divided fibres of the longitudinal muscular coat ; *c*, transverse muscular fibres ;
d, submucous or areolar layer ; *e*, mucous
membrane ; *f*, its papillæ ; *g*, laminated
epithelial lining ; *h*, opening of a mucous
gland, of which the cellular part is seen
embedded deeply in the mucous membrane ;
i, fat-vesicles.

above by the fibres of the longitudinal fasciculi from those of the
inferior constrictor of the pharnyx.
The rings which they form around
the tube have a horizontal direction
at the upper and lower part of the
œsophagus, but in the intervening
space are slightly oblique.   At the
lower end of the œsophagus, both
layers of fibres become continuous
with those of the stomach.

The muscular coat of the upper
end of the œsophagus is of a well-
marked red colour, and consists of
striped muscular fibres ; but lower
down it becomes somewhat paler, and is principally composed of the plain
muscular fibres.   A few striped fibres, however, are found mixed with the
others, and have been traced throughout its whole length, and even, it is
said, upon the cardiac end of the stomach.   (Ficinus.)

The longitudinal fibres of the œsophagus are observed by Hyrtl to be sometimes
joined by a broad band of smooth muscle, passing upwards from the left pleura, and
sometimes also by another from the left bronchus.

The *areolar* coat is placed between the muscular and mucous coats, and
connects them loosely together.

The *mucous membrane* is of firm texture, and is paler in colour than that
of the pharynx or stomach.   From its loose connections its outer surface is
freely movable on the muscular tunic ; and when the latter is contracted,
as happens when the œsophagus is not giving passage to food, the mucous
lining is thrown into longitudinal folds, the inner surfaces of which are in
mutual contact.   These folds again disappear on distension of the canal.

· Minute papillæ are seen upon this mucous membrane, placed at some distance from each other, and the whole is covered with a thick squamous epithelium, which can be traced as far as the cardiac orifice of the stomach, where it suddenly passes into one of a different character, as will be hereafter noticed.

The gullet is provided with many small compound racemose glands, named *œsophageal glands,* which are especially numerous at the lower end of the tube.

Dilatations occasionally occur in the course of the œsophagus. Diverticular pouches are also sometimes found, but appear in all cases to be of hernial origin. Duplicity of the œsophagus in part of its extent, without other abnormality, has been recorded (Blaes, quoted by Meckel).

THE ABDOMINAL PORTION OF THE DIGESTIVE ORGANS.

As that part of the digestive canal which is found beneath the diaphragm, and consists of the stomach and intestines, is situated within the cavity of the *abdomen,* and occupies, together with the liver (the secretion of which it receives), by far the greater part of that cavity, the topographic relations of the abdominal viscera may here be briefly explained.

THE ABDOMEN.

The abdomen is the largest cavity in the body, and is lined by an extensive and complicated serous membrane, named the peritoneum.

It extends from the diaphragm above to the levatores ani muscles below, and is subdivided into two parts : an upper and larger part, *the abdomen,* properly so called ; and a lower part, named the *pelvic cavity.* The limits between the abdominal and pelvic portions of the cavity are marked by the brim of the pelvis.

· The enclosing walls of this cavity are formed principally of muscles and tendons which have been already described (p. 248). They are strengthened internally by a layer of fibrous tissue lying between the muscles and the peritoneum, the different parts of which are described under the names of fascia transversalis, fascia iliaca, and anterior lumbar fascia (p. 257). These walls are pierced by several apertures, through which are transmitted the great vessels and some other parts, such as the several diaphragmatic apertures for the aorta, vena cava, and œsophagus, and the femoral arches and inguinal canals. In the median fibrous substance of the anterior wall lies the umbilical cicatrix. The cavity of the pelvis is also lined with strong fasciæ (p. 260), and partially by peritoneum, and at its lower part it presents the apertures for the transmission of the rectum and the genito-urinary passages.

For the purpose of enabling precise reference to be made to the situation and condition of the contained organs, the *abdomen proper* has been artificially subdivided into certain regions, the boundaries of which are indicated by lines drawn upon the surface of the body. Thus, two horizontal lines drawn round the body divide the cavity into three zones : viz. an upper, a middle, and a lower. One of these lines, commencing at the most prominent point of the lower costal cartilages of one side, is drawn across to the corresponding point on the opposite side, and thence horizontally round the back to the place at which it began. The other line, proceeding from the crest of the ilium of one side, extends to that of the other, and so round the body, as in the former instance. Each of these zones again is subdivided

into three parts by means of two perpendicular lines, drawn from the cartilage of the eighth rib, on each side, down to the centre of Poupart's ligament.

The upper zone is thus marked off into the right and left *hypochondriac*
regions and the *epigastric* region, the depression in the upper part of which
is called *scrobiculus cordis*, or pit of the stomach. The middle zone is
divided into the *umbilical* region in the centre, and the right and left
lumbar regions; and the inferior zone into the *hypogastric* region in the
centre, and the *iliac* region at each side.

On opening the abdominal cavity from the front, the viscera are seen to
lie in an upper and a lower group, separated by the great omentum, which
overhangs those in the lower part. The surfaces, which are in contact one
with another, and with the wall of the cavity, are rendered glistening by
reflections over them of the lining membrane of the cavity, the *peritoneum ;*
and the various organs are found to be attached by means of folds or duplicatures of that membrane, termed mesenteries and omenta, which include the
blood-vessels, nerves, and lymphatics belonging to each organ. In the
upper group, as seen from before, are comprised the liver, stomach, and
a small part of the intestine ; in the lower group, more or less hidden by

Fig. 577.

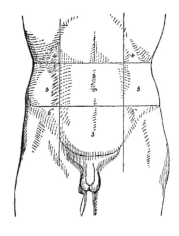

Fig. 577.—Outline of the anterior
surface of the Abdomen, showing
the division into Regions.

1, epigastric region ; 2, umbilical ; 3,
hypogastric ; 4, 4, right and left hypochondriac ; 5, 5, right and left lumbar ;
6, 6, right and left iliac.

the great omentum, are the remaining parts of the alimentary
canal. The spleen, pancreas, and
kidneys constitute a deeper group.

On the right side, projecting
downwards from beneath the diaphragm, is the *liver* with its excretory apparatus, which occupies
the right hypochondrium and part
of the epigastrium and extends a
short way into the left hypochondrium ; to the left, and partly
beneath the liver, is the stomach,
which lies in the epigastric and left
hypochondriac regions ; and closely applied to the left or cardiac end of the
stomach is the *spleen*.

The *stomach* is seen to be connected at its right extremity, named the
pylorus, with the *small intestine*. The first part of the small intestine,
named *duodenum*, forms a deep curve projecting towards the right side,
resting on the posterior wall of the abdominal cavity and right kidney, and
terminating at the left of the middle line, where it emerges from behind the
root of the mesentery, and passes into the second part of the intestine, named
jejunum. The hollow of the curve of the duodenum is occupied by the large
right extremity or head of the pancreas. The remainder of the small intestine, comprising the *jejunum* in its upper two-fifths, and the *ileum* in the

lower three-fifths, is disposed in moveable convolutions, and is attached posteriorly by a broad mesentery to the abdominal wall. It occupies the umbilical and hypogastric regions, from the back part of which the mesentery takes origin, and it extends likewise into the lumbar and iliac regions, besides gravitating into the pelvis. The ileum terminates abruptly in the right iliac region in the *caput cæcum*, a cul-de-sac in which the *great intestine* commences. The cæcum is continued into the *ascending colon*, which lies against the posterior wall of the abdomen, as it passes up through the right lumbar to the right hypochondriac region. The ascending colon is succeeded by the *transverse colon* which passes transversely, or with a pendulous curve, across the abdomen from right to left, resting on the small intestines. Below

Fig. 578.—DIAGRAM OF THE ABDO-
MINAL PART OF THE ALIMENTARY
CANAL. ⅛

1, the stomach; 2, the lower part of the gullet; 3, the left cul-de-sac, and, 4, the pyloric end of the stomach; 5, 6, the duodenum; 7, 8, convolutions of the small intestine; 9, cæcum; 10, the vermiform process; 11, ascending, 12, transverse, and, 13, descending colon; 14, commencement of the sigmoid flexure; 15, rectum.

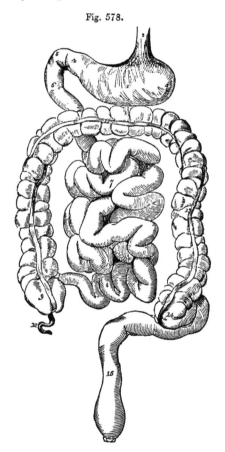

Fig. 578.

the spleen, the transverse colon is continued into the *descending colon* which extends down through the left lumbar to the left iliac region, where it is continued into the more loosely bound *sigmoid flexure*, which occupies that fossa and falls into the *rectum.*

Within the pelvis, the extension downwards of the peritoneal cavity is termed the *recto-vesical fossa:* posteriorly the rectum is observed, and anteriorly the sloping upper wall of the *urinary bladder;* while, in the female, the *uterus* projects upwards between the rectum and bladder, so that a recto-uterine pouch is formed, and the *ovaries* and *Fallopian tubes* are pendant at its sides. The bladder when full, and the uterus in its gravid state, project upwards into the abdomen, and displace more or less of the small intestine.

Subjoined is an enumeration of the organs situated in the different regions of the abdomen.

PARTS SITUATED IN EACH REGION OF THE ABDOMEN.

Epigastric region . .	The middle part of the stomach, with its pyloric extremity, the left lobe of the liver, the hepatic vessels and lobulus Spigelii, the pancreas, the cœliac axis, the semilunar ganglia, part of the vena cava, and also, as they lie between the crura of the diaphragm, part of the aorta, the vena azygos and thoracic duct.
Hypochondriac, right .	The right lobe of the liver, with the gall-bladder, part of the duodenum, the hepatic flexure of the colon, the right suprarenal capsule, and part of the corresponding kidney.
Hypochondriac, left . .	The large end of the stomach, with the spleen and narrow extremity of the pancreas, the splenic flexure of the colon, the left suprarenal capsule and upper part of the left kidney. Sometimes also a part of the left lobe of the liver.
Umbilical	Part of the omentum and mesentery, the transverse part of the colon, transverse part of the duodenum, with some convolutions of the jejunum and ileum.
Lumbar, right . .	The ascending colon, lower half of the kidney and part of the duodenum and jejunum.
Lumbar, left	The descending colon and lower part of the left kidney, with part of the jejunum.
Hypogastric	The convolutions of the ileum, the bladder in children, and, if distended, in adults also ; the uterus when in the gravid state.
Iliac, right	The cæcum, appendix vermiformis, ileo-cæcal valve, ureter, and spermatic vessels.
Iliac, left	The sigmoid flexure of the colon, the ureter and spermatic vessels.

THE PERITONEUM.

The peritoneum or serous membrane of the abdominal cavity is by far the most extensive and complicated of the serous membranes. Like the others it may be considered to form a shut sac, on the outside of which are placed the viscera which it covers. In the female, however, the two Fallopian tubes open at their free extremities into the cavity of the peritoneum. The internal surface is free, smooth, and moist, and is covered by a thin squamous epithelium. The external or attached surface adheres partly to the parietes of the abdomen and pelvis, and partly to the outer surface of the viscera situated within them. The *parietal* portion is connected loosely with the fascia lining the abdomen and pelvis by means of a layer of areolar tissue, distinct from the abdominal fasciæ, and named the *subperitoneal* or *retro-peritoneal* layer ; but it is more firmly adherent along the middle line of the body in front, as well as to the under surface of the diaphragm. The *visceral* portion, which is thinner than the other, forms a more or less perfect investment to most of the abdominal and pelvic viscera.

The folds of the peritoneum are of various kinds. Some of them, constituting the *mesenteries*, connect certain portions of the intestinal canal with the posterior wall of the abdomen ; they are, the mesentery properly so called for the jejunum and ileum, the meso-cæcum, transverse and sigmoid meso-colon, and the meso-rectum. * Other duplicatures, which are called *omenta*, proceed from one viscus to another ; they are distinguished as the great omentum, the small omentum, and the gastro-splenic omentum. Lastly, certain reflexions of the peritoneum from the walls of the abdomen or pelvis to viscera which are not portions of the intestinal canal, are named *ligaments :* these include the ligaments of the liver, spleen, uterus, and bladder.

If the examination of the folds of the peritoneum be commenced on the under surface of the diaphragm, it will be found that on the left side it can be traced back to the posterior wall of the abdomen, and down in front of the upper part of the kidney to the commencement of the descending colon. Further to the right, it is reflected from the diaphragm over the front of the stomach, and from the left of the stomach passes across a very short interval to the spleen, which it completely invests ; and it is continued back from the spleen to the abdominal wall. Still further to the right, the peritoneum is reflected from the diaphragm to the liver, invests the whole superior surface of that organ, and passes round its anterior and lateral margins to invest the whole of its inferior surface, with the exception only of so much as lies behind the portal fissure, viz., the lobule of Spigelius. On the upper surface of the liver the peritoneum is thrown into a right and a left fossa by a vertical antero-posterior fold attaching it to the diaphragm, which is named the *falciform* or *suspensory ligament* of the liver. In the lower margin of this ligament a fibrous cord, consisting of the obliterated remains of the umbilical vein, and named the *round ligament* of the liver, extends upwards from the umbilicus to the longitudinal fissure which divides inferiorly the right lobe of the liver from the left ; and it is the reflection of the peritoneum from this cord which forms the falciform ligament. The thick posterior border of the liver, uninvested with peritoneum, is in contact with the diaphragm ; and the reflexions of peritoneum from the upper and under surfaces of the organ to the parietes above and below this border constitute the *coronary ligament.* Towards the right and left extremities of the liver the superior and inferior layers of the coronary ligament come into contact for a little way, and form the *right* and *left triangular ligaments* of the liver.

The portion of peritoneum reflected from the under surface of the liver,, opposite the portal fissure, passes down over the vena portæ, hepatic artery, and biliary ducts, to the pyloric extremity of the stomach and first part of the duodenum ; while that which invests the part of the liver to the right of the portal fissure is conducted back to the posterior wall of the abdomen. If now, the disposition of the peritoneum be examined in the spot where those two modes of arrangement meet, an aperture sufficiently large to admit a finger, and formed by invagination of the peritoneum, will be found leading, from right to left, behind the hepatic vessels and duct : this aperture is the *foramen of Winslow ;* and the fold of peritoneum in front of it, containing the portal vein, hepatic artery, and biliary ducts, is termed the *small* or *gastro-hepatic omentum.* The foramen of Winslow has above it a portion of the liver; behind it the vena cava inferior; below it the duodenum ; and in front the small omentum. The invagination of peritoneum which takes place at the foramen of Winslow is of great extent, expanding to form a large pouch, which lies behind the stomach, and stretches downwards to a variable degree in front of the transverse colon. This is the *sac of the omentum* or *smaller cavity of the peritoneum,* which will be presently described.

On tracing downwards the peritoneum investing the anterior surface of the stomach, it is seen to be prolonged from the inferior border of that viscus to form a pendulous fold of omentum lying loosely in front of the colon and small intestines, and having a free margin inferiorly. Folding backwards on itself at this margin, the peritoneum passes upwards to the transverse colon and becomes adherent to its surface, whence it is continued back to the abdominal wall.

The sac of the omentum may be laid open by means of an incision a little

Fig. 579.

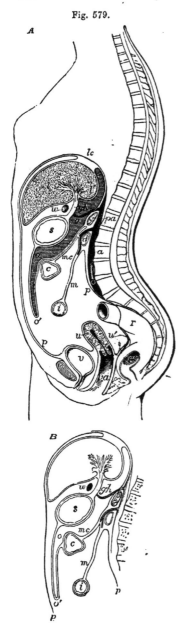

Fig. 579, A.—Diagrammatic Outline of a supposed Section of the Body, showing the Inflections of the Peritoneum in the Female. ⅙

The upper part of the section is a little to the right of the mesial plane of the body, through the quadrate and Spigelian lobes of the liver ; below it is supposed to be mesial : *l c,* placed above the diaphragm opposite to the coronary ligament of the liver ; *l,* the liver ; *l′,* lobe of Spigel ; *s,* stomach ; *c,* transverse colon ; *i,* the small intestine ; *p a,* pancreas ; *a,* the aorta ; *d,* the duodenum ; *v,* urinary bladder ; *u,* uterus ; *r,* rectum ; *r′,* its middle part opened ; *v a,* vagina; *p, p,* the parietal peritoneum lining the front and back of the abdominal cavity ; the line representing the inflections of the greater sac of the peritoneum will be traced from the neighbourhood of *l c,* where it passes on the upper surface of the liver over the upper and lower surfaces of that organ, in the front of *g h,* the gastro-hepatic omentum, over the front of the stomach, down to *o′,* the outer layer of the great omentum, whence it passes back to the vicinity of the pancreas, and re-descends as the upper layer of the transverse meso-colon ; after enclosing the colon it returns on the lower surface of the transverse meso-colon, *m c,* to the root of the mesentery, *m ;* it now forms the mesentery and encloses the small intestine, returning to the duodenum and posterior wall of the abdomen, whence it passes over the rectum, *r,* descends into the recto-vaginal pouch, *u′,* covers the back and front of the uterus and the bladder partially and regains the anterior abdominal wall above the pubes ; as connected with the lesser sac of the peritoneum, *w,* marks the position of the foramen of Winslow as if seen in perspective beyond the section ; the lesser sac with the sac of the omentum is shaded with horizontal lines, and is marked *o o :* round this space the line of the peritoneum may be traced from the diaphragm over the lobe of Spigel, to the back of the gastro-hepatic omentum, thence behind the stomach and down into the sac of the omentum ; it then ascends to the pancreas, which it covers, and thence reaches again the diaphragm.

B, is a sketch of part of a section similar to that of A, but showing the different view frequently taken, according to which the two layers of the meso-colon, after enclosing the colon, descend to form the posterior pair of the layers of the great omentum.

below the great curvature of the stomach. It will then be seen that the inner wall of this sac, having invested the posterior surface of the stomach and commencement of the duodenum, is continued downwards, back to back with the general peritoneum, into the pendulous portion of the omentum, aud, as it returns thence, is applied to the anterior surface of the transverse colon. Passing these parts, it resumes its position of proximity to the peritoneum of the greater sac, and proceeds to the posterior wall of the abdomen. The two layers of peritoneum which thus hang pendulously one within the other, and are derived from the general and the smaller sac, constitute the *gastro-colic* or *great omentum*; while those by which the transverse colon is connected with the abdominal wall are termed the *transverse mesocolon*. In the pendulous great omentum, there being a duplication of both the general peritoneum and the wall of the smaller sac, four layers are to be distinguished, viz., first, an anterior and a posterior layer belonging to the greater sac of the peritoneum, having their smooth surfaces respectively directed forwards to the abdominal wall and backwards to the small intestines; and second, between these, the anterior and posterior layers derived from the lesser sac, lining the omental cavity and gliding one against the other: these four layers are, however, so intimately united and reduced to such extreme tenuity in the adult, that they cannot be separately recognised in the omentum below the colon. In most instances the pendulous part of the omentum presents the appearance of lacework, the interstices of which in corpulent persons are more or less loaded with fat. In some subjects, instead of lying like an apron over the small intestine, it is crumpled into a bundle along the transverse colon, as if displaced by the movements of the intestines against the wall of the abdomen.

The description now given of the great omentum and transverse mesocolon agrees with the appearances most frequently seen in the adult subject, and with the account usually given in English works of Anatomy, the posterior layer of the great omentum being described as separating from the layer within, belonging to the omental sac, when it reaches the transverse colon, so as to pass behind or below that viscus, and from thence as proceeding backwards to the abdominal wall as the posterior or lower layer of the transverse meso-colon. It was, however, long ago pointed out by Haller, and the view has been confirmed by the observations of J. F. Meckel, J. Müller, Hansen, and Huschke, that in the fœtus, and occasionally in the child, or even in the adult, the two posterior layers of the omentum, though adherent to the transverse colon, may be separated from it and from the transverse meso-colon, so as to demonstrate that the transverse meso-colon is really a distinct duplicature of peritoneum. This view has been adopted by Holden and Luschka in their more recent works, and has been verified by Allen Thomson. Figures 579 A, and B, show diagrammatically the difference of the two views. .

The anterior wall of the sac of the omentum invests the whole posterior surface of the stomach; above the small curvature of the stomach it lies back to back with the general peritoneum, completing in conjunction with it a *gastro-phrenic* ligament; and further to the right it forms the posterior layer of the gastro-hepatic omentum, and likewise invests the lobulus Spigelii of the liver, close to the foramen of Winslow. Lying transversely in front of the aorta and in contact with the posterior wall of the abdomen, the pancreas is seen invested anteriorly by the hinder wall of the sac of the omentum. To the left of the stomach the sac extends to the spleen, and usually gives investment to a small portion of that organ at the lower end of its hilus; it thus forms the posterior layer of the *gastro splenic omentum,* the fold by which the spleen is attached to the stomach. The splenic artery lies behind the sac of the omentum in its course to the spleen, but its gastric branches turn round the splenic margin of the sac and reach the stomach by that means. The coronary artery of the stomach reaches the front of the sac by turning round its upper margin ; and the hepatic artery passes round from below, close to the foramen of Winslow.

The disposition of the peritoneum below the level of the transverse mesocolon is comparatively simple. The mesentery of the small intestine, although greatly frilled out in front to correspond in length with the jejunum and ileum to which it affords support, is attached posteriorly by a very short border which extends from the level of attachment of the transverse colon immediately to the left of the middle line, directly down to the right iliac fossa, where the ileum falls into the cæcum. At its widest part the length of the mesentery is from four to six inches between its vertebral and its intestinal border. Between the two layers of serous membrane of which it consists are placed, beside some fat, numerous branches of the superior mesenteric artery and vein, together with nerves, lacteal vessels, and mesenteric glands. In the right and left lumbar region the peritoneum invests the ascending and descending colon usually in less than their whole circumference, and thus binds them closely down to the abdominal parietes, without the intervention of a meso-colon. In some cases the cæcum is suspended at a short distance from the right iliac fossa, by a distinct duplicature of the peritoneum, which is termed the *meso-cæcum ;* but, more commonly, the peritoneum merely binds down this part of the large intestine, and forms a distinct but small mesentery for the vermiform appendix only. The sigmoid flexure is attached to the left iliac fossa by a considerable mesentery, the *sigmoid meso-colon ;* and in the pelvis the rectum is attached by a fold named *meso-rectum.* The other peritoneal folds within the pelvis will be mentioned elsewhere.

Along the colon, and upper part of the rectum, the peritoneum is developed into numerous little projections filled with adipose tissue. These fatty processes are named *appendices epiploicæ.*

At the upper end of the attachment of the mesentery, on its left side, there is always visible a small portion of the terminal part of the duodenum appearing from underneath in about half its breadth ; and on the right side of the mesentery there is often another little angle of duodenum visible between the mesentery and meso-colon. Thus it will be observed that, while the commencement of the duodenum is invested, like the stomach, in front by the general peritoneum and behind by the sac of the omentum, and a succeeding portion is invested only in front, the remainder is crossed by the colon and mesentery, and is only to a small extent in contact with peritoneum.

THE STOMACH.

The stomach is that dilated portion of the alimentary canal which intervenes between the œsophagus and the duodenum, and within which the food is retained for a time to be acted on by the gastric juice, and to be converted into chyme.

This organ is seated in the left hypochondriac and the epigastric regions, extending also into the right hypochondrium. It lies in part against the anterior wall of the abdomen, and in part beneath the liver and diaphragm, and above the transverse colon.

The stomach is of a somewhat conical or pyriform shape. The left extremity is the larger, and is named the cardiac, *great* or *splenic* end. The *right* or *small* end is also named the *pyloric* extremity. Of its two orifices, the one by which food enters from the œsophagus is named the *cardiac* orifice, the other, by which it passes into the duodenum, and which is placed on a somewhat lower level, and more forwards, is the *pyloric* orifice.

The œsophagus terminates in the stomach two or three inches from the great extremity, which projects beyond the place of union to the left, forming the *great cul-de-sac* or *fundus.*

Between the cardiac and the pyloric orifices, the outline of the stomach is curved along its upper and lower borders. The upper border, about three or four inches in length, is concave, and is named the *lesser* curvature; while the lower border, which is much longer, and, except towards the pylorus, convex, forms the *greater* curvature.

Fig. 580.—DIAGRAMMATIC OUTLINE OF
    THE STOMACH. ⅓

a, great curvature; *b*, lesser curvature; *c*, left end, great cul-de-sac, or fundus; *d*, small cul-de-sac or antrum pylori; *o*, œsophageal orifice or cardia; *p*, duodenal orifice or pylorus.

Fig. 580.

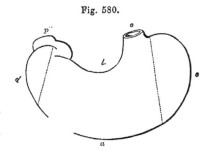

Towards the pylorus, the small end of the stomach describes a double bend, opposite to the first turn of which is a prominence or bulging, sometimes named the *small cul-de-sac* or *antrum pylori.*

Division of the stomach, by constriction, into a right and left pouch, is frequently observed as a temporary condition resulting from spasm. More rarely it is of a permanent character (Struthers, Monthly Med. Journ., 1851).

Dimensions.—These vary greatly, according to the state of distension of the organ. When moderately filled, its length is about ten or twelve inches; and its diameter at the widest part, from four inches to four inches and a half. According to Clendinning, it weighs, when freed from other parts, about four ounces and a half in the male, and somewhat less in the female.

Connections.—The borders of the stomach are connected with folds of peritoneum in their whole extent. Thus, to the superior border is attached the gastro-phrenic ligament and gastro-hepatic omentum; to the inferior border is attached the gastro-colic omentum; and to the left extremity the gastro-splenic omentum. The blood-vessels and lymphatics of the stomach pass within these duplicatures of the peritoneum, and reach the organ along its two curvatures. Its anterior and posterior surfaces are free, smooth, and covered with peritoneum. The anterior surface, which is directed upwards as well as forwards, is in contact above with the diaphragm and the under surface of the liver, and lower down with the abdominal parietes opposite to the epigastric region, which is hence named the *pit* of the *stomach.* The posterior surface is turned downwards and backwards, and rests upon the transverse meso-colon, and farther back, upon the pancreas and great vessels of the abdomen.

At its cardiac orifice it is continuous with the œsophagus, and is, there-fore, fixed to the œsophageal opening in the diaphragm. The pyloric ex-tremity, situated lower down, nearer to the surface, and having greater freedom of motion, is continuous with the duodenum. It is covered by the concave surface of the liver, and in some cases touches the neck of the gall-bladder.

When the stomach is distended, its position and direction are changed. The œsophageal end, being fixed to the back part of the diaphragm, cannot undergo any considerable change ; but the duodenal extremity has more liberty of motion. The lesser curvature is, also, somewhat fixed to the liver by the small omentum, while the great curvature is the most movable part : accordingly, when the stomach is distended, this curvature is elevated and at the same time carried forwards, whilst the anterior surface is turned upwards, and the posterior surface downwards.

Structure.—The walls of the stomach consist of four distinct coats, held together by fine areolar tissue. They are named, in order from without in-wards, the serous, muscular, areolar or submucous, and mucous coats. By some the areolar coat is not reckoned as a separate tunic. Taking all the coats together, the walls of the stomach are thinner than those of the œsophagus, but rather thicker than those of the intestines generally. They are thickest at the pyloric end, and thinnest in the great cul-de-sac.

The *external* or *serous* coat, derived from the peritoneum, is a thin, smooth, transparent, and elastic membrane which closely covers the entire viscus, excepting along its two curvatures. Along the line of these curvatures the attachment is looser, leaving an interval occupied by the larger blood-vessels.

The second or *muscular* coat is composed of three sets of unstriped fibres, disposed in three layers, and named, from their direction, the longitudinal, the circular, and the oblique fibres.

The first or outermost layer consists of the *longitudinal* fibres, which are in direct continuity with those of the œsophagus. They spread out in a radiating manner from the cardiac orifice, and are found in greatest abun-dance along the curvatures, especially on the lesser one. On the anterior and posterior surfaces they are very thinly scattered, or are scarcely to be found. Towards the pylorus they are arranged more closely together and form a thicker uniform layer, which, passing over the pylorus, becomes con-tinuous with the longitudinal fibres of the duodenum.

The second set consists of the *circular* fibres, which form a complete layer over the whole extent of the stomach. They commence by small and thinly scattered rings at the left extremity of the great cul-de-sac, describe larger and larger circles as they surround the body of the stomach concentric to its curved axis, and towards the pyloric end again form smaller rings, and at the same time become much thicker and stronger than at any other point. At the pylorus itself, they are gathered into an annular bundle, which pro-jects inwards into the cavity, and forms, together with areolar tissue and the lining mucous membrane, the pyloric sphincter and valve. Some of the circular fibres appear to be continued from those of the œsophagus, spreading from its right side. According to Pettigrew the fibres of this layer are not, as usually described, mere rings or circles, but rather double loops in the form of the figure of eight, the two parts of which cross each other very obliquely.

The innermost muscular layer is incomplete, and consists of the *oblique* fibres. These oblique fibres are continuous with the layer of circular fibres of the gullet ; they embrace the cardiac orifice on the left, where they form a considerable stratum, and from that point descend obliquely upon the

Fig. 580*.—Sketch of the Distribution of Muscular Fibres in the Stomach (after Pettigrew and from nature). ~ ⅓

A, external layer of longitudinal fibres, as seen from the outside ; B, middle layer of circular fibres as seen on removing the longitudinal layer ; C, deepest layer of oblique fibres as seen from within, after inverting the stomach and removing the mucous membrane : c, the cardiac end ; p, the pyloric end ; in A, the stronger longitudinal fibres passing along the lesser and greater curvatures, and all round the pyloric end, are shown, and the radiating fibres spreading from the root of the gullet over the front and back of the stomach ; in B; the nearly uniform layer of circular fibres in two sets crossing each other very obliquely, and becoming concentric at the cardiac end to the centre of the great culde-sac ; in C, the very oblique bands of fibres which form a continuation of the circular fibres of the gullet and spread in two sets, o, o', one from the right and the other from the left side of the cardia (also partially represented in B), passing over the front and back of the stomach (except its lesser curvature) as far as the pyloric end.

Fig. 580*.

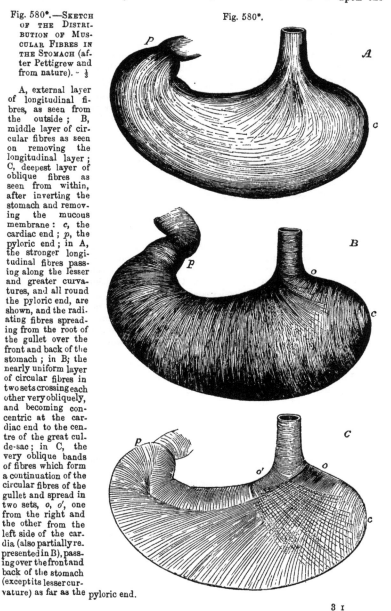

3 I

anterior and posterior surfaces of the stomach, where they spread out from one another, and most of them gradually disappear ; some, however, reach as far as the pylorus. A similar set of fibres, noticed by Henle, and more fully described by Pettigrew, proceed from the right side of the cardia and spread over the front and back of the great cul-de-sac : these are in part continuous with the circular layer. The oblique fibres are best seen from the inside of the stomach, after removing the mucous membrane. In this, as in the circular layer of fibres, Pettigrew believes the figure-of-8 arrangement to prevail. (From unpublished Notes of Researches on the Muscular Fibres of the Stomach, by James Pettigrew, M.D.)

Fig. 581.

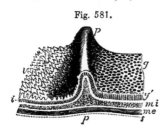

Fig. 581.—DIAGRAMMATIC VIEW IN PERSPECTIVE OF A PORTION OF THE COATS OF THE STOMACH AND DUODENUM, INCLUDING THE PYLORUS. $\frac{2}{1}$

g, the alveolar surface of the stomachal mucous membrane; g', section of the mucous membrane with the pyloric gastric glands; v, the villous surface of the mucous membrane of the duodenum ; i, section of the same with the intestinal glands or crypts of Lieberkühn ; p p, the ridge of the pyloric valve, with a section of its component parts ; m i, deep or circular layer of muscular fibres : these are seen in the section to form a part of the pyloric valve ; m e, external or longitudinal layer of muscular fibres ; s, the serous covering.

The *areolar* or submucous coat of the stomach is a distinct layer placed between the muscular and mucous coats, and connected with both : it consists essentially of a dense filamentous areolar tissue, in which occasional fat-cells may be found ; and it is the seat of division and passage of the larger branches of the blood-vessels.

The internal or *mucous* coat is a smooth, soft, rather thick and pulpy membrane, which has generally a somewhat pink hue owing to the blood in its capillary vessels, but which, after it has been well washed, is of a greyish white or pale straw colour. In some cases, however, it presents this pale aspect without any previous washing. In infancy the vascular redness is more marked, the surface having then a rosy hue, but it becomes paler in childhood, and in aged persons is often of an ash-grey colour. During digestion its vessels become congested, and when examined in that condition it is always of a much brighter pink than at other times.

After death a few hours often suffice to change its colour to a dirty brown tint, mottled and streaked in some cases with dull red lines, corresponding with the course of the veins. This alteration is owing to the exudation of the colouring matter of the blood, and is especially met with in old subjects, in whom the mucous membrane is always thin. In acute inflammation, or after the introduction of irritating substances or of strong acrid poisons, it becomes of a bright red, either all over or in spots, patches or streaks of variable sizes. Corrosive poisons, the gastric juice, and sometimes regurgitating bile, may stain it variously, black, brown, yellow, or green ; and the effect of chronic inflammation is to leave the membrane of a slate-grey colour. Independently of all these modifying circumstances connected with the stomach itself, as was pointed out by Yelloly and others, the colour of the gastric mucous surface is liable to be influenced by causes of a more general nature. Thus, it has been found that in cases of obstructed venous circulation, as when death occurs from hanging or from drowning, and also in certain diseases of the heart, the internal surface of the stomach is reddened to a greater or less extent ; but the amount of vascularity may vary from circumstances which are not well understood, and may be found greatly increased in cases in which none of those now named exist. **Trans. of** Med. Chir. Soc., vol. iv. p. 371.

The gastric mucous membrane is thickest in the pyloric portion of the stomach, and thinnest in the great cul-de-sac. It always becomes thinner in old age.

The outer or *adherent* surface of the mucous membrane is connected with the muscular coat by means of the intervening submucous layer so loosely as to allow of considerable movement or displacement. In consequence of this, and of the great extent and want of elasticity of the mucous membrane as compared with the other coats, the internal surface of the stomach, when that organ is in a contracted state, is thrown into numerous convoluted ridges, *rugæ*, which are produced by the wrinkling of the mucous, together with the areolar coat, and are entirely obliterated by distension of the stomach. These folds of the mucous coat are most evident along the great curvature, and have a general longitudinal direction.

On examining the gastric mucous membrane closely with the aid of a simple lens, it is seen to be marked throughout, but more plainly towards the pyloric extremity, with small depressions or cells named *alveoli*, which have a polygonal figure, and vary from about $\frac{1}{200}$th to $\frac{1}{100}$th of an inch across, being larger and more oblong near the pylorus.

Fig. 582.—Enlarged View of a small part of the Surface of the Mucous Membrane of the Stomach (from Ecker). $\frac{30}{1}$

Fig. 582.

This specimen shows the shallow alveoli, in each of which the smaller dark spots indicate the orifices of a variable number of the gastric glands.

Towards the pyloric region of the stomach, where the mucous membrane is thicker than elsewhere, the margins of these alveoli are elevated into pointed processes or fringes, which may be compared to rudimentary *villi*, the

Fig. 583.—Vertical Transverse Section of the Coats of a Pig's Stomach (from Kölliker). $\frac{30}{1}$

Fig. 583.

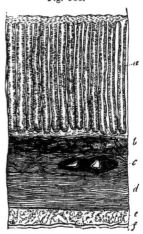

a, gastric glands; *b*, muscular layer of the mucous membrane; *c*, submucous or areolar coat; *d*, circular muscular layer; *e*, longitudinal muscular layer; *f*, serous coat.

perfect forms of those appendages existing only in the small intestine, and making their appearance in the duodenum, immediately beyond the pylorus.

At the bottom of the alveoli, and also in the intervals between them, are seen small round apertures, which are the mouths of minute tubes, placed perpendicularly to the surface, closed at their attached or deep extremity, which rests on the submucous areolar tissue, and opening at the other on the inner surface of the stomach. On making a vertical section of the membrane, and submitting it to microscopic examination, it is seen to consist almost entirely of these small *tubuli*, arranged close to and parallel with each other. Their diameter varies from $\frac{1}{500}$th to $\frac{1}{300}$th of an inch, and their length from $\frac{1}{60}$th to $\frac{1}{20}$th of an inch. At the cardiac end of the stomach, where the membrane is thinnest, they are shorter and are simply tubular; but, in approaching the pyloric portion, they gradually become longer

Fig. 584.

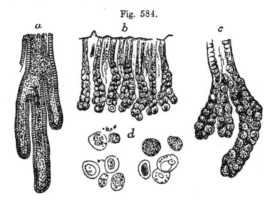

Fig. 584.—THE GASTRIC GLANDS OF THE HUMAN STOMACH (magnified).

a, the deep part of a pyloric gastric gland (from Kölliker); the cylindrical epithelium is traceable to the cæcal extremities.

b, *c*, and *d*, cardiac gastric glands (from Allen Thomson). *b*, vertical section of a small portion of the mucous membrane with the glands magnified 30 diameters; *c*, deeper part of one of the glands, magnified 65 diameters, showing a slight division of the tubes, and a sacculated appearance produced by the large glandular cells within them; the change of the prismatic epithelium into spherical gland-cells within the tube is apparent; *d*, cellular elements of the cardiac glands, magnified 250 diameters.

and assume a more complicated form, for, though quite straight near their orifices, they are curved, clavate, or irregularly sacculated towards their

Fig. 585.

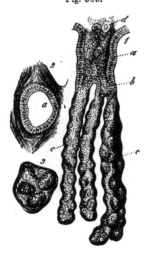

Fig. 585.—PEPTIC GASTRIC GLANDS FROM THE DOG'S STOMACH, MAGNIFIED (from Frey).

1, longitudinal view; *a*, the main duct; *b*, one of the first tubular divisions of the gland; *c*, the single tubes occupied by the gastric or peptic cells; *d*, some of the cells pressed out; 2, cross section of the main duct, showing the epithelial lining; 3, cross section of the simple tubes.

deep or closed extremity. Some are cleft, first into two or three, and finally into six or eight tubular branches. These characters are most perfect near the pylorus. They exist at all parts of the stomach, even where the alveoli are indistinct or absent; they contain a colourless fluid, with granular matter, and appear to be the secreting organs of the gastric mucus and the gastric juice. The tubuli, generally, are formed of a simple homogeneous membrane; fusiform cells supposed to be muscular lie between them on their contiguous or attached surfaces ; and their inner surface is lined with cells. At the pyloric end of the stomach these cells appear to be entirely lined with a simple layer of columnar epithelium ; but in other portions of the organ, only the upper fourth of the tubuli is occupied by epithelium of that character ; the lower three-fourths containing finely granular nucleated cells, which are polygonal or oval in form, are much larger than the columnar, and do not form a stratum on the surface, but completely fill the cavity : these have been termed peptic cells. It has been supposed that only those glands which possess the last-mentioned form of epithelium secrete the gastric juice, and they have accordingly been named *peptic glands*, and distinguished from the *mucous glands*, in which the epithelium is columnar throughout.

A marked distinction has been made out by various observers between peptic and mucous glands of the stomach in the lower animals. Not only have their anatomical characters been found to be different, but likewise their physiological properties, as it has been ascertained that the gastric secretion only possesses its peculiar solvent properties when proceeding from those parts of the stomach which contain glands of the peptic kind. An abrupt separation, however, between the two varieties of gland does not appear to exist in the human subject. (Henle, Syst. Anat. d. Mensch., vol. ii., p. 158, where also other works are referred to.)

Lenticular follicles, similar to those of Peyer's glands, are found in the mucous membrane of the stomach, sometimes studding the greater part of its surface, and giving occasionally a granular or mammillated appearance to it. They are found in greater or less numbers all over the stomach, but are most numerous towards the pylorus. They are best seen in the stomachs of infants and children. Around the cardiac orifice they assume the character of multilocular crypts. They are more frequently found open than shut, the membrane which covers them being extremely delicate. (Allen Thomson, in Goodsir's Annals, i. p. 36.)

A distinct but delicate epithelium exists all over the stomach, covering the margins and floors of the alveoli, and lining the tubuli also. It belongs for the most part to the columnar variety, alternating in some parts with the squamous, which is composed of very minute polygonal scales.

In animals, there is a more or less distinct layer of muscular fibres in intimate relation with the simple basement-membrane. These fibres are of the plain or unstriped variety, and are quite distinct from those which constitute the true muscular coat, being separated from them by the submucous areolar layer.

Vessels and nerves.—The stomach is a highly vascular organ. Its *arterial* branches, derived from all three divisions of the cœliac axis, reach the stomach between the folds of the peritoneum, and form, by anastomosing together, two principal arterial arches, which are placed along its two curvatures. After ramifying between the several coats and supplying them with blood, and especially after dividing into very small vessels in the submucous areolar tunic, the ultimate arterial branches enter the mucous membrane, and ramifying freely, pass to its surface between the tubuli ; here they form a plexus of very fine capillaries upon the wells of the tubules ; and from this plexus larger vessels pass into a coarser capillary network upon the hexagonal borders of the alveoli. The *veins*, corresponding with the arteries, arise from the latter network (Brinton, "Stomach and Intestine" in Cyclop. of Anat.), and, after forming a wide venous plexus in the submucous tissue, return the residual blood into the splenic and superior mesenteric veins, and also directly into the vena portæ. By the breaking up of the arteries into capillaries on the walls of the glands, these are furnished with pure blood for the elaboration of their secretion ; while it is the blood from which that secretion has been drawn which passes on to the capillaries of the free surface, and has added to it whatever materials may be taken into the circulation from the contents of the stomach.

The *absorbents* are very numerous; arising from a very fine superficial plexus immediately underlying the tubular glands, they form a coarser deeply situated network, between the areolar and muscular coats; the vessels proceeding from this network pierce the muscular coats, then follow the direction of the blood-vessels beneath the peritoneal investment, and traverse lymphatic glands found along the two curvatures of the stomach. No trace of lymphatics has been found between the tubuli, therefore the whole depth of the secreting structure intervenes between the layer of lymphatics and the contents of the stomach, whereas capillary blood-vessels are distributed close to the surface; an arrangement which seems favourable for the interchange of material between the contents of the stomach and those of the blood-vessels rather than of the lymphatics.

The *nerves*, which are large, consist of the terminal branches of the two pneumo-gastric nerves, belonging to the cerebro-spinal system, and of offsets from the sympathetic system, derived from the solar plexus. Numerous small ganglia have been found by Remak and others on both the pneumo-gastric and sympathetic twigs. The nerves may be traced through the submucous coat, but no farther, as they then lose their tubular character, and cannot be distinguished from other tissues. (Kölliker.) The left pneumo-gastric nerve descends on the front, and the right upon the back of the stomach.

The pylorus.—While there is no special apparatus at the cardiac orifice of the stomach for closing the passage from the œsophagus, the opening at the pyloric end, leading from the stomach into the duodenum, is provided with a sphincter muscle. On looking into the pyloric end of the stomach, the mucous membrane is seen projecting in the form of a circular fold, called the *pylorus*, leaving a correspondingly narrow opening. Within this fold are circular muscular fibres, belonging to the general system of circular fibres of the alimentary canal, which are here accumulated in the form of a strong band, whilst the longitudinal muscular fibres and the peritoneal coat pass over the pyloric fold to the duodenum, and do not enter into its formation. Externally the pylorus may be easily felt, like a thickened ring, at the right end of the stomach. Internally its opening is usually circular, and less than half an inch across, so that it is the narrowest part of the whole alimentary canal. (See figures 581 and 586.)

Occasionally the orifice is oval, and it is often placed a little to one side. Sometimes the circular rim is imperfect, and there are found instead two crescentic folds, placed one above and the other below the passage (Huschke); and, lastly, there is occasionally but one such crescentic fold.

THE SMALL INTESTINE.

The *small intestine* reaches from the pylorus to the ileo-cæcal valve, at which it opens into the large intestine. It consists of a convoluted tube, measuring on an average about twenty feet in length in the healthy adult, and becoming gradually slightly narrower from its upper to its lower end. Its numerous convolutions occupy the middle regions of the abdomen, and are surrounded by the large intestine. They are connected with the back of the abdominal cavity, and are held in their position by a covering and fold of the peritoneum, named the mesentery, and by numerous blood-vessels and nerves.

The small intestine is arbitrarily divided into three portions, which have received different names; the first ten or twelve inches immediately succeeding to the stomach, and comprehending the widest and most fixed part of the tube, being called the *duodenum*, the upper two-fifths of the remainder being named the *jejunum*, and the lower three-fifths the *ileum*. There are no distinct lines of demarcation between these three parts, but there are certain peculiarities of connection and certain differences of internal structure to be observed in comparing the upper and lower ends of the entire tube, which will be pointed out after it has been described as a whole.

DUODENUM.—This is the shortest and widest part of the small intestine. In length it measures 10 or 12 inches, or nearly the breadth of *twelve* fingers ; hence its name.

It is the widest part of the small intestine, varying in diameter between an inch and a half and two inches. In its course it describes a single large curve somewhat resembling a horse-shoe, the convexity of which is turned towards the right, whilst the concavity embraces the head of the pancreas.

It has no mesentery, and is covered only partially by the peritoneum. Its muscular coat is comparatively thick, and its mucous membrane towards

Fig. 586.—VIEW OF THE DUODENUM FROM BEFORE (slightly altered from Luschka). ⅓

Fig. 586.

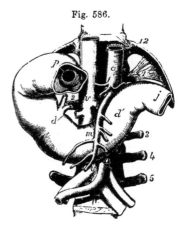

12, the twelfth dorsal vertebra and rib ; 1, 3, 4, 5, transverse processes of the first, third, fourth, and fifth left lumbar vertebræ ; 2, that of the second on the right side ; *a, a,* the abdominal aorta above the cœliac axis and near the bifurcation ; *m,* superior mesenteric artery ; *v, v,* the vena cava above the renal veins and near the bifurcation ; *p,* placed on the first part of the duodenum, points to the pyloric valve seen from the side next the stomach, of which a small part is left connected with the intestine ; *d,* on the descending or second part of the duodenum, indicates the termination of the common bile-duct and the pancreatic duct ; *d',* the third or oblique part of the duodenum ; *j,* the commencement of the jejunum.

the pylorus is the seat of the compound glands of Brunner, to be subsequently described. The common bile-duct and the pancreatic duct open into this part of the intestinal canal.

Three portions of the duodenum, differing from each other in their course and connections, are separately described by anatomists ; viz., the superior, descending, and transverse portions.

The first, or *superior* portion, which is between two and three inches long, commences at the pylorus, and passing upwards, backwards, and to the right side, reaches as far as beneath the neck of the gall-bladder, where the intestine bends suddenly downwards. The first portion of the duodenum is for the most part free, and invested both in front and behind by the peritoneum. Above, and in front of it, are the liver and gall-bladder, and it is commonly found stained by the exudation of bile from the latter a few hours after death. Behind it is the hepatic duct, with the blood-vessels passing up to the liver.

The second, or *descending* portion, commencing at the bend below the neck of the gall-bladder passes vertically downwards in front of the right kidney, as low as the second or third lumbar vertebra, where the bowel turns across to the left to form the third portion. This part of the

duodenum is invested by the peritoneum on its anterior surface only,—
the posterior surface being connected to the right kidney and the vertebral column by areolar tissue. In front is the transverse colon and meso-
colon, the upper layer of which is continuous with the peritoneal covering
of the duodenum. To the left is the head of the pancreas, which adapts
itself to the shape of the intestine on that side. The common bile-duct
descends behind the left border of this part of the duodenum, and, together with the pancreatic duct, which accompanies it for a short distance,
perforates the coats of the intestine obliquely near the lower part of its left
or concave border. In the interior of this part of the intestine the valvulæ conniventes appear numerously; and an eminence or papilla found
about four inches below the pylorus, on the inner and back part of the
intestine, marks the situation of the common orifice of the biliary and pancreatic ducts.

The third or *transverse* or oblique portion, somewhat the longest and narrowest, beginning on the right of the third lumbar vertebra, crosses in front
of the second obliquely from right to left, and, continuing to ascend obliquely
for an inch or more, ends in the jejunum at the left side of the vertebral
column. It is placed immediately behind the root of the transverse meso-
colon, and the commencement of the mesentery, as has been already described, and has the vena cava inferior and the aorta behind it. At its
termination it forms an abrupt angle with the commencement of the jejunum. This is due to its being maintained, at that point, in its position, by
a strong fibrous band descending from the left crus of the diaphragm and
the tissue round the cœliac axis. According to Treitz, muscular fibres
come from both these sources to this part of the duodenum. In subjects
in which the intestines are large and dilated, the curve of the duodenum
may descend to the level of the iliac crest, but, owing to the support given
by the band alluded to, its terminal extremity maintains a uniform position.
Close to this point the superior mesenteric vessels pass from beneath the
pancreas to enter the mesentery on the surface of the duodenum.

JEJUNUM AND ILEUM.—The *jejunum*, originally so called from its having
been supposed to be empty after death, follows the duodenum, and includes the upper two-fifths of the remainder of the small intestine, while
the succeeding three-fifths constitute the *ileum*, so named from its numerous coils or convolutions. Both the jejunum and the ileum are attached
and supported by the mesentery. The convolutions of the jejunum are
situated in part of the umbilical and left iliac regions of the abdomen;
while the ileum occupies part of the umbilical and right iliac regions,
together with the hypogastric, and descends into the pelvis, from which its
lower end, supported by the mesentery, which is here very short, ascends
obliquely to the right and somewhat backwards, over the corresponding
psoas muscle, and ends in the right iliac fossa, by opening into the inner
side of the commencement of the large intestine. There is no defined
limit between the jejunum and the ileum, but the character of the intestine gradually changes from its upper to its lower end, so that portions
of the two intestines, remote from each other, present certain well-
marked differences of structure, which may be here enumerated. Thus,
the jejunum is wider, and its coats are thicker; it is more vascular, and
therefore it has a deeper colour; its valvulæ conniventes are long, wide,
and numerous; its villi are well developed; and the patches of Peyer's
glands are smaller, less frequent, and sometimes confined to its lower
part. The ileum, on the other hand, is narrower; its coats are thin-

ner and paler; the valvulæ conniventes are small, and gradually disappear towards its lower end; the villi are shorter; and the groups of Peyer's glands are larger and more numerous. The diameter of the jejunum is about one inch and a half, that of the ileum about one inch and a quarter. A given length of the jejunum weighs more than the same of the ileum.

At a point in the lower part of the ileum it is not very uncommon to find a cul-de-sac or *diverticulum* given off from the main tube. The origin of these diverticula is explained by reference to the history of development, from which it appears that they arise in connection with the ductus vitello-intestinalis, uniting the intestine with the umbilical vesicle. They are not to be confounded with hernial protrusions of the mucous membrane, which may occur at any point. (See Meckel's Handbook of Anatomy, French edition, vol. ii. p. 431.)

Structure of the Small Intestine.

Structure.—The walls of the small intestine, like those of the stomach, are composed of four coats, viz., the serous, muscular, areolar, and mucous.

The external or *serous* coat, derived from the peritoneum, almost entirely surrounds the intestinal tube in the whole extent of the jejunum and ileum, leaving only a narrow interval along one border of the intestine, where it is reflected from it and becomes continuous with the two layers of the peritoneal duplicature named the mesentery. The line at which this reflexion takes place is named the *attached* or *mesenteric border* of the intestine. The opposite border and sides of the tube, which are covered by the peritoneum, are quite free and movable upon the adjacent parts. The upper part, however, of the small intestine, named the duodenum, is but partially covered by the peritoneum, as has been already more particularly described.

The *muscular* coat consists of two layers of fibres; an outer longitudinal, and an inner or circular set. The *longitudinal* fibres constitute an entire but comparatively thin layer, and are most obvious along the free border of the intestine. The *circular* layer is much thicker and more distinct; its fibres are placed closely together, and run in a circular direction around the bowel, but it does not appear that they individually form perfect rings.

This muscular tunic becomes gradually thinner towards the lower part of the small intestine. It is pale in colour, and is composed of plain muscular fibres. The progressive contraction of these fibres, commencing in any part of the intestine, and advancing in a downward direction, produces the peculiar *vermicular* or *peristaltic* movement by which the digestive mass is forced onwards through the canal. In this movement the circular fibres are mainly concerned; but the longitudinal fibres also aid in it; and those found along the free border of the intestine may have the effect of straightening or unfolding, as it were, its successive convolutions.

The *areolar* or *submucous* coat of the small intestine is a tolerably distinct and whitish layer, of a loose texture, which is connected more firmly with the mucous than with the muscular coat, between which two it is placed. By turning a portion of the intestine inside out, and then blowing forcibly into the cavity, the areolar tunic may be inflated, the air being driven into its areolar tissue, through the part at which the peritoneal investment is wanting. It supports the mucous membrane, and forms a layer of loose substance in which the vessels divide and subdivide into smaller branches, preparatory to entering the mucous tissue. It consists of filamentous areolar tissue, mixed with fine elastic fibres.

The internal or *mucous* coat is characterised by presenting all over its inner surface a finely flocculent or shaggy appearance, like the pile upon velvet, owing to its being covered with multitudes of minute processes, named *villi ;* hence it is also named the *villous* coat. It is one of the most vascular membranes in the whole body, and it is naturally of a reddish colour in the upper part of the small intestine, but becomes paler, and at the same time thinner towards the lower end. The mucous tissue contains beneath its basement membrane, a thin muscular layer, demonstrated easily in animals, but scarcely recognisable in man. It presents for consideration, 1, the large *folds* called *valvulæ conniventes ;* 2, the *villi* and *epithelium ;* 3, the *glands.*

1. *Valvulæ Conniventes.*—The folds and wrinkles found upon the inner surface of the œsophagus and stomach may be completely obliterated by full distension of those parts of the alimentary canal. In the lining membrane of the small intestine, however, there exist, besides such effaceable folds, other permanent ones, which cannot be obliterated, even when the tube is forcibly distended. These permanent folds are the *valvulæ conniventes,* or valves of Kerkring. They are crescentic projections of the mucous membrane, placed transversely to the course of the bowel, each of them reaching about one-half or two-thirds of the distance round the interior of the tube, and they follow closely one upon another along the intestine.

The largest of these valves are about two and a half inches long and one-third of an inch wide at the middle or broadest part; but the greater number are under these dimensions. Large and small valves are often found to alternate with each other. Some of them are bifurcated at one end, and others terminate abruptly, appearing as if suddenly cut off. Each valve consists of a fold of the mucous membrane, that is, of two layers placed back to back, united together by the submucous or areolar tissue. They contain no part of the circular and longitudinal muscular coats. Being extensions of the mucous membrane, they serve to increase the absorbent surface to which the food is exposed, and at the same time they contribute to delay its passage along the intestine.

There are no valvulæ conniventes quite at the commencement of the duodenum ; about an inch or somewhat more from the pylorus they begin to appear ; beyond the point at which the bile and pancreatic juice are poured into the duodenum they are very large, regularly crescentic in form, and placed so near to each other that the intervals between them are not greater than the breadth of one of the valves : they continue thus through the rest of the duodenum and along the upper half of the jejunum ; below that point they begin to get smaller and farther apart ; and finally, towards the middle of the ileum, having gradually become more and more irregular and indistinct, sometimes even acquiring a very oblique direction, they altogether disappear.

2. *Villi.*—The *villi,* peculiar to the small intestine, and giving to its internal surface the velvety or villous appearance already spoken of, are small, elongated, and highly vascular processes, which are found situated closely together on every part of the mucous membrane, over the valvulæ conniventes, as well as between them. They are best displayed by placing a piece of intestine, well cleansed from its mucus, under water, and examining it with a simple lens. The prevalent form of the villi is that of minute, flattened, bell-shaped membranous processes ; others are conical or cylindrical, or even clubbed, or filiform at the free extremity. A few are compound as if two or three villi were connected together at their base.

Their *length* varies from $\frac{1}{4}$th to $\frac{1}{3}$rd of a line, or even more ; and the broad flattened kinds are about $\frac{1}{6}$th or $\frac{1}{8}$th of a line wide, and $\frac{1}{20}$th or $\frac{1}{24}$th of a line thick. They are largest and most numerous in the duodenum and jejunum, and become gradually shorter, smaller, and fewer in number in the ileum. In the upper part of the small intestine Krause has estimated their number at from 50 to 90 in a square line; and in the lower part at from 40 to 70 in the same space : he calculates their total number to be at least four millions.

The *structure* of the villi is complicated : each consists of a prolongation of the proper mucous layer, covered by epithelium and enclosing blood-ves-

Fig. 587.—Magnified View of the Blood-vessels of the Intestinal Villi.

The drawing was taken from a preparation injected by Lieberkühn, and shows in each villus a small artery and vein with the intermediate capillary network.

Fig. 587.

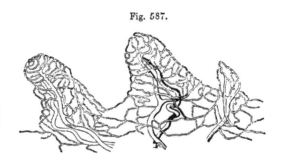

sels, one or more lacteal vessels, and fine muscular fibres, with a greater or less number of small granular corpuscles and fat-globules, of various sizes.

Fig. 588.—Injected Lacteal Vessels in the Villi of the Human Intestine.

A, two villi in which the lacteals are represented as filled with white substance and the blood-vessels with dark (from Teichmann) $\frac{100}{1}$: *a, b,* the lacteal vessels, single in one villus and double with cross loops in the other ; *c,* the horizontal lacteal vessels with which those of the villi communicate ; *d,* the blood-vessels, consisting of small arteries and veins with capillary network between.

B, injected lacteal (shaded dark) in a villus, showing an example not very common of a looped network *a,* which is connected by a single vessel with the horizontal lacteal vessel *b* : the preparation was made from the intestine of a young man who died suddenly while digestion was going on (from W. Krause). $\frac{80}{1}$

Fig. 588.

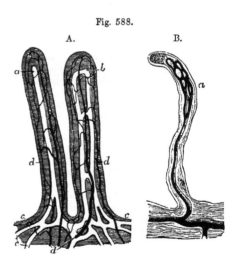

Nerves have not yet been demonstrated in the villi, though they are probably not wanting. Each villus receives one or more small *arterial* twigs, which divide, and form upon its surface, beneath the epithelium and limiting membrane, a fine capillary network, from which the blood is returned for the most part by a single *vein*.

The lacteal lies in the centre of the villus, and is in the smaller villi usually a single vessel, with a somewhat expanded extremity, and of considerably larger diameter than the capillaries of the blood-vessels around. According to the observations of Teichmann, there are never more than two intercommunicating lacteals in a single villus in the human subject; but both he and Frey find a copious network of them in the villi of the sheep. Considerable difference of opinion exists as to the nature of the wall of the lacteal in the villus, and even as to whether or not any wall exists, and this point must be considered as still undetermined. The epithelium of the villi is of the columnar kind ; the cell-wall is delicate, and the nucleus distinct. The nature of both the free and the attached extremities of the cells is involved

Fig. 589.

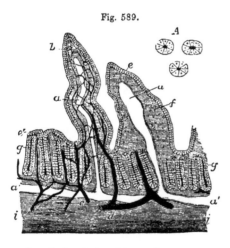

Fig. 589.—Vertical Section of the Intestinal Mucous Membrane of the Rabbit (slightly altered from Frey). $\frac{150}{1}$

Two villi are represented, in one of which the dilated lacteal alone is represented, in the other the blood-vessels and lacteal are both shown injected, the lacteal white, the blood-vessels dark : the section is carried through the tubular glands into the submucous tissue : *a*, the lacteal vessels of the villi ; *a'*, below the glands, the horizontal lacteal, which they join ; *b*, the capillary blood-vessels shown only in one of the villi ; *c*, a small artery ; *d*, a vein ; *e*, the epithelium covering the villi ; *f*, the substance of the villi, presenting interstices which contain lymph-cells ; *g*, tubular glands or crypts of Lieberkühn, some divided in the middle, others cut irregularly ; *i*, the submucous layer.

A, cross section of three tubular glands more highly magnified.

in some doubt. At the free extremity, they present to view a thick layer of substance with vertical striæ, which, on treatment with water, swells out and loses its striated appearance. This layer was first recognised by Kölliker and by Funke, who both consider the striæ to be minute perforating canals ; while Brettauer and Steinach, and likewise Henle, maintain that they are rods comparable with cilia. Brücke, previous to the discovery of the striated body, advanced the opinion that the epithelium-cells were altogether open at their free extremities, and that each communicated likewise with the interior of the villus by a foramen at the deep extremity. Brettauer and Steinach support Brücke's view, in respect that they consider the striated body as continuous with the cell-contents, and not with the cell-wall. With regard to the deep extremities of the epithelial cells, Heidenhain believes that he has observed them prolonged into fine threads, which communicate with branches of anastomosing connective-tissue-corpuscles, and considers that, by means of deep branches of these anastomosing cells opening into the cavity of the lacteal, a channel of communication is established between the lacteal and the surface of the villus. This view has met with some acceptance from its seeming to offer an explanation of the mode in which particles of oil are conveyed from the intestines into the lacteals ; but it cannot

Fig. 590.

A B

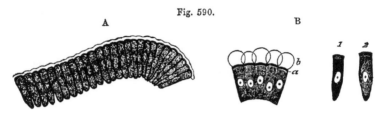

Fig. 590.—EPITHELIUM OF THE INTESTINAL VILLUS OF A RABBIT (from Kölliker). A, $\frac{300}{1}$; B, $\frac{350}{1}$.

A, series of the cylindrical epithelial cells separated from a villus ; a limiting or cuticular membrane or border is seen passing over the free ends of the cells.

B, some of the same cells treated with water; in 1 and 2, and at a, in the left hand series of cells, the striated or porous border is seen ; and at b, in the latter, pellucid drops of mucus which have escaped from the cells.

at present be considered as satisfactorily established. The muscular tissue within the villi was first discovered by Brücke: it consists of a thin stratum of smooth fibres disposed longitudinally round the commencement of the lacteals. Although not always discernible in man, these fibres are distinct in animals ; and in them, on being stimulated, they produce, according to Brücke, a very obvious retraction of the villi.

During digestion, the epithelial cells become turbid with minute oil drops in their interior, which obscure their nuclei. The tissue of the villus itself becomes turbid in like manner ; and clear globules may also be observed, both in the epithelial cells and deeper tissue, which, however, there seems reason to believe, are formed by the running together of smaller particles after death. Kölliker and Donders have both observed minute particles of oil in their passage through the striated body.

A full bibliography, on the subject of the villi, is given by Teichmann in his work "das Saugadersystem," (1861), pp. 77 et seq. ; and the questions at issue are fully discussed in Kölliker's Gewebelehre, 4th edition, and Henle's System. Anatomie. See also Frey, in Zeitsch. f. Wissensch. Zoologie, vol. xiii. Heidenhain's paper is in Moleschott s Untersuchungen z. Naturlehre, vol. iv. Peculiar epithelial cells with deeply hollowed cup-shaped extremities, have been pointed out by Henle, interspersed among the others. It is yet uncertain whether they are a distinct kind of cell, or only a peculiar condition of the ordinary sort.

3. *Glands.*—The glandular structures found in the mucous coat of the small intestine are the crypts or follicles of Lieberkühn, the solitary glands, the patches of Peyer's glands, and Brunner's glands, the last being peculiar to the duodenum.

The *crypts* of Lieberkühn, the smallest of these glandular structures, are found in every part of the small intestine, between the villi, and surrounding the larger glands. They consist of minute tubes, closed at their attached extremity, and placed more or less perpendicularly to the surface, upon

which they open by small orifices.   They appear to be analogous to the
tubuli of the stomach, but they are placed farther apart from each other, and
are sometimes bulged inferiorly, but are hardly ever divided.   Similar
tubules also occupy the whole mucous membrane of the large intestine.
The crypts of Lieberkühn vary in length from the $\frac{1}{30}$th to the $\frac{1}{10}$th of a
line, and their diameter is about $\frac{1}{50}$th of a line.   The walls of the tubes are
thin, and lined with a columnar epithelium : their contents are fluid and
transparent, with granules interspersed, and they never contain fat.   These
crypts are sometimes filled with a whitish substance, which most probably
consists chiefly of desquamated epithelium and mucus.

The *agminated glands*, or *glands of Peyer* (who discovered and described
them in 1677), are found in groups or patches, having an oblong figure, and

Fig. 591 A.

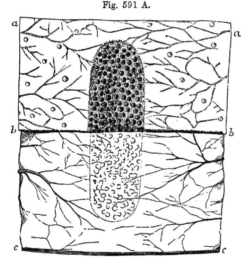

Fig. 591 A.—PATCH OF PEYER'S
GLANDS IN THE ILEUM.

This figure represents, somewhat diagrammatically, and of
the natural size, a patch of
Peyer's glands from near the
middle of the ileum of a young
subject : in the lower half of the
figure the mucous membrane and
the glands have been removed by
dissection, showing the impression left by the patch of glands
by the condensation of the submucous tissue : the piece of
intestine having been opened
along its mesenteric border,
the blood-vessels are seen advancing from the separated
margins towards the centre.

varying from half an inch
to two or even four inches
in length, and being about
half an inch, or rather
more, in width.   These
patches are placed length-
ways in the intestine at that part of the tube most distant from the mesentery ; and hence, to obtain the best view of them, the bowel should be
opened by an incision along its attached border.

The patches of Peyer's glands consist of groups of small, round, flattened
vesicles or capsules composed of a tolerably thick and firm wall of connective
tissue, usually filled with a whitish or rather greyish semi-fluid matter, consisting of round nucleated cells and free nuclei, and situated beneath the
mucous membrane, the surface of which is depressed into little shallow pits,
at or rather under the bottom of which the capsules are placed.   The intermediate surface of the membrane is beset with villi and Lieberkühn's crypts :
the villi are also sometimes found even over the capsules, and the crypts are
collected in circles around the capsules, but do not communicate with them.
Opposite to the patches of Peyer's glands, the mucous and areolar coats of
the intestine adhere more closely together than elsewhere, so that in those
situations it is impossible to inflate the areolar coat.   Fine blood-vessels are
distributed abundantly on the walls of the capsules, and give off still finer
capillary branches, which, supported by a delicate network of connective

tissue, spread through the cavity of each capsule among its semifluid contents, and are disposed principally in lines converging to the centre. In some subjects these small capsules are found almost empty, and then they are

Fig. 591 B.— ENLARGED VIEW OF A PART OF A PATCH OF PEYER'S GLANDS (from Boehm). $\frac{1}{9}$

Fig. 591 B.

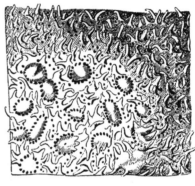

The shaded part of the figure shows the surface of the intestinal mucous membrane in the vicinity of the patch occupied by villi, and between them the orifices of the crypts of Lieberkühn; the lighter part of the figure, in which about a dozen of Peyer's vesicles may be seen, is also beset with villi, and in this part the crypts of Lieberkühn are arranged chiefly in circles round the vesicles.

difficult of detection. They are usually entirely closed; but the elder Krause observed that in the pig they were occasionally open, and a similar observation was made by Allen Thomson, not only in the pig, but in the human intestine also.

The lacteal plexuses, which are abundant in the whole extent of the intestine, are especially rich and composed of wide vessels, where they

Fig. 592 —TRANSVERSE SECTION OF INJECTED PEYER'S GLANDS (from Kölliker). $\frac{80}{1}$

Fig. 592.

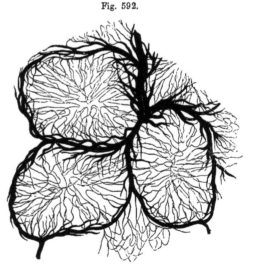

The drawing was taken from a preparation made by Frey: it represents the fine capillary network spreading from the surrounding blood-vessels into the interior of three Peyer's capsules from the intestine of the rabbit.

surround the closed follicles, so closely indeed that these may be said to be imbedded in them; but the lacteals do not penetrate the capsules as the capillary blood-vessels do.

It was formerly presumed without question that Peyer's and the other closed follicles in the alimentary tract constituted a peculiar capsular form of secreting glands; but since the discovery of capillaries in their interior, and of the rich supply of absorbents around them, it has been supposed that they might be more immediately connected with the lymphatic system. This, however, is by no means proved; for, although the interior of the capsules can no longer be compared with the cavities of open glands, there is not sufficient evidence to show whether their contents pass into the intestinal tube or into the lacteals, from which they are as completely separated by intervening texture. The facts which have been ascertained as to their minute structure, and the nature of their contents, seem to bring them rather under the description of vascular glands. It may farther be stated as a point of analogy between them and those structures, that the glands of Peyer belong chiefly to youth. After middle life they become more or less flaccid and empty, and have generally completely disappeared in advanced age.

Fig. 593.

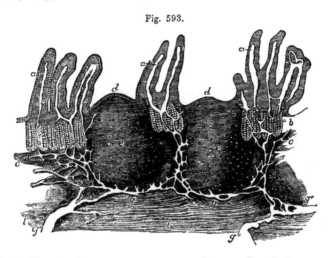

Fig. 593.—Vertical Section of a portion of a Patch of Peyer's Glands, with the Lacteal Vessels injected (from Frey). $\frac{32}{1}$

The specimen from which the drawing was made was obtained from the body of a man of twenty years of age who died suddenly from an injury, and is from the lower part of the ileum ; the epithelium, not represented in the original, is introduced diagrammatically in one part : a, villi, with their lacteals left white ; b, some of the tubular glands ; c, the muscular layer of the mucous membrane ; d, the cupola or projecting part of Peyer's vesicles ; e, their central cavities or substance ; f, the reticulated lacteal vessels occupying the "lymphoid" tissue between the vesicles, joined above by the lacteals from the villi and mucous surface, and passing below into g, the reticulated lacteals under the vesicles of Peyer, which pass into g', the larger lacteals of the submucous layer i.

The observations of Frey and His have further shown that in the intervals between the glands of Peyer and those of Lieberkühn, and also in the substance of the villi, the interstices of the retiform tissue (see Histology, p. lxxix), are everywhere occupied by granular cells of the size and appearance of lymph-cells, and very similar to those contained in the capsules of Peyer's glands.

In all, from twenty to thirty of these oblong patches may in general be found ; but in young persons dying in health, as many as forty-five have been observed. They are larger and placed at shorter distances from each other, in the lower part of the ileum ; but in the upper portion of that intestine and in the lower end of the jejunum, the patches occur less and less frequently, become smaller, and are of a nearly circular form ; they may, however, be discovered occasionally in the lower portion of the duodenum.

Still smaller irregularly shaped clusters of these capsules are found scattered throughout the intestine, and may be regarded as transitions to the next form of glands named *solitary*.

Fig. 594.—Lymphoid or Retiform Tissue of the Intestinal Mucous Membrane of the Sheep (from Frey). $\frac{400}{1}$

Fig. 594.

The figure represents a cross section of a small fragment of the mucous membrane, including one entire crypt of Lieberkühn and parts of several others: *a*, cavity of the tubular glands or crypts; *b*, one of the lining epithelial cells; *c*, the lymphoid or retiform spaces, of which some are empty, and others occupied by lymph-cells, as at *d*.

The *solitary glands* (glandulæ solitariæ) are soft, white, rounded, and slightly prominent bodies, about the size of a millet-seed, which are found scattered over the mucous membrane in every part of the small intestine. They are found on the mesenteric as well as on the free border, between and upon the valvulæ conniventes, and are rather more numerous in the lower portion of the bowel. These small glands have no orifice, but consist of closed vesicles or capsules, exactly resembling those forming the clusters of Peyer's glands, having rather thick but easily destructible walls, and usually

Fig. 595.—Solitary Vesicular Gland of the Small Intestine (from Bœhm). $\frac{40}{1}$

Fig. 595.

The lighter part of the figure represents the elevation produced by the gland ; on this a few villi are seen, and on the surrounding surface of the mucous membrane numerous villi and crypts of Lieberkühn.

containing in their interior an opaque, semifluid substance, which abounds in cells and fine granules. The free surface of the capsules, which is slightly elevated when they are full, is beset with the intestinal villi ; and, placed around them very irregularly, are seen the open mouths of the crypts of Lieberkühn.

Brunner's glands are small rounded compound glands, first pointed out by Brunner, which exist in the duodenum, where they are most numerous at the upper end, in general occupying thickly a space of some inches in extent from the pylorus. According to Huschke, a few of them are also found quite at the commencement of the jejunum. They are imbedded in the areolar tunic, and may be exposed by dissecting off the muscular coat from the outside of the intestine. They are true compound racemose glands, consisting of minute lobules, and containing branched ducts, which open upon the inner surface of the intestine. Their secretion is an alkaline mucus, in which there are no formed elements ; and it has no digestive action upon coagulated albumen. (Kölliker.)

Vessels and Nerves.—The branches of the mesenteric artery, having reached the attached border of the intestine, pass round its sides, dividing into numerous ramifications and frequently anastomosing at its free border. Most of the larger branches run immediately beneath the serous tunic; many pierce the muscular coat, supplying it with vessels as they pass, and having entered the submucous areolar

Fig. 596.

Fig. 596.—ENLARGED VIEW OF ONE OF BRUNNER'S GLANDS FROM THE HUMAN DUODENUM (from Frey).

The main duct is seen superiorly; its branches are elsewhere hidden by the bunches of opaque glandular vesicles.

layer, ramify in it, so as to form a close network, from which still smaller vessels pass on into the mucous coat, and terminate in the capillary network of the folds, villi, and glands of that membrane, which is the most vascular of all the intestinal tissues. The fine capillaries of the muscular coat are arranged in two layers of oblong meshes, which accompany and correspond in direction with the longitudinal and circular muscular fibres. The veins accompany the arteries.

The *absorbents* of the intestine may be conveniently distinguished as those of the mucous membrane and those of the muscular walls. Those of the mucous membrane form a copious plexus which pervades both the mucous and submucous layers, the largest vessels being those which are in the latter layer; but there is not, in the human subject at least, the same distinct division into two strata which has been found in the stomach (Teichmann). With regard to the absorbents of the muscular walls, it has been stated in a former part of this work (p. 491) that, according to the concurrent accounts of the various investigators of this subject, the absorbents of the intestine are in two strata, viz., those of the submucous layer already mentioned, and a subserous set, following principally a longitudinal direction beneath the peritoneum, and having only an interrupted communication with the other through intervening trunks; but more recently, a paper by Auerbach has appeared, in which it is stated as the result of transparent injections, that the only truly subperitoneal plexus which exists is confined to a strip in the immediate neighbourhood of the mesentery; that the longitudinal plexus seen by previous observers is really situated between the circular and longitudinal muscular coats; and that, besides this, there are likewise copious and close minute capillary plexuses, threading the whole thickness of the muscular walls, in complete continuity with the mucous absorbents, and throwing their contents into those larger vessels the position of which had been misunderstood. To the whole of this series of absorbents Auerbach gives the name of "interlaminar plexus." (Virchow's Archiv., vol. xxxiii., p. 340.)

The *nerves* of the small intestine are chiefly derived from the superior mesenteric plexus (see p. 702). This plexus is formed superiorly by nervous branches, of which those in the middle come from the cœliac plexus, and the lateral ones proceed directly from the semilunar ganglion. The plexus and plexiform branches into which it divides cling at first very closely to the larger divisions of the superior mesenteric artery, especially on their anterior surface, and, dividing similarly with the ramifications of the arteries, the branches of the nerves, retaining still a wide plexiform arrangement, pass onwards to the different parts of the intestine between the two folds of the mesentery, and finally, separating somewhat from the blood-vessels, reach the intestine in very numerous branches.

In regard to the nervous distribution in the coats of the intestine, two recent discoveries of considerable interest have been made. One of them, for which we are indebted to Auerbach, consists in the observation of a peculiar nervous plexus, rich in ganglion-cells, which is situated between the circular and longitudinal muscular fibres of the intestine, and to which he has therefore given the name of "plexus myentericus." For the other observation we are indebted to Meissner, who has dis-

Fig. 597 A. Fig. 597 B.

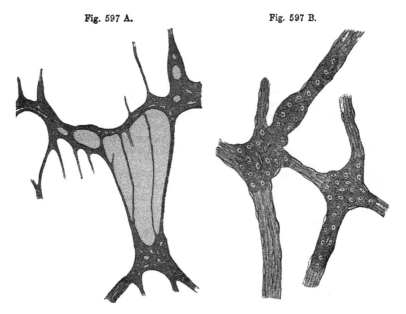

Fig. 597 A.—Nervous Plexus of Auerbach, from the Muscular Coat of a Child's Intestine (from Kölliker). $\frac{30}{1}$

The drawing represents three perforated ganglionic masses united by several nervous cords, of which the thickest is also perforated, forming the "plexus myentericus."

Fig. 597 B.—Small Portion of Meissner's Submucous Nervous Plexus from the Intestine of a Child (from Kölliker). $\frac{350}{1}$

Two ganglia are represented, of which the cells are seen spreading into the nerve-twigs connected with the ganglia: the fusiform particles in the nerve-twigs are small connective tissue corpuscles.

covered a second richly gangliated plexus of nerves situated in the submucous layer, and which is found to communicate freely with the plexus myentericus of Auerbach by means of the larger branches. Both plexuses extend through the whole length of the intestine, from the pylorus to the anus. (Kölliker, Op. cit., pp. 430 and 432.)

THE LARGE INTESTINE.

The large intestine extends from the termination of the ileum to the anus. It is divided into the cæcum (including the vermiform appendix), the colon, and the rectum; and the colon is again subdivided, according to its direction, into four parts, called the ascending, transverse, and descending colon, and the sigmoid flexure.

The length of the large intestine is usually about five or six feet; being about one-fifth of the whole length of the intestinal canal. Its diameter, which greatly exceeds that of the small intestine, varies at different points from two inches and a half to about one inch and a half. It diminishes gradually from its commencement at the cæcum to its termination at the anus; excepting that there is a well-marked dilatation of the rectum just above its lower end.

In outward form, the greater part of the large intestine differs remarkably from the small intestine; for, instead of constituting an even cylindrical tube, its surface is thrown into numerous sacculi, marked off from each other by intervening constrictions, and arranged in three longitudinal rows, separated by three strong flat bands of longitudinal muscular fibres. This sacculated structure is not found in the rectum.

For the sake of convenience, the description of the rectum will be reserved till that of the rest of the great intestine is completed.

The Cæcum. The intestinum cæcum, or caput cæcum coli, is that part of the large intestine which is situated below the entrance of the ileum. Its length is about two inches and a half, and its diameter nearly the same: it is the widest part of the large intestine.

The cæcum is situated in the right iliac fossa, immediately behind the anterior wall of the abdomen. It is covered by the peritoneum in front, below, and at the sides: but behind it is usually destitute of peritoneal covering, and is attached by areolar tissue to the fascia covering the right iliacus muscle. In this case the cæcum is comparatively fixed; but in other instances the peritoneum surrounds it almost entirely, and forms a duplicature behind it, called *meso-cæcum.*

Proceeding from the inner and back part of the cæcum, at its lower end, is a narrow, round, and tapering portion of the intestine, named the *appendix cæci,* or *appendix vermiformis.* The width of this process is usually about that of a large quill or rather more, and its length varies from three to six inches, these dimensions differing much in different cases. Its general direction is upwards and inwards behind the cæcum; and after describing a few slight turns it ends in a blunt point. It is retained in its position by a small fold of peritoneum, which forms its mesentery. The cæcal appendix is hollow as far as its extremity: and its cavity communicates with that of the cæcum by a small orifice, sometimes guarded by a valvular fold of mucous membrane.

This appendix is peculiar, as far as is known, to man and certain of the higher apes, and to the wombat; but in some animals, as in the rabbit and hare, the distal part of the cæcum, being diminished in diameter and highly glandular, may represent a condition of the appendix.

Ileo-cæcal or *ileo-colic valve.*—The lower part of the small intestine, ascending from left to right, and from before backwards, enters the commencement of the large intestine, with a considerable degree of obliquity, about two inches and a half from the bottom of the cæcum, and opposite the junction of the latter with the ascending colon. The opening leading from the ileum into the large intestine is guarded by a valve composed of two segments or folds. This is the *ileo-cæcal* or *ileo-colic valve:* it is also called the valve of Bauhin and the valve of Tulpius, though Fallopius had described it before either of those anatomists.

The entrance between the two segments of the valve is a narrow elongated aperture, lying nearly transverse to the direction of the great intestine. The anterior end of this aperture, which is turned forwards and slightly to the left, is rounded, but the posterior end is narrow and pointed. It is bounded above and below by two prominent semilunar folds, which project

inwards towards the cæcum and colon. The lower fold is the larger of
the two; the upper is placed more horizontally. At each end of the
aperture these folds coalesce, and are then prolonged as a single ridge

Fig. 598.—View of the Ileo-colic Valve
    from the Large Intestine. ½

Fig. 598.

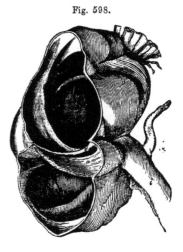

The figure shows the lowest part of the
ileum, *i*, joining the cæcum, *c*, and the
ascending colon, *o*, which have been opened
anteriorly so as to display the ileo-colic
valve ; *a*, the lower, and *c*, the upper segment of the valve.

for a short distance round the cavity
of the intestine, forming the *fræna*
or *retinacula* of the valve. The opposed surfaces of the marginal folds
which look towards the ileum, and
are continuous with its mucous surface, are covered like it with villi ;
while their other surfaces, turned toward the large intestine, are smooth
and destitute of villi. When the
cæcum is distended, the fræna of the
valve are stretched, and the marginal folds brought into apposition,
so as completely to close the aperture and prevent any reflux into the ileum,
while at the same time no hindrance is offered to the passage of additional
matters from thence into the great intestine.

Each segment of the valve consists of two layers of mucous membrane,
continuous with each other along the free margin, and including between
them, besides the submucous areolar tissue, a number of muscular fibres, continued from the circular fibres of the ileum and from those of the large intestine also. The longitudinal muscular fibres, and the peritoneal coat take no
part in the formation of the valve, but are stretched across it uninterruptedly
from one intestine to the other.

The ASCENDING COLON, situated in the right lumbar and hypochondriac
regions, commencing at the cæcum opposite to the ileo-cæcal valve, ascends
vertically to the under surface of the liver, near the gall-bladder, where it
proceeds forwards and then turns abruptly to the left, forming what is
named the *hepatic* flexure of the colon. The ascending colon is smaller than
the cæcum, but larger than the transverse colon. It is overlaid in front by
some convolutions of the ileum, and is bound down firmly by the peritoneum,
which passes over its anterior surface and its sides, and generally leaves an
interval in which its posterior surface is connected by areolar tissue with the
fascia covering the quadratus lumborum muscle, and with the front of the
right kidney. In some cases, however, the peritoneum passes nearly round
it, and forms a distinct though very short right meso-colon.

The TRANSVERSE COLON passes across from the right hypochondrium,
through the upper part of the umbilical region, into the left hypochondrium.
Sometimes it is found as low as the umbilicus or even lower. At each
extremity it is situated deeply towards the back part of the abdominal
cavity, but in the middle it curves forwards, and lies close to the anterior
wall of the abdomen. Hence it describes an arch, the concavity of which is

turned towards the vertebral column ; and it has accordingly been named
the *arch* of the colon.

Above, the transverse colon is in contact with the under surface of the
liver, the gall-bladder, the great curvature of the stomach, and the lower
end of the spleen.   Below it are the convolutions of the small intestine, the
third portion of the duodenum being behind it.   It is invested behind by the
general peritoneum, and in front it adheres to the sac of the omentum.

The DESCENDING COLON is continuous with the left extremity of the transverse colon by a sudden bend named the *splenic flexure*.   It then descends
almost perpendicularly through the left hypochondriac and lumbar regions to
the left iliac fossa, where it ends in the sigmoid flexure.   The peritoneum
affords a covering to it only in front and at the sides, whilst behind it is
connected by areolar tissue to the left crus of the diaphragm, the quadratus
lumborum, and the left kidney.   It is usually concealed behind some convolutions of the jejunum.

The SIGMOID FLEXURE of the colon, situated in the left iliac fossa, consists
of a double bending of the intestine upon itself in the form of the letter S,
immediately before it becomes continuous with the rectum at the margin of
the pelvis opposite to the left sacro-iliac articulation.   It is attached by a
distinct meso-colon to the iliac fossa, and is very movable.   It is placed immediately behind the anterior parietes, or is concealed only by a few turns
of the small intestine.   The sigmoid flexure is the narrowest part of the
colon.

Structure of the large intestine.—The walls of the large intestine consist
of four coats, like those of the stomach and small intestine, namely, the
serous, muscular, areolar, and mucous.

The *serous* and *areolar* coats require no further description here.

The *muscular* coat, like that of the other parts of the intestinal canal, con-

Fig. 599.

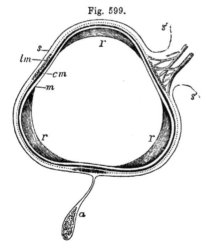

Fig. 599.—OUTLINE SKETCH OF A SEC-
TION OF THE ASCENDING COLON. ¾

s, the serous or peritoneal covering ;
s', *s'*, reflection of this at the attached
border forming a short wide mesentery,
between the folds of which the blood-
vessels are seen passing to the colon ; *a*,
one of the appendices epiploicæ hanging
from the inner border ; *l m*, indicates at
the free border one of the three bands
formed by the thickening of the longitudinal muscular coat; the dotted line
continued from the margins of these
bands represents the remainder of the
longitudinal muscular coat, and the
thick line within it, marked *c m*, represents the circular muscular layer ; *m*,
the mucous membrane at the flattened
part ; *r*, the crescentic bands or indentations which divide the sacculi.

sists of external longitudinal and
internal circular fibres.   The *longitudinal* fibres, though found in a
certain amount all around the intestine, are, in the cæcum and colon,
principally collected into three remarkable flat longitudinal bands.   These
bands, sometimes called the ligaments of the colon, are about half an

inch wide, and half a line thick ; they commence upon the extremity of the cæcum, at the attachment of the vermiform appendix, and may be traced along the whole length of the colon as far as the commencement of the rectum, where they spread out, so as to surround that part of the intestinal tube with a continuous layer of longitudinal muscular fibres. One of these bands, named the *posterior*, is placed along the attached border of the intestine ; another corresponds with its *anterior* border, and, in the transverse colon, is situated at the attachment of the great omentum ; whilst the third band (*lateral*) is found along the free side of the intestine, that is, on the inner border of the ascending and descending colon, and on the under border of the transverse colon. It is along the course of .this. third band that the appendices epiploicæ are. most of them attached. Measured from end to end, these three bands are shorter than the intervening parts of the tube ; and the latter are thus thrown into the sacculi already mentioned : accordingly, when the bands are removed by dissection, the sacculi are entirely effaced, and the colon, elongating considerably, assumes the cylindrical form. The transverse constrictions seen on the exterior of the intestine, between the sacculi, appear on the inside of the intestine as sharp ridges separating the cells, and are composed of all its coats. In the vermiform appendix the longitudinal muscular fibres constitute an uniform layer.

The *circular* muscular fibres form only a thin layer over the general surface of the cæcum and colon, but are accumulated in larger numbers between the sacculi. In the rectum, especially towards its lower part, the circular fibres form a very thick and powerful muscular layer.

Fig. 600.

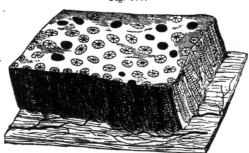

Fig. 600.—SEMI-DIAGRAMMATIC VIEW OF A SMALL PORTION OF THE MUCOUS MEMBRANE OF THE COLON. $\frac{60}{1}$

A small portion of the mucous membrane cut perpendicularly at the edges is shown in perspective ; on the surface are seen the orifices of the crypts of Lieberkühn or tubular glands, the most of them lined by their columnar epithelium, a few divested of it and thus appearing larger ; along the sides the tubular glands are seen more or less equally divided by the section ; these are resting on a wider portion of the submucous tissue, from which the blood-vessels are in a part represented as passing into the spaces between the glands.

The *mucous* membrane differs from the lining membrane of the small intestine in having no folds, like the valvulæ conniventes, as also in being quite smooth and destitute of villi. Viewed with a lens, its surface is seen to be marked all over by the orifices of numerous tubuli, resembling those of the stomach and the crypts of the small intestine. These follicles are arranged perpendicularly to the surface of the membrane ; they are longer and more numerous, and are placed more closely together and at more regular intervals than those of the small intestine. Their orifices are circular, and, when widened by the loss of their epithelial lining, they give the mucous membrane a cribriform aspect.

Besides these, there are scattered over the surface of the whole large intestine numerous *closed follicles*, similar to the solitary glands of the small intestine, but marked by a depression passing down to them between the surrounding tubules (Kölliker). They are most abundant in the cæcum and in its vermiform appendix ; being placed closely all over the latter.

The epithelium, which covers the general surface of the mucous membrane, and lines the tubuli and follicles, is of the columnar kind.

Vessels and Nerves.—In the great intestine of the rabbit, Frey figures the same arrangement of capillary plexuses and venous radicles as has been described in the stomach. He finds also in the rabbit clavate lacteals in rudimentary villi. (Zeitsch. f. Wissensch. Zoologie, vol. xii.) ; but Teichmann's injections in the human subject show no absorbents more superficial than the bases of the tubular follicles.

Nervous plexuses similar to those of the small intestine have also been found in the walls of the large intestine.

THE RECTUM.

The lowest portion of the large intestine, named the *rectum*, extends from the sigmoid flexure of the colon to the anus, and is situated entirely within the true pelvis, in its back part.

Commencing opposite to the left sacro-iliac articulation, it is directed at first obliquely downwards, and from left to right, to gain the middle line of the sacrum. It then changes its direction, and curves forwards in front of the lower part of the sacrum and the coccyx, and behind the bladder, vesiculæ seminales and prostate in the male, and at the back of the cervix uteri and vagina in the female. Opposite to the prostate it makes another turn, and inclines downwards and backwards to reach the anus. The *intestinum rectum*, therefore, so called from its original description being derived from animals, is far from being straight in the human subject. Seen from the front, the upper part of the rectum presents a lateral inclination from the left to the median line of the pelvis, sometimes passing beyond the middle to the right ; and when viewed from the side it offers two curves, one corresponding with the hollow front of the sacrum and coccyx, and the other at the lower end of the bowel, forming a shorter turn in the opposite direction.

Unlike the rest of the large intestine, the rectum is not sacculated, but is smooth and cylindrical ; and it has no separate longitudinal bands upon it. It is about six or eight inches in length ; and is rather narrower than the sigmoid flexure at its upper end, but becomes dilated into a large ampulla or reservoir, immediately above the anus.

The upper part of the rectum is in contact in front with the back of the bladder (or uterus in the female), unless some convolutions of the small intestine happen to descend into the interval between them. This part is surrounded by peritoneum, which attaches it behind to the sacrum by a duplicature named the *meso-rectum*. Lower down, the peritoneum covers the intestine in front and at the sides, and at last its anterior surface only ;

still lower, it quits the intestine altogether, and is reflected forwards to ascend upon the back of the bladder in the male, and of the upper part of the vagina and the uterus in the female. In passing from the rectum to the bladder, the peritoneum forms a cul-de-sac, or recto-vesical pouch, which extends downwards between the intestine and the bladder to within an inch or more from the base of the prostate, and is bounded on the sides by two lunated folds of the serous membrane.

Fig. 601.

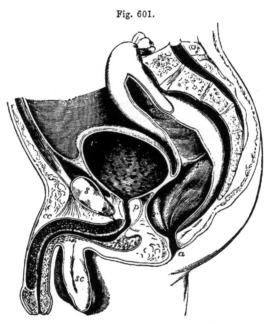

Fig. 601.—Vertical Section of the Pelvis and its Viscera in the Male (from Houston). ⅓

This figure is introduced to illustrate the form, position, and relations of the rectum; it also shows the bladder and urethra with the pelvic inflection of the peritoneum over these viscera: r, r, r, the upper and middle parts of the rectum, and at the middle letter the fold separating the two; $r\ a$, the lower or anal portion; v, the upper part of the urinary bladder; v', the base at the place where it rests more immediately on the rectum; p, the prostate gland and prostatic portion of the urethra; b, the bulb; $c\ c$, the corpus cavernosum penis and suspensory ligament; $s\ c$, the divided tissue within the scrotum.

Below the point where the peritoneum ceases to cover it, the rectum is connected to surrounding parts by areolar tissue, which is mostly loaded with fat. In this way it is attached behind to the front of the sacrum and the coccyx, and at the sides to the coccygei and levatores ani muscles. In front, it is in immediate connection with a triangular portion of the base of the bladder; on each side of this, with the vesiculæ seminales; and farther forwards, with the under surface of the prostate. Below the prostate, where the rectum turns downwards to reach the anus, it becomes invested by the fibres of the internal sphincter, and embraced by the levatores ani muscles, by which it is supported. Lastly, at its termination it is surrounded by the external sphincter ani muscle. In the female, the lower portion of the rectum is firmly connected with the back of the vagina.

Structure. —The rectum differs in some respects from the rest of the large intestine, in the structure of both its muscular and its mucous coats.

The muscular coat is very thick : the external or longitudinal fibres form a uniform layer round it, and cease near the lower end of the intestine ; the internal or circular fibres, on the contrary, become more numerous in that situation, where they form what is named the internal sphincter muscle. The longitudinal fibres are paler than the circular fibres, but both layers become darker and redder towards the termination of the bowel.

The mucous membrane of the rectum is thicker, redder, and more vascular than that of the colon ; and it moves freely upon the muscular coats ; —in that respect resembling the lining membrane of the œsophagus. It presents numerous folds of different sizes, and running in various directions, nearly all of which are effaced by the distension of the bowel. Near the anus these folds are principally longitudinal, and seem to depend on the contraction of the sphincter muscles outside the loosely connected mucous membrane. The larger of these folds were named by Morgagni the *columns* of the rectum (*columnæ recti*). Treitz states that these columns consist of longitudinal muscular fibres, which terminate both superiorly and inferiorly in elastic tissue. Higher up in the intestine, the chief folds are transverse or oblique. Three prominent folds, larger than the rest, being half an inch or more in depth, and having an oblique direction in the interior of the rectum, have been pointed out specially by Houston. One of these projects backwards from the upper and fore part of the rectum, opposite the prostate gland ; another is placed higher up, at the side of the bowel ; and the third still higher. From the position and projection of these folds, they may more or less impede the introduction of instruments. (Houston, Dublin Hospital Reports, vol. v.)

Vessels and Nerves.—The *arteries* of the rectum spring from three sources, viz. the superior hæmorrhoidal branches from the inferior mesenteric; the middle hæmorrhoidal branches from the internal iliac directly or indirectly; and, lastly, the external or inferior hæmorrhoidal branch from the pudic artery. The arrangement of the vessels is not the same throughout the rectum. Over the greater part the arteries penetrate the muscular coat at short intervals, and, at once dividing into small branches, form a network by their communication. Towards the lower end, for four or five inches, the arrangement differs. Here the vessels, having penetrated the muscular coat at different heights, assume a longitudinal direction, passing in parallel lines towards the end of the bowel. In their progress downwards they communicate with one another at intervals, and they are very freely connected near the orifice, where all the arteries join by transverse branches of considerable size. (Quain, Diseases of the Rectum.)

The *veins* are very numerous, and form a complex interlacement resembling that of the arteries just described, and named the hæmorrhoidal plexus. After following a longitudinal course upwards similar to that of the arteries which they accompany, they end partly in the internal iliac vein by branches which accompany the middle hæmorrhoidal artery, and partly in the inferior mesenteric vein. Hence, the blood from the rectum is returned in part into the vena cava, and in part into the portal system. (See Fig. 325.)

The *lymphatics* enter some glands placed in the hollow of the sacrum, or those of the lumbar series.

The *nerves* are very numerous, and are derived from both the cerebro-spinal and the sympathetic systems. The former consist of branches derived from the sacral plexus; and the latter, of offsets from the inferior mesenteric and hypogastric plexuses.

THE ANUS AND ITS MUSCLES.

The *anus*, or lower opening of the alimentary canal, is a dilatable orifice, surrounded internally by the mucous membrane, and externally by the skin, which two structures here become continuous with' and pass into each other. The skin around the borders of the anus, which is thrown into wrinkles or folds during the closed state of the orifice, is covered with numerous sensitive papillæ, and is provided with hairs and sebaceous follicles.

The lower end of the rectum and the margin of the anus are, moreover, embraced by certain muscles, which serve to support the bowel, and to close its anal orifice. These muscles, proceeding from within outwards, are, the internal sphincter, the levatores ani, the coccygei, and the external sphincter. The three last muscles have already been described (pp. 262, 263).

The *internal sphincter* muscle (sphincter ani internus) is a muscular ring or rather belt, surrounding the lower part of the rectum, an inch above the anus, and extending over about half an inch of the intestine. It is two lines thick, and is paler than the external sphincter. Its fibres are continuous above with the circular muscular fibres of the rectum, and, indeed, it consists merely of those fibres more numerously developed than elsewhere, and prolonged farther down than the external longitudinal fibres.

Kohlrausch describes a thin stratum of fibres between the mucous membrane and the internal sphincter, these fibres having a longitudinal direction. Henle thinks this is nothing more than the stratum of fibres belonging to the proper mucous coat ; but Kohlrausch gives it a distinct name, the sustentator tunicæ mucosæ. (Kohlrausch, Anat. and Phys. d. Beckenorgane, Leipzig. 1854.)

DEVELOPMENT OF THE ALIMENTARY CANAL AND PERITONEAL CAVITY.

It has been already casually stated (p. 15) that the epithelial lining of the alimentary canal is derived from the deepest of the three layers into which the germinal membrane divides, while the rest of its walls are derived from a part of the middle layer. To make this clear, it is necessary to state that, while those parts of the middle layer of the embryo which lie next to the chorda dorsalis form the dorsal plates from which the bones, nerves, and muscles of the trunk are derived, the lateral parts lying beyond form, as described by Remak, the *visceral plates,* which on each side divide into a deep and a superficial part, and, at the same time growing inwards, unite together on the ventral aspect of the chorda dorsalis, forming by their union the *mesial plate.* The superficial divisions of the two visceral plates, remaining in contact with the outer epithelial layer of the embryo, form the cutis ; the deep division is the *musculo-intestinal layer,* which forms the walls of the alimentary canal, with the exception of its epithelial lining ; and the space between the superficial and deep divisions is the common pleuro-peritoneal cavity, which becomes separated into the pleural and peritoneal cavities in a subsequent stage of development.

The alimentary canal commences in the form of a groove which opens towards the yelk-cavity of the ovum ; and the internal epithelial and musculo-intestinal layers in which this groove is formed, are continued round the yelk, constituting the walls of the vitelline sac. The open groove is soon changed into a tube at each end, but is left open in the middle upon the ventral aspect, and communicates at first by a wide aperture, but later by means of a tube, named the omphalo-enteric canal or vitelline duct, with the vitelline sac. This duct is soon obliterated, and the vitelline sac becomes the umbilical vesicle, which is thereafter connected for a time with the embryo only by a slender elongated pedicle, which enters at the umbilicus and is accompanied by the omphalo-mesenteric vessels ; this pedicle is finally atrophied and disappears.

The *alimentary canal,* when it first assumes the tubular form, constitutes a simple straight cylinder closed at each end, and placed along the front of the vertebral column, to which it is closely attached at each extremity, whilst in the middle of its course it is connected to the rest of the embryo by a median membranous fold, or rudimental

Fig. 602.

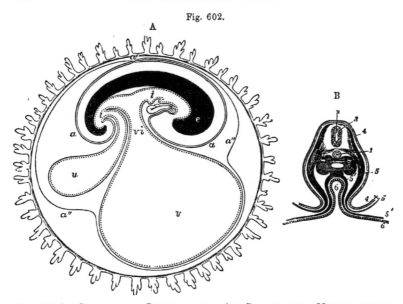

Fig. 602, A.—Diagrammatic Section showing the Relation in a Mammal and in Man between the Primitive Alimentary Canal and the Membranes of the Ovum.

The stage represented in this diagram corresponds to that of the fifteenth or seventeenth day in the human embryo, previous to the expansion of the allantois : *c*, the villous chorion ; *a*, the amnion ; *a'*, the place of convergence of the amnion and reflection of the false amnion *a" a"*, or outer or corneous layer ; *e*, the head and trunk of the embryo, comprising the primitive vertebræ and cerebro-spinal axis ; *i, i,* the simple alimentary canal in its upper and lower portions ; *v*, the yolk-sac or umbilical vesicle ; *v i*, the vitello-intestinal opening ; *u*, the allantois connected by a pedicle with the anal portion of the alimentary canal.

Fig. 602, B.—Transverse Section of the Body of an Embryo, with a Part of the adjacent Membranes, showing the Relation of the Alimentary Cavity to the Layers of the Germinal Membrane (from Remak and Kölliker). $\frac{15}{1}$

1, chorda dorsalis ; 2, 3, spinal marrow ; 4, cuticular layer, and within it the primordial vertebral segments ; 5, the ventral or visceral plates, consisting of the cuticular layer and the outer lamina of the middle germinal layer, passing at 4×5 from the umbilicus into the amnion ; 5', within the embryo, is placed in the peritoneal cavity, below one of the Wolffian bodies and close to the musculo-intestinal lamina ; 6, cavity of the intestine lined by the epithelial or epithelio-glandular layer, which, along with the musculo-intestinal, is continued by the ductus vitello-intestinalis into the yelk-sac, 5' 6.

mesentery. Soon, however, the intestine, growing in length, advances from the spine, and forms a simple loop in the middle of the body, with a straight portion at its upper and lower end, and at the same time becomes slightly dilated in the part destined to form the stomach. The middle of the loop is connected with the umbilical vesicle by the pedicle, and also by the omphalo-mesenteric vessels. The upper extremity of the primitive alimentary tube reaches to the base of the skull and forms the œsophagus and pharynx ; but the mouth is developed by depression of the outer surface of the embryo, above the first branchial arch, and together with the tongue is at first separated from the throat by a partition, which soon gives way. In like manner, the anal orifice does not exist at first, but is formed by invagination of the outer surface, and the opening of a communication between it and the intestine.

Fig. 603.

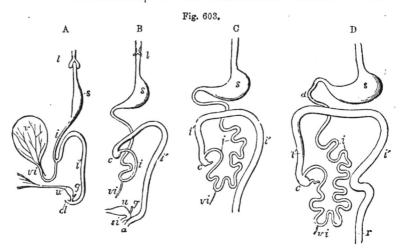

Fig. 603.—OUTLINES OF THE FORM AND POSITION OF THE ALIMENTARY CANAL IN SUCCESSIVE STAGES OF ITS DEVELOPMENT.

A, alimentary canal, &c., in au embryo of four weeks; B, at six weeks; C, at eight weeks; D, at ten weeks; *l*, the primitive lungs connected with the pharynx; *s*, the stomach; *d*, duodenum; *i*, the small intestine; *i'*, the large; *c*, the cæcum and vermiform appendage; *r*, the rectum; *c l*, in A, the cloaca; *a*, in.B, the anus distinct from *s i*, the sinus uro-genitalis; *v*, the yolk sac; *v i*, the vitello-intestinal duct; *u*, the urinary bladder and urachus leading to the allantois; *g*, the genital ducts.

The dilated portion of the tube which forms the stomach turns over on its right side, so that the border, which is connected to the vertebral column by the membranous fold (or true mesogastrium) comes to be turned to the left,—the position of the tube being still vertical, like the stomach of some animals. By degrees it becomes more dilated, chiefly on what is now the left border but subsequently the great curvature, and assumes first an oblique and finally a transverse position, carrying with it the mesogastrium, from which the great omentum is afterwards produced. A slight indication of the pylorus is seen at the third month. Upon the surface of the part of the canal which immediately succeeds the stomach, and which forms the duodenum, the rudiments of the liver, pancreas, and spleen are simultaneously deposited: in connection with the two former, protrusions of the mucous membrane grow into their blastemic mass and form the commencement of their principal ducts.

The place of distinction between the small and the large intestine, which is soon indicated by the protrusion of the cæcum, is at a point just below the apex or middle of the simple loop already mentioned. As the *small* intestine grows, the part below the duodenum forms a coil which at first lies in the commencing umbilical cord, but retires again into the abdomen about the tenth week; afterwards it continues to elongate, and its convolutions become more and more numerous.

The *large* intestine is at first less in calibre than the small. In the early embryo there is at first no cæcum. This part of the bowel gradually grows out from the rest, and in the first instance forms a tube of uniform calibre, without any appearance of the vermiform appendix: subsequently the lower part of the tube ceases to grow in the same proportion, and becomes the appendix, whilst the upper portion continues. to be developed with the rest of the intestine. The cæcum now appears as a protrusion a little below the apex of the bend in the primitive intestinal tube, and, together with the commencing colon, and the coil of small intestine, is at first lodged in the wide part of the umbilical cord which is next the body of the embryo. The ileocæcal valve appears at the commencement of the third month. When the coils of intestine and cæcum have retired from the umbilicus into the abdomen, the colon is at first entirely to the left of the convolutions of the small intestines, but subsequently the first part of the large intestine, together with the meso-colon, crosses over the upper part of the small intestine, at the junction of the duodenum and jejunum.   The cæcum and transverse colon are then found just below the liver; finally, the cæcum descends to the right iliac fossa, and at the fourth or fifth month the parts are in the same position as in the adult.   At first, villous processes or folds of various lengths are formed throughout the whole canal.   After a time these disappear in the stomach and large intestine, but remain persistent in the intermediate portions of the tube.   According to Meckel, the villous processes are formed from larger folds, which become serrated at the edge and divided into separate villi.

The mode of development of the alimentary canal accounts, in some measure, for the principal complication in the folds of the peritoneum.   The stomach being originally straight in form and mesial in position, the small omentum and gastro-phrenic ligament must be regarded as an originally mesial fold with the free edge directed forwards, which afterwards forms the anterior boundary of the foramen of Winslow. Thus the anterior wall of the sac of the omentum, as far as the great curvature of the stomach, may be considered as formed by the right side of a mesial fold, while the peritoneum in front of the stomach belongs to the left side of the same, and a sac of the omentum is a natural consequence of the version and disproportionate growth of the tube between the duodenum and the cardiac orifice of the stomach.   It is obvious that the view of the omental sac, according to which its posterior layers are held to return to the duodenum and posterior wall of the body before proceeding to form the transverse meso-colon (P. 829) is more consistent with the phenomena of development now described, than that which would make them directly enclose the colon. On the other hand, the further elongation of the omental sac and the whole disposition of the peritoneum, with respect to the colon, must be regarded as having taken place after the assumption by the great intestine of its permanent position.

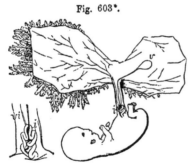

Fig. 603*.

Fig. 603*.—Sketch of the Human Embryo of the Eighth or Ninth Week, showing the Coil of Intestine in the Umbilical Cord.

The amnion and villous chorion have been opened and the embryo drawn aside from them ; *v*, the umbilical vesicle or yolk-sac placed between the amnion and chorion, and connected with the coil of intestine, *i′*, by a small or almost linear tube ; the figure at the side represents the first part of the umbilical cord magnified ; *i*, coil of intestine ; *vi*, vitello-intestinal duct, alongside of which are seen omphalo-mesenteric blood-vessels.

The occurrence of umbilical hernia in its various degrees may be referred to the persistence of one or other of the fœtal conditions in which a greater or less portion of the intestinal canal is contained in the umbilical cord ; and it has been shown that the most common diverticulum of the small intestine is connected with the original opening of the ductus vitello-intestinalis into the ileum (p. 841).

THE LIVER.

The liver is an important glandular organ, very constant in the animal series, being found in all vertebrate, and, in a more or less developed condition, in most invertebrate tribes.   It elaborates and secretes the bile, and otherwise acts, in a manner as yet imperfectly understood, as an elaborator and purifier of the blood.   In the exercise of this latter function, there is formed in its texture an amyloid substance, very easily converted into sugar.

The liver is the largest gland in the body, and by far the most bulky of the abdominal viscera. It measures about ten or twelve inches transversely from right to left, between six and seven inches from its posterior to its anterior border, and about three and a half inches from above downwards at its thickest part, which is towards the right and posterior portion of the

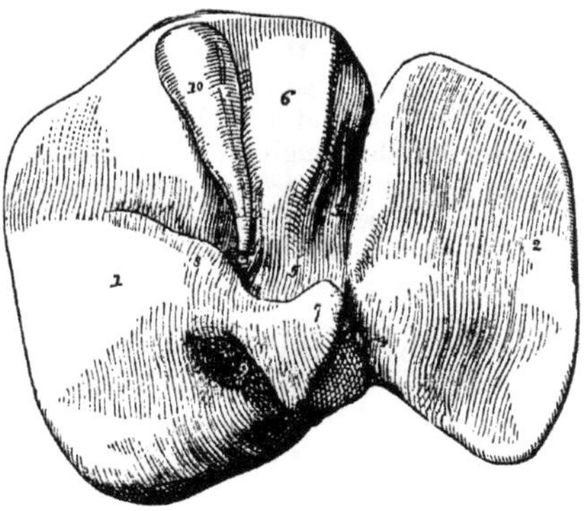

Fig. 604.—SKETCH OF THE UNDER SURFACE OF THE LIVER. ¼

Fig. 604.

The anterior border is turned upwards, and the blood-vessels and ducts have been removed : 1, the right lobe ; 2, the left lobe ; 3, 4, the longitudinal fissure ; 3, its umbilical part ; 4, part containing the ductus venosus ; 5, transverse or portal fissure ; 6, lobulus quadratus ; 7, lobulus Spigelii ; 8, lobulus caudatus ; 9, fissure or fossa of the vena cava ; 10, the gall-bladder in its fossa.

gland. The average bulk, according to Krause, is eighty-eight cubic inches ; according to Beale, one hundred. The ordinary weight in the adult is stated to be between three and four pounds, or more precisely from fifty to sixty ounces avoirdupois.

According to the facts recorded by Reid, the liver weighed, in 43 cases out of 82, between 48 and 58 ounces in the adult male ; and in 17 cases out of 36, its weight in the adult female ranged between 40 and 50 ounces. It is generally estimated to be equal to about 1-36th of the weight of the whole body ; but in the fœtus, and in early life, its proportionate weight is greater. (Reid, in Lond. and Edin. Monthly Journal of Med. Science, April, 1843.)

The specific gravity of the liver, according to Krause and others, is between 1·05 and 1·06 : in fatty degeneration this is reduced to 1·03, or even less.

The parenchyma of the liver has an acid reaction (Kölliker). Beale gives the following results of his analysis of the liver of a healthy man, who was killed by a fall.

Water	68·58
Solid matters	31·42
Fatty matters	3·82
Albumen	4·67
Extractive matters	5·40
Alkaline salts	1·17
Vessels, &c. insoluble in water	16·03
Earthy salts	·33
	100·00

The liver is a solid organ, of a dull reddish-brown colour, with frequently a dark-purplish tinge along the margin. It has an upper smooth and convex surface, and an under surface which is uneven and concave ; the circumference is thick and rounded posteriorly and towards the right extremity, but becomes gradually thinner towards the left and in front, where it forms the sharp anterior and left lateral margins.

The *upper surface* is convex, smooth, and covered with peritoneum. It is marked off into a right portion, large and convex, and a left portion, smaller and flatter, by the line of attachment of the fold of peritoneum named the falciform ligament.

The *under surface*, looking downwards and backwards, is concave and uneven, invested with peritoneum everywhere except where the gall-bladder is adherent to it, and at the portal fissure and fissure of the ductus venosus, which give attachment to the small omentum, the fold of peritoneum which passes round the blood-vessels and ducts of the viscus. On this surface the lobes and fissures of the liver are observed.

The *lobes* of the liver, five in number, are named the right and the left, the lobe of Spigelius, the caudate or tailed lobe, and the square lobe.

The *right* and *left* lobes are separated from each other on the under surface by the longitudinal fissure, and in front by the interlobular notch: on the convex surface of the liver there is no other indication of a separation between them than the line of attachment of the broad ligament. The right lobe is much larger and thicker than the left, which constitutes only about one-fifth or one-sixth of the entire gland.

The other three lobes are small, and might be said to form parts of the right lobe, on the under surface of which they are situated.

The *lobulus quadratus* is that part which is situated between the gall-bladder and the great longitudinal fissure, and in front of the fissure for the portal vein. Its greatest diameter is from before backwards.

The *lobulus Spigelii*, more prominent and less regular in shape than the quadrate lobe, lies behind the fissure for the portal vein, and is bounded on the right and left by the fissures which contain the inferior vena cava and the remains of the ductus venosus.

The *lobulus caudatus* is a sort of ridge which extends from the base of the Spigelian lobe to the under surface of the right lobe. This, in the natural position of the parts, passes forwards above the foramen of Winslow, the Spigelian lobe itself being situated behind the small omentum, and projecting into the omental sac.

The *fissures.*—These are likewise five in number, and are seen on the under surface only. They have all been already incidentally referred to.

The *transverse* fissure, or *portal* fissure, is the most important, because it is here that the great vessels and nerves enter, and the hepatic duct passes out. It lies transversely between the lobulus quadratus and lobulus Spigelii, and meets the longitudinal fissure nearly at right angles. At the two extremities of this fissure, the right and left divisions of the hepatic artery and portal vein, together with the nerves and deep lymphatics enter the organ, while the right and left hepatic ducts emerge.

The *longitudinal* fissure, which separates the right and the left lobes of the liver from each other, is divided into two parts by its meeting with the transverse fissure. The anterior part, named the *umbilical* fissure, contains the umbilical vein in the fœtus, and the remnant of that vein in the adult, which then constitutes the round ligament. It is situated between the square lobe and the left lobe of the liver, the substance of which often forms a bridge across the fissure, so as to convert it partially or completely into a canal. The posterior part is named the *fissure* of the *ductus venosus* (fossa ductus venosi); it continues the umbilical fissure backwards between the lobe of Spigelius and the left lobe; and it lodges the ductus venosus in the fœtus, and in the adult a slender cord or ligament into which that vein is converted.

The *fissure* or *fossa* of the *vena cava* is situated at the back part of the liver, between the Spigelian lobe on the left and the right lobe, and is separated from the transverse fissure by the caudate lobe. It is prolonged upwards in an oblique direction to the posterior border of the liver, and may be said to join behind the Spigelian lobe with the fissure for the ductus venosus. It is at the bottom of this fossa that the blood leaves the liver by the hepatic veins, which end here in the vena cava. The substance of the liver in some cases unites around the vena cava, and encloses that vessel in a canal.

The last remaining fissure, or rather *fossa* (fossa cystis felleæ), is that for the lodgment of the gall-bladder ; it is sometimes continued into a slight notch on the anterior margin of the liver.

Two shallow impressions are seen on the under surface of the right lobe ; one in front (*impressio colica*), corresponding with the hepatic flexure of the colon ; and one behind (*impressio renalis*), corresponding with the right kidney.

The *anterior border* of the liver, a thin, free, and sharp margin, is the most movable part of the gland. Opposite the longitudinal fissure the anterior border presents a notch, and, to the right of this, there is often another slight notch opposite the fundus of the gall-bladder.

The *posterior border* of the liver, which is directed backwards and upwards, is thick and rounded on the right side, but becomes gradually thinner towards the left. It is the most fixed part of the organ, and is firmly attached by areolar tissue to the diaphragm. This border of the liver is curved opposite to the projection of the vertebral column, and has a deep groove for the reception of the ascending vena cava.

Of the two *lateral borders* of the liver, the *right* is placed lower down, and is thick and obtuse ; whilst the *left* is the thinnest part of the gland, is raised to a higher level, and reaches the cardiac part of the stomach.

Ligaments.—The ligaments of the liver, like its lobes and fissures, are commonly described as five, but it seems scarcely necessary to give distinct names to so many parts which are only folds of membrane. One of these, the *coronary ligament*, is the fold of peritoneum by which the posterior border of the liver is attached to the diaphragm : this border lies in contact with the diaphragm, in the greater part of its extent, between the upper and under layers of the peritoneal fold ; but toward the two extremities of the organ these layers come into contact, and form two short mesenteries—the *right* and *left triangular ligaments*, of which the left is the longer and more distinct. Another of these so-called ligaments is the *broad, falciform,* or *suspensory ligament,* a wide thin membrane, composed of two layers of peritoneum, closely united together. By one of its margins it is connected with the under surface of the diaphragm, and with the posterior surface of the sheath of the right rectus muscle of the abdomen as low as the umbilicus ; by another it is attached along the convex surface of the liver, from its posterior border to the notch in its anterior border : the remaining margin is free, and contains between its layers the *round ligament,* a dense fibrous cord, the remnant of the umbilical vein of the fœtus, which ascends from the umbilicus, within the lower edge of the broad ligament, and enters the longitudinal fissure on the under surface. These structures have been already referred to (p. 827).

Position with regard to neighbouring parts.—Occupying the right hypochondriac region, and extending across the epigastric region into a part of the left hypochondrium, the liver is accurately adapted to the vault of the diaphragm above, and is covered, to a small extent in front, by the abdominal parietes. The right portion reaches higher beneath the ribs than the left, .

corresponding thus with the elevated position of the diaphragm on the right side. By means of the diaphragm, the liver is separated from the concave base of the right lung, the thin margin of which descends so as to intervene between the surface of the body and the solid mass of the liver—a fact well known to the auscultator.

The convex surface of the liver is protected, on the right, by the six or seven lower ribs, and in front by the cartilages of the same and by the ensiform cartilage—the diaphragm, of course, being interposed. Being suspended by ligaments to the diaphragm above, and supported below, in common with the rest of the viscera, by the abdominal muscles, the situation of the liver is modified by the position of the body, and also by the movements of respiration ; thus, in the upright or sitting posture, the liver reaches below the margin of the thorax ; but in the recumbent position, the gland ascends an inch or an inch and a half higher up, and is entirely covered by the ribs, except a small portion opposite the substernal notch. Again, during a deep inspiration, the liver descends below the ribs, and in expiration retires upwards behind them. In females the liver is often permanently forced downwards below the costal cartilages, owing to the use of tight stays ; sometimes it reaches nearly as low as the crest of the ilium ; and, in many such cases, its convex surface is indented from the pressure of the ribs.

To the left of the longitudinal fissure the liver is supported on the pyloric

Fig. 605.

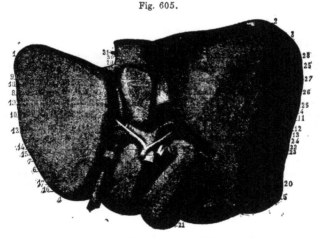

Fig. 605.—LOWER SURFACE OF THE LIVER WITH THE PRINCIPAL BLOOD-VESSELS AND DUCTS (from Sappey). ⅓

The liver has been turned over from left to right so as to expose the lower surface. 1, left lobe ; 2, 3, 4, 5, right lobe ; 6, lobulus quadratus ; 7, pons hepatis ; 8, 9, 10, lobulus Spigelii ; 11, lobulus caudatus ; 12, 13, transverse or portal fissure with the great vessels ; 14, hepatic artery ; 15, vena portæ ; 16, anterior part of the longitudinal fissure, containing 17, the round ligament or obliterated remains of the umbilical vein ; 18, posterior part of the same fissure, containing 19, the obliterated ductus venosus ; 20, 21, 22, gall-bladder ; 23, cystic duct ; 24, hepatic duct ; 25, fossa containing 26, the vena cava inferior ; 27, opening of the capsular vein ; 28, small part of the trunk of the right hepatic vein ; 29, trunk of the left hepatic vein ; 30, 31, openings of the right and left diaphragmatic veins.

extremity and anterior surface of the stomach, on which it moves freely. When the stomach is quite empty, the left part of this surface of the liver may overlap the cardiac end of that viscus. To the right of the longitudinal fissure the liver rests and moves freely upon the first part of the duodenum, and upon the hepatic flexure of the colon, at the junction of the ascending and transverse portions of that intestine. Farther back it is in contact with the fore part of the right kidney and suprarenal capsule.

Vessels.—The two vessels by which the liver is supplied with blood are the hepatic artery and the vena portæ. The *hepatic artery* (p. 408), a branch of the cœliac axis, is intermediate in size between the other two branches of that trunk, being larger than the coronary artery of the stomach, but not so large as the splenic artery. Its size is, therefore, small in comparison with the organ to which it is distributed. It enters the transverse fissure, and there divides into a right and a left branch, for the two principal lobes of the liver.

By far the greater part of the blood which passes through the liver,—and in this respect it differs from all other organs of the body,—is conveyed to it by a large vein, the *vena portæ* (p. 479). This vein is formed by the union of nearly all the veins of the chylopoietic viscera, viz., those from the stomach and intestines, the pancreas and spleen, the omentum and mesentery, and also those from the gall-bladder. It enters the *porta*, or transverse fissure, where, like the hepatic artery, it divides into two principal branches.

The hepatic artery and portal vein, lying in company with the bile-duct, ascend to the liver between the layers of the gastro-hepatic omentum, above the foramen of Winslow, and thus reach the transverse fissure together. The relative position of the three structures is as follows,—The bile-duct is to the right, the hepatic artery to the left, and the large portal vein is behind the other two. They are accompanied by numerous lymphatic vessels and nerves. The branches of these three vessels accompany one another in their course through the liver nearly to their termination ; and in this course are surrounded for some distance by a common investment (Glisson's capsule), which is prolonged into the interior of the organ.

The *hepatic veins*, which convey the blood away from the liver, pursue through its substance an entirely different course from the other vessels, and pass out at its posterior border, where, at the bottom of the fossa already described, they end by two or three principal branches, besides other smaller ones, in the vena cava inferior.

The *lymphatics* of the liver, large and numerous, form a deep and a superficial set, already described (p. 493).

Nerves.—The *nerves* of the liver are derived partly from the cœliac plexus, and partly from the pneumogastric nerves, especially from the left pneumogastric. They enter the liver supported by the hepatic artery and its branches ; along with which they may be traced a considerable way in the portal canals, but their ultimate distribution is not known.

EXCRETORY APPARATUS.—The excretory apparatus of the liver consists of the hepatic duct, the cystic duct, the gall-bladder, and the common bile-duct.

The *hepatic* duct, formed by the union of a right and left branch, which issue from the bottom of the transverse fissure and unite at a very obtuse angle, descends to the right, within the gastro-hepatic omentum, in front of the vena portæ, and having the hepatic artery to its left side. Its diameter is about two lines, and its length nearly two inches. At its lower end it meets with the cystic duct, descending from the gall-bladder ; and the two ducts uniting together at an acute angle, form the common bile-duct.

- The *gall-bladder* is a receptacle or reservoir for such bile as is not immediately required in digestion. It is a pear-shaped membranous sac, three or four inches long, about an inch and a half across at its widest part, and capable of containing from eight to twelve fluid-drachms. It is lodged obliquely in a fossa on the under surface of the right lobe of the liver, with its large end or *fundus*, which projects beyond the anterior border of the gland, directed downwards, forwards, and to the right, whilst its *neck* is inclined in the opposite direction.

The *upper surface* of the gall-bladder is attached to the liver by areolar tissue and vessels, along the fossa formed between the quadrate lobe and the remainder of the right lobe. Its *under surface* is free and covered by the peritoneum, which is here reflected from the liver, so as to include and support the gall-bladder. Sometimes, however, the peritoneum completely surrounds the gall-bladder, which is then suspended by a kind of mesentery at a little distance from the under surface of the liver. The *fundus* of the gall-bladder, which is free, projecting, and always covered with peritoneum, touches the abdominal parietes immediately beneath the margin of the thorax, opposite the tip of the tenth costal cartilage. Below, it rests on the commencement of the transverse colon ; and, farther back, it is in contact with the duodenum, and sometimes with the pyloric extremity of the stomach. The *neck* of the gall-bladder, gradually narrowing, forms two curves upon itself like the letter S, and then, becoming much constricted, and changing its general direction altogether, it bends downwards and terminates in the cystic duct.

The gall-bladder is supplied with blood by the *cystic* branch of the right division of the hepatic artery, along which vessel it also receives nerves from the cœliac plexus. The cystic veins empty themselves into the vena portæ. Beale states that two large veins always accompany one artery.

The *cystic duct* is about an inch and a half in length. It runs downwards and to the left, thus forming an angle with the direction of the gall-bladder, and unites with the hepatic duct to form the common duct.

The *common bile-duct, ductus communis choledochus*, the largest of the ducts, being from two to three lines in width, and nearly three inches in length, conveys the bile both from the liver and the gall-bladder into the duodenum. It continues downwards and backwards in the course of the hepatic duct, between the layers of the gastro-hepatic omentum, in front of the vena portæ, and to the right of the hepatic artery. Having reached the descending portion of the duodenum, it continues downwards on the inner and posterior aspect of that part of the intestine, covered by or included in the head of the pancreas, and, for a short distance, in contact with the right side of the pancreatic duct. Together with that duct, it then perforates the muscular wall of the intestine, and, after running obliquely for three quarters of an inch between its several coats, and forming an elevation beneath the mucous membrane, it becomes somewhat constricted, and opens by a common orifice with the pancreatic duct on the inner surface of the duodenum, at the junction of the second and third portions of that intestine, and three or four inches below the pylorus.

Varieties.—The liver is not subject to great or frequent deviation from its ordinary form and relations. Sometimes it retains the thick rounded form which it presents in the fœtus ; and it has occasionally been found without any division into lobes. On the contrary, Sœmmerring has recorded a case in which the adult liver was divided into twelve lobes ; and similar cases of subdivided liver (resembling that of some animals) have been now and then observed by others. A detached portion, forming a sort of *accessory* liver, is occasionally found appended to the left extremity of the gland by a fold of peritoneum containing blood-vessels.

The gall-bladder is occasionally wanting ; in which case the hepatic duct is much dilated within the liver, or in some part of its course. Sometimes the gall-bladder is irregular in form, or is constricted across its middle, or, but much more rarely, it is partially divided in a longitudinal direction. Direct communications by means of small ducts (named hepato-cystic), passing from the liver to the gall-bladder, exist regularly in various animals; and they are sometimes found, as an unusual formation, in the human subject.

The right and left divisions of the hepatic duct sometimes continue separate for some distance within the gastro-hepatic omentum. Lastly, the common bile-duct not unfrequently opens into the duodenum, apart from the pancreatic duct.

STRUCTURE OF THE LIVER.

Coats.—The liver has two coverings, viz., a serous or peritoneal investment, already sufficiently referred to, and a proper areolar coat.

The *areolar* or *fibrous* coat invests the whole gland. Opposite to the parts covered by the serous coat, it is thin and difficult to demonstrate ; but where the peritoneal coat is absent, as at the posterior border of the liver, and in the portal fissure, it is denser and more evident. Its inner surface is attached to the hepatic glandular substance, being there continuous

Fig. 606.

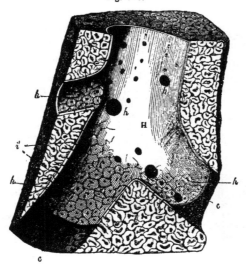

Fig. 606.—SECTION OF A PORTION OF LIVER PASSING LONGITUDINALLY THROUGH A
CONSIDERABLE HEPATIC VEIN, FROM THE PIG (after Kiernan). ⁶⁄₁

H, hepatic venous trunk, against which the sides of the lobules (*l*) are applied ; *h, h, h,* sublobular hepatic veins, on which the bases of the lobules rest, and through the coats of which they are seen as polygonal figures ; *i,* mouth of the intralobular veins, opening into the sublobular veins ; *i′,* intralobular veins shown passing up the centre of some divided lobules ; *l, l,* cut surface of the liver ; *c, c,* walls of the hepatic venous canal, formed by the polygonal bases of the lobules.

with the delicate areolar tissue which lies between the small lobules of the gland. At the transverse fissure it becomes continuous with the *capsule of Glisson,* by which name is designated a sheath of areolar tissue which surrounds the branches of the portal vein, hepatic artery, and hepatic duct, as they ramify in the substance of the liver, and which becomes more delicate as the vascular branches become smaller.

Lobules.—The proper substance of the liver, which has a reddish brown colour and a mottled aspect, is compact, but not very firm. It is easily cut or lacerated, and is not unfrequently ruptured during life from accidents in which other parts of the body have escaped injury. When the substance of the liver is torn, the broken surface is not smooth but coarsely granular, the liver being composed of a multitude of small *lobules,* which vary from half a line to a line in diameter.

These lobules are closely packed polyhedral masses, and in some animals, as in the pig, are completely isolated one from another by areolar tissue continuous with the fibrous coat of the liver and with the capsule of Glisson ; but in the human subject, and in most animals, although they are very distinguishable on account of the disposition both of vessels and parenchyma, they are not distinctly separated, but exhibit continuity through their capillary networks and cellular constituents. Notwithstanding this, however, we may consider the lobules of the human liver as being marked out by slight interlobular intervals.

Fig. 607.

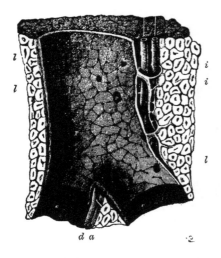

d a   ·2·

Fig. 607.—Longitudinal Section of a Portal Canal, containing a Portal Vein, Hepatic Artery, and Hepatic Duct, from the Pig (after Kiernan). ⁴⁄₁

P, branch of vena portæ, situated in *c, c,* a portal canal, formed amongst the lobules of the liver (*l, l*); *p, p,* vaginal branches of portal vein, giving off smaller ones (*i, i*), named interlobular veins; there are also seen within the large portal vein numerous orifices of the smallest interlobular veins arising directly from it; *a,* hepatic artery; *d,* hepatic duct.

The lobules of the liver have throughout its substance in general the polyhedral form of irregularly compressed spheroids; but on the surface they are flattened and angular. They are all compactly arranged round the sides of branches of the hepatic veins, each lobule resting by a smooth surface or *base,* upon the vein, and being connected with it by a small venous trunk, which arises in the centre of the lobule, and passes out from the middle of its base to end in the larger subjacent vessel. The small veins proceeding from the centre of the lobules are named the *intralobular* veins, and those on which the lobules rest, the *sublobular* veins. If one of these sublobular veins be opened, the bases of the lobules may be seen through the coats of the vein, which are here very thin, giving a tesselated appearance, each little polygonal space representing the base of a lobule, and having in its centre a small spot, which is the mouth of the intralobular vein. When divided in the direction of a sublobular vein, the attached lobules present a foliated appearance, for that part of their surface which is not in contact with the vein is itself slightly lobulated. Cut in a transverse direction, the lobules present a polyhedral form.

The hepatic substance, as exhibited in the arrangement of each lobule, consists of masses of cells and a copious vascular network, closely interwoven, with the intervention of little other tissue. For the sake of convenience, the vascular structure of the liver may be considered first.

Blood-vessels.—The *hepatic veins* commence in the centre of each lobule by the union of its capillary vessels into a single independent intralobular

. Fig. 603.

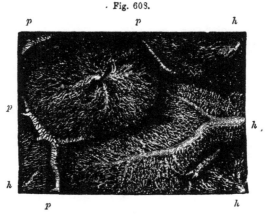

Fig. 608.—CAPILLARY NETWORK OF THE LOBULES OF THE RABBIT'S LIVER (from Kölliker). $\frac{45}{1}$

The figure is taken from a very successful injection of the hepatic veins made by Harting : it shows nearly the whole of two lobules, and parts of three others : *p*, portal branches running in the interlobular spaces; *h*, hepatic veins penetrating and radiating from the centre of the lobules.

vein, as already stated. These minute intralobular veins open at once into the sides of the adjacent sublobular veins. The sublobular veins are of various sizes, and anastomose together. Uniting into larger and larger vessels, they end at length in hepatic venous trunks, which receive no intralobular veins. Lastly, these venous trunks, converging towards the posterior border of the liver, and receiving in their course other small sublobular veins, terminate in the vena cava inferior, as already described. In this course the hepatic veins and their successive ramifications are unaccompanied by any other vessel. Their coats are extremely thin ; the sublobular branches adhere immediately to the lobules, and even the larger trunks have but a very slight areolar investment connecting them to the substance of the liver. Hence the divided ends of these veins are seen upon a section of the liver as simple open orifices, the thin wall of the vein being surrounded closely by the solid substance of the gland.

The vena portæ and hepatic artery, which, together with the biliary ducts,·

enter the liver at the transverse fissure, have a totally different course, arrangement and distribution from those of the hepatic vein. Within the liver the branches of these three vessels lie together in certain canals, called

Fig. 609.

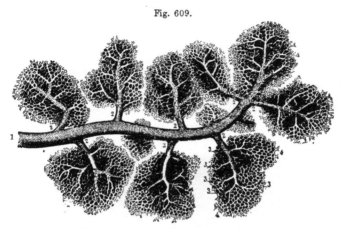

Fig. 609.—INJECTED TWIG OF A HEPATIC VEIN WITH SUBLOBULAR VEINS PASSING INTO THE HEPATIC LOBULES (from Sappey). $\frac{30}{1}$

1, small sublobular hepatic vein; 2, intralobular veins passing into the base of the lobules; 3, their smaller subdivisions; 4, capillary network of communication with the extreme ramifications of the vena portæ.

Fig. 610.

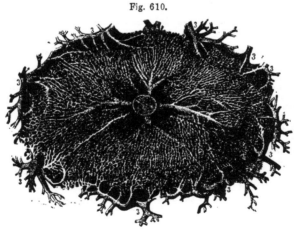

Fig. 610.—CROSS SECTION OF A LOBULE OF THE HUMAN LIVER, IN WHICH THE CAPILLARY NETWORK BETWEEN THE PORTAL AND HEPATIC VEINS HAS BEEN FULLY INJECTED (from Sappey). $\frac{90}{1}$

1, section of the intralobular vein; 2, its smaller branches collecting blood from the capillary network; 3, interlobular branches of the vena portæ with their smaller ramifications passing inwards towards the capillary network in the substance of the lobule.

portal canals, which are tubular passages formed in the substance of the gland, commencing at the transverse fissure, and branching upwards and outwards from that part in all directions. Each portal canal (even the smallest) contains, as shown in a longitudinal section, one principal branch of the vena portæ, of the hepatic artery, and of the biliary duct ; the whole being invested within the larger portal canals by the areolar tissue of the capsule of Glisson.

The *portal vein* subdivides into branches which ramify between the lobules, anastomosing freely around them, and are named *interlobular veins.* The twigs from these penetrate the lobules at their circumference, and end in the capillary network from which the intralobular (hepatic) veins take origin. Within the portal canals the branches of the portal veins receive small tributaries called "vaginal veins," which return to them the blood which has circulated in the capsule of Glisson, and also "venæ advehentes capsulares," from the fibrous coat of the liver.

The *hepatic artery* terminates in three sets of branches, termed vaginal, capsular, and interlobular. The *vaginal* branches ramify within the portal canals, supplying the walls of the ducts and Glisson's capsule. The *capsular* branches appear on the surface of the liver spread out on the fibrous sheath, and are accompanied by the veins which return their blood to the portal branches. The *interlobular* branches accompany the interlobular veins, but are of much smaller diameter. It has been supposed by Kiernan, Ferrein, and Theile, that the blood which they convey is entirely taken up by the portal veins before reaching the capillaries from which the hepatic veins take origin ; but the view, which has been held by other anatomists, that the hepatic arteries transmit blood directly to the capillary network between the portal and hepatic veins, is supported by the experiments of Chrzonszczewsky, mentioned further on.

The *capillary network* is very close, and, in specimens in which it has been filled with transparent injection, can be seen to be continued uninterruptedly from one lobule to another.

The distribution of the portal and hepatic veins within the lobules, as just described, has suggested an explanation of the mottled aspect of the liver, an appearance which formerly led to the erroneous idea of there being two substances in each lobule, one darker than the other. The colour of the hepatic substance itself is pale yellow, and would be uniform throughout, were it not varied according to the quantity of blood contained in its different vessels. Thus, if the system of hepatic veins be congested, the centre of each lobule is dark, and its margin pale : this is the common case after death, and is named by Kiernan *passive* congestion. In what is considered an *active* state of hepatic congestion, the dark colour extends to the portal system, across the interlobular fissures, leaving intermediate spaces, which remain as irregular pale spots : this state occurs especially in diseases of the heart. When, on the other hand, the portal system is congested, which is rare, and occurs generally in children, the margins of the lobules are dark, and their centres pale.

The Hepatic Cells.—The principal part of the secreting substance of the liver, and that which seems to form nearly the whole bulk of the lobules when unprepared sections are examined with the microscope, consists of nucleated cells. The hepatic cells are of a spheroidal, compressed, or polyhedral form, having a mean diameter of from $\frac{1}{1080}$th to $\frac{1}{840}$th of an inch : according to Henle some of them are only $\frac{1}{1716}$th of an inch in diameter. They present some colour even when highly magnified, being of a faint yellowish hue. They usually include a very clear bright vesicular nucleus of a rounded form, within which again one or two nucleoli may be seen. The cells also contain very fine granular or vesicular molecules. In many cases, too, the cells of the human liver and of that of quadrupeds have larger and smaller semi-transparent fat-globules in their interior. The nucleus is frequently quite indistinguishable; and not unfrequently, on the other hand, cells are observed which are provided with two separate nuclei. They are massed in rows or streaks between the vessels, and, in sections made at right angles to the intra-lobular veins, appear as if

Fig. 611.

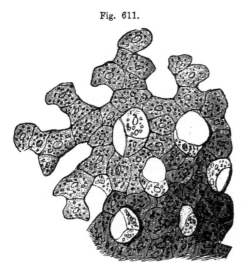

Fig 611.—A SMALL PORTION OF A LOBULE OF THE HUMAN LIVER HIGHLY MAGNIFIED, SHOWING THE HEPATIC CELLS IN CONNECTION AND THE CAPILLARY SPACES BETWEEN THEM (from Kölliker). $\frac{450}{1}$

Fig. 612.—SEPARATE HEPATIC CELLS (from Kölliker). $\frac{400}{1}$

Fig. 612.

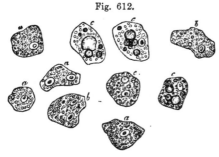

a, most usual form of cells; b, cells containing colour-granules; c, cells containing fat-globules.

radiating from the centre of the lobules towards their circumference. When examined with a higher magnifying power, they are observed to form a continuous web, or solid network, the more obvious openings in which are the spaces occupied by the capillaries with which the cells are interlaced.

Beale states that these cells often appear to be collections of viscid matter around central nuclei, without any distinct cell-wall; but this is open to question. According to Schiff (quoted by Henle), the molecules contained within the hepatic cells consist of the amyloid substance, which is formed in the liver, and from which the glucose obtained from this organ is derived.

The hepatic cells may be washed away from thin sections by dilute solutions of caustic potash, and then the spaces which they occupied are emptied, and the network of capillaries with which they were interlaced is brought more clearly into view, as was pointed out by Rainey; and likewise, according to Henle, narrow bands, which he regards as formed of connective tissue, are to be seen crossing the spaces.

Commencements of the Ducts.—The smallest bile-ducts which are satisfactorily known, ramify between the lobules along with the interlobular blood-vessels : but it is still a matter of discussion how the bile enters these ducts, and what is the mode of its secretion.

Fig. 613.

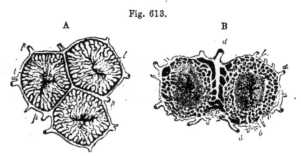

Fig. 613.—DIAGRAMS SHOWING THE ARRANGEMENT OF THE BLOOD-VESSELS AND DUCTS WITHIN AND BETWEEN THE LOBULES, ACCORDING TO KIERNAN. $\frac{20}{1}$

In A, *p, p,* interlobular branches of the portal vein ; *l, l,* intralobular venous plexus, connecting the portal veins (*p, p*) with the intralobular vein (*h*) in the centre, which is the commencing branch of the hepatic vein ; in B, *d, d* are two branches of the hepatic duct, which is supposed to commence in a plexus situated towards the circumference of the lobule marked *b, b,* called by Kiernan the biliary plexus. Within this is seen the central part of the lobule, containing branches of the intralobular (hepatic) vein.

Kiernan described the smallest biliary ducts as commencing within the lobules by numerous ramifications in the form of a close network, which he was only able to inject in the outer part of each lobule. Since the discovery of the hepatic cells, however, it has been very generally supposed

Fig. 614.

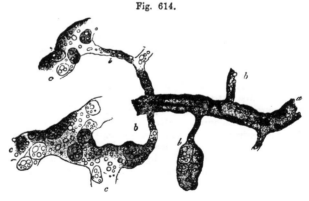

Fig. 614.—VIEW OF SOME OF THE SMALLEST BILIARY DUCTS ILLUSTRATING BEALE'S VIEW OF THEIR RELATION TO THE BILIARY CELLS (from Kölliker after Beale). $\frac{215}{1}$

The drawing is taken from an injected preparation of the pig's liver.

a, small branch of an interlobular hepatic duct ; *b,* smallest biliary ducts ; *c,* portions of the cellular part of the lobule in which the cells are seen within tubes which communicate with the finest ducts.

that, the cells being placed closely together, the bile either passes by irregular interstices between them, or from one cell to another, to reach the smallest ducts at the circumference of the lobules.

Beale, who has investigated the subject with great care, believes that he has succeeded in demonstrating the existence of a basement membrane inclosing, in a plexus of tubes, the intralobular rows or columns of hepatic cells. This basement membrane, he conceives, lines the interstices between the capillary blood-vessels forming the intralobular plexus ; and he states that, although in the adult it becomes so closely incorporated with the walls of the blood-vessels that it is scarcely to be demonstrated as a distinct structure, yet, in the fœtus, the walls of these tubes and those of the vessels are quite distinct. The minute ducts are considered by Beale to be directly continuous with this tubular network : but the tubules containing the hepatic cells, being $\frac{1}{1000}$th of an inch in diameter, and the smallest ducts $\frac{1}{3000}$th or less, there is a great difference in their size ; this difference, he holds, however, is only similar to that which is found, in some other glandular organs, between the proper secreting cavity and the ductal passages.

Kölliker has become convinced of the correctness of Beale's account from an examination of his preparations. Henle believes that the interlobular bile-ducts are completely shut off from the cellular substance of the lobules, which was the theory proposed by Handfield Jones ; and he suggests that the hepatic cells are entirely engaged in the amyloid function of the liver, and unconnected with the biliary secretion.

According to such views as those before stated, the anastomosing network of ducts described by Kiernan would be regarded as artificial passages between the cells, formed by the force of injection ; and there is no doubt that passages of that sort may be made. In recent years, however, Budge, Andrejewic, Hyrtl, Frey, and other observers have succeeded in displaying by injection of the cellular substance of the lobules a network of fine canals of cylindrical form and regular diameter, and having therefore a character which cannot be explained on the supposition that they are irregular interspaces of accidental origin. The apparent improbability of the ducts of a secreting gland taking origin in minute tubes destitute of epithelium, and external to secreting cells, has led to great opposition to the view that the ducts in question are really bile-ducts ; and Reichert has suggested that they are lymphatics : but a set of researches have subsequently been published which invest the theory that the bile-ducts begin within a fine intralobular plexus, with additional weight. Chrzonszczewsky, pursuing a method of experimenting by what may be called the natural injection of colouring matter into the vessels of living animals, by which he had previously succeeded in colouring the tubes and vessels of the kidney, sought for a colouring matter which, when introduced into the blood, would be eliminated in part by the bile without dyeing the textures indiscriminately ; and, after numerous failures, he at last found one substance with the requisite properties,—viz., the sulpho-indigotate of soda (in use under the name of "indigo-carmine"). A saturated watery solution of this substance was introduced, in repeated doses, into the circulation of dogs and sucking-pigs, by the jugular vein ; and in an hour and a half afterwards, while the animals were still living, the blood-vessels were either washed out with chloride of potassium introduced by the portal vein, or they were injected with gelatine and carmine. In specimens prepared in this way, Chrzonszczewsky obtained an extremely fine network of gall-ducts throughout each lobule, injected blue, while the intervening cells remained free from colour. These canals he describes as of regular diameter, without increase of size where they anastomose, and by teasing he obtains portions of them with distinct walls standing·out free from the cells : by warming the section to 113° Fah., the blue colour is destroyed while the canals still remain visible. By killing the animals sooner after the injection, the blue colouring matter was found within the hepatic cells, thus demonstrating that it was through their agency that the canals were filled. Further experiments were made in animals in which the portal vein and hepatic artery had been tied, and the result obtained was that, when the hepatic artery

Fig. 615.—Two small Fragments of Hepatic Lobules, of which the smallest Intercellular Biliary Ducts were filled with Colouring Matter during Life, highly magnified (from Chrzonszczewsky).

Fig. 615.

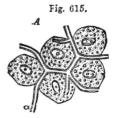

In A, the hepatic cells have been separated, and the intercellular ducts, *a*, are seen not only passing between them, but also in part projecting free ; in this preparation the colour was discharged by heat : in B, the colouring matter remains in the ducts, and the cells are more closely connected together.

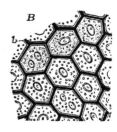

had been tied, the peripheral parts of the lobules showed the blue canals, while the centre of each was left colourless ; and that, when the portal vein had been tied, the reverse effect was produced ; the centre of each lobule showing blue canals, while in the intervening spaces only larger ducts were seen. It is worthy of remark that the appearance of fibres crossing the capillary spaces observed by Henle in sections washed with alkali, might very well be due to such canals·as those described by Budge, Chrzonszczewsky, and others.

Structure of the ducts.—The bile-ducts have strong distensible areolar coats, containing abundant elastic tissue, and their mucous membrane is lined with columnar epithelium. The minute ramifications between the lobules have walls of a more homogeneous nucleated tissue, but the lining of columnar epithelium is still found in them (Henle). The mucous membrane of those which are less minute presents numerous openings, which are scattered irregularly in the larger ducts, but in the subdivisions are arranged in two longitudinal rows, one at each side of the vessel. These openings were formerly supposed to be the orifices of mucous glands ; but, while the main ducts are studded with true mucous glands of lobulated form and with minute orifices, the larger openings now referred to belong, as was pointed out by Theile, to sacs and ramified tubes which occasionally anastomose, and may be studded all over with cæcal projections. Sappey and Henle, who have made these processes the subject of special investigation, find that they are so numerous as sometimes to conceal the parent tube, and on this Henle bases his suggestion (System. Anat.) that they are engaged in the secretion of the bile.

Aberrant biliary ducts.—In the duplicature of the peritoneum forming the· left lateral ligament of the liver, and also in the two fibrous bands which sometimes convert the fossa for the vena cava and the fissure of the umbilical vein into canals, there have been found biliary ducts of considerable size which are not surrounded with lobules.   These aberrant ducts, as they might be called, were described by Ferrein and by Kiernan; they anastomose together in form of a network, and are accompanied by branches of the vena portæ, hepatic artery, and hepatic vein.

Structure of the gall-bladder.—Besides the peritoneal investment, the walls of the gall-bladder are formed of two distinct layers of tissue constituting its areolar or fibrous and its mucous coats.

The *areolar* coat is strong, and consists of bands of dense shining white fibres, which interlace in all directions.   These fibres resemble those of areolar tissue.   In quadrupeds recently killed the gall-bladder contracts on the application of a stimulus ; and in the larger species, as well as in man, muscular fibres of the plain variety have been found mingled with those of the areolar coat.   These fibres have principally a longitudinal direction, but some run transversely.   Their nuclei are indistinct.   The areolar coat forms the framework of the organ, and supports the larger bloodvessels and lymphatics.

The *mucous* coat, which is generally strongly tinged of a yellowish-brown colour with bile, is elevated upon its inner surface into very numerous small ridges, which, uniting together into meshes, leave between them depressions of different sizes and of various polygonal forms. This structure gives to the interior of the gall-bladder an areolar aspect, which is similar to what is seen on a smaller scale in the vesiculæ seminales.   These areolar intervals become smaller towards the fundus and neck of the gall-bladder ; and at the bottom of the larger ones, other minute depressions, which may be seen with a magnifying lens, apparently lead into numerous mucous recesses or follicles.   The whole of the mucous membrane is covered by columnar epithelium, and it secretes an abundance of viscid mucus.

At the places where the neck of the gall-bladder curves on itself, there are strong folds or projections of its mucous and areolar coats into the interior.

In the *cystic duct*, the mucous membrane is elevated internally in a singular manner into a series of crescentic folds, which are arranged in an oblique direction, and succeed closely to each other, so as to present very much the appearance of a continuous spiral valve.   When distended, the outer surface of the duct appears to be indented in the situation of these folds, and dilated or swollen in the intervals, so as to present an irregularly sacculated or twisted appearance.   It is of importance to note the influence of this valve in causing the retention of biliary concretions in the gall-bladder.

Among the monographs which give an account of the structure of the liver, the following may be specially mentioned :—Kiernan, in Phil. Transactions, 1833 ; Theile, in Wagner's Handwörterbuch d. Physiologie, p. 308 ; Rainey on the Capillaries of the Liver, in Microsc. Journal, I. p. 231 ; Handfield Jones, in Phil. Transactions, 1849 and 1853 ; Budge, in Müller's Archiv, 1850 ; Beale, Lectures in Medical Times and Gazette, 1856, and "On some points in the Anatomy of the Liver," in Philos. Trans. 1856 ; Chrzonszczewsky, in Virchow's Archiv, XXXV. p. 153, 1866 ; Frey, in Zeitsch. f. Wissensch. Zoologie, March, 1866.

THE BILE.

The bile, as it flows from the liver, is a thin greenish yellow fluid ; but that which remains in the gall-bladder becomes darker, more viscid, and ropy.   It contains as adventitious particles mucus and epithelium corpuscles, but no hepatic cells.   The specific gravity of the bile is from 1·026 to 1·030. It has a sweetish bitter taste, and an alkaline reaction. It is a saponaceous compound, containing the following ingredients :—water, mucus, colouring matters (composed, according to Berzelius, of a yellow substance named cholepyrrhine, a brown substance named bilifulvine, and a green matter or biliverdine), fatty acids, viz., the margaric and oleic, combined with soda, free fat, cholesterine, salts, and, lastly, the most important ingredient of the bile, namely, the proper biliary matter, which consists, according to Strecker and Lehmann, of two "conjugated acids," formed by the union of one acid, the Choleic, in two isomeric forms (the cholic and choloidic acids of some authors) with Glycin and Taurin respectively. Thus the Glycocholic and Taurocholic acids are formed, each of them consisting principally of carbon and hydrogen, but both containing nitrogen, and the latter a considerable quantity of sulphur. They are combined with soda, but are very readily decomposed, giving rise to ammoniacal and other compounds. Of these two acids the glycocholic is the most important. The bile-pigment affords the most characteristic tests for the detection of biliary matter.

DEVELOPMENT AND FŒTAL PECULIARITIES OF THE LIVER.

The liver begins to be formed at a very early period of fœtal life. Both in the mammal, as seen by Bischoff, and in the bird, as Remak's researches show, it begins in the form of two blind processes from the intestinal tube, immediately beneath the dilatation for the stomach. According to Remak, these processes involve both the epithelial and the fibrous layers of the intestine; and the fibrous layer, rapidly growing, involves the omphalo-mesenteric vein and forms the outline of the liver: meanwhile the internal structure takes origin by the growth of anastomosing cylinders of cells from the epithelial layer, and of tufts of blood-vessels.

Fig. 616.—EARLY CON-
DITION OF THE LIVER
IN THE CHICK ON THE
THIRD DAY OF INCU-
BATION (from J. Mül-
ler). $\frac{10}{1}$

1, the heart, as a simple curved tube ; 2, 2, the intestinal tube ; 3, conical protrusion of the coat of the commencing intestine, on which the blastema of the liver (4) is formed; 5, portion of the layers of the germinal membrane, passing into the yolk-sac.

Fig. 616.

The gall-bladder, according to some authors, is developed as a branch or diverticulum from the bile-duct outside the liver; but Meckel describes it as arising in a deep notch in the substance of the gland. It is at first tubular, and then has a rounded form. The alveoli in its interior appear about the sixth month. At the seventh month it first contains bile. In the fœtus its direction is more horizontal than in the adult.

Size.—In the human fœtus, at the third or fourth week, the liver is said to constitute one-half of the weight of the whole body. This proportion gradually decreases as development advances, until at the full period the relative weight of the fœtal liver to that of the body is as 1 to 18.

In early fœtal life, the right and left lobes of the liver are of equal, or nearly equal, size. Later, the right preponderates, but not to such an extent as after birth. Immediately before birth the relative weight of the left lobe to the right is nearly as 1 to 1·6.

Position.—In consequence of the nearly equal size of the two lobes, the position of the fœtal liver in the abdomen is more symmetrical than in the adult. In the very young fœtus it occupies nearly the whole of the abdominal cavity; and at the full period it still descends an inch and a half below the margin of the thorax, overlaps the spleen on the left side, and reaches nearly down to the crest of the ilium on the right.

Form, Colour, &c.—The fœtal liver is considerably thicker from above downwards than that of the adult. It is generally of a darker hue. Its consistence and specific gravity are both less than in the adult.

Blood-Vessels.—The blood-vessels of the fœtal liver present several important peculiarities, with which, indeed, those previously mentioned are more or less connected. Up to the moment of birth, the greater part of the blood returned from the placenta by the umbilical vein passes through the liver of the fœtus before it reaches the heart ; while a smaller part is transmitted more directly to the right auricle. During fœtal life, the umbilical vein runs from the umbilicus along the free margin of the suspensory ligament towards the anterior border and under surface of the liver, beneath which it is lodged in the umbilical fissure, and proceeds as far as the transverse fissure. Here it divides into *two* branches ; one of these, the smaller of the two, continues onward in the same direction, and joins the vena cava ; this is the *ductus venosus*, which occupies the posterior part of the longitudinal fissure, and gives to it the name of the fossa of the ductus venosus. The other and larger branch (the trunk of the umbilical vein) turns to the right along the transverse or portal fissure, and ends in the vena portæ, which, in as much as it proceeds from the veins of the digestive organs, is in the fœtus comparatively of small dimensions. Moreover, the umbilical vein, as it lies in the umbilical fissure, and before it joins the vena portæ, gives off some lateral branches, which enter the left lobe of the liver. It also sends a few branches to the square lobe and to the lobe of Spigelius.

Fig. 617.

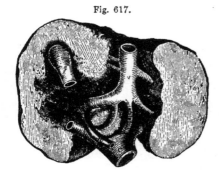

Fig. 617.—UNDER SURFACE OF THE FŒTAL LIVER, WITH ITS GREAT BLOOD-VESSELS, AT THE FULL PERIOD. $\frac{2}{3}$

The rounded outline of the organ, and the comparatively small difference of size between its two lobes, are seen : *a*, the umbilical vein, lying in the umbilical fissure, and turning to the right side at the transverse fissure (*o*), to join the vena portæ (*p*) : the branch marked *d*, named the ductus venosus, continues straight on to join the vena cava inferior (*c*) : a few branches of the umbilical vein enter the substance of the liver at once ; *g*, the gall-bladder.

The blood of the umbilical vein may therefore be considered as reaching the ascending vena cava in three portions. Some is carried into it by the more direct passage of the ductus venosus ; another, the principal portion, passes first through the portal veins, and then through the hepatic veins ; whilst a third portion, supplied by direct branches to the liver, is also returned to the cava by the hepatic veins.

Changes after birth.—Immediately after birth, at the cessation of the current which previously passed through the umbilical vein, and on the establishment of an increased circulation through the lungs, the supply of blood to the liver is diminished by nearly two-thirds. The umbilical vein and ductus venosus become empty and contracted, and soon afterwards they begin to be obliterated, and are ultimately converted into the fibrous cords already described,—that one which represents the umbilical vein constituting the round ligament of the liver. At the end of six days the ductus venosus has been found to be closed ; but it sometimes continues open for several weeks. That portion of the umbilical vein which supplied direct branches to the liver remains open, though diminished in size, and, being in communication with the left branch of the vena portæ, continues afterwards to transmit blood to a part of the liver from that vessel.

Concurrently with, and doubtless in some measure dependent on, the sudden diminution in the quantity of blood supplied to the liver after birth, this organ appears at first to become absolutely lighter; and, according to some data, this decrease of weight is not recovered from until the conclusion of the first year. After that period, the liver, though it increases in size, grows more slowly than the body, so that its relative weight in proportion to the body, which was 1 to 18 just before birth, becomes gradually less and less. At about five or six years of age it has reached the proportion maintained during the rest of life, viz. 1 to 36.

The relative weight of the left lobe to that of the right (which, as above stated, is about 1 to 1.6 immediately before birth) undergoes a subsequent diminution. Thus, at a month old, it has been found to be as 1 to 3, and in after-life the proportion is generally 1 to 4 or 5.

THE PANCREAS.

The *pancreas* is a long, narrow, flattened gland, larger at one end than at the other, and lying deeply in the cavity of the abdomen, immediately behind the stomach, and opposite the first lumbar vertebra. Its larger end, the *head*, turned to the right, is embraced by the curvature of the duodenum, whilst its left or narrow extremity, the *tail*, reaches to a somewhat higher level, and is in contact with the spleen. It extends across the epigastric into both hypochondriac regions.

The right or large end of the pancreas is bent from above downwards, and accurately fills the curvature of the duodenum, to which it is closely adherent. The lower extremity of this curved portion passes to the left, behind

Fig. 618.

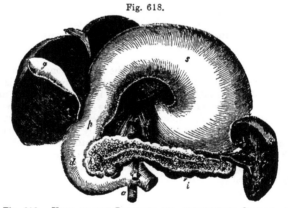

Fig. 618.—VIEW OF THE PANCREAS AND SURROUNDING ORGANS. ⅓

In this figure, which is altered from Tiedemann, the liver and stomach are turned upwards to show the duodenum, the pancreas, and the spleen : *l*, the under surface of the liver ; *g*, gall-bladder ; *f*, the common bile-duct, formed by the union of the cystic duct from the gall-bladder, and the hepatic duct coming from the liver ; *o*, the cardiac end of the stomach, where the œsophagus enters ; *s*, under surface of the stomach ; *p*, pyloric end of the stomach ; *d*, duodenum ; *h*, head of the pancreas ; *t*, tail, and *i*, body of that gland ; the substance of the pancreas is removed in front, to show the pancreatic duct (*e*) and its branches ; *r*, the spleen ; *v*, the hilus, at which the blood-vessels enter ; *c, c,* crura of the diaphragm ; *n*, superior mesenteric artery ; *a*, aorta.

the superior mesenteric vessels, forming the posterior wall of a sort of canal in which they are enclosed. This part of the gland is sometimes marked off from the rest, and is then named the *lesser pancreas.*

The pancreas varies considerably, in different cases, in its size and weight. It is usually from 6 to 8 inches long, about 1½ inch in average breadth, and from half an

3 M

inch to an inch in thickness, being thicker at its head and along its upper border than elsewhere. The weight of the gland, according to Krause and Clendenning, is usually from 2¼ oz. to 3½ oz. ; but Meckel has noted it as high as 6 oz., and Sœmmerring as low as 1½ oz.

The anterior surface of the pancreas is covered with the posterior wall of the sac of the omentum, and is concealed by the stomach, which glides upon it. The posterior surface is attached by areolar tissue to the vena cava, the aorta, the superior mesenteric artery and vein, the commencement of the vena portæ, and the pillars of the diaphragm, all of which parts, besides many lymphatic vessels and glands, are interposed between it and the upper lumbar vertebræ : to the left of the vertebral column it is attached similarly to the left suprarenal capsule and kidney and to the renal vessels. Of the large vessels situated behind the pancreas, the superior mesenteric artery and vein are embraced by the substance of the gland, so as sometimes to be enclosed in a complete canal, through which they pass downwards and forwards, and then emerge from beneath the lower border of the pancreas, between it and the termination of the duodenum. The cœliac axis is above the pancreas ; and in a groove along the upper border of the gland are placed the splenic artery and vein, the vein pursuing a straight, and the artery a tortuous course, and both supplying numerous branches to the pancreas, the narrow extremity of which is thus attached to the inner surface of the spleen. The head of the pancreas, embraced by the inner curved border of the duodenum, is attached more particularly to the descending and transverse portions of that intestine, beyond which it projects somewhat both in front and behind. The ductus communis choledochus passes down behind the head of the pancreas, and is generally received into a groove or canal in its substance.

Structure.—The pancreas belongs to the class of conglomerate glands. In its general characters, and also in its intimate structure, which is racemose, it closely resembles the salivary glands, but it is somewhat looser and softer in its texture. It consists of numerous lobes and lobules, of various sizes, held together by areolar tissue, blood-vessels, and ducts. The lobules, aggregated into masses, are rounded or slightly flattened at the sides, so as to be moulded or adjusted compactly to each other ; their substance is of a reddish cream-colour, and the arrangement of the commencing ducts and vessels is similar to that in the lobules of the parotid gland, which has been already described (p. 816).

The principal excretory duct, called the *pancreatic duct* or *canal of Wirsung* (by whom it was discovered in the human subject in 1642), runs through the entire length of the gland, from left to right, buried completely in its substance, and placed rather nearer its lower than its upper border. Commencing by the union of the small ducts derived from the groups of lobules composing the tail of the pancreas, and receiving in succession at various angles, and from all sides, the ducts from the body of the gland, the canal of Wirsung increases in size as it advances towards the head of the pancreas, where, amongst other large branches, it is usually joined by one derived from that portion of the gland called the lesser pancreas. Curving slightly downwards, the pancreatic duct [then comes into contact with the left side of the ductus communis choledochus, which it accompanies to the back part of the descending portion of the duodenum. Here the two ducts, placed side by side, pass very obliquely through the muscular and areolar coats of the intestine, and terminate, as already described (p. 868), on its internal mucous surface, by a common orifice, situated at the junction of the descending and horizontal portions of the duodenum, between three and four inches below the pylorus. The pancreatic duct, with its branches, is readily distinguished from the glandular substance, by the very white appearance of its thin fibrous walls. Its widest part, near the duodenum, is from 1 line to $1\frac{1}{2}$ line in diameter, or nearly the size of an ordinary quill; but it may be easily distended beyond that size. It is lined by a remarkably thin and smooth mucous membrane, which near the termination of the duct occasionally presents a few scattered follicles.

Varieties.—Sometimes the pancreatic duct is double up to its point of entrance into the duodenum; and a still farther deviation from the ordinary condition is not unfrequently observed, in which there is a *supplementary* duct, derived from the lesser pancreas or some part of the head of the gland, opening into the duodenum by a distinct orifice, at a distance of even one inch or more from the termination of the principal duct. It sometimes occurs that the pancreatic duct and the common bile duct open separately into the duodenum.

Vessels and Nerves.—Like the salivary glands, the pancreas receives its blood-vessels at numerous points. Its arteries are derived from the splenic and from the superior and inferior pancreatico-duodenal branches of the hepatic and superior mesenteric. Its blood is returned by the splenic and superior mesenteric veins. Its lymphatics terminate in the lumbar vessels and glands. The nerves of the pancreas are derived from the solar plexus.

Development.—In its origin and development, the pancreas altogether resembles the salivary glands. It appears a little earlier than these glands, in the form of a small bud from the left side of the intestinal tube, close to the commencing spleen.

Secretion.—Like the saliva, the pancreatic juice is a clear colourless fluid, which has diffused in it a few microscopic corpuscles; it has an alkaline reaction, and coagulates in white flakes when heated. The coagulum is caused by the presence of an albuminoid substance—*pancreatin*—which, like salivin, has the property of converting starch into sugar. The pancreatic juice contains likewise chlorides of sodium and potassium, and phosphates of lime, soda, and magnesia. It readily undergoes decomposition on exposure.

THE SPLEEN.

The spleen is a soft, highly vascular, and easily distensible organ, of a dark bluish or purplish grey colour. It is situated in the left hypochondrium, at the cardiac end of the stomach, between that viscus and the diaphragm, and is protected by the cartilages of the ribs. It is the largest of the structures termed ductless glands, and it is now generally admitted to be intimately connected with the process of sanguification, and is most probably the seat of the formation of blood-corpuscles.

The shape of the spleen is irregular and somewhat variable: it forms a compressed oval mass, placed nearly vertically in the body, and having two faces, one external, convex, and free, which is turned to the left, the other internal and concave, which is directed to the right, and is applied to the cardiac end or great cul-de-sac of the stomach: it also presents an anterior sharper and a posterior blunter margin.

The convex face of the spleen, smooth and covered by the peritoneum, is in contact with the under surface of the left side of the diaphragm, and corresponds with the ninth, tenth, and eleventh ribs. The internal concave face is divided by a vertical fissure, named the *hilus*, into an anterior and posterior portion, both covered with peritoneum, continued round the borders from the convex surface. The anterior of these two portions is the larger, and is closely applied to the stomach; the posterior is in apposition with the left pillar of the diaphragm and left suprarenal capsule. The anterior border of the spleen is thinner than the posterior, and is often

3 M 2

slightly notched, especially towards the lower part. The lower end is pointed, and is in contact with the left end of the arch of the colon, or with the transverse meso-colon. The position of the hilus corresponds with the line of attachment of the gastro-splenic omentum. Along the bottom of this fissure are large openings or depressions, which transmit blood-vessels, with lymphatics and nerves, to and from the interior of the organ. In some cases there is no distinct fissure, but merely a row of openings for the vessels ; and in others the situation of the hilus is occupied by a longitudinal ridge, interrupted by the vascular orifices. The peritoneal connections of the spleen have been already described (pp. 827 and 830). A portion of variable extent behind the hilus, and towards its lower end, will usually be observed deriving its peritoneal covering from the sac of the omentum.

The spleen varies in magnitude more than any other organ in the body ; and this not only in different subjects, but in the same individual, under different conditions, sometimes appearing shrunk, and at others being much distended. On this account it is difficult or impossible to state what are its ordinary weight and dimensions : in the adult it measures generally about 5 or 5½ inches from the upper to the lower end, 3 or 4 inches from the anterior to the posterior border, and 1 or 1½ inch from its external to its internal surface ; and its usual volume, according to Krause, is from 9¾ to 15 cubic inches. In the greater number of a series of cases examined by Reid, its weight ranged from 5 to 7 oz. in the male, and was somewhat less in the female ; but even when perfectly free from disease, it may fluctuate between 4 and 10 ounces. Gray states that the proportion of the spleen to the weight of the adult body varies from 1 : 320 to 1 : 400. In the fœtus the proportion is as 1 : 350. After the age of forty the average weight gradually diminishes, so that in old age the weight of the spleen is to that of the body as 1 : 700 (H. Gray). The specific gravity of this organ, according to Haller, Sœmmerring, and Krause, is about 1·060. In intermittent and some other fevers the spleen is much distended and enlarged, reaching below the ribs, and often weighing as much as 18 or 20 lbs. In enlargement and solidification it has been known to weigh upwards of 40 lbs. ; and it has been found reduced by atrophy to the weight of two drachms.

Small detached roundish nodules are occasionally found in the neighbourhood of the spleen, similar to it in substance. These are commonly named *accessory* or *supplementary* spleens (splenculi ; lienculi). Of these one or two most commonly occur, but a greater number, and even up to twenty-three, have been met with. They are small rounded masses, varying from the size of a pea to that of a walnut. They are usually situated near the lower end of the spleen, either in the gastro-splenic omentum, or in the great omentum. These separate splenculi in the human subject bring to mind the multiple condition of the spleen in some animals, and also the deeper notching of the anterior margin of the organ which sometimes occurs in man.

Structure.—The spleen has two membranous investments—a serous coat derived from the peritoneum, and a special albugineous fibro-elastic tunic. The substance of the organ, which is very soft and easily lacerated, is of a dark reddish-brown colour, but acquires a bright red hue on exposure to the air. Sometimes, however, the substance of the spleen is paler, and has a greyish aspect. It also varies in density, being occasionally rather solid, though friable. The substance of the organ consists of a reticular framework of whitish elastic bands or *trabeculæ*, of an immense proportion of blood-vessels, the larger of which run in elastic canals, and of a peculiar intervening pulpy substance, besides lymphatic vessels and nerves.

The *peritoneal* coat is thin, smooth, and firmly adherent to the elastic tunic beneath, but it may be detached by careful dissection, commencing at the borders of the hilus. It closely invests the surface of the organ, except at the places of its reflection to the stomach and diaphragm, and at the hilus.

The *proper* tunic, much thicker and stronger than the serous, is whitish in colour and highly elastic. It cannot be raised from the spleen, because it is bound down by the trabeculæ of the substance, with the superficial bands of which it is continuous. Along the hilus this coat is reflected into the interior of the spleen, in the form of elastic sheaths or canals, which sur_round or include the large blood-vessels and nerves, and their principal branches. These sheaths ramify with the vessels which they include, and their finer subdivisions are continuous with the trabecular structure. The arrangement of the elastic sheaths and trabeculæ may be easily displayed by pressing and washing out the pulp from a section of the spleen ; and then they are seen to form a close reticulation through the substance. Thus,

Fig. 619.—VERTICAL SEC-TION OF A SMALL SUPER-FICIAL PORTION OF THE HUMAN SPLEEN (from Kölliker). $\frac{1}{38}$

Fig. 619.

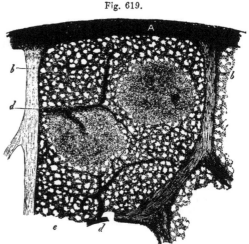

A, peritoneal and fibrous covering ; b, trabeculæ ; cc, Malpighian corpuscles, in one of which an artery is seen cut transversely, in the other longitudinally ; d, injected arterial twigs ; e, spleen-pulp, with the venous spaces and the finest spleen - tissue, the venous spaces all completely full of blood and somewhat wider than na-tural.

the proper coat, the sheaths of the ves-sels, and the trabeculæ, all of a highly elastic nature, constitute a distensible framework, which contains in its inter-stices or areolæ the vessels and the red pulpy substance of the spleen.

Fig. 620.—SECTION OF HARDENED SPLEEN-SUBSTANCE (from Köl-liker after a preparation by Billroth). $\frac{1}{38}$

Fig. 620.

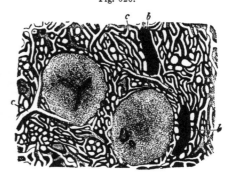

The spleen had been hardened in chromic acid and alcohol. a, Malpighian bodies, one of them having an artery dividing into three twigs within, the other with two arteries cut through ; b, tra-beculæ ; c, artery ; in the re-maining parts the clear spaces are the capillary veins of Billroth, the darker ones are the smallest trabeculæ of the spleen-substance.

These fibrous structures are all composed of interlaced bundles of areolar tissue mixed with a large amount of fine elastic tissue. In addition to these elements, in the spleen of the pig, the dog, and the cat, and to a smaller extent in that of the ox and sheep, there has been found an abundant admixture of plain muscular fibres, resembling those of the middle coat of arteries. Meissner and W. Müller affirm that they also find muscular fibres in the trabeculæ and fibrous coat of the human spleen ; but the existence of such fibres is denied by other observers. The elasticity of the fibrous coat and trabeculæ, together with whatever amount of muscularity they may possess, renders the spleen capable of the great and sudden alterations in size to which it is subject.

The *pulp* of the spleen is of a dark reddish-brown colour : when pressed out from between the trabeculæ it resembles grumous blood, and, like that fluid, it acquires a brighter hue on exposure to the air. This pulpy substance lies altogether outside the veins, between the branches of the venous plexus. As shown by the microscope, it consists chiefly of numerous rounded granular bodies, which have a reddish colour, and are about the size of the blood corpuscles. Their cohesion is very slight, and the terminal tufts of the arterial system of vessels are spread out amongst them. In examining the substance of the spleen, elongated caudate corpuscles are seen in rather large numbers. And besides these there are round nucleated cells, and free nuclei. There are also large cells, some of which are nucleated, and others not, but both of which contain blood-corpuscles in various states of change.

The splenic artery and vein, alike remarkable for their great proportionate size, having entered the spleen by six or more branches, ramify in its interior, enclosed within the elastic sheaths already described. The smaller branches of the arteries run along the trabeculæ, and terminate in the proper substance of the spleen in small tufts of capillary vessels arranged in pencils. These are supported by fine microscopic trabeculæ which run through the pulp in all directions. The main branches of artery which enter the spleen appear to have few or no anastomoses within the substance of the organ, for it has been justly remarked that, if one of them be injected, the material of injection will return by the corresponding vein before spreading to other parts of the spleen : and it only returns by the vein after injection of the pulp. The veins, which greatly exceed the arteries in size, anastomose frequently together, so as to form a close venous plexus, placed in the intervals between the trabeculæ, and supported by them. There is still great difference of opinion as to the manner in which

Fig. 621.

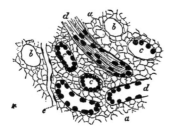

Fig. 621.—A Small Fragment of a Prepared Spleen unravelled (from Kölliker). $\frac{1}{250}$

a, finest reticulum ; *b*, transversely cut capillary veins which have lost their epithelium ; *c*, veins in which the epithelium is more or less preserved ; *d*, longitudinal section of the same ; *e*, a capillary vessel lying in the finest splenic tissue.

the arteries and veins are connected. According to Gray, the capillaries traverse the pulp in all directions, and terminate either directly in the veins, or open into lacunar spaces, from which the veins originate. Billroth and Kölliker admit only the direct termination in veins ; Stieda and W. Müller maintain that a network of intercellular passages intervenes.

Malpighian corpuscles.—On closely inspecting the surface of a section of the spleen, a number of light-coloured spots of variable size are generally seen. For the most part these are evidently the ends of divided trabeculæ or blood-vessels ; but in the ox, pig, sheep, and some other animals, and also in the human subject, there are found distinct whitish vesicular-looking bodies, attached to the trabeculæ which support the small arteries, and imbedded, in groups of six or eight together, in the dark red substance of the spleen. These small vesicular bodies, the *Malpighian corpuscles of the spleen*, are capsules, varying in diameter from $\frac{1}{23}$th to $\frac{1}{60}$th of an inch, and consisting of two coats, the external of which is apparently continuous with the trabecular tissue supporting the arteries. They are filled with a soft, whitish, semi-fluid matter, which contains microscopic globules, resembling, except in colour, those composing the red pulp of the spleen. It may be remarked, that both these kinds of globules are very like the chyle-corpuscles. The vesicular bodies are attached in groups to minute arterial branches ; in some of the lower animals they are sessile, but in the human spleen they are pedunculated. In all cases it is established that they are expansions of the sheaths of the arteries ; those which are sessile are placed usually in the angle of division of two arteries, and are formed by expansion on one side

Fig. 622.—MALPIGHIAN CORPUSCLES OF THE DOG'S SPLEEN (from Kölliker). $\frac{1}{10}$

Fig. 622.

The figure shows a portion of a small artery, to one of the twigs of which the Malpighian corpuscles are attached.

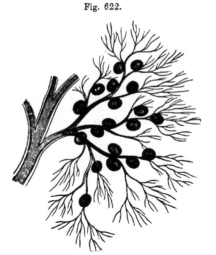

only of the vessels on which they are placed ; those which are pedunculated are pierced by the artery on which they are placed, the expanded sheath having been diffused, as it were, in the capsule, round the vessel (Stieda and Henle). Their walls pass gradually into the pulp on the outside, and on the inside into the contents of the follicle (Busk and Huxley). Capillaries likewise are found within them.

The lymphatic vessels of the spleen, according to Cruikshank and Mascagni, form a superficial and a deep set. The superficial set appear as a network beneath the serous coat, receive occasional branches from the substance of the spleen, and run towards the hilus. The deep lymphatics accompany the blood-vessels, and emerge with them at the hilus, whence, communicating with the superficial set, they proceed along the gastro-splenic omentum to the neighbouring lymphatic glands : the mode in which they commence in the spleen is unknown. The lymphatics of the human spleen, at least the superficial set, are allowed by all to be very difficult to inject. But even in the domestic animals, in which the process is usually more successful, recent observers have not been so fortunate as Cruikshank and Mascagni.

The splenic *nerves*, derived from the solar plexus, surround and accompany the splenic artery and its branches. They have been traced by Remak deeply into the interior of the organ.

The following works on the spleen may be referred to :—Gray, Structure and Use of the Spleen, 1854; Sanders, in Goodsir's Annals; Busk and Huxley on the Malpighian Bodies, in the Sydenham Society's translation of Kölliker's Histology; also Huxley in Micr. Jour., ii. p. 74; Billroth, in Zeitschr. f. Wissensch. Zool., xi. p. 325; W. Müller, Ueber d. fein. Bau der Milz, 1865. Stieda, in Virchow's Archiv, xxiv. p. 540, an abstract of which is given in Medico-Chir. Rev., October, 1862.

Development.—The spleen appears in the fœtus, about the seventh or eighth week, on the left side of the dilated part of the alimentary tube or stomach, and close to the rudiment of the pancreas. By the tenth week it forms a distinct lobulated body placed at the great end of the stomach. After birth it increases rapidly in size; and in a child a few weeks old, it has attained the same proportional weight to the body as in the adult. This organ is peculiar to vertebrate animals.

ORGANS OF RESPIRATION.

THE organs of respiration consist of the thorax (already described), the larynx, the trachea, and the lungs. The larynx, affixed to the upper end of the windpipe, is not only the entrance for air into the respiratory organs from the pharynx, but also the organ of voice, and will be described later.

THE TRACHEA AND BRONCHI.

The *trachea* or *windpipe*, the common air-passage of both lungs, is an open tube which commences at the larynx above, and divides below into two smaller tubes, the right and the left bronchus, one for each lung.

The trachea is placed in the middle plane of the body, and extends from the lower border of the cricoid cartilage of the larynx, on a level with the fifth cervical vertebra in the neck, to a place opposite the third dorsal vertebra in the thorax, where it is crossed in front by the arch of the aorta, and at or immediately below that point it bifurcates into the two bronchi. It usually measures from four inches to four inches and a half in length, and from three-quarters of an inch to one inch in width; but its length and width are liable to continual variation, according to the position of the larynx and the direction of the neck; moreover, it usually widens a little at its lower end, and its diameter is always greater in the male than in the female. In front and at the sides the trachea is rendered cylindrical, firm, and resistant by a series of cartilaginous rings; these, however, are deficient behind, so that the posterior portion is flattened and entirely membranous.

The trachea is nearly everywhere invested by a loose areolar tissue, abounding in elastic fibres, and it is very movable on the surrounding parts. Both in the neck and in the thorax, it rests behind against the œsophagus, which intervenes between it and the vertebral column, and towards its lower part projects somewhat to its left side. The recurrent laryngeal nerves ascend to the larynx on each side in the angle between these two tubes.

In the neck the trachea is situated between the common carotid arteries; at its upper end it is embraced by the lateral lobes of the thyroid body, the middle part or isthmus of which lies across it just below the larynx. It is covered in front by the sterno-thyroid and sterno-hyoid muscles, between which, however, there is left an elongated lozenge-shaped interval in the middle line : this interval is covered in by a strong process of the deep cervical fascia, while, more superficially, another layer not so strong crosses between the sterno-mastoid muscles. The inferior thyroid veins and the *arteria thyroidea ima*, when that vessel exists (p. 340), also lie upon its anterior surface; whilst at the root of the neck, in the episternal notch, the innominate artery and the left carotid pass obliquely over it as they ascend to gain its sides.

In the thorax, the trachea is covered by the first piece of the sternum, together with the sterno-thyroid and sterno-hyoid muscles; lower down, by the left innominate vein, then by the commencement of the innominate artery and left carotid, which pass round to its sides; next by the arch of the aorta and the deep cardiac plexus of nerves, and, quite at its bifurcation, by the extremity of the pulmonary artery, where it divides into its right and left branches. Placed between the two pleuræ, the trachea is contained in the posterior mediastinum, and has on its right side the pleura and pneumo-gastric nerve, and on the left, the left carotid artery, the pneumo-gastric and its recurrent branch, together with some cardiac nerves.

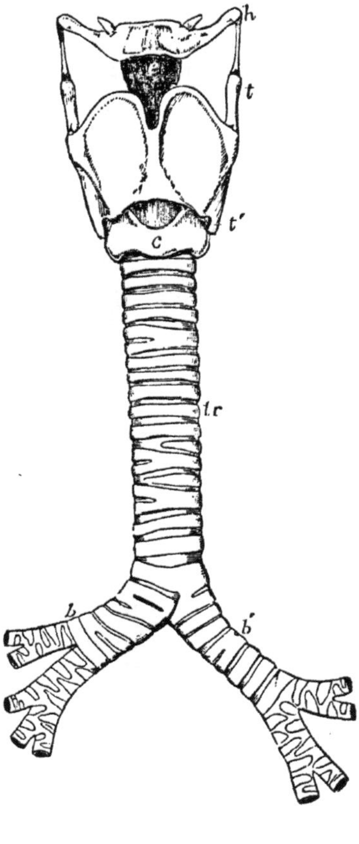

Fig. 623.—Outline showing the General Form of the Larynx, Trachea,
and Bronchi, as seen from before. ½

h, the great cornu of the hyoid bone; *e*, epiglottis; *t*, superior, and *t'*, inferior cornu of the thyroid cartilage; *c*, middle of the cricoid cartilage; *t r*, the trachea, showing sixteen cartilaginous rings; *b*, the right, and *b'*, the left bronchus.

Fig. 623.

The *right and left bronchi* commence at the bifurcation of the trachea and diverge to reach the corresponding lungs. They differ from each other in length, width, direction, and connection with other parts. The *right* bronchus, wider but shorter than the left, measuring about an inch in length, passes outwards almost horizontally into the root of the right lung on a level with the fourth dorsal vertebra: it is embraced above by the vena azygos, which hooks forwards over it, to end in the vena cava superior; the right pulmonary artery lies at first below it and then in front of it. The *left* bronchus, smaller in diameter, but longer than the right, being nearly two inches in length, inclines ob-liquely downwards and outwards beneath the arch of the aorta to reach the root of the left lung, which it enters on a level with the fifth dorsal vertebra, that is, about an inch lower than the right bron-chus. The left bronchus crosses in front of the œsophagus and descend-ing aorta: the arch of the aorta turns backwards and to the left over it, and the left pulmonary artery lies first above it and then on its anterior surface. The remaining connections of each bronchus, as it lies within the root of the corresponding lung, and the mode in which it sub-divides there into bronchia, will be hereafter described.

In form the bronchi exactly resemble the trachea on a smaller scale; they are rounded and firm in front and at the sides, where they are provided with imperfect cartilaginous rings, and they are flattened and membranous behind.

Structure of the Trachea.

The trachea presents for consideration the elastic framework of incomplete cartilaginous rings, layers of fibrous, muscular, and elastic substance, and the lining mucous membrane, with glands.

The *cartilages* and *fibrous membrane.*—The cartilages are from sixteen to twenty in number. Each presents a curve of rather more than two-thirds of a circle, resembling the letter C. The depth from above downwards is from one line and a half to two lines, and the thickness half a line. The outer surface of each is flat, but the inner surface is convex from above downwards, so as to give greater thickness in the middle than at the upper and lower edge. The cartilages are held together by a strong fibrous mem-

Fig. 624.

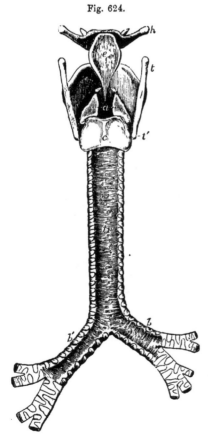

Fig. 624.—Outline showing the General Form of the Larynx, Trachea, and Bronchi as seen from behind. ½

h, great cornu of the hyoid bone; *t*, superior, and *t'*, the inferior cornu of the thyroid cartilage; *e*, the epiglottis; *a*, points to the back of both the arytenoid cartilages, which are surmounted by the cornicula; *c*, the middle ridge on the back of the cricoid cartilage; *t r*, the posterior membranous part of the trachea; *b*, *b'*, right and left bronchi.

brane. This membrane is elastic and extensible in a certain degree, and not only occupies the intervals between the cartilages, but is prolonged over their outer and inner surfaces, so that they are, as it were, imbedded in it.

The layer covering the outer side of the rings is stronger than that within them; and from this circumstance, together with the roundness of their inner surfaces, they may be felt more prominently on the interior than on the exterior of the trachea.

The cartilages terminate abruptly behind. At the back of the trachea, where they are altogether wanting, the fibrous membrane is continued across between their ends, but it is here looser in its texture.

The first or highest cartilage, which is connected by the fibrous membrane with the lower margin of the cricoid cartilage, is broader than the rest, and is often divided at one end. Sometimes it coalesces in a greater or less extent with the cricoid or with the succeeding cartilage. The lowest cartilage, placed at the bifurcation of the trachea into the bronchi, is peculiar in shape ; its lower border being prolonged downwards, and at the same time bent backwards so as to form a curved projection between the two bronchi. The cartilage next above this is slightly widened in the middle line. Sometimes the extremities of two adjacent cartilages are united together, and not unfrequently a cartilage is divided at one end into two short branches, the opposite end of that next it being likewise bifurcated so as to maintain the parallelism of the entire series. The use of these cartilaginous hoops is to keep the trachea open, a condition essential for the free passage of air into the lungs.

Muscular fibres.—Between the fibrous and the mucous membrane at the posterior flattened part of the trachea, there is found a continuous pale reddish layer, consisting of unstriped muscular fibres which pass across, not only between the posterior extremities of the cartilages, but opposite the intervals between the rings also, and have the power of diminishing the area of the tube by approximating the ends of the cartilages. Those which are placed opposite the cartilages are attached to the ends of the rings, and encroach also for a short distance upon the adjacent part of their inner surface.

Outside the transverse fibres are some small fasciculi having a longitudinal direction. These arise, by minute tendons of elastic tissue, in part from the inner surface of the end of the tracheal rings, and in part from the external fibrous membrane.

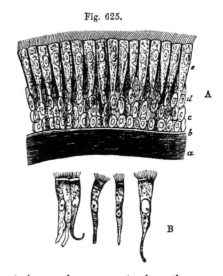

Fig. 625.—CILIATED EPITHELIUM OF THE RESPIRATORY MUCOUS MEMBRANE.

Fig. 625.

A, vertical section of the epithelial lining of the human windpipe (from Kölliker). $\frac{350}{1}$ *a*, *b*, subjacent membrane ; *c*, lowest or spheroidal cells ; *d*, middle or oval cells ; *e*, superficial elongated and ciliated cells.

B, separate columnar and ciliated epithelial cells from the human nasal membrane. $\frac{300}{1}$

Elastic fibres.—Situated immediately beneath the tracheal mucous membrane, and adhering intimately to it, are numerous longitudinal fibres of yellow elastic tissue. They are found all round the tube, internal to the cartilages and the muscular layer, but are much more abundant along the posterior membranous part, where they are principally collected into distinct longitudinal bundles, which produce visible elevations or flutings of the mucous membrane. These bundles are particularly strong and numerous opposite the bifurcation of the trachea. The elastic longitudinal fibres serve to restore the windpipe to its ordinary size after it has been stretched in the direction of its length.

The *glands.*—The trachea is provided with very numerous mucous glands, the constant secretion from which serves to lubricate its internal surface. The largest of these glands are small roundish lenticular bodies, situated at the back part of the tube, lying close upon the outer surface of the fibrous layer, or occupying little recesses formed between its meshes : these are compound glands ; their excretory ducts pass forwards between the muscular fibres and open on the mucous membrane, where multitudes of minute orifices are perceptible. Other similar but smaller glands are found upon and within the fibrous membrane between the cartilaginous rings. Lastly, there appear to be still smaller glands lying close beneath the mucous coat.

The *mucous membrane.*—This is smooth and of a pale pinkish white colour in health, though when congested or inflamed, it becomes intensely purple or crimson. It is covered with a ciliated columnar epithelium, the vibratile movements of which, as may be best seen at the back of the trachea of an animal, tend to drive the mucous secretion upwards towards the larynx. The epithelium is stratified, oval nucleated cells being disposed several rows deep, beneath the columnar cells which bear the cilia.

Vessels and Nerves.—The arteries of the trachea are principally derived from the inferior thyroid. The larger branches run for some distance longitudinally, and then form a superficial plexus with rounded meshes. The veins enter the adjacent plexuses of the thyroid veins. The nerves come from the trunk and recurrent branches of the pneumo-gastric, and from the sympathetic system.

Structure of the Bronchi.

The general structure of the bronchi corresponds with that of the trachea in every particular. Their *cartilaginous* rings, which resemble those of the trachea in being imperfect behind, are, however, shorter and narrower. The number of rings in the right bronchus varies from six to eight, whilst in the left the number is from nine to twelve.

The bronchi are supplied by the bronchial arteries and veins, and the nerves are from the same source as those of the trachea.

THE LUNGS AND PLEURÆ.

The *lungs,* placed one on the right and the other on the left of the heart and large vessels, occupy by far the larger part of the cavity of the chest, and during life are always in accurate contact with the internal surface of its wall. Each lung is attached at a comparatively small part of its inner or median surface by a part named the *root,* and by a thin membranous fold which is continued downwards from it. In other directions the lung is free and its surface is closely covered by a serous membrane, belonging to itself and to the corresponding side of the thorax, and named accordingly, the right or left *pleura.*

THE PLEURÆ.

The *pleuræ* are two independent serous membranes forming two shut sacs, quite distinct from each other, which line the right and left sides of the thoracic cavity, form by their approximation in the middle line the mediastinal partition, and are reflected each upon the root and over the entire free surface of the corresponding lung.

Each pleura consists of a *visceral* and a *parietal* portion : the visceral portion, *pleura pulmonalis*, covers the lung ; and the parietal portion lines the ribs and intercostal spaces, *pleura costales*, covers the upper convex surface of the diaphragm, enters into the formation of the mediastinum, and is reflected on the sides of the pericardium.

The *mediastinum*, or partition between the two pleural cavities, is formed by the reflection of each pleura from the anterior wall of the chest backwards on the pericardium to the root of the lung, and from the back of the root of the lung to the vertebral column. Its division into anterior, middle,

Fig. 626.

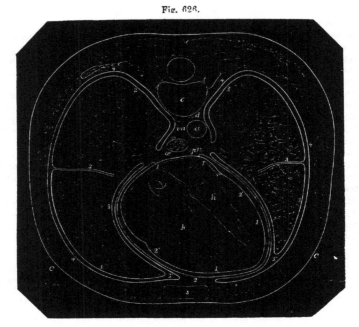

Fig. 626.—Transverse Section of the Chest of a Fœtus, illustrating the Inflections of the Pericardium and Pleuræ (after Luschka and from nature).

The sketch represents the upper surface of the lower section ; the division is carried nearly in a horizontal plane on a level with the interval in front between the fifth and sixth ribs. *s*, the sternum ; *c*, the body of the seventh dorsal vertebra ; *h*, the right, and *h'*, the left ventricle ; *œ*, the œsophagus ; *p n*, the left pneumogastric nerve ; near these letters respectively, the root of the right and left lungs; the right pneumogastric nerve is behind the œsophagus ; *a*, the aorta ; *v a*, the vena azygos ; *d*, thoracic duct ; 1, the cardiac pericardium ; 2, the external pericardium ; 2', the cavity of the pericardium ; 3, the pulmonary pleura passing over the surface, and reflected at the roots of the lungs ; 3', their cavity, and on the right side, the reflection at the mediastinum to the surface of the pericardium ; 4, the external or costal pleuræ; *c, c*, the walls of the chest inclosing the ribs, pectoral muscles, &c.

and posterior mediastina, and the position and contents of each, have been already described (p. 297).

At the root of each lung which is enclosed by its own pleura, the visceral and parietal portions of this membrane are continuous with each other ; and, commencing immediately at the lower border of the root, there is found a triangular fold or duplicature of the serous membrane, extending vertically between the inner surface of the lung and the posterior mediastinum, and reaching down to the diaphragm, to which it is attached by its extremity ; this fold, which serves to attach the lower part of the lung, is named *ligamentum latum pulmonis.*

The upper part of each pleura, which receives the apex of the corresponding lung, projects in the form of a cul-de-sac through the superior aperture of the thorax into the neck, reaching an inch or even an inch and a half above the margin of the first rib, and passing up under cover of the scaleni muscles,—a small slip of which, arising from the transverse process of the last cervical vertebra, is described by Sibson as expanding into a dome-like aponeurosis or fascia, which covers or strengthens the pleural cul-de-sac, and is attached to the whole of the inner edge of the first rib. The right pleura is generally stated to reach higher in the neck than the left ; but, in twenty observations recorded by Hutchinson, the right lung was higher in ten cases, and the left in eight, whilst in two the height was equal on the two sides. Anteriorly the pleural sacs of opposite sides come nearly or altogether into contact behind the second piece of the sternum, and continue so for some distance ; but opposite the lower end of the sternum the right pleura passes beyond the middle line or remains close to it, while the left recedes to a variable distance from the sternum. Inferiorly the pleuræ do not pass quite down to the attachments of the diaphragm, but leave a portion of its circumference in contact with the costal parietes. Owing to the height of the diaphragm on the right side (corresponding with the greater convexity of the liver), the right pleural sac is shorter than the left ; it is at the same time wider. According to Luschka the right pleura, opposite a line descending from the axilla, reaches down to the lower border of the ninth rib, while the left pleura in the same transverse vertical plane reaches to the lower border of the tenth rib.

Structure.—The pleura possesses the usual characters of serous membrane. The costal part of the membrane is the thickest, and may be easily raised from the ribs and intercostal spaces. It is strengthened in these situations by a layer of subserous areolar tissue of considerable thickness. On the pericardium and diaphragm the pleura is thinner and more firmly adherent; but it is thinnest and least easily detached upon the surface of the lungs.

Luschka has described nerves in this membrane, with fine and coarse fibres, which are traceable to the phrenic and sympathetic. Kölliker states that, in the pleura pulmonalis of man, branches of nerves may be seen accompanying the ramifications of the bronchial arteries.

THE LUNGS.

Form.—Each lung is shaped somewhat like a cone, having its base turned downwards, and its inner side much flattened. The *base* is broad, concave, and of a semilunar form, and rests upon the arch of the diaphragm. It is bounded by a thin margin, which is received in the angle between the ribs and the diaphragm ; and it reaches much lower down behind and at the outer side than in front and towards the middle line. The *apex* forms a blunted point, and, as already mentioned, reaches into the root of the neck, above the margin of the first rib, where it is separated from the first portion of the subclavian artery by the pleural membrane.

The *outer surface* of the lung, which moves upon the thoracic parietes, is smooth, convex, and of great extent, corresponding with the arches of the ribs and costal cartilages. It is of greater depth behind than in front. The *posterior* border is obtuse or rounded, and is received into the deep groove formed by the ribs at the side of the vertebral column; measured from above downwards, it is the deepest part of the lung. The *anterior* border is thin and overlaps the pericardium, forming a sharp margin, which touches the sides of the anterior mediastinum, and, opposite the middle of the sternum, is separated during inspiration from the corresponding margin of the opposite lung only by the two thin and adherent layers of the mediastinal septum. The *inner surface* is concave, and in part adapted to the convex pericardium. Upon this surface, somewhat above the middle of the lung, and considerably nearer to the posterior than the anterior border, is the *root*, where the bronchi and great vessels join the lung. Each lung is traversed by a long and deep fissure, which is directed from above and behind, downwards and forwards. It commences upon the posterior border of the lung, about three inches from the apex, and extends obliquely downwards to the anterior and inferior angle, penetrating from the outer surface to within some inches of the root of the organ. The *upper lobe*, the portion of lung which is situated above this fissure, is smaller than the portion below, and is shaped like a cone with an oblique base, whilst the *lower* and larger lobe is more or less quadrilateral. In the right lung there is a second and shorter fissure, which runs forwards and upwards from the principal fissure to the anterior margin, thus marking off a third small portion, or *middle lobe*, which appears like an angular piece separated from the anterior and lower part of the upper lobe. The left lung, which has no such middle lobe, presents a deep notch in its anterior border, into which the apex of the heart (enclosed in the pericardium) is received. This notch is formed by the rapid retreat of the anterior margin of the upper lobe from the middle line, opposite the lower half of the sternum; while inferiorly a tongue-like process of the lower lobe usually projects slightly inwards towards the middle line. Besides these differences in form which distinguish the lungs, it is to be noted that the right lung is shorter, but at the same time wider, than the left, the perpendicular measurement of the former being less, owing to the diaphragm rising higher on the right side to accommodate the liver, whilst the breadth of the left lung is narrowed, owing to the heart and pericardium encroaching on the left half of the thorax. On the whole, however, as is seen on a comparison of their weights, the right is the larger of the two lungs.

At the apices and posterior borders the extent of the lungs accurately corresponds with that of the pleural sacs which contain them, but at the anterior and inferior margins it is not so: the anterior margins pass forwards most completely between the mediastinal and costal pleura during inspiration, and retire to a variable degree from between them in expiration; and in like manner the inferior margins descend, during inspiration, between the costal and diaphragmatic pleuræ, while probably at no time do they ever descend completely to the line of reflection between those membranes.

Weight, Dimensions, and Capacity.—The lungs vary much in size and weight according to the quantity of blood, mucus, or serous fluid they may happen to contain, which is greatly influenced by the circumstances immediately preceding death, as well as by other causes. The weight of both lungs together, as generally stated, ranges from thirty to forty-eight ounces, the more prevalent weights being found between thirty-six and forty-two ounces. The proportion borne by the right lung to the left is nearly that of

and lymphatic glands, are placed on a plane posterior to the great blood-vessels ; the pulmonary artery lies more forward than the bronchus, and to a great extent conceals it, whilst the pulmonary veins are placed still farther in advance. The pulmonary plexuses of nerves lie on the anterior and posterior aspect of the root beneath the pleura, the posterior being the larger of the two.

The order of position of the great air-tube and pulmonary vessels from above downwards differs on the two sides ; for whilst on the right side the bronchus is highest and the pulmonary artery next, on the left, the air-tube, in passing obliquely beneath the arch of the aorta, is depressed below the level of the left pulmonary artery, which is the highest vessel. On both sides the pulmonary veins are the lowest of the three sets of tubes.

Before entering the substance of the lung, the bronchus divides into two branches, an upper and a lower, one for each lobe. The lower branch is the larger of the two, and on the right side gives off a third small branch which enters the middle lobe of that lung.

The pulmonary artery also divides, before penetrating the lung to which it belongs, into two branches, of which the lower is the larger and supplies the inferior lobe. The upper of these two branches gives the branch to the middle lobe. A similar arrangement prevails in regard to the right pulmonary veins, the upper one of which is formed by branches proceeding from the superior and middle lobes of the right lung.

STRUCTURE OF THE LUNGS.

Coverings.—Beneath the serous covering, already noticed, there is placed a thin layer of *subserous* areolar membrane mixed with much elastic tissue. It is continuous with the areolar tissue in the interior of the lung, and has been described as a distinct coat under the name of the second or deeper layer of the pleura. In the lungs of many animals, such as the lion, seal, and leopard, this subserous layer forms a very strong membrane, composed principally of elastic tissue.

Lobules.—The *substance* of the lung is composed of numerous small lobules which are attached to the ramifications of the air-tubes, and are held together by those tubes, by the blood-vessels, and by an interlobular areolar tissue. These lobules are of various sizes, the smaller uniting into larger ones ; they are bounded by flattened sides, and compactly fitted to each other and to the larger air-tubes and vessels of the lungs, those on the surface of the organ having bases, turned outwards, from half a line to a line in diameter. Though mutually adherent by means of fine areolar tissue, they are quite distinct one from the other, and may be readily separated by dissection in the lungs of young animals, and in those of the human fœtus. They may be regarded as lungs in miniature, the same elements entering into their composition as form the lung itself. The structure of a single lobule represents in fact that which is essential in the entire organ, each lobule, besides its investment of areolar membrane, being made up of the following constituents : the *air-tubes* and their *terminating cells,* the *pulmonary* and *bronchial* blood-vessels, with lymphatics, nerves, and interstitial areolar tissue.

Air-tubes.—The principal divisions of the bronchi, as they pass into the lungs, divide into tubes of less calibre, and these again subdivide in succession into smaller and smaller *bronchial* tubes, or *bronchia*, which, diverging in all directions, never anastomose, but terminate separately in the pulmonary parenchyma.· The prevailing form of division is dichotomous ; but sometimes three branches arise together, and often lateral branches are given off at intervals from the sides of a main trunk.   The larger branches diverge at rather acute angles, but the more remote and smaller ramifications spring less and less acutely.   After a certain stage of subdivision each bronchial tube is reduced to a very small size, and, forming what has been termed a *lobular bronchial tube*, enters a distinct pulmonary lobule, within which it undergoes still farther division, and at last ends in the small cellular recesses named *air-cells* or *pulmonary cells*.

Within the lungs the air-tubes are not flattened behind like the bronchi and trachea, but form completely cylindrical tubes.   Hence, although they contain the same elements as the larger air-passages, reduced gradually to a state of greater and greater tenuity, they possess certain peculiarities of structure.   Thus, the *cartilages* no longer appear as imperfect rings running only upon the front and lateral surfaces of the air-tube, but are scattered over all sides of the tube in the form of irregularly shaped plates of various sizes.   These are most developed at the points of division of the bronchia, where they form a sharp. concave ridge projecting inwards into the tube. They may be traced, becoming rarer and rarer and more reduced in size, as far as bronchia only one-fourth of a line in diameter, beyond which the tubes are entirely membranous.   The *fibrous* coat extends to the very smallest tubes, becoming thinner by degrees and degenerating into areolar tissue. The *mucous membrane*, which extends throughout the whole system of air passages, and is continuous with that lining the air-cells, is also thinner than in the trachea and bronchus, but it retains its ciliated columnar epithelium.   The yellow longitudinal bundles of *elastic* fibres are very distinct in both the large and small bronchia, and may be followed by dissection as far as the tube can be laid open, and by the microscope into the smallest

Fig. 627.—PORTION OF THE OUTER SURFACE OF THE Cow's LUNG (from Kölliker after Harting. $\frac{30}{1}$

a, pulmonary vesicles filled artificially with wax ; *b*, the margins of the smallest lobules or infundibula.

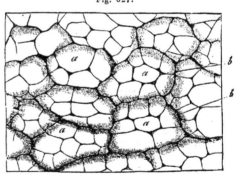

Fig. 627.

tubes.   The *muscular* fibres, which in the trachea and bronchi are confined to the back part of the tube, surround the bronchial tubes with a continuous layer of annular fibres, lying inside the irregular   cartilaginous plates ; they are found, however, beyond the place where the cartilages cease to exist, and appear as irregular annular fasciculi even in the smallest tubes.

Air-cells or *Pulmonary vesicles.*—These cells, in which the finest ramifications of each lobular bronchial tube ultimately terminate, are in the natural state always filled with air.   They are readily seen on the surface and in a section of a lung which has been inflated with air and dried ;

also upon portions of fœtal or adult lung injected with mercury. In the lungs of some animals, as of the lion, cat, and dog, they are very large, and are distinctly visible on the surface of the organ. In the adult human lung their most common diameter is about $\frac{1}{100}$th of an inch, but it varies from $\frac{1}{120}$th to $\frac{1}{70}$th of an inch; they are larger on the surface than in the interior, and largest towards the thin edges of the organ: they are also said to be very large at the apex of the lung. Their dimensions go on increasing from birth to old age, and they are larger in men than in women. In the infant their diameter is usually under $\frac{1}{200}$th of an inch.

The small bronchial tube entering each lobule divides and subdivides from four to nine times, according to the size of the lobule; its branches, which diverge at less and less acute angles, at first diminish at each subdivision, but afterwards continue stationary in size, being from $\frac{1}{30}$th to $\frac{1}{30}$th of an inch in diameter. They lose at last their cylindrical form, and are converted into irregular *lobular passages*, beset, at first sparingly, but afterwards closely and on all sides, with numerous little recesses or dilatations, and ultimately terminate near the surface of the lobule in a group of similar recesses. These small recesses, whether seated along the course or at the

Fig. 628.

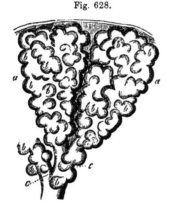

Fig. 628. — SEMIDIAGRAMMATIC REPRESENTATION OF TWO SMALL LOBULI FROM NEAR THE SURFACE OF THE LUNG OF A NEW-BORN CHILD (from Kölliker). $\frac{25}{1}$

a, exterior of the two lobuli or infundibula; *b*, pulmonary vesicles or alveoli on these and on *c*, the smallest bronchial ramifications.

extremity of an air passage, are the *air-cells* or *alveoli ;* and each group of alveoli with the comparatively large passage between them constitutes an *infundibulum*, so called from the manner in which it dilates from the extremity of the bronchial tube. The arrangement of these finest air-passages and air-cells closely resembles, though on a smaller scale, the reticulated structure of the tortoise's lung, in which large open passages lead in all directions to clusters of wide alveoli, separated from each other by intervening septa of various depths.

At the point where the small bronchial tubes lose their cylindrical character, and become covered on all sides with the cells, their structural elements also undergo a change. The *muscular* layer disappears, the longitudinal elastic bundles are broken up into an interlacement of *areolar* and *elastic tissue*, which surrounds the tubes and forms the basis of their walls. The *mucous* membrane becomes exceedingly delicate, consisting merely of a thin transparent membrane, covered by a stratum of squamous instead of ciliated cylindrical epithelium.

The walls of the alveoli, their orifices, and the margins of the septa, are supported and strengthened by scattered and coiled elastic fibres, in addition to which, according to Moleschott, Gerlach, and Hirschmann, there is likewise an intermixture of muscular fibres. It was stated by Rainey, and corroborated by Todd and Bowman, and it is still maintained by Henle, Luschka, and others, that the alveoli are destitute of all epithelium. The presence of nuclei, however, situated in the capillary meshes, and of larger size than those which belong to the capillary walls, is allowed on all hands ; and the majority of recent observers declare the existence of exceedingly

Fig. 629.

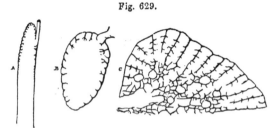

Fig. 629.—DIAGRAMS ILLUSTRATING THE PROGRESSIVE ADVANCE IN THE CELLULAR STRUCTURE OF THE LUNGS OF REPTILES.

A, the upper portion of the lung of a serpent : the summit has cellular walls, the lower part forms merely a membranous sac. B, lung of the frog, in which the cellular structure extends over the whole internal surface of the lung, but is more marked at the upper part. C, lung of the turtle : the cells here have extended so as to occupy nearly the whole thickness of the lung.

delicate squamous epithelial cells. These, according to Eberth, lie in the capillary meshes, from one to three in each, but leave the surfaces of the capillary vessels uncovered. According to others, they join each other over the capillary blood-vessels.

Fig. 630.

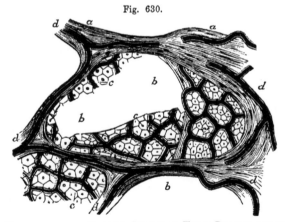

Fig. 630.—FRAGMENT OF THE INJECTED LUNG OF A YOUNG PIG, SHOWING THE MINUTE STRUCTURE OF THE VESICLES (from Hirschmann and Chrzonszczewsky).

a, the areolar and elastic tissue supporting the vesicles ; b, the cavities of two of the vesicles partially cut through ; c, the meshes of the pulmonary capillaries, the latter being filled with dark colouring matter, and the meshes being occupied by regular hexagonal epithelial cells, which in various places are seen to meet each other over the capillary vessels; d, the intervesicular pulmonary vessels. In this instance, the cells observed in each mesh have been more numerous than they are said by Eberth to be in the human subject.

The following writers, among others, maintain the existence of epithelium in the air-cells: Addison, in Phil. Trans., 1842; Rossignol, Recherches sur la Structure intime du Poumon, 1846; Waters, Anatomy of the Human Lung, 1860; Kölliker, in his Gewebelehre; Eberth, in Virchow's Archiv. xxiv., p. 503; and Julius Arnold, in Virchow's Archiv. xxviii., p. 433; Hirschmann, in the same, xxxvi. with addition and drawings by Chrzonszczewsky. The following are among those who deny the existence of epithelium :—Rainey, in Med.-Chir. Trans., vol. xxviii., 1845; and in Brit. and For. Med.-Chir. Review, 1855; Radclyffe Hall, in Med.-Chir. Review, July, 1857; Luschka and Henle, in their works on Human Anatomy; Badoky, in Virchow's Archiv., xxxiii. p. 264.

Fig. 631.

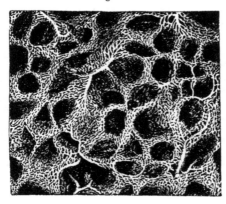

Fig. 631.—Capillary Network of the Pulmonary Blood-vessels in the Human Lung (from Kölliker). $\frac{60}{1}$

The *capillary* network of the pulmonary vessels is spread beneath the thin transparent mucous membrane of both the terminal and lateral air-cells, and is found wherever the finest air-tubes have lost their cylindrical character, and become beset with cells. Around the exterior of each cell there is an arterial circle, which communicates freely with similar neighbouring circles, the capillary systems of ten or twelve cells being thus connected together, as may be seen upon the surface of the lung. From

Fig. 632.

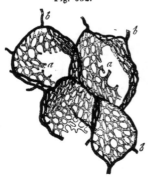

Fig. 632.—Capillary Network on the Pulmonary Vesicles of the Horse (from Frey after a preparation by Gerlach). $\frac{100}{1}$

a, the capillary network; *b*, the terminal branches of the pulmonary artery passing towards and surrounding in part each pulmonary vesicle.

these circular vessels, which vary in diameter from $\frac{1}{1270}$th to $\frac{1}{846}$th of an inch, the capillary network arises, covering the bottom of each cell, ascending also between the duplicature of mucous membrane in the intercellular septa, and surrounding the openings of the cells. As was pointed out by Rainey, the capillary network, where it rises into the intercellular partitions, although it forms a double layer in the lungs of reptiles, is single in the lungs of man and mammalia.

The capillaries are very fine, the smallest measuring, in injected specimens, from $\frac{1}{2540}$th to $\frac{1}{3000}$th of an inch; the network is so close that the meshes are scarcely wider than the vessels themselves. Those which lie nearest to the mouths of the alveoli are observed arching and coiled over and through the elastic fibres found in the interalveolar septa (Luschka and Badoky). The coats of the capillaries are also exceedingly thin, and thus more readily allow the free exhalation and absorption of which the pulmonary cells are the seat.

The *branches of the pulmonary artery* accompany the bronchial tubes, but they subdivide more frequently, and are much smaller, especially in their remote ramifications. They ramify without anastomoses, and at length terminate upon the walls of the air-cells and on those of the bronchia in a fine and dense *capillary network*, from which the radicles of the *pulmonary veins* arise. The smaller branches of these veins, especially near the surface of the lung, frequently do not accompany the bronchia and arterial branches (Addison, Bourgery), but are found to run alone for a short distance through the substance of the organ, and then to join some deeper vein which passes by the side of a bronchial tube, uniting together, and also forming, according to Rossignol, frequent lateral communications. The veins coalesce into large branches, which at length accompany the arteries, and thus proceed to the root of the lung. In their course through the lung, the artery is usually found above and in front of a bronchial tube, and the vein below.

The pulmonary vessels differ from the systemic in regard to their contents, inasmuch as the artery conveys dark blood, whilst the veins carry red blood. The pulmonary veins, unlike the other veins of the body, are not more capacious than their corresponding arteries; indeed, according to Winslow, Santorini, Haller, and others, they are somewhat less so. These veins have no valves. Lastly, it may be remarked that, whilst the arteries of different lobules are independent, their veins freely anastomose together.

The bronchial vessels.—The bronchial *arteries* and *veins*, which are much smaller than the pulmonary vessels, carry blood for the nutrition of the lung, and are doubtless also the principal source of the mucous secretion found in the interior of the air-tubes, and of the thin albuminous fluid which moistens the pleura pulmonalis.

The *bronchial arteries*, from one to three in number for each lung, arise from the aorta, or from an intercostal artery, and follow the divisions of the air-tubes through the lung. They are ultimately distributed in three ways: (1) many of their branches ramify in the bronchial lymphatic glands, the coats of the large blood-vessels, and in the fibrous and muscular walls of the large and small air-tubes, and give supply to a copious capillary plexus in the bronchial mucous membrane, which in fine bronchial tubes is continuous with that supplied by the pulmonary artery; (2) others form plexuses in the interlobular areolar tissue; (3) branches spread out upon the surface of the lung beneath the pleura, forming plexuses and a capillary network, which may be distinguished from those of the pulmonary vessels of the superficial air-cells by their tortuous course and open arrangement, and also by their being outside the investing membrane of the lobules, and by ultimately ending in the branches of the *superficial* set of *bronchial veins*.

The *bronchial veins* have not quite so large a distribution in the lung as the bronchial arteries, since part of the blood carried by the bronchial arteries is returned by the pulmonary veins. The superficial and deep bronchial veins unite at the root of the lung, opening on the right side into the vena azygos, and on the left usually into the superior intercostal vein.

The *absorbent vessels* of the lungs have been already sufficiently described (p. 496).

Nerves.—The lungs are supplied with nerves from the anterior and posterior pulmonary plexuses (pp. 623, 693). These are formed chiefly by branches from the pneumogastric nerves, joined by others from the sympathetic system. The fine nervous cords enter at the root of the lung, and follow the air-tubes. Their final distribution requires further examination. According to Remak, whitish filaments from the par vagum follow the bronchia as far nearly as the surface of the lung, and greyish filaments, proceeding from the sympathetic, and having very minute ganglia upon them in their course, pass both to the bronchia and pleura. Julius Arnold has discovered remarkable bell-shaped ganglionic corpuscles terminating the pulmonary nerves of the frog. (Virchow's Archiv., vol. xxviii. p. 453.)

DEVELOPMENT OF THE LUNGS AND TRACHEA.

The lungs first appear as two small protrusions upon the front of the œsophageal portion of the alimentary canal, completely hid by the rudimentary heart and liver. These primitive protrusions or tubercles are visible in the chick on the third day of incubation, and in the embryos of mammalia and of man at a corresponding stage of advancement. Their internal cavities communicate with the œsophagus, and are lined by a prolongation of its inner layer. At a later period they are connected with the œsophagus by means of a long pedicle, which ultimately forms the trachea, whilst the bronchia and air-cells are developed by the progressive ramification of the internal cavity in the form of cæcal tubes, after the manner of the ducts of glands. According to Kölliker, the human lung in the latter half of the second month presents a granular appearance on the surface, produced by the primitive air-cells placed at the extremities of ramified tubes, which occupy the whole of the interior of the organ; the ramification of the bronchial twigs and multiplication of air-cells goes on increasing, and this to such an extent that the air-cells in the fifth month are only half the size of those which are found in the fourth month.

Fig. 633.

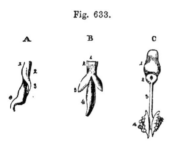

A   **B**   **C**

Fig. 633.—Sketch illustrating the Development of the Respiratory Organs (from Rathke).

A, œsophagus of a chick, on the fourth day of incubation, with the rudimentary lung of the left side, seen laterally; 1, the front, and 2, the back of the œsophagus; 3, rudimentary lung protruding from that tube; 4, stomach. B, the same seen in front, so as to show both lungs. C, tongue and respiratory organs of embryo of the horse; 1, tongue; 2, larynx; 3, trachea; 4, lungs seen from behind.

For a long time the lungs are very small, and occupy only a limited space at the back part of the chest. In an embryo, 16 lines in length, their proportionate weight to the body was found by Meckel to be 1 to 25; in another, 29 lines long, it was 1 to 27; in another, 4 inches in length, 1 to 41; and at the full period, 1 to 70. Huschke found that the lungs of still-born male children were heavier in proportion to the weight of the body than those of female children; the ratio being, amongst females, 1 to 76, and in males, 1 to 55.

Changes after birth.—The lungs undergo very rapid and remarkable changes after birth, in consequence of the commencement of respiration: these affect their size, position, form, consistence, texture, colour, and weight, and should be carefully studied, as furnishing the only means of distinguishing between a still-born child and one that has respired.

1. *Position, size, and form.*—In a fœtus at the full period, or in a still-born child, the lungs, comparatively small, lie packed at the back of the thorax, and do not entirely cover the sides of the pericardium; subsequently to respiration, they expand, and completely cover the pleural portions of that sac, and are also in contact with almost the

whole extent of the thoracic wall, where it is covered with the pleural membrane. At the same time, their previously thin sharp margins become more obtuse, and their whole form is less compressed.

2. *Consistence, texture, and colour.*—The introduction of air and of an increased quantity of blood into the fœtal lungs, which ensues immediately upon birth, converts their tissue from a compact, heavy, granular, yellowish-pink, gland-like substance, into a loose, light, rose-pink, spongy structure, which, as already mentioned, floats in water. The changes thus simultaneously produced in their consistence, colour, and texture, occur first at their anterior borders, and proceed backwards through the lungs : they, moreover, appear in the right lung a little sooner than in the left.

3. *Weight.*—The *absolute weight* of the lungs, having gradually increased from the earliest period of development to birth, undergoes at that time, from the quantity of blood then poured into them, a very marked addition, amounting to more than one-third of their previous weight : for example, the lungs before birth weigh about one and a half ounce, but, after complete expansion by respiration, they weigh as much as two and a half ounces. The *relative weight* of the lungs to the body, which at the termination of intra-uterine life is about 1 to 70, becomes, after respiration, on an average, about 1 to 35 or 40; a proportion which is not materially altered through life. Their *specific gravity* is at the same time changed from 1·056 to about ·342.

4. *Changes in the trachea after birth.*—In the fœtus the trachea is flattened before and behind, its anterior surface being even somewhat depressed ; the ends of the cartilages touch ; and the sides of the tube, which now contains only mucus, are applied to one another. The effect of respiration is at first to render the trachea open, but it still remains somewhat flattened in front, and only later becomes convex.

THE LARYNX, OR ORGAN OF VOICE.

The upper part of the air-passage is modified in its structure to form the *organ of voice.* This organ, named the *larynx,* is placed at the upper and fore part of the neck, where it forms a considerable prominence in the middle line. It lies between the large vessels of the neck, and below the tongue and os hyoides, to which bone it is suspended. It is covered in front by the cervical fascia along the middle line, and on each side by the sterno-hyoid, sterno-thyroid, and thyro-hyoid muscles, by the upper end of the thyroid body, and by a small part of the inferior constrictor of the pharynx. Behind, it is covered by the pharyngeal mucous membrane, and above it opens into the cavity of the pharynx.

The larynx consists of a framework of cartilages, articulated together and connected by proper ligaments, two of which, named the *true vocal cords,* are more immediately concerned in the production of the voice. It also possesses muscles, which move the cartilages one upon another, and modify the form and tension of its apertures, a mucous membrane lining its internal surface, numerous mucous glands, and lastly, blood-vessels, lymphatics, and nerves, besides areolar tissue and fat.

Cartilages of the Larynx.

The cartilages of the larynx consist of three single and symmetrical pieces, named respectively. the *thyroid cartilage,* the *cricoid cartilage,* and the *cartilage of the epiglottis,* and of six others, which occur in pairs, namely, the two *arytenoid cartilages,* the *cornicula laryngis,* and the *cuneiform cartilages.* In all there are nine distinct pieces, the two cornicula and two cuneiform cartilages being very small. Of these, only the thyroid and cricoid cartilages are seen on the front and sides of the larynx ; the arytenoid cartilages, surmounted by the cornicula laryngis, together with the back of the cricoid cartilage, on which they rest, form the posterior wall of the larynx, whilst the epiglottis is situated in front, and the cuneiform cartilages on each side of the upper opening.

The *thyroid* cartilage is the largest of the pieces composing the framework of the larynx. It is formed by two flat lamellæ, united in front at an acute angle along the middle line, where they form a projection at the upper part. This angular projection is subcutaneous, and is much more prominent in the male than in the female, being named in the former the *pomum Adami.* The two symmetrical halves or lamellæ, named the *alæ*, are somewhat quadrilateral in form : the anterior border where they are joined is the shortest, the pomum Adami being surmounted by a deep notch; the posterior free border of each, thickened and vertical, is prolonged upwards and downwards into two processes or cornua, and gives attachment to the stylo-pharyngeus and palato-pharyngeus muscles ; the superior and inferior borders are both of them concave immediately in front of the cornua, while the superior is convex in its anterior half, and the inferior is nearly straight.

The *external* flattened surface of each ala is marked by an indistinct *oblique line* or ridge, which, commencing at a tubercle situated at the back part of the upper border of the cartilage, passes downwards and forwards, so as to mark off the anterior three fourths of the surface from the remaining posterior portion. This line gives attachment below to the sterno-thyroid, and above to the thyro-hyoid muscle, whilst the small smooth surface behind it gives origin to part of the inferior constrictor of the pharynx, and affords attachment, by means of areolar tissue, to the thyroid body. On their *internal* surfaces, the two alæ are smooth and slightly concave, and by their

Fig. 634.

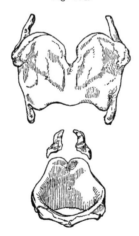

Fig. 634.—Cartilages of the Larynx seen from
before. $\frac{3}{4}$

1 to 4, thyroid cartilage; 1, vertical ridge or pomum Adami ; 2, right ala ; 3, superior, and 4, inferior cornu of the right side ; 5, 6, cricoid cartilage ; 5, inside of the posterior part ; 6, anterior narrow part of the ring ; 7, arytenoid cartilages.

union in front, form a narrow angle within. Of the four *cornua*, all of which bend inwards, the two *superior* or *great* cornua, pass backwards, upwards, and inwards, and terminate each by a blunt extremity, which is connected, by means of the lateral thyrohyoid ligament, to the tip of the corresponding great cornu of the os hyoides. The *inferior* or *smaller* cornua, which are somewhat thicker but shorter, are directed forwards and inwards, and present each, on the inner aspect of the tip, a smooth surface, for articulation with a prominence on the side of the cricoid cartilage.

The *cricoid cartilage*, so named from being shaped like a ring, is thicker in substance and stronger than the thyroid cartilage. It is deep behind, where the thyroid cartilage is deficient, measuring in the male about an inch from above downwards ; but in front its vertical measurement is diminished to a fourth or a fifth of an inch. This diminution is caused by the direction of the superior border, which rises in a convex elevation behind, and descends with a deep concavity in front below the thyroid cartilage ; while the inferior border is horizontal, and connected by membrane to the first ring of the trachea. The posterior elevated part of the upper border is slightly notched in the middle line ; and on the sides of this notch are two convex oval articular facets, directed upwards and outwards, with which the arytenoid cartilages are articulated. The external surface of the cartilage is convex and smooth in front and at the sides, where it affords attachment to the crico-thyroid muscles, and behind these to the inferior constrictor muscle on each side : posteriorly it presents in the middle line a slight vertical ridge, to which some of the longitudinal fibres of the œsophagus are attached. On each side of this ridge is a broad depression occupied by the posterior crico-arytenoid muscle, and externally and anteriorly to that a small rounded and slightly raised surface for articulation on either side with the inferior cornu of the thyroid cartilage. The internal surface is in contact throughout with the mucous membrane of the larynx. The lower border of the cricoid cartilage is circular, but higher up it is somewhat compressed laterally, so that the passage through it is elliptical.

The *arytenoid cartilages* are two in number, and are of a symmetrical form. They may be compared to three-sided pyramids recurved at the summit, resting by their bases on the posterior and highest part of the cricoid cartilage, and approaching near to one another. Each measures from five to six lines in height, about three lines in width, and, in the middle of its inner surface, more than a line from before backwards. Of the three faces, the *posterior* is broad, triangular, and excavated from above downwards, lodging part of the arytenoid muscle. The *anterior*

Fig. 635. — OUTLINE SHOWING THE POSITION AND FORM OF THE ARYTENOID CARTILAGES FROM BEHIND. $\frac{4}{5}$

Fig. 635.

h, hyoid bone ; *t*, the superior, and *t'*, the inferior cornu of the thyroid cartilage ; *c*, placed on the median ridge of the back of the cricoid cartilage ; *a*, placed between the two arytenoid cartilages, to which the letter points by two dotted lines ; the cartilages of Santorini or cornicula are shown above the upper angles ; *tr*, the trachea.

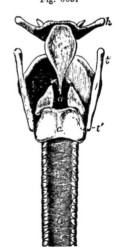

surface, convex in its general outline, and somewhat rough, gives attachment to the thyro-arytenoid muscle, and, by a small tubercle, to the corresponding superior or false vocal cord. The *internal* surface, which is the narrowest of the three, and slightly convex, is nearly parallel with that of the opposite cartilage, being covered by the laryngeal mucous membrane. The anterior and posterior borders, which limit the internal face, ascend nearly in the same vertical plane, whilst the external border, which separates the anterior from the posterior surface, is directed obliquely upwards and inwards.

The base of each arytenoid cartilage is slightly hollowed, having towards its inner part a smooth surface for articulation with the cricoid cartilage. Two of its angles are remarkably prominent, viz., one *external*, short, and rounded, which projects backwards and outwards, and into which the posterior and the lateral crico-arytenoid muscles are inserted ; the other *anterior*, which is more pointed, and forms a horizontal projection forwards, to which the corresponding true vocal cord is attached.

The *apex* of each arytenoid cartilage curves backwards and a little inwards, and terminates in a blunt point, which is surmounted by a small cartilaginous appendage named corniculum laryngis.

The *cornicula laryngis*, or *cartilages of Santorini*, are two small yellowish cartilaginous nodules of a somewhat conical shape, which are articulated with the summits of the arytenoid cartilages, and serve as it were to prolong them backwards and inwards. They are sometimes continuous with the arytenoid cartilages.

The *cuneiform cartilages*, or *cartilages of Wrisberg*, are two very small, soft, yellowish cartilaginous bodies, placed one on each side of the larynx in the fold of mucous membrane, which extends from the summit of the arytenoid cartilage to the epiglottis. They have a conical form, their base or broader part being directed upwards. They occasion small conical elevations of the mucous membrane in the margin of the superior aperture of the larynx, a little in advance of the cartilages of Santorini, with which, however, they are not directly connected.

The *epiglottis* is a median lamella of yellow cartilage, shaped somewhat like an ovate or obcordate leaf, and covered by mucous membrane. It is placed in front of the superior opening of the larynx, projecting, in the ordinary condition, upwards immediately behind the base of the tongue ; but during the act of swallowing it is carried downwards and backwards over the entrance into the larynx, which it covers and protects.

The cartilage of the epiglottis is broad and somewhat rounded at its upper free margin, but inferiorly it becomes pointed, and is prolonged by means of a long, narrow, fibrous band (the thyro-epiglottic ligament) to the deep angular depression between the alæ of the thyroid cartilage, to which it is attached, behind and below the median notch. Its *lateral* borders, which are convex and turned backwards, are only partly free, being in part concealed within the folds of mucous membrane, which pass back on each side to the arytenoid cartilages. The *anterior* or *lingual* surface is free only in the upper part of its extent, where it is covered by mucous membrane. Lower down, the membrane is reflected from it forwards to the base of the tongue, forming one median fold and two lateral frænula, or glosso-epiglottidean ligaments. The adherent portion of this surface is also connected with the posterior surface of the os hyoides by means of a median elastic tissue named the hyo-epiglottic ligament, and is moreover in contact with some glands and fatty tissue. The posterior or *laryngeal* surface of the epiglottis, which is free in the whole of its extent, is concavo-convex from above downwards, but concave from side to side : the lower convexity projecting backwards into the larynx is named the tubercle or cushion. The epiglottis is closely covered by mucous membrane, on removing which, the yellow cartilaginous lamella is seen to be pierced by numerous little pits and perforations, in which are lodged small glands which open on the surface of the mucous membrane.

. The *structure of the cartilages of the larynx.*—The epiglottis, together with the cornicula laryngis and cuneiform cartilages, are composed of what is called yellow or spongy cartilage, which has little tendency to ossify. The structure of all the other cartilages of the larynx resembles that of the costal cartilages, like which, they are very prone to ossification as life advances.

Ligaments of the larynx.—The epiglottidean ligaments and the union of the cricoid cartilage with the trachea have been already mentioned : the other ligaments of the larynx may be divided into thyro-hyoid, crico-thyroid, and arytenoid groups.

Thyro-hyoid ligaments.—The larynx is connected with the os hyoides by a broad membrane and by two round lateral ligaments. The *thyro-hyoid membrane*, or *middle thyro-hyoid ligament*, is a broad, fibrous, and somewhat elastic membrane, which passes up from the whole length of the superior border of the thyroid cartilage to the os hyoides, where it is attached to the posterior and upper margin of the obliquely inclined inferior surface of the bone. Owing to this arrangement, the top of the larynx, when drawn upwards, is permitted to slip within the circumference of the hyoid bone, between which and the upper part of the thyroid cartilage there is occasionally found a small synovial bursa. The thyro-hyoid membrane is thick and subcutaneous towards the middle line, but on each side becomes thin and loose, and is covered by the thyro-hyoid muscles. Behind it is the epiglottis with the mucous membrane of the base of the tongue, separated, however, by much adipose tissue and some glands. It is perforated by the superior laryngeal artery and nerve of each side.

The *lateral thyro-hyoid ligaments*, placed at the posterior limits of the thyro-hyoid membrane, are two rounded yellowish cords, which pass up from the superior cornua of the thyroid cartilage, to the rounded extremities of the great cornua of the hyoid bone. They are distinctly elastic, and frequently enclose a small oblong cartilaginous nodule, which has been named *cartilago triticea :* sometimes this nodule is bony.

Crico-thyroid ligaments.—The thyroid and cricoid cartilages are connected together by a membranous ligament and synovial articulations. The *crico-thyroid membrane* is divisible into a mesial and two lateral portions. The mesial portion, broad below and narrow above, is a strong triangular yellowish ligament, consisting chiefly of elastic tissue, and is attached to the contiguous borders of the two cartilages. Its anterior surface is convex and is partly covered by the crico-thyroid muscles, and is crossed horizontally by a small anastomotic arterial arch, formed by the junction of the crico-thyroid branches of the right and left superior thyroid arteries. The lateral portions are fixed on each side to the inner lip of the upper border of the cricoid cartilage, between the deep muscles and the mucous membrane : they become much thinner as they pass upwards and backwards, and are continuous with the lower margin of the inferior or true vocal cords, becoming blended with them firmly in front.

The *crico-thyroid joints*, between the inferior cornua of the thyroid cartilage and the sides of the cricoid, are two small but distinct articulations, having each a ligamentous capsule and a synovial membrane. The prominent oval articular surfaces of the cricoid cartilage are directed upwards and outwards, while those of the thyroid cartilage, which are slightly concave, look in the opposite direction. The capsular fibres form a stout band behind the joint. The movement allowed is of a rotatory description, the thyroid cartilage revolving on its inferior cornua, and the axis of rotation passing transversely through the two joints.

Arytenoid ligaments.—The arytenoid cartilages are connected below with the cricoid cartilage, above with the cornicula, and in front, by means of fibres contained within the true and false vocal cords, with the thyroid cartilage.

The *crico-arytenoid* articulations are surrounded by a series of thin capsular fibres, which, together with a loose synovial membrane, serve to connect the convex elliptical articular surfaces on the upper border of the cricoid cartilage with the concave articular depressions on the bases of the arytenoid cartilages. There is, moreover, a strong *posterior crico-arytenoid* ligament on each side, arising from the cricoid, and inserted into the inner and back part of the base of the arytenoid cartilage.

The summits of the arytenoid cartilages and the cornicula laryngis have usually a fibrous and synovial capsule to connect them, but it is frequently indistinct.

The *superior thyro-arytenoid* ligaments consist of a few slight fibrous fasciculi, contained within the folds of mucous membrane forming the false vocal cords hereafter to be described, and are fixed in front to the depression between the alæ of the thyroid cartilage, somewhat above its middle, and close to the attachment of the epiglottis : behind, they are connected to the tubercles on the rough anterior surface of the arytenoid cartilages. They are continuous above with scattered fibrous bundles contained in the arytenoepiglottidean folds.

The *inferior thyro-arytenoid ligaments*, placed within the lips of the glottis, and forming the true vocal cords, are two bands of elastic tissue which are attached in front to about the middle of the depression between the alæ of the thyroid cartilage, below the superior cords ; and are inserted behind into the elongated anterior processes of the base of the arytenoid cartilages. These bands are of considerable strength, and consist of closely-arranged parallel fibres. They are continuous below with the thin lateral portions of the crico-thyroid membrane.

Interior of the larynx.—The cavity of the larynx is divided into an upper and a lower compartment by the comparatively narrow aperture of the glottis, or *rima glottidis*, the margins of which constitute in their two anterior thirds the lower or *true vocal cords ;* and the whole laryngeal cavity, viewed in transverse section, thus presents the appearance of an hour-glass, or of two funnels meeting together by their narrower ends. The upper compartment communicates with the pharynx by the *superior aperture* of the larynx, and contains immediately above the rima glottidis the *ventricles* and the upper or *false vocal cords.* The lower compartment passes inferiorly into the tube of the windpipe without any marked constriction or limitation between them (Fig. 638).

The *superior aperture* of the larynx, by which it communicates with the pharynx, is a triangular opening, wide in front and narrow behind, the lateral margins of which slope obliquely downwards and backwards. It is bounded in front by the epiglottis, behind by the summits of the arytenoid cartilages and cornicula laryngis with the angular border of mucous membrane crossing the median space between them, and on the sides by two folds of mucous membrane named the *aryteno-epiglottidean folds*, which, enclosing a few ligamentous and muscular fibres, pass forwards from the tips of the arytenoid cartilages and cornicula to the lateral margins of the epiglottis (Fig. 637).

In studying the form of the laryngeal cavity and its apertures, it is proper to become acquainted with the appearances which they present on examination during life by means of the laryngoscope, and with the relations of these to the anatomical structure. On thus examining the superior aperture, there are seen on each side two rounded elevations, corresponding

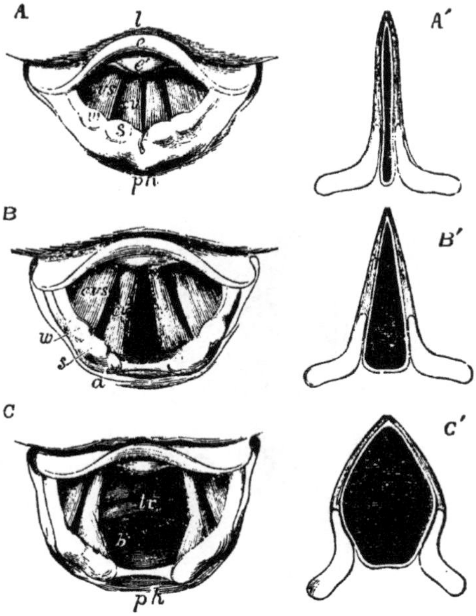

Fig. 636.— THREE LARYN-
GOSCOPIC VIEWS OF THE
SUPERIOR APERTURE OF
THE LARYNX AND SUR-
ROUNDING PARTS IN DIF-
FERENT STATES OF THE
GLOTTIS DURING LIFE (from
Czermak).

Fig. 636.

A, the glottis during the emission of a high note in singing. B, in easy or quiet inhalation of air. C, in the state of widest possible dilatation as in inhaling a very deep breath. The diagrams A', B', and C', have been added to Czermak's figures to show in horizontal sections of the glottis the position of the vocal ligaments and arytenoid cartilages in the three several states represented in the other figures. In all the figures, so far as marked, the letters indicate the parts as follows, viz. : *l*, the base of the tongue ; *e*, the upper free part of the epiglottis ; *e'*, the tubercle or cushion of the epiglottis ; *p h*, part of the anterior wall of the pharynx behind the larynx ; in the margin of the aryteno-epiglottidean fold *w*, the swelling of the membrane caused by the cartilages of Wrisberg ; *s*, that of the cartilages of Santorini ; *a*, the tip or summit of the arytenoid cartilages ; *cv*, the true vocal cords or lips of the rima glottidis ; *cvs*, the superior or false vocal cords ; between them the ventricle of the larynx ; in C, *tr* is placed on the anterior wall of the receding trachea, and *b* indicates the commencement of the two bronchi beyond the bifurcation which may be brought into view in this state of extreme dilatation.

respectively to the cornicula and the cuneiform cartilages ; while in the middle line in front there is a tumescence of the mucous membrane of the lower part of the epiglottis, enabling that structure to close the aperture more accurately when it is depressed, and named the *tubercle* or *cushion of the epiglottis*. The mucous membrane between the arytenoid cartilages is stretched when they are separated, and folded double when they are approximated. (Czermak on the Laryngoscope, translated by the New Sydenham Society.)

On looking down through the superior opening of the larynx, the *glottis* or *rima glottidis* is seen at some distance below, in the form of a long narrow fissure running from before backwards. It is situated on a level with the lower part of the arytenoid cartilages, and is bounded by the *true vocal cords*, two smooth, strong, and straight folds of membrane projecting inwards, with their free edges directed towards the middle line. Above the glottis, another pair of projecting folds is seen, the superior or false vocal cords, which are much thinner and weaker and less projecting than the inferior, and are arched in form. Bounded by the superior and inferior vocal cords are two deep oval depressions, one on each side of the glottis, named the *sinuses*, or *ventricles*, of the larynx; and leading upwards from the anterior parts of these depressions, external to the superior vocal cords, are two small culs-de-sac, named the *laryngeal pouches* or *sacculi.*

Fig. 637.

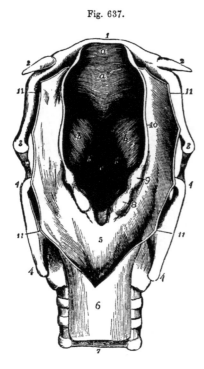

Fig. 637.—PERSPECTIVE VIEW OF THE PHARYNGEAL OPENING INTO THE LA-RYNX FROM ABOVE AND BEHIND.

The superior aperture has been much dilated; the glottis is in a moderately dilated condition; the wall of the pharynx is opened from behind and turned to the two sides. 1, body of the hyoid bone; 2, small cornua; 3, great cornua; 4, upper and lower cornua of the thyroid cartilage; 5, membrane of the pharynx covering the posterior surface of the cricoid cartilage; 6, upper part of the gullet; 7, membranous part of the trachea; 8, projection caused by the cartilage of Santorini; 9, the same belonging to the cartilage of Wrisberg; 10, aryteno-epiglottidean fold; 11, cut margin of the wall of the pharynx; *a*, free part of the epiglottis; *a′*, its lower pointed part; *a″*, the cushion; *b*, eminence on each side over the sacculus or pouch of the larynx; *b′*, the ventricles; *c*, the glottis: the lines on each side point to the margins or vocal cords.

The *superior vocal cords*, also called the *false* vocal cords, because they are not immediately concerned in the production of the voice, are two folds of mucous membrane, each of which forms a free crescentic margin, bounding the corresponding ventricle of the larynx, the hollow of which is seen on looking down into the laryngeal cavity, from the superior vocal cords being separated farther from each other than the inferior cords.

The *inferior* or *true vocal cords*, the structures by the vibration of which the sounds of the voice are produced, occupy the two anterior thirds of the aperture of the glottis. These cords are not mere folds of mucous membrane, but are strengthened near their free margins by the elastic thyro-arytenoid ligaments, and further out by the thyro-arytenoid muscles. The mucous membrane covering them is so thin and closely adherent as to show the light colour of the ligaments through it. Their free edges, which are sharp and straight, and directed upwards, form the lower boundaries of the ventricles, and are the parts thrown into vibration during the production of the voice. Their inner surfaces are flattened, and look towards each other.

The *rima glottidis*, an elongated aperture, situated, anteriorly, between the inferior or true vocal cords, and, posteriorly, between the bases of the arytenoid cartilages, forms when nearly closed a long narrow slit, slightly wider in the centre ; when moderately open, as in easy respiration, its shape is that of a long triangle, the pointed extremity being directed forwards, and the base being placed behind between the arytenoid cartilages ; and in its fully dilated condition it has the figure of an elongated lozenge, the posterior sides of which are formed by the inner sides of the bases of the arytenoid cartilages, while the posterior angle is truncated. This aperture is the narrowest part of the interior of the larynx ; in the adult male it measures about eleven lines or nearly an inch in an antero-posterior direction, and three or four lines across at its widest' part, which may be dilated to nearly half an inch. In the female, and in males before the age of puberty, its dimensions are less, its antero-posterior diameter being about eight lines, and its transverse diameter about two. The vocal ligaments measure about seven lines in the adult male, and five in the female.

The *ventricles*, or *sinuses* of the larynx, situated between the superior and inferior vocal cords on each side, are narrower at their orifice than in their interior. The upper margin of each is crescentic, and the lower straight : the outer surface is covered by the upper fibres of the corresponding thyro-arytenoid muscle.

Fig. 638.

Fig. 638.—ANTERIOR HALF OF A TRANS-VERSE VERTICAL SECTION THROUGH THE LARYNX NEAR ITS MIDDLE.

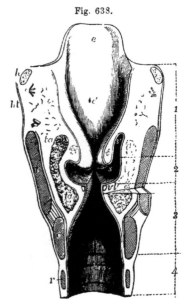

In order to bring the deepest part of one of the sacculi into view, the section is carried somewhat farther forward on the right side : the space between the horizontal dotted lines marked 1, comprises the upper division of the laryngeal cavity ; that marked 2, corresponds to the middle cavity or that of the ventricles ; that marked 3, indicates the lower division of the laryngeal cavity, continued into 4, a part of the trachea ; e, the free part of the epiglottis ; e', its cushion ; h, the divided great cornua of the hyoid bone ; h t, thyro-hyoid membrane ; t, cut surface of the divided thyroid cartilage ; c, that of the cricoid cartilage ; r, first ring of the trachea ; t a, superior and inferior parts of the thyro-arytenoid muscle ; v l, thyro-arytenoid ligament in the true vocal cord covered by mucous membrane at the rima glottidis ; s, the ventricle ; above this, the superior or false cords or margin of the folds above the ventricles ; s', the sacculus or pouch opened on the right side.

The small culs-de-sac named the *laryngeal pouches* lead from the anterior part of the ventricles upwards, for the space of half an inch, between the superior vocal cords on the inner side, and the thyroid cartilage on the outer side, reaching as high as the upper border of that cartilage at the sides of the epiglottis. The pouch is conical in shape, and curved slightly backwards. Its opening into the ventricle is narrow, and is generally limited by two folds of the lining mucous membrane. Numerous small glands, sixty or seventy in number, open into its interior, and it is surrounded by a quantity of fat. Externally to the fat, this little pouch receives a fibrous investment, which is continuous below with the superior vocal cord. Over.

its laryngeal side and upper end is a thin layer of muscular fibres (compressor sacculi laryngis, arytæno-epiglottideus inferior, Hilton) connected above with those found in the aryteno-epiglottidean folds. The upper fibres of the thyro-arytenoid muscles pass over the outer side of the pouch, a few being attached to its lower part. The laryngeal pouch is supplied abundantly with nerves, derived from the superior laryngeal.

Muscles of the Larynx.

Besides certain extrinsic muscles already described—viz., the sterno-hyoid, omo-hyoid, sterno-thyroid, and thyro-hyoid muscles, together with the muscles of the suprahyoid region, and the middle and inferior constrictors of the pharynx, all of which act more or less upon the entire larynx—there are certain *intrinsic muscles* which move the different cartilages upon one another, and modify the size of the apertures and the state of tension of the soft parts of the larynx. These intrinsic muscles are the *crico-thyroid*, the *posterior* and *lateral crico-arytenoid*, the *thyro-arytenoid*, the *arytenoid*, and the *aryteno-epiglottidean*, together with certain other slender muscular fasciculi. All these muscles, except the arytenoid, which crosses the middle line, are in pairs.

The *crico-thyroid muscle* is a short thick triangular muscle, seen on the front of the larynx, situated on the fore part and side of the cricoid cartilage. It arises by a broad origin from the cricoid cartilage, reaching from the

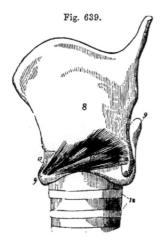

Fig. 639.

Fig. 639.—LATERAL VIEW OF THE CARTILAGES OF THE LARYNX WITH THE CRICO-THYROID MUSCLE (after Willis).

8, thyroid cartilage ; 9, cricoid ; 10, cricothyroid muscle ; 11, crico-thyroid ligament or membrane ; 12, upper rings of the trachea.

median line backwards upon the lateral surface, and its fibres, passing obliquely upwards and outwards and diverging slightly, are inserted into the lower border of the thyroid cartilage, and into the anterior border of its inferior cornu. The lower portion of the muscle, the fibres of which are nearly horizontal, and are inserted into the inferior cornu, is usually distinct from the rest. Some of the superficial fibres are almost always continuous with the inferior constrictor of the pharynx. The inner borders of the muscles of the two sides are separated in the middle line by a triangular interval, broader above than below, and occupied by the crico-thyroid membrane.

The *posterior crico-arytenoid muscle*, situated behind the larynx, beneath the mucous membrane of the pharynx, arises from the broad depression on the corresponding half of the posterior surface of the cricoid cartilage. From this broad origin its fibres converge upwards and outwards to be inserted into the outer angle of the base of the arytenoid cartilage, behind the attachment of the lateral crico-arytenoid muscle. The upper fibres are short and almost horizontal ; the middle are the longest, and run obliquely ; whilst the lower or external fibres are nearly vertical.

Fig. 640.—VIEW OF THE LARYNX AND PART OF THE TRACHEA FROM BEHIND, WITH THE MUS- CLES DISSECTED.

Fig. 640.

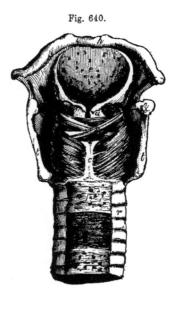

h, the body of the hyoid bone ; *e*, epiglottis ; *t*, the posterior borders of the thyroid cartilage ; *c*, the median ridge of the cricoid ; *a*, upper part of the arytenoid ; *s*, placed on one of the oblique fasciculi of the arytenoid muscle ; *b*, left posterior crico-arytenoid muscle ; *r*, ends of the incomplete cartilaginous rings of the trachea ; *l*, fibrous membrane crossing the back of the trachea ; *n*, muscular fibres exposed in a part.

In connection with the posterior crico-arytenoid muscle, may be mentioned an occasional small slip in contact with its lower border, viz., the *kerato-cricoid* muscle of Merkel. It is a short and slender bundle, arising from the cricoid cartilage near its lower border, a little behind the inferior cornu of the thyroid cartilage, and passing obliquely outwards and upwards to be inserted into that process. It usually exists on only one side. Turner found it in seven out of thirty-two bodies. It is not known to be of any physiological significance. (Merkel, Anat. und Phys. des Menschl. Stimm-und-Spräch-organs, Leipzig, 1857 ; Turner in Month. Med. Journal, Feb. 1860.)

The *lateral crico-arytenoid* muscle, smaller than the posterior, and of an oblong form, is in a great measure hidden by the ala of the thyroid carti-lage. It arises from the upper border of the side of the cricoid cartilage, its

Fig. 641.—DIAGRAMMATIC VIEW FROM ABOVE OF THE DISSECTED LARYNX (after Willis).

Fig. 641.

1, aperture of the glottis ; 2, arytenoid carti-lages ; 3, vocal cords ; 4, posterior crico-arytenoid muscles ; 5, right lateral crico-arytenoid muscle, that of the left side being removed ; 6, arytenoid muscle ; 7, thyro-arytenoid muscle of the left side, that of the right side being removed ; 8, upper border of the thyroid cartilage ; 9, back of the cricoid cartilage ; 13, posterior crico-arytenoid ligament.

origin extending as far back as the articular surface for the arytenoid cartilage. Its fibres, passing obliquely backwards and up-wards, and the anterior or upper ones being the longest, are attached to the external process or outer side of the base of the arytenoid cartilage and to the adjacent part of its anterior surface, in front of the insertion of the posterior crico-arytenoid muscle.

This muscle lies in the interval between the ala of the thyroid cartilage

3 o 2

and the interior of the larynx, being lined within by the mucous membrane of the larynx. Its anterior part is covered by the upper part of the crico-thyroid muscle. The upper part is in close contact and indeed is sometimes blended with the thyro-arytenoid muscle.

The *thyro-arytenoid* is a broad flat muscle situated above the lateral crico-arytenoid. It is thick below and in front, and becomes thinner above and behind. It consists of several muscular fasciculi, which arise in front from the internal surface of the thyroid cartilage, adjacent to the lower two-thirds of the angle formed by the junction of the two alæ. They extend almost horizontally backwards and outwards to reach the base of the arytenoid cartilage. The *lower portion* of the muscle, which forms a thick fasciculus, receives a few additional fibres from the posterior surface of the crico-thyroid membrane, and is inserted into the anterior projection on the base of the arytenoid cartilage and to the adjacent part of the surface close to the insertion of the lateral crico-arytenoid muscle. The *thinner portion* of the thyro-arytenoid muscle is inserted higher up on the anterior surface and

Fig. 642.

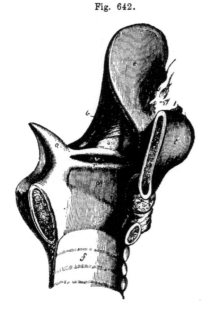

Fig. 642.—VIEW OF THE INTERIOR OF THE LEFT HALF OF THE LARYNX (after Hilton).

a, left arytenoid cartilage ; *c, c,* divided surfaces of the cricoid cartilage ; *t,* thyroid cartilage ; *e,* epiglottis; *v,* left ventricle of the larynx; *r,* left inferior or true vocal cord ; *s,* placed on the inner wall of the laryngeal pouch ; *b,* aryteno-epiglottidean muscle ; *f,* interior of the trachea.

outer border of the arytenoid cartilage. The lower portion of the muscle assists in the formation, or at least contributes to the support, of the true vocal cord, lying parallel with the rima glottidis, immediately on the outer side of the inferior thyro-arytenoid ligament, with which it is intimately connected, and into the outer surface of which some of its fibres are inserted. The upper thin portion, external to the lower, lies upon the laryngeal pouch and ventricle, close beneath the mucous membrane. The entire muscle may be dissected indeed from the interior of the larynx, by raising the mucous membrane of the sinus and vocal cord. Fibres from this muscle pass round the border of the arytenoid cartilage, and become continuous with some of the oblique fibres of the arytenoid muscle, to be presently described.

Santorini described three thyro-arytenoid muscles, an *inferior* and a *middle,* which are constant, and a *superior,* which is sometimes present. The fibres of the superior fasciculus, when present, arise nearest to the notch of the thyroid cartilage, and are attached to the upper part of the arytenoid cartilage. This is named by Sœmmerring the *small* thyro-arytenoid, whilst the two other portions of the muscle constitute the *great* thyro-arytenoid of that author.

Arytenoid and aryteno-epiglottidean muscles.—When the mucous membrane is removed from the back of the arytenoid cartilages, a thick band of transverse fibres constituting the arytenoid muscle is laid bare, and on the surface of this are seen two slender decussating oblique bundles, formerly described as portions of the arytenoid muscle (arytænoideus obliquus), but now more generally considered as parts of the aryteno-epiglottidean muscles, with which they are more closely associated both in the disposition of their fibres and in their action. The *arytenoid* muscle passes straight across, and its fibres are attached to the whole extent of the concave surface on the back of each arytenoid cartilage. The *aryteno-epiglottidean* muscles arising near the inferior and outer angles of the arytenoid cartilages, decussate one with the other, and their fibres are partly attached to the upper and outer part of the opposite cartilage, partly pass forwards in the aryteno-epiglottidean fold, and partly join the fibres of the thyro-arytenoid muscle.

A few fibres associated with the anterior and upper part of the thyro-arytenoid muscle have been described as constituting a *thyro-epiglottidean* muscle.

Fig. 643.—OUTLINE OF THE RIGHT HALF OF THE CARTILAGES OF THE LARYNX AS SEEN FROM THE INSIDE, WITH THE THYRO-ARYTENOID LIGAMENT, TO ILLUSTRATE THE ACTION OF THE CRICO-THYROID MUSCLE.;

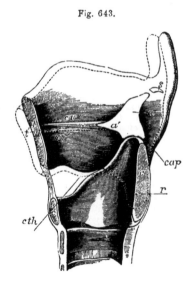

Fig. 643.

t, cut surface of the thyroid cartilage in the middle anteriorly ; *c, c*, the same of the cricoid cartilage before and behind ; *a*, the inner surface of the right arytenoid cartilage ; *a'*, its anterior process ; *s*, the right cartilage of Santorini ; *c v*, the thyro-arytenoid ligament ; the position of the lower cornu of the thyroid cartilage on the outside of the cricoid is indicated by a dotted outline, and *r* indicates the point or axis of rotation of the one cartilage on the other ; *c t h*, indicates a line in the principal direction of action of the crico-thyroid muscle ; *c a p*, the same of the posterior crico-arytenoid muscle ; the dotted line, of which *t'* indicates a part, represents the position into which the thyroid cartilage is moved by the action of the crico-thyroid muscle ; if the arytenoid cartilages are fixed by muscles acting in the direction of *c a p*, the vocal ligaments will be elongated and rendered tense by contraction of the crico-thyroid muscles, as indicated by *c v'*.

Actions of the intrinsic muscles of the larynx.—The *crico-thyroid* muscles produce the rotation forwards and downwards of the thyroid cartilage on the cricoid, which is permitted by the crico-thyroid articulations. In this movement the arytenoid cartilages, being attached to the cricoid cartilage at a level considerably above the axis of rotation, have their distance from the fore part of the thyroid cartilage increased, and therefore, the crico-thyroid muscles increase the tension of the vocal cords. The *thyro-arytenoid* muscles are, in their lower parts, the opponents of the crico-thyroid, raising the fore part of the thyroid cartilage and decreasing the tension of the vocal cords; the upper parts of these muscles, being attached higher up on the arytenoid cartilages, depress them.

The *lateral crico-arytenoid* muscles, by pulling forwards the outer angles of the arytenoid cartilages, approximate the vocal cords to the middle line. The *posterior crico-arytenoid* muscles pull backwards the outer angles of the arytenoid cartilages, and thus draw asunder the posterior extremities of the vocal cords, and dilate the glottis to its greatest extent; they are likewise the elevators of the arytenoid cartilages, being inserted above the articulation.

The *arytenoid* muscle draws the arytenoid cartilages together, and, from the structure of the crico-arytenoid joints, this approximation when complete is necessarily accompanied with depression. The *aryteno-epiglottidean* muscles at once depress and approximate the arytenoid cartilages, which they include in their embrace, and draw down the epiglottis, so as to contract the whole superior aperture of the larynx.

With the aid of the laryngoscope it may be seen that in ordinary breathing the rima glottidis is widely open, and that in vocalisation the vocal cords come closely together; which is effected principally, no doubt, by the action of the lateral crico-arytenoid muscles, assisted by the arytenoid and perhaps by the thyro-arytenoid, and accompanied with a varying amount of contraction of the crico-thyroid muscles. The regulation of the tension of the vocal cords and of the width of the aperture of the glottis, in the production of high and low pitched notes, is probably accomplished by the crico-thyroid and thyro-arytenoid muscles. The movement of the thyroid on the cricoid cartilage, effected by these muscles during the passage of the voice from one extreme of the scale to the other, may be detected by placing the tip of a finger over the crico-thyroid ligament. The arytenoid and aryteno-epiglottidean muscles come into action in spasmodic closure of the upper aperture of the larynx; the complete descent of the epiglottis, however, can only take place when the tongue is retracted as in the act of swallowing.

The manner in which the larynx is affected by the extrinsic muscles, in the acts of deglutition and vocalisation, has been mentioned at pages 191 and 193.

It is remarked by Henle that, with the exception of the crico-thyroid and posterior crico-arytenoid, the muscles of the larynx, namely, those "which lie in the space enclosed by the laminæ of the thyroid cartilage, and above the cricoid, the fibres of which are substantially horizontal, may be regarded in their totality as a kind of sphincter. Such a sphincter is found in its simple form embracing the entrance of the larynx in reptiles; and the complication which it attains in the higher vertebrates arises, like the complication of the muscles generally, from the fibres finding various points of attachment in their course, by which means they are broken up and divided."

The *mucous membrane and glands of the larynx.*—The laryngeal mucous membrane is thin and of a pale pink colour. In some situations it adheres intimately to the subjacent parts, especially on the epiglottis, and still more in passing over the true vocal cords, on which it is extremely thin and most closely adherent. About the upper part of the larynx, above the glottis, it is extremely sensitive. In or near the aryteno-epiglottidean folds it covers a quantity of loose areolar tissue, which is liable in disease to infiltration, constituting œdema of the glottis. Like the mucous membrane in the rest of the air-passages, that of the larynx is covered in the greater part of its extent with a columnar ciliated epithelium, by the vibratory action of which the mucus is urged upwards. The cilia are found higher up in front than on each side and behind, reaching in the former direction as high as the widest portion of the epiglottis, and in the other directions to a line or two above the border of the superior vocal cords: above these points the epithelium loses its cilia, and gradually assumes a squamous form, like that of the pharynx and mouth. Upon the vocal cords also the epithelium is squamous, although both above and below them it is ciliated.

Glands.—The lining membrane of the larynx is provided with numerous *glands*, which secrete an abundant mucus; and the orifices of which may be seen almost everywhere, excepting upon and near the true vocal cords. They abound particularly upon the epiglottis, in the substance of which are found upwards of fifty small compound glands, some of them perforating the cartilage. Between the anterior surface of the epiglottis, the hyoid bone, and the root of the tongue, is a mass of yellowish fat, erroneously named the epiglottidean gland, in or upon which some real glands may exist. Another collection of glands, named *arytenoid*, is placed within the fold of mucous membrane in front of each arytenoid cartilage, from which a series may be traced forwards, along the corresponding superior vocal cord. The glands of the laryngeal pouches have already been described.

Vessels and Nerves of the Larynx.

The *arteries* of the larynx are derived from the superior thyroid (p. 343), a branch of the external carotid, and from the inferior thyroid (p. 371), a branch of the subclavian. The *veins* join the superior, middle and inferior thyroid veins. The *lymphatics* are numerous, and pass through the cervical glands. The *nerves* are supplied from the superior laryngeal and inferior or recurrent laryngeal branches of the pneumogastric nerves, joined by branches of the sympathetic. The superior laryngeal nerves supply the mucous membrane, and also the crico-thyroid muscles, and in part the arytenoid muscle. The inferior laryngeal nerves supply, in part, the arytenoid muscle, and all the other muscles, excepting the crico-thyroid.

The superior and inferior laryngeal nerves of each side communicate with each other in two places, viz., at the back of the larynx, beneath the pharyngeal mucous membrane, and on the side of the larynx, under the ala of the thyroid cartilage (p. 622).

DEVELOPMENT AND GROWTH OF THE LARYNX.

Development.—The rudimentary larynx consists, according to Valentin, of two slight enlargements having a fissure between them, and embracing the entrance from the pharynx into the trachea. According to Reichert, the rudiments of the arytenoid cartilages are the first to appear. Rathke, however, states that all the true cartilages are formed at the same time, and are recognisable together as the larynx enlarges, the epiglottis only appearing later. In the human embryo, Fleischmann could not detect the cartilages at the seventh week, though the larynx was half a line in length, but at the eighth week there were visible the thyroid and cricoid cartilages, consisting at that period of two lateral halves, which are afterwards united together in the sixth month. Kölliker, however, states that Fleischmann had been deceived by the presence of a deep groove, and that by making transverse sections he ascertained that those cartilages are single from the first.

Growth.—During childhood the growth of the larynx is very slow. Richerand found that there was scarcely any difference between the dimensions of this organ in a child of three and in one of twelve years of age. Up to the age of puberty the larynx is similar in the male and female, the chief characteristics at that period being the small size and comparative slightness of the organ, and the smooth rounded form of the thyroid cartilage in front. In the female these conditions are permanent, excepting that a slight increase in size takes place. In the male, on the contrary, at the time of puberty, remarkable changes rapidly occur, and the larynx becomes more prominent and more perceptible at the upper part of the neck. Its cartilages become larger, thicker, and stronger, and the alæ of the thyroid cartilage project forwards in front so as to form at their union with one another, with an acute angle, the prominent ridge of the *pomum Adami*. At the same time, the median notch on its upper border is considerably deepened. In consequence of these changes in the thyroid cartilage, the distance between its angle in front and the arytenoid cartilages behind becomes greater, and the chordæ vocales are necessarily lengthened. Hence the dimensions of the glottis, which, at the time of puberty, are increased by about one-third only in the female, are nearly doubled in the male, and the adult male larynx becomes altogether one-third larger than that of the female.

Towards the middle of life the cartilages of the larynx first show a tendency to ossification ; this commences first in the thyroid cartilage, then appears in the cricoid, and lastly in the arytenoid cartilages. In the thyroid cartilage the ossification usually begins at the cornua and posterior borders; it then gradually extends along the whole inferior border, and subsequently spreads upwards through the cartilage. The cricoid cartilage first becomes ossified at its upper border upon each side, near the two posterior articular eminences, and the ossification invades the lateral parts of the cartilage before encroaching on it either in front or behind. The arytenoid cartilages become ossified from below upwards.

DUCTLESS GLANDS ON THE LARYNX AND TRACHEA.

1. THE THYROID BODY.

The *thyroid body* or *gland* is a soft reddish and highly vascular organ, situated in the lower part of the neck, embracing the front and sides of the upper part of the trachea, and reaching up to the sides of the larynx. It belongs, like the spleen, to the series of structures known as ductless glands; and, although its precise function is unascertained, there is reason to believe that it is in some way connected with the elaboration of the blood.

The thyroid body is of an irregular, semilunar form, consisting of two *lateral lobes*, united together towards their lower ends by a transverse portion named the *isthmus*. Viewed as a whole, it is convex on the sides and in front, forming a rounded projection upon the trachea and larynx. It is covered by the sterno-hyoid, sterno-thyroid, and omo-hyoid muscles, and behind them it comes into contact with the sheath of the great vessels of the neck. Its deep surface is concave where it rests against the trachea

Fig. 644.

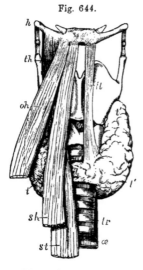

Fig. 644.—SKETCH SHOWING THE FORM AND POSITION OF THE THYROID BODY. ½

The larynx and surrounding parts are viewed from before; on the right side the muscles covering the thyroid body are retained, on the left side they are removed ; *h*, hyoid bone ; *th*, right thryo-hyoid muscle ; *o h*, omo-hyoid ; *s h*, sterno-hyoid ; *s t*, sterno-thyroid ; *c*, on the crico-thyroid membrane above the cricoid cartilage, points by a dotted line to the right crico-thyroid muscle ; *tr*, the trachea ; *œ*, the œsophagus appearing behind and slightly to the left of the trachea ; *t*, the right lobe of the thyroid body partially seen between the muscles ; *t'*, the left lobe entirely exposed; *i*, the isthmus ; *lt*, the fibrous or muscular band termed levator thyroideæ, which is occasionally found in the middle line or to the left side, and which existed in the case from which the figure was taken.

and larynx. It usually extends so far back as to touch the lower portion of the pharynx, and on the left side the œsophagus also.

Each *lateral lobe* measures usually two inches or upwards in length, an inch and a quarter in breadth, and three-quarters of an inch in thickness at its largest part, which is below its middle : the right lobe is usually a few lines longer and wider than the left. The general direction of each lobe is, from below, obliquely upwards and backwards, reaching from the fifth or sixth ring of the trachea to the posterior border of the thyroid cartilage, of which it covers the inferior cornu and adjoining part of the ala. The upper end of the lobe, which is thinner, and sometimes called the *cornu*, is usually connected to the side of the thyroid and cricoid cartilages by areolar tissue.

. The *transverse* part, or *isthmus*, which connects the two lateral lobes together a little above their lower ends, measures nearly half an inch in breadth, and from a quarter to three-quarters of an inch in depth ; it commonly lies across the third and fourth rings of the trachea, but is very inconstant in size, shape, and position, so that the portion of trachea which is covered by it is subject to corresponding variations. From the upper part of the isthmus, or from the adjacent portion of either lobe, but most frequently the left, a conical portion of the thyroid body, named, from its shape and position, the *pyramid*, or *middle lobe*, often proceeds upwards to the middle of the hyoid bone, to which its apex is attached by loose fibrous tissue. Commonly this process lies somewhat to the left ; occasionally it is thicker above than below, or is completely detached, or is split into two parts : sometimes it appears to consist of fibrous tissue only. In many cases, muscular fasciculi, most frequently derived from the thyro-hyoid muscle, but occasionally independent, descend from the hyoid bone to the thyroid gland or its pyramidal process. They are known as the *levator glandulæ thyroideæ.* It sometimes, though rarely, happens that the isthmus is altogether wanting, the lateral lobes being then connected by areolar or fibrous tissue only.

The *weight* of the thyroid body varies ordinarily from one to two ounces. It is always larger in the female than in the male, and appears in many of the former to undergo a periodical increase about the time of menstruation. The thyroid body, moreover, is subject to much variation of size, and is, occasionally, the seat of enormous enlargement, constituting the disease called goitre. The *colour* of the thyroid body is usually of a dusky brownish red, but sometimes it presents a yellowish hue.

Fig. 645.—Magnified View of several Vesicles from the Thyroid Gland of a Child (from Kölliker). $\frac{250}{1}$

a, connective tissue between the vesicles ; *b*, capsule of the vesicles ; *c*, their epithelial lining.

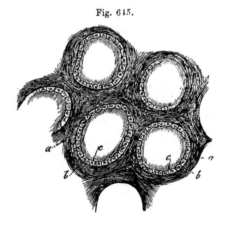

Fig. 645.

Structure.—The texture of this organ is firm, and to the naked eye appears coarsely granular. It is invested with a thin transparent layer of dense areolar tissue, which connects it with the adjacent parts, surrounds and supports the vessels as they enter it, and imperfectly separates its substance into small masses of irregular form and size. The interstitial areolar tissue is free from fat, and contains elastic fibres.

When the thyroid body is cut into, a yellow glairy fluid escapes from the divided substance, which is itself found to contain multitudes of closed vesicles, composed of a simple external capsular membrane, and containing a yellow fluid, with corpuscles resembling cell-nuclei and sometimes nucleated cells floating in it. These vesicles are surrounded by capillary vessels, and are held together in groups or imperfect lobules by areolar tissue. Their size varies from $\frac{1}{850}$th of an inch to that of a millet-seed, so as to be visible to the naked eye,—the size varying, however, in different individuals, more than in the same thyroid body. The *vesicles* are spherical, oblong, or flattened, and are perfectly distinct from each other ; the *corpuscles* within them are in the fœtus and young subject disposed in close apposition and like a single epithelial layer on the inner side of the vesicles, but for the most part detach themselves in the progress of growth. The fluid coagulates by the action of heat or of alcohol, preserving, however, its transparency. According to recent analyses, the substance of the thyroid body consists principally of albumen with traces of gelatine, stearine, oleine, and extractive matter, besides alkaline and earthy salts and water. The salts are chloride of sodium, a little alkaline sulphate, phosphates of potash, lime and magnesia, with some oxide of iron.

Fig. 646.

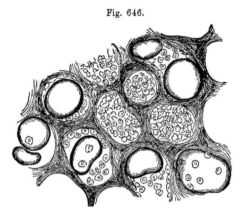

Fig. 646.—VESICLES OF THE THYROID GLAND ENLARGED AND CONTAINING COLLOID MATTER (from Kölliker). $\frac{50}{1}$

One of the most frequent pathological changes to which the thyroid body is subject consists in the accumulation within its vesicles of colloid substance : this may occur without giving rise to very great enlargement of these vesicles, but in certain forms of goître it distends them to an enormous degree.

Vessels.—The *arteries* of the thyroid body (pp. 346 and 371) are the superior and inferior thyroids of each side, to which is sometimes added a fifth vessel named the *lowest* thyroid of Neubauer and Erdmann (p. 340). The arteries are remarkable for their large relative size, and for their frequent and large anastomoses; they terminate in a capillary network, upon the outside of the closed vesicles. The *veins*, which are equally large, ultimately form plexuses on the surface, from which a superior, middle, and inferior thyroid vein (pp. 453 and 460) are formed on each side. The superior and middle thyroid veins open into the internal jugular; the inferior veins emanate from a plexus formed in front of the trachea, and open on the right side into the superior cava, and on the left into the brachio-cephalic vein. The *lymphatics* of the thyroid body are extremely numerous and large, and are supposed to convey into the blood the products formed within the organ.

Nerves.—The *nerves* are derived from the pneumogastric, and from the middle and inferior cervical ganglia of the sympathetic.

Development.—Remak states that the thyroid body is developed from the anterior wall of the pharynx. In a human embryo at the third month, Kölliker found the thyroid body consisting of isolated vesicles, with rounded cells in their interior. The multiplication of these vesicles takes place, according to Kölliker, either by constriction and subsequent division of one vesicle into two, or by a process of gemmation. The transverse part of the gland is said to be developed subsequently to the two lateral lobes. In the fœtus, and during early infancy, this organ is relatively larger than in after life; its proportion to the weight of the body in the new-born infant being that of 1 to 240 or 400, whilst at the end of three weeks it becomes only 1 to 1160, and in the adult 1 to 1800 (Krause). In advanced life the thyroid body becomes indurated, and frequently contains earthy deposit; its vesicles also attain a very large size.

2. THE THYMUS GLAND.

The *thymus gland* or *body* (glandula thymus, corpus thymicum) is a temporary organ which reaches its greatest size at about the end of the second year, after which period it ceases to grow, and is gradually reduced to a mere vestige. Its function, like that of the thyroid body, is unknown, although it is probable that it is in some way connected with the elaboration of the blood in infancy. When examined in its mature state in an infant under two years of age, it appears as a narrow elongated glandular-looking body, situated partly in the thorax, and partly in the lower region of the neck : below, it lies in the anterior mediastinal space, close behind the sternum, and in front of the great vessels and pericardium ; above it reaches upwards upon the trachea in the neck. Its colour is greyish, with a pinkish tinge ; its consistence is soft and pulpy, and its surface appears distinctly lobulated. It consists of *two lateral parts*, or *lobes*, which touch each other along the middle line, and are nearly symmetrical in form, though generally unequal in size, sometimes the left, and at other times the right lobe being the larger of the two. An *intermediate lobe* often exists between the two lateral ones, and occasionally the whole body forms a single mass. The forms of the smaller lobules also differ on the two sides.

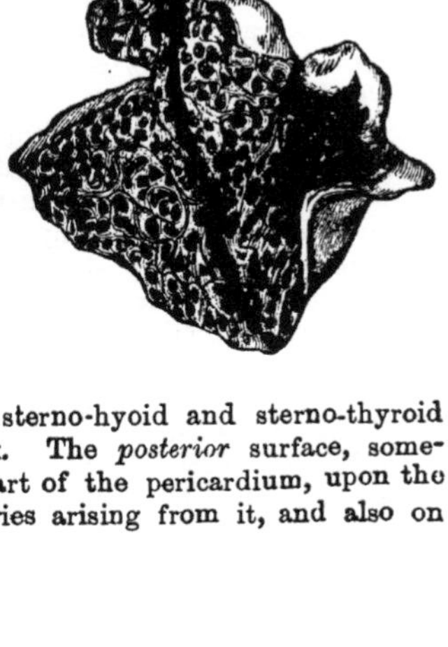

Fig. 647.—ONE LOBE OF THE HUMAN THYMUS GLAND (from Kölliker).

The lower part presents a large cavity which has been opened, and within it are seen numerous apertures leading into the smaller lobes.

Fig. 647.

Each lateral lobe is of an elongated triangular form, its base being directed downwards. The *summit*, or upper extremity, usually mounts up into the neck, reaching as high as to the lower border of the thyroid body. The *base* rests on the upper part of the pericardium, to which it is connected by areolar tissue. The *anterior* surface, slightly convex, is covered by the first and the upper part of the second piece of the sternum, reaching, in the infant at birth, as low down as the level of the fourth costal cartilage. It is attached to the sternum by loose areolar tissue, but opposite the upper part of that bone is separated from it by the origins of the sterno-hyoid and sterno-thyroid muscles, which also cover it in the neck. The *posterior* surface, somewhat concave, rests, in the thorax, upon part of the pericardium, upon the front of the aortic arch and the large arteries arising from it, and also on the left innominate vein, some areolar tissue being interposed between it and these parts. In the neck, it lies upon the front and corresponding side of the trachea. Its *external border* is in contact with the corresponding layer of the mediastinal pleura, near the internal mamma_{ry} artery, and higher up (in the neck), with the carotid artery, or its sheath. The *internal* border is in close contact with that of the opposite lateral lobe. The dimensions of the thymus vary according to its stage of development. At birth it measures above two inches in length, an inch and a half in width at its lower part, and about three or four lines in thickness. Its weight at that period is about half an ounce. Its specific gravity, which is at first about 1·050, diminishes as the gland continues to waste.

Chemical Composition.—The substance and fluid of the thymus contain nearly eighty per cent. of water. Its solid animal constituents are composed essentially of albumen and fibrin in large quantities, mixed with gelatine and other animal matter. The salts are principally alkaline and earthy phosphates, with chloride of potassium.

Structure.—The lateral halves or lobes of the thymus gland are each surrounded by a proper investment of thin areolar tissue, which encloses in a common envelope the smaller masses composing it. This tissue being removed, the substance of the gland is found to consist of numerous compressed lobules, the most of them from two to five lines in diameter, connected by a more delicate intervening areolar tissue. These primary lobules, as they may be called, are each made up externally of smaller or secondary lobules, of a compressed pyriform shape, placed close together with their bases outwardly, and are arranged round an elongated *central stem* (reservoir of the thymus, Cooper), running through each lateral half of the gland, and more or less spirally twisted.

On making a section of the thymus, there is obtained a milky substance consisting of fluid rich in nuclei and small nucleated cells. The walls both of the lobules and the larger stems are limited by a fine homogeneous membrane (Simon); the substance in the interior of this appears at first sight to be entirely composed of corpuscles of the kind just mentioned, varying in diameter from $\frac{1}{2500}$th to $\frac{1}{3000}$th of an inch, and having the appearance of free nuclei ; but, on closer examination, according to Kölliker and His, seen to be mostly contained within delicate cells. The substance contained within the limiting membrane is not, however, a mere fluid with corpuscles, but possesses a delicate reticulum of connective tissue, and, as was first pointed out by Kölliker, likewise capillary blood-vessels, resembling in this respect the substance which occupies the interior of Peyer's glands.

According to Astley Cooper the central stem of the thymus presents a continuous cavity, the ramifications of which pass into both primary and secondary lobules. The existence of a central cavity has been since doubted by some and affirmed by others ; the difficulty, however, may now be regarded as cleared up, since the discovery of connective tissue and blood-vessels within the lobulated structure; for it is admitted that, in the centres of the lobules, sublobules, and central stem, the capillaries and reticulum of connective tissue are deficient and the corpuscles most abundant, while on the other hand it is equally certain that there is no cavity bounded by epithelial lining. Considered in relation to development, Cooper's view is correct; for it has been shown by Simon that the primitive form of the thymus gland is a linear tube, from which, as it grows, lateral branched diverticula subsequently bud out, but that in the mature thymus this tube becomes obscure.

He is of opinion that the central cavity described and figured by Cooper is preternaturally enlarged, owing to over-distension; but that, nevertheless, all the parts of each lateral mass of the thymus are connected with a single

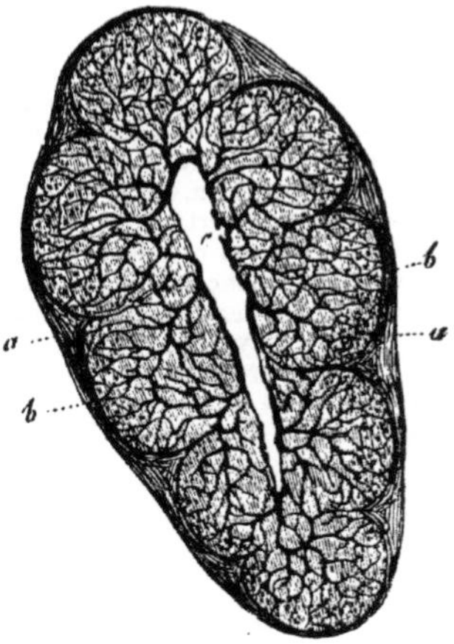

Fig. 648.—Transverse Section of a Lobule of an Injected Infantile Thymus Gland (from Kölliker). $\frac{30}{1}$

Fig. 648.

a, capsule of connective tissue surrounding the lobule; *b*, membrane of the glandular vesicles; *c*, cavity of the lobule, from which the larger blood-vessels are seen to extend towards and ramify in the spheroidal masses of the lobule.

common cavity. (Astley Cooper, Anatomy of the Thymus Gland, Lond., 1832; Simon, Physiological Essay on the Thymus Gland, Lond., 1845; His, on the Lymphatics of the Thymus, in Zeitsch. f. wissensch. Zoologie, X. and XI.; Kölliker and Henle in their respective Handbooks.)

Vessels.—The *arteries* of the thymus are derived from various sources, viz., from the internal mammary arteries, the inferior and superior thyroid, the subclavian and carotid arteries. They terminate in capillary vessels, which form a vascular envelope around and within each vesicle.

The *veins* pursue a different course from the arteries, and, for the most part, open into the left innominate vein.

The *lymphatics* are large. According to the observations of His on the calf, the larger blood-vessels passing to the central canal are each accompanied by two or more lymphatic stems. He finds that these arise from an interlobular plexus of lymphatic spaces destitute of walls, and that this plexus receives its roots from the interior of the lobules; and he advances the opinion that they communicate directly with the central spaces of the lobules; he has not, however, actually observed such a connection.

The *nerves* are very minute. Haller thought they were partly derived from the phrenic nerves, but according to Cooper, no filaments from these nerves go into the gland, though they reach the investing capsule, as does also a branch from the descendens noni. Small filaments, derived from the pneumogastric and sympathetic nerves, descend on the thyroid body, to the upper part of the thymus. Sympathetic nerves also reach the gland along its various arteries.

Development.—The early development of the thymus has been carefully studied by Simon, whose researches were chiefly conducted in the embryos of swine and oxen. In embryos about half an inch in length, it may be seen with the aid of a high magnifying power; and in those of one and a half inch, with the aid of a simple lens. When first distinguishable, it consists of a simple tube closed in all directions, lying along the carotid vessels. The contents of this tube are granular, but do not show regular corpuscles; its walls are delicate and homogeneous. The tube has no connection with the respiratory mucous membrane, as was supposed by Arnold; and so soon as it is discoverable, it is found to be perfectly distinct from the thyroid body. At intervals along the sides of this tube small vesicles bud out, so as to form lateral diverticula, which contain nucleated corpuscles, and which go on subsequently branching out into groups of two or four,—the formation of the permanent vesicles being merely the last repetition of this process. In the human fœtus at the seventh week, Kölliker has seen the thymus lobate at its lower end, and single above; at about the ninth week, the thymus consists of two minute elongated parallel parts lying chiefly on the upper part of the pericardium, and presenting under the microscope a distinct tubulo-vesicular structure filled with polygonal cells; at the twelfth week the thymus is broad, and its surface is entirely covered with lobules; it then increases rapidly until birth, but not with uniform rapidity, for it grows especially during the seventh, eighth, and ninth months of intra-uterine existence.

After birth, the thymus, as already stated, continues to grow to near the end of the second year. According to the observations of Haugstedt and Simon upon the weight of this organ in young animals, it appears for a short time after birth to increase not merely absolutely, but even faster than the rest of the system, and during the next period only to keep pace with the increase of the body. After the second year it ceases to grow, and becomes gradually converted by the eighth or twelfth year into a fatty mass. In this condition the corpuscles of the thymus disappear, forming, according to Simon's opinion, the nuclei of cells which become developed into the cells of adipose tissue. At puberty the thymus is generally reduced to a mere vestige which has entirely lost its original structure, and consists of brownish tissue occupying the upper part of the anterior mediastinum. Occasionally it is still found in good condition at the twentieth year; but generally only traces of it remain at that time, and these are rarely discoverable beyond the twenty-fifth or thirtieth year.

The thymus gland presents no difference in the two sexes. It exists, according to Simon, in all animals breathing by lungs, and is persistent in those which hybernate, though only as a mass of fat.

THE URINARY ORGANS.

THE urinary organs consist of the *kidneys*, the glandular organs by which the urine is secreted, and of the *ureters, bladder*, and *urethra*, which are the organs of its excretion and evacuation. As being locally connected, the *suprarenal* capsules are usually described along with these organs, though they have no relation, as far as is known, to the secretion of urine.

THE KIDNEYS.

The *kidneys*, two in number, are deeply seated in the lumbar region, lying one on each side of the vertebral column, at the back part of the abdominal cavity, and behind the peritoneum. They are situated on a level with the last dorsal and the two or three upper lumbar vertebræ, the right kidney, however, being placed a little lower down than the left, probably in consequence of the vicinity of the large right lobe of the liver. They are maintained in this position by their vessels, and also by a quantity of surrounding loose areolar tissue, which usually contains much dense fat (tunica adiposa). The *size* of the kidneys varies in different instances. Ordinarily, they measure about four inches in length, two and a half inches in breadth, and an inch and a quarter or more in thickness. The left kidney is usually of a longer and thinner shape, whilst the right is shorter and wider.

Weight.—The *average* weight of the kidney is usually stated to be about $4\frac{1}{2}$ oz. in the male, and somewhat less in the female. According to Clendinning, the two kidneys of the male weigh on an average $9\frac{1}{2}$ oz., and those of the female 9 oz. The estimate of Rayer is $4\frac{1}{4}$ oz., for each organ in the male, and $3\frac{2}{3}$ oz. in the other sex. Reid's observations (made on sixty-five males and twenty-eight females, between the ages of twenty-five and fifty-five) would indicate a higher average weight, viz., rather more than $5\frac{1}{4}$ oz. in the former, and not quite 5oz. in the latter,—the difference between the two sexes being therefore upwards of half an ounce. The *prevalent* weights of the kidney, as deduced from the tables of Reid, are, in the adult male (160 observations) from $4\frac{1}{2}$ oz. to 6 oz., and in the adult female (74 observations) from 4 oz. to $5\frac{1}{2}$ oz. The tables more recently published by Peacock give still higher average results as to the weight of these organs. The two kidneys are seldom of equal weight, the left being almost always heavier than the right. The difference, according to Rayer, is equal to about one-sixth of an ounce. The actual average difference was found by Reid in ninety-three cases (male and female), to be rather more than one-fourth of an ounce. The *proportionate* weight of the two kidneys to the *body* is about 1 to 240.

The *specific gravity* of the renal substance is, on an average, 1·052.

The surface of the kidney is smooth and has a deep red colour. Its *form* is peculiar : it is compressed before and behind, convex on its outer and concave on its inner border, and somewhat enlarged at its upper and lower ends.

The *anterior* surface, more convex than the posterior, is directed some-what outwards, and is partially covered at its upper end by the peritoneum, which is separated from it lower down by loose areolar tissue. The duo-denum and ascending colon, both destitute of peritoneum behind, are in contact with the anterior surface of the right kidney, and the descending colon with that of the left. The front of the right kidney, moreover, touches the under surface of the liver, and that of the left the lower ex-tremity of the spleen. The *posterior* surface, flatter than the anterior, and imbedded in areolar tissue, rests partly upon the corresponding pillar of the diaphragm, in front of the eleventh and twelfth ribs; partly on the anterior layer of the lumbar fascia, covering the quadratus lumborum muscle ; and, lastly, on the psoas muscle. The *external border*, convex in its general outline, is directed outwards and backwards towards the wall of the abdo-men. The *internal border*, concave and deeply excavated towards the middle, is directed a little downwards and forwards. It presents in its middle a longitudinal *fissure* bounded by an anterior and posterior lip, and named the *hilus of the kidney*, at which the vessels, the excretory duct, and the nerves enter or pass out. In this hilus, the renal vein lies in front, the artery and its branches next, and the expanded excretory duct or ureter behind and towards the lower part of the hilus. The *upper end* of the kidney, which is larger than the lower, is thick and rounded, and supports the suprarenal capsule, which descends a little way upon its anterior sur-face. This end of the kidney reaches, on the left side, to about the upper border of the eleventh rib, and, on the right, half a rib's breadth lower. It is moreover directed slightly inwards, so that the upper ends of the two kidneys are nearer to each other than the lower ends, which are smaller and somewhat flattened, diverge slightly from the spine, and reach nearly as low as the crest of the ilium. It may here be remarked that, by placing the larger end of a kidney upwards and its flatter surface backwards, or by noticing the relation of the parts in the hilus, the side of the body to which the organ belongs may be determined.

Varieties.—The kidneys present varieties in form, position, absolute and relative size, and number. Thus, they are sometimes found longer and narrower, and some-times shorter and more rounded than usual. Occasionally one kidney is very small, whilst the other is proportionately enlarged. The kidneys may, one or both, be situated lower down than usual, even in the pelvis. Instances are now and then met with in which only one kidney is present, the single organ being sometimes, though not always, formed by the apparent junction of the two kidneys across the front of the great blood-vessels and vertebral column. The united organ has usually the form of a crescent, the concavity of which is directed upwards,—hence the appellation of the *horse-shoe* kidney. Sometimes two united kidneys are situated on one or other side of the vertebral column, in the lumbar region, or, but much more rarely, in the cavity of the pelvis. In other very rare cases, three distinct glandular masses have been found, the supernumerary organ being placed either in front or on one side of the vertebral column, or in the pelvic cavity.

Structure.—The kidney is surrounded by a proper fibrous coat, which forms a thin, smooth, but firm investment, closely covering the organ. It consists of dense fibro-areolar tissue, together with numerous fine elastic fibres, and can easily be torn off from the substance of the gland, to which it adheres by minute processes of connective tissue and vessels.

On splitting open the kidney by a longitudinal section, from its outer to its inner border, the fissure named the *hilus* is found to extend some distance into the interior of the organ, forming a cavity called the *sinus* of the kidney, into the bottom of which the fibrous coat is prolonged. In such a section, also, the commencement of the excretory duct and the disposition of the substance of the organ are seen to the greatest advantage.

The *ureter*, or excretory duct of the gland, which is dilated at its upper end as it approaches the hilus, is seen to expand within the sinus into a funnel-shaped cavity, compressed from before backwards, named the *pelvis* of the kidney. Within the sinus, partly concealed by the vessels, the pelvis divides usually into three, or sometimes only two, principal tubes, which subdivide into several smaller tubes named the *calyces* or *infundibula.* These calyces, which vary in number from seven to thirteen or more, form short funnel-shaped tubes, into each of which a *papilla* of the renal substance projects. A single calyx often surrounds two, sometimes even three papillæ, which are in that case united together: hence, the calyces are in general not so numerous as the papillæ. The spaces between the calyces are occupied by a considerable amount of fat, imbedded in which are seen the main branches of the renal vessels.

Fig. 649.

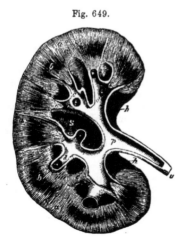

Fig. 649.—Plan of a Longitudinal Section through the Pelvis and Substance of the Right Kidney. ½

a, the cortical substance; *b, b*, broad part of two of the pyramids of Malpighi; *c, c,* the divisions of the pelvis named calyces, or infundibula, laid open; *c'*, one of these unopened; *d*, summit of the pyramids or papillæ projecting into calyces; *e, e,* section of the narrow part of two pyramids near the calyces; *p*, pelvis or enlarged divisions of the ureter within the kidney; *u*, the ureter; *s*, the sinus; *h*, the hilus.

Like the rest of the ureter, the pelvis and greater part of the calyces consist of three coats, viz., a strong external fibro-areolar and elastic tunic, which becomes continuous round the bases of the papillæ with that part of the proper coat of the kidney which is continued into the sinus; secondly, a thin internal mucous coat, which is reflected over the summit of each papilla ; and thirdly, between these two, a double layer of muscular fibres, longitudinal and circular. The longitudinal fibres are lost near the base of the calyx, but the circular fibres, according to Henle, form a continuous circular muscle round the papilla where the wall of the calyx is attached to it.

The *substance* of the kidney consists of two parts, the medullary and cortical, differing in colour, consistence, and structure.

The *internal* or *medullary substance* does not form a continuous structure, but is collected into a series of conical masses called the *pyramids of Malpighi*, the bases of which, directed towards the surface of the kidney, are imbedded in the cortical substance, whilst their apices are turned towards the sinus, and, projecting into the calyces, form the *papillæ* already mentioned. There are generally more than twelve pyramids, but their number is inconstant, varying from eight to eighteen. The medullary portion of the kidney is more dense than the cortical, and is distinctly striated, owing to its consisting of small diverging uriniferous tubes, and to its blood-vessels being arranged in a similar manner. Towards the papillæ the pyramids are of lighter colour than the cortical substance, but at their base they are usually purplish and darker.

The *external* or *cortical* substance is situated immediately within the fibrous capsule, and forms the superficial part of the organ throughout its whole extent to the depth of about two lines, and moreover sends prolongations inwards (septula renum, or columnæ Bertini) between the pyramids. It is of a nearly uniform light crimson brown appearance, and is soft and easily lacerated in directions vertical to the surface. When so lacerated, its torn surface exhibits a columnar appearance, coarser than that of torn medullary substance, and more rough and irregular ; the columnar appearance arising from the alternation of groups of straight and convoluted tubules, and the roughness being caused by the convoluted tubules and the interspersion of small round bodies of a deeper colour, the Malpighian corpuscles. The groups of straight tubules in the cortical substance are continued from those of the medullary substance, and are surrounded by the convoluted tubes into which they pass, not only on their sides but likewise at their outer extremities, so that no straight tubules reach the surface of the organ : they are termed *pyramids of Ferrein*. The Malpighian corpuscles are imbedded among the convoluted tubes, and appear disposed in double rows between the pyramids of Ferrein, and likewise more superficially, but nowhere reach quite to the surface.

The pyramidal masses found in the adult kidney indicate the original separation of this gland into lobules in the earlier stages of its growth (fig. 661). Each of these primitive lobules is in fact a pyramid surrounded by a proper investment of cortical substance, and is analogous to one of the lobules of the divided kidneys, seen in many of the lower animals. As the human kidney continues to be developed, the adjacent surfaces of the lobules coalesce and the gland becomes a single mass, and the contiguous parts of the originally separate cortical investments, being blended together, form the partitions between the pyramids already described. Moreover, upon the surface of the kidney even in the adult, after the removal of the fibrous capsule, faintly marked furrows may be traced on the cortical substance, opposite the intervals in the interior between the several papillæ with their calyces ; and not unfrequently instances occur in which a deeper separation of the original lobules by grooves remains apparent in the adult kidney.

Tubuli uriniferi.—On examining the summit of one of the papillæ carefully, especially with the aid of a lens, a number of small orifices may be seen, varying in diameter from $\frac{1}{300}$th to $\frac{1}{200}$th of an inch ; they are frequently collected in large numbers at the bottom of a slight depression or

3 P

foveola found near the summit of the papilla, but most commonly the surface is pitted over with about a score of minute depressions of this sort. On tracing these minute openings into the substance of the pyramids, they are discovered to be the mouths of small tubes or *ducts*, called the *uriniferous tubes* (tubuli uriniferi), which thus open upon the surface of the several papillæ into the interior of the calyces.

As these tubuli pass up into the pyramidal substance, they bifurcate again and again at very acute angles, their successive branches running close together in straight and slightly diverging lines, and continuing thus to divide and subdivide until they reach the sides and bases of the pyramids, whence they pass, greatly augmented in number, into the cortical substance, where they enter the pyramids of Ferrein. These straight tubules continued up from the orifices in the papillæ are sometimes called *ducts of Bellini*: they are largest near their orifices, at a short distance from which, within the papillæ, their diameter varies, according to Huschke, from $\frac{1}{330}$th to $\frac{1}{240}$th of an inch. Further on in the pyramid they become smaller, measuring about $\frac{1}{600}$th of an inch in diameter, and then do not diminish as they continue to bifurcate, but remain nearly of the same uniform average diameter.

The *convoluted tubes*, tubuli contorti, which form the greater part of the cortical substance, and, together with vessels and connecting stroma, the whole of its outermost portion, vary considerably in diameter, but they maintain commonly the same average width as the straight tubes, namely $\frac{1}{600}$th of an inch. The epithelium in the convoluted tubules may be termed cubical ; it does not present any marked contrasts in thickness, but in some of the smaller tubules it is clear, while in the majority it is turbid, and with its cells ill-defined.

Besides these tubes, long well-known to anatomists, attention has more recently been called by Henle to the presence in the Malpighian pyramids of a number of tubes, which may be roughly estimated as having only a third or a fourth of the diameter of the others, and which, after descending between the larger tubes a variable distance towards the papillæ, then turn abruptly and reascend. The tubes in question have been designated *looped tubes* of Henle. According to this author, the small differ from the large tubes not only in size but in the greater thickness of their walls. By the action of dilute hydrochloric acid, the epithelium of the large tubes is destroyed and that of the looped tubes brought into view. The epithelium of the looped tubes, Henle also states, is clear and squamous towards the papillæ, but towards the bases of the pyramids it becomes turbid, like that of the convoluted tubules.

Chrzonszczewsky, while he both figures and describes looped uriniferous tubes, considers that the merit of having discovered them rests with Ferrein, and that those described by Henle as having squamous epithelium are really blood-vessels. Although, however, it is admitted that loops are formed by blood-vessels very similar to the looped tubes of Henle, it must be regarded as certain that loops of the uriniferous tubules are much more numerous than Chrzonszczewsky is willing to admit, and for a knowledge of them as constant and regularly disposed elements of the renal structure science is indebted to Henle.

Imbedded among the convoluted tubules are the Malpighian corpuscles, the structure and connections of which must be taken into consideration, before the disputed course of the uriniferous tubes can be discussed.

The Malpighian corpuscles are small bodies of a rounded or slightly oblong shape, which have an average diameter of $\frac{1}{120}$th of an inch, but sometimes of only $\frac{1}{200}$th or $\frac{1}{270}$th of an inch. They consist each of a membranous capsule, containing a tuft of blood-vessels. The vascular tuft or

Fig. 650.—DIAGRAMMATIC RE-PRESENTATION OF A PART OF THE STRAIGHT AND CONVO-LUTED URINIFEROUS TUBES WITH THE GLOMERULI (from Frey after a drawing by Müller).

b, b, two large straight tubes in the medullary substance of the pyramid ; *c,* convoluted tubes with several of their terminations in the Malpighian capsules as in *d* ; *a,* three arteries passing up the pyramid and dividing into branches to the glomeruli : the efferent vessels are also represented and the network of capillaries between them and the veins.

glomerulus is formed by a small *afferent* artery breaking up at once into a number of minute branches, which possess simple nucleated walls, form convoluted loops, and are reunited in a single *efferent* vessel, placed close to the afferent : the further history of the afferent and efferent vessels will be described later. The *capsule,* by which the glomerulus is surrounded, is formed of homogeneous membrane. The capsule receives the two vessels at one part ; and at another it is continued into a convoluted uriniferous tubule, as was first pointed out by Bowman. Gerlach and others have considered that it may be formed on one side ; or may be so placed at the extreme point of a looped tubule, that it appears to be continuous with two tubuli ; but it is now generally admitted, as the result of filling the tubes both by injections from the ureter and by extravasation from the glomeruli, that, although in certain amphibia they may be placed

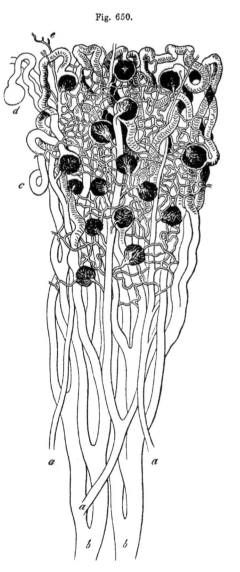

Fig. 650.

laterally, in mammals they are always terminal, communicating with one
tubule only.   The interior of the capsule is lined by a transparent delicate
squamous epithelium ; but there is still much difference of opinion as to the
exact relation of the glomerulus to the epithelium within the capsule.   Bowman has described the glomerulus as hanging naked in the interior of the
capsule, and presenting thus the greatest possible facility for the filtration
of water from its vessels into the tubule : a view which has been supported by
Ecker, Henle, and others.   Kölliker, on the other hand, has observed epithelium on the free extremity of the glomerulus, looking towards the commencing tubule, while at the sides he can find only a single layer, which he

Fig. 651.

Fig. 652.

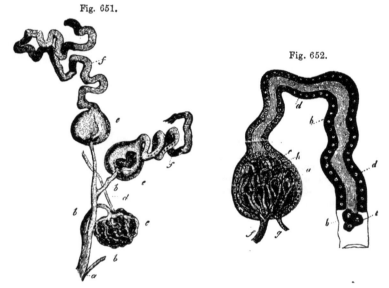

Fig. 651.—Three Malpighian Capsules in connection with the Blood-vessels
and Uriniferous Tubes of the Human Kidney (from Kölliker after Bowman). ⁹⁄₁

a, termination of an interlobular artery ; b, afferent arteries ; c, a denuded vascular
glomerulus ; d, efferent vessel ; e, two of the glomeruli enclosed by the Malpighian
capsules ; f, uriniferous tubes connected with them.

Fig. 652.—Semidiagrammatic representation of a Malpighian Body in its
relation to the Uriniferous Tube (from Kölliker). ²⁰⁰⁄₁

a, capsule of the Malpighian body continuous with b, the membrana propria of the
coiled uriniferous tube ; c, epithelium of the Malpighian body ; d, epithelium of the
uriniferous tube ; e, detached epithelium ; f, afferent vessel ; g, efferent vessel ; h,
convoluted vessels of the glomerulus.

represents as adherent on one side to the glomerulus, and on the other to the
capsular wall. Lastly, Isaacs, Moleschott, and Chrzonszczewsky maintain that
the glomerulus and the capsule have each a separate coating of epithelium,
and they agree in stating that the cells of the layer covering the glomerulus
are considerably larger than those lining the capsule : Chrzonszczewsky re-
commends sections of frozen kidneys as showing very perfectly the two layers
in situ.

Origin, course, and connections of the uriniferous tubules.—When the tubuli uriniferi are followed in their apparent course, the straight tubes are

Fig. 653.—Portion of the Uriniferous Tubes, MAGNIFIED (from Baly).

Fig. 6 3.

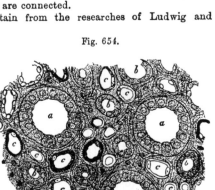

A, portion of a convoluted tube from the cortical substance ; B, epithelial cells from the interior of the tube, magnified 700 diameters.

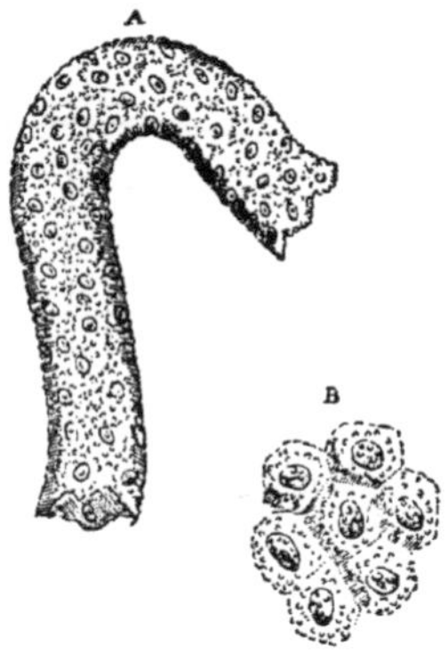

easily traceable from the Malpighian pyramids into the pyramids of Ferrein, and from these into the tubuli contorti. Thus, after the observations of Bowman had demonstrated the connection of the Malpighian corpuscles with the tubuli contorti, it appeared natural to believe that the tubuli contorti at one extremity commenced in capsules of Malpighian corpuscles, and at the other were continued into straight tubules, which opened at the summits of the papillæ. It appears, however, from the concurrent testimony of recent writers, that considerable complexity of arrangement intervenes between the terminations of the straight tubes and the commencements of the tubuli contorti, with which the Malpighian corpuscles are connected.

It may be considered as certain from the researches of Ludwig and

Fig. 654.—Transverse Section of a Renal Papilla (from Kölliker). $\frac{300}{1}$

a, larger tubes or papillary ducts ; *b*, smaller tubes of Henle ; *c*, blood-vessels, distinguished by their flatter epithelium ; *d*, nuclei of the stroma.

Fig. 654.

Zawarykin conducted by means of injections, and from those of Schweigger-Seidel by means of isolation of the tubules in small animals, that the tubuli contorti which commence in the Malpighian corpuscles are continued into the looped tubes of Henle, and that these open into the straight tubules or ducts of Bellini, either directly or through the intervention of convoluted tubes of junction of larger size, said by Schweigger-Seidel to be always present. According to Henle, the straight tubules turn rather sharply round near the surface of the kidney, and again course inwards ; and this appearance, which has been corroborated by other observers, he believes to result from anastomoses of the tubules in arches, two and two. It must be regarded as a question still

open to discussion whether all the smaller tubules which open into these
arches belong to the convoluted and looped tubules already described, as is
believed by the greater number of recent observers, or whether there is not
likewise, as is held by Henle and Chrzonszczewsky, a set of small anastomosing tubules, some of which may have blind extremities.

Fig. 655.

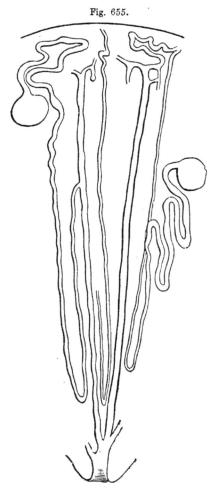

Fig. 655.—DIAGRAM OF THE LOOPED
URINIFEROUS TUBES AND THEIR
CONNECTION WITH THE CAPSULES OF
THE GLOMERULI (from Southey after
Ludwig).

In the lower part of the figure one
of the larger branching tubes is shown
opening on a papilla; in the middle
part three of the looped small tubes
are seen descending to form their
loops, and reascending in the medullary substance; while in the upper
or cortical part two of these tubes,
after some enlargement, are represented as becoming convoluted and
dilated in the capsules of glomeruli.

According to Schweigger-
Seidel the limbs of the looped
tubes are always of unequal size,
that which is continued into the
intermediate tubes being the
larger of the two; and he divides
the loops into two sorts, one in
which the narrowest portion
forms the loop, and another in
which, at the loop, the tube is
of the diameter of the larger of
its two limbs. He likewise
points out the existence of occasional capsular dilatations, where
the looped tubes meet the intermediate portions, which, as he
remarks, may explain the statement of Moleschott, that he had
found in mammals capsules communicating with two tubules.

The investigations of Chrzonszczewsky deserve special mention, on account of the novel
method to which he resorted for
the verification of his views, and
which promises to throw much
light on some of the physiological processes as well as on the minute structure of animals. This method consists in the introduction of the colouring
matter with which the observed vessels are to be filled into the system of a
living animal, either by direct infusion into the blood, or along with food
into the alimentary canal, or by absorption from any of the larger serous
cavities. The results of this mode of colouring the vessels and ducts

have already been noticed.in its application to the liver. For the kidney Chrzonszczewsky made use of the carminate of ammonia, which is freely eliminated with the urine. In order to obtain a full colouring of the blood-vessels, first the renal veins, and afterwards the arteries, are tied soon after a certain portion of the coloured fluid has been introduced into the jugular vein of the living animal. To obtain a coloured injection of the uriniferous tubes, the animal is allowed to live for about an hour after the introduction of the carmine liquid, and then the ureters are tied, while the renal blood-vessels are carefully washed out with a weak solution of common salt; and, to preserve the specimens from after infiltration of the colour, they are immersed in absolute alcohol acidulated with glacial acetic acid.

Fig. 656.

Fig. 657.

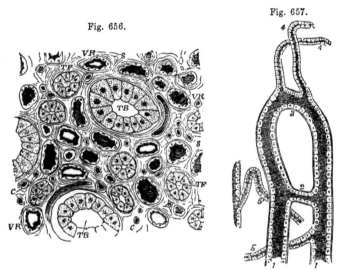

Fig. 656.—Transverse Section of the Medullary Substance of the Pig's Kidney (from Chrzonszczewsky). $\frac{300}{1}$

The drawing represents a small portion of the kidney of an animal into which colouring matter had been infused during life so as to fill the blood-vessels, by which means the distinction between them and the uriniferous tubes both larger and smaller is established, as well as by the different character of the lining epithelium ; the section is made near the papilla ; TB, the larger uriniferous tubes or tubes of Bellini ; TF, the smaller uriniferous tubes or looped tubes of Henle, named by Chrzonszczewsky tubes of Ferrein ; VR, the vasa recta or larger blood-vessels ; c, the small vessels and capillaries ; s, the stroma.

Fig. 657.—Larger and Smaller Uriniferous Tubes from the Medullary Substance of the Pig's Kidney (from Chrzonszczewsky).

1, 1, two of the larger tubes, connected by a transverse tube at 2, and presenting a looped arrangement at 3 ; from this place two smaller uriniferous tubes, 4, 4, are seen taking their origin, as well as at the other places, 5, 5.

Blood-vessels.—The kidneys are highly vascular, and receive their blood from the right and left renal arteries (p. 414), which are very large in proportion to the size of the organs they supply. Each renal artery divides into four or five branches, which, passing in at the hilus, between the vein and ureter, may be traced into the sinus of the kidney, where they lie amongst the infundibula, together with which they are usually imbedded in a quantity of fat. Penetrating the substance of the organ between the papillæ, the arterial branches enter the cortical substance found in the intervals between the medullary cones, and go on, accompanied by a sheathing of areolar tissue derived from the proper coat, and dividing and subdividing, to reach the bases of the pyramids, where they form arches between the cortical and medullary parts, which however are not complete, and in this respect differ from the freely anastomosing venous arches which accompany them. From the arches smaller *interlobular*

Fig. 658.

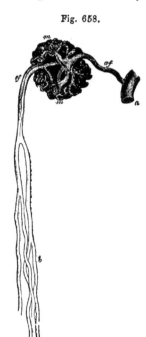

Fig. 658.—Injeoted Glomerulus from the inner part of the Cortical substance of the Horse's Kidney (from Kölliker after Bowman). $\frac{70}{1}$

a, interlobular artery; *af*, afferent artery; *m m*, convoluted vessels of the glomerulus; *ef*, efferent or straight arteriole; *b*, its subdivision in the medullary substance.

arteries are given off, which pass outwards between the double layers of Malpighian capsules which intervene between the pyramids of Ferrein; and from these interlobular arteries are derived the afferent arteries of the glomeruli. The renal arteries give branches likewise to the capsule of the kidney which anastomose with branches of the lumbar arteries, and that so freely that Ludwig was able partially to inject the kidneys of a dog from the aorta after the renal arteries had been tied. (See also Turner as cited at p. 417.) Within the glomerulus the afferent artery breaks up into convoluted branches of very small size, which are gathered together again to form the efferent vessel. The efferent vessel is so far comparable with the vena portæ of the liver that it breaks up again into capillaries, which form a close honeycomb network surrounding the convoluted tubules, and a less copious network with elongated meshes round the straight tubes of the cortical substance. Within the medullary substance are found numbers of straight vessels, *vasa recta*, which lie between the uriniferous tubes, and at the bases of the Malpighian pyramids are arranged in bundles extending inwards from between the pyramids of Ferrein. These vessels partly break up into capillaries, from which returning veins arise, and partly, as has been already noticed, form loops similar to those of the looped tubules of Henle. The mode in which the vasa recta take origin has been made the subject of considerable discussion. According to Bowman, Kölliker, and Ludwig and Zawarykin, the vasa efferentia from the innermost glomeruli are larger than the others, and break into brushes of these vasa recta. Arnold, Virchow, Beale, and others maintain the direct origin of vasa recta from the renal arteries without intervention of the glomeruli. Huschke, Henle, and Hyrtl consider that they take origin in the capillary network of the zone at the base of the pyramids ("neutral zone").

Fig. 659.—Diagram showing the Relation of the Malpighian Body to the Uriniferous Ducts and Blood-vessels (after Bowman).

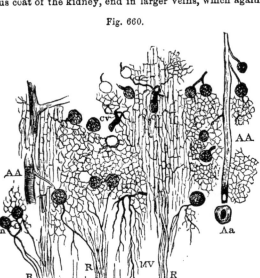

Fig. 659.

a, one of the interlobular arteries; *a'*, afferent artery passing into the glomerulus; *m*, vascular tuft formed within the glomerulus; *c*, capsule of the Malpighian body, forming the termination of and continuous with *t*, the uriniferous tube; *e'*, *e'*, efferent vessels which subdivide in the plexus *p*, surrounding the tube, and finally terminate in the branch of the renal vein *e*.

Small veins, arising by numerous venous radicles from the capillary network of the kidney, are seen near the surface of the gland, arranged so as to leave between them minute spaces, which appear nearly to correspond with the bases of the so-called pyramids of Ferrein. These vessels, some of which have a stellate arrangement (*stellulæ*, Verheyen), joined by numerous branches from the fibrous coat of the kidney, end in larger veins, which again

Fig. 660.—Longitudinal Section of a Part of the Tubular Substance and the adjacent Cortical Substance of the Kidney (from Southey).

The blood-vessels have been minutely injected, and the figure is designed principally to show the origin of the vasa recta. A A, ascending arteries divided longitudinally; C V, cortical veins; A *a*, transverse section of anastomotic arch; *m*, Malpighian bodies; R, vasa recta; M V, medullary veins.

unite into arches round the bases of the pyramids of Malpighi. Here they receive the veins of the pyramids, which commence in a beauti-

Fig. 660.

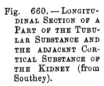

ful plexus round the orifices of the tubuli on the surface of the papillæ. Venous trunks then proceed, in company with the arteries, through the cortical envelope between the pyramids, to the sinus of the kidney. Joining together, they escape from the hilus, and ultimately form a single vein, which lies in front of the artery, and ends in the inferior vena cava (p. 474).

Nerves.—The nerves which have been traced into the kidneys are small. They come immediately from the renal plexus and the lesser splanchnic nerve, and contain filaments derived from both the sympathetic and cerebro-spinal systems. They may be traced accompanying the arteries to their finer branches, but it is uncertain how they end.

Intertubular Stroma.—Between the tubules and vessels of the kidney, although they are disposed closely together, a certain very small amount of interstitial matrix exists, first described by Goodsir, then by Bowman and others, and to which attention has latterly been paid by a number of observers, and especially by Beer. This matrix is for the most part nearly homogeneous, but has a more fibrous character in the vicinity of the ramifications of the blood-vessels. Fibres are likewise described by Ludwig and Zawarykin as passing round the Malpighian corpuscles, and others have been seen by Henle, coiling round the tubes of the medullary substance. The stroma is more abundant in the cortical substance than in the greater part of the medullary; but according to Henle it is very abundant towards the apices of the papillæ. Nuclei and connective tissue corpuscles are scat-tered through its substance. It is much more abundant in animals than in man, and in the human kidney it is more apparent in the young than in the adult, and is also much richer in corpuscles; in this respect resembling the connective tissue generally.

Absorbents.—The lymphatics of the kidney are numerous, consisting of a superficial set, and of deep lymphatics which issue from the hilus with the blood-vessels. According to the researches of Ludwig and Zawarykin, the stroma of the kidney forms a thick network of freely intercommunicating lymphatic spaces, guided to the surface along the tissue round the blood-vessels. These spaces are similar to those previously found by Ludwig and Tomsa in the testicle, and held by His to possess epithelial walls. They are most abundant in the cortical substance.

Among *writings on the kidney,* the following may be here referred to :—Bowman, in Philos. Trans. 1842 ; Toynbee, in Medico-Chir. Trans. 1846; Gerlach, in Müller's Archiv, 1845 ; Johnson, article Ren, in Clyclopædia of Anat. and Phys. ; Isaacs, in Trans. New York Acad. of Medicine, vol. i., 1857 ; Henle, Zur Anatomie der Niere, Göttingen, 1862, and in Handbuch; Ludwig and Zawarykin, in Wiener Kais. Acad. Sitzungsbericht, vol. xlviii. 1864 ; Chrzonszczewsky, in Virchow's Archiv, xxxi. 1864; Schweigger-Seidel, Die Niere des Menschen und der Saügethiere, Halle, 1865 ; Southey, in St. Bartholomew's Hosp. Reports, 1865 ; also, on the stroma, Goodsir, in Lond. and Edin. Journ. of Med. Science, May, 1842 ; and Beer, Die Bindesubstanz d. Menschlichen Niere, Berlin, 1859.

Development.—The development of the kidneys, and also that of the suprarenal cap-sules will be described later with that of the genito-urinary organs.

The Urine.—This is a complex and somewhat variable fluid, containing in solution animal substances characterised by having a large amount of nitrogen in their composition, and derived, it would seem, from the waste of the tissues; also saline substances, and adventitious matters which have been introduced into the blood. The average quantity secreted daily is from 30 to 40 fluid ounces. Its specific gravity varies in health from 1·015 to 1·030, the average standard being 1·020. It is slightly acid in its reaction, and contains some mucus and epithelium. A thousand parts of ordinary urine usually contain 933 parts of water, and 67 of solid matter. The researches of Bowman upon the structure of the kidney in man and animals, render it probable that the solid urinary constituents are secreted by the tubuli, and that the watery part of the urine simply transudes through the vessels of the glomeruli.

The following analysis of the solid contents of the urine is from Lehmann, but it must be considered approximative only, since the proportion of the ingredients is liable to great variation in dependence upon food, exercise, and other conditions:—

Urea	49·68
Uric acid	1·61
Extractive matters, ammoniacal salts, and chloride of sodium	28·95
Alkaline sulphates	11·58
Alkaline phosphates	5·96
Phosphates of lime and magnesia	1·50
	99·28

Among the extractive matters are kreatine, kreatinine, and hippuric acid.

SUPRARENAL BODIES.

The *suprarenal bodies* or *capsules*, or suprarenal glands, (capsulæ atrabiariæ seu renes succenturiati of old anatomists), are two flattened bodies, each of which has a somewhat crescentic or bent triangular shape, and surmounts the corresponding kidney. The *upper* border, convex and thin, is often considerably elevated in the middle so as to form two sides of a triangle. The lower border is concave, and rests upon the anterior and inner part of the summit of the kidney, to which it is connected by loose areolar tissue : it is thick, and almost always deeply grooved. The *posterior* surface rests upon the diaphragm. Its *anterior* surface is covered on the right side by the liver, and on the left by the pancreas and spleen : it presents an irregular fissure named the *hilus*, from which the suprarenal vein emerges. The right capsule, like the right kidney, is placed lower down than the left.

The suprarenal capsules vary in size in different individuals, and the left is usually somewhat narrower at its base, but longer from above downwards, and larger than the right. They measure from an inch and a quarter to an inch and three-quarters in height, and about an inch and a quarter in width ; their thickness is from two to three lines. The *weight* of each in the adult is from one to two drachms.

Besides a covering of areolar tissue mixed frequently with much fat, the suprarenal capsules have a thin fibrous investment. Externally, they have a yellowish or brownish-yellow colour. When divided, they are seen to consist of two substances : one, *external* or *cortical*, is of a deep yellow colour, firm and striated, and forms the principal mass of the organ ; the other, *internal* or *medullary*, is in the adult of a dark brownish-black hue, and so soft and pulpy that some anatomists have erroneously described a cavity within it.

The *fibrous investment* is so intimately connected with the deeper parts that it cannot be removed without lacerating the subjacent structure. Its deeper layers are destitute of elastic fibres, and are particularly rich in nuclei : they are continuous with the septa which enter into the formation of the substance of the organ.

The *cortical part* of the suprarenal body, examined with a low magnifying power, is seen to consist of stroma, in which are imbedded columnar and reticulated masses measuring on an average $\frac{1}{700}$th of an inch in diameter, arranged vertically to the surface of the organ, and containing cellular constituents. In the deepest part of the cortex, however, the colour is darker, and the columnar arrangement is lost, the stroma being more equally scattered ; and immediately beneath the fibrous coat there is another very narrow layer in which the stroma forms oval spaces, of which it is difficult to say whether they communicate with the extremities of the columns or not. These inner and outer layers have been named by J. Arnold respectively zona reticularis and zona glomerulosa, while he applies the term zona fasciculata to the main part ; but as the transition from one of these parts to another is not sudden nor indicated by any line of demarcation, they

Fig. 661.   Fig. 662.

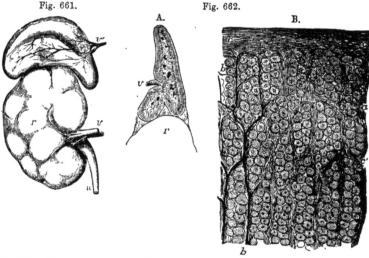

Fig. 661.—FRONT VIEW OF THE RIGHT KIDNEY AND SUPRARENAL BODY OF A FULL GROWN FŒTUS.

This figure shows the lobulated form of the fœtal kidney r ; v, the renal artery and vein ; u, the ureter ; s, the suprarenal capsule, the letter is placed near the sulcus in which the large veins (v') are seen dividing and dipping into the interior of the organ.

Fig. 662.—SECTIONS OF THE SUPRARENAL BODY.

A, vertical section of the suprarenal body of a fœtus twice the natural size, showing the lower notch by which it rests on the summit of the kidney, and the anterior notch by which the veins penetrate, together with the distinction between the medullary and cortical substance. $\frac{2}{1}$

B, longitudinal section of the cortical substance, showing the capsules containing nucleated cells and intervening blood-vessels. $\frac{250}{1}$ The figure represents a small fragment of a section made perpendicularly to the surface in a suprarenal body of which the bloodvessels were partially injected. a, one of the superficial masses of cells (in the zona glomerulosa of J. Arnold) ; a', one of the longer masses slightly deeper (zona fasciculata) ; b, bloodvessels running in the septa of connective tissue between the cell-masses in a part of the specimen ; c, connective tissue and sheath substance on the surface ; c', connective tissue of the septa : this figure, though true to nature in the representation of the several textures, is so far diagrammatic that the space occupied by the shorter masses of cells towards the surface is proportionally too small.

are probably only modifications of the same structure. The contents of the stroma consist of nucleus-like bodies from $\frac{1}{6000}$th to $\frac{1}{3000}$th of an inch in diameter, mixed with minute yellowish granules, and oily particles with granular matter adhering to them, together with large groups of closely set nucleated cells containing granular matter and oily molecules. The cells vary from $\frac{1}{2400}$th to $\frac{1}{1350}$th of an inch in size, and their opposing sides are somewhat flattened, giving them the form of irregular polyhedra : the larger cells are most loaded with oil-globules. In many instances, probably, the appearance of free nuclei and oil-globules is to be explained by cell-walls being ruptured or remaining unrecognised.

According to Simon, the columns consist of distinct tubes with a limiting membrane ; Ecker and others affirm that no continuous tubular cavities exist, but that rows of closed vesicles, many of them oval in shape, and overlapping each other, are placed in such manner as to resemble tubes ; while Gray believes that the walls of adjoining vesicles are sometimes removed by absorption, so that tubular cavities are formed by the coalescence of neighbouring vesicles. Kölliker, however, and other observers, maintain with more correctness, that the so-called vesicles are merely loculi or cavities in the stroma of the organ, possessing no distinct limitary membrane, and producing the appearance of a tubular structure by their apposition in linear series. The small arteries, entering from the surface, run parallel to these columns, frequently anastomose together between them, and surround each row of vesicles with a fine capillary network. Small bundles of nerves pass inwards in the septa between the columns to reach the medullary part of the organ, and their fibres begin to spread out in the zona reticularis, but do not appear to be distributed to the cortical substance.

The *medullary part* of the suprarenal capsule is separated from the cortical part by a layer of connective tissue, the fibres of which are parallel to the two parts, and allow them to be easily separated one from the other in sections prepared for the microscope. In the thinner parts of the adult organ there is no medullary part, or it has shrunk away, and the layer of connective tissue referred to is found separating the deep surfaces of two opposed portions of the cortical part ; but in the young state the distinction of cortical and medullary probably extends throughout the whole. The medullary part is traversed in the centre by venous trunks, which receive the whole of the blood which has passed through the organ. The stroma is delicate, arranged in a reticular manner ; the pulpy substance which lies in it is difficult of examination, but consists of cells, differing from those of the cortex in being destitute of oil globules, and some of them branched. The bundles of nerves which pass through the cortical substance run between it and the medullary substance, and then form a copious interlacement which extends through the whole of the medullary stroma. According to Leidig and Luschka, the cells of the medullary substance are ganglionic ; and Luschka states that he has found them both connected one with another and with nerve-fibres ; but this view still requires confirmation. Moers, while he denies that the cells of the medullary parenchyma are nervous, describes ganglia on the nerves where their bundles begin to break up. The medullary substance receives its blood by the continuation inwards of the capillary network of the cortex, the blood from which is collected by venous radicles which open into the stems in the centre of the organ.

Vessels.—The suprarenal bodies receive *arteries* from three sources, viz., from the aorta, the phrenic, and the renal arteries. The distribution of their capillary vessels has already been mentioned.

The *veins*, which pass out from the centre, are usually united into one for each organ. The right vein enters the vena cava inferior immediately, whilst the left, after a longer course, terminates in the left renal vein.

The *lymphatics* are imperfectly known. Kölliker has seen a few small trunks upon the surface ; and Luschka has, in addition, observed others emerging from the interior in company with the vein.

Nerves.—The nerves are exceedingly numerous. They are derived from the solar plexus of the sympathetic, and from the renal plexuses. According to Bergmann, some filaments come from the phrenic and pneumogastric nerves. They are made up mainly of dark-bordered white fibres, of different sizes, and they have many small ganglia upon them before entering the organ. The nerves are especially numerous in the lower half, and inner border.

Accessory suprarenal capsules are occasionally met with, attached by connective tissue to the main bodies; and varying from a small size up to that of a pea. According to Duckworth, they possess no medullary part.

On the subject of the suprarenal capsules may be consulted,—Ecker, Der feinere Bau der Nebennieren, Braunschweig, 1846; Simon on the Thymus Gland; Frey, article "Suprarenal Capsules," in Cyclop. of Anat. and Phys.; Harley, in the Lancet, June, 1858; Duckworth, in St. Bartholomew's Hosp. Reports, 1865; Moers, in Virchow's Archiv, 1864, vol. xxix. p. 336; J. Arnold, Virchow's Archiv, 1866, vol. xxxv. p. 64; Leidig, Kölliker, Luschka, and Henle, in their Handbooks.

Function.—Nothing is known positively with regard to the functions of the suprarenal capsules. The opinion which has met with most acceptance among physiologists is that these bodies belong to the class of blood-vascular glands, and exert some influence upon the elaboration or disintegration of nutritive material. Bergmann, however, who was the first to point out the richness of their nervous supply, suggested that they were parts of the sympathetic nervous system, and in this opinion he has been followed by Leidig and Luschka; while Kölliker states that, upon anatomical grounds, he is inclined to consider the cortical and medullary portions as functionally different; the former belonging to the group of vascular or ductless glands, the latter appearing to be an apparatus appertaining to the nervous system. Brown-Séquard found that injuries to the spinal cord in its dorsal region produced congestion and subsequent hypertrophy of the suprarenal bodies. Addison has shown that a bronzed tint of skin, together with progressive emaciation and loss of strength, is to be found in conjunction with various forms of disease more or less involving and altering the structure of these bodies.

THE URETERS.

The *ureters* are two tubes which conduct the urine from the kidneys into the bladder. The upper, dilated, funnel-shaped commencement of each in the pelvis of the kidney, into which the calyces pour their contents, has already been described. Towards the lower part of the hilus of the kidney the pelvis becomes gradually contracted, and opposite the lower end of the gland, assuming the cylindrical form, receives the name of ureter. These tubes extend downwards to the posterior and under part or base of the bladder, into which they open, after passing obliquely through its coats.

The ureters measure from fourteen to sixteen inches in length, and their ordinary width is about that of a goose-quill. They are frequently, however, dilated at intervals, especially near their lower end. The narrowest part of the tube, excepting its orifice, is that contained in the walls of the bladder.

Each ureter passes, at first, obliquely downwards and inwards, to enter the cavity of the true pelvis, and then curves forwards, and inwards, to reach the side and base of the bladder. In its whole course, it lies close behind the peritoneum, and is connected to neighbouring parts by loose areolar tissue. Superiorly, it rests upon the psoas muscle, and is crossed, very obliquely from within outwards, below the middle of the psoas, by the spermatic vessels, which descend in front of it. The right ureter is close to the inferior vena cava. Lower down, the ureter passes over the common iliac, or the external iliac vessels, behind the termination of the ileum on the right side and the sigmoid flexure of the colon on the left. Descending into the pelvis, it enters the fold of peritoneum forming the corresponding posterior false ligament of the bladder, and, reaching the side of the bladder near the base, runs downwards and forwards in contact with it, below the obliterated hypogastric artery, and is crossed upon its inner side, in the male, by the vas deferens, which passes down between the ureter and the bladder. In the female, the ureters run along the sides of the cervix uteri and upper part of the vagina before reaching the bladder.

Having reached the base of the bladder, about two inches apart from one another, the ureters enter its coats, and running obliquely through them for about three-quarters of an inch, open at length upon the inner surface by two narrow and oblique slit-like openings, which are situated, in the male, about an inch and a half behind the prostate, and about the same distance from each other. This oblique passage of the ureter through the vesical walls, while allowing the urine to flow into the bladder, has the effect of preventing its return up the ureter towards the kidney.

Structure.—The walls of the ureter are pinkish or bluish white in colour. They consist externally of a dense, firm, areolar, and elastic coat, which in quadrupeds decidedly contracts when artificially irritated. According to Huschke, it possesses two layers of longitudinal fibres : Henle finds only an inner longitudinal and an outer circular layer ; while Kölliker, who formerly described the circular and outer longitudinal layers as the only layers found except in immediate proximity to the bladder, now admits the inner longitudinal and circular as the principal layers, on Henle's authority, and states that the longitudinal fibres external to the circular layer are absent at the upper part of the tube.

Fig. 663.

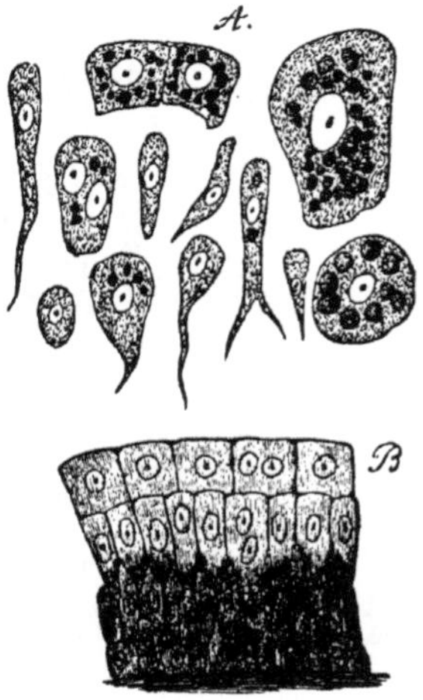

Fig. 663.—Epithelium from the Pelvis of the Human Kidney (from Kölliker). $\frac{350}{1}$

A, different kinds of epithelial cells separated ; B, the same in situ.

Internally, the ureter is lined by a thin and smooth *mucous membrane,* which presents a few longitudinal folds when the ureter is laid open. It is prolonged above upon the papillæ of the kidney, and below becomes continuous with the lining membrane of the bladder. The epithelium is of the spheroidal or transitional character, stratified, and containing, besides rounded cells, others cylindrical and branched (Kölliker and Luschka).

Vessels.—The ureter is supplied with blood from small branches of the renal, the spermatic, the internal iliac, and the inferior vesical arteries. The veins end in various neighbouring vessels.

The *nerves* come from the inferior mesenteric, spermatic, and hypogastric plexuses.

Varieties.—Sometimes there is no funnel-shaped expansion of the ureter at its upper end into a pelvis, but the calyces unite into two or more narrow tubes, which afterwards coalesce to form the ureter. Occasionally, the separation of these two tubes continues lower down than usual, and even reaches as low as the bladder, in which case the ureter is double. In rare cases, a triple ureter has been met with.

In instances of long-continued obstruction to the passage of the urine, the ureters occasionally become enormously dilated, and their opening into the bladder becomes direct, so as to lose its valvular action.

The *urinary bladder* (vesica urinaria) is a hollow membranous and muscular viscus, which receives the urine poured into it through the ureters, retains it for a longer or shorter period, and finally expels it through the urethra.

During infancy it is pyriform, and lies chiefly in the abdomen, but in the adult it is situated in the pelvic cavity behind the pubes, and in the male, in front of the rectum ; in the female, it is separated from the rectum by the uterus and vagina.

The size and shape of the bladder, its position in the abdomino-pelvic cavity, and its relations to surrounding parts, vary greatly, according to its state of distension or collapse.   When quite empty, the bladder lies deeply in the pelvis, and in a vertical antero-posterior section presents a triangular appearance, being flattened before and behind, having its base turned downwards

Fig. 664.

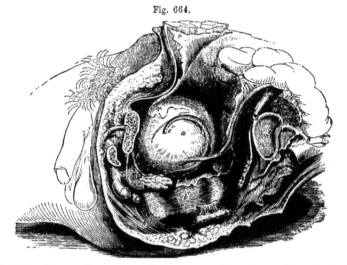

Fig. 664.—LATERAL VIEW OF THE VISCERA OF THE MALE PELVIS (after Quain). ¼

The left os ilium has been disarticulated from the sacrum, the spinous process of the ischium cut through, and the pubes divided to the left of the symphysis ; *a,* the bladder; *b, b',* the rectum ; *c,* membranous part of the urethra ; *d,* section of the left crus or corpus cavernosum ; *e,* bulb of the spongy body of the urethra ; *f,* Cowper's gland ; *g,* section of the body of the pubes ; *h,* sphincter ani muscle ; *i,* part of the left vas deferens ; *m,* articular surface of the sacrum ; *n,* divided spine of the ischium ; *o,* coccyx ; *p,* prostate gland ; *r, r,* peritoneum ; *r',* recto-vesical pouch ; *u,* left ureter ; *v,* left vesicula seminalis.

and backwards, whilst its apex reaches up behind the symphysis pubis (fig. 601).   The surfaces named anterior and posterior have thus a considerable inclination.   When moderately full, it is still contained within the pelvic cavity, and has a rounded form ; but when completely distended, it rises above the brim of the pelvis, and becomes egg-shaped ; its larger end, which is called the *base* or *inferior fundus,* being directed towards the rectum in the male and the vagina in the female ; and its smaller end, or *summit*, resting against the lower part of the anterior wall of the abdomen. Immediately in front of the base is the thickened portion named the *cervix*, or *neck*, which bounds the outlet of the bladder, and connects it below with the urethra.

The long axis of the distended bladder is inclined obliquely upwards and forwards from the base to the summit, in a line directed from the coccyx to a point between the pubes and the umbilicus. In being gradually distended, the bladder curves slightly forwards, so that it becomes more convex behind than in front, and its upper end is by degrees turned more and more towards the front of the abdomen. Lastly, the bladder, when filled, appears slightly compressed from before backwards, so that its diameter in that direction is less than from side to side. Kohlrausch states that, when the bladder is filled during life, it has the shape of a flattened spheroid ; and that, owing to pressure of the intestines from above, and the gravitation of fluid in its interior, its vertical diameter is the shortest. In its ordinary state, the longest diameter in the male is from base to summit ; but in the female its breadth is often greater than its height. The average capacity of the bladder is often stated to be greater in the female than in the male ; and, no doubt, instances of very large female bladders are not unfrequent, but these have probably been the result of unusual distension : in the natural condition, according to Luschka and Henle, the female bladder is decidedly smaller than that of the male.

While freely movable in all other directions upon surrounding parts, the bladder is fixed below to the walls of the pelvis by the neck, and by reflections of the recto-vesical fascia, named the *true ligaments* of the bladder. It is supported, moreover, by strong areolar connections with the rectum or vagina, according to the sex, also in a slighter degree by the two ureters, tho obliterated hypogastric arteries and the *urachus*, by numerous blood-vessels, and, lastly, by a partial covering of the peritoneum, which, in being reflected from this organ in different directions, forms certain folds or duplicatures, named the *false ligaments* of the bladder.

The *anterior* surface is entirely destitute of peritoneum, and is in apposition with the triangular ligament of the urethra, the subpubic ligament, the symphysis and body of the pubes, and, if the organ be full, the lower part of the anterior wall of the abdomen. It is connected to these parts by loose areolar tissue, and to the back of the pubes by two strong bands of the vesical fascia, named the *anterior* true ligaments. This surface of the bladder may be punctured above the pubes without wounding the peritoneum.

The *posterior surface* of the bladder is entirely free, and covered everywhere by the peritoneum, which in the male is prolonged also for a short distance upon the base of the bladder. In the male, this surface is in contact with the rectum, and in the female with the uterus, some convolutions of the small intestine descending between it and those parts, unless the bladder be very full. Beneath the peritoneum, in the male, a part of the vas deferens is found on each side of the lower portion of this surface.

The *summit* (sometimes named the *superior fundus*) is connected to the anterior abdominal wall by a tapering median cord, named the *urachus*, which is composed of fibro-areolar tissue, mixed at its base with some muscular fibres which are prolonged upon it from the bladder. This cord, becoming narrower as it ascends, passes upwards from the apex of the bladder between the linea alba and the peritoneum, to reach the umbilicus,

The *third stratum* of fibres, still more deeply situated, and which might be termed internal longitudinal, was first described by Ellis, who distinguished it as "submucous." It is very delicate, and its fibres, directed longitudinally, are scattered in a regular manner round the cavity of the bladder.

The researches of Pettigrew (including an elaborate series of dissections preserved in the Museum of the Royal College of Surgeons of England) have led him to the conclusion that, with few exceptions, the muscular fibres of all the strata are arranged in figure-8 loops. These loops are directed towards the apex and base, and he regards them as disposed in four sets ; an anterior and a posterior set largely developed, and a right and a left lateral set accessory and less fully developed ; and they are so arranged that at any one spot on the bladder there are to be found decussating groups of fibres, which may be distinguished as longitudinal, horizontal or transverse, and oblique. The extremities of each figure-8 are placed on one aspect of the bladder, and the point of decussation on the opposite aspect ; the arrangement being thus similar to that of a string wound in figure-8 loops round a cylinder. In each set the most superficial loops are compressed laterally and elongated from above downwards, but the succeeding loops as they become more deeply placed are more and more drawn out transversely until those which are nearly circular are reached ; and on passing more deeply than these, the loops become again gradually more and more elongated until those which have been alluded to as internal longitudinal are arrived at. The figure-8 arrangement, stated by Pettigrew to exist in all the groups, is most distinctly seen in the anterior set, which may here be more particularly alluded to. The most superficial fibres of this set form a narrow band some of which are prolonged on the urachus, while others pass round close behind its insertion ; their decussation takes place about midway between the summit of the bladder and the urethra ; and inferiorly they pass forwards to be inserted into the capsule of the prostate, the posterior surface of the pubes, the inner border of the levator ani, and the fascia covering the constrictor urethræ muscle. The points of decussation of the deeper fibres

Fig. 667.

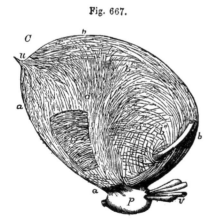

Fig. 667.—VIEW OF THE MUSCULAR FIBRES OF THE BLADDER FROM THE LEFT SIDE (after Pettigrew and from Nature). ⅓

The anterior and posterior superficial fibres are seen in profile running from below upwards, crossing each other by their divergence on the sides of the bladder, and are indicated by the same letters as in the previous figures ; at *c,* a portion of the anterior longitudinal fibres has been removed so as to expose the deeper circular fibres.

as they become more horizontal, are placed lower and lower down. The fibres which cross obliquely are most expanded, and embrace the larger parts of the bladder, taking part, on the posterior wall, in the formation of the so-called circular layer, while the fibres which at their decussation are more nearly horizontal are confined to the regions of the base and neck. The whole of the muscular fibres around the prostate and prostatic portion of the urethra are supposed by Pettigrew to be formed by the lower extremities of the various figures-8.   The general idea of this figure-8 arrangement was first suggested by Sabatier, by whom the more marked examples of it are described ; but it has been fully elaborated by the researches of Pettigrew.

A strong muscular bundle already alluded to passes, as shown by Ellis, with its convexity forwards between the terminations of the ureters, continuous with the longitudinal fibres of these tubes.   Other fibres mentioned by Morgagni, and more fully described under the name of the "muscles of the ureters" by Sir C. Bell, pass forwards from the ureters towards the urethra : they are considered by Pettigrew not as special structures, but as a part of the general arrangement of fibres in that part of the bladder.

On the muscular arrangements of the bladder, see Pettigrew, in Phil. Trans. for 1866 ; Sabatier, Rech. Anat. et Phys. sur les Appareils musculaires correspondants à la vessie et à la prostate dans les deux sexes, 1864 ; and in Henle's Jahresbericht ; Ellis, in Trans. Med. Chir. Society, 1856, and Demonstrations, 1861.

The muscular coat of the bladder forms so irregular a covering, that, when the organ is much distended, intervals arise in which the walls are very thin; and, should the internal or mucous lining protrude in any spot through the muscular bundles, a sort of hernia is produced, which may go on increasing, so as to form what is called a vesical sacculus, or *appendix vesicæ*, the bladder thus affected being termed *sacculated*. Hypertrophy of the muscular fasciculi, which is liable to occur in stricture of the urethra or other affections impeding the issue of the urine, gives rise to that condition named the *fasciculated* bladder, in which the interior of the organ is marked by strong reticulated ridges or columns, with intervening depressions.

Next to the muscular coat, between it and the mucous membrane, but much more intimately connected with the latter, is a well-marked layer of areolar tissue, frequently named the *cellular* or *vascular* coat.   This submucous areolar layer contains a large quantity of very fine coiled fibres of elastic tissue.

The mucous membrane of the bladder is soft, smooth, and of a pale rose colour.   It is continuous above with the lining membrane of the ureters and kidneys, and below with that of the urethra.   It adheres loosely to the muscular tissue, and is thus liable to be thrown into wrinkles, except at the trigone, where it is consequently always more even.   It is covered with a stratified epithelium, the particles of which are intermediate in form between those of the columnar and squamous varieties.   There are no villi upon the vesical mucous membrane, but it is provided with minute follicles, and small racemose glands lined with columnar epithelium, which are most abundant in the vicinity of the neck of the bladder.   The vesical mucus (according to Mandl) is alkaline, and appears to contain alkaline and earthy phosphates.

Vessels.—The *superior vesical* arteries proceed from the remaining pervious portions of the hypogastric arteries; in the adult they appear as direct branches of the internal iliac.   The *inferior* vesical arteries are usually derived from the anterior division of the internal iliac.   In the female the uterine arteries also send branches to the bladder.   The neck and base of the organ appear to be the most vascular portions. The *veins* form large plexuses around the neck, sides and base of the bladder; they eventually pass into the internal iliac veins.   The *lymphatics* follow a similar course.

The *nerves* are derived partly from the hypogastric plexus of the sympathetic, and partly from the sacral plexus of the cerebro-spinal system.   The former are said to be chiefly distributed to the upper part of the bladder, whilst the spinal nerves may be traced more directly to its neck and base.

THE URETHRA.

The urethra is a membranous tube directed in the median line, first vertically and then from behind forwards, beneath the arch of the pubes, in
which situation it opens in the female into the vulva, while in the male it is
enclosed in the spongy substance and prolonged beneath the corpora cavernosa penis. In the female, it serves simply as the excretory passage for the
urine; in the male, it conducts also the seminal fluid. The detailed
anatomy of the male and female urethra will be given with that of the
organs of generation of the respective sexes.

ORGANS OF GENERATION.

THE MALE ORGANS OF GENERATION.

THE male organs of generation include, together with the *testes* and their
proper *excretory apparatus*, a series of structures which for convenience
may be considered first, as they are closely connected with the urethra.
Thus, at its commencement the urethra passes through the *prostate gland*,
and there it receives the excretory ducts of the testes and vesiculæ seminales; emerging from the prostate, it traverses the layers of the subpubic
fascia supported by muscles, and, becoming copiously surrounded with
the erectile tissue of the corpus spongiosum, is pierced by the ducts of
Cowper's glands, and afterwards, in conjunction with the corpora cavernosa,
enters into the formation of the *penis*.

THE PROSTATE GLAND.

The *prostate gland* is a firm glandular body, somewhat resembling a
chestnut in shape and size, which supports the neck of the bladder and
encloses the commencement of the urethra: it is placed in the pelvic cavity,
on the deep aspect of the subpubic fascia, and rests upon the rectum. It
has the form of a flattened cone with its base in contact with the bladder,
and cut obliquely, so that its posterior or rectal surface is much larger than
its anterior or pubic surface. It usually measures about an inch and a
half across at its widest part, an inch and a quarter from its base to its
apex, and nearly an inch in depth or thickness. Its ordinary weight is
about six drachms.

The anterior or pubic surface of the prostate is flattened and marked
with a slight longitudinal furrow; it is about half an inch or rather more
from the pubic symphysis, and there, as well as the sides of the gland, is con-

Fig. 668.

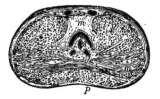

P

Fig. 668.—TRANSVERSE SECTION OF THE PROS-
TATE GLAND THROUGH THE MIDDLE.

u, the urethra, into which the eminence of the
caput gallinaginis rises from below; *s*, the sinus
pocularis, cut through; *d e*, the ejaculatory ducts;
m, superiorly, the deep sphincter muscular fibres;
m, lower down, intersecting muscular bands in the
lateral lobes of the prostate; *p, p*, glandular substance.

nected to the pubic arch by the reflexion
of the pelvic fascia, which forms the
pubo-prostatic *ligaments* or *anterior ligaments* of the bladder. The posterior or rectal surface is smooth, and is marked by a slight depression,

or by two grooves, which meet in front, and correspond with the course of the seminal ducts, as well as mark the limits of the lateral lobes in this situation : it is in close apposition with the rectum, immediately in front of the bend from the middle to the lower or anal part of that viscus, where the surface and posterior border of the gland can be felt by the finger introduced into the intestine. The sides are convex and prominent, and are covered by the anterior portions of the levatores ani muscles, which pass back on each side, from the symphysis pubis and anterior ligament of the bladder, and embrace the sides of the prostate. This part of each levator ani is occasionally separated from the rest of the muscle by areolar tissue, and has been named *levator prostatæ*. The base of the gland is of considerable thickness, and is notched in the middle : its apex is turned towards the triangular ligament. As already stated, the prostate encloses the commencement of the urethra. The canal runs nearer to the upper than to the under surface of the gland, so that in general it is about three lines distant from the former and four or five from the latter ; but it frequently varies greatly in this respect. The prostatic portion of the urethra is about an inch and a quarter long, and is dilated in the middle ; it contains the verumontanum and the openings of the seminal and prostatic ducts, and will be afterwards more particularly described. The common seminal ducts, which pass forwards from the vesiculæ seminales, also traverse the lower part of the prostate, enclosed in a special canal, and open into the urethra.

Fig. 669.

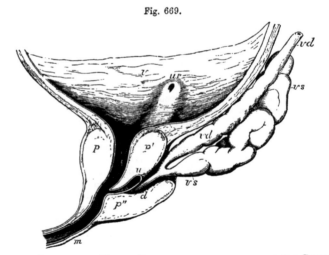

Fig. 669.—Longitudinal Median Section of the lower part of the Bladder and Prostate Gland (after E. H. Weber).

v, inner surface of the urinary bladder ; *u r*, opening of the right ureter, from which a slight elevation runs down to the neck of the bladder ; *p*, upper part of the prostate ; *p′*, the so-called middle lobe ; *p″*, the right lateral lobe ; *u*, the utricle or sinus pocularis ; *d*, the right ejaculatory duct ; *v d*, vas deferens ; *v s*, vesicula seminalis.

This gland is usually described as consisting of three lobes, two of which, placed laterally and separated behind by the posterior notch, are of equal size ; the third, or *middle* lobe, is a smaller rounded or triangular mass, intimately connected with the other two, and fitted in between them on the under side, lying immediately beneath the neck of the bladder and the adjacent part of the urethra. This third lobe is exposed by turning down the seminal vesicles and ducts, between which and the cervix vesicæ it is placed ; being in fact the part of the gland contained between and behind the grooves or fissures by which the ejaculatory ducts reach the urethra. The separation between these lobes, which is little marked in the natural state, becomes often much more apparent in disease.

Structure.—The prostate is enclosed in a dense fibrous coat, which is continuous with the recto-vesical fascia, and with the posterior layer of the triangular ligament, and is rather difficult to tear or cut. Adams describes the fibrous capsule as divisible into two layers, between which the prostatic plexus of veins is enclosed. The prostate is a highly muscular organ ; its external coat contains numerous plain fibres ; within the proper glandular structure, which lies somewhat superficially, there is a strong layer of circular fibres continuous posteriorly with the sphincter vesicæ. Ellis finds that these muscular fibres not only join behind with the circular fibres of the bladder, but are continuous in front with the thin layer hereafter described around the membranous part of the urethra (p. 962). According to Pettigrew, the muscular fibres of the prostate are the lower parts of figure-8 loops, which spread superiorly on the bladder. The substance of the gland is spongy and more yielding ; its colour is reddish grey, or sometimes of a brownish hue. It consists of numerous small follicles or terminal vesicles opening into elongated canals, which unite into a smaller number of excretory ducts. These appear either as pores or as whitish streaks, according to the way in which they are exposed in a section. The epithelium in the vesicular terminations is thin and squamous, whilst in the canals it is columnar. The capillary blood-vessels spread out as in other similar glands on the ducts and clusters of vesicles, and the different glandular elements are united by areolar tissue, and supported by processes of the deep layer of the fibrous capsule (Adams). The ducts open by from twelve to twenty or more orifices upon the floor of the urethra, chiefly in the hollow on each side of the verumontanum (p. 963).—(Adams, Cyclop. of Anat., vol. iv., p. 147 ; Ellis and Pettigrew, referred to at p. 951.)

Vessels and Nerves.—The prostate is supplied by branches of the vesical, hæmorrhoidal, and pudic *arteries*. Its *veins* form a plexus round the sides and base of the gland, which is highly developed in old subjects. These veins communicate in front with the dorsal vein of the penis, and behind with branches of the internal iliac vein. According to Adams, the lymphatics, like the veins, are seen ramifying between the two layers of the fibrous capsule. The *nerves* are derived from the hypogastric plexus.

Prostatic fluid.—This is mixed with the seminal fluid during emission; as obtained from the human prostate soon after death, it has a milky aspect, which is ascribed by Adams to the admixture of a large number of epithelial cells, and he thinks it probable that, as discharged during life, it is more transparent. According to the same observer, the prostatic fluid has an acid reaction, and presents, under the microscope, numerous molecules, epithelial particles both squamous and columnar, and granular nuclei about $\frac{1}{3300}$ inch in diameter. As age advances, this gland is disposed to become enlarged; and its ducts often contain small round concretions of laminated appearance, and varying from a small size up to that of a millet-seed; they sometimes contain carbonate of lime, but are principally composed of animal matter, which in some of them appears to be entirely amylaceous, in others albuminous, and more frequently is of a mixed character. (Virchow's Cellular Pathology, by Chance, p. 369.)

THE PENIS.

The penis, which supports the greater part of the urethra in the male, is composed principally of an erectile tissue, arranged in masses which occupy three long and nearly cylindrical compartments. Of these, two, named *corpora cavernosa penis*, placed side by side, form the principal part of the organ, whilst the other, situated beneath the two preceding, surrounds the canal of the urethra, and is named *corpus cavernosum urethræ* or *corpus spongiosum.*

The penis is attached behind to the front of the pubes, and to the pubic arch, by what is termed the *root ;* in front it ends in an enlargement named the *glans*, which is structurally continuous with the corpus spongiosum. The intermediate portion or *body* of the penis, owing to the relative position of its three compartments, has three somewhat flattened sides, and three rounded borders ; its widest side is turned upwards and forwards, and is named the *dorsum.* The *glans* penis, which is slightly compressed above and below, presents at its extremity a vertical fissure, the external orifice of the urethra ; its base, which is wider than the body of the penis, is hollowed out below to receive the narrowing extremities of the corpora cavernosa ; its border is rounded and projecting, and is named the *corona glandis*, behind which is a constriction named the *cervix ;* the posterior boundary of the glans thus marked off passes obliquely down on each side of the under surface, and ends behind the urethral opening, in a median fold of skin, named the *frœnum.*

The Integuments.—The integument of the penis, which is continued from that of the pubes and scrotum, forms a simple investment as far as the neck of the glans. At this part it leaves the surface and is doubled up in a loose cylindrical fold, constituting the *prepuce* or *foreskin.* The inner layer of this fold returns to the penis behind the cervix, where it is firmly attached ; and the integument, becoming thus again adherent, is continued forwards over the corona and glans, as far as the orifice of the urethra, where it meets with the mucous membrane of the urethra, and behind that orifice forms the *frœnum of the prepuce.* Upon the body of the penis the skin is very thin, entirely free from fat, and, excepting at the root, from hairs also ; in these respects differing remarkably from that on the pubes, which is thick, covers a large cushion of fat, and, after puberty, is beset with hairs : the skin of the penis is moreover very movable and distensible, and is dark in colour. At the free margin of the prepuce the integument changes its character, and approaches to that of a mucous membrane, being red, thin and moist. Numerous sebaceous glands are collected round the cervix of the penis and corona ; they are named the *glands of Tyson* (glandulæ ordoriferæ). Their secretion has a peculiar odour, and was formerly supposed to constitute the white *smegma præputii*, which tends to collect beneath the foreskin ; but that substance consists principally of epithelial cells cast from the opposed cuticular surfaces.

Upon the surface of the glans penis the integument again changes its character ; it ceases to contain glands, but its papillæ are highly developed and extremely sensitive, and it adheres most intimately and immovably to the spongy tissue of the glans.

Beneath the skin, on the body of the penis, the ordinary superficial fascia is very distinct ; it is continuous with that of the groin, and also with the dartoid tissue of the scrotum. Near the root of the organ there is in front a dense band of fibro-elastic tissue, named the *suspensory ligament*, lying amongst the fibres of the superficial fascia ; it is triangular in form ; one edge is free, another is connected with the fore part of the pubic symphysis, and the third with the dorsum of the penis.

The integuments of the penis are supplied with blood by branches of the dorsal artery of the penis and external pudic ; the veins join the dorsal and external pudic veins. Their nerves are entirely derived from the dorsal branches of the pudic nerves.

THE CORPORA CAVERNOSA.

The *corpora cavernosa* form the principal part of the body of the penis, and chiefly determine its form and consistence. They are two cylindrical bodies, placed side by side, flattened on their median aspects, and closely united and in part blended together along the middle line for the anterior three-fourths of their length ; whilst at the back part, in contact with the symphysis pubis, they separate from each other in form of two bulging and

Fig. 670.

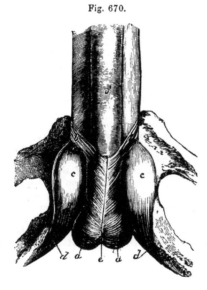

Fig. 670.—Root of the Penis attached to the Rami of the Pubes and Ischium (from Kobelt). $\frac{2}{3}$

a, a, accelerator urinæ muscle covering the bulb of the spongy body of the urethra, which presents at *e,* posteriorly, a median notch ; *b, b,* anterior slips of the muscle or bulbo-cavernosi ; *c, c,* crura of the penis, presenting an oval dilatation, *g,* or bulb of the corpus cavernosum ; *d, d,* erectores penis muscles ; *f,* corpus spongiosum urethræ.

then tapering processes named *crura,* which extend backwards attached to the pubic and ischial rami, and invested by the erectores penis or ischio-cavernosi muscles. Immediately behind their place of union, they are slightly enlarged, so as to form what are named by Kobelt the *bulbs of the corpora cavernosa,* parts which attain a much greater proportionate development in some quadrupeds. In front, the corpora cavernosa are closely bound together into a single rounded extremity, which is covered by the glans penis and firmly connected to its base by fibrous tissue.

The under surface of the united cavernous bodies presents a longitudinal groove, in which is lodged the corpus spongiosum, containing the greatest part of the canal of the urethra. The upper or anterior surface is also marked with a slight median groove for the dorsal vein of the penis, and near the root is attached to the pubes by the suspensory-ligament.

Structure.—The median septum between the two corpora cavernosa is thick and complete behind ; but farther forward it becomes thinner, and imperfectly separates their two cavities, for it presents, particularly towards the anterior extremity, numerous clefts, extending from the dorsal to the urethral edge, and admitting of a free communication between the erectile tissue of the two sides. From the direction of these slits, the intermediate white portions of the septum are made to resemble in arrangement the teeth of a comb, and hence it is named *septum pectiniforme.*

The external fibrous investment of the cavernous structure is white and dense, from half a line to a line thick, and very strong and elastic. It is composed for the most part of longitudinal bundles of shining white fibres, with numerous well-developed elastic fibres, enclosing the two corpora cavernosa in a common covering ; but internal to this, in each compartment, is a layer of circular fibres, which enter into the formation of the septum. (J. Wilson and Ellis.)

From the interior of the fibrous envelope, and from the sides of the septum, numerous lamellæ, bands, and cords, composed of an extensible fibrous tissue, and named *trabeculæ,* pass inwards, and run through and across the cavity in all directions, thus subdividing it into a multitude of interstices, and giving the entire structure a spongy character.

Fig. 671.

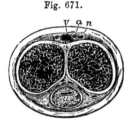

Fig. 671.—TRANSVERSE SECTION OF THE PENIS IN THE DISTENDED STATE.

The integument is represented as surrounding the deeper parts ; the erectile tissue occupying the corpora cavernosa and the septum pectiniforme descending between these bodies ; *u,* placed on the section of the spongy body, marks the urethra in the form of a transverse slit ; *v,* the single dorsal vein ; *a,* the dorsal artery, and *n,* the nerve, of one side.

The trabeculæ, whether lamelliform or cord-like, are larger and stronger near the circumference than along the centre of each cavernous body, and

Fig. 672.—PORTION OF THE ERECTILE TISSUE OF THE CORPUS CAVERNOSUM MAGNIFIED, SHOWING THE AREOLAR STRUCTURE AND THE VASCULAR DISTRIBUTION (from J. Müller).

Fig. 672.

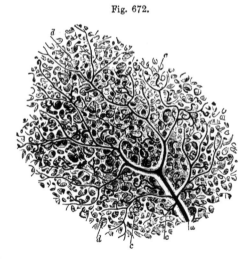

a, a small artery supported by the larger trabeculæ, and branching out on all sides ; *c,* the tendril-like arterial tufts or helicine arteries of Müller ; *d,* the areolar structure formed by the finer trabeculæ.

they also become gradually thicker towards the crura. The interspaces, conversely, are larger in the middle than near the surface ; their long diameter is, in the latter situation, placed transversely to that of the penis ; and they become larger towards the forepart of the penis. They are lined by a layer of squamous epithelium. The trabeculæ contain the ordinary white fibrous tissue and fine elastic fibres, together with pale muscular fibres, arteries, and nerves. The muscular tissue is much more abundant in the penis of some animals than in man.

The intertrabecular spaces form a labyrinth of intercommunicating venous areolæ divided by the trabecular tissue. The spaces of the two sides communicate freely through the septum, especially in front. They return their blood partly by a series of branches which escape between the corpora cavernosa and the corpus spongiosum, and which, accompanied by veins from the latter, mount on the sides of the penis to the vena dorsalis, partly by short veins issuing at the upper surface, and immediately joining the dorsal vein, but principally by veins passing out near the root of the penis and joining the prostatic plexus and pudendal veins. According to Kobelt, there are also communications with the cutaneous veins on the abdomen.

The arteries of the corpora cavernosa are branches of the pudic artery. The proper cavernous arteries (profundæ penis), right and left, supply them chiefly ; but the dorsal artery of the penis also sends twigs through the fibrous sheath, along the upper surface, especially in the fore part of the penis. Within the cavernous tissue, the numerous branches of arteries are supported by the trabeculæ, in the middle of which they run, and terminate in two modes ; some of them subdividing into branches of capillary minuteness which open into the intertrabecular spaces ; while others form tendril-like twigs which project into the spaces, and end in curling dilated extremities—the *helicine arteries* of J. Müller, sometimes singly and sometimes in tufts. The extremity of each curled dilatation would appear to be bound down by a small fibrous band, which according to Henle is usually solid, but is said by Kölliker to contain a capillary continuation of the blood-

Fig. 673.

Fig. 673.—One of the Tufts containing a Helicine Artery more highly magnified (from J. Müller).

The tuft is represented as projecting into the cavity of a vein.

vessel. The helicine arteries are most abundant in the posterior part of the corpora cavernosa, and are found in the corresponding part of the corpus spongiosum also ; but they have not been seen in the glans penis. They are most distinct in man, but are not constant in animals, so that, whatever may be their use, they do not appear to be essential to the process of erection.

CORPUS SPONGIOSUM.

The *corpus spongiosum urethræ* commences in front of the triangular ligament of the perineum, between the diverging crura of the corpora cavernosa, and somewhat behind their point of junction, by an enlarged and rounded extremity named the *bulb*. It extends forwards as a cylindrical, or slightly tapering body, lodged in the groove on the under side of the united cavernous bodies, as far as their blunt anterior extremity, over which it expands so as to form the glans penis already described.

The posterior bulbous extremity, or *bulb of the urethra*, varies in size in different subjects. It receives an investment from the triangular ligament in which it rests, and is embraced by the accelerator urinæ, or bulbo-cavernosus muscle. The posterior extremity of the bulb exhibits, more or less distinctly, a subdivision into two lateral portions or lobes, separated by a slight furrow on the surface, and by a slender fibrous partition within, which extends for a short distance forwards ; in early infancy this is more marked.   It is above this part that the urethra, having pierced the triangular ligament, enters the bulb, surrounded obliquely by a portion of the spongy tissue, named by Kobelt the *colliculus bulbi*, from which a layer of venous erectile tissue passes back upon the membranous portion of the urethra, and also upon the prostatic part, to the neck of the bladder, lying closely beneath the mucous membrane.   At first the urethra is nearer the upper than the lower part of the corpus spongiosum, but it soon gains and continues to occupy the middle of that body.

Structure.—This is essentially the same as that of the corpora cavernosa, only more delicate, or with a much less quantity of the fibrous trabecular structure.   Like the corpora cavernosa, it is distended with blood during erection ; but never acquires the same hardness.   The outer fibrous tunic is much thinner, is less white in colour, and contains more elastic tissue ; the areolæ are smaller, and directed for the most part with their long diameter corresponding to that of the penis ; the trabeculæ are finer and more equal in size ; and the veins form a nearly uniform plexus between them ; in the glans, the meshes of this plexus are smallest and most uniform.   Immediately surrounding the canal of the urethra, and, again, forming part of the external coat of the spongy substance, there are plain muscular fibres, which are continuous posteriorly with those of the bladder.   The helicine arteries are found in the spongy body, excepting in the part which forms the glans penis.   A considerable artery derived from the internal pudic enters the bulb on each side, and supplies the greater part of the spongy body, sending branches as far as the glans penis, which, however, is chiefly supplied by the arteria dorsalis.   Besides these, Kobelt describes, as constantly present, another but much smaller branch of the pudic artery, which, he says, enters the bulb on the upper surface, about an inch from its posterior extremity, and runs forwards in the corpus spongiosum to the glans.   *Veins* issue from the glans and adjoining part of the spongy body, to end in the vena dorsalis penis ; those of the rest of the spongy body for the most part pass out backwards through the bulb, and end in the prostatic and pudic venous plexuses : some emerge from beneath the corpora cavernosa, anastomose with their veins, and end partly in the cutaneous venous system of the penis and scrotum, and partly in the pudic and obturator veins.

The *lymphatics* of the penis form a dense network on the skin of the glans and prepuce, and also underneath the mucous lining of the urethra. They terminate chiefly in the inguinal glands.   Deep-seated lymphatics are also described as issuing from the cavernous and spongy bodies, and passing under the pubic arch with the deep veins, to join the lymphatic plexuses in the pelvis.

The *nerves* of the penis are derived from the pudic nerve and from the hypogastric plexus of the sympathetic (pp. 671 and 703).   They terminate by frequent division, and present indistinct traces of the so-called corpuscula tactûs ; on the glans and bulb of the urethra, some fibres of the cutaneous nerves end in Pacinian bodies.

URETHRA OF THE MALE.

The *male urethra* extends from the neck of the bladder to the extremity of the penis.   Its total length is about eight inches and a half, but varies much according to the length of the penis, and the condition of that organ. Its diameter varies at different parts of its extent, as will be stated more particularly hereafter. The tube consists of .a continuous mucous **mem-**

Fig. 674.

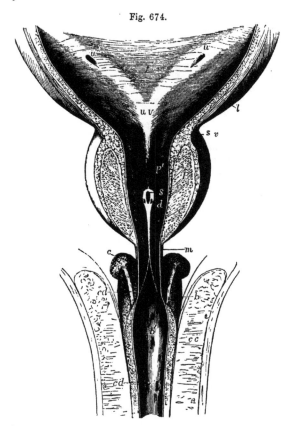

Fig. 674.—The lower part of the Bladder and the Prostatic, Membranous, and Bulbous parts of the Urethra opened from above.

A portion of the wall of the bladder and the upper part of the prostate gland have been removed, the corpora cavernosa penis have been separated in the middle line and turned to the side, and the urethra has been slit up; the bulb is left entire below, and upon and behind it the glands of Cowper with their ducts have been exposed. *t*, placed in the middle of the trigonum vesicæ; *u, u*, oblique apertures of the ureters; from these an elevation of the wall of the bladder is shown running down to *u v*, the uvula vesicæ; *l*, the longitudinal muscular fibres of the bladder passing down upon the prostate; *s v*, the circular fibres of the sphincter surrounding the neck; *p*, the glandular part of the prostate; *p'*, the prostatic portion of the urethra; from the uvula vesicæ a median ridge is seen descending to the caput gallinaginis, in which *s*, indicates the opening of the sinus pocularis, and *d*, that of one of the ductus ejaculatorii; *m*, the commencement of the membranous portion of the urethra ; *b*, the bulb of the spongy body ; *b'*, the bulbous part of the urethra ; *c*, one of Cowper's glands ; *c d, c d*, course and orifice of its duct lying upon the bulb, and passing forward between the spongy body and the urethra, into which along with its fellow it opens ; *c c*, one of the corpora cavernosa.

brane, supported by an outer layer of submucous tissue connecting it with
the several parts through which it passes.   In the submucous tissue there
are, throughout the whole extent of the urethra, two layers of plain muscular fibres, the innermost disposed longitudinally, and the other in a
circular direction.   In accordance with the name or character of those
parts through which it passes, three divisions of the urethra are separately
described as the *prostatic, membranous,* and *spongy* portions.

1. The first, or *prostatic portion,* is the part which passes through the
prostate gland.   It is from 12 to 15 lines in length, is the widest part
of the canal, and is larger in the middle than at either end : at the
neck of the bladder its diameter is nearly 4 lines, then it widens a little,
so as to be rather more than 4 lines, and in old persons 5 or 6, after which
it diminishes like a funnel, until, at its anterior extremity, it is smaller
than at its commencement.   It passes through the upper part of the prostate, above the middle lobe, so that there is more of the gland below it
than above.   Though enclosed in the firm glandular substance, it is more
dilatable than any other part of the urethra ; but immediately at the neck
of the bladder, it is, as elsewhere stated, much more resistant.   The transverse section of the urethra, as it lies in the prostate, is widened from side
to side and somewhat folded upwards in the middle, the upper and under
surface being in contact.

The lining membrane of the prostatic portion of the urethra is thrown
into longitudinal folds, when no fluid is passing along it ; it forms no proper
valve at the neck of the bladder, unless the elevation named the uvula
vesicæ is to be regarded as such.   Somewhat in advance of·this, and continued from it along the floor of the passage, projects a narrow median
ridge, about 8 or 9 lines in length, and $1\frac{1}{2}$ line in its greatest height ; this
ridge gradually rises into a peak, and sinks down again at its anterior or
lower end, and is formed by an elevation of the mucous membrane and
subjacent tissue.   This is the crest of the urethra (crista urethræ), more
generally called *caput gallinaginis* and *verumontanum.*   On each side of this
ridge the surface is slightly depressed, so as to form a longitudinal groove,
named the *prostatic sinus,* the floor of which is pierced by numerous foramina, the orifices of the prostatic ducts.   Through these a viscid fluid oozes
out on pressure ; the ducts of the middle lobe open behind the urethral
crest, and some others open before it.

Sinus pocularis.—At the fore part of the most elevated portion of the
crest, and exactly in the middle line, is a recess, upon or within the margins
of which are placed the slit-like openings of the common seminal or ejaculatory ducts, one at each side.· This median depression, named *sinus
pocularis, vesica prostatica,* or *utricle,* was first described by Morgagni, and
has more lately attracted renewed attention, as corresponding with the
structure which in the female is developed into the uterus.

The utricle forms a cul-de-sac running upwards or backwards, from three
to five lines deep, and usually about one line wide at its entrance and for
some distance up, but acquiring a width of at least two lines at its upper end
or fundus.   The prominent walls of the narrow portion form the urethral crest,
and its fundus appears to lie behind and beneath the middle lobe, and
between the two lateral lobes of the prostate.   Its parietes, which are distinct, and tolerably thick, are composed of fibrous tissue and mucous membrane, together with a few muscular fibres, and enclose on each side the ejaculatory duct ; numerous small glands open on its inner surface. According to
Kobelt and others, the caput gallinaginis contains some well-marked erectile

and muscular tissue, and it has been supposed that this eminence, when dis-
tended with blood, may offer an obstacle to the passage of the semen back-
wards into the bladder. (E. H. Weber, Zusätze zur Lehre vom Baue und
Verrichtungen der Geschlechts-Organe, 1846 ; Huschke in Sœmmerring's
Anatomie, vol. v. ; Leuckart, " Vesicula Prostatica," in Cyclop. of Anat. &
Phys.)

2. The *membranous portion* of the urethra comprises the part between
the apex of the prostate, and the bulb of the corpus spongiosum.　It measures three quarters of an inch along its anterior, but only about half an
inch on its posterior surface, in consequence of the projection upwards on it
of the bulb.　This is the narrowest division of the urethra.　In the middle
its circumference is 0·6 of an inch ; at the end 0·5.　(H. Thompson.)　It is
placed beneath the pubic arch, the anterior concave surface being distant
nearly an inch from the bone, leaving an interval, occupied by the dorsal
vessels and nerves of the penis, by areolar tissue, and some muscular fibres.
Its lower convex surface is turned towards the perinæum, opposite to the
point of meeting of the transverse muscles : it is separated by an interval
from the last part of the rectum.　About a line in front of the prostate, it
emerges from between the anterior borders of the levatores ani, and passes
through the deep layer of the subpubic fascia (p. 260) ; it is then placed
between that and the anterior layer or triangular ligament through which
it passes some way farther forwards, but both of these fibrous membranes
are prolonged upon the canal, the one backwards and the other forwards.
Between these two layers the urethra is surrounded by a little erectile tissue,
by some veins, and also by the fibres of the *compressor urethræ* muscle ;
beneath it, on each side, are Cowper's glands.　The proper or plain muscular
fibres of this portion of the urethra are continued over the outer and inner
surfaces of the prostate into the muscular coat of the bladder posteriorly,
and into those of the spongy portion of the urethra anteriorly.　(Hancock.)

3. The *spongy portion* of the urethra, by far the longest and most variable in length and direction, includes the remainder of the canal, or that
part which is surrounded by the erectile tissue of the corpus spongiosum.
Its length is about six inches.　The part contained within the bulb, sometimes distinguished as the *bulbous portion*, is somewhat dilated ; its circumference being equal to seven-tenths of an inch (Thompson).　The succeeding
portion, as far as the glans, is of uniform size, being intermediate in this
respect between the bulbous and membranous portions.　The cross section
of its canal appears like a transverse slit.　The canal of the urethra situated
in the glans has, on the contrary, when seen in a cross section, the form of
a vertical slit : in this part the canal is again considerably dilated, forming
what is named the *fossa navicularis,* which is from four to six lines in length,
and is most evident in the form of a depression on the floor of the urethra.

Lastly, at its orifice, which is a vertical fissure from two and a half to
three lines in extent, and bounded by two small lips, the urethra is again
contracted and reaches its narrowest dimensions.　In consequence of its
form, and also of the resistant nature of the tissues at its margin, this opening does not admit so large an instrument as even the membranous portion
of the canal.

The *mucous membrane* of the urethra possesses a lining of stratified
epithelium, of which the superficial cells are columnar, except for a short
distance from the orifice, where they are squamous, and where the subjacent
membrane exhibits papillæ.

The whole lining membrane of the urethra is beset with small mucous glands and follicles, commonly named the glands of Littré, the ducts of which pass obliquely forwards through the membranes. They vary much in size and in the degree of loculation and ramification of their cavities. Besides these there are larger recesses or lacunæ, opening by oblique orifices turned forwards or down the canal. These are most abundant along the floor of the urethra, especially in its bulbous part. One large and conspicuous recess, situated on the upper surface of the fossa navicularis, is named the *lacuna magna.*

Cowper's Glands.—In the bulbous portion of the urethra, near its anterior end, are the two openings of the ducts of *Cowper's glands.* These little glands themselves are seated farther back than the bulb, beneath the fore part of the membranous portion of the urethra, between the two layers of the subpubic fascia, the anterior layer supporting them against the urethra. The arteries of the bulb pass above, and the transverse fibres of the compressor urethræ beneath these glands. They are two small firm rounded bodies, about the size of peas, and of a deep yellow colour. They are compound vesicular or racemose glands, composed of several small lobules held together by a firm investment. This latter, as well as the walls of the ducts, contains muscular tissue. The branched ducts, which commence in cellular crypts, unite to form a single excretory duct for each gland, which runs forwards with its fellow for about an inch or an inch and a half beneath the mucous membrane, and the two terminate in the floor of the bulbous part of the urethra by two minute orifices opening obliquely. These glands secrete a viscid fluid, the use of which is not known; their existence is said not to be constant, and they appear to diminish in old age : sometimes there is only one.

Occasionally there is a third glandular body in front of and between Cowper's glands; this has been named the *anterior prostate* or *anti-prostatic gland.*

The muscles in connection with the urethra and penis have been already described (p. 263).

THE TESTES, AND THEIR EXCRETORY APPARATUS.

The *testicles* or *testes*, the two glandular organs which secrete the seminal fluid, are situated in the scrotum, each being suspended by a collection of structures termed the spermatic cord.

The *spermatic cord.*—The parts which enter into this cord are the excretory duct of the testicle, named the vas deferens, the spermatic artery and veins, lymphatics, nerves, and connecting areolar tissue. Besides this, both the cord and the testis have several coverings. The structures mentioned come together to form the cord at the internal or deep abdominal ring (p. 258), and, extending through the abdominal wall obliquely downwards and towards the middle line, escape at the superficial or external abdominal ring (p. 250), whence the cord descends over the front of the pubes into the scrotum.

COVERINGS OF THE TESTIS AND CORD.

The *inguinal canal.*—By the term inguinal canal is understood the space occupied by the spermatic cord as it passes through the abdominal wall. It extends from the deep to the superficial abdominal ring, and is about an inch and a half in length. In the upper part of this course, the cord has placed behind it the fascia transversalis, and is covered in front by the lower fibres of the internal oblique and transversalis muscles ; lower down, it lies in front of the conjoined tendon of these muscles, the fibres of which have arched inwards over it, and its cremasteric covering, is in contact anteriorly with the aponeurosis of the external oblique muscle. The inguinal canal is therefore said to be bounded posteriorly by the fascia transversalis above and the conjoined tendon below, and anteriorly by fibres of the transversalis and internal oblique muscles above, and the aponeurosis of the external oblique muscle below ; while its floor is formed by the curving backwards of Poupart's ligament, and its roof by the apposition of the layers of the abdominal wall.

As it enters the inguinal canal, the cord receives a covering from the infundibuliform fascia, a thin layer continuous with the fascia transversalis, and prolonged down from the margins of the deep abdominal ring ; within the canal it receives a covering from the cremaster muscle and fascia connected with it ; and as it emerges from the canal there is added, superficially to this, the intercolumnar fascia prolonged from the pillars of the superficial abdominal ring.

The *scrotum.*—The *scrotum* forms a purse-like investment for the testes and part of the spermatic cords. Its condition is liable to certain variations according to the state of the health and other circumstances : thus, it is short and corrugated in robust persons and under the effects of cold, but becomes loose and pendulous in persons of weak constitution, and under the relaxing influence of heat. Its surface is marked off into two lateral halves by a slight median ridge, named the *raphe*, extending forwards to the under side of the penis, and backwards along the perinæum to the margin of the anus.

Within the scrotum, the coverings of the cord and testis, as enumerated from without inwards, are the *skin*, superficial fascia and *dartos tissue* of the scrotum, the *intercolumnar fascia*, the *cremaster muscle* and cremasteric fascia, and the *infundibuliform fascia*, which is united to the cord by a layer of loose areolar tissue ; lastly, the testicle has a special serous tunic, named the *tunica vaginalis*, which forms a closed sac, and covers the tunica albuginea or proper *fibrous coat* of the gland.

1. The *skin* in this situation is very thin, and is of a darker colour than elsewhere ; it is generally thrown into rugæ or folds, which are more or less distinct according to the circumstances already mentioned. It is furnished with sebaceous follicles, the secretion from which has a peculiar odour, and it is covered over with thinly scattered crisp and flattened hairs, the bulbs of which may be seen or felt through the skin when the scrotum is extended. The superficial blood-vessels are also readily distinguished through this thin integument.

2. Immediately beneath the skin of the scrotum there is found a thin layer of a peculiar loose reddish-brown tissue, endowed with contractility, and named the *dartos tunic*. This subcutaneous layer is continuous with the superficial fascia of the groin, perinæum, and inner side of the thighs, but acquires a different structure, and is perfectly free from fat. The dartoid tissue is more abundant on the fore part of the scrotum than behind, and, moreover, it forms two distinct sacs, which contain the corresponding testes, and are united together along the middle line so as to establish a median partition between the two glands, named the *septum scroti*, which is adherent below to the deep surface of the raphe and reaches upwards to the root of the penis. The dartos is very vascular, and owes its contractile properties to the presence of a considerable amount of unstriped muscular tissue. Its contractility is slow in its action ; it is excited by the application of cold and of

mechanical stimuli, but, apparently, not by electricity.   By its action the testes are drawn up or sustained, and at the same time the skin of the scrotum is more or less corrugated.

3. The *intercolumnar* or *spermatic fascia*, a very thin and transparent but relatively firm layer, derived from the tendon of the external oblique muscle of the abdomen, is attached above to the margins of the external ring, and is prolonged downwards upon the cord and testicle.   It lies at first beneath the superficial fascia, but lower down beneath the dartos, and it is intimately connected with the layer next in order.

4. The cremasteric layer is composed of scattered bundles of muscular tissue, connected together into a continuous covering by intermediate areolar membrane.   The red muscular portion, which is continuous with the lower border of the internal oblique muscle of the abdomen, constitutes the *cremaster muscle* (p. 251), or *tunica erythroïdes*, and the entire covering is, named the *cremasteric fascia*.

5. The *infundibuliform fascia*, continuous above with the *fascia transversalis* and the subperitoneal areolar membrane, and situated immediately beneath the cremasteric fascia, invests the cord completely, and is connected below with the posterior part of the testicle and the outer surface of its serous tunic.

On forcing air beneath the infundibuliform fascia, a quantity of loose and delicate areolar tissue is seen to connect its internal or deep surface with the vas deferens and spermatic blood-vessels, and to form lamellæ between them. This areolar tissue is continuous above with the subserous areolar tissue found beneath the peritoneum on the anterior wall of the abdomen ; below, it is lost upon the back of the testicle.   Together with the infundibuliform fascia just described, it forms the *fascia propria* of A. Cooper.

Lying amongst this loose areolar tissue, in front of the upper end of the cord, there is often seen a fibro-areolar band, which is connected above with the pouch of peritoneum found opposite the upper end of the inguinal canal, and which reaches downwards for a longer or shorter distance along the spermatic cord.   Occasionally it may be followed as a fine cord, as far as the upper end of the tunica vaginalis ; sometimes no trace of it whatever can be detected.   It is the vestige of a tubular process of the peritoneum, which in the foetus connects the tunica vaginalis with the general peritoneal membrane.   The testicle is placed in the abdomen during the greater part of foetal life ; but at a period considerably prior to its escape from the abdominal cavity, a pouch of peritoneum already extends down into the scrotum.   Into this pouch or *processus vaginalis peritonei* the testicle projects from behind, supported by a duplicature of the serous membrane, named the *mesorchium*.   Sooner or later after the gland has reached the scrotum, the upper part or neck of this pouch becomes contracted and finally obliterated, from the internal abdominal ring down nearly to the testicle, leaving no trace but the indistinct fibrous cord already described, whilst the lower part remains as a closed serous sac, into which the testicle depends, and which is thenceforth named the tunica vaginalis.

In the female an analogous pouch of peritoneum descends, in the foetus, for a short distance along the round ligament of the uterus, and has received the appellation of the *canal of Nuck*.   Traces of it may almost always be seen in the adult.

The neck of the processus vaginalis sometimes becomes closed at intervals only, leaving a series of sacculi along the front of the cord ; or a long pouch may continue open at the upper end, leading from the abdominal cavity into the inguinal canal.

In other instances, the peritoneal process remains altogether pervious, and the cavity of the tunica vaginalis is continuous with that of the peritoneum. In such a case of congenital defect, a portion of intestine or omentum may descend from the abdomen into the inguinal canal and scrotum, and constitute what is named a congenital hernia. Lastly, one or both testes may remain permanently within the abdomen, or their descent may be delayed till after puberty, when it may occasion serious disturbance. Retention of the testes in the abdomen (cryptorchismus) is, in many instances, the accompaniment of arrested development of the glandular structure; it is, however, a peculiarity which is often present without impotence.

In a few mammals, as the elephant, the testes remain permanently within the abdomen; in a much larger number, as the rodentia, they only descend at each period of rut. The complete closure of the tunica vaginalis is peculiar to man, and may be considered as connected with his adaptation to the erect posture.

6. The *tunica vaginalis.*—This tunic forms a shut sac, the opposite walls of which are in contact with each other. Like the serous membranes in general, of which it affords one of the simplest examples, it may be described as consisting of a *visceral* and a *parietal* portion. The visceral portion closely invests the greater part of the body of the testis, as well as the epididymis, between which parts it recedes in the form of a pouch (digital fossa), and lines their contiguous surfaces, and it adheres intimately to the

Fig. 675.

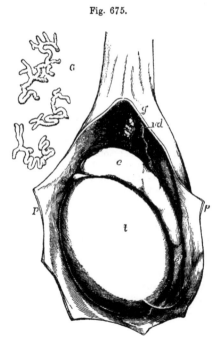

Fig. 675.—The Left Tunica Vaginalis opened. showing the Testis, Epididymis, &c.

p, p, the cut edges of the parietal tunica vaginalis drawn aside laterally, as well as above and below ; *t,* the body of the testicle ; *e,* the globus major of the epididymis ; *e',* the globus minor, near which, *f,* a fold of the tunica vaginalis (or ligament) passes from the body of the testis to the side ; in the upper part of the figure the tunica vaginalis has been slightly dissected off at the place of its reflection on the cord to show *v d,* the vas deferens, and *g,* the organ of Giraldès ; G, the three small nodules of this organ enlarged about ten times, and showing the remains of tubular structure within them.

proper fibrous tunic of the gland. Along the posterior border of the gland, where the vessels and ducts enter or pass out, the serous coat, having been reflected, is wanting.

The parietal or scrotal portion of the tunica vaginalis is more extensive than that which covers the body of the testis ; it reaches upwards, sometimes for a considerable distance, upon the spermatic cord, extending somewhat higher on the inner than on the outer side. It also reaches downwards below the testicle, which, therefore, appears to be suspended at the back of the serous sac, when this latter is distended with fluid.

VESSELS AND NERVES OF THE COVERINGS OF THE TESTIS AND CORD.

The *arteries* are derived from several sources. Thus, the two external pudic arteries (p. 437), branches of the femoral, reach the front and sides of the scrotum, supplying the integument and dartos ; the superficial perineal branch of the internal pudic artery (p. 426) is distributed to the back part of the scrotum ; and, lastly, more deeply seated than either of these, is a branch given from the epigastric artery, named cremasteric, because it is chiefly distributed to the cremaster muscle ; it also supplies small branches to the other coverings of the cord, and its ultimate divisions anastomose with those of the other vessels. The *veins*, which, owing to the thinness of the integuments, are apparent on the surface of the scrotum, follow the course of the arteries. The *lymphatics* pass into the inguinal lymphatic glands.

The *nerves* also proceed from various sources. Thus, the ilio-inguinal, a branch of the lumbar plexus (p. 660), comes forwards through the external abdominal ring, and supplies the integuments of the scrotum ; this nerve is joined also by a filament from the ilio-hypogastric branch of the same plexus : sometimes two separate cutaneous nerves come forward through the external ring. The two superficial perineal branches of the internal pudic nerve accompany the artery of the same nerve and supply the inferior and posterior parts of the scrotum. The inferior pudendal, a branch of the small sciatic nerve (p. 675), joins with the perineal nerves, and is distributed to the sides and fore part of the scrotum. Lastly, the branch of a deeper nerve, springing from the lumbar plexus, and named genito-crural (p. 661), comes into contact with the spermatic cord at the internal abdominal ring, passes with it through the inguinal canal, and supplies the fibres of the cremaster, besides sending a few filaments to the other deep coverings of the cord and testicle.

THE TESTES.

The *testes* are suspended obliquely in the scrotum by means of the cord and membranes already described : they are usually placed at unequal heights, that of the left side being lower than the other. They are of an oval form, but are slightly compressed laterally, so that they have two somewhat flattened sides or faces, an upper and a lower end, an anterior and a posterior border. They are from an inch and a half to two inches long, about an inch and a quarter from the anterior to the posterior border, and nearly an inch from side to side. The weight of each varies from three-quarters of an ounce to an ounce, and the left is often a little the larger of the two.

The front and sides of the testicle, together with the upper and the lower ends, are free, smooth, and closely invested by the tunica vaginalis. The posterior border, however, is attached to the spermatic cord, and it is here that the vessels and nerves enter or pass out. When the testis is suspended in its usual position, its upper end is directed obliquely forwards and outwards, as well as upwards, whilst the lower, which is rather smaller, has the opposite direction. It follows from this that the posterior or attached border is turned upwards and inwards, and the outer flattened face slightly backwards.

Along the outer edge of the posterior border of the gland, and resting also on the neighbouring portion of its outer face, is placed a long narrow body, the *epididymis*, which forms part of the excretory apparatus of the testicle, and is principally composed of the convolutions of a long tortuous canal or efferent duct, to be presently described. Its upper extremity, which is enlarged and obtuse, projecting forwards on the upper end of the testis, is named the *head* of the epididymis, or *globus major ;* the lower, which is more pointed, is termed the *tail*, or *globus minor ;* whilst the intervening and narrower portion is named the *body.* The outer convex surface of the epididymis and the thin anterior border are free, and covered by the tunica vaginalis. The inner surface, except at the upper and lower ends, is also free, and invested by the same tunic, which here forms the digital pouch between the epididymis and the outer face of the testicle, and nearly surrounds the epididymis, except along its posterior border, which is held to the gland by a duplicature of the serous membrane, containing numerous blood-vessels. At its upper and lower extremity, the inner surface of the epididymis is attached to the testicle,—the lower end, or globus minor, by fibrous tissue and a reflection of the tunica vaginalis, the globus major also by the efferent ducts of the testicle.

At the back of the testis and epididymis, beneath the fascia propria, there is found, opposite the lower two-thirds of the testis, a considerable amount of unstriped muscular tissue, the inner muscular tunic of Kölliker.

Situated on the front of the globus major, somewhat to the outer side, there are found in the majority of cases one or more short processes of the tunica vaginalis, containing fine blood-vessels. They are called *corpora Morgagni*, or *hydatids of Morgagni ;* that anatomist having been the first to describe them. One of these, more dilated than the rest, and pyriform in shape, lies closely between the head of the epididymis and the testicle, and appears to be the remains of the fœtal structure, termed Müller's duct : they are without any known physiological importance.

The testis proper, exclusive of the epididymis, is enclosed in a strong capsule, the *tunica albuginea.* This is a dense unyielding fibrous membrane, of a white colour, and about half a line thick, which immediately invests the soft substance of the testicle, and preserves the form of the gland. It is composed of bundles of fibrous tissue, which interlace in every direc-

Fig. 676.

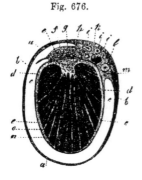

Fig. 676.—Transverse Section through the Right Testicle and the Tunica Vaginalis (from Kölliker).

a, connective tissue enveloping the parietal layer of the tunica vaginalis ; *b,* this layer itself ; *c,* cavity of the tunica vaginalis ; *d,* reflected or visceral layer adhering to *e,* the tunica albuginea ; *f,* covering of epididymis (*g*) on the right or outer side ; *h,* mediastinum testis ; *i,* branches of the spermatic artery ; *k,* spermatic vein ; *l,* vas deferens ; *m,* small artery of the vas deferens ; *n,* lobules of the testis ; *o,* septa or processes from the mediastinum to the surface.

tion. The surface is for the most part covered by the tunica vaginalis, except along the posterior border of the testicle, where the spermatic vessels pass through, and except also at the parts to which the two extremities of the epididymis are attached.

Viewed from the interior, the fibrous tissue of the tunica albuginea is seen to be prolonged forwards, at the posterior and upper border of the testis, for a few lines into the substance of the gland, so as to form within it an incomplete vertical septum, known as the *corpus Highmorianum*, and named by Astley Cooper *mediastinum testis.* Projecting inwards from the back of the testis, it extends from the upper nearly to the lower end of the gland, and it is wider above than below. The firm tissue of which it is composed is traversed by a network of seminal ducts, and by the larger blood-vessels of the gland, which are lodged in channels formed in the fibrous tissue.

From the front and sides of the corpus Highmorianum numerous slender fibrous cords and imperfect septa of connective tissue are given off in radiating directions, and are attached by their outer ends to the internal surface of the tunica albuginea at different points, so as to assist in maintaining the general shape of the testicle, and enclose the several lobes into which the substance of the testis is divided. The whole internal surface of the tunica albuginea is covered by a multitude of fine blood-vessels, which are branches of the spermatic artery and veins, and are held together by a delicate areolar web. Similar delicate ramifications of vessels are seen on the various fibrous offsets of the mediastinum, upon which the blood-vessels are thus supported in the interior of the gland. This vascular network, together with its connecting areolar tissue, constitutes the *tunica vasculcsa* of Astley Cooper.

The proper *glandular substance* of the testicle is a soft but consistent mass of a reddish-yellow colour, which is divided into numerous small lobes of conical form, with the larger ends turned towards the surface of the testicle, and the smaller towards the mediastinum. The number of these *lobes* (lobuli testis) has been estimated at 250 by Berres, and at upwards of 400 by Krause. They differ in size according to their position, those which occupy the middle of the gland and reach its anterior border being longer and larger than the rest. They consist almost entirely of small convoluted tubes, named *tubuli seminiferi, vascula serpentina,* in the interior of which the seminal fluid is secreted. Each lobe contains one, two, three, or even more of these con-

Fig. 677.—PLAN OF A VERTICAL SECTION OF THE TESTICLE, SHOWING THE ARRANGEMENT OF THE DUCTS.

Fig. 677.

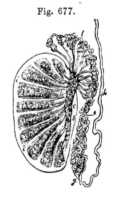

The true length and diameter of the ducts have been disregarded. *a, a,* tubuli seminiferi coiled up in the separate lobes ; *b,* vasa recta ; *c,* rete vasculosum ; *d,* vasa efferentia ending in the coni vasculosi ; *l, e, g,* convoluted canal of the epididymis ; *h,* vas deferens ; *f,* section of the back part of the tunica albuginea ; *i, i,* fibrous processes running between the lobes ; *s,* mediastinum.

voluted tubules, the coils of which, being but loosely held together, may be more or less successfully unravelled by careful dissection under water. Lauth estimates their mean number to be 840, and the average length two feet and a quarter. Their diameter, which is uniform throughout their whole course, is from $\frac{1}{200}$th to $\frac{1}{150}$th of an inch. They present two kinds of convolutions, each tube having a fine and regular undulation, which gives a granular appearance to the whole mass, and this undulating tube being again thrown into complicated folds, which are compressed so as to be elongated in the direction of the lobule. The lobules are never quite distinct, for here and there tubules are always to be found passing from each to those around ; and it sometimes happens that tubules which are divided by a straight plane of

rather tortuous, but afterwards becoming straight, it ascends upon the inner side of the epididymis, and along the back of the testicle, separated from both, however, by the blood-vessels passing to and from the gland. Continuing, then, to ascend in the spermatic cord, the vas deferens accompanies the spermatic artery, veins and nerves, as far as the internal abdominal ring. . Between the testicle and the external ring its course is vertical : it lies behind the spermatic vessels, and is readily distinguished by its hard cord-like feel. Having passed obliquely upwards and outwards along the inguinal canal, and reached the inner border of the internal abdominal ring, it leaves the spermatic vessels (which extend to the lumbar region), and turns suddenly downwards and inwards into the pelvis, crossing over the external iliac vessels, and turning round the outer or iliac side of the epigastric artery. Running beneath the peritoneum, it reaches the side of the bladder, curves backwards and downwards to the under surface of that viscus, and then runs forwards to the base of the prostate gland. In its course within the pelvis, it crosses over the cord of the obliterated hypogastric artery, and to the inner side of the ureter. Beyond this point, where it ceases to be covered by the peritoneum, it is found attached to the coats of the bladder, in contact with the rectum, and gradually approaches its fellow of the opposite side. Upon the base of the bladder, the two vasa deferentia are situated between two elongated receptacles, named the seminal vesicles; and, close to the base of the prostate, each vas deferens ends by joining with the duct from the corresponding seminal vesicle, which is placed on its outer side to form one of the two common seminal or ejaculatory ducts.

The vas deferens measures nearly two feet in length. In the greater part of its extent it is cylindrical or slightly compressed, and has an average diameter of about one line and a quarter ; but towards its termination, beneath the bladder, it becomes enlarged and sacculated, approaching thus in character to the seminal vesicle. Previously to its junction with the duct of that vesicle, it again becomes narrowed to a fine cylindrical canal. The walls of the vas deferens are very dense and strong, measuring one-third of a line in thickness ; whilst, on the other hand, the canal is comparatively fine, its diameter being only from one-fourth to one-half a line. In the sacculated portion the passage is much wider, and the walls are thinner in proportion.

Besides an external areolar investment, and an internal lining mucous membrane, the vas deferens is provided with an intermediate thick tunic, which is dense in structure, somewhat elastic, and of a deep yellowish colour. This coat consists principally of longitudinal muscular fibres, mixed with some circular ones. Huschke describes two longitudinal layers with intermediate circular fibres. The external and middle layers are thick and strong ; but the internal longitudinal stratum is extremely thin, constituting not more than ⅕th of the muscular coat. The vasa deferentia of the dog, cat, and rabbit were found by E. Weber to exhibit lively peristaltic contractions when stimulated by means of electricity.

The surface of the mucous membrane is pale ; it is thrown into three or four fine longitudinal ridges, and, besides this, in the sacculated portion of the duct, is marked by numerous finer rugæ which enclose irregular polyhedral spaces, resembling in this the lining membrane of the vesiculæ seminales. The epithelium is of the columnar kind, not ciliated.

Vas aberrans.—This name was applied by Haller to a long narrow tube, or diverticulum, discovered by him, and almost invariably met with, which leads off from the lower part of the canal of the epididymis, or from the commencement of the vas deferens, and extends upwards in a tortuous manner for one or two inches amongst the vessels of the spermatic cord, where it ends by a closed extremity. Its length, when it is unravelled, ranges from one inch and a half to fourteen inches; and its breadth increases towards its blind extremity. Sometimes this diverticulum is branched, and occasionally there are two or more such aberrant ducts. Its structure appears to be similar to that of the vas deferens. Its origin is probably connected with the Wolffian body of the fœtus, but the exact mode of its formation and its office are unknown. Luschka states that occasionally it does not communicate with the canal of the epididymis, but appears to be a simple serous cyst.

Organ of Giraldès.—This is a minute structure situated in the front of the cord, and in contact with the caput epididymis. It consists usually of several small irregular masses containing convoluted tubules lined with squamous epithelium, and is scarcely to be recognised until the surrounding connective tissue has been rendered transparent by reagents. Its tubules appear to be persistent elements of the Wolffian body. (Giraldès, in Bulletin de la Soc. Anat. de Paris, 1857, and in Journal de la Physiologie, 1861.)

THE SEMINAL VESICLES AND EJACULATORY DUCTS.

The *vesiculæ seminales* are two membranous receptacles, situated, one on each side, upon the base of the bladder, between it and the rectum. When distended, they form two long-shaped sacculated bodies, somewhat flattened above, where they are firmly attached to the bladder, but convex below ; they are widened behind and narrow in front. Their length is usually about two inches and a half, and their greatest breadth from four to six lines ; but they vary in size in different individuals, and also on opposite sides of the same subject.

Their posterior obtuse extremities are separated widely from each other, but anteriorly they converge so as to approach the two vasa deferentia, which run

Fig. 680.

Fig. 680.—Dissection of the Base of the Bladder and Prostate Gland, showing the Vesiculæ Seminales and Vasa Deferentia (from Haller).

a', lower surface of the bladder at the place of reflection of the peritoneum ; *b,* the part above covered by the peritoneum ; *i,* left vas deferens, ending in *c,* the ejaculatory duct ; *s,* left vesicula seminalis joining the same duct ; *s, s,* the right vas deferens and right vesicula seminalis, which has been unravelled ; *p,* under side of the prostate gland ; *m,* part of the urethra ; *u, u,* the ureters, the right one turned aside.

forwards to the prostate between them. The small triangular portion of the base of the bladder, which is marked off by the two vesiculæ seminales at the sides with the vasa deferentia interposed, and behind by the line of reflexion of the peritoneum from the bladder to the rectum, rests mmediately on that intestine.   The seminal vesicles themselves are also supported by the sides of the rectum, but they are separated from the bowel by a layer of the recto-vesical fascia, which holds them to the base of the bladder.

The sacculated appearance of the vesiculæ seminales is owing to their peculiar formation.   Each consists of a tube somewhat coiled and repeatedly doubled on itself, and firmly held in that condition by a dense fibrous tissue.   When unrolled, this tube is found to be from four to six inches long, and about the width of a quill.   Its posterior extremity is closed, so that it forms a long cul-de-sac ; but there are generally, if not always, several longer or shorter branches or diverticula developed from it, which also end by closed extremities.   Its anterior extremity, which forms the fore part of the vesicula, becomes straight and narrowed, and ends opposite the base of the prostate by uniting on its inner side, at a very acute angle, with the narrow termination of the corresponding vas deferens to form a single canal, which is the common seminal or ejaculatory duct.

In structure, the vesiculæ seminales resemble very closely the adjoining sacculated portions of the vasa deferentia.   Besides an external fibro-areolar investment, connected with the recto-vesical fascia, they have a proper coat, which is firm, dense, and somewhat elastic, and consists of rigid white fibres and of others of a deep yellowish-brown hue.   Muscular tissue is found in their walls, at least on the posterior aspect, where longitudinal and transverse bands pass over them from the muscular wall of the bladder (Ellis and Kölliker).   The mucous membrane is pale, or has a dirty brownish-white colour.   It is traversed by multitudes of fine rugæ, which form an areolar structure resembling that seen in the gall-bladder, but deeper and composed of much finer meshes.   The epithelium of the vesiculæ is of the squamous kind ; its particles have a granular character.

The seminal vesicles serve as receptacles or reservoirs for the semen, as is easily proved by a microscopic examination of their contents ; but, besides this, it is supposed by some that they secrete a peculiar fluid which is incorporated with the semen.

The *common seminal* or *ejaculatory* ducts, two in number, are formed on each side by the junction of the narrowed extremities of the corresponding vas deferens and vesicula seminalis, close to the base of the prostate gland. From this point they run forwards and upwards, at the same time approaching each other, and then pass side by side through the prostate between its middle and two lateral lobes.   After a course of nearly an inch, during which they become gradually narrower, they end in the floor of the prostatic portion of the urethra by two small slit-like orifices placed one on each prominent margin of the verumontanum close to the opening of the sinus pocularis.   For a short distance the ejaculatory ducts run in the substance of the walls of this sinus.

The coats of the common seminal duct, as compared with those of the vas deferens and vesicula, are very thin.   The strong outer tunic almost entirely disappears after the entrance of the ducts between the lobes of the prostate, but muscular fibres may be traced into the prostatic portion ; and the mucous membrane becomes gradually smoother, and at length passes into that of the urethra.   The muscular fibres of the duct, according to Henle, are separated in the prostate to form the trabeculæ of a layer of cavernous tissue.

These ejaculatory ducts serve to convey the fluid contained in the seminal vesicles and vasa deferentia into the urethra.

VESSELS AND NERVES OF THE TESTIS.

The testicle and its excretory apparatus receive blood-vessels and nerves from different sources from those which supply the coverings of those parts.

The *spermatic artery*, or proper artery of the testicle (p. 414), is a slender and remarkably long branch, which arises from the abdominal aorta, and passing down the posterior abdominal wall reaches the spermatic cord, and descends along it to the gland. In early fœtal life its course is much shorter, as the testis is then situated near the part of the aorta from which the artery arises. As the vessel approaches the testicle, it gives off small branches to the epididymis, and then divides into others which perforate the tunica albuginea at the back of the gland, and pass through the corpus Highmorianum ; some spread out on the internal surface of the tunica albuginea, whilst others run along between the lobes of the testis, supported by the fibrous processes of the mediastinum. The smallest branches ramify on the delicate membranous septa between the lobes, before supplying the seminiferous tubes.

The vas deferens receives from the superior vesical artery a long slender branch which accompanies the duct, and is hence named the *deferent artery* or *artery of the vas deferens* (p. 421). It ramifies on the coats of the duct, and reaches as far as the testis, where it anastomoses with the spermatic artery.

The *spermatic veins* (p. 473) commence in the testis and epididymis, pass out at the posterior border of both, and unite into larger vessels, which freely communicate with each other as they ascend along the cord, and form a plexus, named the *pampiniform plexus.* Ultimately two or three veins follow the course of the spermatic artery into the abdomen, where they unite into a single trunk, that of the right side opening into the vena cava, and that of the left into the left renal vein.

The *lymphatics* (p. 495) accompany the spermatic vessels and terminate in the lumbar lymphatic glands, which lie about the large blood-vessels in front of the vertebral column. According to Ludwig and Tomsa, they begin in a large network of spaces which completely fills the intervals between the tubuli seminiferi. According to Henle, these intervals are occupied by granular matter, which he compares with the contents of nerve-cells, and which abounds in rounded nuclei. (Ludwig and Tomsa, in Sitzungsberichte d. Kaiserl. Akad. Vienna, 1862.)

The *nerves* are derived from the sympathetic system. The spermatic plexus (p. 701) is a very delicate set of nervous filaments, which descend upon the spermatic artery from the aortic plexus. Some additional filaments, which are very minute, come from the hypogastric plexus, and accompany the artery of the vas deferens.

The vesiculæ seminales receive branches from the inferior vesical and middle hæmorrhoidal arteries. The veins and lymphatics correspond. The nerves belong to the sympathetic system, and come from the hypogastric plexus.

The *semen* is a thick whitish fluid, which consists of a liquor seminis, and of certain solid particles. It is the combined product of the testes and accessory generative glands, the former secreting spermatozoa, and little more fluid than is necessary to convey these ; the latter diluting this secretion with additional fluid.

The *liquor seminis* is colourless, transparent, and of an albuminous nature. It contains floating in it, besides squamous and columnar epithelium cells, oil-like globules and minute granular matter, *seminal granules* (Wagner), and the *spermatozoa* or spermatic filaments.

The *seminal granules* are rounded colourless corpuscles, having a granular aspect. They average about $\frac{1}{4000}$ th of an inch in diameter, and may be allied to mucus-corpuscles.

Fig. 681.

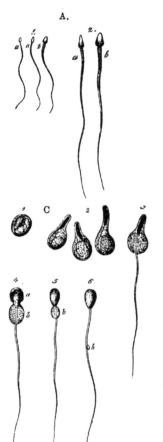

A.

B.

C.

Fig. 681 A.—Spermatic Filaments from the Human Vas Deferens (from Kölliker).

1, magnified 350 diameters; 2, magnified 800 diameters; *a*, from the side; *b*, from above.

Fig. 681 B.—Spermatic Cells and Spermatozoa of the Bull undergoing development (from Kölliker). $\frac{450}{1}$

1, spermatic cells with one or two nuclei, one of them clear; 2, 3, free nuclei with spermatic filaments forming; 4, the filaments elongated and the body widened; 5, filaments nearly fully developed.

Fig. 681 C.—Escape of the Spermatozoa from their Cells in the same Animal.

1, spermatic cell containing the spermatozoon coiled up within it; 2, the cells elongated by the partial uncoiling of the spermatic filament; 3, a cell from which the filament has in part become free; 4, the same with the body also partially free; 5, spermatozoon from the epididymis with vestiges of the cell adherent; 6, spermatozoon from the vas deferens, showing the small enlargement, *b*, on the filament.

The *spermatozoa* are peculiar particles, which, during life and for some hours after being removed from the testicle, perform rapid vibratory or lashing movements. Each consists of a flattened oval part or so-called body, and of a long slender filiform tail. The body is about $\frac{1}{6000}$th of an inch in width, and the entire spermatozoon is from $\frac{1}{500}$th to $\frac{1}{400}$th of an inch in length. The body often contains a minute spot, and, at its junction with the narrow filament or tail, there is frequently a slight projecting fringe or collar. The spermatozoa are developed like nuclei in the interior of the spermatic cells, the cells subsequently becoming enlarged into transparent vesicular bodies of considerable size, in which one or several spermatozoa may be seen. Sometimes a group of cells, each containing a single spermatozoon, is seen enclosed within a parent cell. The spermatozoa are not normally found free until they reach the rete testis.

(Wagner and Leuckart, Article "Semen" in Cyclop. of Anat. and Phys.; Kölliker in Handbuch).

ORGANS OF GENERATION IN THE FEMALE.

The generative organs in the female consist of the ovaries, uterus, and Fallopian tubes, which are named the *internal*, and the vagina and vulva, named the *external* organs of generation.

THE VULVA.

The *vulva*, or *pudendum*, is a general term, which includes all the parts perceptible externally, viz., the mons Veneris, the labia, the hymen or carunculæ, the clitoris, and the nymphæ. The urethra also may be described in connection with these parts.

The integument on the fore part of the pubic symphysis, elevated by a quantity of areolar and adipose substance deposited beneath it, and covered with hair, is termed the *mons Veneris*. The *labia pudendi* (labia externa v. majora) extend downwards and backwards from the mons, gradually becoming thinner as they descend. They form two rounded folds of integument so placed as to leave an elliptic interval (*rima*) between them, the outer surface of each being continuous with the skin, and covered with scattered hairs, whilst the inner is lined by the commencement of the genito-urinary mucous membrane. Between the skin and mucous membrane there is found, besides fat, vessels, nerves, and glands, some tissue resembling that of the dartos in the scrotum of the male. The labia majora unite beneath the mons and also in front of the perineum, the two points of union being called commissures. The posterior or inferior commissure is about an inch distant from the margin of the anus, and this interval is named the perineum. Immediately within the posterior commissure, the labia are connected by a slight transverse fold (*frænulum pudendi*), which has also received the name of *fourchette*, and is frequently torn in the first parturition. The space between it and the commissure has been called *fossa navicularis*.

Beneath the anterior commissure, and concealed between the labia, is the *clitoris*, a small elongated body corresponding in conformation and structure to a diminutive penis, but differing in having no corpus spongiosum nor urethra connected with it below. It consists of two *corpora cavernosa*, which are attached by *crura* to the rami of the ischium and pubes, and are united together by their flattened inner surfaces which form an incomplete pectiniform septum. The body of the clitoris, which is about an inch and a half long, but is hidden beneath the mucous membrane, is surmounted by a small *glans*, consisting of spongy erectile tissue. The glans is imperforate, but highly sensitive, and covered with a membranous fold, analogous to the prepuce. There is a small suspensory ligament, like that of the penis ; and the two ischio-cavernous muscles, named in the female *erectores clitoridis*, have the same connections as in the male, being inserted into the crura of the corpora cavernosa.

From the glans and preputial covering of the clitoris two narrow pendulous folds of mucous membrane pass backwards for about an inch and a half, one on each side of the entrance to the vagina. These are the *nymphæ* (labia interna v. minora). Their inner surface is continuous with that of the vagina ; the external insensibly passes into that of the labia majora. They contain vessels between the laminæ of tegumentary membrane, but, according to Kobelt, no erectile plexus ; indeed, they would seem to correspond to the cutaneous covering of the male urethra (supposed to be split

3 s

open), whilst the erectile structure corresponding to the bulb and spongy body (in two separate right and left halves) lies deeper, as will be presently explained. (Kobelt, Die männlichen und weiblichen Wollustorgane, 1844.)

Fig. 681.*

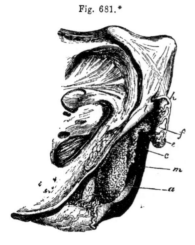

Fig. 681.*—LATERAL VIEW OF THE EREC-TILE STRUCTURES OF THE EXTERNAL ORGANS IN THE FEMALE (from Kobelt). ⅔

The blood-vessels have been injected, and the skin and mucous membrane have been removed; *a*, bulbus vestibuli; *c*, plexus of veins named pais intermedia; *e*, glans clitoridis; *f*, body of the clitoris; *h*, dorsal vein; *l*, right crus clitoridis; *m*, vestibule; *n*, right gland of Bartholin.

Between the nymphæ is the angular interval called the *vestibule*, in which is situated the circular *orifice* of the *urethra*, or *meatus urinarius*, about an inch below the clitoris and just above the entrance to the vagina. The membrane which surrounds this orifice is rather pro-minent ·in most instances, so as readily to indicate its situation.

Immediately below the orifice of the urethra is the *entrance* to the *vagina*, which, in the virgin, is usually more or less narrowed by the *hymen*. This is a thin duplicature of the mucous membrane, placed at the entrance to the vagina ; its form varies very considerably in different persons, but is most frequently semilunar, the concave margin being directed forwards towards the pubes. Sometimes it is circular, and is perforated only by a small round orifice, placed usually a little above the centre ; and occa-sionally it is cribriform, or pierced with several small apertures ; and it may in rare instances completely close the vagina, constituting "imperforate hymen." On the other hand, it is often reduced to a mere fringe, or it may be entirely absent. After its rupture, some small rounded elevations remain, called *carunculæ myrtiformes*.

The *mucous membrane* may be traced inwards from the borders of the labia majora, where it is continuous with the skin : it forms a fold over the vascular tissue of the nymphæ, and is then prolonged into the urethra and vagina. It is smooth, reddish in colour, is covered by a scaly epithelium, and is provided with a considerable number of mucous crypts and follicles and with glands which secrete an unctuous and odorous substance. The mucous crypts and follicles are especially distinct on the inner surface of the nymphæ, and near the orifice of the urethra. The sebaceous glands are found beneath the prepuce, and upon the labia majora and outer surface of the nymphæ.

The *glands of Bartholin* (or of Duverney), corresponding to Cowper's glands in the male, are two reddish yellow round or oval bodies, about the size of a large pea or small bean, lodged one on each side of the commencement of the vagina, between it and the erectores clitoridis muscles, beneath the superficial perineal fascia, and in front of the trans-verse muscles. Their ducts, which are long and single, run forward and open on the inner aspect of the nymphæ, outside the hymen or carunculæ myrtiformes.

Erectile tissue.—All the parts of the vulva are supplied abundantly with blood-vessels, and in certain situations there are masses composed of venous plexuses, or erectile tissue, corresponding to those found in the male. Besides the corpora cavernosa and glans clitoridis, already referred to, there

Fig. 682.

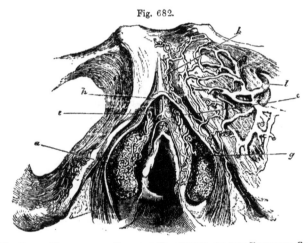

Fig. 632.—Front View of the Erectile Structures of the External Organs in the Female (from Kobelt). ⅔

a, bulbus vestibuli ; *b*, sphincter vaginæ muscle; *e, e*, venous plexus or pars intermedia ; *f*, glans clitoridis ; *g*, connecting veins ; *h*, dorsal vein of the clitoris; *k*, veins passing beneath the pubes ; *l*, the obturator vein.

are two large leech-shaped masses, the *bulbi vestibuli*, about an inch long, consisting of a network of veins, enclosed in a fibrous membrane, and lying one on each side of the vestibule, a little behind the nymphæ. They are rather pointed at their upper extremities, and rounded below : they are suspended, as it were, to the crura of the clitoris and the rami of the pubes, covered internally by the mucous membrane, and embraced on the outer side by the fibres of the coustrictor vaginæ muscle. They are together equivalent to the bulb of the urethra in the male, which, it will be remembered, presents traces of a median division. In front of the bipartite bulb of the vestibule is a smaller plexus on each side, the vessels of which are directly continuous with those of the bulbus vestibuli behind, and of the glans clitoridis before. This is the *pars intermedia* of Kobelt, and is regarded by him as corresponding with the part of the male corpus spongiosum urethræ which is in front of the bulb : it receives large veins coming direct from the nymphæ.

Vessels.—The outermost parts of the vulva are supplied by the superficial pudendal arteries; the deeper parts and all the erectile tissues receive branches from the internal pudic arteries, as in the male. The veins also in a great measure correspond ; there is a vena dorsalis clitoridis, receiving branches from the glans and other parts as in the male ; the veins of the bulbus vestibuli pass backwards into the vaginal plexuses, and are connected also with the obturator veins ; above they communicate with the veins of the pars intermedia, those of the corpora cavernosa and the glans of the clitoris, and also with the vena dorsalis. The lymphatics accompany the blood-vessels.

Nerves.—Besides sympathetic branches, which descend along the arteries, especially for the erectile tissues, there are other nerves proceeding from the lumbar and sacral plexuses ; those from the former being the branches of the genito-crural (p. 660), and those from the latter of the inferior pudendal and internal pudic nerves (P. 675), which last sends comparatively large branches to the clitoris. The mode of termination is not known with certainty ; tactile corpuscles have been seen in the human clitoris, and Pacinian bodies in that of some animals.

THE FEMALE URETHRA.

The *female urethra*, as compared with that of the other sex, is short, representing only the upper half of the prostatic part of the male passage. It is about an inch and a-half in length, and is wide and capable of great distension ; its ordinary diameter is about three or four lines, but it enlarges towards its vesical orifice. The direction of this canal is downwards and forwards, and it is slightly curved and concave upwards. It lies imbedded in the upper or rather the anterior wall of the vagina, from which it cannot be separated.

The external orifice, or *meatus urinarius*, opens in the vulva, beneath the symphysis pubis, nearly an inch below and behind the clitoris, between the nymphæ, and immediately above the entrance to the vagina. From its orifice, which is its narrowest part, the canal passes upwards and backwards between the crura of the clitoris and behind the pubes, gradually enlarging into a funnel-shaped opening as it approaches and joins the neck of the bladder. There is also a dilatation in the floor of the canal, just within the meatus.

The mucous membrane is whitish, except near the orifice ; it is raised into longitudinal plicæ, which are not entirely obliterated by distension, especially one which is particularly marked on the lower or posterior surface of the urethra. Near the bladder the membrane is soft and pulpy, with many tubular mucous glands. Lower down these increase in size and lie in groups, between the longitudinal folds, and immediately within and around the orifice, the lips of which are elevated, are several larger and wider crypts.

The lining membrane is covered with a scaly epithelium, but near the bladder · the · particles become spheroidal. The submucous areolar tissue contains elastic fibres. Outside this there is a highly vascular structure, in which are many large veins. Between the anterior and posterior layers of the triangular ligament, the female urethra is embraced by the fibres of the compressor urethræ muscle.

The *vessels* and *nerves* of the female urethra are very numerous, and are derived from the same sources as those of the vagina.

THE VAGINA.

The *vagina* is a membranous and dilatable tube, extending from the vulva to the uterus, the neck of which is embraced by it. It rests below and behind on the rectum, supports the bladder and urethra in front, and is enclosed between the levatores ani muscles at the sides. Its direction is curved backwards and upwards : its axis corresponding below with that of the outlet of the pelvis, and higher up with that of the pelvic cavity. In consequence of being thus curved, its length is greater along the posterior than along the anterior wall, being in the latter situation about four inches, while in the former it amounts to five or six. Each end of the vagina is somewhat narrower than the middle part: the lower, which is continuous with the vulva, is the narrowest part, and has its long diameter from before

Fig. 683.

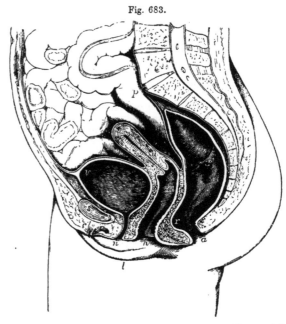

Fig. 683.—Sectional View of the Viscera of the Female Pelvis (after Houston and from nature). ¼

p, promontory of the sacrum ; *s*, symphysis of the pubes ; *v*, the upper part of the urinary bladder ; *v'*, the neck ; *v'*, *n*, the urethra ; *u*, the uterus ; *va*, the vagina ; *r*, the point of union of the middle and lower parts of the rectum : *r'*, the fold between the middle and upper parts of the rectum ; *a*, the anus ; *l*, the right labium ; *n*, the right nympha ; *h*, the hymen ; *cl*, the divided clitoris with the prepuce. The pelvic viscera, having been distended and hardened with alcohol previously to making the section, appear somewhat larger than natural.

backwards ; the middle part is widest from side to side, being flattened from before backwards, so that its anterior and posterior walls are ordinarily in contact with each other : at its upper end it is rounded, and expands to receive the vaginal portion of the neck of the uterus, which is embraced by it at some distance from the os uteri.   The vagina reaches higher up on the cervix uteri behind than in front, so that the uterus appears, as it were, to be let into its anterior wall.

On the *inner surface* of the vagina, along the anterior and the posterior walls, a slightly elevated ridge extends from the lower end upwards along the middle line, similar to the raphe in other situations : these ridges are named the *columns* of the vagina, or *columnæ rugarum*.   Numerous dentated transverse ridges, called *rugæ*, will also be observed, particularly in persons who have not borne children, running at right angles from the columns. These columns and rugæ are most evident near the entrance of the vagina and on the anterior surface, and gradually become less marked, and disappear towards its upper end.

Structure and connections.—The walls of the vagina are thickest in front, in the vicinity of the urethra, which indeed may be said to be imbedded in the anterior wall of the vaginal passage ; in other situations they are thinner. The vagina is firmly connected by areolar tissue to the neck of the bladder, and only loosely to the rectum and levatores ani muscles ; at the upper end, for about a fourth part of its length, it receives a covering behind from the peritoneum, which descends in the form of a cul-de-sac thus far between the vagina and the rectum.

Externally the vagina presents a coat of dense areolar tissue, and beneath this its walls are composed of unstriped muscle, which is not distinctly separable into strata, but is composed chiefly of fibres internally circular and externally longitudinal. Round the tube a layer of loose erectile tissue is found, which is most marked at the lower part.

At its lower end, the vagina is embraced by muscular fibres, which constitute the *sphincter vaginæ*, already described (p. 266).

The mucous membrane, besides the columns and rugæ, is provided with conical and filiform papillæ, numerous muciparous glands and follicles, especially in its upper smoother portion and round the cervix uteri. This membrane, which is continuous with that of the uterus, is covered with a squamous epithelium.

The vagina is largely supplied with vessels and nerves. The arteries are derived from branches of the internal iliac, viz., the vaginal, internal pudic, vesical, and uterine (pp. 422, 428). The veins correspond ; but they first surround the vagina with numerous branches, and form at each side a plexus named the *vaginal* plexus. The *nerves* are derived from the hypogastric plexus of the sympathetic, and from the fourth sacral and pudic nerves of the spinal system ; the former are traceable to the erectile tissue (p. 704).

THE UTERUS.

The *uterus, womb*, or *matrix*, is a hollow organ, with very thick walls, which is intended to receive the ovum, retain and support it during the development of the fœtus, and expel it at the time of parturition. The ova, discharged from the ovaries, reach the uterus by the Fallopian tubes, which open, one at each side, into the upper part of that organ. During pregnancy, the uterus undergoes a great enlargement in size and capacity, as well as other important changes. In the fully developed virgin condition, which is that to which the following description applies, it is a pear-shaped body, flattened from before backwards, situated in the cavity of the pelvis, between the bladder and rectum, with its lower extremity projecting into the upper end of the vagina. It does not reach above the brim of the pelvis. Its upper end is turned upwards and forwards, whilst the lower is in the opposite direction ; so that its position corresponds with that of the axis of the inlet of the pelvis, and forms an angle or curve with the axis of the vagina, which corresponds with that of the outlet of the cavity. The uterus projects, as it were, upwards into a fold of the peritoneum, by which it is covered behind and above, and also in front, except for a short distance towards the lower end, where it is connected with the base of the bladder. Its free surface is in contact with the other pelvic viscera, some convolutions of the small intestine usually lying upon and behind it. From its two sides the peritoneum is reflected in the form of a broad duplicature, named the ligamentum latum, which, together with the parts contained within it, will be presently described.

Fig. 684.

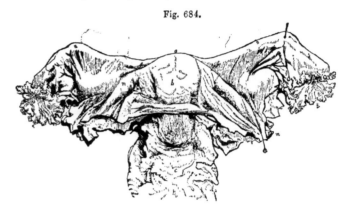

Fig. 684.—ANTERIOR VIEW OF THE UTERUS AND ITS APPENDAGES. ⅓

a, fundus; *b*, body; *c*, cervix; *e*, front of the upper part of the vagina; *n, n*, round ligaments; *r, r*, broad ligaments; *s, s*, Fallopian tubes; *t*, fimbriated extremity; *u*, ostium abdominale; the position of the ovaries is indicated through the broad ligaments, and the cut edge of the peritoneum is shown along the side of the broad ligaments and across the front of the uterus.

The average dimensions of the uterus are about three inches in length, two in breadth at its upper and wider part, and nearly an inch in thickness : its weight is from seven to twelve drachms. It is usually described as possessing a fundus, body, and neck.

The *fundus* is the broad upper end of the body, and projects convexly upwards from between the points of attachment of the Fallopian tubes. During gestation, its convexity is greatly increased, and it surrounds a large part of the uterine cavity. The *body* gradually narrows as it extends from the fundus to the neck ; its sides are straight ; its anterior and posterior surfaces are both somewhat convex, but the latter more so than the former. At the points of union of the sides with the rounded superior border or fundus, are two projecting angles, with which the Fallopian tubes are connected, the round ligaments being attached a little before, and the ovarian ligaments behind and beneath them : these three parts are all included in the duplicature of the broad ligaments. The *neck*, or *cervix uteri*, narrower and more rounded than the rest of the organ, is from six to eight lines long ; it is continuous above with the body, and, becoming somewhat smaller towards its lower extremity, projects into the upper end of the tube of the vagina, which is united all round with the substance of the uterus, but extends upwards to a greater distance behind than in front. The projecting portion is sometimes named the *vaginal part*. The lower end of the uterus presents a transverse aperture, by which its cavity opens into the vagina ; this is named variously *os uteri, os uteri externum*, and (from some supposed likeness to the mouth of the tench fish) *os tincæ*. It is bounded by two thick lips, the posterior of which is the thinner and longer of the two, while the anterior, although projecting less from its vaginal attachment, is lower in position, and, when the tube is closed, rests on the posterior wall of the vagina. These borders or lips are generally smooth, but, after parturition, they frequently become irregular, and are sometimes fissured or cleft.

Fig. 685.

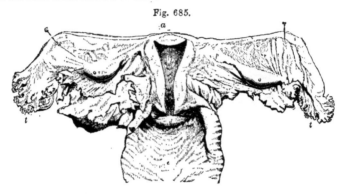

Fig. 685.—POSTERIOR VIEW OF THE UTERUS AND ITS APPENDAGES.

The cavity of the uterus has been opened by the removal of the posterior wall, and the upper part of the vagina has been laid open; *a*, fundus; *b*, body; *c*, cervix; *d*, on the anterior lip of the os uteri externum; *e*, the interior of the vagina; *f*, section of the walls of the uterus; *i*, opening of the Fallopian tube; *o*, ovary; *p*, ligament of the ovary; *r*, broad ligament; *s*, Fallopian tube; *t*, its fimbriated extremity.

Owing to the great thickness of its walls, the *cavity* of the uterus is very small in proportion to the size of the organ. The part within the body of the organ is triangular, and flattened from before backwards, so that its anterior and posterior walls touch each other. The base of the triangle is directed upwards, and is curvilinear, the convexity being turned towards the interior of the uterus. This form is owing to the prolongation of the cavity through the substance of the organ towards its two superior angles, where two minute foramina lead into the Fallopian tubes. At the point where the body is continuous below with the neck, the cavity is slightly constricted, and thus forms what is sometimes named the *internal orifice (os uteri inter-num, isthmus vel ostium uteri)*; it is often smaller than the os externum, and is a circular opening. That portion of the cavity which is within the *neck*, resembles a tube slightly flattened before and behind; it is some-what dilated in the middle, and opens inferiorly into the vagina by the os tincæ. Its inner surface is marked by two longitudinal ridges or columns, which run, one on the anterior, the other on the posterior wall, and from both of which rugæ are directed obliquely upwards on each side, so as to present an appearance which has been named *arbor vitæ uterinus*, also *palmæ plicatæ*: this structure is most strongly marked anteriorly.

Structure.—The walls of the uterus consist of an external serous cover-ing, an internal mucous membrane, and an intermediate proper tissue. The *peritoneal* layer covers the fundus and body, except at the sides and for about half an inch of the lower part of the body in front, which is attached to the base of the bladder.

The *proper tissue* of the uterus constitutes much the greater part of its walls, which are thickest opposite the middle of the body and fundus, and are thinnest at the entrances of the Fallopian tubes. The tissue is very dense : it is composed of bundles of muscular fibres of the plain variety, of small size in the unimpregnated uterus, interlacing with each other, but disposed in bands and layers, intermixed with much fibro-areolar tissue, a large number of blood-vessels and lymphatics, and some nerves. The areolar tissue is more abundant near the outer surface. The arrangement of the muscular fibres is best studied in the uterus at the full period of gestation, in which the bands and layers formed by them become augmented in size, and much more distinctly developed. They may be referred to three sets or orders, viz., external, internal, and intermediate. Those of the *external* set are arranged partly in a thin superficial sheet, immediately beneath the peritoneum, and partly in bands and incomplete strata, situated more deeply. A large share of these fibres arch transversely over the fundus and adjoining part of the body of the organ, and converge at either side towards the commencement of the round ligaments, along which they are prolonged to the groin. Others pass off in like manner to the Fallopian tubes, and strong transverse bands from the anterior and posterior surfaces are extended into the ovarian ligaments. A considerable number of thinly scattered fibres also pass at each side into the duplicature of the broad ligament, and others are described as running back from the cervix of the uterus into the recto-uterine folds or phcæ semilunares. The fibres of the subperitoneal layer are much mixed with areolar tissue, especially about the middle of the anterior and posterior surfaces of the uterus, in which situation many of the superficial fibres appear to have as it were a median attachment from which they diverge. The fibres on the *inner* surface of the uterus are disposed with comparative regularity in its upper part, being arranged there in numerous concentric rings round the openings of the two Fallopian tubes, the outermost and largest circles of the two series meeting from opposite sides in the middle of the uterus. Towards the cervix the internal fibres run more transversely ; elsewhere they take various directions. The *intermediate* fibres, between the external and internal set, pass in bands among the blood-vessels, following less regular courses.

The *mucous membrane* which lines the uterus is thin and closely adherent to the subjacent substance, especially in the body of the organ. It is continued from the vagina, and into the Fallopian tubes. Between the rugæ of the cervix, already described, it is provided with numerous mucous follicles and glands. There are also occasionally found in the same situation certain small transparent vesicular bodies, which, from an erroneous opinion as to their nature, were named the *ovula Nabothi.* They appear to be closed and obstructed mucous follicles, distended with a clear viscous fluid. In the inferior third or half of the cervix, the mucous membrane presents papillæ covered with ciliated epithelium.

In the body of the uterus the mucous membrane is thin, smooth, soft, and of a reddish-white colour. When viewed with a magnifying lens, it is found to be marked with minute dots, which are the orifices of numerous simple tubular glands, somewhat like those of the intestine. Some of these tubular glands are branched, and others are slightly twisted into a coil. These glands can be distinctly seen in the unimpregnated and in the virgin uterus, but they become enlarged and more conspicuous after impregnation (fig. 686). The epithelium is columnar and ciliated.

Ligaments of the uterus.—Where the peritoneum is reflected from the uterus to the bladder in front, and to the rectum behind, it forms, in each position, two semilunar folds, which are sometimes called respectively, the *anterior* and the *posterior ligaments* of the uterus. The former are also named the *vesico-uterine*, and the latter, which are more marked, the *recto-uterine folds.*

Fig. 686.

A.

B

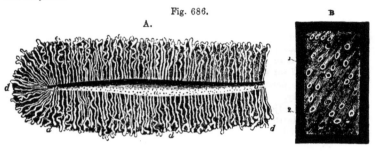

Fig. 686, A.—Section of the Glandular Structure of the Human Uterus at the commencement of Pregnancy (from E. H. Weber). ²⁄₁

a, part of the cavity of the uterus showing the orifices of the glands ; *d*, a number of the tubular glands, some of which are simple, others slightly convoluted and divided at the extremities.

Fig. 686, B.—Small Portion of the Uterine Mucous Membrane after Recent Impregnation, seen from the inner surface (from Sharpey). ¹²⁄₁

The specimen is represented as viewed upon a dark ground, and shows the orifices of the uterine glands, in most of which, as at 1, the epithelium remains, and in some, as at 2, it has been lost.

The *broad ligaments* (ligamenta lata) are formed on each side by a fold of the peritoneum, which is directed laterally outwards from the anterior and posterior surfaces of the uterus, to be connected with the sides of the pelvic cavity. Between the two layers of the serous membrane are placed, first, the Fallopian tube, which, as will be more particularly described, runs along the upper margin of the broad ligament ; secondly, the round ligament, which is in front ; thirdly, the ovary and its ligament, which lie in a special offshoot of the ligamentum latum, behind ; and, lastly, blood-vessels, lymphatics, and nerves, with some scattered fibres from the superficial muscular layer of the uterus. The *ligament of the ovary* is merely a dense fibro-areolar cord, containing also, according to some authorities, uterine muscular fibres, and measuring about an inch and a half in length, which extends from the inner end of the ovary to the upper angle of the uterus, immediately behind and below the point of attachment of the Fallopian tube ; it causes a slight elevation of the posterior layer of the serous membrane, and, together with the ovary itself, forms the lower limit of a triangular portion of the broad ligament, which has been named the *ala vespertilionis* or bat's wing.

The *round ligaments* are two cord-like bundles of fibres, about four or five inches in length, attached to the upper angles of the uterus, one on either side (ligamentum teres uteri), immediately in front of the Fallopian tube. From this point each ligament proceeds upwards, outwards, and forwards, to gain the internal inguinal ring ; and after having passed, like the spermatic cord in the male, through the inguinal canal, reaches the fore part of the pubic symphysis, where its fibres expand and become united with the substance of the mons Veneris. Besides areolar tissue and vessels, the round ligaments contain plain muscular fibres, like those of the uterus, from which, indeed, they are prolonged. Each ligament also receives a covering from the peritoneum, which, in the young subject, is prolonged under the form of a tubular process for some distance along the inguinal canal: this, which resembles the processus vaginalis originally existing in the same situation in the male, is named the canal of Nuck : it is generally obliterated afterwards, but is sometimes found even in advanced life.

Blood-vessels and Nerves.—The *arteries of the uterus* are four in number, viz., the right and left *ovarian* (which correspond to the spermatic of the male) and the two *uterine.* Their origin, as well as the mode in which they reach the uterus and ovaries, has been already described (pp. 414, 422). They are remarkable for their frequent anastomoses, and also for their singularly tortuous course; within the substance of the uterus they seem to be placed in little channels or canals. The *veins* correspond with the arteries; they are very large, and form the uterine plexuses, and their thin walls are in immediate contact with the uterine tissue. The course of the lymphatics is described at p. 495; they are very large and abundant in the gravid uterus. The *nerves* have been fully described at p. 704. They are derived from the inferior hypogastric plexuses, the spermatic plexuses, and the third and fourth sacral nerves.

The *changes which take place in the uterus* from age, menstruation, and gestation, and the characters presented by this organ after it has once performed the latter function, can only be very generally indicated here.

For some time after *menstruation* first commences, the uterus becomes rounder and slightly enlarged at each period, its os externum becomes more rounded, and its lips swollen ; subsequently these periodical alterations are not so marked. The organ itself, however, always becomes more turgid with blood, and the mucous membrane appears darker, softened, and thickened.

In *gestation* more extensive alterations ensue, which necessarily affect the size, shape, and position of the organ, the thickness and amount of substance in its walls, the dimensions and form of its cavity, and the character of its cervix and of its os externum and os internum. Its weight increases from about one ounce to one pound and a half or even three pounds. Its colour becomes darker, its tissue less dense, its muscular bundles more evident, and the fibres more developed. The principal increase is in the muscular tissue, and this increase takes place not only by the enlargement of already existing elements, the fibre-cells becoming enlarged from seven to eleven times in length, and from two to five times in breadth (Kölliker), but also by new formation. The former process is general ; the latter occurs mainly in the innermost layers, and continues until the sixth month of pregnancy, when new formation ceases. The round ligaments become enlarged, and their muscular structure more marked ; the broad ligaments are encroached upon by the intrusion of the growing uterus between their layers. The mucous membrane and the glands of the body of the uterus become the seat of peculiar changes, which lead to the formation of the decidual membrane ; whilst the membrane of the cervix loses its columns and rugæ. The blood-vessels and lymphatics are greatly enlarged, and it is observed that the arteries become exceedingly tortuous as they ramify upon the organ. The condition of the nerves in the gravid uterus has been previously fully referred to (p. 704).

After parturition, the uterus again diminishes, its enlarged muscular fibres undergoing oleaginous degeneration and becoming subsequently absorbed, while a new set of minute fibre cells are developed. The organ, however, never regains its original virgin character. Its weight usually remains from two to three ounces in those who have had children ; its cavity is larger; the os externum is wider and more rounded, and its margins often puckered or fissured ; the arteries continue much more tortuous than they are in the virgin, and its muscular fibres and layers remain more defined.

Age.—In the infant, the neck of the uterus is larger than the body; and the fundus is not distinguished either by breadth or by convexity of outline. These parts afterwards enlarge gradually, until, at puberty, the pyriform figure of the womb is fully established. The arbor vitæ is very distinct, and indeed at first reaches upwards to the highest part of the cavity.

From the gradual effects of more advanced .*age* alone, independent of impregnation, the uterus shrinks, and becomes paler in colour, and harder in texture; its triangular form is lost; the body and neck become less distinguishable from each other; the orifices also become less characteristic.

For further details with regard to uterine changes, the reader is referred to Farre on "Uterus and its Appendages" in Cyclop. of Anat. and Phys.

THE OVARIES AND FALLOPIAN TUBES.

The *ovaries*, the parts corresponding to the testicles of the male (ovaria, testes muliebres), are two somewhat flattened oval bodies, which are placed one on each side, nearly horizontally, at the back of the broad ligament of the uterus, and are enveloped by its posterior membranous layer. The ovaries are largest in the virgin state; their weight is from three to five scruples, and they usually measure about one inch and a half in length, three-quarters of an inch in width, and nearly half an inch in thickness; but their size is rather variable. Each ovary is free on its two sides, and also along its posterior border, which has a convex outline; but it is attached by its anterior border, which is straighter than the other, and along the line of its attachment exhibits a deep *hilus* by which the vessels and nerves enter. Its inner end is generally narrow, and is attached to the dense cord already described as the *ligament* of the *ovary*, which connects it with the uterus. Its outer extremity is more obtuse and rounded, and has attached to it one of the fimbriæ of the Fallopian tube.

Structure.—The colour of the ovaries is whitish, and their surface is either smooth, or more commonly irregular, and often marked with pits or clefts resembling scars. Beneath the peritoneal coat, which covers it everywhere except along its attached border, the ovary is enclosed in a *proper fibrous coat* (tunica albuginea), of a whitish aspect and of considerable thickness, which adheres firmly to the tissue beneath, being in structural continuity with it. When the deeper ovarian substance is divided, it is seen to consist of a firm reddish-white vascular structure called the

Fig. 687.

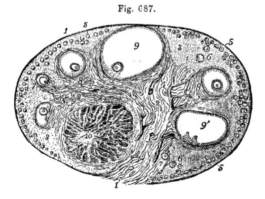

Fig. 687.—VIEW OF A SECTION OF THE PREPARED OVARY OF THE CAT (from Schrön). ♀

1, outer covering and free border of the ovary; 1', attached border; 2, the ovarian stroma, presenting a fibrous and vascular structure; 3, granular substance lying external to the fibrous stroma; 4, bloodvessels; 5, ovigerms in their earliest stages occupying a part of the granular layer near the surface; 6, ovigerms which have begun to enlarge and to pass more deeply into the ovary; 7, ovigerms round which the Graafian follicle and tunica granulosa are now formed, and which have passed somewhat deeper into the ovary and are surrounded by the fibrous stroma; 8, more advanced Graafian follicle with the ovum imbedded in the layer of cells constituting the proligerous disc; 9, the most advanced follicle containing the ovum, &c.: 9', a follicle from which the ovum has accidentally escaped; 10, corpus luteum presenting radiated columns of cellular structure.

stroma, the fibres of which, although forming a felted tissue, have, with the blood-vessels, principally a radiating direction from the hilus to the rest of the surface. It contains numerous spindle-shaped cells, and also, according to some writers, unstriped muscular tissue. Towards the surface, the ovarian tissue, which in this part has been distinguished as *cortical*, presents, especially in children, a different appearance from the deeper or *medullary* part, from being granular, and having within it great numbers of small vesicles, the Graafian vesicles or follicles, which are absent from the deep part. After the period of puberty, a certain number of the Graafian vesicles, varying from twelve to thirty or more, have attained a larger size, some having a diameter of from $\frac{1}{20}$th to $\frac{1}{6}$th of an inch, or even more. The great majority however, remain much smaller : thus Henle estimated the number of vesicles of about $\frac{1}{50}$ to $\frac{1}{80}$th of an inch in diameter, in the two ovaries of a girl of eighteen, at 72,000 (Syst. Anat., II. 483).

The *vesicles of De Graaf*, when dilated, are filled in part with a clear, colourless, albuminous fluid, the larger ones approaching the surface of the ovary, on which they may sometimes be distinguished as semi-transparent elevations. Each of these vesicles includes, besides its fluid contents, the *ovum*—a small round vesicular body, imbedded in a layer of cellular substance. Sometimes, though rarely, two ova have been found in one vesicle.

Fig. 688.

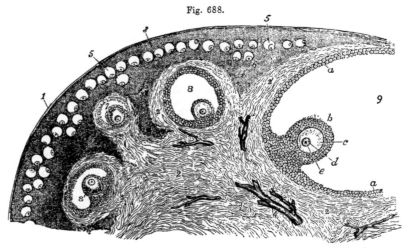

Fig. 688.—VIEW OF A PORTION OF THE SECTION OF THE PREPARED CAT'S OVARY, REPRESENTED IN THE PRECEDING FIGURE, MORE HIGHLY MAGNIFIED (from Schrön).

1, outer covering of the ovary; 2, fibrous stroma ; 3, cortical layer of granular substance towards the surface; 3′, deeper parts of the granular substance ; 4, blood-vessels ; 5, ovigerms forming a layer near the surface ; 6, one or two of the ovigerms sinking a little deeper and beginning to enlarge ; 7, one of the ovigerms farther developed, now enclosed by a prolongation of the fibrous stroma, and consisting of a small Graafian follicle, within which is situated the ovum covered by the cells of the discus proligerus ; 8, a follicle farther advanced ; 8′, another which is irregularly compressed ; 9, the greater part of the largest follicle, in which the following indications are given ; *a*, epithelial or cellular lining of the follicle constituting the membrana granulosa ; *b*, the portion reflected over the ovum named discus proligerus ; *c*, vitellus or yelk part of the ovum surrounded by a vesicular membrane, which becomes afterwards the zona pellucida ; *d*, germinal vesicle ; *e*, germinal spot or nucleus.

The developed vesicle has two coats, viz., an external *vascular tunic*, and an internal tunic named the *ovicapsule*, which is lined with a cellular or epithelial layer, the *membrana granulosa.* At first the ovum appears near the centre of the vesicle, while the latter is still very small, but, in the mature condition, it lies towards the internal surface of the ovi-capsule, imbedded in the *discus proligerus,* a small flattened heap of granular cells, continuous with the membrana granulosa.

The *ovum* itself, first discovered in mammals by Baër, is a spherical body, very constant in size, being about $\frac{1}{120}$th of an inch in diameter ; it consists of a thick, colourless, and transparent envelope (zona pellucida), which surrounds the substance of the yelk. Within the yelk, which is made up of granular matter, is situated a still smaller vesicular body, named the *germinal vesicle,* which is about $\frac{1}{720}$th of an inch in diameter ; and in this again is an opaque spot, having a diameter only of $\frac{1}{3500}$th to $\frac{1}{2500}$th of an inch, and named the *germinal spot* (macula germinativa).

The ova make their appearance in the ovary at so early a period that even at the time of birth it has been found too late to observe their mode of origin. It has been ascertained that the ovum makes its appearance before the ovisac, and that the germinal vesicle is the first part of the ovum to be formed, the granular substance of the yelk being gradually deposited round it. Around the ovum a circle of cells becomes visible, grows thicker, and divides into two layers, the outer of which becomes the membrana granulosa, while the inner adheres to the ovum, forming the discus proligerus. The precise nature and mode of origin of .the inner tunic of the Graafian vesicle is matter of dispute, and indeed Henle denies that there is any homogeneous membrane distinct from the outer cells of the membrana granulosa.

According to Schrön's observations on the cat, the ova make their first appearance near the surface of the ovary, and the vesicles become deeper placed as they grow larger : it is only in the later stages of growth, when the great expansion of the vesicles presses aside the surrounding tissues, that they are again brought into contact with the surface. From observations by Gröhe it appears that the process is similar in the human ovary. A beaded arrangement of the ova, as if developed in anastomosing primitive gland tubes, has been observed by Valentin in the ovaries of young animals, and more recently by Pflüger in the adult cat. Spiegelberg finds similar appearances in the human fœtus. But the existence of such tubular glandular structure and its relation to the commencing ova is still under discussion, and requires farther observation for its determination.

Fig. 689.

Fig. 689. — OVUM OF THE SOW REMOVED FROM THE GRAAFIAN VESICLE, WITH ITS CELLULAR COVERING (from M. Barry). $\frac{100}{1}$

1, germinal spot or nucleus ; 2, germinal vesicle ; 3, the yelk ; 4, the zona pellucida or external covering of the ovum ; 5, part of the tunica granulosa or proligerous disc ; 6, some adherent granules or cells.

The Graafian vesicle, as it becomes more fully dilated, approaches the surface of the ovary. By rupture of the vesicle the ovum, covered by the cells of its proligerous disc, escapes into the Fallopian tube, and is thus conveyed into the womb, while the ruptured vesicle becomes converted, by hypertrophy of its walls, into a yellow mass named *corpus luteum,* which after persisting for a time, dwindles down into a small fibrous cicatrix.

On the subject of the ovum the following works may be mentioned :—
Martin Barry's Researches on Embryology, in Phil. Trans., 1838 and 1839 ;
Allen Thomson, Article "Ovum," in Cyclop. of Anat. and Phys., where
also the literature will be found referred to ; Farre, "Uterus and Appendages," in the same ; Pflüger, Die Eierstöcke der Säugethiere und des
Menschen, Leipzig, 1863; Schrön, in Zeitsch. f. Wissensch. Zoologie, vol.
xji. p. 409 ; Gröhe, in Virchow's Archiv, vol. xxvi. p. 271 ; also in
Virchow's Archiv, vol. xxix. p. 450 ; Spiegelberg, in Virchow's Archiv,
vol. xxx. p. 466 : and Henle, in his Handbuch.

The *Fallopian tubes.*—These tubes, which may be considered as ducts
of the ovaries, or oviducts, and which serve to convey the ovum from
thence into the uterus, are inclosed in the free margin of the broad ligaments. They are ·between three and four inches in length. Their inner
or attached extremities, which proceed from the upper angles of the uterus,
are narrow and cord-like ; but they soon begin to enlarge, and proceeding outwards, one on each side, pursue an undulatory course, and at
length, having become gradually wider, they bend backwards and downwards towards the ovary, about an inch beyond which they terminate in
an expanded extremity, the margin of which is divided deeply into a
number of irregular processes named *fimbriæ ;* one of these, somewhat
longer than the rest, is attached to the outer end of the corresponding
ovary. The wide and fringed end of the Fallopian tube, or rather *trumpet,*
as the term "tuba·" literally signifies, is turned downwards and towards
the ovary, and is named the *fimbriated extremity* (morsus diaboli). In
the midst of these fimbriæ, which are arranged in a circle, the tube itself
opens by a round constricted orifice, *ostium abdominale,* placed at the

Fig. 690.

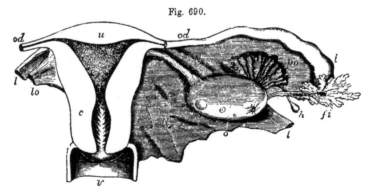

Fig. 690.—Diagrammatic View of the Uterus and its Appendages, as seen from
behind. ½

The uterus and upper part of the vagina have been laid open by removing the posterior
wall ; the Fallopian tube, round ligament, and ovarian ligament have been cut short and
the broad ligament removed on the left side ; *u,* the upper part of the uterus ; *c,* the
cervix opposite the os internum ; the triangular shape of the uterine cavity is shown and
the dilatation of the cervical cavity with the rugæ termed arbor vitæ ; *v,* upper part of the
vagina ; *od,* Fallopian tube or oviduct ; the narrow communication of its cavity with
that of the cornu of the uterus on each side is seen ; *l,* round ligament; *lo,* ligament of
the ovary ; *o,* ovary ; *i,* wide outer part of the right Fallopian tube ; *f i,* its fimbriated
extremity ; *po,* parovarium ; *h,* one of the hydatids frequently found connected with the
broad ligament.

bottom of a sort of fissure leading from that fringe which is attached to the ovary. It is by this orifice that an ovum is received at the time of its liberation from the ovary, and is thence conveyed along the tube to its uterine extremity, which opens into the womb by a very minute orifice, admitting only a fine bristle, and named *ostium uterinum*. The part of the canal which is near the uterus is also very fine, but it becomes gradually larger towards its abdominal orifice, and there it is again somewhat contracted : hence, the term *isthmus*, given by Henle to the uterine half, and *ampulla* to the outer half of the Fallopian tube.

Beneath the external or peritoneal coat the walls of the tube contain, besides areolar tissue, plain muscular fibres like those of the uterus, arranged in an external longitudinal and an internal circular layer. The mucous membrane lining the tubes is thrown into longitudinal plicæ, which are broad and numerous in the wider part of the tube, and in the narrower part are broken up into very numerous arborescent processes : it is continuous, on the one hand, with the lining membrane of the uterus, and at the outer end of the tube with the peritoneum ; presenting an example of the direct continuity of a mucous and serous membrane, and making the peritoneal cavity in the female an exception to the ordinary rule of serous cavities, *i.e.* of being perfectly closed. The epithelium in the interior of the Fallopian tube is, like that of the uterus, columnar and ciliated ; the inner surface of the fimbriæ is also provided with cilia, and Henle has even detected ciliated epithelium on their outer or serous surface, but it here soon passes into the scaly epithelium of the peritoneal membrane.

Vessels and nerves of the ovaries and Fallopian tubes.—The ovaries are supplied by the ovarian *arteries*, analogous to the spermatic in the male, which anastomose freely by an internal branch with the termination of the uterine arteries. Sometimes this anastomotic branch is so large that the ovary seems to be supplied almost entirely by the uterine artery. The ovarian artery always sends numerous branches to the Fallopian tube. The smaller arteries penetrate the ovary along its attached border, pierce the proper coat, and run in flexuous parallel lines through its substance. The *veins* correspond, and the ovarian veins form a plexus near the ovary named the pampiniform plexus. The *nerves* are derived from the spermatic or ovarian plexus ; and also from one of the uterine nerves, which invariably send an offset to the Fallopian tube.

The *parovarium* (Kobelt), or Organ of Rosenmüller, is a structure which can usually be brought plainly into view by holding against the light the fold of peritoneum between the ovary and Fallopian tube. It consists of a group of scattered tubules lying transversely between the Fallopian tube and ovary, lined with epithelium, but having no orifice. The tubules converge, but remain separate at their ovarian end, and at the other are more or less distinctly united by a longitudinal tube. The parovarium consists of a few tubules formed in connection with the Wolffian body, which, partaking in the growth of the surrounding textures, have remained persistent during life. The duct which unites them is sometimes of considerable size, and is prolonged for some distance downwards, in the broad ligament. Its more developed form in some animals constitutes the duct of Gaertner, afterwards referred to as arising from a persistent condition of the Wolffian duct.

DEVELOPMENT OF THE URINARY ORGANS.

The *Wolffian Bodies* and their *Excretory ducts*.—The development of the genitourinary organs in reptiles, birds, and mammalia, including man, is preceded by the formation of two *temporary* glands, named after their discoverer, C. F. Wolff, the

Wolffian Bodies. In the embryos of the higher mammalia these organs are proportionally smaller, and disappear earlier than in those of the lower mammalia, birds, or reptiles. In the human subject, accordingly, the Wolffian bodies are relatively small, and are found only in an early stage of fœtal development. In the mammalian embryo, at a period when the intestinal canal still communicates with the umbilical vesicle by a wide orifice, the Wolffian bodies appear in the form of two slight ridges of blastema, placed one on each side of the line of attachment of the intestine to the vertebral column. On reaching their full size, which in man seems to take place about the fifth week of embryonic life, they have the appearance of two oblong reddish masses placed on the sides of the vertebral column, and extending from the lower end of the abdomen to the vicinity of the heart. Their structure is glandular; clear pedunculated vesicles may be early discovered in them, opening into an excretory

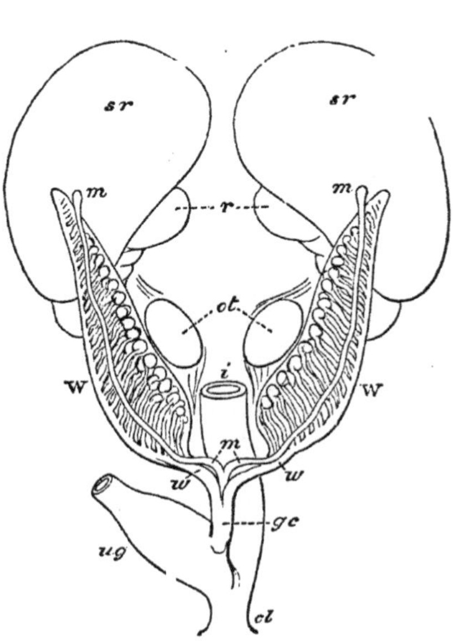

Fig. 691. — DIAGRAM OF THE WOLFFIAN BODIES, MÜLLERIAN DUCTS, AND ADJACENT PARTS, PREVIOUS TO SEXUAL DISTINCTION, AS SEEN FROM BEFORE.

s r, the suprarenal bodies; *r*, kidneys; *o t*, common blastema of ovaries or testicles; W, Wolffian bodies; *w*, Wolffian ducts; *m, m*, Müllerian ducts; *g c*, genital cord; *u g*, sinus urogenitalis; *i*, intestine; *c l*, cloaca.

Fig. 691.

duct which runs along the outer side of each organ. These vesicles subsequently become lengthened into transverse and somewhat tortuous cœcal tubes, which still retain a dilatation, like the capsule of a Malpighian body, at their inner extremity. The Wolffian bodies are highly vascular, their larger blood-vessels running between and parallel with the transverse tubules. In the embryo of the coluber natrix, Rathke first observed vascular tufts, which he compared to the Malpighian corpuscles of the kidneys; and since the time of his discovery, Malpighian tufts have been found in the Wolffian bodies of birds and mammals. The ducts of the two bodies open into the sac of the allantois, to be presently described.

A whitish secretion has been seen in the ducts of the Wolffian bodies of birds and serpents resembling the urine of those animals, and as the fluid of the allantois also has been found to contain uric acid, it is reasonable to think that the Wolffian bodies perform the office of kidneys during the early part of fœtal life. They are accordingly sometimes named the *primitive* or *primordial kidneys*.

As development advances, the Wolffian bodies rapidly become proportionally shorter and thicker: they shrink towards the lower part of the abdominal cavity, and soon become almost entirely wasted. By the middle of the third month only traces of them are visible in the human embryo. They take no part in the formation of the kidneys or suprarenal capsules, nor in that of the ovaries or body of the testes, but are connected with the origin of a part of the seminal passages in the male sex.

The *Kidneys* and *Ureters.*—The *kidneys* commence subsequently to and independently of the Wolffian bodies. They already exist about the seventh week, as two small

3 T

dark oval masses, situated behind the upper part of the Wolffian bodies, which are still large and completely hide the kidneys. Though at first smooth and oval, the kidneys soon assume their characteristic general outline, and about the tenth week are distinctly lobulated. The separate lobules, generally about fifteen in number, gradually coalesce in the manner already described; but at birth, indications of the original lobulated condition of the kidney are still visible on the surface, and the entire organ is more globular in its general figure than in the adult. The kidneys are then also situated lower down than in after-life.

The formative blastema of the kidney, as observed by Rathke in the fœtal calf, soon contains a series of club-shaped bodies which have their larger ends free and turned outwards, and their smaller ends or pedicles directed inwards towards the future hilus, where they are blended together. As the organ grows these bodies increase in number, and finally, becoming hollow, form the *uriniferous tubes*. At first, short, wide, and dilated at their extremities, the tubuli soon become elongated, narrow, and flexuous, occupying the whole mass of the kidney, which then appears to consist of cortical substance only. At a subsequent period, the tubuli nearest the hilus become straighter, and thus form the medullary substance. The tubuli, as shown by Valentin, are absolutely, as well as relatively, wider in the early stages of formation of the kidney. The Malpighian corpuscles have been seen by Rathke in a sheep's embryo, the kidneys of which measured only two and a half lines in length.

With regard to the mode of the first appearance of the pelvis and ureter, the statements of embryologists are very conflicting. The *ureters*, it is stated by Rathke, commence *after* the kidneys, and then become connected with the hilus of each organ, and with the narrow ends of the club-shaped bodies in its interior. At first, according to him, the growing tubuli do not seem to communicate with the cavity of the ureter; but, subsequently, when the wide upper portion of this canal or *pelvis* of the kidney has become divided to form the future *calyces*, the pencil-like bundles of the tubuli open into each subdivision of the ureter, and give rise at a later period to the appearance of the *papillæ* and their numerous orifices. The lower ends of the ureters soon come to open into that part of the sac of the allantois which afterwards becomes converted into the bladder. The researches of Müller and Bischoff are in general confirmatory of Rathke's account. Valentin believes that the ureter (which he has seen at the earliest periods), the pelvis of the kidney, and the uriniferous tubules, are formed in a general blastema, independently of one another; and that, each part first becoming separately hollowed out, their cavities afterwards communicate with each other. Bischoff states that the ureters appear at the same time as the kidneys, and are formed in continuity with the uriniferous tubules, and moreover that all these parts, which are at first solid, are excavated, not separately, but in common, in the farther progress of development. Lastly, according to Remak's observations on the chick, the kidneys of that animal commence as two hollow projections from the cloaca, internal to the ducts of the Wolffian bodies, which afterwards elongate and ramify so as to form both the ureters and kidneys.

In the advanced fœtus and in the new-born infant, the kidneys are relatively larger than in the adult, the weight of both glands, compared with that of the body, being, according to Meckel, about one to eighty at birth.

The *Suprarenal Bodies.*—These organs have their origin from blastema, independent both of the kidneys and of the Wolffian bodies. Valentin describes them as originating in a single mass, placed in front of the kidneys, and afterwards becoming divided. Meckel has also seen them partially blended together, Müller has found the suprarenal bodies in contact, but not united. Bischoff has always seen them separate, and in early conditions closely applied to the upper end of the Wolffian bodies. Kölliker has also observed them united by a bridge of substance, in which the splanchnic nerves were lost. From all this it is plain that the solar plexus and suprarenal capsules are closely united in the early fœtal state; but it by no means follows that they have a common origin.

In quadrupeds the suprarenal bodies are at all times smaller than the kidneys; but in the human embryo they are for a time larger than those organs, and quite conceal them. At about the tenth or twelfth week, the suprarenal bodies are smaller than the kidneys; at birth the proportion between them is 1 to 3, whilst in the adult it is about 1 to 22. They diminish in aged persons.

· *The Allantois, Urinary Bladder, and Urachus.*—The name of Allantois was originally given to a membranous sac which is appended to the umbilicus of various quadrupeds in the fœtal state, and which communicates with the urinary bladder by means of a canal passing through the umbilical aperture, and named the urachus. These several parts are formed out of one original saccular process, which passes out from the cloacal termination of the intestine, and which subsequently becomes distinguished into the bladder, the urachus, and the allantois strictly so called; but modern embryologists employ the term allantois also to signify the original common representative of the several parts referred to. In this sense, an allantois may be said to exist not only in mammalia, but also in birds and reptiles, subject, however, to great differences in its subsequent development and relative importance. Thus, in Batrachians it never extends beyond the abdominal cavity; in scaly reptiles, on the other hand, as well as in birds and in some quadrupeds, it ultimately surrounds the body of the fœtus, and spreads itself over the inner surface of the chorion or outer covering of the ovum, whilst in other quadrupeds, its extra-abdominal portion is of small extent. In man, the allantois proper is not only very insignificant in point of size, but also extremely limited in duration, for it vanishes at a very early period in the life of the embryo; and, whilst in many animals it serves both as a receptacle for the secretion of the fœtal urinary organs, and as a vehicle to conduct the umbilical vessels from the body of the embryo to the chorion to form the placenta (or some equivalent vascular structure), it seems in the human species to serve merely for the latter purpose. The allantoid process communicates below with the intestinal canal, and receives the wide excretory ducts of the Wolffian bodies, the ureters, and the Fallopian tubes or vasa deferentia. By Baër, Rathke, and others, the allantois has been regarded as formed from the intestinal tube, and by Reichert as developed upon the excretory ducts of the Wolffian bodies. Bischoff states that, in the embryos of the rabbit and dog, it commences before the appearance of either the Wolffian bodies or the intestine, as a solid mass projecting forwards from the posterior extremity of the body. This mass soon becomes hollowed into a vesicle, which is covered with blood-vessels, and communicates with the intestine. Continuing rapidly to enlarge, it protrudes between the visceral plates, and, when these close together, through the opening of the umbilicus, forming in the rabbit a pear-shaped sac, which conveys blood-vessels (soon recognised as the umbilical vessels) to the chorion, to form the fœtal part of the placenta.

In the human embryo, the allantois ceases, at a very early period, to be found beyond the umbilicus, and in the lower part of its extent, within the abdomen, it becomes widened to form the bladder, whilst the upper part, or urachus, becomes constricted, and is at length completely closed, and remains only in the form of a ligament, with minute traces of its original hollow structure already described along with the urinary bladder.

The lower part of the allantois, or rudimentary bladder, receiving, as already mentioned, the efferent canals of the Wolffian bodies, as well as those of the kidneys and of the ovaries or testes, at first communicates freely with the lower end of the intestinal tube, and when this becomes opened to the exterior, there is formed a sort of cloaca, as in adult birds and reptiles. Soon, however, a separation takes place, so as to produce for the genito-urinary organs a distinct passage to the exterior: this is named the *sinus urogenitalis,* and is situated in front of the termination of the intestine.

DEVELOPMENT OF THE ORGANS OF GENERATION.

The development of the parts concerned in the reproductive function does not begin until after the rudiments of the principal organs of the body have appeared. The internal organs of generation first make their appearance, and for a brief period no sexual difference is perceptible in them. The external organs, which subsequently begin to be formed, are also identical in appearance in the two sexes, as late as the fourteenth week.

The *internal Organs of Generation.*—The *Ovaries* and *Testes.*—The rudiments of the ovaries or testes, for it cannot at first be determined which are ultimately to be produced, appear after the formation of the allantois and Wolffian bodies, but a little sooner than the kidneys. They consist of two small whitish oval masses of blastema placed on the inner border of the Wolffian bodies. At first, they are placed near to one another, and parallel; the Wolffian bodies being at that time large, and occupying the whole posterior part of the abdomen. But as the kidneys grow, above and internal to the Wolffian bodies, the latter are displaced outwards, and with them the reproductive organs. At this time the sex becomes distinguishable; for in the female the ovary becomes elongated and flattened, and it assumes at first an oblique and then a nearly transverse direction; whereas, in the male, the testis becomes

Fig. 692.

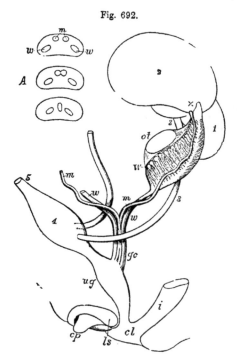

Fig. 692. — DIAGRAM OF THE PRIMITIVE URINARY AND SEXUAL ORGANS IN THE EMBRYO PREVIOUS TO SEXUAL DISTINCTION.

The parts are shown chiefly in profile; the kidney and suprarenal body of the right side are omitted, and the Müllerian and Wolffian ducts are shown from the front. 1, left kidney; 2, suprarenal body; 3, ureter; 4, urinary bladder; 5, urachus; *o t*, the mass of blastema from which ovary or testicle is afterwards formed; W, left Wolffian body; ×, part at the apex from which the coni vasculosi are afterwards developed in new blastema; *w, w,* right and left Wolffian ducts; *m, m,* right and left Müllerian ducts uniting together and with the Wolffian ducts in *g c,* the genital cord; *u g,* sinus uro-genitalis; *i,* lower part of the intestine; *c, l,* common opening of the intestine and urogenital sinus; *c p,* elevation which becomes clitoris or penis; *l s,* ridge from which the labia majora or scrotum is formed.

rounder and thick, and, together with the Wolffian body, retains its vertical position, although displaced downwards and outwards. Subsequently the tubuli seminiferi are developed within the testis, and ova in the superficial strata of the ovary.

Uterus and Fallopian tubes: Epididymis and Vasa deferentia.—The excretory duct of each Wolffian body lies from the first along its outer border, and in the succeeding part of its course is continued down from the extremity of the Wolffian body to the sinus urogenitalis. As the Wolffian body begins to change its position, and at the same time decrease in size, a white thread of blastema appears on the front of that body, and runs along the inner side of the Wolffian duct in its whole course; this forms the *Müllerian duct.* At the upper extremity of the Wolffian body, and close to the white thread, there is likewise developed a pyramidal mass of blastema, occupying the position originally held by the upper tubules of that gland, which seem to be absorbed to give place to it. The Müllerian duct, commencing by a slightly dilated extremity, descends in front of the excretory duct, to the lower end of the Wolffian body, where it dips down in front of that body, then turns over that duct so as to get behind it, and on arriving at the middle line comes in contact with its fellow of the opposite side, with which, and with the lower parts of the Wolffian ducts, it is united into a single cord, named the *genital cord.* The Müllerian ducts become as it were fused together, first at the upper and lower parts of the genital cord, and subsequently through its whole extent, while the Wolffian ducts remain separate to their ends.

Another structure is likewise seen at this stage in connection with the Wolffian body in both sexes, viz. an elevation of peritoneum, with tissue enclosed, which extends from the lower end of the testis or ovary to the point where the excretory and Müllerian ducts quit their contact with the lower extremity of the Wolffian body, and, there becoming stronger, extends onwards from this point to the processus vaginalis or canal of Nuck. The further development of these parts in each of the two sexes requires a separate description.

Fig. 693. — DIA-
GRAM OF THE FE-
MALE TYPE OF
SEXUAL ORGANS.

Fig. 693.

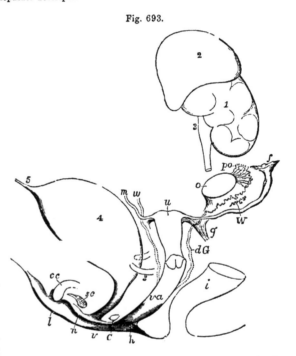

This and the following figure represent diagrammatically a state of the parts not actually visible at one time; but they are intended to illustrate the general type in the two sexes, and more particularly the relation of the two conducting tubes to the development of one as the natural passage in either sex, and to the natural occurrence of vestiges of the other tube, as well as to the persistence of the whole or parts of both tubes in occasional instances of hermaphroditic nature.

1, the left kidney; 2, suprarenal body; 3, ureter, of which a part is removed to show the parts passing within it; 4, urinary bladder; 5, urachus; o, the left ovary nearly in the place of its original formation; p o, parovarium; W, scattered remains of Wolffian tubes near it; d G, remains of the left Wolffian duct, such as give rise to the duct of Gaertner, represented by dotted lines; that of the right side cut short is marked w; f, the abdominal opening of the left Fallopian tube; u, the upper part of the body of the uterus, presenting a slight appearance of division into cornua; the Fallopian tube of the right side cut short is marked m; g, round ligament, corresponding to gubernaculum; i, lower part of the intestine; v a, vagina; h, situation of the hymen; C, gland of Bartholin (Cowper's gland), and immediately above it the urethra; c c, corpus cavernosum clitoridis; s c, vascular bulb or corpus spongiosum; n, nympha; l, labium; v, vulva.

In the female, the vagina, uterus, and Fallopian tubes are formed out of the Müllerian ducts. That portion of the ducts in which they become fused together is developed into the vagina, the cervix, and part of the body of the uterus; and the peculiarity of the mode of fusion accounts for the occurrence, as a rare anomaly, not only of double uterus, but of duplicity of the vagina, coincident with communication between two lateral halves of the uterus. The part of the Müllerian duct extending from the base of the Wolffian body, to the point where the two ducts meet, constitutes in animals with horned uteri, the cornu of the uterus; but in the human subject it remains comparatively short, entering into the formation of the upper part of the organ. The remaining upper portion of the Müllerian duct constitutes the Fallopian tube—becoming at first open and subsequently fringed at its upper extremity. In the peritoneal elevation between the ovary and the base of the Wolffian body the fibrous ligament of the ovary is developed, while that part which proceeds onwards to the canal of Nuck becomes the round ligament of the uterus. The excretory ducts of the Wolffian bodies. disappear in the human female, but in the pig and some ruminants they persist as the canals of Gaertner. The parovarium is generally believed to consist of the vestiges of some of the tubules of the Wolffian body, but it is held by Banks to

Fig. 694.

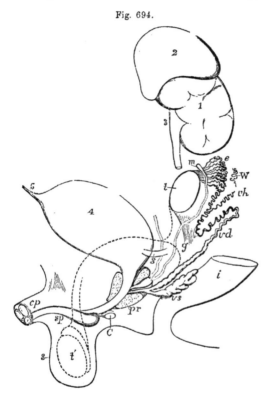

Fig. 694.—Diagram of the Male Type of Sexual Organs.

1, 2, 3, 4, and 5, as in the preceding figure ; t, testicle in the place of its original formation ; e, caput epididymis ; v d, vas deferens ; W, scattered remains of the Wolffian body constituting the organ of Giraldés ; v h, vas aberrans ; m, Müllerian duct, the upper part of which remains as the hydatid of Morgagni, the lower part represented by a dotted line as descending to the sinus pocularis constitutes the cornu and tube of the uterus masculinus ; g, the gubernaculum ; v s, the vesicula seminalis ; p r, the prostate gland ; C, Cowper's gland of one side ; c p, corpora cavernosa penis cut short ; s p, corpus spongiosum urethræ ; s, scrotum ; t', together with the dotted lines above, indicates the direction in which the testicle and epididymis change place in their descent from the abdomen into the scrotum.

owe its origin to a development of tubes in the whitish blastema previously mentioned, which appears in connection with the upper part of that body when it begins to shrink, and which, in the male, gives rise to the upper part of the epididymis ; and in this view Allen Thomson is disposed to concur.

In the male, the Müllerian ducts are destined to undergo little development and are of no physiological importance, while the ducts of the Wolffian bodies, and probably also some part of their glandular substance, form the principal part of the excretory apparatus of the testicle. The united portion of the Müllerian ducts remains as the vesicula prostatica, which accordingly not only corresponds with the uterus, as was shown by Weber, but likewise, as pointed out by Leuckart, contains as much of the vagina as is represented in the male. In some animals the vesicula prostatica is. prolonged into cornua and tubes; but in the human subject the whole of the ununited parts of the Müllerian ducts disappear, excepting, as suggested by Kobelt, their upper extremities, which seem to be the source of the hydatids of Morgagni. The excretory duct of the Wolffian body, from the base of that body to its orifice, is converted into vas deferens and ejaculatory duct, the vesicula seminalis being formed as a diverticulum from its lower part.

With respect to the formation of the epididymis, our information is not altogether complete. According to the greater number of the most recent observations, it appears certain that the larger convoluted seminal tube, which forms the body and globus minor of the epididymis, arises by a change or adaptation of that part of the Wolffian duct which runs along the outer side of the organ. The vas aberrans or vasa aberrantia of Haller appear to be the remains also in a more highly convoluted form of one or more of the tubes of the Wolffian body still adhering to the excretory duct of the organ, and their communication with the main tube of the epididymis receives an explanation from that circumstance. But there are no direct observations on record of the process of conversion of these fœtal structures into the permanent forms. As to the coni vasculosi in the upper part of the epididymis, still more doubt has prevailed. Since Müller's discovery in birds of the collateral duct named after him, and the extension of this discovery to mammals, it has been customary to regard the upper part of the epididymis as produced by a transformation of the tubes and duct in the upper part of the Wolffian body, according to the views most fully given by Kobelt; but doubts have been entertained by some as to the correctness of this view, and more recent observations by Banks appear to prove that it must, in some degree, be modified.

, According to Banks, the origin of the coni vasculosi is due to a process of development occurring in a new structure or mass of blastema which had been previously observed by Cleland, and which is deposited at the upper end of the Wolffian body, and close to the Müllerian duct. Within this blastema Cleland showed that the tubes of the efferent seminal vessels and the coni vasculosi, together with the tube which counects them, are formed anew, while the tubes of the adjacent part of the Wolffian body are undergoing an atrophic degeneration. This has been confirmed by the detailed observations of Banks, who has further shown the continuity of their uniting tube with the Wolffian excretory duct.

Should this view prove to be correct, the caput epididymis must be regarded not simply as a conversion of the upper part of the Wolffian body, but rather as a new formation or superinduced development in blastema connected with it.

The coni vasculosi, so formed, become connected with the body of the testicle by means of a short straight cord, which is afterwards subdivided into the vasa efferentia. The peritoneal elevation descending from the testis towards the lower extremity of the Wolffian body, is the upper part of the plica gubernatrix, and becomes shortened as the testicle descends to meet the lower end of the epididymis; the peritoneal elevation which passes down into the scrotum, and is continuous with the other, is the more important part of the plica gubernatrix, connected with the gubernaculum testis. The spermatic artery is originally a branch of one of those which go to the Wolffian body, and ascends from the surface of the Wolffian body to the upper part of the testis, along the ligaments connecting them; but, as the testis descends, the artery lies entirely above it, and the secreting substance of the Wolffian body remains adherent to it; and hence it is that the organ of Giraldès, which consists of persistent Wolffian tubules, is found in a position superior to the epididymis. (For a fuller account of this complicated subject the student is referred to Banks " On the Wolffian Bodies." Edin. 1864.)

The *descent of the testicles* is a term applied to the passage of the testes from the abdominal cavity into the scrotum. The testicle enters the internal inguinal ring in the seventh month of fœtal life: by the end of the eighth month it has descended into the scrotum, and, a little time before birth, the narrow neck of the peritoneal pouch, by which it previously communicated with the general peritoneal cavity, becomes closed in the manner elsewhere described (p. 965), and the process of peritoneum, now entirely shut off from the abdominal cavity, remains as an independent serous sac. The peritoneal pouch, or processus vaginalis, which passes down into the scrotum, precedes the testis considerably in its descent, and into its posterior part there projects a considerable columnar elevation already alluded to, which is filled with soft tissue, and is termed *plica gubernatrix*. There is likewise a fibrous structure attached inferiorly to the lower part of the scrotum, and surrounding the peritoneal pouch above, which may be distinguished as the *gubernacular cord*, both this and the plica gubernatrix being included in the general term *gubernaculum testis* (J. Hunter). The gubernacular cord consists ef fibres which pass downwards from the sub-peritoneal fascia, others which pass upwards from the superficial fascia and integument, and others again which pass both upwards and downwards from the internal oblique muscle and the aponeurosis of the external oblique; it exhibits, therefore, a fusion of ·the layers of the abdominal wall. Superiorly, it surrounds the processus vaginalis, without penetrating the plica gubernatrix; and the processus vaginalis, as it grows, pushes its way down through the gubernacular cord and disperses its fibres. By the time that the testis enters the internal abdominal ring, the processus vaginalis has reached a considerable way into the scrotum; and, as the testis follows, the plica gubernatrix becomes shorter, till it at last disappears; but it cannot be said that the shortening of the plica is the cause of the descent of the testicle, and much less (as has been held by some) that the muscular fibres of the gubernacular cord are the agents which effect this change of position. The arched fibres of the cremaster muscle make their appearance on the surface of the processus vaginalis as it descends, while its other fibres are those which descend in the gubernacular cord. (See for a further account of this process, and the various views which have been held with regard to the descent of the testicles, Cleland, "Mechanism of the Gubernaculum Testis." Edinburgh, 1856.)

The *External Organs of Generation.*—In the human subject, these have for some time the same form in both sexes; but, in animals in which the penis is prolonged to the umbilicus, that circumstance forms one of the very earliest sexual distinctions, inasmuch as the clitoris hangs free.

Fig. 695.

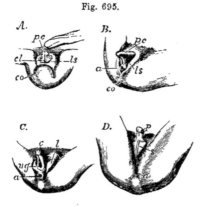

Fig. 695.—DEVELOPMENT OF THE EXTERNAL SEXUAL ORGANS IN THE MALE AND FEMALE FROM THE INDIFFERENT TYPE (from Ecker).

A, the external sexual organs in an embryo of about nine weeks, in which sexual distinction is not yet established, and the cloaca still exists : B, the same in an embryo somewhat more advanced, and in which, without marked sexual distinction, the anus is now separated from the urogenital aperture : C, the same in an embryo of about ten weeks, showing the female type : D, the same in a male embryo somewhat more advanced. Throughout the figures the following indications are employed; *p c*, common blastema of penis or clitoris ; to the right of these letters in A, the umbilical cord ; *p*, penis ; *c*, clitoris ; *cl*, cloaca ; *ug*, urogenital opening ; *a*, anus ; *l s*, cutaneous elevation which becomes labium or scrotum ; *l*, labium ; *s*, scrotum ; *c o*, coccygeal elevation.

Up to the fifth week, according to Tiedemann, there is no separate genito-urinary orifice, and indeed no anus. Previous to this period, or about the beginning of the fourth week, there is a common opening, for the intestine, the generative, and the urinary organs, *i. e.*, a *cloacal* aperture. In front of this simple opening, there soon appears a small recurved projecting body, which, as it enlarges, becomes grooved along the whole of its under surface. This is the rudimentary *clitoris*, or *penis*, at the summit ·of which an enlargement is formed which becomes the *glans*. The margins of the groove seen on its under surface are continued backwards on each side of the common aperture, which is now elliptical, and is bounded laterally by two large cutaneous folds. Towards the tenth or eleventh week a transverse band, the commencing *perinæum*, divides the anal orifice from that of the genito-urinary passage, which latter now appears as a rounded aperture, placed below the root of the rudimentary clitoris or penis, and between the prolonged margins of the groove beneath that organ. This opening, but not the clitoris or penis, is concealed by the large cutaneous folds already mentioned. In this condition, which continues until the twelfth week, the parts appear alike in both sexes, and resemble very much the perfect female organs. The rudiments of *Cowper's glands* are, it is said, seen at an early period, near the root of the rudimentary clitoris or penis, on each side of the genito-urinary passage.

In the female, the two lateral cutaneous folds enlarge, so as to cover the clitoris and form the *labia majora*. The clitoris itself remains relatively smaller, and the groove on its under surface less and less marked, owing to the opening out and subsequent extension backwards of its margins to form the *nymphæ*. The *hymen* begins to appear about the fifth month. Within the nymphæ, the urethral orifice, as already mentioned, becomes distinct from that of the vagina.

In the male, on the contrary, the *penis* continues to enlarge, and the margins of the groove along its under surface gradually unite from the primitive urethral orifice behind, as far forward as the glans, so as to complete the long canal of the male *urethra*. This is accomplished about the fifteenth week. When this union remains incomplete, the condition named hypospadias is produced. In the meantime the *prepuce* is formed, and, moreover, the lateral cutaneous folds also unite from behind forwards, along the middle line or *raphé*, and thus complete the *scrotum*, into which the testicles do not descend until the last month of fœtal existence.

The following tabular scheme of the corresponding parts of the genitourinary organs in the two sexes, and of their relation to the formative rudiments of the common embryonic type, may be useful in fixing attention on the more important points of the subject.

FEMALE.	COMMON EMBRYONIC.	MALE.
Ovary	I.—Common reproductive gland	Body of Testicle.
	II.—Wolffian body.	
	1. New blastema at upper part	Coni vasculosi.
Parovarium	2. Tubular substance of the gland	Organ of Giraldés. Vasa aberrantia.
Irregular vestiges near parovarium (?)	3. Excretory duct along the gland	Convoluted tube of epididymis.
Duct of Gaertner in some animals	4. Duct below the gland	Vas deferens.
	III.—Duct of Müller.	
Fimbriæ and Fallopian tube	1. Upper end and part along the Wolffian gland	Hydatid of Morgagni and other vestigial vesicles.
Cornu uteri	2. Free part of duct	Cornu of sinus pocularis in some animals.
Uterus and vagina	3. Fused part of both ducts in the genital cord	Sinus pocularis.
Female urethra	IV.—Pedicle of the allantois	Upper portion of prostatic part of urethra.
Vestibule	V.—Sinus urogenitalis	Lower portion of prostatic with membranous part of urethra.
Glands of Bartholin	Common blastema	Glands of Cowper.
	VI.—Common sexual prominence and integumental folds.	
Corpora cavernosa clitoridis	1. Common blastema	Corpora cavernosa penis.
Labia majora	2. Outer integumental folds	Scrotum.
Nymphæ	3. Inner integumental folds	Integument of lower surface of penis.
Vestibular bulbs and other erectile tissue	4. Common blastema	Bulb and corpus spongiosum urethræ.
	VII.—Peritoneal folds and gubernacular bands.	
Canal of Nuck	1. Inguinal peritoneal pouch	Processus vaginalis.
Ovarian ligament	2. Band from genital gland to base of Wolffian body	Tissue connecting testicle and globus minor.
Round ligament of uterus	3. Band descending from base of Wolffian body	Gubernaculum testis.

MAMMARY GLANDS.

The mammary glands (mammæ), the organs of lactation in the female, are accessory parts to the reproductive system. They give a name to a large class of animals (Mammalia), which are distinguished by the possession of these organs. When fully developed in the human female, they form, together with the integuments and a considerable quantity of fat, two rounded eminences (the breasts) placed one at each side on the front of the thorax. These extend from the third to the sixth or seventh rib, and from the side of the sternum to the axilla. A little below the centre of each breast, on a level with the fourth rib, projects a small conical body named the *nipple* (mammilla), which points somewhat outwards and upwards. The surface of the nipple is dark, and around it there is a coloured circle or *areola*, within which the skin is also of a darker tinge than elsewhere. In the virgin, these parts are of a rosy pink colour, but they are always darker in women who have borne children. Even in the second month of pregnancy, the areola begins to enlarge and acquire a darker tinge ; these changes go on increasing as gestation advances, and are regarded as reliable signs in judging of suspected pregnancy. After lactation is over, the dark colour subsides, but not entirely. The skin of the nipple is marked with many wrinkles, and is covered with papillæ ; besides this, it is perforated at the tip by numerous foramina, which are the openings of the lactiferous ducts : and near its base, as well as upon the surface of the areola, there are scattered rounded elevations, which are caused by the presence of little glands with branched ducts, four or five of which open on each elevation. The tissue of the nipple contains a large number of vessels, together with much plain muscular tissue, and its papillæ are highly sensitive ; it is capable of a certain degree of erection from mechanical excitement, which may be partly caused by turgescence of its vessels, but is probably due, in greater part, to contraction of the muscular fibres.

The base of the mammary gland, which is nearly circular, is flattened, or slightly concave, and has its longest diameter directed upwards and outwards towards the axilla. It rests on the pectoral muscle, and is connected to it by a layer of areolar tissue. The thickest part of the gland is near the centre, opposite the nipple, but the full and even form of the breasts depends chiefly on the presence of a large quantity of fat, which lies beneath the skin, covers the substance of the gland, and penetrates the intervals between its lobes and lobules. This fatty tissue, which is of a bright yellow tinge and rather firm, is divided into lobulated masses by numerous laminæ of fibrous or very dense areolar tissue, which are connected with the skin on the one hand, and on the other with the firm areolar investment of the gland itself, which investment is connected behind by similar laminæ with the areolar membrane covering the pectoral muscle ; these laminæ serve to support the gland. Beneath the areola and the nipple there is no fat, but merely the firm areolar tissue and vessels surrounding the lactiferous ducts.

Structure.—The mammary gland consists of a number of distinct glandular masses or lobes, each having a separate excretory duct, held together by a very firm intervening fibrous or areolar tissue, and having some adipose tissue penetrating between them. Each of these divisions of the gland is again subdivided into smaller lobes, and these again into smaller and smaller lobules, which are flattened or depressed, and held together by areolar tissue, blood-vessels, and ducts. The substance of the lobules, especially as contrasted with the adjacent fat, is of a pale, reddish cream-colour, and is rather firm. It is composed principally of the vesicular commencements of the lactiferous ducts, which appear like clusters of minute rounded cells, having a diameter from ten to thirty times as great

Fig. 696.

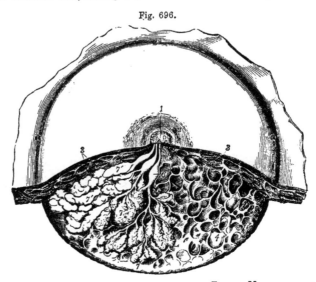

Fig. 696.—Dissection of the lower half of the Female Mamma during the period of Lactation (from Luschka). ⅔

In the left-hand side of the dissected part the glandular lobes are exposed and partially unravelled ; and in the right-hand side, the glandular substance has been removed to show the reticular loculi of the connective tissue in which the glandular lobules are placed : 1, upper part of the mammilla ; 2, areola ; 3, subcutaneous masses of fat ; 4, reticular loculi of the connective tissue which support the glandular substance and contain the fatty masses ; 5, one of three lactiferous ducts shown passing towards the mammilla where they open ; 6, one of the sinus lactei or reservoirs ; 7, some of the glandular lobules which have been unravelled ; 7', others massed together.

as that of the capillary vessels by which they are surrounded. These cells open into the smallest branched ducts, which, uniting together to form others of larger size, finally end in a single excretory canal corresponding to one of the chief subdivisions of the gland. The canals proceeding thus from the principal lobes composing the gland are named the *galactophorous ducts*, and are from fifteen to twenty in number ; they converge towards the areola, beneath which they become considerably dilated, especially during lactation, so as to form *sacs* or *sinuses* two or even three lines wide, which serve as temporary though small reservoirs for the milk. At the base of the nipple all these ducts, again reduced in size, are assembled together, those in the centre being the largest, and then proceed side by side, surrounded by areolar tissue and vessels, and without communicating with each other, to the summit of the mammilla, where they open by separate orifices ; these orifices are seated in little depressions, and are smaller than the ducts to which they respectively belong. The walls of the ducts are composed of areolar tissue, with longitudinal and circular elastic filaments.

The mucous membrane is continuous with the common integument at the orifices of the ducts; its epithelium is scaly or tesselated, and in the smallest ducts and their ultimate vesicles consists of cells having a diameter very little exceeding that of their nuclei.

Fig. 697.

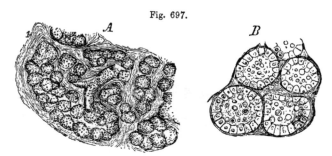

Fig. 697.—MAGNIFIED VIEWS OF THE GLANDULAR SUBSTANCE OF THE MAMMA DURING THE PERIOD OF LACTATION (from Henle).

A, section of a small lobule of the gland, magnified 60 diameters; 1, stroma of connective tissue supporting the glandular tissue; 2, terminal ramuscule of one of the gland-tubes; 3, glandular vesicles.

B, four glandular vesicles magnified 200 diameters, showing the lining epithelial cells and some milk-globules within them.

Blood-vessels and Nerves.—The *arteries* which supply the mammary glands are the long thoracic and some other branches of the axillary artery, the internal mammary, and the subjacent intercostals. The *veins* have the same denomination. Haller described a sort of anastomotic venous circle surrounding the base of the nipple as the *circulus venosus*. The *nerves* proceed from the anterior and middle intercostal cutaneous branches.

In the *male*, the mammary gland and all its parts exist, but quite in a rudimentary state, the gland itself measuring only about six or seven lines across, and two lines thick, instead of four inches and a half wide and one and a half thick, as in the female. Occasionally the male mamma, especially in young subjects, enlarges and pours out a thin watery fluid; and, in some rare cases, it has secreted milk.

Varieties.—Two or even three nipples have been found on one gland. An additional mamma is sometimes met with, and even four or five have been observed to co-exist; the supernumerary glands being most frequently near the ordinary pair, but sometimes in a distant part of the body, as the axilla, thigh, or back.

DIVISION II.

SURGICAL ANATOMY.

I.—SURGICAL ANATOMY OF THE ARTERIES.

In the description of the several blood-vessels, the points bearing on operative surgery have been indicated in detail. The leading facts to be attended to by the surgeon in the operation of placing a ligature on the chief arterial trunks will be now collectively considered.*

SURGICAL ANATOMY OF THE COMMON CAROTID ARTERY.

The common carotid artery does not furnish any branch, save in very rare instances. In a practical or surgical point of view, the branches arising sometimes close to its upper end may be disregarded, so that a ligature can be applied to any part of the vessel, except immediately at its commencement or termination. When the case is such as to allow a choice, the point which combines the most favourable circumstances for the operation, is opposite the lower end of the larynx. Here a large space would, in ordinary cases, intervene between the ligature and the ends of the vessel; and at the same time this part is free from the difficulties offered by the muscles lower down, and by the superior thyroid veins, if the artery be secured near its bifurcation. But it has been shown (p. 345) that the carotid artery occasionally bifurcates below the usual position—opposite the lower margin of the larynx, and even, however rarely, lower than this. In such cases, should the artery be laid bare at the point of division, it would be best to tie the two parts separately, close to their origin, in preference to tying the common trunk near its end. If, in consequence of very early division of the common carotid or its entire absence (cases which, however, are of extremely rare occurrence), two arteries (the external and internal carotids) should happen to come into view in the operation supposed, the most judicious course would doubtless be to place the ligature on that artery, which, upon trial, as by pressure, should prove to be connected with the disease.

In performing the operation, the direction of the vessel and the inner margin of the sterno-mastoid muscle are the surgeon's guides for the line of incision. Before dividing the integument it is well to ascertain whether the anterior jugular vein be in the line of incision. Should the operation be performed at the lower part of the neck, some fibres of the muscles will require to be cut across in order to lay the artery bare with facility; and the necessity for this step increases in approaching towards the clavicle. After the super-

---

* The plates referred to in this section are those of Richard Quain "On the Arteries."

ficial structures have been divided, assistance will be derived from the trachea or the larynx, as well as from the pulsation, in determining the exact situation of the artery. The trachea, from its roughness, may be

Fig. 698.

Fig. 698.—View of the Right Common Carotid and Subclavian Arteries, with the Origin of their Branches and their Relations (from R. Quain)   ⅓

. *e*, front of the hyoid bone ; *f*, thyroid cartilage; *g*, isthmus of the thyroid gland ; *h*, the trachea above the inter-clavicular notch of the sternum ; *i, i′*, the sawn ends of the clavicle, the portion between them having been removed ; *k*, the first rib ; *m*, scalenus medius; *p*, on the longus colli muscle, pointing to the pneumogastric nerve ; IV, the uppermost of the nerves of the axillary plexus ; A, the innominate artery ; 1, right common carotid artery ; 1′, placed on the left sterno-thyroid muscle, points to a part of the left common carotid ; 2, internal carotid ; 2′, upper part of the internal jugular vein, which has been removed between *i*, and 2′; 3, and 4, external carotid ; 3, is placed at the origin of the superior thyroid artery ; 4, at that of the lingual ; 5, the superior thyroid artery ; 5′, the thyroid or glandular branch ; 8, the first part, 8′, the third part of the arch of the subclavian artery ; 8″, the subclavian vein separated from the artery by the scalenus anticus muscle ; 9, is placed on the scalenus anticus muscle in the angle between the transversalis colli and suprascapular branches of the thyroid axis ; 10, outer part of the supra-scapular artery ; 10′, transverse cervical branches passing into the deep surface of the trapezius ; 10″, the posterior scapular artery, represented as rising directly from the third part of the subclavian artery, and passing through the axillary plexus of nerves and under the levator anguli scapulæ ; 11, on the scalenus anticus muscle, points to the inferior thyroid artery near the place where the ascending muscular artery of the neck is given off ; the phrenic nerve lies on the muscle to the outside ; at *i*, the suprasternal twig of the suprascapular artery is shown.

readily felt in the wound, even while the parts covering it have still some thickness. The sheath of the vessels is to be opened over the artery—near the trachea—for thus the jugular vein is most easily avoided. This vein, should it lie in front of the artery, as it sometimes does on the left side, and especially at the lower part of the neck on that side, will be a source of much difficulty in completing the operation, i. e., in passing the aneurism needle with the ligature about the artery. To surmount the difficulty much caution is required. The operator will find it advantageous to have the circulation in the vein (which in such operations becomes turgid and very large) arrested at the upper end of the wound by means of an assistant's finger. In most cases, if not in all, it is best to insert the aneurism needle conveying the ligature on the outer side of the artery, for thus the vagus nerve and the jugular vein will be most effectually avoided.

SURGICAL ANATOMY OF THE SUBCLAVIAN ARTERIES.

The subclavian artery is so deeply placed, its connections with important parts are so intimate and varied, and the branches are so large in proportion to the length of the trunk, that operations on this vessel present, in most cases, considerable difficulties to the surgeon. But the difficulties, it will be found, vary in different cases.

The last division of the artery, that beyond the anterior scalenus muscle (p. 366), is the part which is most favourably circumstanced for the application of a ligature in the case in which such an operation is most frequently called for, namely, aneurism affecting the artery in the axilla. This part is preferable chiefly because the vessel is here nearest to the surface, and most remote from the origin of the large branches. But, though the subclavian artery appears to be easy of access above the clavicle while the parts are in their natural position, it is to be remembered that, when an aneurism exists in the axilla, the clavicle may be so much elevated in consequence of the presence of the tumour, as to be placed in front of the vessel, or even above it. In such circumstances, the artery lies at a great depth, and at the same time the structures in front and behind it (the clavicle on the one hand, the vertebræ with the muscles covering them on the other hand), cannot, in any degree, be drawn asunder to facilitate the steps of the operation. It is when the outer part of the clavicle is thus raised from the ordinary horizontal position, that the height to which the artery arches above the bone becomes a point of importance. In most cases it happens that a portion of the artery is a short distance (about an inch) above the clavicle [plate 3] ; but occasionally, as before mentioned (p. 367), it rises much higher [plate 20, fig. 3]; or it may be lower than usual, lying close behind the bone [fig. 2]. If, in a case rendering the operation necessary, the clavicle should be unusually raised, the accessibility of the vessel in the neck will differ in these several conditions : in one, the artery could be arrived at only by proceeding from above downwards behind the bone : in another, a part of it would still be higher than the bone. This will serve, in part at least, to account for differences in the time which the operation for tying the subclavian artery has occupied in the hands of different surgeons, and even in the hands of the same surgeon in different cases.*

The principal facts bearing on the actual performance of an operation on the third part of the subclavian artery, will now be briefly recalled. The most prominent or convex part of the clavicle, the part of the bone opposite which the vessel lies, will serve as a guide for the middle of the first incision, which is to be made a little above the clavicle, and parallel with it. If (after noting with the eye, or marking on the surface the line at which it is desired to make the incision), the integument be drawn downwards over the clavicle, the parts covering the bone may be divided with freedom.

: With the integument, the platysma and several nerves are divided in this incision, but no vessel is endangered, except in those rare cases in which the cephalic vein or the external jugular crosses over the clavicle [plate 25, figs. 4, 5]. It will, in most cases, be an advantage to add a short vertical incision, directed downwards to the middle of the horizontal one. Should the sterno-mastoid muscle be broad at its lower end, or should the interval between that muscle and the trapezius be insufficient for the farther steps of the operation, a portion of the former muscle, or even of both muscles, must be divided [plate 25, fig. 7].

The external jugular vein next presents itself with the veins joining it from the shoulder, and, as this vein is usually over the artery, it must be held aside, or it may be necessary to divide it. If divided, the lower end of the vessel requires the application of a ligature as well as the upper one, in consequence of the reflux of blood from the subclavian vein. The omo-hyoid muscle will be turned aside if necessary ; and now must be determined the exact position at which the artery is to be sought by division of the deeper fascia. If the clavicle have its usual horizontal direction, the first rib is the best guide to the vessel. The brachial nerves are here, it is to be remembered, close to the vessel,—so much so, that the ligature has in several cases been passed in the first instance round one of them instead of the artery. But if, in consequence of the disease rendering the operation necessary, the outer end of the clavicle be much raised, then it will, in many cases, be more easy to place the ligature on the artery above the insertion of the scalenus muscle, or even behind that muscle. Above the first rib, the situation of the vessel may be ascertained by means of the brachial nerves and the scalenus muscle ; and, before the membrane covering them is divided, the position of these structures may be ascertained by the difference they offer to the touch. The cord-like nerves and the smooth flat muscle may thus be readily distinguished. At the same time the influence of pressure at a particular point in controlling the pulsation in the aneurism, will in this, as in other operations on the arteries, assist the surgeon.

* This statement will be illustrated by reference to two cases which occurred at nearly the same time in the practice of the same surgeon. In March, 1819, M. Dupuytren tied the subclavian artery for axillary aneurism, and the result was in all respects favourable. —See "Leçons orales," &c., t. iv.; and M. Marx in "Repert. général d'anatomie," &c. 1826.

Two or three weeks afterwards the same surgeon, being engaged in performing an operation of the same kind, was compelled to discontinue it for a time in consequence of the sufferings of the patient, and an hour and forty-eight minutes elapsed before the operation was concluded. The patient died of hæmorrhage in four days ; and, on examination after death, it was found that the artery had been perforated with the aneurism needle. One of the large nerves and half the artery had been included in the ligature. This case is reported by Dr. Rutherford, R.N., who was present at the operation, in "Edinburgh Med. and Surg. Journal," vol. xvi. 1820.

Before concluding the remarks on the third division of the artery, it should be mentioned that the suprascapular or transverse cervical artery may be met with in the operation, which in other cases may be complicated by the occurrence of a branch, or, however rarely, of branches, taking rise beyond the scalenus muscle.

The *second division of the subclavian artery* is the part which rises highest in the neck, and on this account it may be advantageously selected for the application of a ligature when the vessel is difficult of access beyond the muscle. The chief objection to operating on the artery in this situation arises from the contiguity of the large branches. Care is necessary in dividing the scalenus muscle to avoid the phrenic nerve and the internal jugular vein. Moreover, the fact of the entire of the subclavian artery being in apposition with the pleura, except where it rests on the rib, must be borne in mind.

Some difficulty may arise from a change in the position of the artery, as when it lies between the fibres of the anterior scalenus, or when it is in front of that muscle ; but such cases are of very rare occurrence, and the knowledge of the fact that the vessel may be thus displaced, will assist the surgeon in the event of difficulty arising from this cause.

Before it reaches the scalenus muscle the left subclavian artery [plate 2] may be said to be inaccessible for the application of a ligature, in consequence of its depth and its close connection with the lung and other structures calculated to create difficulty in an operation, among which may be mentioned the internal jugular and left innominate veins. To the difficulties resulting from the manner of its connection with the parts now named, must be added the danger of performing an operation in the neighbourhood of the large branches.

On the right side, though deeply placed and closely connected with important parts, the first division of the subclavian artery may be tied without extreme difficulty. But inasmuch as the length of the vessel, between its three large branches on the one hand, and the common carotid on the other, ordinarily measures no more than an inch, and often less, there is little likelihood of the operation in question being successfully performed in any case ; and the probability of success must be held to be still farther diminished when it is considered that the length of the free part of the artery is sometimes lessened by one of the large branches arising nearer than usual to its commencement.

In order to place a ligature on the portion of the right subclavian artery here referred to, it is necessary to divide by horizontal incisions the three muscles which cover it, together with the layers of fascia between and beneath them [plate 17, fig. 1]. While the muscles are being divided, a branch of the suprascapular artery will probably require to be secured [plate 16]. The position of the inner end of the clavicle and of the trachea, and the effect of pressure with the finger on the circulation in the aneurism or in the limb, will assist the surgeon in finding the artery without dissecting the surrounding parts to an unnecessary and injurious extent—a precaution of importance in all cases. In the farther steps of the operation, the exact position of the internal jugular vein, the vagus nerve, and the pleura, are to be well remembered.

The right subclavian artery is occasionally somewhat more deeply placed than usual in the first part of its course : and this occurs when it springs from the left side of the arch, or, more frequently, when it separates from the innominate behind the carotid [plate 20, fig. 4].

3 U

SURGICAL ANATOMY OF THE BRACHIAL ARTERY.

In the operation for tying the brachial artery, the known direction of the vessel, and the inner margin of the biceps muscle, chiefly aid in determining its position (p. 382). In consequence of the thinness of the parts which cover the artery, and the position of the basilic and median basilic veins with respect to it, even the integuments must be divided with care. After turning aside the superficial vein, should that be necessary, and dividing the fascia, the median nerve will probably come into view, and the artery will then be readily found. This is the course required under ordinary circumstances. But it may happen that, after dividing the fascia, it will be necessary to cut through a layer of muscular fibres in order to bring the artery into view [plate 37, figs. 3, 4, 5]. The influence of pressure with the finger, in controlling the circulation, will enable the surgeon to determine if

Fig. 699.

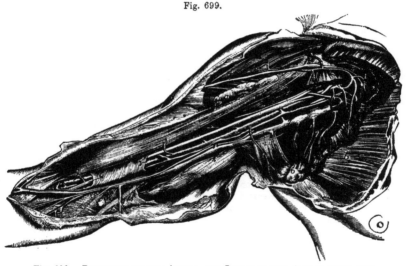

Fig. 699.—DISSECTION OF THE AXILLA AND INSIDE OF THE ARM TO SHOW THE AXILLARY AND BRACHIAL VESSELS (from R. Quain). ¼

The greater and lesser pectoral muscles have been divided so as to expose the axillary vessels : *a*, the inserted portion of the pectoralis major ; *b*, the pectoral portion ; 1, 1, axillary artery ; +, +, the median nerve formed by the two portions of the plexus which surround the artery; 1′, placed on a part of the sheath of the brachial vessels, and 1″, on the lower part of the biceps muscle, point to the brachial artery surrounded by its venæ comites ; 2, 2, axillary vein ; 3, 3, the basilic vein ; the upper figure is placed on the triceps muscle, the lower on the fascia near the junction of the ulnar vein : on the basilic vein are seen the ramifications of the internal cutaneous nerve ; 4, on the deltoid and 4′, on the clavicular part of the great pectoral muscle, mark the cephalic vein joining the acromio-thoracic and through it the axillary vein ; 5, 5, placed on the divided portions of the pectoralis minor, point to the origin and branches of the acromio-thoracic artery ; 6, placed on a group of axillary glands, indicates the alar thoracic and subscapular vessels ; 7, placed on the trunk of the axillary vein, points by a line to one of the venæ comites of the brachial vein, which being joined by the other higher up passes into the axillary vein ; the ulnar nerve is seen passing from below the basilic vein towards the inner condyle ; near 1, placed on the coraco-brachialis muscle is seen the musculo-cutaneous nerve before it passes through that muscle.

the vessel be behind the muscular fibres, and will guide him to the place at which they ought to be divided.

Again, as the brachial artery occasionally deviates from its accustomed place in the arm, it is prudent, before beginning an operation on the living body, to be assured of its position by the pulsation. Should the vessel be thus displaced, it has the ordinary coverings of the brachial artery, except at the lower part of the arm, where some fibres of the pronator teres will require to be divided in an operation for securing the vessel.

When the brachial artery is double, or when two arteries are present in the arm, both being usually placed close together, they are accessible in the same operation. The circumstance of one being placed over the fascia (should this very unfrequent departure from the usual arrangement exist) will become manifest in the examination which ought to be made in all cases before an operation is begun. And, as regards the occasional position of one of the two arteries beneath a stratum of muscular fibres, or its removal to the inner side of the arm (in a line towards the inner condyle of the humerus), it need only be added that a knowledge of these exceptional cases will at once suggest the precautions which are necessary, and the steps which should be taken when they are met with.—The foregoing observations have reference to operations on the brachial artery, above the bend of the elbow ; the surgical anatomy of the vessel opposite that joint requires a separate notice.

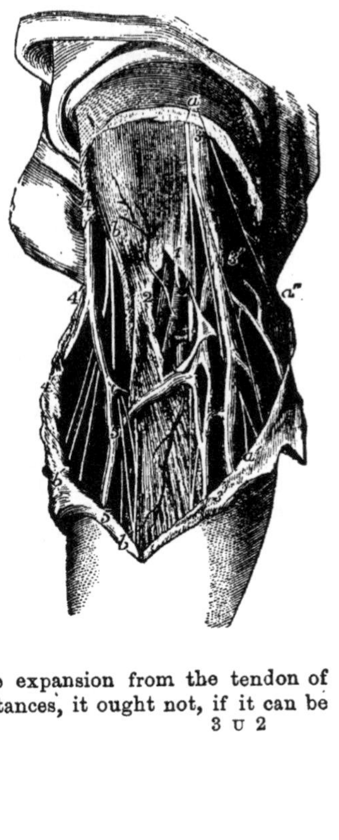

Fig. 700.—SUPERFICIAL DISSECTION OF THE BLOOD-VESSELS AT THE BEND OF THE ARM (from R. Quain). ⅓

Fig. 700.

a, two branches of the internal cutaneous nerve ; *a′*, *a′*, the descending twigs of the same nerve ; *b*, placed over the biceps near its insertion and close to the external cutaneous nerve ; *b′*, anterior twigs of the same nerve accompanying the median vein ; 1, placed on the fascia near the bend of the arm, above the place where it has been opened to show the lower part of the brachial artery with its venæ comites, of which one is entire, marked 2, and the other has been divided ; +, is placed between this and the median nerve ; 3, basilic vein ; 3′, 3′, ulnar veins ; 4, cephalic vein ; 4′, radial vein ; 5, 5, median vein ; 3′, 5, median basilic vein ; 4′, 5, median cephalic vein.

At the *bend of the elbow* the disposition of the brachial artery is chiefly, or, at least, most commonly, of interest in a surgical point of view, because of its connection with the veins from which blood is usually drawn in the treatment of disease. The vein (median basilic) which is generally the most prominent and apparently best suited for venesection is commonly placed over the course of the brachial artery, separated from it only by a thin layer of fibrous structure (the expansion from the tendon of the biceps muscle) ; and under such circumstances, it ought not, if it can be avoided, to be opened with a lancet, except in a part which is not contiguous to the artery.

When two arteries are present, instead of the ordinary single trunk, they are commonly close together ; but it now and then happens that an interval exists between them—one being in the usual situation of the brachial, the other nearer, in different degrees in different cases, to the inner condyle of the humerus.   There is on this account an additional reason for precaution when venesection is to be performed ; and care is the more necessary, as the second artery may be immediately under the vein without the interposition of fascia [plate 41].

SURGICAL ANATOMY OF THE COMMON ILIAC ARTERIES.

The common iliac artery (p. 418), extending in a line from the left side of the umbilicus towards the middle of Poupart's ligament, and being placed at its commencement on a level with the highest part of the iliac crest, may be approached in an operation, by dividing the abdominal muscles to a sufficient extent in the iliac region, and a little above this part of the abdomen.   The incision may be made, beginning about Poupart's ligament, to the outer side of its middle, and running parallel with that structure towards the anterior superior spine of the hip-bone, thence curving for a couple of inches towards the umbilicus.   In this way the artery will be approached from below, but, if a tumour extends along the external iliac artery, this plan of operation will be objectionable, for the swelling itself, and, it may be, the adhesion of the peritoneum to its surface, will be sources of serious difficulty.   Should the aneurism extend upwards in the abdomen it will be best to approach the artery from the side, or rather from above,—not from below.   The essential part of the operation, so far as the abdominal muscles are concerned, is, that they should be divided to the extent of five or six inches at the side of the abdomen, beginning about two inches above the level of the umbilicus and ending lower than the iliac spine, the incision being curved outwards towards the lumbar region. Sir P. Crampton, in an operation to tie this artery, divided the muscles from the end of the lowest rib, straight down nearly to the iliac crest, and thence forward a little above the border of the bone as far as its spine.*   This plan is well devised for the object.

The fascia behind the muscles (fascia transversalis) is to be cut through with care, and the peritoneum is to be raised from that and the iliac fascia, as well as from the subjacent membrane (sometimes containing fat) which is interposed between the serous and the fibrous membranes.   With the peritoneum the ureter will be raised, as this adheres to it.

The artery will be seen on the last lumbar vertebra ; and, on the right side of the body, large veins will be in view in close connection with it, viz., both common iliac veins, and the commencement of the lower vena cava [plate 55].   It will be remembered, that in some cases (without transposition of the viscera, as well as with that condition) the iliac veins are joined on the left instead of the right side ; and that in another small class of cases the junction of those veins is delayed, so to say [plate 58, figs. 1, 2, 3].   The effect of either of these conformations of the venous system would be to give to the artery on the left side much more than the usual complication with veins.   Lastly, the thin subserous membrane covering the artery is divided without any difficulty, to admit the passage of the ligature.

* Med. Chir. Trans., vol. xvi.

The common iliac artery is in most cases of sufficient length to admit the application of a ligature without much apprehension of secondary hæmorrhage occurring in consequence of insufficiency in this respect. But

Fig. 701.—VIEW OF THE RIGHT EXTERNAL AND INTERNAL ILIAC ARTERIES OF THE MALE. ⅓

Fig. 701.

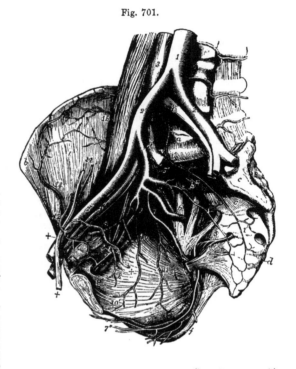

The viscera of the pelvis have been removed as well as the internal iliac veins. 1, lower part of the abdominal aorta; 1', middle sacral artery; 2, 2, common iliac arteries; 2', right external iliac; 3, lower part of the vena cava inferior; 4, 4, common iliac veins; the number on the right points by a line to the right internal iliac artery; 4', right external iliac vein; 5, placed on the ilio-lumbar nervous trunk, points to the posterior division of the internal iliac artery giving off the gluteal; 5', ilio-lumbar artery; 5", lateral sacral artery with branches passing into the anterior sacral foramina; 6, placed on the anterior division of the first sacral nerve, points to the sciatic artery coming from the anterior division of the internal iliac; 7, pudic artery; 7', the same artery passing behind the spine of the ischium, and proceeding within the ischium and obturator internus muscle, accompanied by the pudic nerve towards the perinæum; towards f, inferior hæmorrhoidal branches are given off; 7", superficial perineal artery and nerve; 8, hypogastric artery, with the obliterated remains of the umbilical artery cut short, and 8', superior vesical branches rising from it; 9, obturator artery with the corresponding nerve and vein; 9', the pubic twigs which anastomose with descending twigs of the epigastric artery, and from which, by the enlargement of one of them, the aberrant obturator artery may proceed; 10, inferior vesical; 11, middle hæmorrhoidal vessels rising in this instance from the pudic; 12, epigastric artery winding to the inside of +, +, the vas deferens and spermatic cord; 13, circumflex iliac artery; 14, spermatic artery and vein divided superiorly; 15, twigs of the ilio-lumbar artery proceeding to anastomose with the circumflex iliac.

it has been shown (p. 420) to be in some instances very short—so short that the operation would be inadmissible. In any case in which the common trunk is thus short, it would probably be more prudent to place a ligature on the external iliac and another on the internal iliac, at the origin of each, than to tie the common iliac artery, or the external iliac alone near its commencement.

The surgeon has it in his power to judge of the length of the artery during the operation, and to determine as to the propriety of tying the one vessel or the other, for the iliac arteries are under his view almost as fully as if dissected. Arteries in other parts of the body are, on the contrary, only seen at the point at which it has been beforehand determined to place the ligature.

SURGICAL ANATOMY OF THE INTERNAL ILIAC ARTERY.

This artery has been tied for aneurism affecting one of its large branches on the back of the pelvis—the gluteal or sciatic (p. 420). It is arrived at by dividing the abdominal muscles before the iliac fossa to a greater extent than is required for exposing the external iliac—in the manner of the operation first mentioned for the common iliac artery. The vein, a large one, is, it will be borne in mind, behind the artery and in contact with it [plate 55]; it is occasionally double [plate 58, fig. 6].

There is some difference in the degree of difficulty that would be experienced in securing the internal iliac artery in different cases. This is owing to the fact that, when short, (and, as stated before, it often is so,) the artery is placed deeply in the pelvis; whereas, when the length is more considerable, it is accessible above that cavity.

Again, when the artery is very short, [as represented, for instance, in plate 58, fig. 1], it would probably be more safe to tie the common iliac, or both the external and the internal iliacs at their origin, than to place a ligature on the latter only, close to a strong current of blood.

SURGICAL ANATOMY OF THE EXTERNAL ILIAC ARTERY.

The external iliac artery (p. 431) admits of being tied in a surgical operation at any part except near its upper and lower end; the near neighbourhood of the upper end being excepted on account of the circulation through the internal iliac, and the lower end on account of the common position of the branches (epigastric and circumflex iliac). Occasional deductions from this statement occur in consequence of a branch or branches taking origin near or at the middle of the artery; and, as the operator may see such a branch, he will avoid placing a ligature very near it.

The incision through the muscles to reach the artery, commencing a little above the middle of Poupart's ligament, may be directed parallel with the ligament upwards and outwards as far as its outer end, where the incision may be curved with advantage for a short space (about an inch) upwards.

This and the other iliac arteries might be operated on by means of straight incisions in a line from the umbilicus to the middle of Poupart's ligament, or a little to the outer side of this line. But the division of the muscles on the fore part of the abdomen is liable to the objection that the peritoneum must be disturbed in front as well as behind; and, moreover, a curved incision has the advantage of giving more room laterally than one which is merely straight.

The muscles and the fascia transversalis being divided, and the peritoneum (to which the spermatic vessels adhere) being raised, the artery is found where the finger of the surgeon, introduced into the wound, begins to descend into the true pelvis, along the border of the psoas muscle.

In contact with the artery will be seen the following structures, each occupying the position already mentioned, viz., lymphatic glands, the circumflex iliac vein, and the external iliac vein [plate 55].

In order to pass the ligature, it is necessary to divide the thin and some-times resistent subserous membrane, which binds the vessel down to the fascia iliaca.

The femoral artery (p. 434) is accessible to the surgeon for the application of a ligature without serious difficulty in its entire length ; but, as the lower half is deeply placed, the difficulty of reaching this part is greatest, and renders it necessary to divide and disturb the surrounding structures to a greater extent than where the vessel is nearer the surface. For these reasons the upper part of the artery is to be preferred for the performance of the operation adverted to, in all cases in which other circumstances do not control the choice of the surgeon. But the upper part of the femoral artery is not equally eligible for the application of a ligature at all points, in consequence of the position of the branches—an important consideration in the surgical anatomy of this vessel.

Close to the commencement of this artery are two considerable branches (epigastric and circumflex iliac) ; and between one and two inches lower down the deep femoral branch ordinarily takes its rise. A ligature placed on the arterial trunk in the interval between those branches, that is to say, on the common femoral artery, is in the near neighbourhood of two dis-turbing causes,—two sources of danger, so near that the prospect of a favour-able issue to the operation is, under ordinary circumstances, very small.

Moreover, it has been shown amid the facts detailed before (p. 441), that the origin of the deep femoral is often less than the average distance from Poupart's ligament ; and that, not unfrequently, a considerable branch (one of the circumflex arteries) takes its rise from the common femoral artery. When these circumstances are considered, the operation of tying the common femoral artery, or the femoral artery within two inches of its commencement, must be regarded as very unsafe. And it may be added, that the conclusion to which the anatomical facts would lead is fully confirmed by the results of cases in which the operation has been actually performed.

It remains to determine where a ligature applied to the main artery shall be sufficiently distant from the origin of the deep femoral below it, to be free from the disturbing influence of the circulation through that great branch. It has been shown that now and then a case occurs in which the profunda is given off at the distance of from two to three inches below Pou-part's ligament—in only a single instance out of a large number of observa-tions did the space referred to amount to four inches.

From the foregoing remarks the inference to be deduced is, that the part of the femoral artery to be preferred for the operation supposed, is at the distance of between four and five inches below the lower margin of the abdominal muscles.

Remarks on the operation.—The position of the artery being determined, and the integument and fat divided, a vein may be met with lying on the fascia, over the course of the artery. The saphenous vein, being nearer to the inner side of the limb than the line of incision, is not seen in the opera-tion. The fascia lata, which is now to be divided, has a more opaque ap-pearance over the vessels than over the muscles, for the colour of the latter appears through the membrane. After dividing the fascia, the edge of the sartorius muscle will, in many cases, require to be turned aside ; and occasionally this muscle crosses the thigh so directly, that it must be drawn considerably outwards in order to reach the artery [plate 74, fig. 4]. To the exact point at which the sheath of the vessels, and even the fascia should be

Fig. 702.

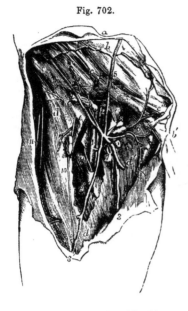

Fig. 702.—SUPERFICIAL DISSECTION OF THE FEMORAL VESSELS, WITH THEIR SMALLER BRANCHES IN THE RIGHT GROIN (from R. Quain). ¼

a, the integument of the abdomen ; *b*, the superficial abdominal fascia ; *b'*, the part descending on the spermatic cord ; *c, c*, the aponeurosis of the external oblique muscle ; *c'*, the same near the external abdominal ring ; *c''*, the inner pillar of the ring ; *d*, the iliac part of the fascia lata ; *d'*, the pubic part, *e, e*, the sheath of the femoral vessels laid open, the upper letter is immediately over the crural aperture ; *e'*, placed on the sartorius muscle partially exposed, points to the margin of the saphenic opening ; 1, femoral artery, having the femoral vein 2, to its inner side, and the septum of the sheath shown between the two vessels ; 3, the principal saphenous vein ; 3', its anterior branch ; 4, the superficial circumflex iliac vein and arterial branches to the glands of the groin ; 5, the superficial epigastric vein ; 6, the external pudic arteries and veins ; 7 to 8, some of the lower inguinal glands receiving twigs from the vessels ; 9, internal, 10, middle, and 11, external cutaneous nerves.

cut through, the pulsation of the artery will guide the operator. A small nerve may present itself in this part of the operation. The immediate investment of the artery should be opened to the smallest possible extent, and the knife or other instrument should be sparingly used at this stage of the operation : the object being to disturb the artery from its connections, including its nutrient vessels (vasa vasorum), as little as possible, and likewise to avoid wounding any of the small muscular branches which spring from most arteries at irregular intervals. The division of an artery of the size of those last referred to at a distance from the source from which it springs is of little importance. It contracts, and soon ceases to bleed. But when it is divided close to the trunk, blood issues from it as it would if an opening equal in size to the calibre of the little branch were made in the trunk itself.

In order to avoid injuring the vein, which is separated from the artery only by a thin partition of areolar tissue, the point of the aneurism-needle, which conveys the ligature, is to be kept close to the artery.

Other veins of occasional occurrence may render increased care necessary, for example, those small branches which cross the artery or course along its surface ; or it may be a larger vein—a division of the femoral vein when it is double, or the deep femoral vein when the ligature is applied a little higher than usual [plate 75].

To reach the femoral artery in the middle of the thigh, the depth of the vessel being considerable, the incision through the integuments must be proportionally long. As the sartorius is directly over the vessel, the operation may be performed by turning the muscle either towards the outer or the inner side of the limb ; and the incision would be made according to the plan adopted, at the inner or the outer margin of the muscle. The

Fig. 703.—Deep View of the Femoral Artery and its Branches on
the Left Side (from R. Quain). ¼

Fig. 703.

The sartorius muscle has been removed in part, so as to expose the artery in the middle third of the thigh ; *a*, the anterior superior iliac spine ; *b*, the aponeurosis of the external oblique muscle near the outer abdominal ring, from which the spermatic cord is seen descending towards the scrotum ; *c*, the upper part of the rectus femoris muscle ; *d*, abductor longus ; *e*, fibrous sheath of Hunter's canal covering the artery ; 1, femoral artery ; 1', femoral vein divided and tied close below Poupart's ligament ; 2, profunda femoris artery ; 3, anterior crural nerves ; 4, internal circumflex branch ; 5, superficial pudic branches ; 6, external circumflex branch, with its ascending transverse and descending branches separating from it ; 6', twigs to the rectus muscle ; 7, branches to the vastus internus muscle ; 8, and 9, some of the muscular branches of the femoral.

preferable mode appears to be to divide the integument on or over the muscle, near its inner margin, so as to arrive directly upon the muscle and draw it outwards, after cutting freely through the investing fascia. The fibrous structure stretched over the vessels from the adductors to the vastus internus muscle being divided, the position of the femoral vein and saphenous nerve are to be kept in view in completing the operation. In the first steps of the operation in this part of the thigh, injury to the long saphenous vein is to be guarded against.

Before concluding the observations on the femoral artery, a very small class of cases claims a word of notice. It has happened (in Sir Charles Bell's case) that the application of a ligature to a femoral artery has not been followed by the usual consequence of cessation of the pulsation in the aneurism ; and the uninterrupted continuance of the circulation was found, on examination after death, to be attributable to the circumstance of the artery being double where the ligature was applied, while the two parts became re-united above the tumour. If such a case should again be met with in an operation, the surgeon instructed by the case alluded to, and by other examples of the same arrangement of the arteries which have since been observed, might at once, under the guidance of the pulsation, or of the effect of pressure in controlling the circulation through the aneurism, divide the covering of areolar tissue over the second part of the artery, and tie it likewise.

II. SURGICAL ANATOMY OF THE PARTS CONCERNED IN CERTAIN ABDOMINAL HERNIÆ.

Besides the surgical anatomy of the principal arteries, certain parts of the walls of the abomen and pelvis are to be now considered with reference to surgical operations in which the viscera of those cavities are from time to time concerned.

The walls of the abdomen, when in a healthy state, unaffected by injury, disease, or malformation, retain the viscera within the cavity under all circumstances ; but where certain natural openings exist for the passage of blood-vessels, protrusions of the viscera, constituting the disease named " hernia " or " rupture " are liable to occur under the influence of the compression to which the organs are subjected during the production of efforts. For the replacement of the viscus so protruded, an accurate acquaintance with the structure of the part through which the protrusion takes place is required by the surgeon ; and, on this account, an examination of the seat of the hernia as a surgical region becomes necessary.

Two of the openings by which herniæ escape from the abdomen are situate close together at the groin.　One is the canal in the lower part of the broad abdominal muscles, which gives passage in the male to the duct and vessels of the testis (spermatic cord), and in the female to the round ligament of the womb.　The second opening exists at the inner side of the large femoral blood-vessels.

Hernial protrusions are likewise found to escape at the umbilicus, in the course of the blood-vessels which occupy that opening in the fœtus, or in the immediate neighbourhood of the opening ; and at the thyroid foramen, where the obturator vessels and nerve pass downwards to the adductor muscles of the thigh.　According to the situation they occupy these herniæ are named respectively inguinal, femoral, umbilical, and obturator.　They will now be separately noticed ; but, inasmuch as the structure of the parts connected with the umbilical and obturator herniæ is by no means intricate, and as, moreover, it is noticed with sufficient detail in text-books of practical surgery, it will be unnecessary to refer farther in this work to those forms of hernia.

OF THE PARTS CONCERNED IN INGUINAL HERNIA.

The inguinal hernia, it has been stated above, follows the course of the spermatic cord from the cavity of the abdomen.　We shall therefore, before adverting to the hernial protrusions, examine the structure of the abdominal walls in the neighbourhood of the canal in which the cord is placed ; and for this purpose it will be supposed that the constituents of those walls are successively laid bare and everted to such an extent as would be permitted by two incisions made through them, and reaching, one along the linea alba for the length of three or four inches from the pubes, the other, from the upper end of the vertical incision outwards to the superior spine of the hip-bone.

The *superficial fascia* (p. 257) is connected along the fold of the groin with Poupart's ligament and the upper end of the fascia lata ; and, after descending over the spermatic cord into the scrotum, it becomes continuous with the membrane of the same kind which covers the perinæum. Its thickness varies much in different persons, on account of the different quantity of fat contained within its meshes ; but in the scrotum the fascia is devoid of fat ; as it also is elsewhere towards the internal surface, where its density is at the same time augmented. From the varying thickness of this structure on the abdomen and the scrotum, as well as in different persons, it will be inferred that the depth of incision required to divide it in an operation must vary considerably.

The *superficial vessels* of the groin are encased by the fascia, and are held to separate it into two layers. The vessels which ramify over the inguinal canal and the scrotum are the external pudic and epigastric arteries and veins (p. 437 and 475). The veins, especially the epigastric, are considerably larger than the arteries they accompany. Some of these vessels are wounded in operations performed for the relief of strangulated hernia ; but the bleeding from them is small in quantity and rarely requires the application of a ligature or other means to arrest it. The lymphatic glands of the groin (p. 489) admit of being arranged in two sets—one being placed over Poupart's ligament and parallel with that structure ; while the other series is upon the upper part of the thigh at its middle, about the saphenous opening in the fascia lata.

When the superficial fascia is removed, the aponeurosis of the *external oblique muscle* (p. 249) is in view, together with, in the male body, the spermatic cord, in the female body the round ligament of the uterus, which emerge from an opening close to the outer side of the pubic spine. The lowest fibres of the aponeurosis, as they approach the pubes, become separated into two bundles which leave an interval between them for the passage of the cord or round ligament. One of the bands, the upper one and the smaller of the two, is fixed in front of the symphysis of the pubes ; and the lower band, which forms the lower margin of the aponeurosis, being stretched between the anterior superior iliac spine and the pubes, is named Poupart's ligament, or the femoral arch. This latter tendinous band has considerable breadth. It is fixed at the inner end to the spine of the pubes, and, for some space outside that process of the bone, to the pectineal ridge. In consequence of the position of the pectineal ridge at the back part of the bone, the ligament is tucked backwards ; and its upper surface affords space for the attachment of the other broad muscles, at the same time that it supports the spermatic cord. Poupart's ligament does not lie in a straight line between its two fixed points ; it curves downwards, and with the curved border the fascia lata is connected. It is owing to the last-mentioned fact that the so-named ligament, together with the rest of the aponeurosis of the external oblique, is influenced by the position of the thigh, being relaxed when the limb is bent, and the converse. Moreover, the change of the position of the limb exercises a corresponding influence on the state of the other structures connected with Poupart's ligament.

The interval left by the separation of the fibres of the aponeurosis above referred to, is named the *external abdominal ring*, and the two bands by which it is bounded are known as its *pillars* or *columns*. The space is triangular in shape, its base being the crest of the pubes, while the apex is at the point of separation of the two columns. The size of the ring varies considerably in different bodies ;—in one case its sides will be found closely applied to the spermatic cord ; while, in another, on the contrary, the space is so considerable as to be an obvious source of weakness to the abdominal parietes. It is usually smaller in the female than in the male body.

Fig. 704.

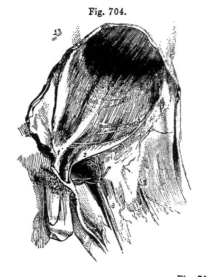

Fig. 704.—The Aponeurosis of the External Oblique Muscle and the Fascia Lata.

1, the internal pillar of the abdominal ring; 2, the external pillar of the same (Poupart's ligament); 3, transverse fibres of the aponeurosis; 4, pubic part of the fascia lata; 5, the spermatic cord; 6, the long saphenous vein; 7, fascia lata.

Between the pillars of the abdominal ring is stretched a thin fascia, named from that circumstance, "intercolumnar;" and a thin diaphanous membrane prolonged from the edges of the opening affords a covering (fascia spermatica) to the spermatic cord and the tunica vaginalis testis. The cord, in passing through the ring, lies over the outer pillar.

Fig. 705.

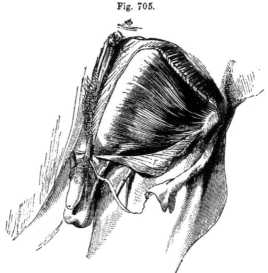

Fig. 705.—Deeper Dissection of the Abdominal Wall in the Groin.

The aponeurosis of the external oblique muscle having been divided and turned down, the internal oblique is brought into view with the spermatic cord escaping beneath its lower edge; 1, aponeurosis of the external oblique; 1', lower part of the same turned down; 2, internal oblique muscle; 3, spermatic cord; 4, saphenous vein.

Internal oblique muscle (p. 250)—After removing the aponeurosis of the external oblique, this muscle is laid bare. The lower fibres are thin and often of a pale colour. Immediately above Poupart's ligament the outer part is muscular, the inner part tendinous. The spermatic cord, when about to escape at the external abdominal ring, passes beneath the fleshy part of the muscle. The fibres in this situation varying considerably in direction from those of the rest of the muscle, pass inwards from Poupart's ligament at first nearly parallel with that structure ; and, becoming tendinous, they join with the tendon of the transversalis.

Fig. 706.

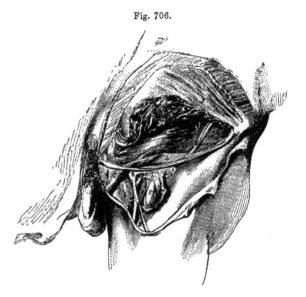

Fig. 706.—The Inguinal Canal and Femoral Sheath fully exposed.

After the removal of the lower part of the external oblique (with the exception of a small slip including Poupart's ligament), the lower portion of the internal oblique has been raised, and thereby the transversalis muscle and fascia have been brought into view. The femoral artery and vein are seen to a small extent, the fascia lata having been turned aside and the sheath of the blood-vessels laid open. 1, external oblique muscle; 2, internal oblique; 2', part of same turned up; 3, transversalis muscle. Upon the last-named muscle is seen a branch of the circumflex iliac artery, with its companion veins ; and some ascending tendinous fibres are seen over the conjoined tendon of the two last-named muscles; 4, transversalis fascia ; 5, spermatic cord covered with the infundibuli-form fascia from the preceding. 6, upper angle of the iliac part of fascia lata ; 7, the sheath of the femoral vessels ; 8, femoral artery ; 9, femoral vein ; 10, saphenous vein ; 11, a vein joining it.

Transversalis muscle.—This muscle (p. 253) does not, in general extend down as far as the internal oblique ; so that, the latter being removed, an interval is observable between the edge of the transversalis and Poupart's ligament, in which the transversalis fascia comes into view ; and in which the spermatic cord is seen after having penetrated that fascia. The lower edge of the muscle is commonly close above the opening for the cord in the subjacent membrane, while its tendon curves to the inner side ; so that the margin of the muscle ·with its tendon has a semicircular direction with respect to the aperture.

The tendinous fibres in which the fleshy parts of the two preceding muscles end, are connected together so as to form one layer, which is named the "conjoined tendon of the internal oblique and transverse muscles." This tendon is fixed to the crest of the pubes in front of the rectus muscle, and likewise to the pectineal ridge. It is thus behind the external abdominal ring, and serves to strengthen the wall of the abdomen where it is weakened by the presence of that opening.

A band of tendinous fibres, directed upwards and inwards over the conjoined tendon in a triangular form, gives additional strength to the abdominal wall in the same situation, but the fibres of this structure are often very indistinct.

Where the spermatic cord is in apposition with the preceding muscle, the cremaster muscle of the testis descends over it. The fibres which compose this muscle are, from their colour, more easily distinguished than the other investments of the cord ; and this is especially the case in robust persons ; or when they are hypertrophied, as sometimes happens in cases of long-standing hernia. The outer part of the cremaster is much larger than the portion connected with the pubes ; and the latter is sometimes absent (p. 253).

When observed in different bodies the lower parts of the internal oblique and transverse muscles present some differences in their physical characters as well as in the manner in which they are disposed with respect to the spermatic cord. Thus :—

a. The transversalis, in some cases, is attached to but a small part of Poupart's ligament, and leaves, therefore, a larger part of the abdominal wall without its support. On the other hand, that muscle may be found to extend so low down as to cover the internal abdominal ring together with the spermatic cord, for a short space. Not unfrequently the fleshy fibres of the two muscles are blended together as well as their tendons.

b. Cases occasionally occur in which the spermatic cord, instead of escaping beneath the margin of the internal oblique, is found to pass through the muscle, so that some muscular fibres are below as well as above it. And examples of the transversalis being penetrated by that structure in the same manner are recorded.*

c. In his latest account of the structure of these parts Sir A. Cooper described the lower edge of the transversalis as curved all round the internal ring and the spermatic cord. " But the lower edge of the transversalis has a very peculiar insertion, which I have hinted at in my work on Hernia. It begins to be fixed in Poupart's ligament, almost immediately below the commencement of the internal ring, and it continues to be inserted behind the spermatic cord into Poupart's ligament as far as the attachment of the rectus."† With this disposition of its fibres, the muscles would, in the opinion of the last-cited authority, have the effect of a sphincter, in closing the internal ring, and would thus tend to prevent the occurrence of hernia. But the principal object with which the attention of surgeons has been fixed on the muscles in this situation, is in order to account for the active strangulation of hernial protrusions at the internal abdominal ring, and in the inguinal canal.

Fascia transversalis.—This membrane is described as part of the general lining of the abdominal walls (p. 258). Closely connected with the transversalis muscle by means of the areolar tissue interposed between the fleshy fibres of the muscle, it is united below to the posterior edge of Poupart's

---

* Recherches Anatomiques sur les Hernies, &c., par J. Cloquet, p. 18 and 23. Paris, 1817. Inguinal and Femoral Herniæ, by G. J. Guthrie, plate I. London, 1833.

† Observations on the Structure and Diseases of the Testis, second edition, p. 36. Ed. by Bransby B. Cooper, F.R.S. London, 1841.

ligament, there joining with the fascia iliaca; and on the inner side it blends with the conjoined tendon of the internal oblique and transversalis muscles, as well as with the tendon of the rectus. The fascia possesses very different degrees of density in different cases; in some being little more than a loose areolar texture, while in others it is so resistant at the groin—towards which part it increases in thickness, and especially at the lower side of the internal abdominal ring—that it is calculated to afford material assistance to the muscles in supporting the viscera. By an oval opening in this membrane the spermatic cord, or the round ligament of the womb, begins its course through the abdominal parietes. This opening, named the *internal abdominal ring*, is opposite the middle of Poupart's ligament, and usually close above that structure, but occasionally at a distance of three or four lines from it. Its size varies a good deal in different persons, and is considerably greater in the male than the female. From the edge of the ring a thin funnel-shaped elongation (infundibuliform fascia; fascia spermatica interna, Cooper), is continued over the vessels of the spermatic cord.

Epigastric Artery.—The position of this vessel is one of the most important points in the anatomy of the inguinal region, from the close connection which it has with the different forms of inguinal hernia and with the femoral hernia. Accompanied by two veins (in some instances by only one) the vessel ascends under cover of the fascia last described obliquely to the rectus muscle, behind which it then proceeds to its ultimate distribution (p. 432). In this course the artery runs along the inner side of the internal abdominal ring—close to the edge of the aperture or at a short interval from it. The vessels of the spermatic cord are therefore near to the epigastric artery; and the vas deferens, in turning from the ring into the pelvis, may be said to hook round it.

The *Inguinal Canal.*—This channel, by which the spermatic cord passes through the abdominal muscles to the testis, begins at the internal abdominal ring, and ends at the external one. It is oblique in its direction, being parallel with and immediately above the inner half of Poupart's ligament; and it measures two inches in length. In front the canal is bounded by the aponeurosis of the external oblique muscle in its whole length, and at the outer end by the fleshy part of the internal oblique also; behind it, is the fascia transversalis, together with, towards the inner end, the conjoined tendon of the two deeper abdominal muscles. Below, the canal is supported by the broad surface of Poupart's ligament, which separates it from the sheath on the large blood-vessels descending to the thigh, and from the femoral canal at the inner side of those vessels.

The spermatic cord, which occupies the inguinal canal, is composed of the arteries, veins, lymphatics, nerves, and excretory duct (vas deferens) of the testis, together with a quantity of loose areolar tissue mixed up with those parts. The direction of the vessels just enumerated requires notice. The artery and vein incline outwards from the lumbar part of the vertebral column to reach the internal abdominal ring, where, after being joined by the vas deferens as it emerges from the pelvis they change their course, inclining inwards along the inguinal canal; at the end of which they become vertical. There are thus repeated alterations in the direction of the vessels; and while at the beginning and ending all are close to the middle line of the body, they are considerably removed from that point where they come together to emerge from the abdominal cavity.

The coverings given from the constituent parts of the abdominal wall to the spermatic cord and the testis, namely, the cremasteric muscular fibres with the two layers of fascia (the infundibuliform and spermatic fasciæ) between which those fibres are placed, are very thin in their natural state ; but they may be readily distinguished in a surgical operation from the investing superficial fascia, by their comparative density and the absence of fat.

In order to examine the *peritoneum* at the groin, it will be best to divide that membrane with the abdominal muscles by two incisions drawn from the umbilicus—one to the hip-bone, the other to the pubes. The flap thus formed being held somewhat outwards, and kept tense, a favourable view will be obtained of the two fossæ (*inguinal fossæ* or *pouches*) with the intervening crescentic fold. This fold is formed by the cord remaining from the obliterated umbilical artery, which being shorter than the outer surface of the serous sac, causes this to project inwards ; and as the length of the cord differs in different cases, so likewise do the size and prominence of the peritoneal fold vary accordingly.

The lowest part of the outer fossa will be generally found opposite to the entrance into the internal abdominal ring and the femoral ring, while the inner one corresponds with the situation of the external abdominal ring. But the cord representing the umbilical artery, which it has been stated causes the projection of the serous membrane into a fold, does not uniformly occupy the same position in all cases. Most frequently it is separated by an interval from the epigastric artery, while in some cases it is immediately behind that vessel. There is necessarily a corresponding variation in the extent of the external peritoneal fossa. This fact will find its practical application when the internal form of inguinal hernia is under consideration.

Between the peritoneum and the fascia lining the abdominal muscles is a connecting layer of areolar structure named the *subserous areolar membrane*. A considerable quantity of fat is in some cases found in this membrane.

The relative position of some of the parts above referred to may be here conveniently stated, by means of measurements, made by Sir A. Cooper, and adopted after examination by J. Cloquet. But, as the distance between given parts varies in different cases, the following measurements must be regarded only as a general average :—

	MALE.	FEMALE.
From the symphysis of the pubes to the anterior superior spine of the ilium . . .	5½ inches. ...	6 inches.
From the same point to the spine of the pubes . .	1⅛ ,, ...	1¾ ,,
,, to the inner part of the external abdominal ring	0⅞ ,, ...	1 ,,
,, to the inner edge of the internal abdominal ring	3 ,, ...	3¼ ,,
,, to the epigastric artery on the inner side of the internal abdominal ring . .	2¾ ,, ...	2⅞ ,,

From the preceding account of the structure of the abdominal wall at the groin, it will be inferred that the defence against the protrusion of the viscera from the cavity is here weaker than at other parts. The external oblique muscle and the fascia transversalis are perforated, while the two intervening muscles are thinner than elsewhere, and more or less defective. To this it must be added that the viscera are impelled towards the same part of the abdomen by the contraction of the diaphragm and the other abdominal muscles, in the production of efforts to overcome resistance; and these are the circumstances under which protrusions actually take place.

INGUINAL HERNIÆ.

The protrusions of the viscera, or herniæ, which occur in the course of the inguinal canal, are named "inguinal." Of this form of the disease two varieties are recognised : and they are distinguished according to the part of the canal which they first enter, as well as by the position which they bear with respect to the epigastric artery. Thus, when the hernia takes the course of the inguinal canal from its commencement, it is named *oblique*, because of the direction of the canal, or *external*, from the position which its neck bears with respect to the epigastric artery. On the other hand, when the protruded part, without following the length of the canal, is forced at once through its termination, *i. e.* through the external abdominal ring, the hernia is named, from its course, *direct*, or, from its relation to the epigastric artery, *internal*. In these, the two principal varieties of inguinal hernia, there are some modifications which will be adverted to in the special notice of each.

Oblique inguinal hernia.—In the common form of this hernia the protruded viscus carries before it a covering of peritoneum (the *sac* of the hernia), derived from the outer fossa of that serous membrane ; and, in passing along the inguinal canal to the scrotum, it is successively clothed with the coverings given to the spermatic vessels from the abdominal parietes. The hernia and its sac lie directly in front of the vessels of the spermatic cord (the intestines and the peritoneum having the same position relatively to those vessels in the abdomen) ; but, when the disease is of long standing, the vessels may be found to be separated from each other, and pressed more or less towards the side or even the fore part of the sac, under the influence of the weight of the tumour. The hernia does not extend below the testis, even when it attains large size. That it does not is owing, doubtless, to the intimate connection which the coverings of the cord have with the tunica vaginalis testis.

When the hernia does not extend beyond the inguinal canal, it is distinguished by the name *bubonocele*: and when it reaches the scrotum, it is commonly named from that circumstance *scrotal* hernia.

There are two other varieties of oblique inguinal hernia, in which the peculiarity depends on the condition of the process of peritoneum that accompanies the testis when this organ is moved from the abdomen. In ordinary circumstances the part of the peritoneum, connected immediately with the testis, becomes separated from the general cavity of that serous membrane by the obliteration of the intervening canal ; and the hernial protrusion occurring after such obliteration has been completed, carries with it a distinct serous investment—the sac. But if the hernia should be formed before the process of obliteration is begun, the protruded part is then received into the cavity of the tunica vaginalis testis, which serves in the place of its sac. In this case the hernia is named *congenital* (hernia tunicæ vaginalis,—Cooper). It is thus designated, because the condition necessary for its formation usually exists only about the time of birth ; but the same variety of the complaint is occasionally found to be first formed in the adult, obviously in consequence of the tunica vaginalis remaining unclosed,—still continuous with the peritoneum. The congenital hernia, should it reach the scrotum, passes below the testis ; and, this organ being imbedded in the protruded

3 x

viscus, a careful examination is necessary in order to detect its position. This peculiarity serves to distinguish the congenital from the ordinary form of the disease.

Fig. 707.
A.   B.

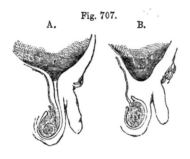

Fig. 707.—DIAGRAMS OF A PART OF THE PERITONEUM AND THE TUNICA VAGINALIS TESTIS.

In the first, A, the serous investment of the testis is seen to be an elongation from the peritoneum ; while in the second, B, the two membranes are shown distinct from each other. 1, the peritoneal cavity ; 2, the testis.

To the second variety of inguinal hernia, in which the distinguishing character depends on the state of the tunica vaginalis testis, the name "infantile " has been applied (Hey). The hernia in this case is covered with a distinct sac, the peculiarity consisting in the circumstance of the rupture with its sac being invested by the upper end of the tunica vaginalis. The relative position of the two serous membranes (the hernial sac and the tunica vaginalis) may be accounted for by supposing the hernia to descend when the process of the peritoneum, which accompanies the testis from the abdomen, has been merely closed at the upper end, but not obliterated for any length. As the tunica vaginalis at this period extends upwards to the wall of the abdomen, the hernia, in its descent, soon meets that membrane and becomes invested by it. The exact mode of the investment has not yet been clearly made out by dissection. It may be that the hernia passes behind the upper end of the large serous tunic of the testis, which then laps round the sac from before, or that the tunica vaginalis is inverted from above so as to receive the hernia in a depression. But the fact most material for the surgeon is fully ascertained—namely, that during an operation in such a case, the hernial sac is met with only after another serous bag (the tunica vaginalis testis) has been divided. The peculiarity here described has been repeatedly found present in the recently formed hernia of grown persons. The term infantile, therefore, like congenital, has reference to the condition of certain parts, rather than to the period of life at which the disease is first formed.

In the female, oblique inguinal hernia follows the course of the round ligament of the uterus along the inguinal canal, in the same manner as in the male it follows the spermatic cord. After escaping from the external abdominal ring, the hernia lodges in the labium pudendi. The coverings are the same as those in the male body, with the exception of the cremaster, which does not exist in the female : but it occasionally happens that some fibres of the internal oblique muscle are drawn down over this hernia in loops, so as to have the appearance of a cremaster (Cloquet).

A strictly congenital inguinal hernia may occur in the female, the protruded parts being received into the little diverticulum of the peritoneum (canal of Nuck), which sometimes extends into the inguinal canal with the round ligament. But as this process of the peritoneum, in such circumstances, would probably not differ in any respect from the ordinary sac, there are no means of distinguishing a congenital hernia in the female body.

Direct inguinal hernia (internal : ventro-inguinal).—Instead of following the whole course of the inguinal canal, in the manner of the hernia above described, the viscus in this case is protruded from the abdomen to the groin directly through the lower end of the canal, at the external abdominal ring ; and at this point the two forms of hernia, if they co-existed, would come together. At the part of the abdominal wall through which the direct inguinal hernia finds its way, there is recognised on its posterior aspect a triangular interval, the sides of which are formed by the epigastric artery, and the margin of the rectus muscle, and the base by Poupart's ligament. It is commonly named the triangle of Hesselbach. Through this space the hernia is protruded, carrying before it a sac from the internal fossa of the peritoneum ; and it is in general forced onwards directly into the external abdominal ring.

Fig. 708.—INTERNAL VIEW OF THE VESSELS RELATED TO THE GROIN.

Fig. 708.

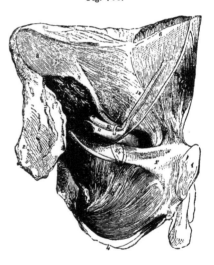

A portion of the wall of the abdomen and the pelvis is here seen on the posterior aspect, the os innominatum of the left side and the soft parts connected with it having been removed from the rest of the body. 1, symphysis of the pubes ; 2, irregular surface of the hip-bone which has been separated from the sacrum ; 3, ischial spine ; 4, ischial tuberosity ; 5, obturator internus ; 6, rectus, covered with an elongation from 7, fascia transversalis ; 8, fascia iliaca covering the iliacus muscle ; 9, psoas magnus cut ; 10, iliac artery ; 11, iliac vein ; 12, epigastric artery and its two accompanying veins ; 13, vessels of the spermatic cord, entering the abdominal wall at the internal ring. The ring was in this case of small size ; 14, two obturator veins ; 15, the obliterated umbilical artery. The cord, it will be remembered, is not naturally in contact with the abdominal parietes in this situation.

The coverings of this hernia, taking them in the order in which they are successively applied to the protruded viscus, are the following :—The peritoneal sac and the subserous membrane which adheres to it, the fascia transversalis, the tendon common to the internal oblique and transverse muscles, and the intercolumnar (external spermatic) fascia derived from the margin of the external abdominal ring, together with the superficial fascia and the integuments.

With respect to one of the structures enumerated, namely, the common tendon of the two deeper muscles, considerable variety exists as to its disposition in different cases. In place of being covered by that tendon, the hernia may be found to pass through an opening in its fibres, or to escape beneath it. Cremasteric muscular fibres are met with (rarely, however,) upon this hernia.

The spermatic cord is commonly placed behind the outer part of the direct inguinal hernia, especially at the external abdominal ring. It is here that the hernia and the cord in most cases first come together ; and

3 x 2

their relative position results from the points at which they respectively pass through the ring, the former being upon the crista of the pubes,

Fig. 709.

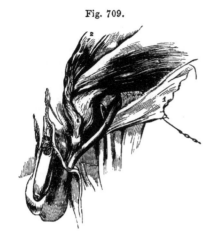

Fig. 709.—A Direct Inguinal Hernia on the Left Side, covered
by the Conjoined Tendon of the
Internal Oblique and Transverse Muscles.

1, aponeurosis of the external oblique; 2, internal oblique turned up;
3, transversalis muscle; 4, fascia
transversalis; 5, spermatic cord; 6,
the hernia. A small part of the epigastric artery is seen through an
opening made in the transversalis
fascia.

while the latter drops over the
outer pillar of the opening.
The hernial sac is not, however, in this case (as the sac
of the external form of the
disease is) in contact with the
vessels of the cord. The investments given from the fascia
transversalis to those vessels and to the hernia respectively, are interposed.

But the point at which the internal inguinal hernia passes through the

Fig. 710.

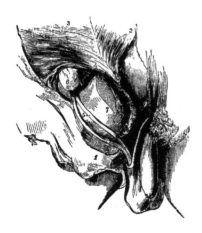

Fig. 710.—A Small Oblique Inguinal
Hernia, and a Direct One on the
Right Side.

A little of the epigastric artery has
been laid bare, by dividing the fascia
transversalis immediately over it. 1,
tendon of the external oblique; 2, internal oblique turned up; 3, transversalis; 4, its tendon (the epigastric artery
is shown below this number); 5, the
spermatic cord (its vessels separated);
6, a bubonocele; 7, direct hernia protruded beneath the conjoined tendon of
the two deeper muscles, and covered by
an elongation from the fascia transversalis.

triangular space above described,
as marked on the posterior aspect
of the abdominal wall, is subject
to some variation.   Instead of
pushing directly through the external abdominal ring, (the most
frequent position), the hernia
occasionally enters the inguinal canal nearer to the epigastric artery, and,
passing through a portion of the canal to reach the external ring, has
therefore a certain degree of obliquity. This change in position may coincide with a change of the peritoneal fossa, which furnishes the hernial sac

—a change, namely, from the internal fossa to the external one. The alteration of the fossa does not, however, in all cases coincide with a change in the position of the hernia ; for the cord remaining from the obliteration of the umbilical artery, (which separates the fossæ,) instead of crossing behind the triangle of Hesselbach so as to leave room at each side of it for a hernia to penetrate that space, lies, it has been already stated, some-times directly behind the epigastric artery :—indeed, according to the observations of Cloquet, it is most frequently in this position ; * and when the cord in question is so placed, the hernia, whatever may be its position in the triangle of Hesselbach, can occupy only the internal peritoneal fossa. The inference, however, most important in a practical or surgical point of view, to be drawn from the varying position of the neck of the internal hernia, has reference not to the cord just alluded to, but to the epigastric artery—i. e. to the greater or less distance of the neck of the sac from that vessel.

The investments of the internal hernia are likewise liable to be influenced by the position at which it penetrates the abdominal wall. It is in all likelihood when the protrusion occurs outside the ordinary situation, that the hernia escapes beneath the conjoined tendon of the two deeper muscles. It is, moreover, under the same circumstances that the hernia is more directly in front of the spermatic cord, and that the cremasteric fibres are among its investments. (Ellis.)

The internal inguinal hernia is very rarely met with in *the female.* In the single example of the disease observed by Richard Quain, as well as in the cases (a very small number) found recorded in books, the hernia, though not inconsiderable in size, was still covered with the tendon of the external oblique muscle.†

Distinctive diagnosis of oblique and direct inguinal herniæ.—The following circumstances, which are brought together from the facts detailed in the preceding pages, or are inferences from those facts, will serve to distin-guish the two forms of the disease from one another. The oblique hernia, when recently formed, is elongated and narrow at its upper part, being restrained by the tendon of the external oblique muscle. It is, however, attended with a degree of fulness in the inguinal canal, as well as tender-ness upon pressure being made over the canal. After passing through the external abdominal ring, it is observed to be directly in front of the spermatic cord. The direct hernia, when of small size, is globular ; it is protruded more immediately over the pubes ; causes no fulness or tender-ness in the canal ; and the spermatic cord is usually behind its outer side. But the distinction between the two herniæ admits of being made only when the disease is recent and the tumour moderate in size ; for, when oblique inguinal hernia is of long standing, and has attained considerable

---

* Recherches, &c., p. 39, note.

† See "Treatise on Ruptures," by Mr. Lawrence, 4th edit. p. 213, and an essay by M. Velpeau in "Annales de Chirurgie Française et étrangère," tom. i. p. 352.

M. Velpeau, in the essay just referred to, proposes to recognise three varieties of internal hernia, viz., 1, the ordinary form which passes straight through the external abdominal ring ; 2, an outer oblique variety, which passes through a part of the inguinal canal ; and 3, an inner oblique one, which entering the abdominal wall close to the edge of the rectus muscle, is directed outwards in order to reach the opening in the external oblique muscle. The first two forms adverted to by M. Velpeau have been described in the text. With respect to the third variety or class sought to be introduced by that surgeon, it should be observed that he seems to have been led to propose it by the observation of a single case—an example of internal hernia in the female.

size, the obliquity of the inguinal canal no longer remains,—the internal ring being enlarged and brought inwards opposite the external one,—while at the same time the epigastric artery, borne inwards by the hernia, curves along the inner side of the sac. Under this change, the oblique hernia assumes the appearance of one primarily direct.

Operations for the relief of inguinal hernia.—This account of the disposition of the parts connected with the different forms of inguinal hernia may be concluded by a brief statement of the application of the anatomical facts in practical surgery, either in simply replacing the hernial protrusion, or in the operation required to attain that object when the hernia is otherwise irreducible. In the efforts to effect the replacement of the protruded parts (the taxis), it is to be borne in mind that the abdominal muscles, which, in most cases, are the sole obstacle to the attainment of that end, become relaxed to some extent by flexing the thigh and inclining the trunk forwards. The direction, too, which the protruded part follows through the abdominal walls, ought to influence the direction given to the pressure required in restoring it.

When the operation required to set free the constriction which prevents the restoration of the protruded viscus to the abdomen is undertaken, the parts covering the hernia or a portion of it at the upper end, are to be divided, so as to allow the introduction of a knife beneath the " stricture"; and this (the stricture) will be found at the external ring, or, more frequently, at the internal one. To accomplish the object, the tendon of the external oblique is to be laid bare by an incision beginning somewhat above the upper end of the hernia, and extending downwards below the external ring. If, on examination, the stricture should be ascertained to be at the last-named opening, the division of a few fibres of its circumference will allow a sufficient dilatation for the replacement of the hernia; but if, as generally happens, the seat of the stricture should prove to be higher up,—in the inguinal canal or at the internal ring,—the aponeurosis of the external oblique is to be cut through over the canal, and the lower edge of the internal muscles, one of which commonly constitutes the stricture, is then to be divided on a director insinuated beneath them.

In the operation indicated in the last paragraph, the sac of the hernia is supposed to be left unopened,—the course which it is best to adopt when the stricture is external to that membrane. Occasionally, however, it happens that the sac itself is the cause of the constriction. When this is the case, or when from some other reason the surgeon is unable, after such an operation as that above noticed, to replace the hernia, it becomes necessary to lay the sac open, in order to divide the constriction at its neck. When the incision required in the last-mentioned step of the operation is being made, the epigastric artery is not to be overlooked. From the position which that vessel holds with respect to the oblique and direct forms of hernia respectively, it necessarily follows that an incision outwards through the neck of the sac, in the former variety of the disease, and inwards in the latter, would be free from risk on account of the artery; but, inasmuch as the oblique hernia is liable, in time, to assume the appearance of one primarily direct, and a want of certainty as to the diagnosis must, on this account, exist in certain cases,—as, moreover, it is advantageous to pursue one course which will be applicable in every case,—the rule generally adopted by surgeons in all operations for inguinal herniæ, is to carry the incision through the neck of the sac directly upwards from its middle.

The hernia distinguished as 'femoral' leaves the abdomen at the groin, under the margin of the broad abdominal muscles, and upon the anterior border of the hip-bone, immediately at the inner side of the large femoral blood-vessels. After passing downwards for about an inch or less, the hernia turns forwards to the fore part of the thigh at the saphenous opening in the fascia lata; and when it has reached this point the swelling may be felt and seen.

The muscles of the abdomen, beneath the edge of which the femoral hernia escapes, are represented by the aponeurotic band of the external oblique muscle, which is commonly known as Poupart's ligament, but which, in connection with the femoral hernia, is named the *femoral* or *crural arch.* Extending from the anterior superior iliac spine to the pubes, this band widens at its inner end, and, inclining or folding backwards, is fixed to a part of the pectineal line, as well as to the pubic spine of the hip-bone. The small triangular portion attached to the pectineal line is known as Gimbernat's ligament (Hey). The outer edge of this part is concave and sharp; with other structures, to be presently described, it forms the inner boundary of the aperture through which the hernia descends. The breadth and strength of Gimbernat's ligament vary in different bodies, and with its breadth the size of the opening which receives the hernia will likewise vary.

The space comprised between the femoral arch and the excavated margin of the pelvis is occupied by the conjoined psoas and iliacus, with the anterior crural nerve between those muscles, and the external iliac artery and vein at their inner side. Upon these structures the fascia which lines the abdomen is so arranged as to close the cavity against the escape of any part of the viscera, except at the inner side of the blood-vessels. But the arrangement of the parts situate thus deeply (towards the cavity of the abdomen) will be most conveniently entered upon after those nearer to the surface shall have been examined. To this examination we now proceed.

The general disposition of the *superficial fascia* met with on removing the common integument from the groin has been described (p. 292). In connection with the present subject it will be enough to mention the following facts. The deeper layer of this structure adheres closely to the edge of the saphenous opening, and the careful removal of it is necessary in order adequately to display that aperture. Where it masks the saphenous opening, the deep layer of the superficial fascia supports some lymphathic glands, the efferent vessels of which pass through it; and the small portion of the membrane so perforated is named the *cribriform fascia.* The superficial and the deep fasciæ adhere together along the fold of the groin likewise; and this connection between the two membranes serves the purpose, at least, of drawing the integument the more evenly into the fold of the groin, when the limb is bent at the hip-joint.

By Scarpa the deep layer of the superficial fascia which covers the abdomen was described as an emanation from the fascia lata, extended upwards over the external oblique muscle.* But different modes of viewing the continuity of such structures depend very much on the manner of conducting the dissection. In the present case, for example, the fascia may be said to proceed from above or from below, according as the parts are dissected from the abdomen downwards, or from the thigh upwards.

* A Treatise on Hernia, translated by Wishart, p. 247.

Such difference, however, is no more than a verbal one, the material fact being merely that the two membranes are connected together along the groin.

The separation of the *fascia lata* into two parts at the saphenous opening, and the position and connections of each part, having been described in detail, only a few points in the arrangement of this membrane will be noticed in this place. At the lower end of the saphenous opening the iliac division of the fascia is continuous with the pubic by a well-defined curved margin immediately above which the saphenous vein ends; above the opening a pointed cornu (falciform process—Burns*) of the same portion of the fascia extending inwards in connection with the femoral arch reaches Gimbernat's ligament; and in the interval between the two points now referred to (*i. e.*, from the upper to the lower end of the saphenous opening), the iliac portion of the fascia lata blends with the subjacent sheath of the femoral vessels as well as with the superficial fascia. The pubic part of the fascia covers the pectineus muscle, and is attached to the pectineal ridge of the hip-bone. Immediately below the femoral arch the iliac and pubic portions lie one before, the other behind, the femoral blood-vessels and their sheath: they occupy the same position with respect to the femoral hernia.

Fig. 711.

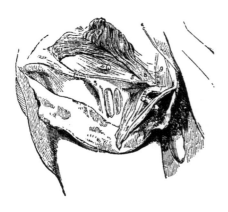

Fig. 711.—THE GROIN OF THE RIGHT SIDE DISSECTED SO AS TO DISPLAY THE DEEP FEMORAL ARCH.

1, the outer part of the femoral arch; 1', part of the tendon of the external oblique muscle, including the femoral arch, and also the inner column of the external inguinal ring, projecting through which is seen a portion of the spermatic cord cut; 2, the femoral arch at its insertion into the spine of the pubes. The fibres outside the numeral are those of Gimbernat's ligament; 3, the outer part of the femoral sheath; 4, the spermatic cord, after having perforated the fascia transversalis; 5, the deep femoral arch—its inner end where it is fixed to the pubes; 6, internal oblique muscle; 7, transversalis. Beneath the lower edge of this muscle is seen the transversalis fascia, which continues into the femoral sheath under the deep femoral arch; 8, conjoined tendon of the internal oblique and transversalis muscles; 9, a band of tendinous fibres directed upwards behind the external abdominal ring.

For an account of the superficial arteries and veins which ramify in the integument in the neighbourhood of the groin, see pp. 437 and 475.

* Edinb. Med. and Surg. Journal, vol. ii. p. 263, and fig. 2.

In the first edition of Hey's Practical Observations in Surgery, the upper end of this process of the fascia was named the "femoral ligament;" and since then several anatomists have distinguished the same part as "Hey's ligament." But Mr. Hey dropped the designation in the subsequent editions of the same work, and there seems no good reason for continuing it. Compare the original edition (1803), p. 151, and plate 4, with the third edition (1814), p. 147, and plates 4, 5, and 6.

The anterior or iliac part of the fascia lata being turned aside, the *sheath of the femoral vessels* will be in view. The sheath is divided by septa, so that each vessel is lodged in a separate compartment, and the vein is separated by a thin partition from the artery on one side and from the short canal for the lymphatics on the other side. Along the thigh the sheath is filled by the artery and vein, but behind the femoral arch it is widened at the inner side. Here it is perforated for lymphatic vessels, and on this account is said to be "cribriform."* This inner wider part of the sheath receives the femoral hernia ; and in connection with the anatomical description of that disease it is designated the femoral canal. At its upper end the sheath of the vessels is continuous with the lining membrane of the abdomen—with ·the fascia transversalis at its fore part, and with the fascia iliaca behind.

When the femoral arch is being removed it will be found that a bundle of fibres springing from its under surface outside the femoral vessels, extends across the fore part of the femoral sheath, and, widening at its inner end, is fixed to the pectineal line behind Gimbernat's ligament. This fibrous band is known as the *deep femoral arch.* Connected with the same part of the bone is the conjoined tendon of the internal oblique and transverse muscles ; the tendon lies before the attachment of the deep femoral arch. In many cases the last-named structure is not strongly marked; and it may be found to blend with the tendon of the muscles just referred to. Not unfrequently it is altogether wanting.

Attention may now be directed to the internal surface of the abdomen. When the peritoneum has been removed, it will be observed that the fasciæ lining the cavity form for the most part a barrier against the occurrence of hernia ; for outside the iliac vessels the fascia iliaca and fascia transversalis are continuous with one another behind the femoral arch. These fasciæ are, in fact, but parts of the same membrane, to which different names are assigned for the convenience of description, just as distinctive names are applied to portions of the same artery. But where the iliac artery and vein occur, the arrangement of the fasciæ is different. The vessels rest upon the fascia iliaca ; and the membranes, instead of joining at an angle as elsewhere, are continued into the sheath of the vessels in the manner above described.†

The sheath is closely applied to the artery and vein, so that in the natural or healthy state of the parts there is no space left for the formation of a hernia in the compartments which belong to those vessels ; but at the inner side of the blood-vessels will be found a depression which is occupied but partly with the lymphatics. This is the femoral ring, the orifice of the femoral canal.

Femoral ring. —After the removal of the peritoneum, this opening is not at first distinctly discernible, being covered with the laminated membrane (subserous) which intervenes between the peritoneum and the walls of the abdomen. That part of the membrane which covers the ring was found by

* The word "cribriform" being applied to this part as well as to the layer of the superficial fascia stretched across the saphenous opening, the two structures are distinguished in the following manner :—the former is known as the cribriform portion of the sheath of the vessels, while to the latter is assigned the name of cribriform fascia.

† Some anatomists describe the sheath of the vessels as continued down from the membranes in the abdomen, while others regard it as an emanation from the fascia of the thigh, but continuous with the abdominal fasciæ. As this difference in the manner of viewing the structure in question does not alter the facts in any way, it is quite immaterial which of the modes of description is adopted.

Cloquet to possess in some cases considerable density; and, from being the only barrier in this situation between the abdomen and the top of the thigh, it was named by that observer the *crural septum* (septum crurale). But this structure is no more than areolar tissue with enclosed fat, and it forms oftentimes but a very slight partition. On clearing it away, the ring is displayed (fig. 346). It is a narrow opening, usually of sufficient size to admit the end of the fore finger; the size, however, varies in different cases, and it may be said to increase as the breadth of Gimbernat's ligament diminishes, and the converse. It is larger in the female than in the male body. On three sides the ring is bounded by very unyielding structures. In front are the femoral arches; behind is the hip-bone covered by the pectineus muscle and the pubic layer of the fascia lata; on the outer side lies the external iliac vein, but covered with its sheath; and on the inner side are several layers of fibrous structure connected with the pectineal line—namely, Gimbernat's ligament, the conjoined tendon of the two deeper abdominal muscles, and the fascia transversalis, with the deep femoral arch. The last-mentioned structures—those bounding the ring at the inner side—present respectively a more or less sharp margin towards the opening.

Femoral canal.—From the femoral ring, which is its orifice, the canal continues downwards behind the iliac part of the fascia lata (its falciform process), in front of the pubic portion of the same membrane, and ends at the saphenous opening. It is rather less than half an inch in length; but in its length the canal varies a little in different cases.

Blood-vessels.—Besides the femoral vein, the position of which has been already stated, the epigastric artery is closely connected with the ring, lying above its outer side. It not unfrequently happens that the obturator artery descends into the pelvis at the outer side of the same opening, or immediately behind it; and in some rare cases that vessel turns over the ring to its inner side. Moreover, an obturator vein has occasionally the same course; and small branches of the epigastric artery will be generally found ramifying on the posterior aspect of Gimbernat's ligament. In the male body, the spermatic vessels are separated from the canal only by the femoral arch.

To the foregoing account of the anatomical arrangement of the parts concerned in femoral hernia, may be added certain measurements, showing the distances of some of the most important from a given point. They are copied from the work of Sir A. Cooper:*—

	MALE.	FEMALE.
From the symphysis pubis to the anterior spine of the ilium	5¾ inches	... 6 inches.
From same point to the middle of the iliac vein	2⅝ ,,	... 2¾ ,,
,, to the origin of the epigastric artery	3 ,,	... 3¼ ,,
,, to the middle of the lunated edge of the fascia lata	3¾ ,,	... 2¾ ,,
,, to the middle of the femoral ring	2¼ ,,	... 2⅜ ,,

Descent of the hernia.—When a femoral hernia is being formed, the protruded part is at first vertical in its course; but at the lower end of the canal, after the passage of about half an inch, it undergoes a change of direction, bending forward at the saphenous opening; and, as it increases in size, it ascends over the iliac part of the fascia lata and the femoral arch. The hernia thus turns round those structures, passing from

* On Crural Hernia, p. 5.

behind them to their anterior surface.   Within the canal the˙hernia is very small, being constricted by the unyielding structures which form that passage ; but when it has passed beyond the saphenous opening, it enlarges in the loose fatty layers of the groin ; and, as the tumour increases, it extends outwards in the groin towards the iliac spine of the hip-bone.   Hence its greatest diameter is transverse.

Fig. 712.—VIEW OF THE RELATION OF THE VESSELS OF THE GROIN TO A FEMORAL HERNIA, &c. (from R. Quain). ¼

Fig. 712.

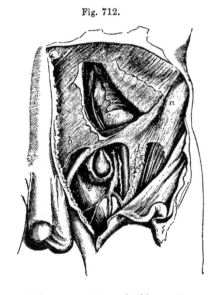

In the upper part of the figure a portion of the flat muscles of the abdomen has been removed, displaying in part the transversalis fascia and peritoneal lining of the abdomen : in the lower the fascia lata of the thigh is in part removed and the sheath of the femoral vessels opened : the sac of the femoral hernial tumour has also been opened.

a, anterior superior spinous process of the ilium ; b, aponeurosis of the external oblique muscle above the external inguinal aperture ; c, the abdominal peritoneum and fascia transversalis ; d, the iliac portion of the fascia lata near the saphenic opening ; e, sac of a femoral hernia ; 1, points to the femoral artery ; 2, femoral vein at the place where it is joined by the saphenic vein ; 3, epigastric artery and vein passing up towards the back of the rectus muscle ; +, placed upon the upper part of the femoral vein close below the common trunk of the epigastric and an aberrant obturator artery ; the latter artery is seen in this case to pass close to the vein and between it and the neck of the hernial tumour.

Coverings of the hernia.—The sac which is pushed before the protruded viscus, is derived from the external fossa of the peritoneum ; except, however, when the cord of the obliterated umbilical artery is placed outside its ordinary position, in which case the serous membrane furnishes the sac from its internal fossa.   After the sac, the hernia carries before it the subserous membrane (septum crurale of Cloquet), which covers the femoral ring, and likewise an elongation from the sheath of the femoral vessels. These two structures combined constitute a single very thin covering, known as the fascia propria of the hernia (Cooper).   It sometimes happens that the hernia is protruded through an opening in the sheath, which therefore in that event does not contribute to form the fascia propria.

Diagnosis.—Passing over the general symptoms of abdominal herniæ and the means of forming the diagnosis between a hernia and several other diseases with which it is liable to be confounded,—subjects which fall within the province of treatises on practical surgery,—the observations to be made in this place may be limited to the anatomical circumstances which characterise femoral hernia, and serve to distinguish it from the inguinal form of the complaint.   When the inguinal hernia descends to the scrotum or to the labium pudendi, and when the femoral hernia extends some distance outwards in the groin, no error in diagnosis is likely to arise. It is only in distinguishing between a bubonocele and a femoral hernia of moderate size that a difficulty occurs. The position of the femoral hernia is, in most cases, characteristic. The tumour is upon the thigh, and a narrowed part, or neck, may be felt sinking into the thigh near its middle. Besides, the femoral arch is usually to be traced above this hernia, while that band is lower than the mass of a tumour lodged in the inguinal canal. At the same time the inguinal tumour covers the femoral arch, and cannot be withdrawn from it like a femoral hernia, when it has turned over that cord. Some assistance will be gained, in a doubtful case, from the greater facility with which the tumour emerging at the saphenous opening admits of being circumscribed, in comparison with the bubonocele, which is bound down by a more resistent structure—the aponeurosis of the external oblique muscle. Other practical applications of the foregoing anatomical observations come now to be considered.

The taxis.—During the efforts of the surgeon to replace the hernia, the thigh is to be flexed upon the abdomen and inclined inwards, with a view to relax the femoral arch ; the tumour is, if necessary, to be withdrawn from over the arch, and the pressure on it is to be directed backwards into the thigh.

The operation.—The replacement of the hernia by the means just adverted to being found impracticable, an operation is undertaken with the view of dividing the femoral canal (or some part of it), thereby widening the space through which the protruding viscus is to be restored to the abdomen, or with the view of relieving strangulation when the restoration of the part is not possible or not desirable. Inasmuch as the manner of conducting the operation chiefly depends on the place at which the constricting structures are to be cut into, it will be convenient in the first instance to determine this point ; and with this object we shall inquire into the practicability and safety of making incisions into the femoral canal at different points of its circumference. As the hernia rests upon the pelvis, the posterior part of the canal may at once be excluded from consideration ; so likewise may its outer side on account of the position of the femoral vein, and also the outer part of its anterior boundary, because of the presence of the epigastric artery in this direction. There remains only the inner boundary with the contiguous part of the anterior one, and through any point of this portion of the ring or canal an incision of the required extent (always a very short one) can be made without danger in nearly all cases. The sources of danger are only occasional ; for the urinary bladder, when largely distended, and the obturator artery when it turns over the femoral ring—a very unusual course—are the only parts at the inner side of the hernia liable to be injured ; while the last-named vessel, when it follows the course just referred to, and in the male the spermatic cord, are the structures in peril when the anterior boundary of the canal is cut into towards the inner side of the hernia (see p. 624 and fig. 291).

Returning now to the steps of the operation :—After it has been ascertained that the urinary bladder is not distended, the skin is to be divided by a single vertical incision made on the inner part of the tumour, and extending over the crural arch. When the subcutaneous fat (the thickness of which is very various in different persons) is cut through, a small bloodvessel or two are divided, and some lymphatic glands may be met with. The hæmorrhage from the blood-vessels seldom requires any means to restrain it ; but the glands, if enlarged, retard the operation in some degree. The fascia propria of the hernia, which succeeds to the sub-cutaneous fat, is distinguished by its membranous appearance and the absence of fat. It is very thin, and caution is required in cutting through it, as the peritoneal sac is immediately beneath : the two membranes are indeed in contact, except in certain cases to be presently noticed. A flat director is now to be insinuated between the hernial sac and the inner side of the femoral canal, space for the instrument being gained by pressing its smooth surface against the neck of the hernia. On the groove of the director so introduced, or under the guidance of the fore finger of the left hand if the use of the director should be dispensed with, the probe-pointed bistoury is passed through the canal, and the dense fibrous structure of which it consists is divided, the edge of the knife being turned upwards and in-wards, or directly upwards. By the former plan of relieving the stricture, the parts divided are the following,—viz., the falciform process of the fascia lata and the structures fixed to the pectineal line of the pubes, namely, Gimbernat's ligament, and, it may be, the tendon of the two deep abdominal muscles, with the fascia transversalis, and the inner end of the deep femoral arch ; while if the incision be directed upwards, the falciform process of the fascia lata and the two femoral arches are divided. The opening being sufficiently dilated, the protruded part is restored to the abdomen as with the taxis.

But it may be found necessary to lay open the hernial sac in order to examine its contents, or in order to relieve the impediment to the return of the hernia if that should happen to reside in the neck of the sac itself. In this case it will probably be required to add to the vertical incision already made through the integuments and superficial fascia, another directed out-wards over the tumour, and parallel with the femoral arch. Such additional incision is readily made, by passing the scalpel beneath the integument and fat, and cutting outwards after the skin has been pierced with the point of the knife. The sac being now opened, the hernia knife is used at the inner side of its neck, while the bowel is guarded with the left hand. During the restoration of the protruded parts, some advantage will be gained if the edges of the divided sac should be held down with a pair or two of forceps in the hands of an assistant.

In the foregoing observations, it has been stated that the fascia propria is in contact with the sac of the hernia, except in certain cases. The exception is afforded by the interposition of fat, and sometimes in consider-able quantity. The adipose substance is deposited in the subserous mem-brane ; it has the peculiarity of resembling the fat lodged in the omentum, and it is occasionally studded with small cysts, containing a serous fluid. The hernia will be most readily found in such circumstances behind the inner part of the adventitious substance ; which should be turned outwards from the inner side, or cut through.

III.—THE PERINÆUM AND ISCHIO-RECTAL REGION.

A connected view of the structures which occupy the outlet of the pelvis becomes necessary, in consequence of the important surgical operations occa-sionally performed on the genito-urinary organs and the rectum, which are contained in that part. In the examination of these structures, which it is proposed to make in this place, attention will be confined to the male body.

The hip-bones as they bound the outlet of the pelvis are already sufficiently described (p. 97). The anterior portion of the space, which is appropriated to the urethra and the penis, is named the *perinæum*. This part is triangular, the sides being formed by the sides of the pubic arch meeting at the symphysis pubis, while a line extended between the two ischial tuberosities represents the base of the triangle. In well-formed bodies the three sides of the space are equal in length ; but cases occur in which, by the approximation of the ischiatic tuberosities, the base is narrowed ; and we may anticipate the practical application of the anatomical facts so far as to state here, that this circumstance exercises a material influence on the operation of lithotomy, inasmuch as the incisions required in that operation, instead of being oblique in their direction, must, in such circumstances, be made more nearly straight backwards.

That portion of the outlet of the pelvis which lies behind the perinæum may be named the ischio-rectal region. It contains the end of the rectum ; and it is defined by the ischial tuberosities, the coccyx, and the great gluteal muscles. We shall now proceed to the detailed examination of the two parts thus mapped out.

The skin of the *perinæum* continued from the scrotum, and partaking of the characters it has on that part, is dark-coloured, thin, and extensible, loosely connected with the subjacent textures, and in the male body studded with crisp hairs. Around the anus, it is thrown into folds, which are necessary to allow the extension of the orifice of the bowel, during the passage of masses of fæcal matter ; and along the middle of the perinæum the median ridge or raphe of the scrotum is continued backwards to the anus. By this mark upon the skin, the large triangle in which is comprised

Fig. 713.

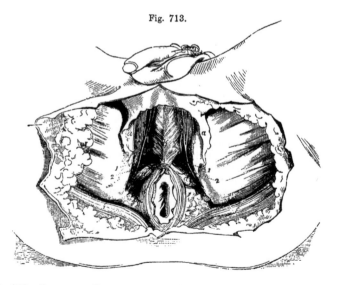

Fig. 713.—Superficial Dissection of the Perinæum and part of the Thighs.

a, superficial fascia ; *b*, accelerator urinæ ; *c*, erector penis ; *d*, transversus perinæi ; *e*, upper point of sphincter ani ; *f*, the edge of the gluteus maximus; 1, superficial perineal artery ; 2, superficial perineal nerve.

the whole perinæum, is subdivided into two equal parts.   To one of these smaller spaces the operations usually performed for gaining access to the urinary bladder are for the most part restricted.   The skin of the perinæum is provided with numerous sebaceous follicles.

From the muscles of the perinæum the skin is separated by areolar tissue and fat, except in the neighbourhood of the anus, where the sphincter of the bowel is immediately in contact with the integument.   The deeper part of the fatty subcutaneous membrane,—the *superficial fascia* (p. 259),— taking on a fibrous appearance, has, in a great measure, the same arrangement and characters as the corresponding structure of the groin.   With that membrane the layer is continuous in front through the scrotum, but at other points it is confined to the perinæum, being fixed laterally to the sides of the pubic arch, while it is continued posteriorly, beneath the sphincter ani and in front of the rectum, into the deep perineal fascia. It is in consequence of these connections of the superficial fascia, that abscesses do not attain a large size in the perinæum, and that urine effused in consequence of rupture of the urethra does not extend backwards to the rectum or outwards to the thigh, but continues forwards, and, if an outlet for its escape should not be afforded by the surgeon, reaches successively the scrotum, the penis, and the groin above Poupart's ligament.   In extreme cases the extravasated fluid would spread from the position last mentioned over the anterior part of the abdomen and even to the thorax, its extension downwards to the thigh being restrained by the attachment of the superficial fascia along the fold of the groin.

Fig. 714.

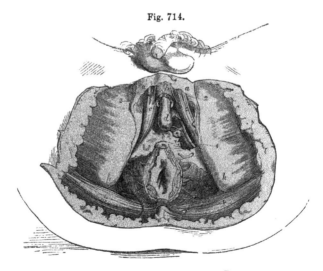

Fig. 714.—Deeper Dissection of the Perinæum.

The perineal muscles have been removed and also the fat in the ischio-rectal fossa ; *a*, superficial fascia ; *b*, accelerator urinæ ; *c*, crus penis ; *d*, the bulb ; *e*, triangular ligament of the urethra ; *f*, levator ani ; *g*, sphincter ; *h*, tuberosity of the ischium ; *k*, gluteus maximus ; *, Cowper's gland of the left side ; 1, pudic artery ; 2, superficial perineal artery and nerve.   The inferior hæmorrhoidal arteries and the artery of the bulb are likewise shown.

of urine, which, after the operation, continues for a time to trickle from the bladder. But the prostate and the neck of the bladder, on the contrary, are to be incised only for a small extent. The reasons for this rule may be stated as follows. By accumulated experience in operations on the living body, it has been found that the structures now under consideration, when slightly cut into, admit of dilatation, so as to allow the passage of a stone of considerable size, and that no unfavourable consequence follows from the dilatation. Moreover, when these parts are freely divided (cut through), the results of lithotomy are less favourable than in the opposite circumstances. The less favourable results adverted to appear to be due to the greater tendency to infiltration of urine in the subserous tissue of the pelvis ; and the occurrence of this calamity probably depends on the fact that, when the prostate has been fully cut through, the bladder is at the same time divided beyond the base of the gland, and the urine then is liable to escape behind the pelvic fascia (which it will be remembered is connected with both those organs at their place of junction) ; whereas, if the base of the gland should be left entire, the bladder beyond it is likewise uninjured, and the urine passes forwards through the external wound.

The steps of the operation by which the foregoing general rules are sought to be carried out are the following. The grooved staff having been passed into the bladder (and this instrument ought to be of as large size as the urethra will admit), and the body or the patient, as the case may be, having been placed in the usual position—by which position the perinæum is brought fully before the operator with the skin stretched out—the first incision is begun about two inches before the anus, a little to the left of the raphe of the skin, and from this point it is carried obliquely backwards in a line about midway between the tuber ischii and the anus, extending a little way behind the level of the latter. During the incision, the knife is held with its point to the surface, and it is made to pass through some of the subcutaneous fatty layer as well as the skin. Now, the edge of the knife is applied to the bottom of the wound already formed, in order to extend it somewhat more deeply ; and the forefinger of the left hand is passed firmly along for the purpose of separating the parts still farther, and pressing the rectum inwards and backwards out of the way. Next, with the same finger passed deeply into the wound from its middle and directed upwards, the position of the staff is ascertained, and the structures still covering that instrument are divided with slight touches of the knife,—the finger pressing the while against the point at which the rectum is presumed to be. When the knife has been inserted into the groove of the staff (and it reaches that instrument in the membranous part of the urethra) it is pushed onwards through the prostatic portion of the canal with the edge turned to the side of the prostate, outwards, or, better, outwards with an inclination backwards. The knife being now withdrawn, the forefinger of the left hand is passed along the staff into the bladder. With the finger the parts are dilated, and with it, after the staff has been withdrawn, the position of the stone is determined and the forceps is guided into the bladder.

In case the calculus is known to be of more than a moderate size, and the knife used is narrow, the opening through the side of the prostate may be enlarged as the knife is withdrawn, or the same end may be attained by increasing the angle which that instrument, while it is being passed onwards, makes with the outer part of the staff. And if the stone should be of large size, it will be best to notch likewise the opposite side of the prostate before the forceps is introduced. The same measure may be resorted to afterwards should much resistance be experienced when the foreign body is being extracted. Lastly, this part of the operation (the extraction of the stone)

Fig. 718.

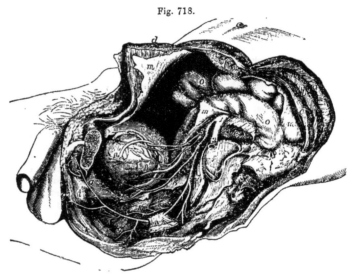

Fig. 718.—View of the Distribution of the Arteries to the Viscera of the Male Pelvis, as seen on the removal of the Left Os Innominatum, &c. (from R. Quain). ¼

a, left external oblique muscle of the abdomen divided ; *b*, internal oblique ; *c*, transversalis ; *d, d*, the parts of the rectus muscle divided and separated ; *e*, psoas magnus muscle divided ; *f*, placed on the left auricular surface of the sacrum, points by a line to the sacral plexus of nerves ; *g*, placed on the os pubis, sawn through a little to the left of the symphysis, points to the divided spermatic cord ; *h*, the cut root of the crus penis ; *i*, the bulb of the urethra ; *k*, elliptical sphincter ani muscle ; *l*, a portion of the ischium near the spinous process, to which is attached the short sacro-sciatic ligament ; *m*, the parietal peritoneum ; *n*, the upper part of the urinary bladder ; *n, n′*, the left vas deferens descending towards the vesicula seminalis ; *n″*, the left ureter ; *o*, the intestines ; 1, the common iliac at the place of its division into external and internal iliac arteries ; 2, left external iliac artery ; 3, internal iliac ; 4, obliterated hypogastric artery, over which the vas deferens is seen passing, with the superior vesical artery below it ; 5, middle vesical artery ; 6, inferior vesical artery, giving branches to the bladder, and descending on the prostate gland and to the back of the pubes ; 7, placed on the sacral plexus, points to the common trunk of the pudic and sciatic arteries ; close above 7, the gluteal artery is seen cut short ; 8, sciatic artery cut short as it is escaping from the pelvis ; 9, placed on the rectum, points to the pudic artery as it is about to pass behind the spine of the ischium ; 9′, on the lower part of the rectum, points to the inferior hæmorrhoidal branches ; 9″, on the perinæum, indicates the superficial perineal branches ; 9‴, placed on the prostate gland, marks the pudic artery as it gives off the arteries of the bulb and of the crus penis ; 10, placed on the middle part of the rectum, indicates the superior hæmorrhoidal arteries as they descend upon that viscus.

should be conducted slowly, so as gradually to dilate the parts without lacerating them ; and the forceps should be held with its blades one above the other.

The Structures divided in the Operation.—In the first incision the integument and the subjacent fatty layer are divided ; afterwards a small part of the accelerator urinæ, and the transversus perinœi with the transverse artery. Then the deep perineal fascia with the muscular fibres between its layers, the membranous part of the urethra, the prostatic part of the canal, and, to a small extent, the prostate itself are successively incised.

The Blood-vessels : their relation to the incisions.—The transverse artery of the perinæum with, it may be, the superficial artery of the perinæum, is the only artery necessarily cut through when the vessels have their accustomed arrangement ; for in such circumstances the artery of the bulb is not endangered if the knife be passed into the staff in a direction obliquely upwards, the artery being anterior to the groove of that instrument ; neither is there a risk of wounding the pudic artery; unless the incisions through the deep parts (the prostate for instance) should be carried too far outwards.*

But in some cases the arteries undergo certain deviations from their accustomed arrangement, whereby they are rendered liable to be wounded in the operation. Thus, the artery of the bulb when it arises, as occasionally happens, from the pudic near the tuber ischii, crosses the line of incision made in the operation.† The arterial branches ramifying on the prostate are in some instances enlarged, and become a source of hæmorrhage,‡ and the veins, too, on the surface of that gland, when augmented in size, may give rise to troublesome bleeding.§ Lastly, it should be added that the occasional artery (accessory pudic), which takes the place of the pudic when defective, inasmuch as it lies on the posterior edge of the prostate, might be divided if the gland were cut through to its base, and only in this event.‖

---

* For reference to some cases in which the pudic artery was divided in lithotomy, see Crosse's "Treatise on Urinary Calculus," p. 21. London, 1835.

† "The Anatomy of the Arteries," &c., by R. Quain, p. 442, and plate 64†, figs. 1 and 2. A case in which death resulted from division of the artery of the bulb is recorded by Dr Kerr, in the "Edinb. Med. and Surg. Journal," July, 1817, p. 155.

‡ See an essay, entitled "Remarks on the Sources of Hæmorrhage after Lithotomy," by James Spence, in the "Edinburgh Monthly Journal of Medical Science," vol. i. p. 166 ; 1841. And "The Arteries," &c., by R. Quain, p. 445.

§ "The Arteries," &c., by R. Quain, p. 446, and plate 65, fig. 3.

‖ Ibid. p. 444, and plate 63. An instance in which fatal consequences resulted from the division of such an artery has been placed on record. See "Case of Lithotomy attended with Hæmorrhage," by J. Shaw, in "The London Medical and Physical Journal," vol. lv. p. 3, with a figure. 1826.

DIVISION III.

DISSECTIONS.

THE object of the following Directions is to serve as a short and simple guide for the display of the structure of the body by students in dissecting-rooms, the various organs and their parts being mentioned in the order in which they may best be exposed, and such methods being indicated as may enable each student to obtain the greatest amount of information from his dissection, and at the same time to prevent interference among the neighbouring dissectors as much as possible.

I. GENERAL MANAGEMENT OF THE DISSECTIONS.

1. In different schools, various plans are pursued in the allotment of portions of the body to different dissectors. According to the method here recommended, the subject is divided into ten parts, five on each side of the body, which are left in connection with one another until the dissection is sufficiently advanced to admit of their being conveniently separated. The boundaries of the parts are so adjusted, that by their due observance interference between the different dissectors may be as much as possible avoided.

2. In the case of a male subject, a day is recommended to be set apart at the commencement for the dissection of the perinæum. Thereafter, and in the case of a female subject, immediately on its being brought into the rooms, the subject is to be placed with the face downwards for four days, during which time the posterior regions are to be dissected, in so far as within reach, in the order afterwards mentioned for each part. It is then to be turned and laid upon its back, when a dissection of the various parts in front is to be made. The whole dissection is supposed to be completed within six weeks—the time fixed by the Anatomy Act.

3. The dissection of the head and neck and of the limbs should be begun at once when the subject is laid upon its face; that of the abdomen as soon as it is turned on the back, and the thorax must not be opened until the upper limbs are removed. The limbs ought not to be removed until the parts which connect them with the trunk have been fully dissected, and an opportunity has been given for the examination of the surgical anatomy of the subclavian artery and the parts concerned in hernia, by the dissectors of the head and the abdomen; all of which may be accomplished before the tenth day. The further dissection of the several parts may then proceed in accordance with the methods suggested in the special directions.

4. It is to be observed that, although in the special directions all the organs mentioned are supposed to be brought under review in one dissection, it may be necessary for the student, in order to obtain a full knowledge of them, to dissect each of the parts more than once. This is especially the case with the head and neck. It is incumbent therefore upon the student to make a selection of different objects in each dissection, under the guidance of the demonstrator, in order that he may progressively obtain a full view of the whole.

5. Those students who have not previously dissected, are recommended to select the limbs for their first and second dissections, after they shall have obtained a sufficient knowledge of the bones and joints ; and for the most part, the junior students ought not, in a first or second dissection, to attempt to expose more than the muscles and the largest vessels and nerves. In their third and subsequent dissections they will gradually come to make a more complete display of all the parts.

6. In the dissection of the limbs, no interference between the dissectors of opposite sides can occur ; but in the head and neck, thorax and abdomen, there is a necessity for the students who are engaged with the parts of opposite sides to act in concert. The viscera must be examined by them together, and it will frequently happen that the dissectors of only one side can work at the same time. When such is the case, the one dissector should give his assistance to the other by reading or otherwise ; and it will sometimes be found advantageous for those having the same parts of opposite sides to make in concert different kinds of dissections on the opposite sides of the body ; as for example, to dissect the muscles chiefly on one side, and the vessels and nerves on the other, or the orbit from above on one side and in a lateral view on the other, etc.

II. SPECIAL DIRECTIONS FOR THE DISSECTION OF EACH PART.

I.—HEAD AND NECK.

THE right and left sides of this region constitute each a part. Its dissection may occupy the full time, or about six weeks ; two hours or more daily being devoted to it. Its inferior boundary extends from the sternum, along the clavicle, to the acromion process ; and thence to the spinous process of the third cervical vertebra. It may be found impossible to follow out in one part the whole of the dissections indicated below ; and therefore the dissector ought rather, if his time is limited, to make a selection for repeated dissections, following, as nearly as possible, the methods described. Many of the smaller points of detail may be passed over by the junior student ; and there are some which can only be observed in a favourable condition of the subject.

1. *Integument of the Cranium.*—The subject being placed with the face downwards, during the first two days, the scalp and the back of the neck (to the third cervical vertebra) are to be dissected ; and while this is being done, only one dissector should work at a time. An incision is to be made along the middle line, from the spinous process of the third cervical vertebra, forwards over the head, to the root of the nose, and another from immediately behind the ear to meet the first at the vertex, care being taken not to cut deeper than through the skin. The flaps of integument thus marked out are to be reflected from above downwards, the posterior one first.

At the back of the neck the posterior and upper parts of the trapezius and sterno-mastoid muscles will be laid bare (pp. 200 and 193) ; and, between these, a part of the splenius muscle, and, when the trapezius is not strongly developed, a small angle of the complexus muscle will be brought into view. These muscles are to be left undivided at present. On the posterior part of the cranium the structures to be examined are the occipital artery and vein, and the great occipital nerve, which pierces the complexus and trapezius muscles (pp. 351 and 634) ; the small occipital nerve, which passes upwards along the posterior border of the sterno-mastoid muscle (p. 638) ; and, beneath these, the occipital part of the occipito-frontalis muscle (p. 169), which passes upwards from the superior curved line of the occipital bone. Behind the ear are the retrahens auriculam muscle and the posterior auricular artery and nerve (pp. 353 and 612) ; above the ear is the attollens auriculam muscle ; and in front of the ear the attrahens auriculam muscle connected with the attollens, the temporal artery and vein, the small temporal branch of the third division of the fifth nerve, and the superior branches of the facial nerve (pp. 170, 353, 612, and 606). Passing upwards on the forehead, are the frontal part of the occipito-frontalis muscle, the frontal vein, the supraorbital and frontal arteries, and the supraorbital and supratrochlear nerves (pp. 360 and 597).

2. *Interior of the Cranium and Brain.*—During the third and fourth days the brain and its membranes are to be removed and studied, and the interior of the base of the skull dissected to show the sinuses, blood-vessels, and nerves ; and, if there is time (as may be the case, should the head have been previously opened), the orbit may be examined from above. To remove the calvarium, the temporal aponeurosis and upper part of the temporal muscle having been dissected, let the scalpel be carried round the cranium from a point a little above the occipital protuberance, so as to pass across the forehead at about an inch above the orbits ; and having cleared a small portion of the bone on the circle so made, let the external table. of the skull be sawn through, leaving the inner table undivided. Let the inner table be cracked completely round by a few smart strokes of the chisel and mallet, and the calvarium may then be pulled away from the dura mater which lines it. The superficial aspect of the dura mater having been observed, and the superior longitudinal sinus laid open and inspected (p. 462), the dura mater is to be divided on a level with the sawn edge of the skull, excepting where it touches the middle line ; this will permit the arachnoid membrane and pia mater to be examined, as well as the cerebral veins entering the superior longitudinal sinus ; and when these veins are divided, the falx cerebri will be seen dipping down between the cerebral hemispheres. The falx cerebri is then to be separated from its attachment to the crista galli and thrown backwards (p. 562).

It will now be in the dissector's option to remove the brain at once from the body, or to examine it in situ as far as the ventricles. If the latter plan, which is generally to be preferred, be adopted, the dissectors ought first to examine the convolutions of the upper aspect of the brain, noticing the anterior and posterior cerebral arteries arching respectively backwards and forwards ; they will then slice away the hemispheres to the level of the corpus callosum, and observe the extent of that structure, its transverse markings, the raphe and the longitudinal lines (p. 540). They will proceed by incisions at the sides of the corpus callosum, to open the lateral ventricles separately, so as to expose their cavities with the anterior and posterior cornua and the parts lying on their floor : they must afterwards cut across the corpus callosum near the forepart, and raising it carefully, divide with scissors the septum lucidum which separates the lateral ventricles, and notice between its layers the fifth ventricle. The lateral ventricles having been thus thrown into one, the structures forming their floor are more fully seen, viz., the corpora striata, tæniæ semicirculares, the optic thalami iu part, the choroid plexus, the upper surface of the fornix, the foramen of Monro, the anterior and posterior cornua, and the hippocampus minor. The descending cornu is now to be exposed, on one side only, by cutting away the cerebral substance above and external to it, and in it will be found the hippocampus major, pes hippocampi, tænia hippocampi, and fascia dentata.

The fornix is to be divided immediately above the foramen of Monro, and reflected ; by which means its inferior surface will be brought into view, as also the upper surface of the velum interpositum. The connections of the anterior extremity of the velum may then be cut across, and that structure likewise turned back so as to lay bare the third ventricle ; but in doing this care should be taken lest the pineal body, which is adherent to the under surface of the velum interpositum near its back part, should be raised out of its place. The objects seen in and near the third ventricle, are now to be studied : viz., the optic thalami, the three smaller commissures, viz., anterior, middle and posterior, the pineal body and its crura, the corpora quadrigemina, and the anterior opening of the iter a tertio ad quartum ventriculum ; also the anterior crura of the fornix should be traced down as far as possible towards the corpora albicantia. The velum having been replaced, the transverse fissure of the cerebrum ought now to be opened by division of the remains of the corpus callosum and fornix in the middle line, and it may be followed in its whole extent to the extremity of the descending cornu. By this proceeding the veins of Galen will be traced back through the velum interpositum to the margin of the tentorium, and, on division of the velum, the valve of Vieussens and the origin of the fourth nerve, as well as that of the optic tract, may be seen ; but if the view of these objects should not be satisfactory, they may be again examined after removal of the brain from the skull.

The remaining part of the brain is to be removed by cutting the tentorium on each side sufficiently to allow the cerebellum to be raised, and dividing the spinal cord and vertebral arteries as low as possible, the spinal accessory and suboccipital nerves, and the cranial nerves in order from behind forward, with the infundibulum and internal carotid arteries ; after which the brain is to be laid on a flat plate with the base uppermost. If, however, it has been decided to remove the brain entire from the body, this may either be done in the manner now described, or, with the subject temporarily placed for the purpose on its back. According to the latter mode of procedure, which is the most customary, the anterior lobes of the brain are gently raised, and the olfactory bulbs lifted from the surface of the ethmoid bone ; the optic nerves, internal carotid arteries, the infundibulum, and the third pair of nerves are successively divided ; the anterior attachment of the tentorium is then to be cut on each side so as to secure the divisions of the fourth pair of nerves before they have any chance of being torn. The tentorium is then to be more extensively divided, and after it, in their order, the remaining nerves, the vertebral arteries, and the spinal cord. In studying the base of the brain, the distribution of the arteries should be first observed, with their union in the circle of Willis (p. 363). After they are removed, and the less adherent portions of the arachnoid membrane and pia mater are stripped off, except from the angle between the cerebellum and medulla oblongata, the principal parts of the brain visible from the base are to be examined. These are :—the fissure of Sylvius separating the anterior from the middle lobe, and contained in it the hidden convolutions or island of Reil ; at the entrance of the fissure the locus perforatus anticus, and terminating in it the inferior part of the transverse fissure of the cerebrum ; also, the crura cerebri emerging from before the pons Varolii, the anterior extremity of the corpus callosum lying in the bottom of the great longitudinal fissure, and below it, proceeding backwards in the middle line, the lamina cinerea, the optic commissure, the tuber cinereum, the infundibulum, the corpora albicantia, and the locus perforatus posticus (p. 536). The principal objects to be noted on the medulla oblongata are the anterior pyramids with their decussation and the olivary bodies on the front, and the restiform bodies on its lateral aspect ; posteriorly are the posterior pyramids, and the calamus scriptorius, and its prolongation downwards into the minute remains of the central canal of the spinal cord (p. 514). The fourth ventricle, situated between the medulla oblongata and cerebellum, is now brought into view, and at its sides will be observed the fringes of pia mater called choroid plexus of the fourth ventricle, the two small lobules of the cerebellum at the sides of the medulla oblongata named the flocculi or subpeduncular lobes, and behind them the amygdalæ ; while above the medulla are the parts belonging to the middle lobe of the cerebellum, afterwards more fully noticed (p. 521).

The origins of the cranial nerves may next be examined. The first pair or olfactory tracts and bulbs are seen on the anterior lobes, and should be traced back to the white striæ by which they arise at the inside of the fissure of Sylvius ; the second and fourth nerves are seen passing round the crura cerebri, the optic tracts from the corpora quadrigemina, optic thalami, and corpora geniculata, the fourth nerve from the valve of Vieussens ; the third pair lying close together on the inner aspects of the crura cerebri ; the fifth pair emerging by two roots from the front of the pons Varolii ; the sixth in front of the anterior pyramids ; the seventh nerve in two parts, the portio dura and portio mollis, in the angle between the medulla oblongata, pons Varolii, and cerebellum ; the eighth pair in three parts, the glosso-pharyngeal, vagus or pneumo-gastric, and spinal accessory, in front of the restiform body ; the ninth pair in front of the olivary body ; and the suboccipital (or first cervical nerve of some authors) close below the ninth (p. 583). The cerebellum is to be separated from the structures to which it is attached by division of its superior, middle and inferior crura. The general disposition of its convolutions and the superior vermiform process will be noted, as also the parts entering into the formation of the inferior vermiform process lying in the vallecula beneath, viz., the pyramid, uvula, and laminated tubercle, together with the posterior velum. Sections of the cerebellum are to be made to exhibit the arbor vitæ and the grey centre known as the corpus dentatum or rhomboideum. In conclusion, sections may be made of the pons Varolii to show its transverse and longitudinal fibres, of the medulla oblongata to show the olivary nucleus or corpus dentatum, and of the crura cerebri to show the locus niger.

The venous sinuses, arteries, and nerves in the base of the skull, ought now to be examined, if there be time, before the subject is turned on its back. The superior longitudinal sinus, the inferior longitudinal and the straight sinus (with the veins of Galen entering it), and the posterior occipital sinus, are to be traced to the torcular Herophili ; and the lateral sinuses from that point to the jugular foramina. The cavernous sinuses, joined together by means of the circular sinus, are then to be opened ; and the superior and inferior petrosal sinuses, and the transverse sinus (p. 461). In the vicinity of the cavernous sinus the relations of the 3rd, 4th, 5th, and 6th nerves are to be exhibited, and also the internal carotid artery and the Gasserian ganglion (pp. 594 and 359) ; after which the nerves are to be replaced in situ and protected with cotton dipped in spirits, that they may be ultimately traced forward in the dissection of the orbit. The pituitary body is to be removed from its position in the sella turcica, and its form and structure examined (p. 539).

If the above examination of the sinuses cannot be accomplished, at this stage of dissection, the interior of the skull must be carefully cleaned, and protected from the air by replacing the skull-cap or otherwise. The dissectors must also attend to the preservation of the parts at the back of the neck before the subject is turned.

3. *Cervical Region superficially, and Posterior Cervical Triangle.*—It is essential that within four days after the subject has been laid upon its back, the dissection of the posterior and inferior triangle of the neck be completed, so that the third part of the subclavian artery may be seen to advantage before the clavicle and the vessels and nerves of the superior extremity are divided.

With this view, a superficial dissection is advised of the whole cervical region. Make an incision in the middle line from the sternum to the chin ; another from the acromion, along the clavicle, to the sternum ; and a third from the chin to the back of the ear ; and let the flaps so obtained be reflected backwards ; care being taken not to injure the fibres of the platysma myoides, nor the nerves which lie in the superficial fascia. The platysma is to be examined and reflected upwards (p. 178); after which, let the external and anterior jugular veins be laid bare, and also the cutaneous branches of the cervical plexus of nerves, viz. :—superiorly, the superficial cervical, great auricular, and small occipital nerves; and, inferiorly, the suprasternal, supraclavicular, and supra-acromial nerves : these will be traced most easily from their line of emergence at the posterior border of the sterno-mastoid muscle (pp. 459 and 638). Let the disposition of the deep cervical fascia also be noted (p. 197).

The dissector will then cut down through the fat at the lower part of the posterior border of the sterno-mastoid muscle, and uncover the omo-hyoid muscle, whose posterior belly emerges from behind the sterno-mastoid, and forms the superior boundary of the inferior division of the posterior triangle. He will remove the fat and lymphatic glands from the inferior triangle, until the scalenus anticus muscle is reached, which will serve as a guide to the third part of the subclavian artery and vein, and the superior trunks of the brachial plexus of nerves (pp. 366 and 643). Besides these structures, the dissector will observe, while engaged with this space, if the sterno-mastoid muscle be narrow, the phrenic nerve upon the surface of the scalenus anticus muscle ; he will find the suprascapular nerve and the small branch to the subclavius muscle both coming from the trunk formed by the fifth and sixth nerves, the transverse cervical and suprascapular arteries, and part of the scalenus medius and posticus muscles, as well as the lower set of the chain of lymphatic glands which lie along the line of the sterno-mastoid muscle (pp. 644 and 499). The superior part of the posterior triangle is next to be dissected by clearing the upper attachments of the scaleni muscles, with the splenius colli and levator scapulæ (p. 177), when the arrangement of the cervical plexus will be seen, together with the origin of the phrenic nerve (p. 636); also the spinal accessory nerve emerging from the substance of the sterno-mastoid muscle, and forming connections with the cervical plexus before it disappears beneath the trapezius muscle (p. 625). The seven cervical and first dorsal nerves are to be cleaned up to their emergence from the intervertebral foramina, the communicating branches of the sympathetic nerve being preserved if possible (p. 691); and the posterior thoracic nerve and the branch to the rhomboid muscles are to be found (p. 643).

4. *Anterior Triangle and Deep parts of the Neck.*—Let a dissection of the deep fascia and of the sterno-hyoid and sterno-thyroid muscles be made in the middle line between the larynx and sternum, to exhibit the relations of the trachea as connected with the operation of tracheotomy (p. 888), in particular noticing the position of the innominate artery, the common carotid arteries, the thyroid body, the inferior thyroid veins, and the arteria thyroidea ima, if it be present (pp. 340 and 920). The dissection of the anterior triangle of the neck is now to be proceeded with, by cleaning the whole of the sterno-mastoid, sterno-hyoid and sterno-thyroid muscles, and the anterior belly of the omo-hyoid muscle (p. 191); and in front of the sheath of the great vessels the descendens noni nerve, with its twigs to the three last named muscles, is to be laid bare (p. 626). Let the sheath of the vessels be opened, and the upper part of the common carotid artery exposed, with the pneumo-gastric nerve and internal jugular vein beside it; mark the place of its division into external and internal carotid arteries, and examine the first part of these two vessels, following the external carotid up to the parotid gland. Let the digastric and stylo-hyoid muscles be cleaned, and the parts be exposed in the submaxillary triangle, viz., the superficial part of the submaxillary gland, the submental branch of the facial artery, and the mylo-hyoid muscle, with the nerve that supplies it (pp. 183 and 608); observe also the ninth cranial or hypoglossal nerve lying close to the stylo-hyoid muscle, and dissect out its branch to the thyro-hyoid muscle (p. 627).

The sterno-mastoid muscle is to be divided about three inches from its upper end, and the superior part is to be dissected quite up to the bone, care being taken not to cut the spinal accessory nerve which pierces it. The sterno-hyoid and sterno-thyroid muscles ought now to be divided near their lower end, the thyroid body dissected, and its form and relations noted. The dissector will then direct his attention to the branches of the external carotid artery; he will dissect the superior thyroid artery and note its sterno-mastoid branch (already cut), and the hyoid, laryngeal, and crico-thyroid branches; he will dissect also the commencement of the ascending pharyngeal artery, the occipital artery as far as the occipital groove of the temporal bone, the posterior auricular artery, the lingual artery as far as the border of the hyoglossus muscle, and the facial artery as far as the lower jaw (p. 346); he will also lay bare the pneumo-gastric nerve as far as convenient, tracing the superior and external laryngeal branches (p. 622).

In the lower part of the neck, the subclavian artery is now to be examined in the three parts of its course; and the different relations of the subclavian and common carotid arteries in the first part of their courses on the two sides of the body are to be carefully compared (p. 364). The internal jugular and the subclavian veins, with the branches entering them, are to be dissected, and on the left side the arched part of the thoracic duct descending into the angle of junction of these two veins (pp. 459, 469, and 488). The branches of the subclavian artery are to be displayed, viz., the vertebral and internal mammary arteries, the thyroid axis, from which arise the inferior thyroid giving off the ascending cervical artery, the suprascapular artery, and most frequently the transverse cervical dividing into the superficial cervical and posterior scapular ; lastly, there are the deep cervical and superior intercostal arteries coming off either as a single trunk or separately (p. 366). The frequent origin of the posterior scapular artery from the third part of the subclavian artery and other varieties will here require to be attended to (p. 372). The trunk of the sympathetic nerve is to be dissected, with its three cardiac and its other branches, as high as the first cervical nerve (p. 688) ; and the recurrent laryngeal branch of the vagus nerve is to be found between the gullet and trachea, and traced up to the larynx (p. 622).

5. *Superficial Dissection of the Face.*—In proceeding with this region, the dissectors ought to expose in concert the superficial muscles of the face on one side, keeping only the principal blood-vessels and nerves. They ought likewise to make a more detailed exposure and dissection of these vessels and nerves on the other side, for which purpose the superficial muscles must be in some measure sacrificed. If this method cannot be followed in concert, each dissector must display as much as possible all the parts on his own side, in which case he will do best to begin with the superficial muscles.

To exhibit the superficial muscles of the face, the skin is to be reflected from the middle line, from which one or two such transverse incisions as shall seem necessary are to be directed outwards. It is most convenient to begin with the orbicularis palpebrarum muscle, removing the skin from the circumference to the margin of the eyelids, and dividing it along these margins (p. 171). The muscles between the eye, nose and upper lip may then be exposed, the principal of which are these :—the compressor naris, the levator labii superioris alæque nasi, the levator proprius labii superioris, and the zygomatici, more deeply the corrugator supercilii, the levator anguli oris, the pyramidalis nasi continued from the frontalis, the dilatator naris, &c. Below the mouth the depressor anguli oris and depressor labii inferioris will be seen. A more complete view of the orbicularis oris may be obtained by dissecting it from the inner aspect of the lips ; and the levator menti is best displayed by making an incision down to the bone in the middle line, and dissecting outwards.

To expose the nerves and blood-vessels of the face, the skin may be reflected as stated above from the middle line outwards. The surface of the parotid gland is to be cleaned, and search made for the branches of the facial nerve as they emerge from underneath its upper and anterior margins (p. 614). The duct of the parotid gland, and the transverse facial artery, are also to be dissected (p. 354). The branches of the facial nerve are to be followed forward, and, as far as possible, their connections with the infraorbital, buccal and inferior labial branches of the fifth nerve are to be traced. Let the dissector cut the superior attachment of the levator proprius labii superioris muscle, and, dissecting down upon the infraorbital foramen, follow out the distribution of the infraorbital nerve and artery emerging from it (pp. 602 and 357). Let him also cut carefully down upon the mental foramen, and follow out the inferior labial nerve and artery emerging thence (p. 608).

The facial artery and vein with their branches are to be dissected out from the point to which they have been previously traced at the border of the jaw. The principal branches of the artery, such as the inferior labial, the superior and inferior coronary, the lateral nasal and the angular, are to be exposed (p. 350).

The branches of the facial nerve should be traced backwards through the parotid gland to the emergence of the main trunk from the stylomastoid foramen : while this is being done, the connections of this nerve with the auriculo-temporal branch of the fifth and with the great auricular nerve will be preserved, and the twigs to the posterior belly of the digastric muscle and the stylo-hyoid muscle should be sought for, close to the skull (p. 613). The continuation of the external carotid artery into the superficial temporal will be seen ; and, in dissecting out the remains of the parotid gland, the position and relations of that gland can be studied (p. 814). In this part of the dissection, the student should also observe the connections of the part of the cervical fascia which separates the parotid and submaxillary glands, and which is continuous with the strong band known as the stylo-maxillary ligament (p. 197). Finally, the dissector may clean and examine the tarsal and nasal cartilages (pp. 706 and 771).

6. *Deep Dissection of the Face.*—The masseter muscle, and the nerve and artery which enter its deep surface from the sigmoid notch of the lower jaw are to be examined (p. 181), and the temporal fascia removed, the orbital twig of the superior maxillary nerve being sought between its layers (p. 600). By means of the saw and bone-nippers, the zygomatic arch may then be divided in front and behind in such a manner as exactly to include the origin of the masseter muscle, which should be turned downwards and backwards, the masseteric nerve and artery being in the meantime preserved. Let the coronoid process be divided by a vertical and horizontal incision with the saw and nippers as low down as possible, care being taken not to cut the buccal nerve, which lies in close contact with the temporal muscle. The coronoid process with the temporal muscle attached is to be reflected upwards, and the neck of the jaw is to be divided a little below the condyle, and as much of the ramus of the jaw is to be removed as can be cut away without injury to the inferior dental artery and nerve which enter the foramen. The internal maxillary artery with its branches is to be exposed as far as can be done without injury to the external pterygoid muscle, on whose outer surface it generally lies ; it is frequently, however, covered by it. The gustatory and inferior dental nerves will be seen below the inferior border of the external pterygoid muscle, the latter nerve giving off the mylo-hyoid branch before entering the inferior dental canal, and resting on the fibrous slip commonly known as the internal lateral ligament of the jaw, between which and the jaw the internal maxillary artery likewise passes. Above the superior border of the same muscle will be seen the anterior and posterior deep temporal arteries and nerves, and between the two parts of the same muscle, the buccal nerve and vessels. After the external pterygoid muscle has been examined (p. 182), the temporo-maxillary articulation is to be studied (p. 132), and opened by cutting the external lateral ligament and dividing the capsule of the joint above and below the interarticular fibro-cartilage, and the condyle of the jaw is to be disarticulated ; care being taken not to cut the auriculo-temporal division of the inferior maxillary nerve, which is in close contact with the inner side of the capsule (p. 606). The external pterygoid muscle may now be turned forward along with the head of the jaw, and its nerve found ; after which it may be removed.

The branches of the internal maxillary artery in the vicinity of the pterygoid muscles are thus brought fully into view, viz. : in the first part of its

course, the inferior dental, the middle meningeal giving off the small meningeal artery, the two deep temporal, the pterygoid and other muscular
branches : next, more deeply within the pterygoid muscles, the posterior
superior dental and the infraorbital branches (p. 354). The chorda tympani
nerve is to be dissected upwards to the fissure of Glaser, from its point of
junction with the gustatory nerve under cover of the external pterygoid
muscle, and the branches of the inferior maxillary nerve are to be traced
back to the foramen ovale (p. 605) : the auriculo-temporal nerve will frequently be found embracing the middle meningeal artery. The internal
pterygoid muscle is to be examined as far as it can be laid bare (p. 181).
The auriculo-temporal division of the inferior maxillary nerve is then to be
traced to its distribution, and the pinna of the ear is to be dissected so as to
show the form and extent of its cartilage, the small muscles on its surface,
and the final distribution of its nerves (p. 741).

7. *The Orbit.*—The dissection of the orbit and the parts passing into it
may next be proceeded with. Let a vertical cut be made with the saw
through the frontal bone, near the inner angle of the orbit, immediately
above the fovea trochlearis ; and another from above the ear, downwards
and forwards, through the lateral wall of the skull, towards the sphenoidal
fissure. Remove the outer part of the malar bone with the bone-nippers,
separate carefully with the handle of the knife the periosteum and contents of the orbit from the upper and outer walls, and unite the inner
saw-cut with the sphenoidal fissure, immediately outside the optic foramen,
by means of the chisel ; then, with the bone-nippers, remove the isolated
piece of bone so as to unroof the orbit, and afterwards divide the periosteum
longitudinally, and reflect it. On the upper surface of the contents of the orbit
posteriorly is the fourth nerve, which is to be traced forwards from the cavernous sinus where it enters the orbital surface of the trochlearis muscle, and that
muscle is to be displayed (pp. 594 and 179). The frontal nerve, occupying
the middle of the space, is to be traced back to its origin from the
ophthalmic division of the fifth nerve (p. 597). The lachrymal gland is to
be exposed (p. 709); and from its posterior border the lachrymal nerve is to
be traced back to its origin from the ophthalmic nerve, while at the same
time its malar branch and palpebral distribution may also be seen. The
levator palpebræ muscle, and the inferior, external, and internal recti
muscles are to be displayed (p. 179), and the ocular surface of each
cleared ; when the sixth nerve will be seen ending in the external rectus,
and branches of the third in the other three recti muscles. These nerves
are now to be traced backwards between the two heads of origin of the
external rectus muscle to the cavernous sinus (pp. 593 and 610). Below
the superior rectus muscle the nasal nerve, derived from the ophthalmic,
will be seen crossing the optic nerve ; it will be followed to the anterior
internal orbital foramen, and its infratrochlear branch traced to the lower
eyelid ; it is then to be dissected back to its origin, and the long and delicate root of the lenticular ganglion sought for on the outer side of the
optic nerve. The ophthalmic or lenticular ganglion is on the outside of the
optic nerve, and may be most easily found by tracing the short and thick
twig which runs into it from the inferior division of the third nerve. In
front of the ganglion its ciliary branches may be seen (p. 599). The
remainder of the fat is to be removed from the lower part of the orbit ;
the distribution of the ophthalmic artery is to be displayed (p. 360) ; and
the lower division of the third nerve is to be traced forwards to the
inferior rectus and obliquus muscles. By a slight dissection from the front

of the orbit the insertions of these muscles may be more fully displayed. The contents of the orbit may be afterwards divided behind and turned forward, to admit of the tensor tarsi muscle and the lachrymal sac being dissected. Finally, if the subject be favourable, the nasal nerve may be traced through the ethmoid bone to its distribution in the interior of the nares, and its external twig to the tip of the nose examined.

8. *Deep view of the Fifth Nerve. Spheno-palatine and Otic Ganglia. Internal Ear.*—After the dissection of the orbit has been completed, the foramen rotundum and infraorbital canal are to be laid open, and the superior maxillary nerve and its orbital and dental branches dissected (p. 600). Remove with the saw a further portion of the skull towards the meatus externus, reaching as far as the foramen spinosum, and with the chisel or nippers cut down close to the foramen ovale ; remove also a portion of the bone above the pterygoid processes so as to open up the spheno-maxillary fossa, and the spheno-palatine ganglion will be brought into view. The connection of the ganglion with the superior maxillary nerve may then be made out. Trace the nasal and naso-palatine branches of the ganglion through the spheno-palatine foramen, and the palatine branches passing downwards. Lay open the Vidian canal and dissect the Vidian nerve back to the great superficial petrosal nerve (p. 603). At the same time the infraorbital, spheno-palatine, descending palatine, and Vidian branches of the internal maxillary artery will be noted (p. 356). The otic ganglion may also be in part seen by breaking open the foramen ovale, following upwards the nerve of the internal pterygoid muscle, and slightly everting the trunk of the inferior maxillary nerve (p. 608). The twigs from this ganglion to the tensor palati and tensor tympani muscles may be found. The otic ganglion, however, can only be seen to advantage in dissections made from the inner side of the internal pterygoid muscle and inferior maxillary nerve. The Eustachian tube may be laid bare in the posterior part of its course, and may be opened, and the attachment of the tensor tympani above it shown (p. 747).

By now sawing the wall of the skull down to the margin of the external auditory meatus, and removing with the bone-nippers, cautiously, the anterior wall of the meatus externus, the membrana tympani may be exposed ; and by unroofing the tympanic cavity in continuation of the Eustachian tube backwards, the malleus, incus, and stapes, as well as the tendon of the tensor tympani muscle, will be brought into view (p. 748). The mode of action of the latter on the membrana tympani may be studied ; also the chorda tympani nerve will be seen traversing the cavity. The malleus and incus are to be carefully removed ; then, placing one point of the bone-nippers in the internal auditory meatus, lay open with the other the vestibule and cochlea, and let the relation of the portio mollis and portio dura nerves to these cavities be observed (pp. 610 and 615). The manner in which the stapes fits into the fenestra ovalis may now be seen to advantage, the tendon of the stapedius muscle requiring, however, to be cut across before that ossicle can be removed. With the aid of the bone-nippers, the fleshy part of the stapedius may be laid bare, descending in the mastoid part of the temporal bone, close to the facial nerve ; and, in favourable circumstances, the chorda tympani may be traced back to the facial nerve.

9. *Submaxillary and Sublingual Regions.*—Let the lower jaw be divided in front of the masseter muscle, and let the gustatory and mylo-hyoid nerves be followed from the pterygoid into the submaxillary region. The

3 z

anterior belly of the digastric muscle is to be divided at the chin and turned down. The mylo-hyoid muscle is to be separated from its fellow in the middle line and from the hyoid bone, and reflected toward the jaw, in order to expose the deeper parts. The tongue is to be put on the stretch by fastening it forward ; the lower jaw is to be divided by a vertical saw-cut between the first and second incisor teeth, leaving intact the attachment of the genio-hyoid muscle ; the fragment of loose bone is to be raised, and the mucous membrane of the mouth slit up to the tip of the tongue. The dissector will first trace carefully out the gustatory nerve, where it is in contact with the submaxillary gland, and will exhibit the submaxillary ganglion connected with it (p. 609). He will then isolate the submaxillary and sublingual glands, and will observe the relations of Wharton's duct, the sublingual ducts, and the gustatory and hypoglossal nerves (p. 816). He will examine the hyoglossus muscle, the genio-hyoid, the genio-hyoglossus, stylo-glossus, and stylo-pharyngeus muscles (p. 185) ; also the glosso-pharyngeal nerve (p. 615), and the stylo-hyoid ligament (p. 52). On dividing the hyo-glossus muscle, the subjacent part of the lingual artery may be followed into its sublingual and ranine branches ; its small hyoid branch and its branch to the dorsum of the tongue may also be seen ; as well as those deep branches of the facial artery which have not yet been examined, viz., the ascending palatine and the tonsillar branches (p. 348).

10. *Parts close to the external basis of the Cranium.*—If the styloid process be nipped through at its base and thrown down with the three styloid muscles attached, the dissector will be enabled to examine more particularly the pharyngeal plexus of nerves (p. 690). He may then also examine the relations of the internal carotid artery and internal jugular vein (p. 359) : and he will follow up the hypoglossal, spinal accessory, pneumo-gastric, glosso-pharyngeal, and sympathetic nerves to the basis of the skull ; examining the connexions of the superior ganglion of the latter with the other nerves mentioned, and with the anterior divisions of the upper cervical nerves (p. 620). The jugular foramen and the carotid canal are to be opened into ; and the eighth nerve, and the internal carotid artery with the carotid plexus accompanying it, are to be followed into the interior of the cranium (pp. 619 and 688). Before leaving this part of the dissection, the students dissecting the head ought to make together a complete revision of all the parts in connexion with the basis of the cranium.

11. *Pharynx, Larynx, Palate, Tongue, Nares, &c.*—Let the remains of the carotid arteries be removed, and the pharynx drawn away from its loose connection with the upper cervical vertebræ ; and let the base of the skull be divided between the pharynx and the recti capitis antici muscles ; then, leaving the neck and back part of the skull for a later examination, let the pharynx, with the parts in its vicinity, be prepared for dissection by distending its walls with hair or tow. The constrictor muscles of the pharynx are to be cleaned and examined, as also the origins of the levator and circumflexus palati muscles (p. 187). The next step is to open the pharynx from behind, by an incision in the middle line, and a transverse one close to the base of the skull ; and to examine the apertures of the nares, fauces, glottis, œsophagus, and Eustachian tubes (p. 819). The muscles of the soft palate are then to be dissected ; more particularly the insertions of the levator and circumflexus palati ; the palato-pharyngeus and palato-glossus corresponding in position to the posterior and anterior pillars of the fauces, and in the middle line the azygos uvulæ (p. 189). The Eustachian tube should also be dissected out.

The larynx and tongue are to be separated from the upper jaw, and the surface of the tongue and the tonsils examined, as well as such of the intrinsic muscles of the tongue as may be visible (p. 805). The dissectors will then proceed to the study of the larynx, carefully cleaning it (p. 905) ; and after the glottis and true and false vocal cords have been sufficiently inspected, they may remove the mucous membrane, tracing at the same time the distribution of the superior and inferior laryngeal nerves, and the laryngeal branch of the superior thyroid artery. The muscles of the larynx will then be fully dissected. The crico-thyroid, the arytenoid, the aryteno-epiglottidean, and the posterior crico-arytenoid muscles can be seen without injuring the cartilages ; but to expose the lateral crico-arytenoid and the thyro-arytenoid muscles, it is necessary to remove the upper part of one ala of the thyroid cartilage. Lastly, the ventricles and pouches of the larynx are to be examined, the vocal ligaments are to be dissected out, and, the muscular substance having been removed from the cartilages, their uniting ligaments, and the joints by which they move on one another, are to be studied.

In concluding this stage of the dissection, let a vertical section of the nares and hard palate be made on one side of the septum nasi. Let the meatus of the nose, the nasal duct, and the maxillary antrum be examined (p. 773) ; and, if the subject be in good condition, a view may be obtained of the palatine and naso-palatine branches of the spheno-palatine ganglion, as well as of the distribution of the descending palatine artery in the palate (pp. 603 and 357).

12. *Deep Muscles and articulations of the Neck and Head.*—The muscles attached to the cervical vertebræ are now to be examined. In front of the vertebral column, the student will observe the scaleni, longus colli, recti capitis antici major and minor, and rectus lateralis muscles (p. 193) ; then turning to the posterior aspect, he will dissect the remains of the levator anguli scapulæ (p. 203), splenius, trachelo-mastoid, and complexus muscles to their attachments (p. 234), and notice the portion of the occipital artery covered by the splenius, with its branch the princeps cervicis (p. 351). The recti capitis postici major and minor, and the obliqui capitis superior and inferior, with the suboccipital nerve supplying them, are to be dissected out (pp. 239 and 632), and the course of the vertebral artery displayed as it lies in the groove of the atlas (p. 367). Lastly, the arches of the vertebræ are to be removed, and the joints and ligaments examined, especially those between the atlas, axis, and occipital bone, among which the transverse ligament of the atlas and the crucial and odontoid ligaments require particular attention (p. 125).

II.—UPPER LIMBS OR SUPERIOR EXTREMITIES.

The right and left limbs constitute each a part. Their dissection should extend over a period of not less than four weeks. They each include, along with the limb itself, the axilla or armpit, and the structures which lie between the trunk of the body and the bones of the shoulder and arm. The muscles of the back and the spinal cord are also to be dissected by those having the upper limbs. The omo-hyoid muscle, however, and the upper parts of the trapezius, levator anguli scapulæ, splenius, trachelo-mastoid, and complexus muscles should be left uninjured for the dissectors of the head and neck. The inferior boundary of this part on the trunk of the body is indicated by a line passing along the outer and lower borders of the latissimus dorsi, the serratus magnus, and the pectoralis major muscles.

1. *Muscles of the Back ; Spinal Cord.*—During the first four days, while the subject is lying on its face, the dissection of the back and spinal cord below the level of the third cervical vertebra is to be completed. Let an incision be made in the middle line from the level of the third cervical vertebra to the sacrum, a second from the acromion to the spine of the seventh cervical vertebra, and a third from the point where the fold of the axilla meets the arm to the acromion. If the student be a beginner, let him at once dissect out the trapezius muscle in the direction of its fibres (except the part of it which falls within the boundary of the dissection of the head and neck), and afterwards the latissimus dorsi, following up its fibres as close as possible to the tendon of insertion ; but let him not reflect the skin further than is necessary to exhibit the anterior border of the latissimus dorsi (p. 200). If the student be a senior, he will, previously to the dissection of these muscles, also display the cutaneous branches of the posterior divisions of the spinal nerves, which lie upon their surface (p. 633).

The trapezius muscle is to be divided by a vertical incision at the distance of two inches from its vertebral attachment, and on its deep aspect the spinal accessory nerve and the superficial cervical artery are to be displayed (pp. 625 and 373). The rhomboid and levator anguli scapulæ muscles may then be dissected (pp. 202 and 643), and the nerve to the rhomboids, reaching their deep surface from above. The latissimus dorsi muscle is to be divided by means of an incision carried along its attachment to the lumbar fascia from its superior border, at about three inches from its vertebral attachment, downwards and outwards towards the external border, leaving uncut the slips attached to the lower ribs and crest of the ilium. The rhomboid muscles are also to be divided, and the posterior scapular artery dissected (p. 373). The serrati postici superior and inferior muscles may then be dissected, and the vertebral aponeurosis seen (pp. 233 and 240) ; after which a view may be obtained of the serratus magnus muscle from its internal aspect (p. 207).

The posterior serrati muscles and the vertebral aponeurosis may now be divided, and the dissection of the muscles composing the erector spinæ may be proceeded with (p. 234). Beginning with the ilio-costalis or sacro-lumbalis muscle, the student will dissect first its six or seven slips of direct insertion into the lower ribs, then the slips attached to the upper ribs, constituting the musculus accessorius ad ilio-costalem : he will afterwards turn the muscle outwards and trace the separate heads of origin of the musculus accessorius from the lower ribs into their insertions above ; and also the similar origins of the ascendens cervicis muscle from the upper ribs. He will next treat the longissimus dorsi muscle in the same manner, dissecting first the costal insertions on its outer side, and then, having separated it from the spinalis dorsi muscle (which always requires the division of a tendon running between the two muscles), make out the insertions into the transverse processes of the vertebræ. The issue of posterior branches of spinal nerves, and of intercostal and lumbar vessels, will guide the dissector to the separation of the masses of muscle (pp. 633 and 404). The continuation of the long muscles into the ascendens cervicis and transversalis cervicis, and the origins of the trachelo-mastoid, are then to be traced upwards in the neck. To see the last-named muscle, however, the splenius muscle must be dissected and vertically divided ; and the complexus and semispinalis dorsi and colli muscles may then be examined: Lastly, the deepest muscles, multifidus spinæ, rotatores spinæ, interspinales, and inter-transversales are to be dissected (p. 238).

At this stage of the dissection a good view may be obtained of the posterior margins of the obliquus externus and obliquus internus muscles of the abdomen, and of the posterior and middle layers of the lumbar aponeurosis, which are continuous behind with the transversalis muscle : the dissection of these muscles, however, belongs to the abdomen, and they must not be injured (p. 199).

The next proceeding for a senior dissector is to lay bare the spinal cord ; for this purpose he will straighten as much as possible the lumbar vertebræ, by placing blocks underneath the abdomen, and will let the neck hang slightly downwards. He will then saw through the laminæ of the dorsal and lumbar vertebræ on each side, keeping the edge of the saw directed slightly inwards, and will continue the saw-cuts below on the back of the sacrum, so as to meet each other where the sacral canal becomes incomplete. The part so isolated may easily be raised with the chisel, and with the bone-nippers the whole laminæ of the vertebræ may be removed, attached to one another by their elastic ligaments. In several spaces of the lower dorsal region the articular processes of the vertebræ may be removed, so as to expose one or more of the spinal nerves issuing from the canal, and these, being dissected for a little distance beyond their ganglia, may be afterwards taken out along with the cord. The theca of dura mater ought now to be made as clean as possible by removing the fat from its surface, and, after being examined, should be slit open, that the other membranes and the relations of the cord may be examined in situ ; more particularly, the ligamentum denticulatum, the position of the lower extremity of the cord, the cauda equina, and the filum terminale will be observed (pp. 502 and 565). The spinal cord and its membranes are then to be removed from the body and stretched out upon a table, when the anterior and posterior roots of the nerves and some of the ganglia in connection with the latter may be observed ; also the external form and structure of the cord, with the anterior, middle, and posterior columns, the anterior and posterior fissure, &c. ; and, lastly, several sections of the cord, in different places, may be made to exhibit the relations of the grey and white matter within.

2. *Pectoral Region and Axilla.*—Within four days after the subject has been laid upon its back, the pectoral region and the axilla are to be dissected. Let a median incision be made in front of the sternum, and from its upper end let another be carried along the clavicle to the acromion, and thence downwards to the inside of the middle line of the arm, a little below the fold of the axilla, and a third horizontally outwards from the lower end of the sternum. Then let the skin be reflected from the pectoralis major muscle (p. 203), and let the senior student in doing this preserve the fibres of the platysma myoides and the suprasternal and supra-clavicular branches of the cervical plexus of nerves descending over the clavicle (pp. 170 and 639), the anterior cutaneous branches of the inter-costal nerves, with the accompanying twigs from the internal mammary artery near the middle line, and two or three small anterior twigs of the lateral cutaneous branches of the intercostal nerves appearing round the lower border of the muscle (pp. 656 and 375). If the subject be a female, let him also dissect the mammary gland, and in raising the general integument leave the skin of the nipple, by carrying round it a circular incision of about two inches in diameter (p. 1002). By raising the skin

within this circle the lactiferous ducts and sacculi will be brought into view.

Let the clavicular portion of the pectoralis major muscle now be divided near the clavicle for the examination of the subclavicular space, preserving the external anterior thoracic nerve as it passes to that muscle (p. 645); and let the costo-coracoid membrane and sheath of the axillary vessels be examined (p. 229). Then let the sheath be removed, and let the termination of the cephalic vein and the parts of the axillary artery and vein brought into view be studied, and also the superior or short thoracic, acromio-thoracic, and thoracico-humeral branches (pp. 377 and 468).

For the dissection of the axillary space, the skin and the fascia are to be separately raised from its surface (p. 230), and in the first place the great vessels and nerves of the limb should be carefully exposed as they pass from the axilla into the brachial region, but without much disturbing their position. The axillary artery and vein are then to be followed upwards, and the fat removed from within the space, when the long thoracic vessels will be found chiefly along the anterior border, the subscapular vessels principally along the posterior border, and the alar twigs more in the middle. At this stage there will also be seen on the inner wall of the axilla the intercosto-humeral with other lateral cutaneous branches of intercostal nerves piercing the serratus magnus muscle (p. 657), the posterior thoracic nerve descending on the surface of that muscle to supply it (p. 644), and on the posterior wall the three subscapular nerves (p. 645). When the axilla has been sufficiently studied, the remainder of the pectoralis major muscle is to be divided; the pectoralis minor muscle also is to be dissected and divided, and the internal anterior thoracic nerve, which supplies it, is to be found. By this proceeding the axillary vessels will be exposed in their whole course, and the origins of the branches of the axillary artery may be more fully examined, viz., the acromio-thoracic, the alar thoracic, short and long thoracic, and subscapular arteries, and the anterior and posterior circumflex arteries. Three cords of the brachial plexus will also be seen; the outer one giving off the musculo-cutaneous, the external anterior thoracic, and the outer head of the median nerve; the inner giving off the inner head of the median nerve, the internal cutaneous nerve, the nerve of Wrisberg, and the ulnar; the posterior giving off the three subscapular nerves, the circumflex, and the musculo-spiral nerve. At this time, after removing the costo-coracoid fascia, the subclavius muscle should be cleaned and examined (p. 206).

On the fourth day after the subject has been placed upon its back, the clavicle is to be sawn through the middle, or disarticulated at its sternal end, if this should be recommended by the Demonstrator. The dissector of the arm may then, in company with the dissector of the head and neck, on the same side, obtain a continuous view of the upper part of the brachial plexus, and trace the origins of the suprascapular and posterior thoracic nerves (p. 641). The axillary vessels and the main trunks of the brachial plexus of nerves are afterwards to be securely tied together opposite the outer border of the first rib, and divided above the ligature; the lower parts may subsequently be retained in position, by tying them to the portion of the clavicle left with the arm. The serratus magnus muscle may now be put upon the stretch, and should be fully studied before the removal of the arm (p. 207).

3. *Scapular Muscles, Vessels, and Nerves.*—After the arm has been removed, the first duty of the dissector is to clean the parts which have

been already laid bare, and to dissect all the cut muscles, so as to bring their attachments completely into view ; he may then remove the redundant masses which are no longer required, preserving only such portions of tendons and muscles as may be necessary for subsequent revision of their relations to the joints and their attachments to the bones.   He will then clean the deltoid muscle, beginning from behind, so as to save as much as possible the cutaneous branches of the circumflex nerve (pp. 208 and 645).   He will dissect the teres major muscle, and the quadrangular and triangular intervals which are separated by the long head of the triceps muscle, and lie between the teres muscle and the scapula ; and he will lay bare, as far as can be done without injury to the muscles, the structures which pass through these intervals, viz., in the upper or quadrangular one, the circumflex nerve, with its branch to the teres minor muscle, and the posterior circumflex artery, and in the lower or triangular interval, the dorsal branch of the subscapular artery (p. 380).   The deltoid muscle is next to be removed from the whole of its superior attachment, and beneath it will be seen the bursa that lies between the acromion and shoulder-joint (p. 138), and the branches of the circumflex vessels and nerve.   The teres minor, infraspinatus and supraspinatus muscles are to be dissected and reflected, and the distribution of the suprascapular nerve and artery traced.   While this is done, neither the deltoid ligament nor acromion need be divided.   The subscapular muscle is likewise to be examined, with the two short subscapular nerves which supply it ; and on reflecting this muscle, the subscapular bursa will be observed communicating with the shoulder-joint.   In removing the muscles attached to the scapula, the student should bring into view the anastomoses of the posterior scapular, suprascapular, acromio-thoracic, dorsal branch of the subscapular, and the circumflex arteries.   The scapular muscles may then be cut short at their attachments to the humerus.

4. *Subcutaneous view of the Arm.*—In proceeding with the dissection of the arm, if the part be in a condition favourable for the purpose, the dissector may at once display the cutaneous nerves and veins as far as the wrist (p. 647).   He will, in that case, make an incision all the way down to the wrist in front of the limb ; or, should it be deemed advisable not to remove the integument so far, he may terminate his incision half-way down the fore-arm.   For the easier preservation of the cutaneous nerves, which lie close to the aponeurosis of the limb, he will remove the subcutaneous fat by reflecting it in the direction from above downwards.   The intercosto-humeral nerve is to be traced down to its distribution (p. 657).   The nerve of Wrisberg and the internal cutaneous branch of the musculo-spiral nerve will be most easily traced from their deep origins (pp. 646 and 652). The internal cutaneous nerve will be found piercing the aponeurosis on the inside of the arm in two separate places, a few inches above the elbow ; and on the outer side will be found the two external cutaneous branches of the musculo-spiral nerve, appearing in the line of the external intermuscular septum ; while at the bend of the elbow, towards the outer side, the musculo-cutaneous or external cutaneous nerve will be observed emerging from the deep parts.   Near the elbow, on the inner side, there is a small lymphatic gland, and on the subcutaneous part of the olecranon a small synovial bursa.   Further down, there may be seen on the inner side a cutaneous branch from the ulnar nerve, below the middle of the fore-arm ; on the outer side the radial nerve becoming superficial two or three inches above the wrist ; and in front the palmar cutaneous branch of the

median nerve immediately above the annular ligament. On the fore-arm will be found the radial, median and ulnar veins; in front of the elbow the median-cephalic and median-basilic veins, together with the deep median branch; and in the upper arm the cephalic and basilic veins (p. 466).

5. *Brachial Region more deeply.*—The student will now remove the aponeurosis from the front of the arm. He will first dissect out the brachial artery with the venæ comites clinging to it and intercommunicating round it, and the median nerve crossing in front (p. 381). Arising from the inner side of the artery he will find the superior profunda branch turning backwards with the musculo-spiral nerve, a little further down the inferior profunda branch accompanying the ulnar nerve, and a little above the elbow, the anastomotic resting on the brachialis anticus muscle : while from the outer side of the brachial artery a variety of muscular branches are observed to spring. The inferior profunda sometimes arises from the superior profunda branch. Not unfrequently two large arteries will be found in the arm, in consequence of a high division of the main trunk ; the radial or ulnar artery, most frequently the former, being given off from the brachial at a higher point than usual, and sometimes even as high as the axillary artery. In some of these cases the artery which arises out of place lies superficially to the aponeurosis of the limb. The biceps and coraco-brachialis muscles are next to be dissected, and the deep part of the musculo-cutaneous nerve, which gives them branches (pp. 212 and 648). The dissector will be careful to preserve the aponeurotic slip of insertion of the biceps, which lies superficially to the vessels at the bend of the arm. The aponeurosis is to be removed from the back of the arm, and the intermuscular septa are to be examined (p. 230) : the triceps muscle is to be dissected, and the superior profunda artery and musculo-spiral nerve are to be traced to its outer side (pp. 214 and 652). The musculo-spiral nerve is to be followed to its division into the radial and posterior interosseous trunks, and its branches, to the brachialis anticus, supinator longus and extensor carpi radialis longior displayed. The space in front of the elbow should next be dissected, so as to show the relations in it of the brachial, ulnar, and radial arteries, with the radial recurrent and anterior ulnar recurrent branches, and the median and radial nerves (pp. 389 and 397). The brachialis anticus muscle should also at this time be fully exposed down to its place of insertion.

6. *Shoulder-joint, &c.*—The articulations at the upper part of the arm ought now to be examined (p. 134). The conoid and trapezoid parts of the ligaments uniting the clavicle to the coracoid process are first to be dissected, and their uses studied ; then the acromio-clavicular articulation, and the suprascapular and coraco-acromial ligaments of the scapula ; lastly, the shoulder joint is to be dissected, the capsule is to be cleaned, the coraco-humeral ligament dissected, and the tendons of muscles in close relation with the joint examined. When lastly the capsule is opened, the origin of the long head of the biceps in connection with the glenoid ligament will be seen, and also the prolongations of the synovial membrane round the long head of the biceps and beneath the subscapular muscle.

7. *The Fore-arm in front.*—Let the aponeurosis be removed from the front of the fore-arm, and let the five superficial muscles arising from the inner condyle of the humerus be dissected, beginning with the pronator radii teres ; exhibiting its two heads of origin with the median nerve between

them, and proceeding successively to the flexor carpi radialis, palmaris longus (which, however, is often absent), flexor sublimis digitorum, and flexor carpi ulnaris (p. 215) ; displaying the branches of the median nerve to the first four muscles, and that of the ulnar nerve to the last-mentioned muscle and to the flexor profundus digitorum (pp. 651 and 659). The course of the radial and ulnar arteries and nerves in the fore-arm is also to be studied. From the radial artery (p. 394) will be seen given off the radial recurrent, the muscular branches, the anterior carpal branch and the superficial volar ; while arising from the ulnar artery (p. 388) will be seen the anterior and posterior ulnar recurrent, and the interosseous, dividing into anterior and posterior interosseous, and giving off the branch to accompany the median nerve. This last branch, the comes nervi mediani, derives importance from being not unfrequently developed as a third principal trunk of the fore-arm, which passes down into the superficial palmar arch. The muscular branches of the ulnar artery, and its anterior and posterior carpal branches, are also to be exposed. The deep layer of muscles, consisting of the flexor longus pollicis, flexor profundus digitorum and pronator quadratus, are next to be dissected (p. 219) ; and along with them, lying on the interosseous membrane, and giving twigs to the muscles, the interosseous branch of the median nerve, and accompanying it, the anterior interosseous artery (p. 390).

8. *The Hand in front.*—For the dissection of the front of the hand, let an incision be made down the middle of the palm, a second transversely through the skin above the division of the fingers, and others down the middle of each finger. Let the palmar aponeurosis be exposed (P. 231), preserving the palmaris brevis muscle which is attached to its inner margin (p. 225) ; and let the skin be reflected from the front of the fingers and thumb, so as to exhibit the sheaths for the tendons, and the two digital branches of the artery and nerve on each (p. 218). The palmar aponeurosis is then to be removed, and the trunks of the ulnar and median nerves will be brought into view (pp. 649 and 651), as also the ulnar artery, the superficial volar branch of the radial artery, and the superficial palmar arch (p. 393). The short muscles of the thumb, viz., the abductor, opponens, flexor brevis, and adductor pollicis, are to be dissected, with the twigs of the median nerve supplying the three first, and the insertion of the flexor longus pollicis ; then the abductor, opponens, and flexor minimi digiti, with the twigs of the ulnar nerve supplying them, and its deep branch piercing them (p. 225). The annular ligament is to be cleaned and the synovial sheath behind it examined ; the tendons of the superficial and deep flexors are to be followed to their insertions, and the lumbricales muscles dissected. The deep branch of the ulnar artery may now be traced to the deep palmar arch, and that of the ulnar nerve to its distribution in all the interossei, two of the lumbricales, the adductor pollicis and the inner part of the flexor brevis pollicis muscle. The deep palmar arch and its branches are also to be fully examined (p. 400).

9. *The Fore-arm and Hand Posteriorly.*—For the dissection of the back of the fore-arm and hand let the remainder of the integument and aponeurosis be carefully reflected, and let the distribution of the ulnar and radial nerves to the dorsal aspects of the fingers be traced (p. 653). The muscles are then to be dissected in the following order, viz., the supinator longus, extensores carpi radiales longior and brevior, extensor carpi ulnaris, extensor communis digitorum and extensor minimi digiti, the extensor indicis, the three extensores pollicis, and, lastly, the anconeus and supinator brevis muscles (p. 220). There will be found passing through the fibres

of the last-mentioned muscle, the posterior interosseous nerve ; and on the interosseous membrane the posterior interosseous artery, with its recurrent branch ; they are both to be traced to their distribution (pp. 654. and 391). The lower part of the radial artery which has hitherto been hid from view may also now be studied : its posterior carpal and its metacarpal branch will be seen, together with the dorsal branches of the thumb and index finger (p. 398). The termination on the back of the wrist of the anterior interosseous artery after passing through the interosseous membrane is also to be noticed. Finally, the interossei muscles are to be dissected on both the palmar and dorsal aspects of the hand (p. 227).

10. *Articulations of the Fore-arm and Hand.*—The dissector may now return to an examination of the elbow-joint and other articulations of the upper limb. In connection with the elbow-joint, he will first make a revision of the relations of the soft parts to the joint, such as those of the triceps, brachialis anticus, and supinator brevis muscles, the muscles attached to the outer and inner condyles of the humerus, and the median, musculospiral, and ulnar nerves, together with the anastomoses of the superior and inferior profunda and the anastomotic branches of the brachial, with the two ulnar, the radial, and the interosseous recurrent arteries. The dissector will then proceed to examine in detail the internal and external lateral ligaments, the anterior and the thin posterior ligaments, the orbicular ligament, the synovial membrane, and the cartilaginous surfaces of the bones (p. 138). The dissector should carefully observe the different kinds of motion of which the parts are capable, and the variations in the tightness of the ligaments and in the relations of external parts induced by these motions. In examining the lower radio-ulnar articulation, the dissector will particularly study the relations of the triangular fibro-cartilage, and the nature of the movements in pronation and supination of the hand ; and, in the carpal joints, the extent of the synovial cavities and the position of the cartilage and interosseous ligaments.

III.—THORAX.

The right and left sides of this region constitute each a part. Its dissection may be completed within three weeks. It includes the deep dissection of the thoracic parietes, the viscera of the thoracic cavity, together with the upper surface of the diaphragm. It is indispensable that the dissectors of opposite sides should be present together and act in concert.

1. *Parietes and Pleura.*—The dissection is to be commenced on the fifth day after the subject has been placed upon its back, that is, the tenth day after it has been first placed in the rooms. The external and internal intercostal muscles, and the intercostal arteries and nerves in the anterior part of their course, together with the parietal pleura, are to be first dissected (pp. 240, 402 and 655). Then let the internal mammary artery on the right side be laid bare by the removal of the 2nd, 3rd, 4th, 5th and 6th costal cartilages, in order that its relation to the sternum, and its anterior intercostal and perforating branches may be observed (p. 374). The corresponding costal cartilages on the left side may then be divided close to the ribs, and the ribs belonging to those cartilages on both sides are then to be divided as smoothly as possible about three inches beyond their angles ; in doing which the dissectors must be careful to avoid injuring their hands upon the sharp spicula of the sawn extremities of the ribs. The anterior

limits of the pleural cavities and the position of the anterior mediastinum can now be examined, together with the position of the heart and great vessels in relation to the lungs and the walls of the thorax (p. 299). That this may be done more effectually, the lungs should be inflated through a tube introduced into the throat or wind-pipe, and their different positions and relations in the inflated and collapsed state attentively examined. The body of the sternum is next to be separated from the manubrium, and, together with the adherent costal cartilages of the left side, removed; and on the fragment of the thoracic wall thus separated the triangularis sterni muscle and its relation to the internal mammary artery may be further examined. The dissectors will then complete their examination of the anterior mediastinum, observing in its upper part the remains (if any) of the thymus body, and will carefully study the remaining reflections of the pleura. The heart within the pericardium is also to be observed (p. 313). In making this dissection the student may be required to separate the parietal from the pulmonary pleura, by breaking up with his fingers, or the handle of the knife, the inflammatory adhesions which are often met with. Great care must be taken to clean with a sponge and wash the interior of the chest and the surface of the lungs, first with water, and subsequently with preserving fluid (p. 892).

2. *Parts External to the Pericardium.*—The phrenic nerve will be seen on each side beneath the pleura in front of the root of the lung, and is to be dissected out; when its relation to the internal mammary artery, which it crosses at the upper part of the chest, and the branch of the latter artery which accompanies it, are to be observed (p. 640). The structures above the pericardium are then to be dissected. Foremost will be found the innominate veins and superior vena cava, with the termination of the vena azygos, and several smaller veins, viz., the inferior thyroid, internal mammary, superior intercostal, and bronchial veins (p. 453); and behind the veins, the innominate, left carotid, and left subclavian arteries arising from the arch of the aorta (pp. 340, 341 and 364). The pneumo-gastric nerves will also be found, that of the right side lying external to the innominate artery, and its recurrent branch turning round behind the subclavian artery; and that of the left side passing down in front of the arch of the aorta, with its recurrent branch winding behind the aorta (p. 618). Likewise crossing the arch of the aorta, on their way to the superficial cardiac plexus, will be found the cervical cardiac branch of the left pneumo-gastric nerve, and, usually, the superior cardiac branch from the sympathetic nerve on the left side (p. 690). The other cardiac nerves, viz., the cervical cardiac branch of the right pneumo-gastric nerve, the thoracic cardiac branches of both pneumo-gastric nerves, the three cardiac branches of the sympathetic chain of the right side, and the middle and inferior branches of the left side, are to be sought on the front and sides of the trachea, as they pass down to the deep cardiac plexus. The distribution of the pneumo-gastric nerves is then to be traced to the lungs and œsophagus; and, as far as possible, the posterior and anterior pulmonary plexuses are to be brought into view (p. 623). After that has been done, the roots of the lungs are to be fully dissected, the relations of the pulmonary arteries and veins and the bronchi observed, and the bronchial arteries traced to their origins (pp. 897 and 402).

3. *Interior of the Pericardium and Heart.* — The pericardium having been examined on its outer aspect, is then to be cut open, and its interior carefully inspected (p. 300); after which it is to be removed, its remains

being cleared away from the trunks of vessels entering and emerging from the heart. The arch of the aorta may now be fully studied, and the cord of the ductus arteriosus displayed passing between the commencement of the left pulmonary artery and the arch of the aorta (pp. 331 and 382). The students will then proceed to the dissection of the heart, examining first its external form (p. 302), and afterwards dissecting the right and left coronary arteries and the coronary vein (pp. 338 and 482). They will then make an opening into the right auricle, by means of one incision from the point of entrance of the vena cava superior to near the entrance of the vena cava inferior, and another from the auricular appendage to the middle of the first incision. They will remove and wash out the blood from the right side of the heart, and will particularly observe in the auricle the arrangement of the musculi pectinati, the annulus ovalis, the Eustachian valve guarding the vena cava inferior, the orifice of the coronary vein guarded by the valve of Thebesius, and the foramina Thebesii (p. 308). When the examination of the right auricle has been completed, the dissector will pass the forefinger of the left hand through the auriculo-ventricular orifice, and open the right ventricle by two incisions, one along the anterior border, close to the septum of the heart, prolonged upwards to the commencement of the pulmonary artery, and the other passing from the first, along the superior border of the ventricle, immediately below the auriculo-ventricular sulcus, care being taken not to injure the anterior segment of the tricuspid valve. The principal objects to be noted in this ventricle are the tricuspid valve with the chordæ tendineæ and musculi papillares which act upon it, the other arrangements of columnæ carneæ, the infundibulum, and the semilunar valves of the pulmonary artery on their cardiac aspect. In exposing the latter, the incision into the ventricle should be carried into the pulmonary artery between two of the segments of the valve (p. 310). To examine the left side of the heart, let the inferior vena cava be dissected a little out of its aperture in the diaphragm, and let it be divided, and the heart thrown upwards. The left auricle is then to be opened by a transverse incision near its ventricular margin, and by two short incisions at right angles to the first; and after being carefully sponged out, its cavity and auricular appendage, the remains of the valve of the foramen ovale, and the entrance of the pulmonary veins on each side will be examined (p. 311). The left ventricle is to be opened by a process similar to that employed for opening the right; and after it is carefully cleaned, the mitral valve and its relation to the aortic orifice, and the cardiac aspect of the semilunar valves which guard the latter are to be studied (p. 312).

4. *Deep Cardiac Nerves, Bronchi, &c.* — The aorta is to be divided within an inch above its origin, and the first part of the vessel is to be opened to examine the semilunar valves and the sinuses of Valsalva (p. 307). At this stage of the dissection a fuller examination may be made of the cardiac nerves as they enter the superficial and deep cardiac plexus : the cardiac ganglion will also be found, and the coronary plexus traced a short way along the coronary vessels (p. 698). The dissectors may then divide the trachea an inch or two above its bifurcation, remove the heart and lungs, and examine more in detail the disposition and structure of the bronchi (p. 888).

5. *Parts in the posterior mediastinum, &c.* — Returning to the thoracic cavity, the dissectors will examine the œsophagus (p. 821), the descending aorta with its intercostal branches (p. 401), the main vena azygos, and its left branch (p. 469) and, lying between the vena azygos and aorta, the thoracic

duct (p. 487). The thoracic duct may be followed, with the concurrence of the dissectors of the head and neck, to its termination in the angle of junction of the left internal jugular and subclavian veins ; and, with the assistance of the dissectors of the abdomen, it may be also followed down to its commencement under the crus of the diaphragm. The sympathetic nerve with its chain of ganglia, is now to be traced over the heads of the ribs and the vertebral column : its communications with the intercostal nerves are to be made out, and the splanchnic nerves arising from it dissected (p. 693).

The upper surface of the diaphragm having been cleaned with the knife, the dissectors of the thorax will examine along with those of the abdomen the anatomy of this muscle, directing their attention to its various muscular and tendinous parts, and to the apertures for the passage of the aorta, gullet, and vena cava inferior, and observing the distribution of nerves and blood-vessels in its substance (p. 243).

6. *Articulations.*—When the dissection of the rest of the thorax has been completed, the dissectors will, if the subject be favorable, make an examination of the articulations of the vertebral column and ribs (p. 121). Let them study, in particular, the anterior and posterior common ligaments, the intervertebral substance, the ligamenta subflava of the arches, the form and movements of the articular processes, and the various costo-vertebral, costo-transverse and other ligaments. In doing this, the dissectors should make an attentive examination of the nature and extent of the movements of the different ribs, and the manner in which they are influenced by the movements of the vertebral column.

IV.—ABDOMEN AND PELVIS.

The right and left sides of these regions constitute each a part. Their dissection should not be completed in less than four or five weeks. It comprehends the examination of the perinæum and genital organs, the abdominal parietes over the whole of the external oblique muscles, extending in front to the linea alba and below to Poupart's ligament, the viscera and deeper parts of the abdomen and pelvis, and the lower surface of the diaphragm.

1. *Perinæum.*—If the subject be a male, the first day on which it is in the rooms will be set apart for the dissection of the perinæum ; and of this opportunity the dissectors of the abdomen must be prepared to avail themselves. A lithotomy staff is to be passed into the bladder, and, the hands and feet having been tied together, the subject is to be placed in the same position as for the operation of lithotomy, near the edge of the table. A block is then to be placed below the pelvis, and the scrotum is to be tied up to the handle of the staff. The body may, however, be still more conveniently maintained in the proper position, as is done in some schools, by means of a simple frame with two upright spokes, behind which the limbs are placed while the perinæum is projected forwards between them. A careful incision is to be made in the middle line from the back of the scrotum to the anus, and, being carried round the margin of the anus, is to be prolonged as far as the coccyx ; while a transverse incision is to be directed across the middle line in front of the anus from one ischial tuberosity to the other. Let the dissector reflect the flaps of skin, exposing the external sphincter, and clear out the fat completely from the ischio-rectal fossa of the left side, taking care not to injure the reflection of fascia which bounds it in front in a line with the central point of the perinæum ; and let him study the walls of the fossa (p. 261). On the right side,

enough of fat ought to be left in the ischio-rectal fossa to protect the levator
ani and obturator fascia ; the inferior hæmorrhoidal vessels and nerves may
be dissected towards the border of the sphincter (pp. 426 and 672), and the
hæmorrhoidal branch of the 4th sacral nerve may be seen emerging from between the levator ani and coccygeus muscles (p. 668). The two layers of
the superficial fascia in the part of the perinæum anterior to the anus are to
be distinguished, the most superficial corresponding to what may most
correctly be termed the subcutaneous·adipose tissue, and being continued
over the ischio-rectal fossa, while the deeper layer terminates behind by
dipping deeply in front of that fossa. The most superficial layer having
been removed, the blowpipe may be introduced beneath the deep layer in
the anterior half of the perinæum, so that, by inflating the connective tissue
underneath it, its external limits, its septum in the middle line, and its
continuity forwards may be demonstrated. It may then be slit open, and
will be found to be attached to the arch of the pubes externally, to be
continuous with the dartos in front, and to be reflected backwards to the
triangular ligament behind (p. 259). Underneath it will be found the
three long scrotal nerves, viz., the two superficial perineal branches of the
pudic and the inferior pudendal branch of the small sciatic nerve, which
are to be traced backwards (pp. 670 and 675) : also the superficial and
transverse perineal arteries are to be dissected out (p. 426). The muscles
on which these structures lie are then to be cleaned, viz., the accelerator
urinæ embracing the urethra, the erector penis lying upon the crus penis,
and the transversalis perinæi (p. 264). In the area between these muscles,
subjacent to them, will be observed the triangular ligament or anterior
layer of the subpubic fascia, and its relations, especially to the urethra, are
to be studied (p. 260). It is then to be divided near the bone, and on its
deep aspect the deep transversalis muscle, the constrictor urethræ, and the
artery of the bulb are to be dissected. The deep transversalis muscle is to
be divided, and Cowper's glands are to be sought for in the middle line
beneath (p. 963). On the left side are to be traced out the pudic artery and
nerve ; in doing which the branches of the artery to the bulb and the corpus
cavernosum should be observed (pp. 425 and 670). Lastly, a good view of
the inferior aspect of the prostate gland may be obtained by dividing the
sphincter ani from the accelerator urinæ muscles at the central point of the
perinæum (p. 264). In the dissection of the perinæum, constant reference
should be made to the bearing of its anatomy on the operations of
lithotomy (p. 1039). At this period, the dissectors may remove one of the
testicles for the sake of dissecting it while fresh.

2. *Abdominal wall anteriorly.*—The dissection of the abdominal parietes,
in either sex, is to be commenced on the day on which the subject is laid on
its back, with a careful examination of the fascia of the inguinal region, on
each side, as far as Poupart's ligament. This should, if possible, be undertaken in association with the dissector of the lower limb. An incision is to
be made in the middle line from the xiphoid cartilage to the pubes, avoiding
the umbilicus, and a transverse one meeting the first, inwards from the
anterior superior spine of the ilium. Let the dissector raise the lower of
the two flaps of skin thus marked out, remove the subcutaneous layer of fat
and fascia, and reflect the deeper layer, usually called superficial fascia, in
the same direction as the skin, so as to see the manner in which it is bound
down in the line of Poupart's ligament. Let him at the same time observe
the superficial epigastric and circumflex iliac arteries and veins, and the continuation of the superficial fascia over the region of the spermatic cord

towards the scrotum in the male, or to the labia in the female (pp. 257 and 437). He will also examine the external abdominal ring, its pillars, the intercolumnar fascia, and the emergence of the spermatic cord in the male, or the round ligament of the uterus in the female (pp. 964 and 986); and he will notice the terminal branches of the ilio-inguinal and ilio-hypogastric nerves (p. 660).

The integument is next to be removed from the upper part of the abdomen, and along with it the subcutaneous fat; only a sufficient thickness of superficial fascia being at first left to preserve the cutaneous nerves. These will be found in two ranges, the one situated near the middle line, and consisting of the anterior branches of the lower intercostal nerves, the other range emerging laterally, and consisting of the lateral cutaneous branches of the same nerves (p. 657). Let the external oblique muscle then be fully dissected, its posterior border being brought, if possible, into view (p. 248). The aponeurosis of the external oblique muscle is next to be divided by an incision carried transversely inwards from the anterior superior spine of the ilium, the inferior part of the aponeurosis being left for future examination; and the dissector will proceed to separate successively the attachments of the muscle to the crest of the ilium and each of the eight lower ribs, and will reflect the muscle towards the middle line as far as it admits of it. The internal oblique muscle, having next been examined, is to be reflected in the same way, and the transversalis muscle exposed and examined (pp. 250, 253).

The deeper parts involved in the descent of inguinal hernia are now to be studied. For this end, the remaining part of the aponeurosis of the external oblique muscle is to be divided along its inner attachment, down to the symphysis pubis; the lower border of the internal oblique muscle is to be examined, and, in the male, the cremasteric muscular fibres which are continuous with it are to be followed down to the testicle. The lower parts of the internal oblique and transversalis muscles are to be successively detached from Poupart's ligament and turned inwards, and their conjoined tendon is to be made evident. The fascia transversalis, with the internal abdominal ring, is now brought into view, and the subperitoneal fat may be seen shining through it (p. 258). The student will observe particularly the structures which lie in contact with the spermatic cord in its course from the internal to the external abdominal ring, and which are described as forming the walls of the inguinal canal (p. 963). He will also raise the fascia transversalis, and note the infundibuliform fascia and the circumflex iliac and epigastric arteries (p. 432); and will acquaint himself with the relations of the latter to the direct and oblique varieties of inguinal hernia, and with the coverings which these herniæ receive in their descent (p. 1029).

Poupart's and Gimbernat's ligaments may now be examined from the deep aspect, and, by separating the subperitoneal fat from the junction line of the fascia transversalis and fascia iliaca, the student will obtain a view of the deep crural arch, the crural ring, and the septum crurale (p. 258)— structures which are to be noted in relation to femoral hernia (p. 1033). He will then open the sheath of the rectus muscle; dissect it and the pyramidalis muscle (p. 253); follow the epigastric artery in the substance of the rectus muscle from below, and the abdominal branch of the internal mammary artery from above; and will, at the same time, examine the deficiency in the lower part of the posterior wall of the sheath of the rectus muscle, and the semilunar folds of Douglas (p. 250).

3. *Male Genital Organs.*—If the subject be a male, the penis ought at

this time to be dissected.   On removal of the skin, the dorsal arteries, vein; and nerves, together with the suspensory ligament, will be brought into view (pp. 428, 479, and 671).   The corpora cavernosa, corpus spongiosum, and glans are then to be dissected ; and the glans may, with care, be separated from the corpora cavernosa (p. 956).   The pendulous portion of the penis is to be cut across, the section examined, and the urethra slit open.

· The testicles and spermatic cord will next be dissected.   The fascia cremasterica is to be laid open, and the cremasteric branch of the epigastric artery and genital branch of the genito-crural nerve found (p. 964).   The fascia propria is to be removed, and the elements of the cord examined, viz., the vas deferens and the spermatic artery, veins, and nerves.   The testicle may then be removed, the tunica vaginalis opened, and the appearance and relations of the epididymis and vas deferens noticed (p. 967). The caput epididymis, in front of which will be seen the hydatid of Morgagni, is to be raised from the tunica albuginea, and the epididymis and coni vasculosi are to be dissected out.   The tunica albuginea must then be divided, and the arrangement of the tubuli seminiferi in the lobules made apparent under water, and the mediastinum exhibited.

4. *Abdominal Cavity ; Peritoneum ; Small Intestines, and Colon.*—The cavity of the abdomen is to be opened by a vertical and a transverse incision crossing one another on the left side of the umbilicus ; but the vertical incision is, in the first instance, to be arrested at the umbilicus, in order that the urachus and the fossæ into which the peritoneum is thrown by the obliterated hypogastric arteries may be examined.

The peritoneal cavity, especially the pelvic part, is to be carefully sponged out and all grumous fluid removed from it, and a piece of cotton soaked with spirit is to be laid in the recto-vesical fossa ; on the adoption of these precautions the practicability and comfort of the later parts of the dissection materially depend.   The general arrangement of the viscera is first to be examined, including the position and relations of the stomach, spleen, liver, duodenum, jejunum, ileum, coecum and other parts of the colon, the rectum, and the kidneys (p. 823).   The folds of the peritoneum are next to be studied (p. 826).   This membrane should be followed transversely and vertically throughout the abdominal cavity, and the line of attachment of the mesentery to the wall of the abdomen should be displayed.   The disposition of the foramen of Winslow and the great omentum should then be investigated, and, in order that the interior of the sac of the great omentum may be seen, a transverse cut should be made into it below the arch of vessels which lies along the great curvature of the stomach, and by this means the posterior surface of the stomach and the anterior surface of the pancreas will also be brought into view.   When the disposition of the great omentum has been observed, the small or gastro-hepatic omentum, the gastro-splenic omentum, the meso-colon, and the relations of the duodenum to the peritoneum will be easily followed.

· After the study of the peritoneum has been completed, the transverse colon is to be lifted upwards, and the small intestines turned over to the left side, in such a manner as to display the whole of the upper or right side of the mesentery ; and the distribution of the superior mesenteric artery, from the lower border of the pancreas downwards, with the accompanying vein and plexus of nerves, is then to be brought out by dissection (pp. 410 and 702).

From its right side the artery will be seen to give off the middle colic, right colic, and ileo-colic branches, and from its left side about a dozen branches to the small intestines ; and of these intestinal branches the dissector may trace the primary, secondary and tertiary arches of anastomosis. If the left branch of bifurcation of the middle colic artery be followed, it will lead the dissector to the left colic, and so to the trunk of the inferior mesenteric artery, with its accompanying vein and nerves, situated to the left of the mesentery ; and to study these the intestines must now be turned over to the right side. In addition to the left colic, the sigmoid branch of the inferior mesenteric artery will then also be seen, and the first part of the superior hæmorrhoidal vessels before they descend into the pelvis upon the meso-rectum (p. 412).

The dissector will now tie the intestine, a little below the termination of the duodenum, with two ligatures about an inch and a half distant, and will divide it between the ligatures ; in like manner he will secure and divide the great intestine at the lower extremity of the sigmoid flexure : he will then remove from the body the whole length of intestine between the upper and lower ligatures. To do this properly, he must begin from above, and pulling the ligatured extremity upwards with his left hand, with his right apply the scalpel lightly to the edge of the mesentery, close to the bowel. By this means the whole small intestines may with ease be removed, and the mesentery left in the abdomen. The large intestine may now also be removed as far as the rectum. The intestines are to be taken to the trough, and there they are to be thoroughly cleaned, by having water run through them from the jejunal end. They may then be spread out on a table and inflated, in order that the relative length and diameter of the different parts may be observed, the arrangement of muscular bands on the colon, and other facts as to their structure (pp. 840 and 854). The small intestine is to be separated from the great, several inches above the cæcum. A portion near the upper end may be cut separate, inflated and dried, in order to show the valvulæ conniventes which are thus put upon the stretch. The remainder is to be slit open in its whole extent, which may be best done with a pair of scissors, one of the points of which has been blunted with a small piece of cork ; the appearance of the mucous membrane in the different parts is then to be studied, attention being particularly directed to the distribution of the villi and valvulæ conniventes (p. 842), and to the patches of Peyer's glands (p. 846). The great intestine is next to be divided some inches beyond the cæcal valve, and the remainder is to be slit up and its mucous membrane examined. Lastly, the cæcum is to be very carefully washed, and the structure and action of the cæcal valve studied, by filling the portion of colon with water (p. 852). The water will be returned although the portion of ileum be left untied, and the position of the valve when closed may thus be seen. The cæcum may then be slit open on the side opposite the valve, and the vermiform appendage may also be opened to observe its glandular structure.

5. *Stomach and Duodenum, Pancreas, Spleen, and Liver.*—The duodenum and stomach are to be slightly inflated, and the arteries arising from the cœliac axis are to be dissected (p. 406). The student may begin by dissecting the splenic artery, following its course to the spleen, and observing its branches to the pancreas, to the stomach, the vasa brevia, and the left gastro-epiploic artery. Let him next trace the coronary artery of the stomach along the small curvature of that organ. Then, in following out the hepatic artery to its division into right and left branches, he will find the pyloric branch anastomosing with the coronary artery ; the cystic branch going to the gall-bladder ; and the gastro-duodenal branch dividing

4 A

into the right gastro-epiploic which anastomoses with the left gastro-epiploic, and the superior pancreatico-duodenal which anastomoses with the inferior pancreatico-duodenal branch of the superior mesenteric artery.

The inferior mesenteric vein will be traced upwards behind the pancreas to join the splenic vein, which, passing transversely onwards to meet the superior mesenteric vein, will be seen to form with it the trunk of the vena portæ (p. 479). The position of the common bile-duct with reference to the hepatic artery and portal vein is to be observed, and the duct is to be traced up into the hepatic and cystic ducts and downwards to the duodenum (p. 867). The relations and structure of the pancreas are then to be examined, and the pancreatic duct is to be traced along its posterior aspect to its termination in the duodenum along with the common bile-duct (p. 881). The spleen may now be removed, its blood-vessels dissected, a section made of it, and some of the pulp may be washed away to show the trabecular structure in the interior of the organ (p. 883). The stomach may now be removed along with the duodenum, and a careful examination made of the structure of these organs ; the shape of the stomach, its three layers of muscular fibres, and the construction of the pyloric valve being specially noted (p. 830).

The liver is next to be studied. Its ligaments, viz., the falciform ligament, the round ligament or obliterated umbilical vein, the coronary, and the two lateral or triangular ligaments are first to be examined ; after which the organ may be removed from the body (p. 865). In doing this, the inferior vena cava must be divided both above and below the liver. The dissectors may now observe the division of the liver into a right and left lobe, as also the quadrate, Spigelian, and caudate lobes : they will likewise note the various fissures, viz., the transverse or portal ; the longitudinal or antero-posterior, divided into an anterior part containing the remains of the umbilical vein, and a posterior part in which the remains of the ductus venosus are situated ; the fissure or fossa of the gall-bladder, and the fissure or fossa of the vena cava (p. 862). They will observe the openings of the hepatic veins into the part of the vena cava imbedded in the posterior border of the liver, and follow the divisions of the hepatic arteries, portal vein and hepatic ducts, as far as possible into the substance of the liver. In doing this the capsule of Glisson sheathing these parts is to be observed ; the appearance of the substance of the liver may then be exhibited by minuter dissection ; and the gall-bladder having been opened and washed, the structure of its coats and the peculiar reticulated arrangement of its mucous membrane may be examined.

6. *Deep Posterior part of the Abdominal Cavity.*—On returning to the examination of the parts remaining in the abdomen, the dissectors will begin by tracing out the plexuses of the sympathetic nerves. The superior and inferior mesenteric plexuses, in connection with the aortic plexus, are to be traced upwards into the solar plexus, and the nerves proceeding from the aortic plexus downwards into the hypogastric plexus. The solar plexus will be found surrounding the aorta at the root of the cœliac axis ; also, its semilunar ganglia, one on each side, and the splanchnic nerves passing through the crura of the diaphragm to terminate in it (p. 699). The dissectors will now follow the plexiform nerves which emanate from the solar plexus and surround the arteries in the neighbourhood ; namely, the cœliac plexus subdividing into hepatic, splenic and coronary ; also, the suprarenal and renal, and the spermatic plexuses. In doing this the suprarenal capsules will fall under observation, and care is to be taken in the

removal of the surrounding adipose tissue not to injure their substance, which is easily torn (p. 939) : after they have been carefully cleaned, these bodies may be examined by incisions into their substance.

The aorta and inferior vena cava are then to be dissected, and also the common and external iliac arteries and veins, together with the kidneys and ureters.   The branches of the aorta to be examined are the inferior phrenic, the cœliac axis, the superior mesenteric, the suprarenal, the renal, the spermatic, the inferior mesenteric, the origins of the four pairs of lumbar arteries, and, continuing the direction of the aorta from its point of bifurcation, the middle sacral artery.   The two common iliac arteries and veins must at this time be cleaned, also the ureters ; and the dissection may be carried down along the external iliac vessels, as far as the origin of the epigastric and circumflex iliac arteries ; in doing which the relations of the iliac arteries and veins will be carefully observed (pp. 418, 473, and 477). The position and relations of the kidneys are now to be examined, and more particularly the position of the renal artery, renal vein, and ureter, as they enter the gland (p. 926).   The kidneys having been removed from the body, are to be opened by a transverse vertical section, to exhibit the pelvis, calyces, and pyramids, the cortical and internal tubular substances, and the Malpighian glomeruli : the fibrous tunic which invests the kidney is also to be observed.   The receptaculum chyli or commencement of the thoracic duct will be found beneath the right crus of the diaphragm (p. 487), as also the commencement of the vena azygos in connection with some of the lumbar veins (p. 469).

7. *Upper and Posterior Wall of the Abdomen.*—The diaphragm is now to be dissected (p. 243).   Anteriorly will be found its attachments to the six lower ribs interdigitating with those of the transversalis muscle ; posteriorly will be found the two crura and the ligamenta arcuata externa and interna ; while the fibres passing from all those parts will be traced to their connection with the central tendon ; and the openings for the aorta, œsophagus, and vena cava inferior will be examined.   The surface of the psoas magnus muscle is next to be cleaned, as well as that of the psoas parvus lying superficial to it (if it be present) (p. 272) ; and, emerging from the fibres of the psoas magnus, the genito-crural nerve will be found and followed downwards.   The nerves of the lumbar plexus will be observed principally on the outer and inner aspects of the psoas muscles (p. 660). The fibres of these muscles are to be dissected away from the nerves.   In addition to the communicating branches of the plexus, there will be observed, proceeding from the anterior division of the first lumbar nerve, the ilio-hypogastric and ilio-inguinal nerves, often united into one ; from the second lumbar nerve the external cutaneous and genito-crural nerves ; from the second, third, and fourth lumbar nerves together, the anterior crural and the obturator nerves ; and, lastly, the lumbo-sacral cord, formed by the union of a part of the fourth with the whole of the fifth nerve (p. 658).   On the bodies of the vertebræ will be found the lumbar part of the chain of sympathetic ganglia ; the branches of communication between which and the spinal nerves are to be dissected (p. 696).

At this time the dissectors ought to revert to the arrangement of the posterior part of the transversalis muscle.   This they will find to be continued into an aponeurosis which is connected behind with three layers ; of these the most posterior is the fascia lumborum observed in the dissection of the back, the second lies in front of the erector spinæ muscle, and the foremost is a much thinner membrane placed in front of the quadratus

4 A 2

lumborum muscle. The quadratus lumborum and iliacus muscles are now to be dissected (pp. 255 and 271). On removing the iliacus from the iliac fossa, the distribution of the ilio-lumbar artery will be traced, and its anastomoses with the last lumbar and the circumflex iliac artery exhibited (p. 429).

8. *Dissection of the Pelvis.*—The pelvis with several of the lumbar verte-bræ ought now to be separated from the rest of the trunk, and before proceeding further, the dissector should carefully remove the superfluous masses of muscle and other soft parts adherent to the outer surface of the bones.

Female Genital Organs.—If the subject be a female, the perinæum is first to be dissected. The exact position of the orifice of the urethra is to be examined with reference to the passing of the catheter (p. 980). The fat is to be removed from between the ischium and rectum ; and, as this is being done, the inferior hæmorrhoidal and superficial perineal vessels and nerve will be brought into view (pp. 426 and 670). The sphincter muscles of the rectum and vagina, the levator ani and transversalis muscles, and the obturator fascia will be seen (p. 265). From among the fat on the fore part are to be dissected out the crura of the clitoris and the erector muscles embracing them ; and on the side of the vulva the bulbus vestibuli. The glands of Bartholin are to be sought at the back part of the lower end of the vagina, and the duct of each followed to its orifice by the side of the hymen or carunculæ myrtiformes. Internally to the crus clitoridis, the triangular ligament or subpubic fascia will be found extending from the pubic arch to the vagina (pp. 977 and 260).

The bladder ought now to be partially inflated, and the reflections of the peritoneum in the pelvic cavity examined, especially the posterior, lateral, and anterior false ligaments of the bladder, and in the female the broad ligament of the uterus, with the ovary, Fallopian tube, and round ligament (pp. 947 and 985). Let the peritoneum then be reflected from the walls of the pelvis so as to exhibit the lateral and anterior true ligaments of the bladder, and the whole internal aspect of the pelvic fasciæ (p. 260). In order to have a complete view of these fasciæ, it will be necessary to remove a portion of the os innominatum of the right side. This must be done in such a manner as not to interfere with the attachments of the fasciæ : while, therefore, the anterior and lower part of the bone with the acetabulum is to be removed, the brim of the pelvis and the boundary of its outlet are to be preserved, as well as the sacro-sciatic foramina. With a little care, and preliminary observation of the form of the innominate bone, this may be done by means of a single section with the saw, carried close by the brim of the pelvis, and downwards in such a direction as to remove the greater part of the thick-ness of the ischial tuberosity and pass as near as possible to the sacro-sciatic notches, without breaking into them. By this means the hip-joint may be removed intact ; and, should it not have been dissected along with the leg, to which it properly belongs, the dissectors of the abdomen will now have an opportunity of examining it ; and may especially observe the action of the ligamentum teres, by removing the deep part of the aceta-bulum, while the capsule of the joint is left intact (p. 151).

Returning to the pelvis, the opening in its lateral wall is to be enlarged, if necessary, with the bone-nippers, and the obturator internus muscle is to be carefully removed, and the peculiar arrangement of its tendon remarked (p. 269). On the inner aspect of that muscle will be found superiorly the undivided pelvic fascia, inferiorly the obturator fascia, and between the two the white band stretching from the symphysis pubis to the spine of the ischium, which marks the level at which the pelvic fascia splits into the recto-vesical and obturator fasciæ; while in the upper part of the obturator foramen the obturator vessels and nerve will be seen issuing from the interior. If the ischio-rectal fossa be now thoroughly cleaned, a complete view of the layers of fascia will be obtained, and of their relation to the levator ani muscle (p. 260). The brim of the pelvis is next to be sawn through near the symphysis pubis, on the side on which the dissection has been made, and is to be removed. By this means, if the subject be a male, the relations of the fascia to the prostate gland will be better seen. The ureters and the vasa deferentia are to be followed as far as the bladder; the sympathetic nerves of the hypogastric plexus are to be traced in their distribution to the pelvic viscera (p. 702); and the branches of the internal iliac vessels are to be dissected. The internal iliac artery will be found to give off to the walls of the pelvis and to the external parts, the gluteal, iliolumbar, and lateral sacral arteries, constituting the branches of its posterior division; the obturator, internal pudic, and sciatic arteries in connection with its anterior division: while to the viscera it supplies the superior vesical with the obliterated hypogastric artery, the inferior vesical giving the middle hæmorrhoidal, and, in the female, the uterine and vaginal arteries (p. 420). The first group may perhaps be best seen on the entire side, and the second and third group on the dissected side of the pelvis. On the former side the sacral nerves are to be displayed (p. 268), and the origin of the pyriformis muscle examined. The junction of the lumbosacral cord with the anterior divisions of the three first sacral nerves and a branch of the fourth, to form the sacral plexus, will now be brought into view (p. 669). The gluteal nerve will be found arising from the lumbosacral cord (p. 667); and arising from the sacral plexus will be found the great and small sciatic nerves, the pudic nerve, the nerve to the obturator internus muscle, and other muscular branches (p. 670). The remaining branches of the fourth sacral nerve will be found to aid the hypogastric plexus in the supply of nerves to the viscera: at the same time the small fifth sacral and coccygeal nerves may also be dissected (p. 668). The coccygeus and levator ani muscles are to be cleaned on their upper aspects, when they will be seen to form a continuous muscular floor to the pelvic cavity (p. 262). The chains of sympathetic ganglia are then to be dissected in front of the sacrum, and, if possible, the lowest parts traced to their junction in front of the coccyx (p. 696).

9. *Pelvic Viscera.*—It may be proper to examine the muscular walls of the bladder in the inflated condition of the organ, before its removal from the pelvis (p. 944); after which the viscera are to be separated from their attachments to the walls of the pelvis, and removed in one mass.

The rectum may then be carefully dissected away from the rest of the viscera, the extent of its connection with them being at the same time observed (p. 856). Its muscular coats having been sufficiently examined, it is to be slit open and washed, in order that the general appearance and folds of its mucous membrane may be seen. In the male subject the prostate gland enveloped in its fibrous covering, the vesiculæ seminales, and the vasa deferentia are to be carefully dissected (pp. 952 and 971); the bladder is to be opened from before, the neck being left in the first instance entire; and the openings of the ureters and urethra, with the trigone between them, are to be examined (p. 948). The prostatic, membranous, and bulbous parts of the urethra are then to be slit open from above, the varying diameter of the urethra observed, as also in its prostatic part, the verumontanum or caput gallinaginis, the sinus pocularis, and the orifices of the common ejaculatory ducts (p. 961). The junction of the vas deferens and vesicula seminalis to form the common ejaculatory duct is to be displayed ; and a longitudinal section of the prostate gland may be made to show its thickness, consistence, and structure : the relations of its base to the neck of the bladder should be particularly observed, with the circle of veins of the vesical plexus in the angle between them.

In the female subject the bladder is to be opened and examined as in the male, and the length and diameter of the urethra observed (p. 980). The vagina is then to be cut open a little on one side of the middle line in front, when the rugæ of its mucous membrane will be seen ; also, at its entrance, the carunculæ myrtiformes, and, projecting into it above, the cervix uteri (p. 981). The ovary with its ligament and mesovarium, the Fallopian tube, the round ligament of the uterus, and, between the ovary and Fallopian tube, the tubules termed parovarium or organ of Rosenmüller, are next to be dissected, and the external configuration of the uterus examined (p. 982). The student will then notice the position and appearance of the os uteri externum, and will open the uterus on its anterior aspect by a line of section which, by dividing into two superiorly, is prolonged to both of the cornua (p. 984). He will thus see the size and shape of the triangular cavity of the uterus, the cavity of the cervix, the rugæ of its mucous membrane, and the os uteri internum.

10. *The Pelvic Ligaments.*—At the conclusion, the articulations of the pelvic bones may be examined, if they are still in a condition fit for dissection (p. 147). The symphysis pubis with its concentric laminæ of fibrocartilage is first to be examined ; then the articulation of the pelvis with the fifth lumbar vertebra, especially the sacro-vertebral and ilio-lumbar ligaments : the great and small sacro-sciatic ligaments should be cleaned, and, by removing the remains of the origin of the obturator internus muscle, the obturator membrane.

The anterior and posterior ligaments and the intervertebral disc of the sacro-coccygean articulation are to be observed : lastly, the strong posterior and the thinner anterior sacro-iliac ligaments having been dissected, the last mentioned is to be divided, and the cartilaginous surfaces of the sacro-iliac synchondrosis are to be brought into view by forcing open the articulation.

V.—LOWER LIMBS OR INFERIOR EXTREMITIES.

The right and left limbs constitute each a part, the dissection of which should extend over a period of not less than four weeks. It includes the whole limb below Poupart's ligament and the crest of the ilium, but not the perinæum.

1. *The Gluteal Region.*—The dissection of the gluteal region, the back of the thigh, and the popliteal space is to be completed in the four days during which the subject lies on its face. To remove the integument from the buttock let an incision be carried along the crest of the ilium, brought downwards in the middle line of the sacrum and curved outwards in the fold of the nates, then directed obliquely to the outside of the thigh about five or six inches below the great trochanter. The junior student will at once proceed to clean the gluteus maximus muscle in the direction of its fibres (p. 266). The senior student will examine the arrangement of the

cutaneous nerves in this region.   Of these he will find, descending over the
crest of the ilium, in order from before backwards, the lateral branches of
the last dorsal and ilio-hypogastric nerves (pp. 658 and 660), with several
branches of the lumbar nerves (p. 634) ; and, piercing the gluteus maximus
muscle near its posterior attachment, some small cutaneous twigs from the
posterior divisions of the upper sacral nerves (p. 635) ; lastly, turning
round its inferior border, branches from the small sciatic nerve (p. 675).
It will be observed that the fascia lata, which is strongly developed over
that part of the gluteus medius which lies in front of the gluteus maximus
muscle, on reaching the upper border of the gluteus maximus, divides into
two laminæ, of which one is continued on the superficial, and the other on
the deep aspect of that muscle (p. 292).   Care is to be taken to lay bare
the inferior border of the gluteus maximus in its whole extent ; and a
synovial bursa over the tuberosity of the ischium is to be sought for.   The
muscle is then to be divided close to its iliac and sacral attachment, and in
turning it forward, the sciatic artery and the superficial branch of the
gluteal artery will come into view.   The branches of these arteries and of
the small sciatic nerve which enter the muscle are to be followed out to
some extent, and they may then be divided to permit the complete reflection of the muscle.   While this is being done, a large synovial bursa will be
found between the trochanter major and the insertion of the gluteus maximus into the fascia lata.

The fascia lata is to be removed from the upper part of the gluteus
medius muscle, and the parts exposed by the removal of the gluteus maximus are to be cleaned in their order from above downwards, viz. : the back
part of the gluteus medius muscle, the gluteal vessels (p. 429), the pyriformis
muscle, the sciatic vessels and the great and small sciatic nerves (p. 674), the
gemelli muscles, superior and inferior, with the tendon of the obturator internus muscle between them (p. 268).   The tendon of this muscle may now be
dissected from between the gemelli, divided and turned back, to show the
synovial cavity in which it plays upon the smooth trochlear surface of the
ischium.   The quadratus femoris, the tendon of the obturator externus muscle
situated more deeply, the upper part of the adductor magnus muscle, and the
origin of the hamstring muscles are then to be exposed.   From the small
sciatic nerve the inferior pudendal branches will be seen given off, in addition
to those already mentioned, and from the sciatic artery, besides muscular
branches, the coccygeal branch, the branch to the great sciatic nerve, and
that by which it anastomoses with the internal circumflex artery may be
traced.   On the spine of the ischium also will be seen the pudic vessels and
nerve, and the nerve to the obturator internus muscle (pp. 425 and 670) ;
and descending under cover of the tendon of the obturator internus and the
gemelli is the small nerve to the quadratus femoris.

The gluteus maximus muscle having been entirely removed from its upper
attachment, and the tendon of insertion being left, the gluteus medius is to
be raised from the ileum in three-fourths of its extent ; its anterior border
and that of the gluteus minimus muscle being left for dissection from the
front.   The attachments of the gluteus medius muscle are to be observed,
as also the superior and inferior deep branches of the gluteal artery, and the
distribution of the gluteal nerve (pp. 429 and 667).   The posterior part of
the gluteus minimus may then be raised from the ileum to show the extent
of its attachment to that bone, and its relation to the capsule of the hip-
joint.

2. *The Popliteal Space.*—It is advisable to dissect this space before the

posterior femoral region. In order to open it the integument may be divided by a longitudinal incision of considerable length, which may be crossed if necessary by a transverse one in the middle of the space, sufficient to allow the integument to be thrown freely back. On removal of the superficial fat, the fascia lata, which is strong in this region, will come into view, and, in the lower part of the space, the terminal twigs of the small sciatic nerve (p. 675), and the upper part of the short saphenous vein (p. 476). The fascia lata is to be divided, and the fat carefully removed from the space, its boundaries cleaned, and the vessels and nerves with their branches traced. Superiorly the biceps muscle on the outside, and the semitendinosus and semimembranosus muscle on the inside, and inferiorly the heads of the gastrocnemius muscle with the small belly of the plantaris will thus be exposed.

Lying in the space the dissector will find the external and internal popliteal nerves giving off their articular and sural branches (pp. 676 and 679), and more deeply the popliteal vessels in a common sheath (p. 441). He will follow out the branches of the popliteal artery, viz., its five articular branches, the superior, azygos, and inferior, and its sural branches. On the surface of the popliteal artery, where it enters the space, may be found a twig of the obturator nerve (p. 663).

When the dissection of the popliteal space has been completed, it is to be united to that of the gluteal region by an incision along the posterior part of the thigh. The course of the small and great sciatic nerves will thus be laid bare, together with the biceps, semitendinosus, and semimembranosus muscles, the twigs of the great sciatic nerve supplied to these muscles, and to the adductor magnus, and the four perforating branches of the deep femoral artery (p. 439) ; the posterior aspect of the adductor magnus muscle will also be exposed.

3. *The Front of the Thigh.*—On the day on which the subject is laid upon its back, the student should begin the dissection of the front of the thigh, by studying the fasciæ connected with the descent of femoral hernia. For this purpose an incision is to be made from the neighbourhood of the anterior superior spinous process of the ilium inwards, in the line of the groin, and carried half way down the inside of the thigh. The large flap of integument thus marked out is to be raised and turned outwards. The subcutaneous fascia is then to be laid bare by the removal of any fat, and it will be advantageous if this can be done in concert with the dissector of the abdomen (p. 292). Various small superficial arteries and veins will be seen, viz. : the superficial epigastric, superficial circumflex iliac, and superior and inferior superficial pudic (p. 437). The fascia lata will be laid bare, and the cribriform fascia overlying the saphenous opening. On the surface of the fascia lata will be brought into view the internal or long saphenous vein passing into the saphenous opening, frequently presenting two branches (p. 475) ; nearly in front of the femoral artery, the crural branch of the genito-crural nerve; and, in front of the anterior superior spine of the ilium, the external cutaneous nerve (p. 660). A twig of the ilio-inguinal nerve may also be seen distributed to the skin of a small part of the thigh close to the pubes. The border of the saphenous opening is to be made distinct by removing the cribriform fascia, and in doing this the attachment of the superior cornu or falciform process to the pubic portion of the fascia lata is to be shown (p. 293). This falciform process is then to be separated from the fascia lata and turned to the outside sufficiently to expose the infundi-buliform or crural sheath, investing the femoral vessels, and the dissector

may examine the three compartments into which this sheath is divided, and which contain respectively the artery, the vein, and a lymphatic gland ; the latter blocking up the crural aperture between the femoral vein and Gimbernat's ligament, through which femoral hernia descends. All the relations of these parts are to be carefully studied with special reference to the operations for strangulated femoral hernia (p. 1036).

The incision on the inner side of the thigh is now to be prolonged downwards towards the middle line beyond the knee, and the dissection of the front of the thigh continued. The two middle and the two internal cutaneous branches of the anterior crural nerve, together with the branch from the internal saphenous nerve to the integument of the knee, and the internal saphenous vein, will be dissected out, and the fascia lata in front of the thigh made clean (p. 664). The fascia is then to be removed, and the communications of the internal cutaneous, internal saphenous, and obturator nerves sought in the lower part of the inner aspect of the thigh (p. 666). Scarpa's triangle is now to be cleaned, and the dissection of the femoral vessels both in that space and in the after part of their course is to be studied (p. 434). Towards its termination below the middle of the thigh, the femoral artery will be observed to be covered by a tendinous expansion, which conceals it for a part of its course before it pierces the tendon of the adductor magnus muscle : in the passage so formed, known as Hunter's canal, the femoral artery, which is accompanied by the internal saphenous nerve, will be seen to give off the anastomotic branch (p. 293).

The deep femoral artery should be dissected as far as the upper border of the adductor longus muscle ; and the origins of its first branches are to be brought into view, viz. : the internal circumflex artery, dividing into ascending, transverse, and descending branches. One or both of the circumflex arteries often arise from the femoral artery immediately above the origin of the deep femoral (p. 438). The sartorius muscle is to be cleaned, and likewise the gracilis muscle, and the surface of the other adductors ; the relations of the inferior tendons of the sartorius, gracilis, and semitendinous muscles may also be exposed (pp. 273 and 276). The student will then direct his attention to the outer part of the thigh near the hip. He will there dissect the fascia lata from the remaining part of the gluteus medius muscle, and from the tensor vaginae femoris muscle, leaving at first a strip of the fascia extending down to the knee on the outside of the leg, and he will afterwards expose the deeper band of the fascia which passes inwards to the hip-joint from within the upper part of the muscle (pp. 273 and 292). He will also find the branch of the gluteal nerve to the tensor vaginae femoris by dissecting between it and the gluteus medius muscle (p. 667). Let him divide successively the tensor vaginae femoris and the remains of the gluteus medius and minimus, and dissect the two last muscles down to their inferior attachments, so as to exhibit the bursae between them and the trochanter major, and the connection of the gluteus minimus with the capsule of the hip-joint (p. 268). While engaged with this proceeding, he will be enabled to dissect more particularly the ascending and transverse branches of the external circumflex artery, and to examine their anastomoses with the gluteal artery (p. 438). Let him then clean the rectus muscle, trace its anterior and posterior heads close to their origins, and observe the positions of the limb in which they are respectively tightened (p. 274). The trunk of the anterior crural nerve is now to be cleaned, its branches to the extensor muscles are to be dissected, the internal saphenous nerve laid bare as far as the knee, and the slender twigs to the pectineus muscle seen passing behind the femoral vessels. These last may be most easily found if the common femoral artery be previously divided (p. 664). If the accessory obturator nerve is present, it will now be seen passing over the brim of the pelvis to the outer border of the pectineus muscle which it partly supplies (p. 666). The pectineus and adductor longus muscles are then to be divided, and their attachments carefully dissected. The continuation of the profunda femoris artery behind the adductor longus is to be cleaned ; and its four perforating branches, of which the fourth is the continuation of the artery, will be seen piercing the adductor magnus muscle (p. 439). When the pectineus muscle has been reflected, the accessory obturator nerve may be traced to its communication with the main obturator nerve, to the pectineus muscle, and to the hip-joint. The anterior division of the obturator nerve is to be traced down in front of the adductor brevis muscle, and on division of the pectineus muscle its posterior division to the adductor magnus will come into view. The obturator nerve will be observed to supply all the adductor group of muscles (p. 662). The dissector will now trace the internal circumflex artery ; he will find it dividing into two branches, one of which passes inwards in front of the obturator externus and adductor brevis muscles, while the other is directed backwards to anastomose with the sciatic artery, and gives off a branch to the hip-joint which enters it by the notch of the acetabulum (p. 439). The obturator externus muscle is to be cleaned, and the external and internal divisions of the obturator artery are to be laid bare from among its fibres (pp. 269 and 423).

The adductor magnus muscle is then to be cleaned and examined (p. 277); and after it the conjoined insertion of the psoas and iliacus muscles (p. 271); the vastus externus, vastus internus and crureus muscles, together with the deep fibres of the latter, called subcrureus, which are inserted into the synovial membrane of the knee-joint (p. 275).

4. *Hip-joint.*—When this stage of the dissection has been reached, the student may either saw through the femur and leave the hip-joint to a more convenient opportunity, or dissect the joint at this time, and afterwards disarticulate the femur. The latter plan is usually to be preferred. In that case, the attachments of all the muscles which act upon or are related to the hip-joint are to be reviewed, and those which remain uncut are to be severed ; the capsular ligament is to be cleaned ; its thinness or deficiency on the posterior aspect, and the thick accessory or ilio-femoral ligament, strengthening it in front, are to be noted (p. 151). The relation of the head of the femur to the acetabulum in the various positions of the limb and foot are to be observed. The capsule may then be opened, and the cotyloid, transverse, and round ligaments examined, together with the articular surfaces and synovial membrane : the limb may then be removed from the body.

5. *The Back of the Leg.*—After the separation of the limb from the trunk, and when the divided structures have been cleaned and cut conveniently short, the student will proceed with the dissection of the calf and back of the leg, by directing an incision down the middle of the limb to the heel, and reflecting the skin to each side. He will trace the external and internal saphenous veins as far as the outer and inner ankle (p. 475); accompanying the latter he will find the internal saphenous nerve (p. 666), and along with the former he will find the external saphenous nerve arising from the union of the communicans tibialis and communicans fibularis branches of the internal and external popliteal nerves respectively (p. 677). He will also

find another cutaneous branch of the external popliteal nerve ramifying on the outer side of the leg. The gastrocnemius muscle is then to be cleaned, and the nerves and vessels entering it are to be more particularly dissected (p. 283). Its thin and flat tendon is then to be carefully divided at its lower part from that of the soleus, and the muscle is to be turned upwards. The soleus muscle will thus be brought into view, and, resting upon it the plantaris (which however is sometimes absent). Between the soleus muscle and the knee-joint the popliteus muscle will be seen protected by the popliteal aponeurosis, and, crossing it, the lower part of the popliteal vessels and internal popliteal nerve giving off branches to these muscles (pp. 442 and 676). The popliteus muscle is to be preserved to be dissected more particularly with the knee-joint. The plantaris and soleus muscles are to be separated from their superior attachments, and the nature and connexions of the tendo Achillis examined (p. 285); after which the latter may be divided near its insertion. The deep fascia is then to be divided, and the flexor longus digitorum, tibialis posticus, and flexor longus pollicis muscles, lying in this order from within outwards, are to be dissected (pp. 286 and 288). The anterior tibial artery will be seen perforating the interosseous membrane to arrive at the front of the leg, and the posterior tibial artery, venæ comites, and nerve are to be studied, and the branches of the nerve to the popliteus and other three deep muscles followed; while the peroneal artery is to be traced downwards in the fibres of the flexor longus pollicis muscle, and will be observed to give off the anterior peroneal and a communicating branch to the posterior tibial artery (pp. 444 and 677). The relations of the tendons, artery and nerve behind and below the inner ankle are to be particularly noted.

6. *The Sole of the Foot.*—The skin is to be reflected by means of an incision along the middle line of the heel and sole, and a transverse one across the balls of the toes. The plantar cutaneous branch of the posterior tibial nerve is to be traced to its distribution; and, on removing the fat from the plantar aponeurosis, an outer and inner set of small nerves and vessels will also be found perforating the latter (p. 296). Below the inner ankle the internal annular ligament is to be cleaned, and the tibialis posticus muscle is to be dissected to its insertion (pp. 288 and 295). The skin is to be divided up the middle of the toes; the sheaths for the flexor tendons are to be exhibited, and the digital arteries and nerves on both sides of each of them are to be traced. The plantar aponeurosis is then to be removed by dissection from the muscles which it covers as much as possible, so as to expose the abductor pollicis, flexor brevis digitorum, and abductor minimi digiti muscles (p. 289). The insertions of the tendons of the flexor brevis digitorum are to be followed by dividing the sheaths on the toes; its posterior attachment is to be divided, and the branch of the internal plantar nerve which supplies it sought. This will bring into view the tendons of the flexor longus digitorum and flexor longus pollicis, the union of which will be noted; the flexor accessorius and the lumbricales muscles will now also be dissected (p. 287). Crossing the flexor accessorius muscle are the external plantar artery and nerve; the artery is to be followed to the deeper part of its course where it forms the plantar arch. The branches of the nerve to the flexor accessorius and abductor minimi digiti are to be found, its distribution to the two outer toes is to be traced, as also the origin of its deep branch (p. 679). The flexor accessorius muscle is to be removed from its broad origin, and the tendons of the flexor longus pollicis and flexor longus digitorum are then to be divided. The internal

plantar artery is to be dissected forwards to the inner side of the great toe; and the internal plantar nerve, after giving branches to the abductor pollicis, flexor brevis pollicis and two inner lumbricales muscles, will be traced forwards to its distribution on both sides of the three inner toes and one side of the fourth toe (pp. 446 and 677). The deep branch of the external plantar nerve is to be traced to its distribution in the two outer lumbricales, the transversus pedis, adductor pollicis, and all the interossei muscles, save the outermost two, which, together with the flexor minimi digiti, are supplied by the external digital branch. The arch of the external plantar artery will at the same time be traced to the first inter-osseous space, and its digital and other branches dissected (p. 447). After these parts have been examined, the attachments of the flexor brevis and abductor pollicis, transversus pedis, and flexor brevis minimi digiti muscles are to be fully studied.

7. *The Front of the Leg, and Dorsum of the Foot.*—The remaining in-tegument having been removed from the front of the leg and upper surface of the foot, the dissector will trace the cutaneous veins and nerves in this region.   On the inner border of the foot will be found the small terminal twigs of the internal saphenous nerve, and in front of the inner ankle the commencement of the great saphenous vein (pp. 475 and 666); while on the foot externally, and passing behind the outer ankle, will be observed the external or posterior saphenous vein and nerve (pp. 476 and 677).   On the middle of the leg externally, the musculo-cutaneous nerve will be seen piercing the aponeurosis and becoming superficial, and its distribution is to be traced to the inner side of the great toe and to the adjacent sides of the toes in the three outer interdigital spaces (p. 680); while the first inter-digital space will be found supplied by a branch continued from the anterior tibial nerve.   Immediately above and to the inside of the ankle-joint will be found the upper transverse and the lower oblique parts of the anterior annular ligament or retinaculum binding down the tendons of the extensor muscles (p. 295).   These are to be kept, the rest of the aponeurosis being removed : there will thus be exposed in order from within outwards, the tibialis anticus, extensor pollicis, extensor longus digitorum, and peroneus tertius muscles, which are to be dissected to their insertions (p. 279).   On the dorsum of the foot the extensor brevis digitorum is also to be dissected ; preserving at the same time the anterior tibial vessels and nerves, and the musculo-cutaneous nerves already mentioned.   Arising from the outer aspect of the fibula, the peroneus longus and brevis muscles are then to be cleaned (p. 282): the latter is to be traced to its insertion, but the course of the tendon of the peroneus longus across the sole of the foot will be more fully seen when the ligaments are dissected.   The musculo-cutaneous nerve is to be traced upwards to its origin from the external popliteal or peroneal nerve, and, as it pierces the fibres of the peronei muscles in its course round the fibula, its branches to these muscles will be seen.   The anterior tibial nerve is then to be traced beneath the muscles and round the fibula, and downwards on the front of the interosseous membrane, and will be found to supply in the leg the extensor longus digitorum, tibialis anticus, extensor pollicis, and peroneus tertius muscles, and on arriving at the foot, the extensor brevis digitorum (p. 682).   The anterior tibial artery will at the same time be dissected, and its branches traced, viz., its recurrent branch passing upwards on the tibia through the origin of the tibialis anticus muscle, to anastomose with the articular branches of the popliteal artery ; its muscular branches, and its external and internal malleolar branches ;

here there will generally be seen the anastomoses between the external malleolar and the anterior peroneal arteries (p. 449). The continuation of the anterior tibial artery as the dorsal artery of the foot is to be traced forwards to its junction with the plantar arch in the first interosseous space, and its tarsal and metatarsal branches are to be examined with the branches supplied by the latter to the three outer interosseous spaces (p. 450). Finally, the interossei muscles are to be dissected and examined in their dorsal and plantar aspects (p. 291).

8. *The Knee-Joint, Ankle-Joint, and Articulations of the Foot.*—The tendons passing near the knee-joint are, in the first place, to be cleaned; and the anastomoses of blood-vessels upon the knee are to be more particularly examined, viz., 'the anastomotic branch of the femoral artery, the external and internal superior articular, and external and internal inferior articular branches of the popliteal artery, and the recurrent branch of the anterior tibial artery. The three parts of the insertion of the tendon of the semimembranosus muscle and the posterior ligament are to be exhibited (p. 271): the popliteus muscle is then to be dissected out, and its tendon traced to its origin (p. 285); the tendon of the biceps muscle is also to be dissected to its insertion in connection with the external lateral ligament (p. 270); and at the same time the internal lateral ligament is to be displayed (p. 153). In front the ligamentum patellæ is to be cleaned, and the extension upwards of the synovial sac of the knee-joint carefully examined; the joint may then be opened by cutting into the synovial sac at this place, and reflecting the remains of the quadriceps extensor femoris muscle. Inside will be seen the ligamentum mucosum, the alar ligaments, and the fatty processes of the synovial membrane; the extent of the synovial cavity will be carefully inspected, and with a little dissection the crucial ligaments may then be brought into view. The capsule of the joint ought next to be entirely removed, in order that the form and actions of the lateral and crucial ligaments and the movements of the semilunar cartilages may be better studied. The structure of the latter will be best seen after the femur has been separated from the tibia.

The movements of the ankle-joint ought to be studied in connection with those of the tarsal articulations (p. 158). Its principal ligaments are to be cleaned externally, viz., the external lateral in three distinct parts, the internal lateral, and the transverse or posterior. When the internal examination·of this joint has been completed, the superior and inferior tibiofibular articulations and the interosseous membrane are to be studied. On the dorsum of the foot the numerous short dorsal ligaments of the tarsal and metatarsal bones are to be cleaned. On the sole of the foot the superficial and deep parts of the calcaneo-cuboid ligament, the inserted tendons of the tibialis posticus and peroneus longus muscles, the scaphoido-cuboid, scaphoido-cuneiform, and various other shorter ligaments are to be dissected.

The examination of the remaining joints of the foot may .then be completed in the following order: the posterior articulation of the astragalus and calcaneum, bounded in front by the strong interosseous ligament; the articulation of the astragalus, calcaneum, and scaphoid, in which the inferior calcaneo-scaphoid ligament is especially to be observed; the calcaneo-cuboid articulation; the articulation between the cuboid and fourth and fifth metatarsal bones; the articulation between the scaphoid and cuneiform bones, which passes forwards between the latter; the articulation between the two outer cuneiform bones and the second and third metatarsal bones; and the articulation between the internal cuneiform and first metatarsal bone.

The mode of connection of the metatarsal bones with each other is to be observed ; the interosseous, dorsal and plantar ligaments of their bases, and the transverse metatarsal ligament of their heads.   Lastly, the articulations of the metatarsal bones with the first phalanges, and of the phalanges with each other are to be dissected.   In connection with the great toe, the arrangement of the sesamoid bones deserves particular attention.

INDEX.

ARTERY—*continued.*
perforating, of hand, 400
of thigh, 439
of thorax, 374
pericardiac, 374, 402
perineal, superficial, 426
in females, 428
transverse, 427
peroneal, 444
anterior, 445
petrosal, 752
pharyngeal, ascending, 357
phrenic, inferior, 416
superior, 374
plantar, external, 447
internal, 446
popliteal, 441
dissection of, 1080, 1083
profunda, of arm, inferior, 384
superior, 383
of penis, 428
of thigh, 437
pterygoid, 356
pterygo-palatine, 357
pubic, 434
of obturator, 424
pudic, 425
accessory, 428
external, 437
in female, 428
pulmonary, 331, 898, 902, 903
from abdominal aorta, 406
development of, 326
in fœtus, 328
valves of, 307
pyloric, 408
radial, 394
dissection of, 1065
peculiarities of, 385, 386, 399
of index finger, 399
ranine, 349
recurrent, of deep palmar arch, 400
interosseous posterior, 391
radial, 397
tibial, 449
ulnar, 389
renal, 414, 935
sacral, middle, 418
lateral, 430
scapular, posterior, 367, 373
sciatic, 429
sigmoid, 512
spermatic, 414, 975
spheno-palatine, 357
spinal, anterior, 369
of intercostals, 404
of inferior thyroid, 371
of lumbar, 417
posterior, 369
of vertebral, 368
splenic, 408, 886
sternal, 374
sterno-mastoid, 351
stylo-mastoid, 353, 752, 768
subclavian, 364

ARTERY—*continued.*
subclavian, development of, 327
dissection of, 1053
peculiarities of, 337, 367
surgical anatomy of, 1007
sublingual, 348, 349, 818
of submaxillary gland, 817
submental, 350, 351, 818
subscapular, 379
of suprascapular, 373
supra-acromial, 373
supra-orbital, 360
suprarenal, 413
suprascapular, 366, 371
sural, 442
tarsal, 450
temporal, 353
anterior, 354
deep, 356
middle, 354
posterior, 354
thoracic, acromial, 379
alar, 379
external, 379
long, 379
superior, 379
of thumb, dorsal, 398
large, 399
thymic, 374
thyroid, inferior, 366, 371, 919, 922
lowest, 340, 922
superior, 346, 919, 922
tibial, anterior, 448
dissection of, 1083, 1084
peculiarities of, 451
posterior, 444
dissection of, 1083
peculiarities of, 445
tonsillar, 350, 351, 814
tracheal, 371
transverse, of basilar, 370
cervical, 373
of face, 354
humeral or scapular, 371
of perinæum, 427
tympanic, 356, 752
ulnar, 388
dissection of, 1065
peculiarities of, 385, 386, 391
umbilical, 328, 421
uterine, 422, 987
vaginal, 422
of liver, 873
of vas deferens, 421
vertebral, 366, 367
development of, 327
peculiarities of, 369
vesical, inferior, 421
superior, 421
vesico-prostatic, 421
vestibular, 767
Vidian, 357
volar, superficial, 397
Arthrodia (ἄρθρον, a joint), 120

4 B 2

Muscular contractility or irritability, cxxix
 duration of, after death, cxxx
 stimuli of, cxxx
 current, cxxix
 rigidity, cxxxi
 sense, cxxix
MUSCULAR TISSUE, General Anatomy of, cxv
 blood-vessels of, cxxiii
 chemical composition of, cxxviii
 cleavage into disks, cxx
 connection with tendons, cxxi
 corpuscles of, cxxi
 cross stripes of, cxvii, cxix
 development of, cxxvi
 fasciculi of, cxvi, cxvii
 fibres of, striped, cxvii
  branched, cxxi
  length and ending of, cxxi
 unstriped, cxxiv

THE END.

BRADBURY, EVANS, AND CO., PRINTERS, WHITEFRIARS.

Quain,
    Quain's elements of anat-
omy. 7th ed., ed. by W.
Sharpey, A.Thomson, and
J.Cleland.

Binding

Lightning Source UK Ltd.
Milton Keynes UK
UKHW021353110119
335396UK00015B/1024/P